THE OXFORD HANDBOOK OF

PEACEBUILDING, STATEBUILDING, AND PEACE FORMATION

THE OXFORD HANDBOOK OF

PEACEBUILDING, STATEBUILDING, AND PEACE FORMATION

Edited by

OLIVER P. RICHMOND

and

GËZIM VISOKA

OXFORD
UNIVERSITY PRESS

OXFORD
UNIVERSITY PRESS

Oxford University Press is a department of the University of Oxford. It furthers the University's objective of excellence in research, scholarship, and education by publishing worldwide. Oxford is a registered trade mark of Oxford University Press in the UK and certain other countries.

Published in the United States of America by Oxford University Press
198 Madison Avenue, New York, NY 10016, United States of America.

Library of Congress Cataloging-in-Publication Data
Names: Richmond, Oliver P., editor. | Visoka, Gëzim, editor.
Title: The Oxford handbook of peacebuilding, statebuilding, and peace formation / edited by Oliver P. Richmond and Gëzim Visoka.
Other titles: Handbook of peacebuilding, statebuilding, and peace formation
Description: New York, NY : Oxford University Press, 2021. |
Includes bibliographical references and index.
Identifiers: LCCN 2020058306 (print) | LCCN 2020058307 (ebook) |
ISBN 9780190904418 (hardback) | ISBN 9780197576410 (epub)
Subjects: LCSH: Peace-building. | Nation-building. |
Peaceful change (International relations)
Classification: LCC JZ5538 .O949 2021 (print) | LCC JZ5538 (ebook) |
DDC 327.1/72—dc23
LC record available at https://lccn.loc.gov/2020058306
LC ebook record available at https://lccn.loc.gov/2020058307

DOI: 10.1093/oxfordhb/9780190904418.001.0001

To Rand Engel, a true peacebuilder.

Contents

PART IV POSTLIBERAL PEACE AND PEACE FORMATION

Acknowledgments

This Handbook is part of our ongoing efforts to expand the conceptual and empirical frontiers of academic research on peace and conflict studies. Building on the legacy of previous generations of peace and conflict scholars and practitioners, it attempts to offer an authoritative and critical overview of conceptual foundations and political implications of peacebuilding, statebuilding, and peace formation at the global, regional, and local levels.

We would like to thank all the contributors to this Handbook for their efforts and dedication in providing fresh and comprehensive thinking on the policies, practices, examples, and discourses underlining various segments of peacebuilding, statebuilding, and peace formation. We would also like to express our gratitude to our colleagues in the School of Law and Government and the Faculty of the Humanities and Social Science at Dublin City University and the Department of Politics at the University of Manchester.

At Oxford University Press, we are enormously grateful to Angela Chnapko for her enthusiasm and support and to Alexcee Bechthold for her editorial assistance throughout this project.

We hope that this work will serve both as a teaching guide and a reference for peace and conflict scholars, international relations scholars, political scientists, sociologists, area studies scholars, and others working on subjects related to peace and conflict around the world.

Oliver P. Richmond, Manchester
Gëzim Visoka, Dublin

Contributors

Cecilia M. Bailliet is a professor at the Department of Public and International Law at the University of Oslo.

Alex J. Bellamy is a professor of peace and conflict studies and director of the Asia Pacific Centre for the Responsibility to Protect at the University of Queensland.

Annika Björkdahl is a professor of political science in the Department of Political Science at Lund University.

Roland Bleiker is a professor of international relations in the School of Political Science and International Studies at the University of Queensland.

Volker Boege is an honorary research fellow in the Department of Political Science and International Studies at the University of Queensland.

John D. Brewer is a professor of postconflict studies in the Senator George J. Mitchell Institute for Global Peace, Security and Justice at Queen's University Belfast.

Naji Bsisu is an assistant professor of political science at the Maryville College.

David Chandler is a professor of international relations at the Department of Politics and International Relations, University of Westminster.

Daniel J. Christie is Professor Emeritus of Psychology at Ohio State University.

Timothy Donais is an associate professor in the Department of Global Studies at Wilfrid Laurier University.

John Doyle is director of the Institute for International Conflict Resolution and Reconstruction and professor of international politics in the School of Law and Government at Dublin City University.

Allard Duursma is a senior researcher in the Center for Security Studies at ETH Zurich.

Pamina Firchow is a professor in the Heller School for Social Policy and Management at Brandeis University.

Maria Raquel Freire is a professor in international relations at the Faculty of Economics of the University of Coimbra.

Rachel M. Gisselquist is a senior research fellow at the UN University's World Institute for Development Economics Research (UNU-WIDER), Finland.

Catherine Goetze is a senior lecturer in international relations at the University of Tasmania.

Berit Bliesemann de Guevara is a professor in international politics and founding director of the Centre for the International Politics of Knowledge in the Department of International Politics at the Aberystwyth University.

Aidan Hehir is a reader in international relations in the Department of Politics and International Relations at the University of Westminster.

Emma Hutchison is an associate professor in the School of Political Science and International Studies at the University of Queensland.

Vivienne Jabri is a professor of international politics in the Department of War Studies at King's College London.

Paul Jackson is a professor of African politics and director of research for the social sciences at the University of Birmingham and a research fellow in the Centre for Gender and Africa Studies, University of the Free State.

Richard Jackson is a professor of peace studies and director of the National Centre for Peace and Conflict Studies at the University of Otago.

Beate Jahn is a professor of international relations and head of the Department of International Relations at the University of Sussex and president of the European International Studies Association.

Mary Kaldor is emeritus professor of global governance and director of the Foreign, Commonwealth and Development Office-funded Conflict Research Programme in LSE IDEAS at the London School of Economics and Political Science.

Stefanie Kappler is an associate professor in conflict resolution and peacebuilding in the School of Government and International Affairs at Durham University.

John Karlsrud is a research professor at the Norwegian Institute of International Affairs.

Gráinne Kelly is a lecturer in the School of Applied Social and Policy Sciences, Ulster University.

Kai Michael Kenkel is an associate professor in the Institute of International Relations at Pontifical Catholic University of Rio de Janeiro and an associate researcher at the German Institute of Global and Area Studies.

Walt Kilroy is an assistant professor in the School of Law and Government and associate director of the Institute for International Conflict Resolution and Reconstruction at the Dublin City University.

Denisa Kostovicova is an associate professor in global politics at the European Institute at the London School of Economics and Political Science.

Florian Krampe is a senior researcher at the Stockholm International Peace Research Institute.

SungYong Lee is an associate professor in the National Centre for Peace and Conflict Studies at the University of Otago.

Nicolas Lemay-Hébert is a senior lecturer in the Department of International Relations, Coral Bell School of Asia Pacific Affairs at the Australian National University.

Kristoffer Lidén is a senior researcher at the Peace Research Institute Oslo.

Gearoid Millar is a senior lecturer in peace and reconciliation and head of the Department of Sociology at the University of Aberdeen.

Amanda Murdie is the Thomas P. and M. Jean Lauth Public Affairs Professor and head of the Department of International Affairs at the University of Georgia, USA.

Andries Odendaal is a freelance specialist in local peacebuilding based in South Africa.

Sandra Pogodda in a lecturer in peace and conflict studies in the Department of Politics at the University of Manchester.

Necati Polat is a professor in the Department of International Relations at the Middle East Technical University, Turkey.

Michael Pugh is Professor Emeritus at Bradford University, UK.

Oliver P. Richmond is a research professor in international relations, peace, and conflict studies in the Department of Politics at the University of Manchester. He is also international research professor at Dublin City University, Ireland.

Licínia Simão is an assistant professor at the Faculty of Economics of the University of Coimbra.

Jackie Smith is a professor of sociology in the Department of Sociology at the University of Pittsburgh.

Sarah Smith is a research officer at the LSE Centre for Women, Peace, and Security.

Ashok Swain is a professor of peace and conflict research, UNESCO chair of International Water Cooperation, and the director of research at the School of International Water Cooperation at Uppsala University.

Ioannis Tellidis is an associate professor of international relations at the College of International Studies at Kyung Hee University.

Nathalie Tocci is director of the Istituto Affari Internazionali in Rome and an honorary professor at the University of Tübingen.

Catherine Turner is an associate professor of international law in the Durham Law School at Durham University.

Gëzim Visoka is an associate professor of peace and conflict studies in the School of Law and Government at Dublin City University.

Joanne Wallis is a professor in the Department of Politics and International Relations at the University of Adelaide.

Christoph Zürcher is a professor in the Graduate School of Public and International Affairs at the University of Ottawa.

CHAPTER 1

INTRODUCTION

International, State, and Local Dynamics of Peace in the Twenty-First Century

OLIVER P. RICHMOND AND GËZIM VISOKA

PEACEBUILDING and statebuilding have become major areas of research within international relations (IR), security studies, development studies, economics, and peace and conflict studies (see Richmond 2020). Through them, democracy, capitalism, and human rights as well as the state and international architecture are assumed to be central to liberal or neoliberal forms of peace, respectively. Not only are they well-established academic subdisciplines, but they are also major policy areas within the UN and other international and regional organizations, as well as the international financial institutions such as the World Bank, and they take a significant place in the foreign and security policies of many established and emerging democracies. It is difficult to know whether the policies or the theories came first.

Peacebuilding and statebuilding have become the main approaches for preventing, managing, and mitigating global insecurities, dealing with the humanitarian consequences of civil wars, and expanding democracy and neoliberal economic regimes (UN Secretary-General 2018). They comprise a wide range of interventionary components, such as imposing and incentivizing rules, norms, institutions, and conditions to govern postconflict transitions in the area of elections and institution-building, security sector reform, economic reconstruction, promotion of civil society, rule of law and justice, reconciliation, and transitional justice. They produce a hierarchy within these elements as well as in security terms. Peace formation is a relatively new concept that foregrounds the emancipatory practices constituted via the local and the everyday in view of war, the state, and international architecture. Drawing on previous writings on localized processes of peacebuilding, it examines how local actors attempt to influence their own experience of peace and state, and how these develop in parallel to or in relation with international peacebuilding and statebuilding projects.

The Oxford Handbook of Peacebuilding, Statebuilding, and Peace Formation endeavors to offer an authoritative and comprehensive overview of peacebuilding and associated debates and practices relating to peace processes, peace settlements, peacekeeping, development, statebuilding, and peace formation. It offers a record, a set of contributions that is perhaps the most comprehensive so far, from some of the key thinkers of this era. This is in order to capture the evolution and potential of these theories, doctrines, and practices as they formulate themselves into a perhaps more concrete international peace architecture in the longer term. It does so by examining a range of disciplinary perspectives, situating peacebuilding and statebuilding within broader contemporary social and global politics, and in particular the gradual emergence of an international peace architecture comprising intergovernmental, international, and transnational institutions and organizations as well as social movements aimed at the expansion of rights and the redefinition of the terms of emancipation in the modern world. It also thus examines the emergence and dynamics of transnational peacebuilding and local peace formation actors and their relations with the global peace architecture, which in the past have tended to be framed by concepts like conflict management, resolution, transformation, and peacebuilding in isolation from broader dynamics.

With contributions from many leading authors, the Handbook offers a systematic overview of conceptual foundations, political implications, and tensions at the global, regional, and local levels, as well as key policies, practices, examples, and discourses underlining all segments of peacebuilding and statebuilding both as discursive formulations and as policy practices. It also analyzes peace formation processes in relation to these formulations and practices. Accordingly, we hope the Handbook can serve as an essential guide to this vast intellectual landscape. Part I investigates various disciplinary approaches to peacebuilding, statebuilding, and peace formation. This is followed in part II by a range of chapters covering the different elements of peacebuilding and statebuilding praxis. Part III disaggregates the theoretical elements of these praxes. Part IV turns to peace formation approaches and their implications.

Disciplinary Perspectives

Part I of this Handbook outlines the field's main disciplinary approaches. Efforts to build peace and reconstruct states have developed within the parameters of an international system that appears to be shifting from liberal internationalist and multilateral, with a vibrant global civil society valuing human rights, toward populist, nationalist, and neoliberal, with ineffective multilateral and global civil sets of networks. This shift is undermining the wide range of political, economic, and sociocultural forms of interventions that inevitably require a pluriverse of theoretical and methodological approaches for making sense of them. As such, scholarly perspectives on peacebuilding, statebuilding, and peace formation are far from being monolithic: concepts are used differently, theoretic and methodological premises differ, and even their conceptions of peace range from negative to positive, from everyday to hybrid. Most draw on peace and conflict studies, IR theories, and other social science disciplines (see Richmond, Pogodda, and Ramović 2016).

Despite a long-standing attempt within peace research to develop an understanding of peace, the majority of contemporary debates still remain embedded within the orthodox approaches to IR (namely positivist debates derived from realism, liberalism, and Marxism). Realist and liberal thinking on war and peace tends to remain centered around survival, security, and a balance of power between states, augmented by efforts to forge contested visions of world peace around democracy, global institutions, and capitalism. Marxism, on the other hand, offers a grassroots vision of peace as a struggle for emancipation from determinist structures of the international political economy, perhaps via violent and global revolution (Richmond 2020). Sociology, anthropology, environmentalism, and geography have also offered valuable insights for understanding different scales, actors, and practices of peace across different historical and political settings, much of them pointed to expanded understandings of rights within the states-system, international legal and multilateral architecture.

The contemporary international peace architecture is generally taken to have emerged from liberal forms of hegemony aimed at restraining imperial and later industrial state geopolitics (Richmond 2005). It connected the political form of the state and the international system and legislated basic norms against violence, war, and conflict via institutional arrangements built to prevent or provide a remedy. Thus peace has become associated with cooperation, institutions, law, rights, checks and balances, multilateralism, and international organization (see Jahn 2013). Liberal theory rests upon various assumptions of universal rationality, individual liberty, social contracts, and cooperation in domestic and international affairs, as well as the need for a legitimate domestic government and international governance (Richmond 2002). The international approach, in terms of peacebuilding, is defined by liberal norms, laws, and institutions, harking back to liberal internationalism in the early twentieth century and later liberal institutionalism, which emerged via the UN, humanitarians, donors, and other international agencies (Williams 2006). Statebuilding, however, was connected to nationalism as well as imperial agendas for international order, reinvented for the neoliberal era at the end of the twentieth century. The normative agenda of liberal peace, though, remains embedded in both sets of practices, reflecting realist contours of geopolitics meshed with cosmopolitan narratives and liberal checks and balances. As Beate Jahn (chapter 2, "Liberal Internationalism") illustrates, although the liberal peacebuilding model had largely failed to achieve its goals by the end of the twentieth century, it has become an integral element of the international order. Its main function is not the establishment of emancipatory peace but the management of domestic, international, and transnational conflicts generated by the inherent contradictions of geopolitics, capitalism, and liberalism. Kristoffer Lidén (chapter 3, "The Ethics of Liberal Peacebuilding") illustrates that the limited efficiency and legitimacy of liberal peacebuilding does not necessarily go against the normative foundations of political liberalism—neither in their internationalist nor cosmopolitan manifestations. In a nutshell, the vast majority of peacebuilding and statebuilding theories and practices do not reject the internationalist objectives of promoting peace in accordance with human rights within a state framework. Rather, they tend to criticize the lack of consistency in this objective.

Efforts to institute peace within international law are at the heart of international peace architecture. Cecilia M. Bailliet (chapter 4, "The International Law of Peace") discusses the normative evolution of the concept of peace in international law. Peace through law is promoted widely as the most normatively sustainable approach to resolving disputes and promoting human rights and global justice, but as it rests upon the state it must balance sovereignty with cosmopolitan rights. Bailliet traces two major normative movements in international law: the principle of peaceful coexistence among nations and the right to peace, which demonstrate the struggle between pluralist and universalist understandings of peace in world politics (see Bailliet and Larsen 2015). While the right to peaceful coexistence seeks to recognize but also contain the existing distribution of power and influence among nations, the right to peace seeks to push for a more singular understanding of peace with emancipatory tendencies by globalizing the Western understanding of peace.

While realist and liberal perspectives of peacebuilding tend to have a more static view of peace related to the historical political struggles of the nineteenth and twentieth centuries for independence, territorial integrity, sovereignty, and industrialization, increasingly peacebuilding and statebuilding are now conceptualized as social constructs. In this form they are nurtured by interstate socialization, transnational norms, and knowledge production, scaled up from the local and the social levels to the state and thence the international system. Constructivism features prominently in peacebuilding and statebuilding studies. From a constructivist point of view, peace is achieved by establishing rules and institutions that enable creating of new identities and interests through intersubjective interactions. These transform the identities of parties in conflict toward creating a consensual meaning for peace. Yet peace means different things to different actors in different contexts, and peacebuilding efforts are always contextual, for distinct actors and for specific purposes. Joanne Wallis (chapter 5, "The Social Construction of Peace") looks at the contribution of constructivist social theory in peace and conflict studies. Wallis unpacks how peace is perceived as socially constructed and a product of human agency and ideas and practices constituted and instantiated within intersubjective social contexts. She shows that a constructivist analysis of peace allows us to take a reflexive approach by making less certain, more dynamic, contingent, and flexible ontological assumptions regarding how peace is understood and practiced in conflict-affected societies. This signifies that peace can be constructed based on certain norms and ideas that are reproduced and institutionalized by social practices of actors, which can produce new norms and knowledge that inform other sequential modes of peace scaled up through law and institutions. Yet constructivism also exposes important tensions in conceiving peace both as a process and as a goal, which may follow multiple paths simultaneously, remaining always unfinished and producing unintended outcomes. Constructivism has informed a wide range of peacebuilding theories and practices that tend to focus on identifying key actors, processes, and norms that are more prone to durable peace as well as point out contextual and transnational blockages to peace.

A large body of knowledge on peacebuilding and statebuilding studies draws on different strands of critical theory. Since its inception, peace research in IR has been one of the pioneering subdisciplines that has maintained a normative and practical commitment to peace, justice, security, and emancipation and has endorsed both deconstructive and reconstructive modes of critique (Visoka 2019). In general, critical approaches to peace and conflict studies have emerged in response to fallacies about international interventions aimed at peacebuilding in conflict-affected societies. Their primary goal has been to criticize power relations, values, and the impact of liberal interventionist paradigms. As Vivienne Jabri and Oliver P. Richmond (chapter 6, "Critical Theory and the Politics of Peace") show, the critique of peacebuilding and statebuilding encompasses perspectives of peacebuilding as governmentality, drawing on Michel Foucault's analytics of power, postcolonial critiques that focus on practice as a mode of colonial rationality, and other, alternative perspectives that focus on the agency of the subaltern, the agency of women, and the broader question of how local voices come to impact and resist control by external forces. Compared to other critical debates in IR, most of the critical accounts in peace and conflict studies are deeply rooted in empirical research and observation of discourses and practices. A distinct feature of critical peacebuilding approaches is commitment to nonviolence and emancipatory alternatives.

Richard Jackson (chapter 7, "Pacifism in International Relations") explores the contribution of pacifism to IR and demonstrates its relevance to discussions around peacebuilding, statebuilding, and peace formation. Jackson holds that pacifism is ideally placed to offer insights and suggestions for both practitioners and theorists of peacebuilding, such as criticizing violence, power and just war theory and exploring the contribution of civilian capacity and nonviolent forms of agency to enhance peace, security, and justice. Pacifism's unique contribution to peace and conflict studies is a commitment to the construction of peaceful societies rooted in critical interrogation of dominant forms of knowledge accompanied by real-world political activism. Thus, as Jabri and Richmond argue, critical theory has made very significant contributions in holding peacebuilding and statebuilding praxis to account and creating a platform for its further emancipatory expansion, perhaps via peace formation from below, cognizant of postcolonial renderings of power.

Sociology has played a formative role in different debates of peace and conflict studies, especially in the early work on peace research and conflict resolution, but only recently has it started to play a more prominent role in peacebuilding and statebuilding debates (Goetze 2017). It offers a standpoint from which to view historical, structural, societal, and agential problems and how they relate to the development of the different aspects of international norms, practices, and agency as responses to violence. In other words, sociologically informed approaches concur that conflict, cooperation, and reconciliation are at their most vivid at the local-scale habitus, where the effects of conflict are most viscerally experienced. Catherine Goetze and Berit Bliesemann de

Guevara (chapter 8, "International Political Sociology of Peacebuilding") explore how sociological concepts, theories, and methods have been engaged to study postconflict peacebuilding. They highlight that international political sociology enables studying peacebuilding interventions in conflict situations; it enables studying the configuration of power relations; and it enables the deployment of different methodological lenses for observing social practices, norms, and institutions in conflict-affected societies. Goetze and de Guevara survey the existing literature on peacebuilding and offer important insights on how to read, see, and write about peacebuilding interventions from the point of view of political sociology. Political sociology offers a reflective theorization of peacebuilding interventions and a more detailed empirical investigation of the social forces and interactions involved in postconflict processes. They indicate that the next phase of research should be to listen sociologically and observe reflexively the fluidity of social relations in peacebuilding contexts.

Central to understanding the peace process is to explore locations and spatial features that enable or inhibit peace efforts. Annika Björkdahl and Stefanie Kappler (chapter 9, "Spaces of Peace") offer an overview of the spatial understanding of peace and conflict. They argue that a focus on the spatial dimensions of peace requires a relational and contextual reading in order to appreciate the nonlinear ways in which peace impacts spatiality. By looking at how space transforms social relations Björkdahl and Kappler demonstrate that peace no longer is fully determined by the top-down planning of space, but physical places and their meaning are locally determined by resident communities. They show that the physical environment shapes social interactions, governs movement through space, and enables or hampers peacebuilding efforts and their likelihood for success on the ground. Understanding peace in terms of space and place opens up the possibility for emplaced agencies for peacebuilding to move between different scales, such as from the local to the global. A spatial approach to peace and conflict studies thus requires that geo-epistemologies of peace should be revised as the role and function of a given social space are subject to the agents and communities surrounding it. This requires rethinking how we observe and interact with conflict-affected places as this is where people experience peacebuilding and conflict dynamics in their everyday lives (see Björkdahl and Buckley-Zistel 2016).

Our understanding of peacebuilding, statebuilding, and peace formation is directly influenced by the methodological approaches we rely on. Currently we have two dominant methodological movements associated with positivism and rationalism that focus on states and power and transnational institutions and norms, and those associated with constructivist and postpositivist critiques that focus on societal and everyday aspects of peace. Pamina Firchow (chapter 10, "Peace Methods and Methodologies") offers an overview of the main methods and epistemological approaches across different disciplinary strands of peace and conflict studies. Firchow's analysis shows that different methodologies tend to (re)produce different knowledges about peace. The positivist and rationalist approaches that rely on methodological nationalism and liberalism tend to draw on quantitative methods and reproduce specific modes of peace which privileges the dominant order and state system and their quest for power and desire for stability.

On the other hand, more critical and postpositivist approaches that may contain elements of methodological liberalism and everydayism tend to use qualitative methods and generate more critical, alternative, and emancipatory knowledge that challenges the existing order and its normative foundations. Firchow's chapter offers valuable insights on how to reconcile and combine these distinct methodological approaches to promote more pluriversal and reflexive knowledge. Multiple theoretical, historical, and empirical methods are obviously required in concert to provide a balanced perspective on the evolution of this international peace architecture.

Peace and conflict research tends to rely on IR and related social science disciplines that have bureaucratic and Western-centric understandings of peacebuilding and statebuilding. It is clear that this approach has failed to fully explain the dynamics, agency, and hybridity of human society and institutions when it comes to peace, or that inequality is conflict inducing. Critically unpacking peace architecture means engaging ethnographically and sociologically through a postcolonial lens with local peace processes, networks, relationality, peace, and state formation. So rescuing peacebuilding from neoliberal epistemological frameworks requires an anthropological and ethnographic sensitivity. Gearoid Millar (chapter 11, "Ethnographic Peace Research") provides an introduction to ethnographic peace research (EPR) as an attempt to promote a forceful and rigorous research agenda that explores local experiences, perceptions, concepts, and understandings of conflict, violence, and peace. Millar argues that EPR can accomplish three tasks. First, it can be used to test existing or to generate new theories regarding local experiences of conflict, violence, and peace, largely in order to contribute to academic debate. Second, it can be used to assess the outcomes of the projects implemented by peacebuilding or statebuilding organizations largely to inform the future decisions of policymakers, planners, and funders. Third, it can be used as an action research endeavor to generate knowledge and solutions that contribute to local peace, security, justice, and development.

Although textual analysis is evident across all peace and conflict research methods, images and visuality influence how we view, understand, and respond to violence and peace efforts. Emma Hutchison and Roland Bleiker (chapter 12, "Visuality of Peace and Conflict") offer a conceptual and empirical overview of the importance of visuality in enhancing peace formation and reconciliation in conflict-affected societies and responses to humanitarian crises. Hutchison and Bleiker draw attention to the historical, relational, and performative functions of images, pointing out that there are inevitable power relationships involved in the nexus between visuality and dominant ways of seeing and perceiving society and politics. They look at the role of humanitarian and peace photography in shaping the ethics, policies, emotions, and practices of international interventions. In particular, images enable scholars, practitioners, and affected communities to learn from past injustices and violence in order to conceptualize and initiate political action for building peace and seeking justice.

Knowledge production about conflict-affected societies is dominated by Western epistemologies, which influence how we know peace and what policy solutions are offered to resolve conflicts and build peace (Acharya and Buzan 2010; Visoka and Musliu

2019). As a result of this, non-Western or local knowledge is often marginalized and its usefulness disregarded. Yet there is a growing realization that global and non-Western knowledge on peace and conflict is crucial to understanding the pluriversal dynamics of peace. Necati Polat (chapter 13, "Peace in Non-Western Theory") problematizes different geo-epistemological distinctions of peace and looks at different visions of peace outside the European and Western epistemologies. Polat offers a theoretical and historical overview of some of the key invocations and meanings of peace across different parts of the world, thus questioning the contemporary association of peace with a particular region and modernity.

Since the adoption of the Women, Peace and Security agenda by the UN, gender-based understandings of peace, security, and development have received greater scholarly and policy attention (Davies and True 2019). Sarah Smith (chapter 14, "Gender, Security, and Peacebuilding") examines how gender is relevant to, shapes, and is shaped by security and peacebuilding. Smith argues that mainstreaming gender in peace and conflict studies has contributed to making visible previously marginalized, intersectional experiences and knowledge and exposing the gendered logics that inform this exclusion and are fundamentally entwined with and productive of the priorities and practices of security and peacebuilding. Accordingly, rectifying gender-discriminatory understandings of peace and security requires reconceptualizing what constitutes security and peace, and redesigning the institutions and processes that pursue these goals. Smith argues that an intersectional and nonessential view of gender must come to the forefront of both academic and policy work on security and peacebuilding alike in order to challenge the binary and exclusionary frameworks of thinking about identities and agendas of peace that obscure lived experience and perpetuate hierarchical social and power relations.

Peacebuilding and statebuilding interventions seek to intervene not only at the state and institutional levels but also at the social and individual levels. Daniel Christie (chapter 15, "Peace Psychology") offers insights into how psychosocial approaches to understanding and mitigating conflicts have been deployed since the inception of the Cold War. Christie traces two waves of debates on peace psychology: the first focuses on generating concepts, themes, and perspectives aimed at the prevention of nuclear war and the mitigation of intractable conflicts characterized by repeated cycles of violence; the second focuses on understanding the underlying structures that fueled and sustained violence in civil wars.

Understanding and explaining the international peace architecture thus requires an attempt to bring IR together with political philosophy and history, with empirical anthropology, sociology, and the critical interrogation of the government, society, and economy of stratified societies, in the light of geopolitics and international political economy. A multidisciplinary perspective, relating disciplines and multiple methodologies (as opposed to dogmatic disciplinary theories and methods) provides a more dynamic understanding of peace and order not just in the usual micro, system, and states-system terms but also as an edifice and architecture built on the layers and stages of long-forgotten sediments.

Peacebuilding and Statebuilding in Global Politics

Part II turns to different elements of peacebuilding and statebuilding praxis. Peacebuilding was initially understood as a grassroots and bottom-up process in which a local consensus led to a positive peace. As the concept evolved it came to represent a convergence of the agendas of peace research, conflict resolution, and conflict management approaches (Jabri 2007). This convergence culminated in the contemporary liberal peacebuilding project, which in itself has been partially subsumed within a liberal statebuilding enterprise (Richmond 2010). Forming a much broader international peace architecture than expected, however, it now comprises the UN system; the donor system; international financial institutions; NATO, the European Union, the African Union, and other regional organizations; international law; human rights and humanitarian law; and a global-to-local system of NGOs and social movements. The architectural blueprint of peacebuilding and statebuilding revolves around democracy, human rights, law, free trade, and an international community that promotes associated values.

International peacekeeping after the Second World War began as a modest means of subduing conflicts between states; it grew in importance partly to facilitate processes of decolonization. Despite many controversies, UN peacekeeping remains one of the only multilateral avenues for restoring peace and order in societies affected by violent conflict. Peacekeeping was a major contribution to the twentieth-century project of peace in the sense of providing a tool with which a preliminary, negative peace could be consolidated in a state-centric world. However, integrated missions and peacebuilding interventions since the end of the Cold War have adopted a radically different approach, indicating an ambition to create a liberal state without necessarily receiving local consent. The meaning of peacekeeping has evolved since its inception. Initially it was affiliated with monitoring ceasefires; later it expanded to protecting civilians, using force, and providing security for postconflict peacebuilding. The changing nature of global affairs and the emergence of new transnational security challenges have expanded the doctrine and type of actors engaged in international peacekeeping. Scholarly work on international peacekeeping tends to focus on the historical evolution of peace operations, the politics of peacekeeping at the UN and among member states, the polemics of how to measure the success of peacekeeping, the challenges posed by regionalization and privatization of peacekeeping, and organizational and normative challenges (Weiss and Daws 2008).

There is overwhelming consensus that peacekeeping is a positive instrument for the advancement of a stable international order and for the protection of civilians and their human rights and an example of international cooperation among troop contributors, as Walt Kilroy shows (chapter 18, "Protection of Civilians"). Yet both realist and critical scholars view peacekeeping as a reflection of power politics, centered on the advancement of national interest, and an expression of dominant and rising global powers to expand regional hegemony and assert their influence. Michael Pugh (chapter 17, "Peacekeeping") contends that while peacekeeping had undoubtedly secured many lives, its relationship to peacebuilding has broadly sustained asymmetries and inequalities in the global system. As Pugh argues, in practice a peacekeeping system operates as a double asymmetry: peripheral and former colonial states rebranded as fragile and failing states are objects of intervention, whereas former colonies rebranded as democratic and functioning states use peacekeeping as a foreign policy instrument to retain their power in specific regions of interest. Peacekeeping is clearly vital to maintaining the current order. It has formed part of the interventionist, modernization, and trusteeship project, which has naturalized the current international hierarchy. Aidan Hehir (chapter 16, "International Interventions") looks at the implications of international interventions for state sovereignty and the world order. Hehir traces the evolution of the norms of international intervention and discusses institutional and political ambiguities within the international peace architecture that leave room for abusing with the use of force. The question is whether this evolving system can still aid in the development of a more secure and just order (see Lemay-Hébert 2019).

Moreover, Alex J. Bellamy (chapter 19, "The United Nations and the Responsibility to Rebuild") looks at the relationship between the Responsibility to Protect (R2P) norm and postconflict peacebuilding efforts. Bellamy argues that peacebuilding and statebuilding were integral parts of R2P since its inception, recognizing that states and societies are at their most vulnerable as they emerge out of violent conflict. However, the responsibility to rebuild societies after military interventions has been sidelined in some of the most recent cases, such as Libya, largely because the responsibility to rebuild was separated from the R2P agenda as it was seen as part of other institutional frameworks within the UN. The R2P-peacebuilding gap has thus become a trap, where intervention without a reconstruction agenda is prone to creating fragile states and fears of prolonged and costly reconstruction tend to undermine efforts to prevent atrocities. Bellamy argues that one way of overcoming this gap is to integrate atrocity prevention and non-reoccurrence into peacebuilding activities (see also Bellamy and Dunne 2016).

In addition to the UN's global role, numerous regional organizations have joined the international peace architecture. Notably, the inability of the UN to respond to a wide range of conflicts and the limits of peacekeeping and peacebuilding operations along with the rise of competing organizations and actors have resulted in the regionalization of peace efforts. In the past twenty years the European Union has increased its role in resolving conflicts and building peace in neighboring regions and beyond. Although the existing EU peacebuilding framework relies on the existing UN liberal peacebuilding and statebuilding project, the EU recently has moved toward a more pragmatic approach in its external relations. Nathalie Tocci (chapter 20, "The European Union and Peacebuilding") looks at the evolution of the EU's approach to peacebuilding. Tocci argues that the EU has moved away from a hyperliberal idealist approach to building peace and reconstructing states to a more principled pragmatism centered on supporting states' and societies' resilience and integrating the EU's capabilities to leverage greater impact. The initial approach of the EU, which tended to use civilian capacity, financial incentives, and political conditionality in peacebuilding and statebuilding activities, was unable to resolve protracted disputes and build a sustainable peace due to passive enforcement capacity. This pushed the EU to change its approach so that it no longer disguises its geopolitical interests in the pursuit of security, prosperity, democracy, and rule-based international order. The EU's new approach is thus moving from peacebuilding to resilience-building and from normative to pragmatic frameworks for dealing with conflict and building peace.

The normative claims propagated by peacebuilding organizations have been challenged on historical, ideological, and cultural grounds. In conflict-affected societies, interventions' subject-citizens have intuitively felt that the liberal peace and the neoliberal state, as well the global economy and governance system in which they are embedded, merely naturalize historical power dynamics and the global North-South divide. At the global level, rising powers have also contested the universality and consistency of Western-led interventions as well as utilized its operational flaws to associate it with self-interest and dominance. Kai M. Kinkel (chapter 21, "Rising Powers and Peacebuilding") explores how peacebuilding activities serve as the locus for rising powers to contest the rules underpinning the international order and to stake their claim to influence and produce a post-Western order. Beyond the EU, regional organizations and actors such as China, Brazil, India, South Africa, and Turkey have taken on substantial roles in peacebuilding efforts across the globe. However, Kinkel argues that while rising powers have been able to bring a wealth of experience into the practice of peacebuilding and to a degree enhance its global legitimacy, they have done so largely from within the normative and institutional frameworks set by the major Western states.

Along with shattering global institutions, the international peace architecture is also shaped by networks of nonstate actors and social movements. Global civil society is not just expected to be subcontractors of donors, governments, and institutions but is also supposed to drive the state and international order in a progressive direction following the international standards set by the UN, Western allies, and international financial institutions, as well as developments in international law. Mary Kaldor and Denisa Kostovicova (chapter 23, "Global Civil Society, Peacebuilding, and Statebuilding") look at the role of global civil society in generating but also in undermining emancipatory and civic forms of peacebuilding. They advocate for a broader understanding of civil society as a generator of civility and promoter of legitimate political authority rather than associating it with NGOs and narrow technocratic functions in postconflict societies. Yet, as Naji Bsisu and Amanda Murdie (chapter 24, "Networks of Peace") show, coordination and cooperation among different humanitarian organizations, NGOs, and military interveners remain a challenge that often affects the overall quality of peace. There is also strong evidence that the growing association of peace with various forms of justice has been influenced by networks that have tussled with the state and international system in order to produce a new and more hybrid framework and push for expanded rights. In particular, critical social movements have posed significant challenges for approaches to contemporary peace, especially in exposing their unstable geopolitical and geo-economic framework and revealing the importance of networks, mobility, and new technology as instruments to enhance critical agency for peace. Jackie Smith (chapter 22, "Globalization of Peace") looks at the relationship between globalization and peacebuilding efforts. Smith describes interstate peacebuilding initiatives as hegemonic and inseparable from global capitalism, primarily concerned with stabilization and reduction of violent conflicts that disrupt the operation of global capitalism rather than addressing the structural violence and blockages to peace. As an alternative, Smith argues, there is a long history of counterhegemonic efforts by social movements that advocate fundamental transformations of unequal social realities as a necessary effort to promote human rights, dignity, and basic survival.

Intrinsic to the politics of statebuilding is the gradeability of states according to state capacity, predominantly using Western models of statehood as ranking frameworks. This inevitably creates hierarchies of states, using categories such as stable, fragile, and collapsed states, which in turn elevates different interventionary regimens and techniques. Nicolas Lemay-Hébert (chapter 25, "Peace, Intervention, and State Fragility") explores the relationship between discourses of state fragility and peacebuilding and statebuilding interventions. Lemay-Hébert shows that states ranked as fragile and weak are more prone to international intervention and top-down impositions. Such labels permit extensive statebuilding interventions, which tend to focus on creating capable states—not so much as means for self-sufficiency as to retain hierarchical power relations between states—while making permissible regime change in the name of democracy promotion and neoliberal economic reforms in the name of societal recovery. While statebuilding interventions have mostly been unsuccessful, blame for failure is attributed to incorrigible societal and local conditions in conflict-affected states and difficulty at transforming fragile states through external interventions. Lemay-Hébert shows that this framing of failures has elevated other discourses, such as resilience-building, which in itself encourages new forms of intervention without taking responsibility for the outcomes. Resilience ideology embedded in the diversity of risks and vulnerabilities that lead to fragility permits restructuring of the liberal interventionism agenda to widen the net of states prone to fragility.

Efforts at peacebuilding and statebuilding are also widely overshadowed by the rise of new interventionary approaches to terrorism, which more often than not are confined to efforts that seek to produce a durable and sustainable peace. Ioannis Tellidis (chapter 26, "Terrorism and Peacebuilding") provides an overview of the relationship between terrorism studies and the practices of counterterrorism, and peace and conflict studies and the practices of peacebuilding. Tellidis argues that although terrorism poses serious security concerns, magnifying the risk that terrorism poses allows the state to reinforce its position of control by claiming the urgency to implement extraordinary measures. Terrorism also elevates more extreme forms of securitization, which often fail to address the underlying conditions and historical and socioeconomic triggers of political violence. Similarly, most peacebuilding interventions tend to focus on peace and governance and state security rather than addressing more profound aspects of emancipation, justice, and development. Tellidis adds that the work of critical peace and conflict studies in exploring the everyday and local dynamics of peace and conflict offers valuable grounds for overcoming some of the fallacies of top-down understandings of peacebuilding and terrorism and promoting alternative approaches that tackle the complex blockages to peace.

Most of the debates on peacebuilding and statebuilding tend to focus on civil wars, which are mostly due to ethnic and other identity contentions. However, since the Arab Spring, violent conflicts and revolutions related to regime change are increasingly the focus of peace and conflict studies. These conflicts reveal multilayered contentions that not only determine the scale of political violence but also reveal the state and societal dynamics that shape postrevolution transitions and the prospects for sustaining peace. Sandra Pogodda (chapter 27, "Peace after Revolutions") explores the distinct challenges that contemporary revolutions are posing to domestic reformers and external interveners. Pogodda looks at the dynamics of everyday state formation led by local agency to delineate the political space of the state outside of the formal decision-making processes and thus fundamentally transforms state-society relations seeking greater justice, dignity, and participation.

In recent years, the security environment has changed significantly as new forms of threat and thus of warfare emerge. These new threats have come to be known as hybrid wars and are characterized by a mixture of coercive and subversive activity and the conventional and unconventional methods used by state and nonstate actors to achieve specific objectives while remaining below the threshold of formally declared warfare. Hybrid wars may be proxy and asymmetric conflicts pursued by nonstate, minority, and migrant groups; information warfare and cyberattacks; terrorist attacks and identity-based radicalization; or covert intervention in electoral processes and democratic institutions. Maria Raquel Freire and Licínia Simão (chapter 28, "Peace and Security in the Age of Hybrid Wars") explore how the rise of hybrid threats and wars has shaped the security discourse and practices of peacebuilding. They argue that the concept of hybrid wars, although useful to understand the complexity of contemporary security challenges, is not new. Yet hybrid wars as a new reference point has implications for the international peace and security architecture, in particular in the renewal of great power rivalry and the expansion of more fluid, yet extreme, forms of interventions. In particular, hybrid wars have pushed the UN to change its understanding of peacebuilding and design more flexible and adaptive missions that mesh together elements of peacekeeping, peacemaking, and peacebuilding elements. Freire and Simão conclude that reinforcing resilience of states and societies will most likely remain the central approach to promote peace and security through multilateral and unliteral efforts.

The international peace architecture is currently responding to a shift from analog to digital approaches, incorporating new information and communication technologies in peacekeeping and peacebuilding interventions. In particular, we are noticing an increased use of new technologies as part of UN peacekeeping operations to enhance situational awareness and anticipate risks. Allard Duursma and John Karlsrud (chapter 29, "Technologies of Peace") demonstrate how digital and web-based information and communication technologies can be used to prevent and manage armed violence, foster inclusive societies, and ensure a durable and high-quality peace. Although certain aspects of emerging digital approaches to peacebuilding and statebuilding are promising, they cannot yet bypass or resolve older, analog conflict dynamics revolving around the state, territorialism, and state formation.

Disaggregating Peacebuilding and Statebuilding

Part III disaggregates the theoretical elements of these dominant praxes. The goal of ending violence (both overt and structural) while avoiding using violence (both overt and structural) is critical for a world in which rights, democracy, justice, and independence are equated with more positive, emancipatory, and empathetic forms of peace. International peacebuilding and statebuilding have emerged as mechanisms for promoting and imposing these goals on fragile and conflict-affected societies. Peacebuilding and statebuilding represent both heuristic and political processes of intervention conducted by the Global North's dominant states and its institutions, often aimed at the Global South, conflict-affected, or developing countries. They are aimed at building states in unstable regions around the world and are thought to represent a significant advance on the practices of peacekeeping, mediation, and other forms of intervention that were employed before the end of the Cold War (Doyle and Sambanis 2006). They seek to modernize, maintain, and further develop an international system based on neoliberal states that are embedded in an international architecture of liberal peace. Thus part III delves into the main elements that underline peacebuilding and statebuilding interventions. The international peace architecture now has an array of peacekeeping, statebuilding, and peacebuilding tools.

Statebuilding as a policy-driven theoretical field has taken over discussions of peacebuilding, conflict resolution, and transformation (Chandler and Sisk 2013). These approaches emphasize liberal and neoliberal forms of governance and top-down thinking about peace, rather than bottom-up approaches. They practice an outside-in construction of peace, whereby outside actors import the specialized knowledge, procedures, and structures, with an inside-out approach, whereby disputants attempt to renegotiate this process according to their own interests, culture, and frameworks (see Fukuyama 2004). The state is the architecture that maintains this form of peace, in both its domestic and its international relations and structures. Statebuilding interventions are based upon the belief that state-level and local-scale political institutions, under pressure from the need to modernize and develop both internally and externally, have failed to accommodate necessary change and thus require direction from external actors better equipped to reform the state and its peace system to reflect the democratic will of the local people. Statebuilding interventions tend to start with a liberal or neoliberal institutional design and set of norms and expect compliance from a range of local partners and subjects in conflict-affected settings around the world. They assume recipient societies consent to such imposition, must undertake all-encompassing political and societal transformation, and must adhere to and integrate the international economic, security, and political structures despite these structures' injustices, inequalities, and contradictory assemblages of rights, identities, markets, interests, and institutions (see Wallensteen 2015). Modern forms of statebuilding, organized around the need to respond to the various dynamics of state formation, focus on institutions for security, the separation of powers, democracy, and the rule of law. Human rights and civil society are a rhetorically important flourish but are of less concern. Markets are crucial, even though applied neoliberal ideology largely undermines the developmental and democratizing approach statebuilding proposes.

Thus statebuilding now aims to create prosperous and stable liberal or neoliberal states framed by a "good governance" agenda, in the image of those who intervene. While liberal peace has discursively supported progressive processes through institutionalization and civil society building, concerns with stability have accommodated ethnonationalist groups and ignored other local potentials for peace. In particular, the form of the state that has emerged as the focus of statebuilding is highly problematic from a local and ethical perspective, generally cannot engage with short-term and pressing needs and issues, and its institutions, security, and rights frameworks depend on state-level hegemony and external support. Statebuilding interventions compound these contradictions by creating weak states by design that would fail if not for often opportunistic international support, and which reflect a compromise between international and elite interests and identities rather than those of its general population (Richmond 2014). This form of statehood, as David Chandler (chapter 30, "Statebuilding") shows, is globally governed, with little autonomy for the citizen. The political subject of statebuilding is expected to be resilient rather than autonomous so that it adopts the international projection of the state and works to make both the state and everyday life viable rather than determining institutions, law, rights, security, and needs for itself. The contradictions of this approach have rapidly become clear in former war zones and development settings alike.

Statebuilding represents yet another extension of a very conservative liberal peace, which exchanges emancipatory content and capability for power rather than legitimacy. It has produced states that are failed by design (Richmond 2014). They suffer from a lack of legitimacy and broad acceptance on the ground, play into the hands of exclusive elite projects, tend to be securitized and to follow northern understandings of rationalism and liberal institutionalism, represent a range of categories and hierarchies unsuited to context, and offer a legibility and a legal framework that is distant. They exclude local populations or endorse their relative inequality in the international system, fail to take into account local history and society, follow standard operating procedures regardless of past failures, and prevent any local debate about the type of state that might be built. Statebuilding has emerged as a disciplinary, bureaucratic, and problem-solving process devised to maintain the neoliberal state, current patterns of resource distribution, the liberal normative claim of superiority, global governance, externality and conditionality as well as the framework of rights, intervention, and moderated forms of sovereignty (Richmond 2014, 9). Statebuilding creates a range of processes that may also be conflict-inducing at the local level. The extent of the reforms that may be required, in particular with the adoption of Western standards and norms, have led some to argue that democratization should be delayed while institutions are built (Richmond 2014; Paris 2004). Christoph Zürcher (chapter 31, "Democratization and Peacebuilding") illustrates how all peacebuilding missions launched by the UN after the Cold War have integrated democracy and its contents, such as good governance, rule of law, and free and fair elections, at the core of the peace that is being built. Yet external effort for democratization can accelerate the power struggle among the warring parties in the aftermath of conflict, which can exacerbate political violence rather than political moderation (see Mac Ginty 2013).

The international community has attempted to engineer solutions to ethnonationalist conflicts mainly through power-sharing arrangements wherein all ethnic and social groups are offered democratic representation and rights. John Doyle (chapter 32, "Power-Sharing in Divided Societies") shows that power-sharing models of government as a contribution to peacemaking have dominated constitutional design since the mid-1990s, but they remain highly contested. International peacebuilders operate on the wrong assumption that there is a local commitment to democratic and Western forms of governance and human rights and overlook the competition for power among local and national groups. In practice we have seen that such efforts have failed to erode nationalism and promote sustainable peace. Contrary to what is expected, peacebuilding interventions seem to be complicit in the mutation of new forms of nationalism, which aim to derail and delay prospects for peacebuilding and reconciliation. Within institutions, peace tends to be captured by organized politics of ethnonationalist elites who may be insiders or outsiders to a peace process but are committed to obstructing the conflict transformation and peace consolidation through shadow networks of patronage, nepotism, corruption, sabotage, and political violence. In practice, statebuilding does not generate new types of liberal subjects but instead is part of a political debate over how different subjects and their interests and identities may coexist with externalized forces.

Statebuilding derives from the assumption that when sovereignty is weak and the state lacks capacity in the domestic arena, when it fails to contribute to regional stability or comply with international norms or law, intervention is justified, necessary, and desirable. The establishment of the rule of law and justice system is meant to create orderly societies that enforce law through judicial procedures and public policing and that function on the principles of good governance and respect for human rights. Yet a precondition for society that functions based on the rule of law is the ability of state institutions to enforce the law. Reforming the army, police, and judiciary is seen as a crucial precondition for establishing the rule of law in postconflict societies. As Paul Jackson (chapter 33, "Statebuilding, Security Sector Reform, and the Rule of Law") shows, the rule of law has been subject to the same critique as other aspects of peacebuilding. It imposes on local political arrangements a technocratic and Western model of how to organize a law-abiding society. The rule of law is vital for the construction of viable states, but it also requires a polity that is not controlled and captured by elites. Statebuilding interventions tend to privilege predatory elites as they consider them the main interlocutors in imposing the rule-of-law regimes and partners for preserving peace and stability. Efforts to establish the rule of law and promote justice are prone to failure as long as international interventions promote neoliberal rather than social democratic models of statebuilding, which enables the predatory elites to evade responsibility and accountability to citizens. In other words, the rule of law becomes a mechanism with which the predatory elites capture the state and disguise their power through the law. The rule of law is more likely to succeed if it emerges as part of a consensual agreement between citizens and polity, which relies on the local political, historical, cultural, and social context. Theorists and peacebuilding actors need to move from an institutional peace-as-governance agenda to an alternative or at least additional everyday agenda for producing a hybrid form of peace. Putting communities first entails a rethinking of the priorities of peace, placing them in much broader parameters than the state or regional security from a rational actor or liberal-institutionalist perspective. In terms of peacebuilding this would place human needs, particularly economic and security needs, before free-market reform and in parallel to local custom, context, institutions, democracy, second-generation human rights—and a rule of law that protects the citizen in context, not just their prospective rights and property. This would all have to be placed into the context of a much deeper and broader notion of the international peace architecture, transcending geopolitics, liberal peace, neoliberal states, and the constraining frameworks of the states-system.

Parallel to democratization, security, and the rule of law, transitional justice and reconciliation have been incorporated into the peacebuilding and statebuilding toolbox. Justice, peace, and democracy are seen as mutually reinforcing objectives (se Murphy 2017). Transitional justice entails the set of mechanisms, practices, and issues that seek to confront and deal with past violations of human rights and humanitarian law in order to ensure accountability, bring about justice, and achieve reconciliation (Teitel 2002). While during regime change and postsocialist transitions the role of international assistance for transitional justice is limited, in the case of postconflict situations the scope and nature of international engagement is much broader and more complex. Transitional justice mechanisms include individual prosecutions; international, domestic, and hybrid courts; truth commissions; and vetting and dismissal mechanisms. By aiming to fulfill the transitional justice agenda, the UN and other regional organizations have taken on the responsibility for strengthening domestic law enforcement and judicial institutions, establishing human rights review bodies, and establishing tribunals, truth and reconciliation mechanisms, and victim reparation programs (see UN Secretary-General 2004). Yet the relationship between peacebuilding and justice and the establishment of postconflict social justice remain problematic. This problem revolves around the definition of justice itself and the argument either that justice needs to be incorporated into any self- sustaining peace or that social justice may have to be secondary in the short to medium term to the creation of negative peace. Justice has often remained subservient to stability and a limited notion of peace because so many individuals and organizations in conflict environments are implicated in violence, corruption, or crimes against humanity.

In this context, as Catherine Turner (chapter 34, "Transitional Justice and Peacebuilding") shows, transitional justice processes in conflict-affected societies are received with mixed reactions as a result of the contested track record of international, hybrid, and national mechanisms for bringing justice to victims and contributing to conflict prevention in the future. The international community considers the arrest, prosecution, and trial of war crimes suspects to be a normative necessity to serve justice, truth, and reconciliation. However, the lack of commitment, resources, and capacities, alongside the prioritization of stability over justice, has effectively undermined the prospects for offering a measure of justice to victims' families and survivors of atrocities and gross human rights abuses. In particular, international-led and top-down approaches to dealing with the past and transitional justice are widely criticized for their limited ability to include local perspectives and represent the interests of all affected communities, thus failing to secure local legitimacy and wide public support. The superficial engagement of local actors and limited participation of affected communities seek to legitimize a process with its visibility but delivers little if anything to the families of victims and war survivors. Bottom-up approaches are rarely used, although they would be more appropriate, as they place the affected communities at the very center of the transitional justice process and recognize their right to be consulted (Shaw and Waldorf 2010). A bottom-up and citizen-driven process of dealing with the past can be complex and, most important, depends on the political will of government and society to deal with the past. In many conflict-affected societies political hostilities and the readiness of society to deal with the past tend to determine the viability of bottom-up transitional justice processes.

One of the ultimate goals of peacebuilding is to achieve reconciliation among the groups in conflict. Gráinne Kelly (chapter 35, "Reconciliation and Peacebuilding") examines the pluriversal nature of reconciliation. Conceptual debates tend to consider reconciliation as a process, an outcome, an aspirational end-state (see Lederach 1997). As such it must result in mutual recognition, coexistence, mutual trust, forgiveness, and equality. Yet achieving this aspirational goal is hindered by predatory elites who tend to use conflicting ethnic and identity cleavages as pretext to hold on to their power and prolong ethnic antagonism. Moreover, genuine efforts at reconciliation are hindered by negative forms of peace that rest the understanding of reconciliation and peace upon the bureaucratic and economic rationale of state governmentality rather than on rights, equality, and justice as local community demands. Existing efforts at reconciliation have failed to recognize that sustainable peace rests as much on a social, grassroots, and midlevel process of accommodation and reconciliation as on the technology of the state with the support and normative preferences of international organizations or donors. The state continues to rest on elite politics (which are often predatory) rather than build a local social contract, democratization, and reconciliation or reconstruct the regional pluralism that existed before the violent conflict (Richmond 2014). In turn, dominant approaches have a pragmatic understanding of reconciliation as an ongoing process without an idealistic destination. This entails defining reconciliation as a long-term and generational process, which should be altered and designed in accordance with evolving circumstances and the nature of intergroup hostilities, in hopes that legacies of the past will be addressed and the societies will move forward in one way or another (see Llewellyn and Philpott 2014). While this realistic take on reconciliation tends to be grounded on rights, needs, and local context, processual views of reconciliation risk being instrumentalized for political reasons either to prolong peacebuilding interventions or offer enduring agency to predatory and ethnonationalist and ethnoreligious elites. As John D. Brewer (chapter 36, "Religion and Peacebuilding") shows, the postconflict struggle is not only over the legitimacy of the state and access to its political, economic, and cultural resources, but religious affiliation defines the boundaries of the groups that are in competition (see Appleby, Little, and Omer 2015). Engaging with root causes of conflict and promoting reconciliation became more marginal as bureaucratic and stability-oriented approaches to peacebuilding emerged.

The connection of peacebuilding and statebuilding with liberal institutions, neoliberalism, and individualism rather than context, culture, or needs and welfare ignores the experience of postwar reconstruction and the development of the Western liberal state. The type of security produced by this approach tends to be of an international, regional, and state nature rather than grounded in local experiences or needs. It bypasses issues such as reconciliation and has a rather ambiguous relationship with transitional justice. Contemporary statebuilding is directly connected to neoliberalism, which has direct implications for development, peace, and social equality. Rachel M. Gisselquist (chapter 37, "Foreign Aid and Peacebuilding") looks at the role of aid in conflict-affected societies and the ways aid may support peacebuilding. Gisselquist argues that foreign aid can directly influence peacebuilding processes. First and foremost, by addressing the primary socioeconomic concerns of local communities—poverty, infrastructure, healthcare, and education—it enables broader structural political and societal changes. Foreign aid can expand the opportunities for economic development and foreign direct investment, thus connecting the national economy with global markets (del Castillo 2008). Moreover, foreign aid and its linkage to normative conditionality can directly impact the dynamics of national peacebuilding processes (see Mac Ginty and Peterson 2015). Yet despite the importance of foreign aid, the focus of statebuilding on neoliberal modes of governance to promote self-help appears to lead to widening inequality, mirrored in a circumvention of the institutions of the liberal state by a range of elites and a distancing (or rejection) of the state by society in general. The evidence so far suggests that neoliberal states are ill-equipped to deal with conflict and make peace because their institutions are too weak to prevent competition from spilling over into violence or to redress historical inequality (Duffield 2007). Neoliberal states do not offer comprehensive responses to economic needs and fail to provide public services quickly enough to undercut currents of violence or address the root causes of conflict (see Chandler and Reid 2016).

Across different postconflict settings, the World Bank and International Monetary Fund have focused on economic development, though they have often been accused of weakening the social capacity of the state and the welfare of the population by favoring growth over equality. These dynamics are exaggerated in postconflict and development settings, where inequality is statistically at its highest and self-help is least plausible. This is because of sociohistorical practices, communal forms of social organization, a lack of property rights, and weak state capacities to protect property, provide services, or provide law and order. Thus as inequality is a root of some dimensions of conflict—especially related to power, material resources, and identity—the state and peace being built are already failed by design (Richmond 2015; Visoka, Doyle, and Newman 2019). In turn, inequality exacerbates political competition over the distribution of scarce resources in any society and is particularly problematic in postconflict societies, where inequalities are already acute and weapons are widely available. Peacebuilding and statebuilding explicitly cannot offer a sustainable peace unless both local and global inequalities are on the peace agenda because of their causal role in violence and war. Development and modernization need to be determined locally, in a context of international acceptance of the value of localized or customary systems, different speeds of development and a commitment to mitigate inequality where there is a need for rapid development, and a locally determined peace dividend (Richmond 2015). For instance, Florian Krampe and Ashok Swain (chapter 39, "Environmental Peacebuilding") show that states recovering from violent conflicts might not be genuinely interested in pursuing sustainable development policies rooted in environmental and social justice rather than focus on building a neoliberal state.

Contemporary statebuilding interventions continue to reflect the social and ideological predilections of dominant interveners, whether the United Nations, World Bank, major donors, or minor donors and other agencies (see Philpott and Powers 2010). This has an impact on the generation of local legitimacy. The states being built reflect neither local, sociohistorical frameworks for legitimacy nor identity, and so command little loyalty. However, in practice, statebuilding and peacebuilding interventions have based their legitimacy not on the consent, compliance, and support of local subjects (which peace formation activities would signal, authorize, and legitimate) but on external authority granted by the UN Security Council and reproduced through the technologies of power, donations, and norm and knowledge transfers. In peacebuilding and statebuilding debates, securing local legitimacy is considered crucial for the success of peacebuilding interventions. Aware that the legitimacy of UN peacebuilding operations may erode over time, sustaining the local legitimacy of international peace interventions has emerged as a political necessity to control local expectations, reduce local resentment, and above all save liberal peacebuilding and prolong foreign rule over conflict-affected societies (Richmond and Mac Ginty 2020). Over time, across many postconflict societies, it became obvious that external legitimacy, legal validity, and acceptance among great powers and international bodies are not sufficient for the success of peacebuilding interventions. In the peacebuilding literature, as Timothy Donais (chapter 38, "Local Ownership, Legitimacy, and Peacebuilding") shows, local legitimacy has come to signify a wide range of everyday and institutional attributes that signify the subject's acceptance of and support for local or international authorities based on individual or collective norms and values. Yet statebuilding has failed to recognize that state and society are implicated in each other. Multiple forms of legitimacy are required for statebuilding to contribute to an emancipatory form of peace. Statebuilding is at best ambiguous because it is embedded in conflict between international norms, interests, neoliberalism, and local legitimacy. Local responses are also ambiguous, but autonomy, democracy, rights, local ownership, and the need to avoid harm indicate that they should be prioritized in any statebuilding process. In reality, legitimacy emanating from local, state, and international sites tends not to be aligned in the way that liberal peace thinking imagines. Neither does it align with geopolitical assumptions about the centrality of state power in supporting legitimate authorities. This points to the need to think about fluid and networked forms of legitimacy, agonistically misaligned, which operate from multiple sites of local authority connecting to state agencies and institutions, and onward to the institutions of global governance despite normative and practical differences (again pointing to the lacuna of peace formation).

Promoting local ownership over political processes is increasingly viewed as a means of providing internal legitimacy for international peacebuilding and statebuilding interventions (von Billerbeck 2016). In principle, local ownership is a vital framework for developing an emancipatory form of peace. It predicates local, contextual, cultural, and democratic choices about rights, needs, and institutions, as well as international norms and standards. It also promises to represent contractual responsibility and accountability, negotiated between "internationals" and the "local," and contextualized relationships between them. However, in practice, local ownership and participation have become buzzwords in peacebuilding and statebuilding interventions, partly used to legitimate the role of internationals in postconflict settings and enable local agency to develop in a liberal setting. Local ownership, which resembles colonial indirect rule, seeks to make interventions more acceptable to local subjects. Local ownership falsely indicates that local actors have an influence over the statebuilding process, as well as indicating a contractual relationship between providers and recipients. As the internationals prescribe what ownership means, they disregard local agency, norms, and preferences. International actors are meant to persuade local actors that the liberal peace, developed by a neoliberal state, is what they want and should own. Following this logic, local ownership is impossible unless local actors define what it is they want to "own" and how they develop it through their relationship with international actors. This would require subaltern actors in conflict-affected societies finding a way of overcoming power imbalances that silence them.

Existing approaches to peacebuilding and statebuilding tend to have common characteristics and to suffer from a range of destabilizing unintended consequences (Visoka 2016). First, they prefer the exercise of power or a fudge over sustainable and emancipatory solutions to conflicts reflecting subaltern political claims for emancipation from the roots and dynamics of violence. Second, a limited power-sharing framework tends to be developed more or less always within the confines of modern statehood that empower predatory elites. As a result, they often weaken the need for pluralism with respect to identity problems, preferring instead to use ethnopolitical and territorial meanings of citizenry. Finally, they depend on very limited resources to provide security, development, and rights, preferring efficiency over justice, reconciliation, and contextual needs. In particular, they avoid discussions of justice in historical and contemporary, local and global settings, especially vis-à-vis material aspects of conflict, meaning the issues of the dispute are not addressed. Ultimately, most of the states that have been the subject of developmentalism in modernization form, peacekeeping, mediation, peacebuilding, or statebuilding, have since experienced authoritarian and oligarchical rule, have failed to distribute a material peace dividend even if security is improved, and have frequently experienced renewals of violence and social unrest. Thus what is missing from debates on intervention, peacebuilding, and statebuilding is an understanding of how people and communities act to make peace themselves. Statebuilding itself is unstable, and its limitations from local perspectives—both elite and societal—have necessitated international involvement over and over again. Progressive forms of peace cannot be achieved unless the international architecture itself is progressive across its full spectrum, including in the areas of the environment, rights, material needs, institutions, justice, and law.

Postliberal Peace and Peace Formation

Part IV turns to grassroots and bottom-up peace formation approaches and their implications, pointing to everyday dynamics and the production of hybrid political orders and hybrid peace. Mainstream approaches aimed at ending conflict and making peace through liberal peacebuilding and neoliberal statebuilding have reached an impasse. Both statebuilding and liberal peacebuilding strategies have failed to connect with their local populations in conflict-affected states (Richmond 2011). Failed statebuilding has been the result of externalized states being based on blueprints determined by decontextualized and depoliticized agendas for states that provide security, political rights and institutions, and market access, but little in the way of political rights (Richmond 2014). They end up buttressing problematic elites and their often chauvinistic, nationalistic, or personal interests, and lack a connection in context, on the ground, among populations that have their own understandings of identity, sovereignty, institutions, rights, law, and needs according to their own sociohistorical and cultural traditions and context. They have produced negative forms of peace rather than reconciliation, rights, equality, and justice, as examined by SungYong Lee (chapter 41, "Local Resistance and Hybrid Peace"). Negative hybrid peace entails situations where the root causes of conflict have not been addressed, though the institutional framework of the state exists and is mainly of benefit to elites or existing power structures. Moreover, a hybrid form of state exists that incorporates the components of the liberal and neoliberal peace at a level perhaps not appreciated by international actors but very strongly modified by its local context, social and power configurations, and cultural norms (see Richmond and Mitchell 2012).

States have come into being as a result of a mixture of local and international dynamics and intervention and are effectively failed by design. Such states lack core capacity in many crucial areas, partly because externally driven statebuilding processes dictate they do not play a role in social welfare and redistribution of material resources, or because the imposed standards and norms are ill-suited to specific contexts. Attempts to create a sustainable peace have proceeded without the involvement of the subjects of that peace in its many theaters around the world. Thus part IV looks at alternative ways of thinking about peace processes both in theory and in practice (drawing on a tradition implied in the work of such conflict scholars as John Burton, Elise Boulding, and John Paul Lederach). Without incorporating a better understanding of the multiple and often critical agencies involved in peace formation, the states emerging from statebuilding will remain as they are: failed by design. This is because they are founded on externalized systems, legitimacy, and norms rather than a contextual, critical, and emancipatory epistemology of peace. Engaging with the processes of how peace emerges may aid international actors in gaining a better understanding of the roots of a conflict, how local actors may be assisted, how violence and power-seeking may be ended or managed, and how local legitimacy may emerge, as well as playing a role in revising international order. By looking at the smaller scale and often invisible local attempts related to peace formation, some answers may emerge to the pressing question of how large-scale peacebuilding or statebuilding may be significantly improved and made more representative of the lives, needs, rights, and ambitions of its subjects.

Oliver Richmond (chapter 40, "Peace Formation and the Reshaping of International Peacebuilding") offers an intellectual résumé of peace formation by looking at its promise and paradoxes. Peace formation implies a reconstruction of political community, the state, and international organizations from the bottom up if they are to be representative, democratic, and responsive to the situations of their subjects in local, state, regional, and global contexts (see also Richmond 2016). Local processes of peace formation are often in partial opposition to statebuilding and liberal peace processes, as well as to violent forms of state formation. Peace formation maintains that any peace process and the innovations it has required emerged not merely from better legal, security, state, or international design but from its emancipatory response to everyday political claims of ordinary people made in conflict-affected societies (subaltern political claims, often aimed at basic security and resources, services, and of course human rights). Peace formation processes are driven partly by local, peaceful forms of agency (though often critical and resistant). Peace formation arises through subaltern power/agency in sectors of society where nonviolent, peaceful change is sought, but it also has implications for the evolution of the international peace architecture. It aims to negate local violence and preserve and recondition local identity and political institutions, as demonstrated in Volker Boege's contribution (chapter 42, "Hybrid Political Orders and Customary Peace"). It engages with, and influences, international actors, from donors to the United Nations. It sees political subjects as formative of the state, economy, society, and the international community. Peace formation is limited in scale and scope but is often well versed in the historical, political, economic, social, and cultural dynamics required in any stable and peaceful governance system in context. In other words, it comprehends the dynamics of legitimate political authority necessary to make a sustainable peace in ways external actors cannot, rendering them reliant on extremely problematic blueprints that lack local legitimacy and any real capacity. The peace formation project is also fragile, however, subsisting only because of great ingenuity and local, transnational, and transversal support networks. It offers a clear advance over older conflict management, resolution, transformation, and peacebuilding systems and may well provide internationals with an avenue to enable peace. However, it may also highlight norms, values, and interests that are unacceptable at the international level or seek to reconcile strongly opposed positions from all these quarters.

The concept of peace formation differs from previous work on local peace activities, however, as it offers a preliminary and alternative explanation for the formation of the state from below (rather than via elite-driven violence) and extends the notion that the international peace architecture has emerged from various transnational, transversal, and local peace and social movements. Peace formation does not expect that mobilization, agency, and the institutions that emerge to make and maintain peace are shaped mainly by the knowledge and power of external actors. Instead peace formation arises from within a sociopolitical context, learning from many sources, across transnational, transversal, and international scales. If it is to be widely influential, peace formation needs a physical infrastructure that can provide ad hoc or organized opportunities for meetings, negotiation, mediation, and the dissemination of information. Recently there has been a widespread attempt across postconflict sites around the world to bring localized peace processes into the formal, externally driven, statebuilding and peacebuilding processes and to promote the formation of localized infrastructures for peace. This contributes to a multilevel and multidimensional architecture of peace that includes societies and their own political processes and modes of governance, and this may lead to more localized peace processes. As Andries Odendaal (chapter 43, "Local Infrastructures for Peace") demonstrates, peace networks and informal processes often emerge around problems related to local violence, early warning, everyday needs and services, and political contests. It is thought that they negotiate, mediate, and monitor local or even state-level peace agreements and become actively involved in questions of state reform. They may operate in informal and formal state spheres, aiming to produce a range of collaborative peace processes appropriate to context and culture.

Although they may have only a limited impact, infrastructures for peace offer a high level of local ownership and so a potential basis for the local emergence of formal legitimate authority. Peace infrastructures appear to connect formal with informal sectors, local, state, and often international, through systems of pluralist dialogue that slowly turn into institutions. They create a nexus between the community of peace actors in local contexts, parts of the state, and some internationals that are more concerned about sustainable peace than isolated notions of security or development inherent in negative versions of peace. However, ambiguities pertain as to who owns them and where they actually come from and whether their role is the fundamental reform of the state or merely to persuade society to accept the state as it is. By moving into the public space, peace infrastructures risk being co-opted, either by traditional and predatory centers of power or by internationals intent on their own ideological understandings of peacebuilding and statebuilding. One major problem with peace formation is how easily extinguishable it is and how easily it may drop from view (though sometimes withdrawal is strategic given the concentration of power at state and international levels). Peace formation may risk being stifled by clientelism and neopatrimonialism, which is often not in favor of reform.

Peace formation operates in local contexts and hidden cultural, social, economic, and political spaces and increasingly influences the state and international institutions. These peace formation processes constitute a widening pool of political subjects (see Visoka 2017). They maintain a memory of historical peace practices and institutions related to custom, culture, identity, religion, or Western norms of human rights and representation. They seek to influence and hold government to account, seek international support to modernize and expand their practices, and set an example to wider society. They raise expectations relating to the need for a positive and emancipatory form of peace, slowly insert contextual modes of politics into institution-building processes and legal and constitutional frameworks, and influence donors and other international actors. Peace formation agency is one of several foundations of a legitimate and emancipatory peace, as discussed in Gëzim Visoka's contribution (chapter 44, "Emancipatory Peace"). Peace formation constructs emancipatory projects in contextual scales, reconciling demands for self-determination and autonomy. It also does so in local languages, through translation, while respecting identity. It points to broad questions of justice, resource distribution, and the need for empathy and localized as well as transnational legitimacy. It may be the basis for legitimacy of any polity aimed at an emancipatory and positive hybrid peace. While peace formation offers a progressive rethinking of peace, rights, class, gender, and resource distribution, the state (as well as the international political economy) often is controlled by elites who block progress, indirectly utilizing the resources that internationals provide them with. Despite the potential to produce a positive hybrid peace, in practice a negative hybrid peace emerges from peace formation because of its inevitable encounter with the various forms of power.

References

Acharya, A., and B. Buzan. 2010. *Non-Western International Relations Theory: Perspectives on and beyond Asia*. London: Routledge.

Appleby, R. S., D. Little, and A. Omer, eds. 2015. *The Oxford Handbook of Religion, Conflict, and Peacebuilding*. Oxford: Oxford University Press.

Bailliet, C. M., and K. M. Larsen, eds. 2015. *Promoting Peace through International Law*. New York: Oxford University Press.

Bellamy, A. J., and T. Dunne. 2016. *The Oxford Handbook of the Responsibility to Protect*. Oxford: Oxford University Press.

Björkdahl, A., and S. Buckley-Zistel, eds. 2016. *Spatializing Peace and Conflict: Mapping the Production of Places, Sites and Scales of Violence*. London: Palgrave Macmillan.
Chandler, D., and J. Reid. 2016. *The Neoliberal Subject: Resilience, Adaptation and Vulnerability*. Lanham, MD: Rowman & Littlefield.
Chandler, D., and T. D. Sisk, eds. 2013. *The Routledge Handbook of International Statebuilding*. London: Routledge.
Davies, S., and J. True, eds. 2019. *The Oxford Handbook of Women, Peace, and Security*. New York: Oxford University Press.
del Castillo, G. 2008. *Rebuilding War-Torn States: The Challenge of Post-Conflict Economic Reconstruction*. New York: Oxford University Press.
Doyle, M., and N. Sambanis. 2006. *Making War and Building Peace*, Princeton, NJ: Princeton University Press.
Duffield, M. 2007. *Development, Security and Unending War*. London: Polity.
Fukuyama, F. 2004. *State-Building: Governance and World Order in the Twenty-First Century*. London: Profile Books.
Goetze, C. 2017. *The Distinction of Peace: A Social Analysis of Peacebuilding*. Ann Arbor: University of Michigan Press.
Jabri, V. 2007. *War and the Transformation of Global Politics*. London: Palgrave Macmillan.
Jahn, B. 2013. *Liberal Internationalism: Theory, History, Practice*. London: Palgrave Macmillan.
Lederach, J. P. 1997. *Building Peace: Sustainable Reconciliation in Divided Societies*. Tokyo: United Nations University Press.
Lemay-Hébert, N., ed. 2019. *Handbook on Intervention and Statebuilding*. London: Edward Elgar.
Llewellyn, J. L., and D. Philpott. 2014. *Restorative Justice, Reconciliation, and Peacebuilding*. New York: Oxford University Press.
Mac Ginty, R., ed. 2013. *The Routledge Handbook of Peacebuilding*. London: Routledge.
Mac Ginty, R., and J. H. Peterson, eds. 2015. *The Routledge Handbook to Humanitarian Action*. London: Routledge.
Murphy, C. 2017. *The Conceptual Foundations of Transitional Justice*. Cambridge: Cambridge University Press.
Paris, R. 2004. *At War's End: Building Peace after Violent Conflict*. Cambridge: Cambridge University Press.
Philpott, D., and G. F. Powers, eds. 2010. *Strategies of Peace: Transforming Conflict in a Violent World*. New York: Oxford University Press.
Richmond, O. P. 2002. *Maintaining Order, Making Peace*. London: Palgrave Macmillan.
Richmond, O. P. 2005. *The Transformation of Peace*. London: Palgrave Macmillan.
Richmond, O. P., ed. 2010. *Palgrave Advances in Peacebuilding: Critical Developments and Approaches*. Basingstoke: Palgrave Macmillan.
Richmond, O. P. 2011. *A Post Liberal Peace*. London: Routledge.
Richmond, O. P. 2014. "Jekyll or Hyde: What Is Statebuilding Creating? Evidence from the 'Field.'" *Cambridge Review of International Affairs* 27 (1):1–20.
Richmond, O. P. 2015. "The Impact of Socio-Economic Inequality on Peacebuilding and Statebuilding." *Civil Wars* 16 (4):449–467.
Richmond, O. P. 2016. *Peace Formation and Political Order in Conflict Affected Societies*. New York: Oxford University Press.
Richmond, O. P. 2020. *Peace in International Relations*. 2nd ed. London: Routledge.

Richmond, O. P., and R. Mac Ginty, eds. 2020. *Local Legitimacy and International Peacebuilding*. Edinburgh: Edinburgh University Press.
Richmond, O. P., and A. Mitchell, eds. 2012. *Hybrid Forms of Peace: From Everyday Agency to Post-Liberalism*. London: Palgrave Macmillan.
Richmond, O. P., S. Pogodda, and J. Ramović, eds. 2016. *The Palgrave Handbook of Disciplinary and Regional Approaches to Peace*. London: Palgrave Macmillan.
Shaw, R., and L. Waldorf. 2010. *Localizing Transitional Justice: Interventions and Priorities after Mass Violence*. Sandford, CA: Sandford University Press.
Teitel, R. G. 2002. *Transitional Justice*. Oxford: Oxford University Press.
UN Secretary-General. 2004. *The Rule of Law and Transitional Justice in Conflict and Post-Conflict Societies*, UN Doc. S/2004/616, 23 August.
UN Secretary-General. 2018. *Peacebuilding and Sustaining Peace*. UN Doc. A/72/707-S/2018/43, 18 January.
Visoka, G. 2016. *Peace Figuration after International Intervention: Intentions, Events and Consequences of Liberal Peacebuilding*. London: Routledge.
Visoka, G. 2017. *Shaping Peace in Kosovo: Politics of Peacebuilding and Statehood*. London: Palgrave Macmillan.
Visoka, G. 2019. "Critique and Alternativity in International Relations." *International Studies Review* 21 (4):678–704.
Visoka, G., J. Doyle, and E. Newman, eds. 2019. *Routledge Handbook of State Recognition*. London: Routledge.
Visoka, G., and V. Musliu, eds. 2019. *Unravelling Liberal Interventionism: Local Critiques of Statebuilding in Kosovo*. London: Routledge.
von Billerbeck, S. 2016. *Whose Peace? Local Ownership and United Nations Peacekeeping*. Oxford: Oxford University Press.
Wallensteen, P. 2015. *Quality Peace: Strategic Peacebuilding and World Order*. New York: Oxford University Press.
Weiss, T. G., and S. Daws. 2008. *The Oxford Handbook on the United Nations*. Oxford: Oxford University Press.
Williams, A. 2006. *Liberalism and War*. London: Routledge.

PART I

DISCIPLINARY PERSPECTIVES

CHAPTER 2

LIBERAL INTERNATIONALISM

BEATE JAHN

Peacebuilding came to be closely associated with liberal internationalism following the end of the Cold War. The demise of the Soviet Union left behind a world dominated by powerful liberal states. For many observers, this "triumph" of liberalism seemed to indicate its alignment with the progressive forces of history (Ikenberry 2006, 161, 158, 146). In contrast to other political regimes, Western liberal democracy seemed capable of resolving all "fundamental contradictions" within society and thus constituted, as Francis Fukuyama (1989, 8, 4) famously argued, "the final form of human government."

This basic assumption was widely shared, if often less flamboyantly, by academics, politicians, and publics and gave rise to profound optimism. If liberal principles, practices, and institutions were indeed capable of resolving fundamental contradictions within society, then their worldwide realization offered the solution to problems ranging from poverty to civil wars (Sørensen 2000, 287). Moreover, following the demise of the Soviet Union, liberal actors now had the power and the opportunity to spread these liberal principles to other parts of the world. The 1990s thus witnessed a wave of political, economic, and military policies designed to promote democracy, protect human rights, liberalize markets, and consolidate peace at the end of civil conflict (Paris 2004, 38; Duffield 2001; Jahn 2007; Zaum 2012). Today, however, much of this early optimism is gone (Rampton and Nadarajah 2017, 444; Richmond 2018; Paris 2010). Instead we are confronted with a curious paradox: despite the fact that peacebuilding operations largely failed to achieve their goals, peacebuilding continues unabated and has become an integral element of the international order.

Analyzing these developments, this chapter will first establish the core tenets of the liberal approach to peacebuilding. The second part provides an account of peacebuilding operations in practice. It traces a cycle of failures and reforms that ultimately led to the current paradox: the successful integration of peacebuilding into the fabric of the world order despite its continuing failures. The third section offers an alternative conception of liberalism and shows that the main function of peacebuilding lies in managing the tensions and contradictions inherent in a liberal world order. This analysis suggests, in conclusion, that, in one form or another, peacebuilding policies will continue to play an important role as long as the liberal world order exists.

The Core Tenets of Liberal Peacebuilding

Following the end of the Cold War, peacebuilding operations were intimately associated with liberal foreign and international policies. They were based on a liberal philosophy of history designed to establish liberal principles, practices, and institutions in postconflict societies and pursued by a wide range of liberal actors. The liberal philosophy of history posits a natural and progressive development of humankind and its social and political relations. Within this framework, the power of liberalism after the end of the Cold War was interpreted as evidence for "mankind's ideological evolution" from absolutism through Bolshevism and fascism to liberal democracy (Fukuyama 1989, 3). The economic power of liberalism seemed to prove that capitalism was indeed the most efficient way of creating prosperity, and widespread political support seemed to confirm the validity of the core liberal value of human freedom (Plattner 2008, 29; Jahn 2012).

Conversely, this philosophy of history offered an explanation for continuing conflicts. If liberal principles, practices, and institutions generated domestic and international peace, prosperity, and respect for human rights, then their absence was responsible for conflicts (Fukuyama 1989, 8; Moravcsik 1997, 529, 530, 532). The existence of poverty, autocracy, failing states, and conflict indicated that "the vast bulk of the Third World" was still stuck at earlier stages of development (Fukuyama 1989, 3, 15). On the basis of these assumptions, therefore, consolidating peace required nothing else but the introduction of liberalism in postconflict societies.

Liberalism, in turn, was specifically associated with three core institutions: a market economy, democracy, and human rights. The market economy provided the material foundations for peace because it generated economic growth and prosperity and hence the material satisfaction of the population. Poverty, in contrast, produced social unrest and conflicts over scarce resources. Free-market economies, moreover, relied on free trade and therefore had no interest in international wars disrupting economic exchange. Poor countries, in contrast, had little to lose and much to gain from fights over territory and resources. Poverty produced mass migration, drug trafficking, disease, and even terrorism, thus undermining a stable international order. In short, state failure—and hence the conflicts arising from it—was widely attributed to economic failure (Sachs 2001, 187; David 1992–1993, 135–136). The consolidation of peace in conflict-ridden societies thus required economic development, and this, as the model of liberal states demonstrated, was best pursued through free markets and free trade.

Politically, liberalism was identified with democracy. By providing all individuals and groups in society with peaceful means to express their interests, democracy was seen to foster the development of a social contract on which lasting peace could be built (Boutros-Ghali 1996). It thus prevented civil wars and ethnic cleansing in the domestic sphere as well as international aggression (Diamond 1995, 6–7; Doyle 1996). Dictatorship, in contrast, generated political resistance among the population. Moreover, in nondemocratic societies elites did not have to heed the will of the population and could use wars to distract from domestic problems (David 1992–1993, 131–134, 138–140). The introduction of democracy thus promised to provide the population in postconflict situations with peaceful means to express and resolve their differences.

The third institution with a crucial role in establishing peace was the set of liberal principles of individual and human rights. The recognition of civil and political rights in liberal states ensured respect for all individuals, and the rule of law provided a peaceful avenue for addressing grievances and instilling respect for all human beings (Boutros-Ghali 1996). In contrast, conflicts in the Global South were often fueled by traditional cultures based on religious, ethnic, and political hatred vilifying and dehumanizing enemies and thus paving the way for brutality, violence, and war (David 1992–1993, 136–138, 150; Haass 2003, 145). The consolidation of peace thus required the codification of individual and human rights in constitutions and respect for the rule of law.

Moreover, democracy, capitalism, and human rights were seen to be mutually constitutive: economic development was expected to fuel democratization, and democracy required civil and political rights and the rule of law, which also played a crucial role in the protection of private property and free markets. Peace, development, and democracy were thus inextricably linked (Boutros Ghali 1996). Taking liberal states as a model, therefore, peacebuilding operations during the 1990s "all pursued the same general strategy": "democratization and marketization" (Paris 2004, 19). Based on the liberal philosophy of history, peacebuilding operations were designed as short-term interventions, simply kick-starting the natural development toward liberalism.

Peacebuilding operations thus entailed the introduction of democratic elections, marketization programs, and constitutional and judicial reforms codifying civil and political rights. In order for these new institutions to work, police and judicial officers had to be retrained, markets deregulated, and the development of civil society associations and political parties supported (Paris 2004, 19). In practice, peacebuilding operations are therefore closely associated with peacekeeping, statebuilding, democracy promotion, humanitarian intervention, and development policies.

These policies were pursued by a wide range of liberal actors. They included the development agencies of leading liberal states that made aid conditional on democratization. The International Monetary Fund and the World Bank insisted on the liberalization of trade, the deregulation of markets, and the privatization of state-owned industries in return for assistance. International organizations in the field of peacekeeping—the Organization for Security and Cooperation in Europe, the European Union, the North Atlantic Treaty Organization, the Organization of American States—were themselves largely constituted by liberal states. But even organizations with a more pluralist membership like the United Nations and its specialist organizations (the General Assembly, the UN Commission on Human Rights, and the UN Development Program) subscribed in the climate of the early 1990s to these liberal principles. And though some NGOs criticized the way these principles were implemented in target states, they nevertheless tended to agree with the liberal political and economic goals of peacekeeping operations (Paris 2004, 22–35).

Peacebuilding, in sum, was intimately associated with liberalism because it was based on a liberal worldview, designed to establish liberal institutions in postconflict societies, and pursued by liberal actors.

Peacebuilding in Practice

In practice, peacebuilding policies can broadly be divided into three phases. In each case, the failure of peacebuilding operations led to analysis and reform. Yet despite the fact that none of these reforms enabled peacebuilding operations to achieve their stated goal—to prevent the recurrence of fighting and to consolidate peace—they have nevertheless become an integral feature of the international order. Based on the liberal assumptions set out in the first section of this chapter, the first generation of peacebuilding operations in the 1990s focused on transforming state institutions—but largely failed to achieve their aim. Instead, as Roland Paris's analysis shows, one can distinguish among three different outcomes. In the first group of cases, comprising Angola, Rwanda, Cambodia, Liberia, and Bosnia, the rapid introduction of democratic elections either triggered renewed fighting or exacerbated existing conflicts. In the second, more successful group of cases—Croatia, Namibia, and Mozambique—the end of conflict was precipitated not by the introduction of liberal institutions but by the withdrawal of external parties (Serbia and South Africa) to the conflict. And in a third group of states—Mozambique, Guatemala, Nicaragua, El Salvador—the imposition of economic liberalization led to increased poverty and inequality, which in turn fed an upsurge of criminal violence and thus reproduced the very causes of the original conflict (Paris 2004, 63–147).

Observers attributed these failures to the lack of preconditions for a functioning market democracy in target states. The introduction of elections had not led to a self-perpetuating democracy because the target countries lacked state capacity, rational economic structures, a stable social and political order, the rule of law, accountability, political parties, and a vigorous civil society (Diamond 1996, 33). Similarly, the World Bank argued that its economic liberalization program failed because of the lack of good governance, corruption, and the barriers presented by traditional culture (Williams and Young 1994, 99). Peacebuilders, in short, had overlooked that conflict-ridden countries were in fact much less developed than anticipated (Paris 1997, 2010; Barnett 2006).

Target countries therefore needed a wholesale reconstruction of all aspects of society. Peacebuilders had to provide security, build up administrative structures, constitute legitimate governments, mobilize civil society, build state capacity, provide justice and reconciliation, fight corruption, shift political culture—essentially produce an entirely new population with different ideas and habits (Williams and Young 1994, 96; Diamond

1996; Jahn 2007; Schmidt 2013). In addition, it was now suggested that peacebuilding measures had to be sequenced: economic development and capacity building had to be achieved before political freedom could be introduced. Peacebuilders, in other words, had to take on government functions and "suppress certain forms of political expression" in target countries (Paris 2004, 209). This encompassing program of transformations also required a much longer term commitment than anticipated for the first generation of peacebuilding (Carothers 2002, 16; Paris 2004, 177). In short, peacebuilding had become statebuilding (Richmond 2018, 224).

Elements of this new approach were implemented in Kosovo, East Timor, and Sierra Leone. However, a long-term commitment to state- and nationbuilding abroad is costly and difficult to sell to democratic publics in liberal states. Moreover, following the terrorist attacks in September 2001 the focus of liberal states shifted to national security, and resources were committed to the war in Afghanistan and later in Iraq. In both cases, the war initially focused on the removal of the incumbent regime, which—reverting to the first generation of peacebuilding—was expected to generate automatic development toward market democracy and hence also peace (Packer 2006). Postconflict reconstruction began only when the removal of the Taliban and Saddam Hussein led to insurgency and civil war.

Yet again these efforts failed. Large-scale statebuilding undertaken by external forces was reminiscent of colonialism and generated local resistance as well as an authoritarian peace following hegemonic interests (Mac Ginty 2011; Chandler 2010; Richmond 2018). Meanwhile in Afghanistan foreign troops as well as myriad peacebuilding actors—from NGOs through international organizations to state development agencies—have now been operating for seventeen years. Yet even this long-term commitment never managed to put an end to insurgency and civil war, failed to establish a sustainable market democracy, and steadily increased the influence of the Taliban (Rubin 2006), which now controls substantial parts of the country. Similarly the intervention in Iraq led to insurgency, increased crime and insecurity, and political instability (Nash 2006)—not only in the domestic sphere but for the entire region.

Analyses of these failures of second-generation peacebuilding focused on two issues. First, observers criticized the "one size fits all" approach of imposing the Western model of market democracy onto radically different local contexts (Mac Ginty 2011; Richmond 2011). Second, these top-down approaches dramatized a major contradiction between the aim to spread liberal norms of political freedom and self-determination and a peacebuilding practice that required the denial of political freedom to local actors (Jahn 2007, 223; Paris 2004, 209; Diamond 1996, 35). Second-generation peacebuilding was thus reminiscent of colonialism, imperialism, and trusteeship (Paris 2004, 231; Ollapally 1995, 434; Richmond 2018, 222; Cunliffe 2012).

These analyses have given rise to the third generation of peacebuilding, characterized by a "local turn." Top-down approaches were to be replaced with a grassroots approach: local actors were to participate in the design of particular projects and in their implementation. And such local ownership was expected to lead to a "hybrid," "postliberal," or "post-Westphalian peace" (Richmond 2010, 2011; Mac Ginty 2011;

Verkoren and Leeuwen 2013; Jabri 2013; Chesterman 2007; Bliesemann de Guevara 2010; Randazzo 2017; Paffenholz 2015). This language of local ownership, of engagement with local forces, of national authority, of consensus building, of attention to local priorities has been widely adopted.

In practice, however, policies of local ownership provided opportunities for those whose ideas, priorities, initiatives, and capabilities conformed to the goals and procedures of external peacebuilders and marginalized others (Kušić 2018). International organizations justify their interventions by soliciting pro forma local support; NGOs use local engagement to secure continued funding; local elites become responsive to external funders rather than domestic constituents; contributing to peace operations has become a major source of income for many states in the Global South (Cunliffe 2018). Local and international elites, in sum, captured the opportunities provided by the language of local ownership "in ways that least benefit the marginalised" (Richmond 2018, 225).

Instead of leading to a hybrid peace, the third generation of peacebuilding split domestic and international society into winners and losers—integrating the former into a liberal world order and marginalizing the latter. This development accounts, on the one hand, for growing popular resistance to liberal principles and institutions in postconflict and liberal societies alike. On the other hand, it explains the establishment of an impressive architecture of peacebuilding institutions in the current world order: the UN Peacebuilding Commission, the UN Peacebuilding Fund, and the UN Peacebuilding Support Office, in addition to the development agencies of leading states and other international organizations and NGOs. Peacebuilding, in sum, has become integrated into the liberal world because it offers opportunities to some sections of world society despite its failure to establish peace.

Liberalism and the Paradox of Peacebuilding

Each phase of peacebuilding, as we have seen, was beset by a particular problem. The first generation of peacebuilding revealed a gap between theory and history, the second a tension between freedom and oppression, and the third a false distinction between the local and the global. Both the sequence and the substance of these problems have their roots not in the strategic or tactical design of peacebuilding operations but at a much deeper level: in the contradictions of liberalism itself.

The gap between liberal assumptions about history and its actual development has its roots in the very foundations of liberal thought. Core liberal institutions are derived from the claim that human beings are "naturally" born free. Materially, however, this freedom could be realized only on the basis of private property ensuring the individual's independence from others, as Locke argued. And politically this freedom required the individual's consent to government (Locke 1994, 269–294, 336, 351). Alas, these claims were not based on historical evidence (Jahn 2013, 43–44). In Locke's time and throughout much of history, most individuals did not own private property, most governments were not based on consent, and hence most individuals were not free in this liberal sense. Early liberal thought was thus confronted with the same problem peacebuilders of the first generation encountered in target states: their theoretical assumptions did not match the historical realities.

In light of this disjuncture between theory and history, the establishment of liberal polities first required the constitution of their historical preconditions: the spread of private property leading to the production of free individuals who could then exercise their freedom through participation in government. Historically, therefore, the establishment of liberal polities entailed the protection of private property and the expropriation/privatization of common property (Acemoglu and Robinson 2006, 349–350; McNally 1988, 8–9, 62; Perelman 2000, 175). But since the vast majority of the population who depended for their livelihood on common property could not be expected to consent to their own expropriation, it also required their political oppression (Locke 1994, 384). The establishment of liberalism thus entailed economic appropriation and expropriation, political emancipation and oppression.

Current peacebuilding operations replicate these policies precisely: when it turned out that the preconditions for a "natural" development of liberal society were missing, political oppression became necessary until "effective governmental institutions" and the "transition to market democracy" ensured a constructive vote that advances a stable and lasting (liberal) peace (Paris 2004, 206, 191). Second-generation peacebuilding thus suffered from the contradiction of pursuing the liberal norm of freedom through political oppression. This contradiction, however, does not have its roots in the particular shortcomings of postconflict societies. Instead it is an integral feature of liberalism itself—which for much of its history denied political rights to "unreliable" populations: non-property owners, women, slaves, indigenous populations.

These internal contradictions of liberalism—its practice of appropriation and expropriation, political emancipation and oppression—generate deep social and political divisions between the poor and the rich, the politically free and the disenfranchised. Addressing this contradiction, colonialism has played a constitutive role for the theory and practice of liberalism (Locke 1994, 299; Jahn 2013, 49–52). Through the appropriation of other people's land (and labor), political tensions could be exported and economic benefits imported. The latter were then used to address domestic social and political unrest through economic compromises (such as welfare policies), which provided even the poorer sections of the population with a stake in the liberal system—and thereby paved the way for the introduction of democracy (Jahn 2018, 54–55). However, this constitution of liberal market democracies at home simultaneously deprived communities abroad of property and political rights. The internal contradictions of liberalism were thus ameliorated through external expansion—which required a fundamental distinction between a domestic political sphere in which the rule of law applied and an international sphere characterized by power politics.

The local turn in peacebuilding repeats this move, presenting the local/nonliberal and the global/liberal as the result of endogenous developments (Rampton and Nadarajah 2017, 442; Sabaratnam 2013) and thus obscuring that this distinction is itself constitutive of liberalism (Hutchings 2013). The local/nonliberal and the global/liberal are in fact the product of centuries of liberal power politics in the international sphere. And the conflicts to which liberalism supposedly offers solutions are themselves the result of liberal internationalism, as David Rampton and Suthaharan Nadarajah (2017) show for the case of Sri Lanka. Instead of peace, the local turn produces societies deeply divided between those who benefit from privatization and political participation and those who are expropriated and politically marginalized, as Katarina Kušić (2018) has shown for the case of Serbia. Peacebuilding is therefore the latest instantiation in a long line of liberal internationalist policies—from colonialism through modernization policies to Cold War interventions—that ensure the reproduction of liberalism by projecting its internal tensions into the international sphere (Dillon and Reid 2009).

And it is this function that explains the paradoxical outcome of peacebuilding policies. These policies have become an integral and well-institutionalized part of the international order because they serve the reproduction of liberal power. They provide liberal elites with (one of many) means to reproduce themselves materially—through the introduction of market forces (expropriation) in target states, the provision of jobs for liberal NGOs, development agencies, international organizations, and militaries; politically through the co-optation and integration of elites and the marginalization of "nonliberal" sections of (world) society; and ideationally by "blaming the victims" for "illiberal" liberal policies (Barnett, Fang, and Zürcher 2014; Laffey and Nadarajah 2012).

All three problems of peacebuilding policies, in sum, can be traced back to the inherent contradictions of liberalism, and the order in which they haunt these policies matches the dynamics generated by liberalism: closing the gap between liberal theory and historical reality requires power politics, and such "illiberal" policies are reconciled with liberal principles by projecting them into an international, nonliberal, local sphere fundamentally separated from domestic, liberal, global polities and policies. Peacebuilding thus relies on a romanticization of liberalism itself. It overlooks, first, that liberalism is not aligned with the forces of history but rather is historically constructed; second, that this construction of liberalism systematically requires expropriation and power politics in the domestic and the international sphere; and third, that these liberal policies constitute not peace but rather divided societies and conflict.

Conclusion: The Future of Peacebuilding

Identifying these contradictions within liberalism itself suggests that the core function of peacebuilding does not lie in the establishment of peace—even if that aim surely motivates most of its proponents. Instead, peacebuilding policies serve to manage the domestic, international, and transnational conflicts generated by the inherent contradictions of liberalism. Peacebuilding policies are thus an integral part of the liberal world order and represent only the latest instantiation of policies that played this role for its constitution and reproduction: colonialism, trusteeship, modernization policies.

The fate of these earlier policies thus provides some indications for the prospects of peacebuilding. Neither colonialism nor trusteeship or modernization policies managed to achieve their stated aims. Instead, each of these policies eventually generated conflict and resistance, which undermined their legitimacy. Nevertheless, they were resurrected in revised versions and under new names—but essentially serving the same functions. It is no surprise, therefore, that peacebuilding, peacekeeping, statebuilding, democracy promotion, humanitarian intervention, and development policies today explicitly draw on models of colonialism, trusteeship, and modernization policies (Cooper 2002; Ignatieff 2003).

If peacebuilding policies indeed replicate the dynamics of their predecessors, their propensity to generate conflict and resistance will eventually undermine their legitimacy. But as long as the liberal world order exists, its inherent contradictions need to be managed. Peacebuilding policies will therefore need to be reinvented—if in a different form and under a different name.

References

Acemoglu, D., and J. A. Robinson. 2006. *Economic Origins of Dictatorship and Democracy.* Cambridge: Cambridge University Press.

Barnett, M. 2006. "Building a Republican Peace: Stabilizing States after War." *International Security* 30 (4):87–112.

Barnett, M., S. Fang, and C. Zürcher. 2014. "Compromised Peacebuilding." *International Studies Quarterly* 58 (3):608–620.

Bliesemann de Guevara, B. 2010. "Introduction: The Limits of Statebuilding and the Analysis of State-Formation." *Journal of Intervention and Statebuilding* 4 (2):111–128.

Boutros-Ghali, B. 1996. *An Agenda for Democratization.* New York: United Nations.

Carothers, T. 2002. "The End of the Transition Paradigm." *Journal of Democracy* 13 (1):5–21.

Chandler, D. 2010. "The Uncritical Critique of 'Liberal Peace.'" *Review of International Studies* 36 (S1):137–155.

Chesterman, S. 2007. "Ownership in Theory and in Practice: Transfer of Authority in UN Statebuilding Operations." *Journal of Intervention and Statebuilding* 1 (1):3–26.

Cooper, R. 2002. "The New Liberal Imperialism." http://www.guardian.co.uk/world/2002/apr/07/1/print. Accessed 16 February 2013.

Cunliffe, P. 2012. "Still the Spectre at the Feast: Comparisons between Peacekeeping and Imperialism in Peacekeeping Studies Today." *International Peacekeeping* 19 (4):426–442.

Cunliffe, P. 2018. "From Peacekeepers to Praetorians: How Participating in Peacekeeping Operations May Subvert Democracy." *International Relations* 32 (2):218–239.

David, S. R. 1992–1993. "Why the Third World Still Matters." *International Security* 17 (3):127–159.

Diamond, L. 1995. *Promoting Democracy in the 1990s: Actors and Instruments, Issues and Imperatives.* Report to the Carnegie Commission on Preventing Deadly Violence. New York: Carnegie Corporation.

Diamond, L. 1996. "Is the Third Wave Over?" *Journal of Democracy* 7 (3):20–37.

Dillon, M., and J. Reid. 2009. *The Liberal Way of War: Killing to Make Life Live.* London: Routledge.

Doyle, M. W. 1996. "Kant, Liberal Legacies, and Foreign Affairs." In *Debating the Democratic Peace,* edited by M. E. Brown, S. N. Lynn-Jones, and S. E. Miller, 3–57. Cambridge, MA: MIT Press.

Duffield, M. 2001. *Global Governance and the New Wars: The Merging of Development and Security.* London: Zed Books.

Fukuyama, F. 1989. "The End of History?" *National Interest,* Summer, 3–18.

Haass, R. N. 2003. "Toward Greater Democracy in the Muslim World." *Washington Quarterly* 26 (3):137–148.

Hutchings, K. 2013. "Liberal Quotidian Practices of World Ordering." In *Liberal World Orders,* edited by T. Dunne and T. Flockhart, 157–172. Oxford: Oxford University Press.

Ignatieff, M. 2003. "State Failure and Nation-Building." In *Humanitarian Intervention. Ethical, Legal, and Political Dilemmas,* edited by J. L. Holzgrefe and R. O. Keohane, 299–321. Cambridge: Cambridge University Press.

Ikenberry, G. J. 2006. *Liberal Order and Imperial Ambition.* Cambridge, UK: Polity Press.

Jabri, V. 2013. "Peacekeeping, the Local and the International: A Colonial or a Postcolonial Rationality?" *Peacebuilding* 1 (1):3–16.

Jahn, B. 2007. "The Tragedy of Liberal Diplomacy, Part II." *Journal of Intervention and Statebuilding* 1 (2):211–229.

Jahn, B. 2012. "Critique in a Time of Liberal World Order." *Journal of International Relations and Development* 15 (2):145–157.

Jahn, B. 2013. *Liberal Internationalism. Theory, History, Practice.* Basingstoke, UK: Palgrave.

Jahn, B. 2018. "Liberal Internationalism: Historical Trajectory and Current Prospects." *International Affairs* 94 (1):43–61.

Kušić, K. 2018. "Locating Subjects, Disrupting Intervention: Youth Empowerment and Agricultural Modernisation in Serbia." *PhD* dissertation, Aberystwyth University.

Laffey, M., and S. Nadarajah. 2012. "The Hybridity of Liberal Peace: States, Diasporas and Insecurity." *Security Dialogue* 43 (5):403–420.

Locke, J. 1994. *Two Treatises of Government.* Edited by Peter Laslett. Cambridge: Cambridge University Press.

Mac Ginty, R. 2011. *International Peacebuilding and Local Resistance: Hybrid Forms of Peace.* Basingstoke, UK: Palgrave.

McNally, D. 1988. *Political Economy and the Rise of Capitalism: A Reinterpretation.* Berkeley: University of California Press.

Moravcsik, A. 1997. "Taking Preferences Seriously: A Liberal Theory of *International Politics.*" *International Organization* 51 (4):513–553.

Nash, W. L. 2006. "Iraq: Democracy or Civil War." *Prepared Testimony before the House Subcommittee on National Security, Emerging Threats, and International Relations,* 1–4. Washington, DC: Council on Foreign Relations.

Ollapally, D. 1995. "Third World Nationalism and the US after the Cold War." *Political Science Quarterly* 110 (3):417–434.

Packer, G. 2006. *The Assassins' Gate: America in Iraq.* London: Faber and Faber.

Paffenholz, T. 2015. "Unpacking the Local Turn in Peacebuilding: A Critical Assessment towards an Agenda for Future Research." *Third World Quarterly* 36 (5):857–874.
Paris, R. 1997. "Peacebuilding and the Limits of Liberal Internationalism." *International Security* 22 (2):54–89.
Paris, R. 2004. *At War's End: Building Peace after Civil Conflict*. Cambridge: Cambridge University Press.
Paris, R. 2010. "Saving Liberal Peacebuilding." *Review of International Studies* 36 (2):337–365.
Perelman, M. 2000. *The Invention of Capitalism: Classical Political Economy and the Secret History of Primitive Accumulation*. Durham, NC: Duke University Press.
Plattner, M. F. 2008. *Democracy without Borders? Global Challenges to Liberal Democracy*. Lanham, MD: Rowman and Littlefield.
Rampton, D., and S. Nadarajah. 2017. "A Long View of Liberal Peace and Its Crisis." *European Journal of International Relations* 23 (2):441–465.
Randazzo, E. 2017. *Beyond Liberal Peacebuilding: A Critical Exploration of the Local Turn*. Abingdon, UK: Routledge.
Richmond, O. P. 2010. "Resistance and the Post-Liberal Peace." *Millennium: Journal of International Studies* 38 (3):665–692.
Richmond, O. P. 2011. *A Post-Liberal Peace*. Oxford: Routledge.
Richmond, O. P. 2018. "Rescuing Peacebuilding? Anthropology and Peace Formation." *Global Society* 32 (2):221–239.
Rubin, B. R. 2006. *Afghanistan's Uncertain Transition from Turmoil to Normalcy*. CSR No. 12. New York: Council on Foreign Relations.
Sabaratnam, M. 2013. "Avatars of Eurocentrism in the Critique of the Liberal Peace." *Security Dialogue* 44 (3):259–278.
Sachs, J. D. 2001. "The Strategic Significance of Global Inequality." *Washington Quarterly* 24 (3):187–198.
Schmidt, J. 2013. "Democracy Promotion in a Post-Political World." PhD diss., University of Westminster.
Sørensen, G. 2000. "The Impasse of Third World Democratization: Africa Revisited." In *American Democracy Promotion. Impulses, Strategies, and Impacts*, edited by M. Cox, G. J. Ikenberry, and T. Inoguchi, 287–307. Oxford: Oxford University Press.
Verkoren, W., and M. van Leeuwen. 2013. "Civil Society in Peacebuilding: Global Discourse, Local Reality." *International Peacekeeping* 20 (2):159–172.
Williams, D., and T. Young. 1994. "Governance, the World Bank, and Liberal Theory." *Political Studies* 42 (1):84–100.
Zaum, D. 2012. "Beyond the 'Liberal Peace.'" *Global Governance* 18 (1):121–132.

CHAPTER 3

THE ETHICS OF LIBERAL PEACEBUILDING

KRISTOFFER LIDÉN

SPEAKING to a central problem in ethics and political theory, international assistance to conflict-ridden countries is torn between two key concerns of global justice. On the one hand, it responds to a call for international solidarity across nations. On the other, the international politics of peace reflect a global hierarchy, where powerful countries on top of the hierarchy set the premises of the assistance. This dilemma is familiar from the ethics of development assistance and humanitarian intervention but took on new significance with the intensification of global governance after the end of the Cold War (Dower 2009; Duffield 2001; Hutchings 2010; Weiss 2013; Lidén 2014).

Throughout this period, the primary focus in international ethics remained on topics like distributive justice, humanitarian intervention, and global institutional reform rather than on more mundane forms of global governance within the current international order. There was continued debate on the moral responsibility to assist foreign countries, but not so much on the practicalities of how this ought to be done (Sen 2010). The assumption seems to have been that "where there is a will, there is a way"—and that questions like the operationalization of peacebuilding go beyond the scope of normative political theory (e.g., Rawls 1999, 93; see Wolff 2011). Peacebuilding efforts throughout this period nonetheless demonstrate that assisting foreign countries while respecting norms of equality, self-determination, and cultural diversity is easier said than done.

Today the experiment of "liberal peacebuilding" is generally considered a failure (Richmond and Mac Ginty 2014). Instead of externally driven "rapid liberalization" (first-generation liberal peacebuilding) or top-down and "outside-in" liberal statebuilding (second generation), peacebuilding has turned to "resilience," supporting domestic initiatives and institutions in building peace from the bottom up and "inside out" (Bargués-Pedreny 2018; Chandler 2017). While peacebuilding in its heyday replaced peacekeeping as the baseline of UN mandates for peace operations, the focus has returned from the peacebuilding objective of addressing root causes of war to the security objective of "stabilizing" countries combined with "the protection of civilians" (Karlsrud 2018).

Although the idea of promoting peace through political and economic liberalization has been downplayed as a strategy for peace operations, it nonetheless still lives on in UN policies of conflict prevention, development assistance, prevention of extremism, and the UN Sustainable Development Goal 16 to "promote peaceful and inclusive societies for sustainable development, provide access to justice for all and build effective, accountable and inclusive institutions at all levels" (UN General Assembly 2015; UN 2018). Moreover, despite terrible recent wars, the idea of a liberal democratic peace between and within countries continues to find strong statistical support (summarized in Pinker 2018, ch. 11). This includes a positive correlation between international peacekeeping and a reduction in the number, duration, and lethality of civil war throughout the period of "failed peacebuilding" (Goldstein 2011; Hegre, Hultman, and Nygård 2018). Against this backdrop, the ethics of liberal peacebuilding as a mode of global governance remains highly relevant.

The fall of liberal peacebuilding is often treated as a fall of liberalism as a normative approach to international peace operations, boding for more realist security strategies or the search for emancipatory postliberal, postcolonial, and/or posthumanist alternatives (Chandler 2017; Richmond 2016). However, as argued in this chapter, the lack of efficiency and legitimacy of liberal peacebuilding does not necessarily go against the normative foundations of political liberalism—in neither their internationalist nor their cosmopolitan manifestations. Indeed, peacebuilding practices in the post–Cold War period were less in line with liberal political principles than it may seem from the promotion of liberal market democracy. Furthermore, there is more continuity between the idea of liberal peacebuilding and current, supposedly nonliberal, policies of resilience, stabilization, and counterinsurgency than commonly assumed (Chandler 2017, 32–38). Efforts at drawing theoretical lessons from the demise of liberal peacebuilding therefore need to be more precise about where exactly the policies got it wrong. Unhelpfully, critiques of liberal peacebuilding are often presented as derived from empirical observation, downplaying their reliance on controversial political theoretical presuppositions.

This chapter starts out by outlining the political theory of liberal peacebuilding and situating it in relation to the strands of realism, internationalism, and cosmopolitanism in international ethics. It is demonstrated that the critics of liberal peacebuilding do not necessarily reject its internationalist objectives of promoting peace in accordance with human rights within a state framework—as defined by the UN Charter—but criticize a lack of coherence with this objective. Rather than subjecting the liberal internationalist premises of peacebuilding to standard realist and cosmopolitan arguments, the critical debate therefore primarily plays out *within* the strand of internationalism, adding empirically informed nuance to the ethics of contemporary global governance. Through the writings of John Rawls, Michael Walzer, and Simon Caney, it is demonstrated that empirical findings on the failure of liberal peacebuilding can be harmonized with the liberal internationalism that the critique tends to refute. My argument is not that this proves the critics wrong, but that in order to draw political lessons from practices of liberal peacebuilding more attention must be devoted to how the critique challenges central normative presuppositions of prevalent versions of liberal internationalism. In the last section, some of these theoretical challenges are pointed out, invoking elements of critical, poststructural, and postcolonial theory.

Liberal Peacebuilding between Intervention and Sovereignty

With the end of the Cold War, international attention to civil war made peace operations mandated by the UN a new domain for the global promotion of "the liberal peace" through the advancement of democracy, human rights, private property, and free-market economy (Doyle 1997, 207). This agenda resonated with Immanuel Kant's proposition for a pacific union among liberal states, with the crucial exception that it was based on a form of international political engagement in the domestic affairs of states that had no place in Kant's vision (Doyle 1997, 257; Kant [1795] 1996, First Definitive Article; see also Perreau-Saussine 2010).

In the essay "Kant's Idea of Perpetual Peace: At Two Hundred Years' Historical Remove," Jürgen Habermas (1998) argues that such political interference nonetheless is consistent with Kant's position when the historical transformations of the intermediate centuries are taken into account. In his view, an increased interdependence of nations has made noninterference an outdated and insufficient source of global peace and self-determination. He writes, "A purely negative conception of peace [peace as the absence of war] was sufficient for Kant's purposes. Today this is unsatisfactory not only because of the breakdown of the limits on the conduct of war, but above all because wars have social causes" (185). This is particularly manifest, he argues, in conflict-ridden countries in the Third World where "the state infrastructure and monopoly of the means of violence are so weakly developed (Somalia) or have disintegrated to such an extent (the former Yugoslavia), where the social tensions are so extreme and the threshold of tolerance of political culture so low" that internal order is disrupted, with the threat of national, ethnic, or religious disintegration as a result (184). With this context, Habermas contends that the foundations of Kant's political thought justify policies employing "all means short of military force, including humanitarian intervention, to influence the internal affairs of formally sovereign states with the goal of promoting self-sustaining economies and tolerable social conditions, democratic participation, the rule of law, and cultural tolerance" (185). This is the theoretical rationale of *liberal peacebuilding*.

When the topic of postconflict peacebuilding was addressed in the research volume of the report of the International Commission on Intervention and State Sovereignty (ICISS 2001, 195–203) on "the responsibility to protect," it concentrated on the issue of UN-mandated "transitional administrations." These involve the establishment of temporary sovereign international rule in territories where the sovereignty of a former governmental authority has been suspended but before functions of a new internationally recognized government are fully operative, as in the cases of Kosovo and East Timor. While such interference evidently requires careful considerations of state sovereignty, the great bulk of peacebuilding operations since the 1990s have taken place in situations where a formal agreement was reached with local authorities and some capacity for self-governance was still in place. The ICISS report refers to these types of operations as "control" and "assistance." Control involves a transitional authority exercising the powers of direct control in certain domains of government under a UN mandate, as in Cambodia and Bosnia and Herzegovina, while assistance implies that "the local administration may not be in complete disarray, and the transitional authority provides some overall coherence and an international standard for the development of governance structures" (ICISS 2001, 195–203). In the light of "transitional administration" and its less intrusive version of control, assistance emerges as undeserving of a careful independent discussion in spite of being the most common variant of peacebuilding. It is nonetheless exactly in this gray zone between UN-mandated intervention and situations of regular "development cooperation" that ethical and legal standards for global governance are most wanting (see Orford 2011). In the words of Jean L. Cohen (2012, 320), "These practices [of UN mandated global governance] are Janus-faced: while in some instances they prevent or help rectify humanitarian disasters and protect human rights, in others they lead to the evisceration of the rule of law and constitutionalism on all levels of the international political system, and undermine the foundational Charter principles of sovereign equality and human rights."

The governance exercised by international peace operations for the sake of peacebuilding falls between intervention and state sovereignty in a literal chronological sense: after international interventions to end civil war or genocide, as in Bosnia, Kosovo, and East Timor in the 1990s, the objective of peacebuilding was to build a legitimate sovereign state that could eventually be left in peace. In general, peacebuilding operations tended to start out in a rather coercive mode, backed by military peacekeeping forces and possibly preceded by coercive diplomacy if not outright military intervention. Then the operations gradually slid toward the "sovereignty" side of the scale by seeking to transfer political control to local authorities and becoming as redundant as possible. At least, this is how peacebuilding was supposed to work.

Peacebuilding governance also falls between intervention and sovereignty in a *legal* sense because it involves an unprecedented degree of *legal political interference* in the domestic affairs of conflict-ridden countries. It thereby evokes long-standing debates in the ethics of international law and politics between proponents and opponents of benevolent interference. (e.g., Mill [1867] 1973; Wheeler 2000). In international law, intervention and state sovereignty are mutually exclusive terms, and peacebuilding is categorized as nonintervention as long as it enjoys the formal consent of the authorities of the target country.

There are, however, many reasons why the authorities of war-torn countries may accept such extensive foreign political engagement—even when they disagree with the diagnosis and cure. First, the immediate need for assistance in the aftermath of civil war may outweigh the concern for autonomous self-governance. Second, the military and economic costs of noncooperation with major powers and international organizations may outweigh the disadvantages of the interference. Third, in situations of domestic political struggle, state authorities often have self-interest in international assistance because their power (as against political and military opponents) is reinforced by international resources and support (Hameiri 2010, ch. 7; Hameiri, Hughes, and Scarpello 2017).

In effect, a mandate from sovereign authorities does not guarantee that the political will of their subjects is represented by peacebuilding governance. The reliance of peacebuilding policies upon political strategies that are defined *for* instead of *by* its political subjects corresponds to the dictionary definition of "paternalism": the "policy . . . of controlling people in a paternal way by providing them with what they need but giving them no responsibility or freedom of choice" (Hornby 1989). On this account, Kant's skepticism against political intervention for the "emancipation of others" seems as relevant as ever, and a shadow is cast over the liberal interventionism proposed by Habermas. Yet if the alternative to such interference is civil war and massive preventable suffering, the question that follows is whether these troubles can be addressed in ways that are more representative of the diverse and often conflicting interests of the affected societies.

The initial thrust of liberal peacebuilding was to solve this problem of self-determination by arranging national elections early in the political process. As documented by Paris (2004) in *At War's End*, this turned out to be troublesome or even counterproductive in the absence of effective democratic institutions, requiring a shift to "institutionalisation before liberalisation" through international statebuilding. The reliance of such statebuilding on international actors and support nonetheless reinforced the criticism of liberal peacebuilding as advancing foreign norms, institutions, and interests (Paris and Sisk 2009).

Situating the Problem within International Ethics

Normative positions on sovereignty and intervention in world politics are often divided into *realist*, *internationalist*, and *cosmopolitan* approaches (for an overview, see Brown 2002; Dower 2009). While internationalist approaches are premised on the rights of states, realist approaches deny the possibility of such legal or moral principles in the international sphere. Cosmopolitans, on the other hand, extend international law and ethics from the rights of states to the rights of individuals as "world citizens." Only if the state system is the best way to advance the rights of individuals on a global scale can it be part of the solution from a cosmopolitan perspective. Internationalism does not exclude such a preoccupation with individual/human rights but sees them as a concern for states to manage—both within their own jurisdictions and, potentially, through multilateral cooperation.

Despite the extensive political interference implicated in liberal peacebuilding policies, they do not seek to replace the state system with a cosmopolitan order. Given this internationalist basis, one might expect the critics of liberal peacebuilding to rely either on a political realist rejection of its moral foundations or on a cosmopolitan rejection of its state centrism. Instead, the most common form of critique falls within the internationalist approach broadly defined. That is, it does not tend to reject the moral ambitions of peacebuilding or engage with questions of international legal reform for its achievement but scrutinizes the perils and prospects of liberal peacebuilding on the internationalist premises of current international law.

This point may be spelled out by reconstructing the ethical justification of liberal peacebuilding in UN policies as follows:

A. States have a collective responsibility to promote peace in accordance with human rights within the confines of state sovereignty, as defined in the UN Charter.
B. Civil war is an impediment to peace and human rights. In addition to suffering, death, and devastation for those directly affected, it creates insecurity for whole societies and hinders socioeconomic development. Civil war also tends to spark or reinforce international conflict and is sometimes associated with international terrorism. If peacebuilding alleviates the harmful consequences of civil war and reduces the likelihood of its recurrence within the confines of state sovereignty, it is therefore ethically desirable and in accordance with the basic purposes of the UN Charter.
C. Liberal peacebuilding:
   1. creates peace by transforming war-torn countries into liberal democracies;
   2. is the most legitimate form of peacebuilding, as it actively promotes human rights without violating state sovereignty;
   3. contributes to international peace and security by expanding a zone of peace between liberal states.
D. Liberal peacebuilding is therefore ethically desirable and should be a central concern of states and international organizations.

Most critics of liberal peacebuilding do not primarily address criterion A or B but do address C1. When the "liberalism" of liberal peacebuilding is criticized, it is therefore the particular form of liberalism expressed in premise C1 that is attacked. Most such criticism implicitly *confirms* a commitment to propositions A and B. Thus it is rather the specific political ideas underpinning recent peacebuilding practices that are criticized. This criticism has consequences for the theoretical premises of C2 and C3 as well, and evidently for D. In the debate on liberal peacebuilding, the rejection of proposition C is therefore often presented as a general rejection of proposition D. This rejection is then found morally appalling by proponents of D based on a moral commitment to A and B. Hence the debate is wrongly framed as a classic contestation over the "liberal" norms of human rights and democracy rather than on their controversial application to peacebuilding (C).

This tendency is most explicit in the "revisionist" debate on prescriptive alternatives to liberal peacebuilding on the normative premises of B (see Lidén 2013; Richmond and

Mac Ginty 2014). Here, insights from theoretical strands like realism, Marxism, communitarianism, and cosmopolitanism are integrated, but within the basically liberal internationalist normative framework of A. This may be a pragmatic mode of immanent critique under nonideal circumstances, or it may reflect a genuine commitment to these norms.

In effect, the scope of the debate on liberal peacebuilding primarily spans liberal and different critical forms of internationalism broadly defined, as well as "solidarist" and "pluralist" variants thereof. (These are sometimes associated with the broader notions of cosmopolitanism and communitarianism, respectively; cf. Brown 2002.) *Solidarist* internationalism overlaps with cosmopolitanism in premising state sovereignty on certain universal moral criteria such as human rights (see Wheeler 2000). When these criteria are not met, it is seen as a collective responsibility of states not only to offer assistance to the state in question but to resolve the problem. In its liberal interpretation, making international assistance to troubled countries conditional on political liberalization is a logical consequence of this solidarist position. The *pluralist* strand of internationalism excludes any such political interference in the domestic affairs of states (e.g., Jackson 2000; Walzer 1983). Overlapping with basic realist and communitarian presumptions, pluralists expect solidarist interference both to entail exploitation of troubled countries by their powerful protectors and to undermine local culture and morality.

Solidarism and pluralism are often defined as mutually exclusive positions along these lines. Yet as variations of internationalism they are more convincingly conceived as opposites on a gradual scale (Buzan 2004, 49; Caney 2005, 12; Dower 1998, 25). Indeed this is the scale within which normative debate on the interpretation of the basic principles of current international law is situated, as distinguished from debate on the ethical justifiability of these principles. As exemplified in the next sections, the literature on liberal peacebuilding helps adjust this scale to fit evolving practices of global governance.

Between Solidarism and Pluralism: Rawls, Walzer, and Caney

When John Rawls (1999) derives an international law of *peoples* rather than of *individuals* from his political liberalism in *The Law of Peoples*, it exemplifies efforts at striking a balance between solidarist and pluralist concerns. In a solidarist vein, he requires states not only to respect the sovereignty of other states but to honor human rights: "Although suitable for a society of well-ordered peoples, [the principle of nonintervention] fails in the case of a society of disordered peoples in which wars and serious violations of human rights are endemic" (37–38). In principle, this entails a right of "lawful states" to subject "outlaw states" to "forceful sanctions and even to interventions" for the promotion of human rights (81). It also implies a duty of states to actively promote human rights in "burdened societies" "whose historical, social, and economic circumstances make their achieving a well-ordered regime . . . difficult if not impossible" (108–111). For these purposes, Rawls argues, international institutions and practices should be established "to serve as a kind of confederative center and public forum for their common opinion and policy toward non-well-ordered regimes" (93).

This argument may be expected to justify the premising of peacebuilding on political and economic liberalization. Rawls's liberal solidarist precepts are nonetheless counterbalanced by a pluralist concern for the self-determination of peoples. In order to allow for considerable autonomy and diversity, this makes him define human rights narrowly as the rights to life, liberty, property, and equality: "Among the human rights are the right to life (to the means of subsistence); to liberty (to freedom from slavery, serfdom, and forced occupation, and to a sufficient measure of liberty of conscience to ensure freedom of religion and thought); to property (personal property); and to formal equality as expressed by the rules of natural justice (that is, that similar cases be treated similarly)" (Rawls 1999, 65, references omitted). As long as nonliberal peoples fulfill these basic human rights, their right to self-determination is to be respected. This excludes political sanctions (military, economic, diplomatic) to push for liberalization (59, 110). It even implies that *assistance* to burdened societies ought not to promote a particular political model like liberalism but adapt the promotion of human rights to the local cultural and political context (107–108). Rawls recognizes that this complicates the duty of assistance considerably as compared to a universal recipe like liberal statebuilding. Furthermore, he rejects the alternative of mere financial aid, because the fulfillment of human rights rather relies on a change of political and social institutions. Hence the solution is not to avoid engagement in political matters like peacebuilding but to interfere in a way that promotes basic human rights without violating the principle of self-determination.

While this argument is based on a moral defense of diversity, imposed liberalization is also left as *unrealistic* due to the rooting of social and political institutions in political culture and religious, philosophical, and moral traditions: "[T]he political culture of a burdened society is all-important; and . . . at the same time there is no recipe, certainly no easy recipe, for well-ordered peoples to help a burdened society to change its political and social culture" (Rawls 1999, 108). This argument—from one of the world's leading proponents of political liberalism—is surprisingly similar to the critique of liberal peacebuilding for imposing foreign ideas, norms, and institutions.

In the essay "Governing the Globe," Michael Walzer (2004)—a profiled critic of liberal interventionism—draws a continuum between the extremes of "unity" (global state) and "division" (anarchy) and argues in favor of a pluralist compromise between the two ("third degree pluralism"; 186). While rejecting the cosmopolitan argument for a world federation, this compromise is still more "solidarist" than the current world order on his account. In addition to the lack of capacity of the current international system to alleviate global inequality, the lack of control of war between and among states is a central motivation of his argument (179). The solution, he argues, is "the familiar anarchy of states mitigated and controlled by a threefold set of non-state agents: organizations like the U.N., the associations of international civil society, and regional unions like the EC

[European Community]" (186). It involves building on "the institutional structures that now exist, or are slowly coming into existence, and to strengthen all of them, even if they are competitive with one another" (186–187). In order to address the problem of war, this order requires a UN with a military force of its own capable of humanitarian interventions and a strong version of peacekeeping, he argues. Combined with a vibrant global civil society and effective regional federations, this order "would bring many of the advantages of a global federation but with greatly reduced risks of tyranny from the center" (188).

In Walzer's pluralist argument we find a justification for extensive peace operations that does not rely on liberal political principles. It is instead motivated by a classic internationalist concern for allowing states (as "political communities") to pursue their self-determined goals. While the starkest pluralists see civil war as a merely internal affair, Walzer sees it as a situation where states can benefit from international assistance. This, however, requires that the political terms of the assistance are set by legitimate authorities of the assisted countries and not by foreign actors. If the authorities request assistance for liberal statebuilding as a self-determined choice, it would therefore harmonize with Walzer's argument. Yet as a general approach to peace operations his argument supports the critique of liberal peacebuilding as a universal recipe, and the rhetoric of stabilization and resilience is more in line with his argument.

The task of balancing between a solidarist commitment to human rights on the one hand and a pluralist concern with cultural diversity and egotist foreign policies on the other is not reserved for dedicated internationalists, however. In *Justice beyond Borders* Simon Caney (2005) rejects the element of moral relativism in the arguments of Walzer and Rawls, while retaining a pragmatic approach to the scope and potential of cosmopolitan engagement within the current international system (78–85, 192–196). In principle, he defends a "maximalist" liberal account of civil and political human rights and favors the establishment of cosmopolitan democracy (along the lines outlined by David Held, among others) as the most coherent way of realizing these rights on a global scale (Caney 2005, 156–160; Held 1995, 279). Yet this position does not result in an argument for the imposition of liberal statehood in ways that would disregard the paternalism and practical shortcomings of such imposition emphasized by the liberal peacebuilding critique. As long as the international system remains defined by state-centric international law, the prescriptive implications of his argument are therefore of a liberal solidarist kind in spite of its cosmopolitan foundations.

Central to Caney's argument is the assumption that civil and political rights do not undermine but guarantee autonomy and cultural diversity, although policies violating both norms often are wrongly justified by human rights defenders. This argument significantly raises the bar for the promotion of human rights through global governance, as compared to the idea of coercive liberal statebuilding. If past policies of liberal peacebuilding have tended to ignore the problem of cultural diversity and the norm of self-determination, the ensuing practices have therefore not been consistent with civil and political rights on Caney's premises, even if motivated by the goal of realizing exactly these rights through liberal democracy. Hence, in practice, the challenge Caney leaves peace operations with is quite similar to that of Rawls, except he does not do so on the basis of a valuation of societies, cultures, peoples, or political communities in themselves (due to their value for individuals) but out of concern for the political liberty of the individuals concerned.

Against this backdrop, we see that Paris (2010) was right when arguing that the criticism of liberal peacebuilding for not being sufficiently adapted to local agency and conditions can be integrated in a revised form of liberal peacebuilding. A consideration of political alternatives therefore requires a more nuanced engagement with the theoretical premises of the critique. Without resorting to a full exposition of these alternative perspectives, the final section sketches out some of the theoretical arguments in the critique that challenge the liberal solidarist and pluralist positions represented here by Rawls, Caney, and Walzer.

Critical Theories for the Ethics of Global Governance

To recap: If the theory and practice of liberal peacebuilding were the same, we could argue that the historical problems confronted prove the theories wrong. And if the turn of peace operations from peacebuilding to stabilization represented a turn from "liberal" to "realist" foreign policy, we could argue that the liberal theories of Rawls and Caney are no longer relevant for contemporary practices and that the moderate pluralism of Walzer is as solidarist as it gets. Yet, as we have seen, there is no necessary connection between liberal peacebuilding and theories of liberal internationalism (or cosmopolitanism). Moreover, while liberal peacebuilding turns out to be less liberal than one might expect, we have seen how the turn to resilience, stabilization, and protection is more in line with liberal internationalism than commonly assumed.

Still, as evidenced by several contributions to this Handbook, the critique of (quasi-) liberal peacebuilding involves analyses of and prescriptions for international engagement that depart from prevalent premises of liberal internationalism without resorting to conventional realist, Marxist, or cosmopolitan arguments. Drawing on different strands of critical (Jones 2001), poststructural (Edkins 1999), and postcolonial (Seth 2013) theory (where the last overlaps with the two former while concentrating on marginalized or subaltern people in the Global South), these analyses and prescriptions address the ontologies and epistemologies through which understandings of the ethical objective of advancing peace in accordance with human rights and self-determination are formed. While widely disparate, their challenge to liberal internationalism can be summed up under the broad headings of (1) agency and (2) knowledge in global politics.

(1) The disconnect between theory and practices of liberal peacebuilding could be interpreted as an oversight among otherwise liberal actors or as evidence of a lack of commitment of supposedly liberal actors to liberal principles. In line with the latter interpretation, critics like Chandler and Richmond see the disconnect as symptomatic for

a general absence of truly liberal political agency in world politics—even at the historical "liberal moment" of the post–Cold War era. This does not prove the arguments of Rawls and Caney wrong, as they do not make such empirical claims about the current state of international affairs. Yet it actualizes theories on nonliberal political preconditions of liberal foreign policies. Such preconditions include economic, security, and moral interests and the (inter)subjective ideas through which these interests are conceived (for an overview, see Lidén 2013). This perspective implies that the advancement of peace, human rights, and democracy in world politics requires a different understanding of agency than we find in prevalent versions of liberal internationalism.

The failure of liberal peacebuilding is often attributed to the maladapted nature of liberal institutions in non-Western, nonliberal societies. If international actors would have had the power to impose such institutions through peacebuilding efforts, this criticism of their "foreignness" would imply neocolonial imposition of the kind pluralists reject. However, the history of recent practices demonstrates the limits to the power of international actors in imposing such "foreign solutions" in the first place—be they liberal or not—when these are not aligned with the interests of local political elites or institutions (cf. Autesserre 2010; Barnett and Zürcher 2009). Where liberal internationalists may interpret this as an effect of "illiberal" or "traditional" norms and ideas, critical, poststructuralist, and postcolonial theories tend to see a genuine conflict of interests between "liberal peacebuilders" and the societies they intervene in. This is not conceived as a conflict between liberal and nonliberal positions because (a) the peacebuilders are not seen as truly liberal and (b) liberal principles like human rights and democracy take on a different significance than presumed by liberal peacebuilding policies (e.g., Lizée 2011, 168–192). Literature on norms and practices of gender in liberal peacebuilding presents a rich source of insights into these dynamics (e.g., Boutron 2018; Doeland and Skjelsbaek 2018; O'Reilly 2012; Rey and McKay 2006; Väyrynen 2010).

When Habermas justifies liberal peacebuilding as a response to the disruption of "internal order" in war-torn states, it exemplifies how intervention relies on a certain liberal institutionalist imaginary. Similarly, Rawls's categories of well-ordered, burdened, and outlaw societies equate peoples with their state-based political order—and the absence of a well-functioning modern state makes them a problem to be resolved or "left alone" by liberal actors. This categorization ignores the historical antecedents to ex-colonial states, the impact of global actors and structures on local conditions, as well as the presence of norms and institutions that are not reducible to a liberal/nonliberal dichotomy (e.g., Kapoor 2008; Richmond 2019).

Drawing on Homi Bhabha's conceptualization of the limited power of colonial rule, the result of "failed" peacebuilding was characterized as *hybrids* of international and local norms and practices (Bhabha 1994; Mac Ginty 2011; Richmond 2009). In order to better grasp these dynamics, the state-centric language of sovereignty and intervention is replaced with a focus on the micro-level of peace operations—"turning local" in a theoretical and methodological sense, evoking notions like resistance, co-optation, manipulation, and "the everyday" (Mac Ginty and Richmond 2013). This "local turn" is also associated with a local turn in peacebuilding policy—but there is no straightforward path from one to the other, and a local turn in policies that does not rely on a local turn in theory runs counter to this critical tradition. The local turn has sparked a critical debate on what a local perspective and hybridity really entails—and whether it is the right way forward for critical theories of peace and development (Chandler and Richmond 2015; Debiel, Held, and Schneckener 2016; Hameiri and Jones 2017; Lidén and Jacobsen 2016; Randazzo 2017; Richmond and Mitchell 2011; Sabaratnam 2017; Visoka 2016). Instead of ending the debate on liberal peacebuilding, this turn has thereby sparked a theoretical venture that goes to the heart of the ethical problem of global governance.

In line with these arguments on the international and local dimensions of the agency involved in global governance, Shahar Hameiri characterizes liberal peacebuilding as a mode of state regulation for the management of global risks—played out as a continuation of local struggles for power in the target countries. He argues that instead of concerted global governance or isolated international interference, these practices form semipermanent structures of "transnational governance" through strategic partnerships between international and local actors based on mutual interests rather than on a shared commitment to political liberalism (Hameiri 2010, 2011; Hameiri and Jones 2017). When advancing political and economic liberalization, Hameiri argues, it is not primarily an expression of liberal internationalist normative principles but of prevalent (Western) knowledge and expertise on what it means to maintain peace and security within a state framework (see also Neumann and Sending 2014).

(2) Relating to this point, a second element in the critique of liberal peacebuilding concerns the ability to conceptualize policies of peace and development across the domestic contexts of their application. While the global scope of relations between states, corporations, and individuals makes it relevant to formulate global policies concerning those relations, it is different from formulating global policies for relations that are more specific for individual societies. The emphasis on resilience and local ownership can be a response to this problem, but it can also reinforce the problem if defined in universal terms (Chandler 2012; Mac Ginty and Richmond 2013).

If drawing the boundary between the global and local political spheres at the state level, this is the classic pluralist argument against solidarist engagement. However, the identification of noninterference with the principle of state sovereignty also makes a problematic projection of political institutions upon societies in which the state is only one among several decisive institutions below and across the state level, and where these are fundamentally impacted by international actors and structures. Starting from a limited locality instead of from universal policies when considering the meaning of peace and development therefore does not exclude that international assistance or reforms are part of the solution (Robertson 1995).

With the universal reach of the state, one might argue that at least when it comes to statebuilding or state regulation certain policies may apply across contexts—as with policies of "good governance." The problem of epistemology arises, nonetheless, when these policies are translated into concrete measures on behalf of the societies in question. Speaking to this problem, Nehal Bhuta (2008, 519) examines the resemblance of contemporary international statebuilding with colonial political thought on how to promote the security and welfare of native subjects.

The explicit racism that underpinned colonial rule is no longer commonplace, and it is not backward customs but modern warfare and unconsolidated statehood after independence from colonial rule that incite UN-authorized statebuilding efforts. Yet, Bhuta (2008, 521) writes, the conception of *politics as technology* "remains fundamentally constitutive of disciplines such as economic development and . . . the emergent knowledge-complex of state-building and democratization."

Contrary to many of his contemporary critics of liberal peacebuilding, Bhuta argues that the political response ought not to be the generation of more advanced knowledge on international statebuilding and its adaptation to local culture. It is not the contents of past operations but their very political form that he rejects. Like colonial bureaucracy, such knowledge in the service of rule creates "its own objects and its own experts" (Bhuta 2008, 520). Acting upon this knowledge not only may result in *failure*, he argues, but may reinforce global *injustice* because the knowledge reproduces the ideas and institutions of the powerful in a way that consolidates their power: "The production, consecration and circulation of expertise about how to intervene in the politics of other societies and other peoples, is deeply implicated in the structures of inequality and strategies of domination inherent in any world order" (534).

This worry about the epistemology of global governance is reflected in the essay "Decolonizing Global Ethics" by Kimberly Hutchings (2019), where she addresses the problem of thinking about the ethics of global politics without reducing the world to a homogeneous universal perspective. She argues that the solution is not to come up with better universals or procedures that can comprise diversity but to embrace dissonance and negotiations between different perspectives on the premise that "there is no overriding rule and no shared ideal" (122). Speaking directly to the situatedness of ethics, she sums up her argument as follows: "I am suggesting, therefore, that taking pluriversality seriously means shifting our understanding of global ethics away from seeing it as a route to determining answers to questions of global justice and toward seeing it as an embodied, reflective practice contingently attached to specific goals and contexts" (2019, 123). This approach could be expected to exclude international efforts to assist war-torn countries based on universal norms of peace, human rights, and democracy. However, if the shortcomings of such efforts are acknowledged, including the subjective biases of international actors, Hutchings might argue that her approach is the only way a global ethic of international engagement could still be pursued.

Seen together, these two elements of the critique of liberal peacebuilding regarding the agency and knowledge of global governance represent a challenge to the theoretical presuppositions of liberal solidarist as well as pluralist internationalism, entailing other variations along this solidarist-pluralist continuum. Although Rawls, Walzer, and Caney could reply that they of course recognize the problem of global contestation and cultural diversity and welcome any attempt at adjusting their arguments to local non-Western contexts, we have seen how the critiques question their very ontological and epistemological preconditions. By relating this theoretical debate to concrete practices of international peace operations, the critics of liberal peacebuilding thereby help advance ethical considerations on the global promotion of peace and the role for political and economic liberalization therein.

Conclusion

At the zenith of liberal peacebuilding, the battle for the ethics of global governance stood between cosmopolitan democracy and more conservative modes of liberal internationalist reforms (Brown 2002). The ways in which quasi-liberal objectives of global governance relied on instrumental calculations of states and corporations benefiting from a Western hegemony was left for radical critics to point out. With the "return" of nonliberal history, the pendulum of ethical considerations has swung too far in the opposite direction, ignoring the continued relevance of liberal internationalist norms and institutions as a constraint on war and exploitation in world politics.

The ethics of liberal peacebuilding followed this dialectic, where its interventionism initially drew inspiration from illusionary "post-Westphalian" arguments by cosmopolitans like Habermas. The eventual "fall" of liberal peacebuilding is symptomatic for a fall of the conception of global governance as a concerted global force for good—a transitional phase toward a fully cosmopolitan world order (Gaskarth 2015). The fragility and illusions that were on display in peacebuilding makes empirically informed theory on its justification and effects a source of inspiration for the ethics of the multifaceted practices associated with global governance. Instead of a contestation over cosmopolitan pretensions, we are left with a battle between nuanced variations of internationalism in dire need of theoretical refinement. As with the notion of peacebuilding, a first step in this direction might be to get rid of the very term "global governance" due to its confusing allusions, but doing so also risks disentangling the critique from the policy debates it engages with.

In his review of the recent history of peacebuilding, Chandler (2017) argues that the critique of liberal peacebuilding risks being co-opted by realist and neoconservative agendas due to its lacking a sense of the power and rationales at play. In this chapter, we have seen how liberal internationalist positions may embrace critical arguments for pragmatism, resilience, and a "local turn" in peace operations. Debates on the politics of peace and development therefore need to engage more explicitly with ethical arguments on concrete political alternatives in specific contexts of conflict or disaster instead of assuming that political implications follow from merely descriptive theoretical or empirical accounts of general policies and their implementation.

References

Autesserre, S. 2010. *The Trouble with the Congo: Local Violence and the Failure of International Peacebuilding*. Cambridge: Cambridge University Press.

Bargués-Pedreny, P. 2018. *Deferring Peace in International Statebuilding: Difference, Resilience and Critique*. Milton, UK: Routledge.

Barnett, M., and C. Zürcher. 2009. "The Peacebuilder's Contract: How External Statebuilding Reinforces Weak Statehood." In *The Dilemmas of Statebuilding: Confronting the Contradictions of Postwar Peace Operations*, edited by R. Paris and T. D. Sisk, 23–52. London: Routledge.

Bhabha, H. 1994. *The Location of Culture*. New York: Routledge.
Bhuta, N. 2008. "Against State-Building." *Constellations* 15 (4):517–542.
Boutron, C. 2018. "Engendering Peacebuilding: The International Gender Nomenclature of Peace Politics and Women's Participation in the Colombian Peace Process." *Journal of Peacebuilding & Development* 13 (2):116–121.
Brown, C. 2002. *Sovereignty, Rights and Justice: International Political Theory Today*. Cambridge, UK: Polity Press.
Buzan, B. 2004. *From International to World Society? English School Theory and the Social Structure of Globalisation*. Cambridge: Cambridge University Press.
Caney, S. 2005. *Justice beyond Borders: A Global Political Theory*. Oxford: Oxford University Press.
Chandler, D. 2012. "Resilience and Human Security: The Post-interventionist Paradigm." *Security Dialogue* 43 (3):213–229.
Chandler, D. 2017. *Peacebuilding: The Twenty Years' Crisis, 1997–2017*. New York: Springer.
Chandler, D., and O. P. Richmond. 2015. "Contesting Postliberalism: Governmentality or Emancipation?" *Journal of International Relations and Development* 18: 1–24. doi.org/10.1057/jird.2014.5
Cohen, J. 2012. *Globalization and Sovereignty: Rethinking Legality, Legitimacy, and Constitutionalism*. Cambridge: Cambridge University Press.
Debiel, T., T. Held, and U. Schneckener. 2016. *Peacebuilding in Crisis: Rethinking Paradigms and Practices of Transnational Cooperation*. Abingdon: Routledge.
Doeland, E. M., and I. Skjelsbaek. 2018. "Narratives of Support and Resistance: A Political Psychological Analysis of the Implementation of UNSCR 1325 in Bosnia and Herzegovina." *Political Psychology* 39 (5):995–1011.
Dower, N. 1998. *World Ethics: The New Agenda*. Edinburgh: Edinburgh University Press.
Dower, N. 2009. *The Ethics of War and Peace*. Cambridge, UK: Polity Press.
Doyle, M. W. 1997. *Ways of War and Peace: Realism, Liberalism and Socialism*. New York: Norton.
Duffield, M. 2001. *Global Governance and the New Wars*. London: Zed Books.
Edkins, J. 1999. *Poststructuralism and International Relations: Bringing the Political Back In*. Boulder, CO: Lynne Rienner.
Gaskarth, J., ed. 2015. *Rising Powers, Global Governance and Global Ethics*. London: Routledge.
Goldstein, J. S. 2011. *Winning the War on War: The Decline of Armed Conflict Worldwide*. New York: Dutton.
Habermas, J. 1998. "Kant's Idea of Perpetual Peace: At Two Hundred Years' Historical Remove." In *The Inclusion of the Other: Studies in Political Theory*, edited by Ciaran P. Cronin and Pablo De Greiff, 165–202. Cambridge, MA: MIT Press.
Hameiri, S. 2010. *Regulating Statehood: State Building and the Transformation of the Global Order*. Basingstoke, UK: Palgrave Macmillan.
Hameiri, S. 2011. "A Reality Check for the Critique of the Liberal Peace." In *A Liberal Peace? The Problems and Practices of Peacebuilding*, edited by S. Campbell, D. Chandler, and M. Sabaratnam, 191–208. London: Zed Books.
Hameiri, S., C. Hughes, and F. Scarpello. 2017. *International Intervention and Local Politics: Fragmented States and the Politics of Scale*. Cambridge: Cambridge University Press.
Hameiri, S., and L. Jones. 2017. "Beyond Hybridity to the Politics of Scale: International Intervention and 'Local' Politics." *Development and Change* 48 (1):54–77.

Hegre, H., L. Hultman, and H. M. Nygård. 2018. "Evaluating the Conflict-Reducing Effect of UN Peacekeeping Operations." *Journal of Politics* 81 (1):215–232.

Held, D. 1995. *Democracy and Global Order: From the Modern State to Cosmopolitan Governance*. Cambridge, UK: Polity Press.

Hornby, A. S. 1989. *Oxford Advanced Learner's Dictionary*, edited by A. P. Cowie. Oxford: Oxford University Press.

Hutchings, K. 2010. *Global Ethics: An Introduction*. Cambridge, UK: Polity Press.

Hutchings, K. 2019. "Decolonizing Global Ethics: Thinking with the Pluriverse." *Ethics & International Affairs* 33 (2):115–125.

ICISS. 2001. The Responsibility to Protect: Research, Bibliography, Background. Supplementary Volume to the Report of the International Commission on Intervention and State Sovereignty. Ottawa: International Development Research Centre for ICISS.

Jackson, R. H. 2000. The Global Covenant: Human Conduct in a World of States. Oxford: Oxford University Press.

Jones, R. W., ed. 2001. Critical Theory and World Politics. London: Lynne Rienner.

Kant, I. (1795) 1996. "Toward Perpetual Peace." In Immanuel Kant: Practical Philosophy, edited by M. J. Gregor, 311–352. New York: Cambridge University Press.

Kapoor, I. 2008. The Postcolonial Politics of Development. New York: Routledge.

Karlsrud, J. 2018. The UN at War: Peace Operations in a New Era. Cham, Switzerland: Palgrave Macmillan.

Lidén, K. 2013. "In Love with a Lie? On the Social and Political Preconditions for Global Peacebuilding Governance." *Peacebuilding* 1 (1):73–90.

Lidén, K. 2014. *Between Intervention and Sovereignty: Ethics of Liberal Peacebuilding and the Philosophy of Global Governance*. Oslo: Department of Philosophy, Classics, History of Art and Ideas, University of Oslo.

Lidén, K., and E. K. U. Jacobsen. 2016. "The Local Is Everywhere: A Postcolonial Reassessment of Cultural Sensitivity in Conflict Governance." In *Cultures of Governance and Peace: A Comparison of EU and Indian Theoretical and Policy Approaches*, edited by J. P. Burgess, O. P. Richmond, and R. Samaddar, 132–149. Manchester: Manchester University Press.

Lizée, P. P. 2011. *A Whole New World: Reinventing International Studies for the Post-Western World*. London: Palgrave Macmillan.

Mac Ginty, R. 2011. *International Peacebuilding and Local Resistance: Hybrid Forms of Peace*. Basingstoke, UK: Palgrave Macmillan.

Mac Ginty, R., and O. P. Richmond. 2013. "The Local Turn in Peace Building: A Critical Agenda for Peace." *Third World Quarterly* 34 (5):763–783.

Mill, J. S. (1867) 1973. "A Few Words on Nonintervention." In *Essays on Politics and Culture*, 368–384. Gloucester, MA: Peter Smith.

Neumann, I. B., and O. J. Sending. 2014. *Governing the Global Polity*. Ann Arbor: University of Michigan Press.

O'Reilly, M. 2012. "Muscular Interventionism." *International Feminist Journal of Politics* 14 (4):529–548.

Orford, A. 2011. *International Authority and the Responsibility to Protect*. Cambridge: Cambridge University Press.

Paris, R. 2004. *At War's End: Building Peace after Civil Conflict*. Cambridge: Cambridge University Press.

Paris, R. 2010. "Saving Liberal Peacebuilding." *Review of International Studies* 36: 337–365.

Paris, R., and T. D. Sisk. 2009. "Introduction: Understanding the Contradictions of Postwar Statebuilding." In *The Dilemmas of Statebuilding: Confronting the Contradictions of Postwar Peace Operations*, edited by R. Paris and T. D. Sisk, 1–19. London: Routledge.

Perreau-Saussine, A. 2010. "Immanuel Kant on International Law." In *The Philosophy of International Law*, edited by S. Besson and J. Tasioulas, 53–77. Oxford: Oxford University Press.

Pinker, S. 2018. *Enlightenment Now: The Case for Reason, Science, Humanism, and Progress*. New York: Viking.

Randazzo, E. 2017. *Beyond Liberal Peacebuilding: A Critical Exploration of the Local Turn*. London: Routledge.

Rawls, J. 1999. *The Law of Peoples*. Cambridge, MA: Harvard University Press.

Rey, C., and S. McKay. 2006. "Peacebuilding as a Gendered Process." *Journal of Social Issues* 62 (1):141–153.

Richmond, O. P. 2009. "A Post-liberal Peace: Eirenism and the Everyday." *Review of International Studies* 35 (3):557–581.

Richmond, O. P. 2016. *Peace Formation and Political Order in Conflict Affected Societies*. New York: Oxford University Press.

Richmond, O. P. 2019. "Human Rights and the Development of a Twenty-First Century Peace Architecture: Unintended Consequences?" *Australian Journal of International Affairs* 73 (1):45–63.

Richmond, O. P., and R. Mac Ginty. 2014. "Where Now for the Critique of the Liberal Peace?" *Cooperation and Conflict* 50 (2):171–189.

Richmond, O. P., and A. Mitchell, eds. 2011. *Hybrid Forms of Peace: From the "Everyday" to Post-liberalism*. London: Palgrave.

Robertson, R. 1995. "Glocalization: Time-Space and Homogeneity-Heterogeneity." In *Global Modernities*, edited by M. Featherstone, S. M. Lash, and R. Robertson, 25–44. London: Sage.

Sabaratnam, M. 2017. *Decolonising Intervention: International Statebuilding in Mozambique*. London: Rowman & Littlefield.

Sen, A. 2010. *The Idea of Justice*. London: Penguin.

Seth, S. 2013. *Postcolonial Theory and International Relations: A Critical Introduction*. London: Routledge.

UN. 2018. *Peacebuilding and Sustaining Peace: Report of the Secretary General*. A/72/707-S/2018/43. New York: United Nations.

UN General Assembly. 2015. "Transforming Our World: The 2030 Agenda for Sustainable Development. A/RES/70/1." https://www.refworld.org/docid/57b6e3e44.html.

Väyrynen, T. 2010. "Gender and Peacebuilding." In *Palgrave Advances in Peacebuilding: Critical Developments and Approaches*, edited by O. P. Richmond, 137–153. London: Palgrave Macmillan.

Visoka, G. 2016. *Peace Figuration after International Intervention: Intentions, Events and Consequences of Liberal Peacebuilding*. London: Routledge.

Walzer, M. 1983. *Spheres of Justice: A Defence of Pluralism and Equality*. Oxford: Blackwell.

Walzer, M. 2004. *Arguing about War*. London: Yale University Press.

Weiss, T. G. 2013. *Global Governance: Why? What? Whither?* Cambridge, UK: Polity.

Wheeler, N. J. 2000. *Saving Strangers: Humanitarian Intervention in International Society*. Oxford: Oxford University Press.

Wolff, J. 2011. *Ethics and Public Policy: A Philosophical Inquiry*. London: Routledge.

CHAPTER 4

THE INTERNATIONAL LAW OF PEACE

CECILIA M. BAILLIET

THE articulation of peace as an essential purpose of the international order received much support after both World War I and World War II (Richmond 2007). The aim of peace influenced the establishment of institutions oriented toward implementing peace, such as the failed League of Nations, the United Nations, and the regional organizations, including, inter alia, the Organization of American States, the African Union, and the European Union (which won the Nobel Peace Prize in 2012).[1] "Peace through law" perspectives were promoted within Western scholarship through the work of Hersch Lauterpacht (1946), who linked peace to human rights, and Hans Kelsen (1944), who called for international criminal justice to support peace. Grenville Clark and Lois Sohn (1958, 1960, 1966) spent three decades designing a proposal they called *World Peace through World Law*, suggesting reform of the United Nations framework to address components of peace, including global disarmament, the establishment of an international police force, protection of the environment, compulsory use of the International Court of Justice, increased use of conciliation, and application of equity to disputes. Many of these topics remain timely; however, they are presently being discussed additionally through the lens of other articulated values, including sustainable development, solidarity, security, and cooperation (Botchway and Kwarteng 2018; Etzioni 2018).[2] This development marks a departure from the liberal international order, linked to the Pax Americana, which influenced the evolution of human rights and international law for the past fifty years.[3]

The international law of peace has been enunciated in two forms: as peaceful coexistence and as a human right to peace. Both iterations may be considered to reflect perspectives from the Global South, confirming a shift in international law and human rights marking a post-Western epoch in which changes within the Responsibility to Protect doctrine raise queries as to marginalization of human rights and the risk of security-oriented peace.

The Principles of Peaceful Coexistence

In 1648 the Peace of Westphalia ended the religious wars in Europe, inspiring the international law theorists Grotius, Pufendorf, and Vattel. They articulated the concept of a natural right to peaceful coexistence, in which sovereigns respected the right of each to determine internal matters (such as religion) without interference and to limit aggression. This was not principled pacifism, as these theorists supported just war, such as for self-defense or even to assert trade interests. The duty to pursue peaceful coexistence was owed to the people of the nation, foreign nations, and mankind. In turn, Kant elaborated a vision of perpetual world peace brought about by a world federation of republics that renounced war.

Peaceful Coexistence was further developed during the period of Decolonization. Specifically, Five Principles of Peaceful Coexistence were identified in the Sino-Indian Agreement of 29 April 1954 on Trade and Freedom of Movement: (1) Mutual Respect for Territorial Integrity and Sovereignty, (2) Nonaggression, (3) Noninterference in Internal Affairs, (4) Equality and Mutual Benefit, and (5) Peaceful Coexistence itself.[4] The principles of peaceful coexistence were later adopted within the Bandung Declaration on Promotion of World Peace and Cooperation at the 1959 Asian-African Conference. Peaceful coexistence is more often identified with negative peace rather than positive peace, and its components have been adopted within the UN Charter (Article 2) and the UN Declaration on Principles of International Law concerning Friendly Relations and Cooperation among States (UN General Assembly 1970).

On 25 June 2016 Russia and China issued a joint statement on the promotion of international law that refers to the principles of peaceful coexistence as forming the foundation for the international legal order (Russian Federation, Ministry of Foreign Affairs 2016):

> The Russian Federation and the People's Republic of China reiterate their full commitment to the principles of international law as they are reflected in the United Nations Charter, the 1970 Declaration on Principles of International Law concerning Friendly Relations and Cooperation among States in accordance with the Charter of the United Nations. They are also guided by the principles enshrined in the Five Principles of Peaceful Coexistence. The principles of international law are the cornerstone for just and equitable international relations featuring win-win cooperation, creating a community of shared future for mankind, and establishing common space of equal and indivisible security and economic cooperation.

The joint statement reaffirms fundamental principles relating to peaceful coexistence, which are enshrined within the UN Charter: sovereign equality, prohibition of the threat or use of force in violation of the Charter, nonintervention in the internal or external affairs of states, noninterference by states in the internal affairs of other states with the aim of forcing change of legitimate governments, and the peaceful settlement of disputes. In addition, Russia and China express condemnation of unilateral military interventions, unilateral sanctions, and extraterritorial application of national law by states not in conformity with international law.[5] The acknowledgment of these standards as constitutive elements is significant because it provides a measure to exchange views on state behavior, including their own. These provisions have resulted in criticism of contradictory practice, given Russia's actions in Ukraine and Crimea, as well as China's actions in the South China Sea.

Peaceful Coexistence and the Responsibility to Protect

There may be concern that the principles of peaceful coexistence may impede Responsibility to Protect (R2P) operations. However, China's policy in allegiance with the pursuit of peaceful coexistence appears to support a cautious interpretation of R2P: requiring host state consent for peacekeeping operations, limiting the collective use of force as an absolute last-resort response to mass-atrocity crimes (as opposed to generalized humanitarian situations), after exhaustion of peaceful mechanisms (including diplomacy, mediation, etc.), in accordance with broad consensus within the regional organization, and only pursuant to a UN Security Council authorization (responding to a threat or breach of the peace, not seeking to externally impose regime change). In this manner, the principle of peaceful coexistence would not be in conflict with R2P but rather would serve to ensure that the framework for operations remains in accordance with the sequential requirements of the UN Charter (Teitt 2009; Barelli 2018; Bellamy and Davies 2009; Newman 2013; Howe 2018; Bae, Infanzon, and Abbe 2018).

The normative reemergence of peaceful coexistence offered by Russia and China signals an interest in identifying the policy as forming a pragmatic basis for an international legal order that supports respect for different national traditions and political perspectives. This initiative may be considered a response to concerns about instability given global trends of polarization and ideological divisions weakening the liberal, rules-based order. However, a point of concern is the explicit identification of security and economics as fields of cooperation. Hence there is a need to analyze whether this prioritization will marginalize the role of human rights in the context of peace (UN Human Rights Council 2012).

The Declaration on the Right to Peace

The initiative to promote recognition of a "right to peace" began with its characterization as a solidarity right and was adopted in various resolutions by the UN General

Assembly, dating back to the 1978 Declaration on the Preparation of Societies for Life in Peace. The first wave of collective rights was grounded in the context of decolonization and resistance and considered to be interrelated, identifying peoples as subjects and placing a duty of cooperation on states: the right to self-determination (which is considered to be *jus cogens*), the right to development, and the right to permanent sovereignty over natural resources (UN General Assembly 1978, 944). Renewed engagement by NGOs throughout the years culminated in the establishment of the Working Group on the Right to Peace by the UN Human Rights Council that created a draft Declaration on the Right to Peace in 2016 (UN Human Rights Council 2016). Hence the right to peace is part of a second-wave of "third-generation rights" articulating African, Asian, Islamic, and Latin American concepts of collective rights, responsibility, and social justice, like the right to a healthy environment. Rosa Freedman (2013, 942) has described a trend toward hybridity: "Events within the international arena demonstrate a current revolution regarding the Global South moving away from postcolonial discourses and towards ensuring that their own, heterogeneous, norms, values and identities are recognised within international law." She suggests that certain forums within the UN, in particular the Human Rights Council, provide a space in which regional groups seek to negotiate new norms. Hybrid rights represent pluralism within human rights. What marks the new international law of peace is its characterization as an obligation owed by states to societies instead of individual-claim rights. Hence implementation of peace will be pursued in a broader manner than traditional justiciability.

It would perhaps initially appear to be noncontentious that the international community would be interested in strengthening recognition of the value of peace in the world by classifying it as a human right. This would enable peace to be promoted within the UN and regional and national human rights structures. However, this initiative sparked a division between many Western states and the majority of Eastern and Southern states. The former group exhibited reluctance in articulating peace as a human right, taking issue with its scope, form, and formulation, stating that enforcement would be vague. Peace may be perceived alternatively as a metaright or moral right (unenforceable in court), a basic human need, an aspiration, or an obligation. As Freedman (2013) explains, hybrid rights do not have a tangible substance that can be protected or a tangible victim; nevertheless they impact human rights and form part of the discourse when articulated. This section assesses the text of the Declaration to explain how the language may be interpreted from both liberal and statist/collective perspectives. It represents a new form of international law in the post-Western age which moves away from a focus on individual-claim rights and toward hybrid rights that seek to add collective perspectives to human rights output.

Preamble

The Declaration has a lengthy preamble that refers to both negative and positive elements of peace. It begins by referring to the classic principles of peaceful coexistence: prohibition of the threat or use of force against the territorial integrity or political independence of any state, peaceful settlement of disputes, the obligation of nonintervention, cooperation, equal rights and self-determination of peoples, and sovereign equality (UN General Assembly 2016).

Further, it deplores terrorism but also warns that counterterrorist measures that do not conform with human rights, international humanitarian law (IHL), or refugee law also threaten peace. It recognizes minority rights and hence signals a possible tension with counterterror agendas that target minority groups that are perceived to pose a threat to governments. There are situations where groups allegedly seeking self-determination may not be considered legitimate movements and instead may be characterized as separatist or terrorist groups that violate the right to peace. Newman (2013, 243–244) explains:

> States such as Russia, China, India and Sri Lanka would prefer to address their "internal" separatist and political challenges as domestic—not international—issues, and they do not want to support international scrutiny of "domestic" issues, as a matter of principle, even in cases in which they do not have strong direct interests; hence, Russia and China blocked the imposition of UN sanctions on Zimbabwe in 2008, and a UN resolution condemning Burma in 2007. In both cases Russia and China—whatever other interests they may have had—stated that the situations in Zimbabwe and Burma did not constitute a threat to international peace and security and were therefore domestic issues. This reflects a staunchly pluralistic worldview, and is symptomatic of a broader, fundamental, difference between loosely Western and non-Western approaches to dealing with human rights challenges.

Russia and China took a similar stance in relation to Venezuela in 2019, revealing that concern for destabilization on account of violating sovereignty takes priority over human rights. Yet the Declaration expresses that peace and security, development, and human rights are the pillars of the United Nations system and form the foundation for collective security and well-being, as well as being interlinked and mutually reinforcing. This illustrates that peace may indeed be considered an umbrella concept, like the other terms listed, and therefore requires clarification of its components and how it is to be interpreted. The puzzle is the fact that peace (as well as the other concepts) may be a prerequisite for the enjoyment of human rights, while respect for human rights may also be a necessary foundation for the enjoyment of peace. Hence there is a risk of circular reasoning with respect to these terms, as well as other terms, such as "justice" and "democracy." Indeed, there is an academic debate as to whether one may sacrifice justice in favor of peace, e.g., passing amnesty laws in transitional periods, or whether one cannot have a state of peace unless justice has been pursued through prosecution of violators of human rights (Odio Benito 1998; Grono 2018; Dancy 2018).

The Declaration calls for attention to peaceful communication and appears to adopt a liberal perspective by referring to human dignity as a fundamental criterion that may be interpreted as a potential limit to excessive state action in the name of collective peace. This is a good example of Freedman's mark of hybridity, which builds on rather than rejecting existing human rights, adding an element that was missing.

Operative Paragraphs

The operative paragraphs of the Declaration are short. Article 1 states, "Everyone has the right to enjoy peace such that all human rights are promoted and protected and development is fully realized."[6] This language suggests a negative right that shall not be interfered with by the state. On the other hand, it may also imply a positive obligation upon the state to ensure that the person benefits from the condition of peace, so that peace may be regarded as a type of metaright, like security, that enables the person to enjoy human rights (Lazarus 2007).

It is arguable that there is a liberalistic focus on the individual. Peace is the essential foundation to enable a person to live his or her life in dignity in pursuit of his or her aspirations. Yet a statist or Global South perspective would highlight the articulation of peace as a collective aim that the state has an active role to pursue, so that peace as a value is not exclusively individualistic in nature. From this perspective, this is not an articulation of a claim right but a nonjusticiable obligation. Samantha Besson (2013) has clarified the difference between duties and responsibilities in human rights; the former refer to specific contexts and threats and are held by states and by institutions that have effective control over persons, while responsibilities are abstract moral requirements held by individuals and nonstate institutions. Following this analysis, peace would appear to have hybrid characteristics merging aspects of responsibility and duty.

Article 2 of the Declaration places primary responsibility on states to guarantee the structural framework necessary for individual enjoyment of both internal and external peace: "States should respect, implement and promote equality and non-discrimination, justice and the rule of law, and guarantee freedom from fear and want as a means to build peace within and between societies." This recognizes the importance of strengthening the key components of positive peace policies and actually repeats identification of constituent rights that also pertain to the metaright security/core human security concept: equality and freedom from fear and freedom from want (Burgess and Gräns 2012; see also: UN General Assembly 2005, para. 143). In this sense, the right to peace actually incorporates human security. A liberal interpretation would imply the role of the legislature and judiciary to protect marginalized members of society, thereby connecting to sustainable development goal 16 on peace, justice, and strong institutions. The statist view would emphasize the importance of maintaining the authority of the law as a means to guarantee peace; state action, rather than separation of powers, is the vehicle for attainment of peace on behalf of collective society. The use of the word "should" instead of "shall" implies a non-binding duty that limits the ability to make a legal claim or receive a remedy. The identification of human security in Article 2 coupled with the recognition of development in Article 1 indicates acceptance of Asian, African, and Latin American perspectives on socioeconomic rights as providing the key to ensuring people adequate lives, which in turn will buttress peace.

Article 1 it is unclear as to whether states are expected to protect individuals or the international community from violations by third parties, such as corporate actors, foreign states, or international organizations, nor is it obvious what degree of effort by the state is necessary to fulfill the obligation. The Declaration leaves open the possibility for use of regular accountability mechanisms within the UN, regional, and national human rights mechanisms for reporting, observation, and follow-up, such as the UN Charter–based bodies or treaty bodies (UN Human Rights Office of the High Commissioner n.d.). The duty is oriented toward both the individual and the international community in order to underscore and protect the community value of peace. It is likely that NGOs and other actors will interpret the scope of the right to peace expansively.

Article 3 refers to a transnational framework for broad institutional enforcement of peace rather than identify specific remedies: the UN, specialized agencies, international, regional, national, and local organizations, and civil society are to support and assist the implementation of the Declaration. Universalists and those favoring hybrid approaches to human rights would argue that nonstate actors have a role to play in guaranteeing peace. This is important given that some states are unwilling or unable to fulfill their roles or to complement the state in order to attain a more holistic implementation (see UN Human Rights Office of the High Commissioner 2011). Statist approaches would consider that international organizations and civil society have an obligation to support the state in its pursuit of peace. The language appears in keeping with identification of fulfillment of a responsibility by multiple, transnational actors to promote peace as a prerequisite for enjoyment of human rights, in keeping with hybrid traditions on implementation of collective rights.

Article 4 calls for transnational form for institutional promotion of peace education: "International and national institutions of education for peace shall be promoted in order to strengthen among all human beings the spirit of tolerance, dialogue, cooperation and solidarity." This suggests the creation of a culture of peace to build trust, empathy, and skills for nonviolent dispute resolution. Statist regimes would highlight the values of cooperation and solidarity as strengthening communitarian policies.

Article 5 of the Declaration states that its provisions are to be read in accordance with the UN Charter, indicating that it is an instrument of Realistic Peace, not Principled Peace. This provision suggests that the right to peace is a qualified right, not an absolute right. It is implied that the UN Charter's exceptions to the prohibition of the use of force, i.e., self-defense in accordance with Article 51 and authorization by the UN Security Council under Chapter VII predominate. Thus, it is necessary to reflect on the relationship between the Right to Peace and the Responsibility to Protect doctrine.

The Right to Peace and the Responsibility to Protect

A question arises as to the possible interface between the right to peace and the Responsibility to Protect doctrine (UN General Assembly 2005, para. 138). R2P is a political commitment (not a legal obligation) that aims to prevent the root causes of internal conflict and respond to mass-atrocity crimes with adequate measures. States have the primary responsibility to protect their populations against international crimes, and the international community has the responsibility to assist states in protecting human rights through peaceful means subject to state consent (e.g., preventive diplomacy, development cooperation, dialogue, training, disarmament, demobilization, reintegration, transitional justice, and peacekeeping) or act collectively in accord with the UN Charter. Coercive measures may include sanctions, prosecution, or, in extreme situations, military intervention to protect people from crimes against humanity, war crimes, genocide, or ethnocide only when authorized by the UN Security Council on a case-by-case basis. The articulation of a *responsibility* to act, as opposed to a *right* to intervene, harmonizes with the orientation of the right to peace toward recognizing the common responsibility of the state and international community for the fulfillment of peace. The UN's 2005 World Summit Outcome Document explains that intervention may be appropriate "should peaceful means prove inadequate and national authorities manifestly fail to protect their populations." (UN General Assembly 2005, para. 139). However, in terms of establishing a normative foundation, Monica Hakimi (2018) argues that UN Security Council resolutions authorizing the use of force are particularistic and do not set clear precedents for similar cases, and that academic analysis has resulted in contradictory conclusions; hence their future impact on establishing a customary international standard may be limited.

Furthermore R2P is tainted by accusations that it is selectively applied, as various initiatives have been vetoed within the UN Security Council, abandoning people in Gaza, Syria, Myanmar, Venezuela, Yemen, and elsewhere. Meanwhile the actions that have been authorized—in Darfur, Libya, Côte d'Ivoire, Mali, and the Central African Republic—have been characterized as either promoting regime change, marked by the participation of former colonial powers, or lacking effectiveness in fulfilling their mission. There is vagueness resulting from the blurring of boundaries between war and peace; one may note Osterdahl and van Zadel's (2009, 205) suggestion that one refer to a continuum between the two instead of distinct phases. There is also a tension between respecting the principle of sovereignty and the challenge faced by the international community to respond to mass-atrocity crimes and support the establishment of functioning institutions within territories that have deep societal divisions and a history of dysfunctional state actors. As Lemay-Hébert and Visoka (2017, 147) note, the pursuit of "normal peace" practices is a phenomenon "aimed at disciplining and regulating societies deemed 'abnormal' or 'dysfunctional' through wide-ranging forms of interventions."

Suhrke (2014, 282) confirms the tendency of the international community to ignore local law and instead transplant Western reforms, which are then criticized or ignored by the local population. In fact she concludes that "large, international peace operations tend to undermine the development of effective and legitimate state power and institutions of justice that can defuse tension, address sources of conflict, and restrain violence." Chinkin and Kaldor (2017, 542) concur that "the international presence institutionalizes the causes of conflict, disempowers those who would be the drivers of change, and either freezes an intolerable status quo, and/or sets the scene for the next round of protest and conflict." Richmond (2015, 6–10) suggests that "peace has been narrowed to a focus on security and the state; hence the discourse of statebuilding, which is oriented toward security, rights, and markets but offers little in the way of a peace dividend that might be attractive or necessary at the local level." Suhrke suggests that the international community pursues a stereotype of a postwar state that assumes the predisposition of the state and society toward disorder and violence. She offers the critique that, on the contrary, some postwar states are strong states that use targeted violence, indirect violence (denial of access to basic necessities), detention, and coercion against social groups perceived to be the enemy. In this scenario, Suhrke recommends that the state institutions need "taming" (such as through human rights training) rather than building up the national police. Alternatively, in the scenario of a weak state, she opines that it may be better to establish order and services by supporting traditional dispute-resolution institutions, public-private partnerships, and community policing, thereby bypassing "statebuilding as the central objective" (Suhrke 2014, 274). She credits the human rights regimes and the inclusion of human rights field missions in UN peace operations for monitoring violations and supporting local human rights organizations. Indeed the UN Secretary General's (2019) "Report on Responsibility to Protect" addressing prevention asserted that there is a need to retain human rights engagement: "[T]he field presence of international actors—for instance, on human rights monitoring, investigation and reporting, engagement with civil society and local communities—can be instrumental in contributing to prevention." It cited political, social, economic, gender, and religious discrimination as the roots of conflict (see also UN Secretary General 2013).

The right to peace harmonizes with R2P to the extent that the latter recognizes the responsibility to rebuild society and promote recovery and reconciliation, addressing root causes of violence through the prism of security, justice, and sustainable development, thereby interfacing with positive peace. These measures span support for electoral systems, property registration and dispute resolution mechanisms, judicial and police reform, agricultural markets, access to water, healthcare, etc. Yet in practice challenges to peace consolidation in the rebuilding phase are linked to the common phenomenon of complaints regarding victor's justice that impedes reconciliation, continues gender imbalance in participation within transitional institutions and programs, and maintains impunity and exclusionary power hierarchies that impede the transparent functioning of justice and political institutions. UN and regional peacekeeping missions are increasingly emphasizing security and counterterror operations at the expense of improving governance and participation of different groups within institutions, in spite of being formally encouraged to pursue sustainable peace strategies (International Peace Institute 2018; UN 2016).

The language in UN Security Council (2018) Resolution 2247 on peacekeeping appears to be broad and holistic, addressing long-term governance and development perspectives. It "reiterat[es] the need for a comprehensive approach to conflict prevention and sustainable peace, which comprises operational and structural measures for the prevention of armed conflict and addresses its root causes, including through strengthening the rule of law at international and national levels and promoting sustained economic growth, poverty eradication, social development, sustainable development, national reconciliation, good governance, democracy, gender equality and respect for, and protection of, human rights." Yet the same resolution also emphasizes the primary responsibility of national authorities to define strategies for peacebuilding and sustaining peace, which appear to have a narrow focus. Specifically, the resolution emphasizes the importance of national reform of police, justice, and corrections institutions as a key to building peace and ending impunity. The UN and civil society are identified as playing supportive roles to the state:

> Reaffirming the primary responsibility of national governments and authorities in identifying, driving and directing priorities, strategies and activities for peacebuilding and sustaining peace, emphasizing that sustaining peace is a shared task and responsibility that needs to be fulfilled by the Government and all other national stakeholders, and in this regard, emphasizing that inclusivity is key to advancing national peacebuilding processes and objectives in order to ensure that the needs of all segments of society are taken into account, and stressing that civil society can play an important role in advancing efforts to sustain peace . . .
>
> Reaffirms the importance of national ownership and leadership in peacebuilding, whereby the responsibility for sustaining peace is broadly shared by the Government and all other national stakeholders, and recognizes that United Nations police, justice and corrections components can contribute to building and sustaining peace by supporting host-State police, justice and corrections institutions, as mandated.

The current normative output remains largely state-centric, in spite of recognition of the role of nongovernmental actors, as they are clearly given supplementary roles. Fox (2014, 257) has described postconflict states as kind of "twilight zone" in which "some actors receive Chapter VII mandates but have no authority to supervene local actors or bilateral donors . . . [and others] have authorizations that are limited topically or geographically, [or as third scenario] local actors are created by international actors. . . . Because the nature of each actor dictates the applicable law the result will be a patchwork of obligations following no necessary hierarchy."

As the UN itself does not have an army, R2P is further complicated by a pluralistic, opaque framework for accountability, given the number of UN-authorized missions that are either conducted, buttressed, or, on the contrary, not sufficiently supported by external, regional, and subregional institutions, such as the North Atlantic Treaty Organization, Economic Community of West African States, Association of Southeast

Asian Nations, Organization for Security and Cooperation in Europe, Arab League, Organization of Islamic Cooperation, European Union, or African Union (Kingah and Seiwert 2016). There is often a lack of clarity in the legal mandates drawn up by the UN Security Council or regional or subregional organizations, while the practical implementation of the mandate is conducted by the troop-contributing countries on the ground, often overseen by regional/external actors or stakeholders (including civil society groups and private actors) that may be coordinating or cooperating with missions. In turn, the missions are subject to pressure by host governments which themselves may be held responsible for the loss of lives or displacement of civilians. Jennifer S. Easterday (2014, 384) adds the observation that interinstitutional (international and national) power rivalries also prevent cohesive peacebuilding.

One aspect to take into consideration is the impact of the geographic imbalance regarding roles within peace missions. The majority of troop-contributing countries are from Africa and South Asia, and within the permanent members in the UN Security Council only China is among the top ten Blue Helmets (UN 2018). In terms of funding, the largest donors are the United States, China, and Japan; hence there are different priorities among the actors that can create gaps and inconsistencies. As an example, in spite of the fact that the UN Security Council claims that an integrated, strategic, and coherent approach to peacebuilding requires recognition that security, development, and human rights are closely interlinked and mutually reinforcing, there has been a notable decrease in emphasis on human rights in favor of security in terms of staffing, due to the preferences of China and Russia, thereby weakening the mechanism for "taming" the state favored by Suhrke (see UN Security Council 2018; Gladstone 2018).

Peace Agreements, *Jus Post Bellum*, and the Human Security Concept

Finally, additional normative confusion may result from national peace agreements that impact the interpretation of the international law of peace by their reference to international norms, such as the right to self-determination and the expectation of justice (Sheeran 2011). Peace agreements are strongly influenced by their specific context, and this may result in compromises that run counter to positivistic expectations of retributive accountability (required by international human rights law and international criminal law) in favor of restorative justice, as stated by Hillebrecht, Huneeus, and Borda (2018, 280): "The Colombian solution to the peace/justice dilemma is to offer lighter punishment for the most heinous crimes, in exchange for greater involvement by the wrongdoer in legal processes aimed at repairing the harm and restoring social bonds."

Chinkin and Kaldor (2017) note that peace agreements and transitional justice mechanisms rarely incorporate economic and social rights that lie at the heart of causes of conflict and are the key to increasing equality and stability within society, according to a positive peace approach. Suhrke (2014, 273) is supportive of recognition of the need for flexible perspectives: "While it is expected and often claimed that accountability mechanisms will strengthen rule of law and thereby peace and democracy, the statistical evidence does not support this view. The case-study literature has documented difficult trade-offs and significant costs in terms of social and political conflict. A legal obligation to institute transitional justice mechanisms, e.g. by mandating provisions to this effect in peace agreements approved by the Security Council, would deny the importance of context."

Easterday (2014, 380) further notes the encumbrances of peace agreements, which: "are limited by who sits at the table and can result in counter-productive political arrangements. . . . They also leave gaps and silences with respect to critical issues, such as gender considerations, that can undermine peace efforts. With international involvement, they may reflect neo-colonialist tendencies. . . . [T]here is little proof that constitutional reform through peace agreements contributes to lasting or sustainable peace." As the next step, Easterday credits the R2P doctrine (as well as democratic governance and transitional justice) as one of the conceptual normative sources for the emergence of the subfield *jus post bellum*, which she broadly interprets as being composed of principles, norms, and rules from treaties, customary international law, and soft law from different fields of international law, peace agreements, and informal arrangements and practices from nonstate actors. She argues that "in order to foster sustainable peace, the application of particular *jus post bellum* laws and norms would be measured against peace-oriented policy goals and principles that seek to maximize and reinforce the peace project" (383). Easterday presents the puzzle within *jus post bellum* as fragmentation, as issues may be framed as "human rights" or "security" issues, resulting in battles between institutions linked to the different regimes, thereby inhibiting a cohesive peacebuilding strategy and leading to a high rate of failed peace agreements.

In comparison, Chinkin and Kaldor's (2017) support for recognition of the relevance of the human security concept may be considered an attempt to bridge the gap between the two regimes. They identify human security as "what individuals enjoy in rights-based law-governed societies" (32) and as constructed by Human Rights Law, International Criminal Law, IHL, Refugee Law, and principles with respect to Internally Displaced Persons. However, they suggest that this forms a "new law of peace" (565) which is itself compartmentalized and fragmented and should be joined with the Sustainable Development Goals. People who suffer from human rights violations without legal mechanisms for redress are much more likely to resort to violence, and the absence of accountability for perpetrators of violations often encourages them to continue to violate. In contrast, societies in which human rights are upheld and protected are more resilient to crisis factors and routinely deal with grievances through peaceful systems.

Yet the conceptualization of human rights in conflict settings is challenging; they are norms with political, social, cultural, anthropological, ethical, legal, and economic implications to be interpreted and applied in societies in flux. In addition, human rights vary from conceptions in universal instruments, regional instruments, and national constitutions. This indicates that peace strategies should map the legal and social culture and contexts before designing a human rights–based framework for conflict prevention and sustainable peace. The key challenge is how the international community can legitimize human rights in the local context. Should the focus be on promotion of a culture of peace, support to administrative agencies and civil society for reconciliation activities over traditional accountability mechanisms?

The puzzle is how to address the "conflict continuum" through the lens of human rights? Can the integration of a human rights–based approach lead to better conflict prevention? Initiatives to remove limitations of freedom of expression, assembly, and association are intended to create viable spaces for a pluralistic civil society and reduce the risk of violence; however, the need to restrict hate speech and discrimination that may incite violence results in contradictory tensions. Moreover the focus on civil and political rights needs to be balanced with the right to an adequate standard of living, food, healthcare, and employment.

Thus the international law of peace is malleable, perpetually under construction, and may well be characterized as transnational law, due to the primacy of the local context in measuring its legitimacy and influencing its evolution. The R2P doctrine raises a quandary: while the right to peace incorporates respect for human rights, the current orientation of R2P missions in practice may imply that the UN's purpose of preserving peace and security may de facto take priority over its other purpose, to promote human rights, in spite of the written commitment of the normative instruments to pursue balance. Hence there is potential incongruity between the right to peace and R2P that is due to marginalization of human rights. Engdahl (2019) calls for recognition of the common ground between the three concepts; indeed there is need for clarification in order to ensure effective implementation of the international law of peace. Chinkin and Kaldor (2017, 33) assert that human security means "[r]epresentative and accountable international authority for the use of force, the recasting of war as a humanitarian catastrophe and a massive violation of human rights, the recasting of self-defence as scaled up self-defence of individuals not states, and the application of human rights law, as well as IHL, to any use of force, as well as large scale disarmament." It is a vision with peace at the core but will require a consequential change of orientation within the international community.

Future Evolution of the International Law of Peace

In 2017 the UN Human Rights Council adopted "Promotion of the Right to Peace," which underscored the elements of positive peace, including dialogue for conflict resolution in order to attain development. The United States published its explanation of its vote against the declaration, serving as confirmation of its status as a persistent objector to the evolution of a customary international right to peace. The measure passed nonetheless because of the fundamental will of the states of the Global South to support the elucidation of the collective interest and public good of peace.[7]

This trend mirrors the UN's platform for sustainable development goals to address the global challenges of poverty, inequality, climate, environmental degradation, prosperity, peace, and justice. Indeed, at the regional level, the Inter-American Court of Human Rights has underscored the justiciability of "second- and third-generation" rights, pursuing transformative constitutionalism; hence it is possible to suggest that the right to peace may develop through crystallization into customary international law (Bogdandy et al. 2017). Western states that object to this evolution may be concerned about alleged potential constraining effects of "third-generation" rights upon state action, but their objections reveal a lack of understanding of how to interpret hybrid rights. Furthermore, it may be argued that the modern epoch is characterized by the recognition of the pluralistic responsibility of nonstate actors, including multinational corporations, for respecting human rights, both individual and collective. The new international law of peace represents a call for recognition of values held by diverse societies. It is likely that the international law of peace and human rights will continue to evolve in the post-Western epoch characterized by new iterations of community interests (Besson 2018).

Notes

1. The EU Treaty of Lisbon, Article 2, states, "The Union's aim is to promote peace, its values, and the well-being of its peoples." The Organization of American States was created to achieve peace and justice and to strengthen peace and security within the continent; the OAS Charter recognizes the principles of peaceful dispute resolution; of particular interest, social justice and social security are recognized as forming the bases of lasting peace. Article 3 of the African Union Constitution sets forth as objectives to promote peace, security, and stability on the continent. Its principles include the peaceful coexistence of member states and the right to live in peace and security, as well as peaceful resolution of conflicts.
2. See UN Human Rights Council (2018b): "States have a collective responsibility to uphold the principles of human dignity, equality and equity at the global level"; UN Human Rights Council (2018a): "[I]nternational solidarity is a broader concept and principle that includes sustainability in international relations, especially international economic relations, the peaceful coexistence of all members of the international community, equal partnerships and the equitable sharing of benefits and burdens. . . . [T]he promotion of international cooperation is a duty for States. . . . [I]nternational solidarity is a powerful tool for addressing the structural causes of poverty, inequality and other global challenges: the widening gap between economically developed and developing countries is unsustainable and . . . impedes the realization of human rights in the international community."

3. In 2018 the United States withdrew from the UN Human Rights Council, blocked the re-appointment of a World Trade Organization Appellate Body member, and denounced the legitimacy of the International Criminal Court. In contrast, China has increased its contributions to UN institutions. It has also supported the creation of new institutions, such as the Asian Infrastructure Investment Bank and Belt and Road Tribunals to provide dispute resolution of commercial contracts.
4. India and People's Republic of China: Agreement (with exchange of notes) on trade and freedom of movement between Tibet Region of China and India. Signed at Peking on 29 April 1954 (UN 1958, 57–81). This agreement confirmed that Tibet fell within the territory of China.
5. See Wuerth (2018), explaining how selectivity and politicization occurring in cases involving enforcement of human rights in the context of humanitarian interventions weaken international law and may threaten peace.
6. UN General Assembly, Declaration on the Right to Peace: resolution/adopted by the General Assembly, 2 February 2017, A/RES/71/189, available at: https://www.refworld.org/docid/589c72134.html [accessed 13 November 2020]

   The Declaration is available at: http://www.un.org/ga/search/view_doc.asp?symbol=A/C.3/71/L.29 https://documents-dds-ny.un.org/doc/UNDOC/LTD/N16/352/85/PDF/N1635285.pdf? OpenElement
7. Additional solidarity rights include the right to intergenerational equity and the right to participate in cultural heritage.

References

Bae, S., A. Infanzon, and M. J. Abbe. 2018. "Active in Not Being Active (AINBA): How East Asian Powers Accept R2P." *Asian Journal of Political Science* 26:1, 16–34, doi:10.1080/02185377.2017.1366348.

Barelli, M. 2018. "Preventing and Responding to Atrocity Crimes: China, Sovereignty, and the Responsibility to Protect." *Journal of Conflict & Security Law* 23 (2): 173–201.

Bellamy, A. J., and S. E. Davies. 2009. "The Responsibility to Protect in the Asia-Pacific Region." *Security Dialogue* 40 (6): 547–574. https://doi.org/10.1177/0967010609349907.

Besson, S. 2013. "The Allocation of Anti-Poverty Rights Duties: Our Rights but Whose Duties?" In *Poverty and the International Economic System*, edited by K. Nadakavukaren Schefer, 408–432. Cambridge: Cambridge University Press.

Besson, S. 2018. "Community Interests in the Identification of International Law: With a Special Emphasis on Treaty Interpretation and Customary Law Identification." In *Community Interests across International Law*, edited by Eyal Benvenisti and Georg Nolte, 36–49. Oxford: Oxford University Press.

Bogdandy, A., E. Ferrer Mac-Gregor, M. Morales Antoniazzi, F. Piovesan, and X. Soley. Eds. 2017. *Transformative Constitutionalism in Latin America: The Emergence of a New Jus Commune*. Oxford: Oxford University Press.

Botchway, T. P. and A. H. Kwarteng. 2018. "Developing International Law in Challenging Times." *Journal of Politics and Law* 11: 53–63.

Burgess, P., and J. O. N. A. S. Gräns. 2012. "Human Security." *Contemporary Security and Strategy*, 89–105.

Chinkin, C., and M. Kaldor. 2017. *International Law and New Wars.* Cambridge: Cambridge University Press.

Clark, G., and L. B. Sohn. 1958, 1960, 1966. *World Peace through World Law.* Cambridge, MA: Harvard University Press.

Dancy, G. 2018. "Deals with the Devil? Conflict Amnesties, Civil War, and Sustainable Peace." *International Organization* 72 (2): 387–421.

Easterday, J. S. 2014. "Peace Agreements as a Framework for Jus Post Bellum." In *Jus Post Bellum: Mapping the Normative Foundations*, edited by Carsten Stahn, Jennifer Easterday, and Jens Iverson, 379–415. Oxford: Oxford University Press.

Engdahl, O. 2019. "Protection of Human Rights and the Maintenance of International Peace and Security: Necessary Precondition or a Clash of Interests." In *The Research Handbook on International Law and Peace*, edited by Cecilia Bailliet, 127–145. Cheltenham, UK: Edward Elgar.

Etzioni, A. 2018. "The Rising (More) Nation Centered System." *Fletcher Forum of World Affairs* 42: 29.

Fox, G. H. 2014. "Navigating the Unilateral/Multilateral Divide." In *Jus Post Bellum: Mapping the Normative Foundation*, edited by Carsten Stahn, Jennifer S. Easterday, and Jens Iverson, 229–258. Oxford: Oxford University Press.

Freedman, R. 2013. "Third-Generation Rights: Is There Room for Hybrid Constructs within International Law?" *Cambridge Journal of International and Comparative Law* 2 (4): 935–951.

Gladstone, R. 2018. "China and Russia Move to Cut Human Rights Jobs in UN Peacekeeping." *New York Times*, 27 June.

Grono, N. 2018. *Negotiating Peace and Justice: Considering Accountability and Deterrence in Peace Processes.* International Crisis Group.

Hakimi, M. 2018. "The Jus Ad Bellum's Regulatory Form." *American Journal of International Law* 112 (April): 151–190.

Hillebrecht, C., A. Huneeus, and S. Borda. 2018. "The Judicialization of Peace." *Harvard International Law Journal* 58 (Summer): 279–330.

Howe, B. 2018. "Divergent Interpretations of the R2P and Human Security: Implications for Governance Challenges in Myanmar." *Korean Journal of Security Affairs* 23 (1): 80–101.

International Peace Institute. 2018. *Prioritizing and Sequencing Peacekeeping Mandates: The Case of MINUSMA.* May. New York: International Peace Institute.

Kelsen, H. 1944. *Peace through Law.* Chapel Hill: University of North Carolina Press.

Kingah, S., and E. Seiwert. 2016. "The Contested Emerging Norm and Practice of a Responsibility to Protect: Where Are the Regional Organizations?" *North Carolina Journal of International Law* 42 (Fall): 115–189.

Lauterpacht, H. 1946. "The Grotian Tradition in International Law." *British Yearbook of International Law* 23: 1–51.

Lazarus, L. 2007. "Mapping the Right to Security." In *Security and Human Rights,* edited by B. Goold and L. Lazarus, 325–346. London: Bloomsbury.

Lemay-Hébert, N., and G. Visoka. 2017. "Normal Peace: A New Strategic Narrative of Intervention." *Politics & Governance* 5 (3): 146–156.

Newman, E. 2013. "R2P Implications for World Order." *Global Responsibility to Protect* 5 (3): 235–259.

Odio Benito, E. 1998. "Justice for Peace: No to Impunity." In *Reining in Impunity for International Crimes and Serious Violations of Fundamental Human Rights: Proceedings of the Siracusa Conference 17–21 September 1998 (Association Internationale de Droit Penal)*, edited by Christopher C. Joyner, 149–154. Toulouse, France: Érès.

Osterdahl, I, and E. van Zadel. 2009. "What Will Jus Post Bellum Mean? Of New Wine and Old Bottles." *Journal of Conflict and Security Law* 14: 175–207.

Richmond, O. P. 2007. *The Transformation of Peace*. Palgrave.

Richmond, O. P. 2015. "*The Dilemmas of a Hybrid Peace: Negative of Positive?*" *Cooperation & Conflict* 50: 50–68.

Russian Federation, Ministry of Foreign Affairs. 2016. "The Declaration of the Russian Federation and the People's Republic of China on the Promotion of International Law." 25 June. http://www.mid.ru/en/foreign_policy/position_word_order/-/asset_publisher/6S4RuXfeYlKr/content/id/2331698.

Sheeran, S. P. 2011. "International Law, Peace Agreements, and Self-Determination: The Case of the Sudan." *International and Comparative Law Quarterly* 60 (2): 423.

Suhrke, A. 2014. "Post-War States: Differentiating Patterns of Peace." In *Jus Post Bellum: Mapping the Normative Foundations*, edited by C. Stahn, J. Easterday, and J. Iverson, 267–284. Oxford: Oxford University Press.

Teitt, S. 2009. "Assessing Polemics, Principles and Practices: China and the Responsibility to Protect." *Global Responsibility to Protect* 1 (2): 208–236.

UN. 1958. *Treaty Series*. Vol. 299. https://treaties.un.org/doc/publication/unts/volume%20299/v299.pdf.

UN. 2016. "Security Council Unanimously Adopts Resolution 2282 (2016) on Review of United Nations Peacebuilding Architecture." 27 April. https://www.un.org/press/en/2016/sc12340.doc.htm.

UN. 2018. "Summary of Troop Contributing Countries by Ranking." 31 December. https://peacekeeping.un.org/sites/default/files/2_country_ranking_8.pdf.

UN General Assembly. 1970. Declaration on Principles of International Law concerning Friendly Relations and Cooperation among States in accordance with the Charter of the United Nations, A/RES/2625(XXV). 24 October.

UN General Assembly. 1978. Declaration on the Preparation of Societies for Life in Peace. A/RES/33/73, 15 December. http://www.un-documents.net/a33r73.htm.

UN General Assembly. 2005. World Summit Outcome. A/RES/60/1, 24 October.

UN General Assembly. 2016. Declaration on the Right to Peace. A/C.3/71/L.29, 31 October. http://www.un.org/ga/search/view_doc.asp?symbol=A/C.3/71/L.29.

UN Human Rights Council. 2012. "Open-Ended Intergovernmental Working Group on a Draft United Nations Declaration on the Right to Peace." http://www.ohchr.org/EN/HRBodies/HRC/RightPeace/Pages/WGDraftUNDeclarationontheRighttoPeace.aspx.

UN Human Rights Council. 2016. Draft declaration of the right to peace. A/HRC/32/L.18, 1 June.

UN Human Rights Council. 2017. *Promotion of the Right to Peace*. A/HRC/35/L.4, 16 June.

UN Human Rights Council. 2018a. Resolution 38/2 on Human Rights and International Solidarity. A/HRC/Res/38/2, 16 July.

UN Human Rights Council. 2018b. Resolution 38/3 Enhancement of International Cooperation in the Field of Human Rights. A/HRC/Res/38/3, 16 July.

UN Human Rights Office of the High Commissioner. n.d. "Human Rights Bodies." http://www.ohchr.org/EN/HRBodies/Pages/HumanRightsBodies.aspx.

UN Human Rights Office of the High Commissioner. 2011. *Guiding Principles on Business and Human Rights*. http://www.ohchr.org/Documents/Publications/GuidingPrinciplesBusinessHR_EN.pdf.

UN Secretary General. 2013. "Responsibility to Protect: State Responsibility and Prevention." A/67/929-S/2013/399, 9 July.
UN Secretary General. 2019. "Report on Responsibility to Protect: Lessons Learned for Prevention." A/73/898-S/2019/463, 10 June.
UN Security Council. 2018. Resolution 2247 (2018). S/RES/2447, 13 December.
Wuerth, I. 2018. "International Human Rights Law: An Unexpected Threat to Peace." *Marquette Law Review* 10(3): 803–818.

CHAPTER 5

THE SOCIAL CONSTRUCTION OF PEACE

JOANNE WALLIS

THE word "peace" is littered throughout treaties and declarations of the United Nations and other international institutions, where it is uncritically presented as the ultimate goal of international cooperation. Yet there is no universally valid definition of peace, which has no intrinsic meaning or metaphysical existence. Instead constructivist approaches demonstrate that peace is socially constructed; it is the product of human agency, and ideas and practices relating to peace are constituted and instantiated within intersubjective social contexts. In short, peace means different things to different actors in different contexts. Although constructivist approaches represent a broad church, they share fundamental principles. First, that actors "act towards objects, including other actors, on the basis of meanings that the objects have for them." Second, that "[a]ctors acquire identities—relatively stable, role-specific understandings and expectations about self—by participating in such collective meanings." These "identities are inherently relational"; that is, they exist within a specific "social world." Third, that "identities are the basis of interests," which are themselves socially constructed and, in turn, inform actions. Fourth, that "[a]n institution is a relatively stable set or 'structure' of identities and interests" (Wendt 1992, 396–399). Consequently, ideational structures, that is, systems of shared ideas, beliefs, and values, are as important as material structures in shaping the behavior of the agents who operate in them. Power plays a role in these processes, and constructivism recognizes the "continual contest for control over the power necessary to produce meaning in a social group" (Hopf 1998, 180).

Therefore constructivism allows us to recognize that peace is "mutually constituted by actors employing constitutive rules and social practices" and can have "multiple meanings for different actors based on their own communities of intersubjective understandings and practices" (Hopf 1998, 174). For our analysis of the social construction of peace, we will first consider how peace is constituted and known as an idea or norm, and then how it is reproduced and institutionalized through social practices. Although these two processes are discussed sequentially, there is a recursive loop between them: ideas and norms structure practices, which create ideas and norms (Neumann 2002).

The Agents of Peace

The first question we must consider is who the agents of peace are. Constructivism has traditionally focused on states and has argued that peace resembles a "security community" (Adler 1998, 167). This concept describes the integration of states "to the point that they have a sense of community, which, in turn, creates the assurance that they will settle their difference short of war" (Adler and Barnett 1998, 3). But, reflecting our interest in peacebuilding, we will follow peacebuilding literature and practice to look inside states. Most notably, the "local turn" has advocated "a central role for local people as agents for peace" (Paffenholz 2015, 857). If we assume that actors' identities are socially constructed, this suggests that there are no ideal-type or unitary actors; instead the agents of peace are contested, fragmented, and relational. Agents and structures are also co-constitutive, and there are constant processes of negotiation and renegotiation between them. For example, in the peacebuilding context, the "local" is "partially produced by what internationals find, initiate or are willing to fund" (Heathershaw 2013, 280), as is the "international" by "locals." This reflects that there are power dynamics evident in our decisions about which actors we identify as exercising agency with respect to peace. Indeed while the local turn represents a positive move to consider the day-to-day dynamics of peace, there is a danger in focusing too tightly on the local level and overlooking the influence of the "institutional and geopolitical context" (Pingeot 2018, 365).

The political geography concept of scale has been proposed as a way to avoid idealizing and dichotomizing agents and structures. Scale refers to "hierarchized social, political and economic territorial spaces" (Hameiri and Jones 2017, 56), each denoting "the arena and moment, both discursively and materially, where socio-spatial power relations are contested and compromises are negotiated and regulated" (Swyngedouw 1997, 140). Although this concept has not been explicitly linked to constructivism, it nevertheless reflects that approach by arguing that "scales . . . are not natural; they are (re)produced through strategic agency and socio-political contestation" (Hameiri and Jones 2017, 60). Questions of agency and scale highlight the importance of interrogating ontological assumptions and their epistemological consequences, including their role in constituting social reality. Ontology is concerned with "the claims or assumptions that a particular approach to social enquiry makes about the nature of social reality—claims about what exists, what it looks like, what units make it up and how these units interact with one another" (Blaike 1993, 6). Epistemology refers to "the claims or assumptions made about the ways in which it is possible to gain knowledge of reality" (6–7). An interest in epistemology highlights that human knowledge is situated; that is, it has a "*causal* history, even if it also has a rational structure," and "its status *as* knowledge is dependent upon its possessors being located in a particular kind of situation" (Antony 2006, 60).

Guided by the concept of the "liberal peace" (Doyle 2005), both the discourse and the practice of peacebuilding at the international level have been dominated by ontological individualism, that is, the ontological assumption that individual humans are the sole, unique, and ultimate constituents of social reality to which all else is reducible, including institutions, states, and markets. International peacebuilding interventions in conflict-affected societies have consequently aimed at institutionalizing the main tenets of the liberal peace: democratization, the rule of law, human rights, and free-market economies (Richmond 2006). However, the coherence of the liberal peace should not be overstated, and many international peacebuilding interventions have instead settled for the more "*status quo* aspirations" of stability and security (Chandler 2010, 148).

As constructivism holds that the origins of knowledge and practices are social, this highlights the importance of an ontology that focuses on relational processes of knowing and practicing peace. This draws on the work Lev Vygotsky (1981, 163), who saw all individuals as inextricably related, both to each other and to their physical environment, and argued that "social relations or relations among people genetically underlie all higher (mental) functions and their relationships." This underlines that we need to take a reflexive and contextual approach that provides space for recognizing a more relational ontology in many conflict-affected societies. Such a relational ontology should recognize that "a wider range of—sometimes hidden—inter-related identities, individuals, social collectivities, regimes and systems are involved in determining everyday social reality" (Wallis and Richmond 2017, 427). A relational ontology could focus on relationships, rather than individuals or groups, in order to identify how agency is exercised with respect to peace. However, such an approach should not claim that relationships are free of power relations but instead reveal which power relations and actors are obstacles to peace. Therefore analyses of the social construction of peace need to be alive to "ontological asymmetry," which asks us to pay attention to "whether and how one group's" narratives are marginalized and delegitimized while another's are backed by power and enjoy broader societal appeal (Rumelili and Celik 2017, 285).

A related concept informed by constructivist thought is that of "ontological security," which refers to security of identity, that is, "the need to experience oneself as a whole, continuous person in time—as being rather than constantly changing—in order to realise a sense of agency" (Mitzen 2006, 342). However, this emphasis on individuals reveals the "liberal agendas" of many constructivists (Adler 2012, 134). As liberalism makes the ontological assumption that individuals are autonomous, it does not necessarily account for the more complex and relational ways in which agency may be exercised. For example, in conflict-affected societies social actors may be individual humans, but they may also come from the spiritual realm, such as ancestors and religious deities, and from the natural habitat, including animals, plants, and places. That is, nonhuman beings can be "active members of society" that "directly influence how humans organize themselves" (Watts 2013, 23).

The Structure of Peace

The second question we must consider is what role structure, material factors, and power play in the construction of peace. Although it initially saw structure and agency as co-constitutive (Wendt 1992), over time constructivism became more structuralist, focusing on norms and modes of legitimacy at the expense of highlighting the power of agents (Wendt 1999). This "structure versus agency" debate generated a second "ideationalism versus materialism" debate, rooted in the fact that the exact relationship between ideational and material structures in early formulations of constructivism was unclear. This saw constructivism characterized primarily as an ideational approach yet did not entirely exclude material factors, acknowledging, at a minimum, that ideational factors give meaning to material structures (Abdelal, Blyth, and Parsons 2010) or that material factors are a secondary influence on human interaction (Finnemore and Sikkink 2001). This led to many constructivists emphasizing the role of ideas in social life at the expense of acknowledging the influence of the material world, including power (Adler 1997). Yet power is both ideational and material; it can work either directly in the interactions of particular actors or more diffusely through "underlying social structures and systems of knowledge" (Barnett and Duvall 2005, 42).

These debates reflect the liberal (and, more broadly, Western) epistemological-ontological bases of constructivism, which assume that only humans have agency and that they are separate from and perceive each other and the world. Yet, as noted, in many conflict-affected societies nonhuman beings are social actors and are irreducible from each other. This challenges us to recognize the agency of nonhuman—material and immaterial—beings and objects (including spirits, animals, plants, and rocks) and how they intersubjectively exercise power, relate to each other, and constitute both the social world and one another. This builds on actor-network theory (Latour 1987) and the "radically relational" ontological proposition that "action, intentionality, consciousness, subjectivity and morality all derive from relations between entities," so that "agency and power are themselves relational effects" (Lockie 2004, 50). This has consequences for how we analyze structure, material factors, and power in conflict-affected societies, compelling us to recognize that agency is "not an inherent feature of individuals but an effect of the differential distribution of power, knowledge, and recognition in social topographies" (McCourt 2016, 481), including the "qualities having to do with the human spirit that compel behaviour" (de Leon 2018, 19). Yet while we must broaden our understanding of the relational aspects of structure and agency, we should avoid "reduc[ing] power to relations," in the process overlooking their "material and social character" (Joseph 2018, 427, 433). This is particularly important in conflict-affected societies, in which material factors such as colonization, globalization, neoliberalism, and interventionism are influential.

Knowing Peace

According to constructivist approaches, ideas and knowledge about peace are intersubjective and have constitutive effects on social reality. Based on this understanding, we can examine why actors converge around specific norms and understandings of peace, how they shape their identities, how these generate their interests, and how these create institutions or structures. At the international level, this can explain why the liberal peace has been institutionalized as the dominant norm that has guided many efforts at peacebuilding since the end of the Cold War, often under the auspices of the United Nations. It also underscores the importance of identifying which ideas and knowledge have shaped understandings of peace in conflict-affected societies and the interests that lie behind them.

Therefore constructivism assists us to analyze the consequences of the dominance of liberal ontological assumptions in international peacebuilding efforts for ideas and knowledge about peace. As liberalism assumes that individuals are rational and act according to reason, the idea of the liberal peace assumes the normative universality of liberalism and prioritizes formal, "rational" understandings of peace, focused on supposedly technical, context-free, and transferrable concepts such as democratic institutions, market economies, and "human rights." This has implications when international actors are engaged in peacebuilding interventions. It is broadly agreed that most interveners regard "everyday peace" (Mac Ginty 2014), that is, the bottom-up methods that ordinary people in conflict-affected societies use to exercise their local agency and everyday diplomacy to survive, as "banal, mundane, and unimportant" (Autesserre 2017, 124). This results in a "knowledge hierarchy," whereby the expertise of foreign interveners is more highly valued than that of local populations (125).

Reflecting this knowledge hierarchy, the dominance of the norm of the liberal peace at the international level has obscured more context-specific and relational understandings and knowledge about peace. Constructivism invites us to examine how local knowledge influences how peace is socially constructed in conflict-affected societies. This reveals that, when international peacebuilders seek local knowledge, they often "define, reduce or reify local socio-political practices and institutions as objects onto which Western conceptions of liberal individualism are projected" (Wallis and Richmond 2017, 429). Alternatively, international peacebuilders attempt to "rationalise" or "formalise" this knowledge, "leading to misrecognition, suppression and obscuring power dynamics and divisions" (429). Consequently the idea of peace is often based on incomplete knowledge of the social reality, which can generate misleading and counterproductive conclusions concerning how peace is understood and known (Ayoob 2002).

This suggests that we need to dismantle this knowledge hierarchy by adopting more reflexive and contextual approaches that recognize subaltern positionality and its local knowledge about peace. Local knowledge is hidden, flexible, and contingent, as it is often drawn from the wider human, natural, and spiritual habitat and influenced by social habits, practices, and virtues. That is, local knowledge may be better characterized as both ontological and cosmological; it can represent a "theory of the universe" that describes "what is and what is not, what can be and what cannot" (Beier 2005, 45). It is also highly political, as it is the product of local dynamics of power, exchange, and contestation, and continually evolving due to its contact with external forces. This means that local knowledge cannot usually easily "be discovered, neatly registered and codified into a 'rational,' bureaucratic, econometric, legal, universal, and formal framework"

(Wallis and Richmond 2017, 429). Indeed attempts to "rationalise" or "formalise" local knowledge "can lead to misrecognition, suppression and obscuring power dynamics and divisions" (429).

Reflecting constructivist thinking, the concept of "vernacular security" applies this approach to security (Bubandt 2005), and analogies can be drawn for our discussion of peace. The vernacular security approach seeks to help explain how actors "construct and describe experiences of security and insecurity in their own vocabularies, cultural repertoires of knowledge and categories of understanding" (Croft and Vaughan-Williams 2017, 22). The concept of vernacular security is also based on the constructivist assumption that security is a socially situated and discursively defined practice.

As constructivism holds that ideas and knowledge about peace are socially constructed, this opens up possibilities of changing those ideas and knowledge. For example, just as constructivists recognize how the diffusion of norms influences understandings of what constitutes a legitimate international order (Barnett and Finnemore 1999), the diffusion of liberal norms has influenced contemporary peacebuilding. This recognition has transformative potential, as it provides space to imagine alternative sociopolitical realities and consequently different ontological assumptions from which to start an analysis of peace, which provides opportunities to challenge hegemonic ideas such as "sovereignty" and "liberalism" and for the diffusion of alternative norms.

Constructivist accounts of peace have also drawn on the Copenhagen school's concept of securitization (Buzan and Waever 2003). This concept is based on the constructivist observation that interests are socially constructed, as it says that we should treat security as a "speech act" (Onuf 1998, 66); that is, what determines whether something is a security issue is whether people describe it as a security issue and if those who listen accept that description. Once a matter has been securitized, it is prioritized above normal politics, and extraordinary means are necessary to address the problem. It has been argued that an analogous process occurs with respect to peace, with "pacifization" described as "the process of framing an issue as related to 'peace' to justify taking certain measures" (Lupovici 2013, 205).

However, a challenge arises when differences in the framing and discourses of peace are irreconcilable. The concept of agonistic peace has been proposed to address this (Rumelili and Celik 2017, 284). This concept draws on the literature on agonistic democracy, which recognizes that identity differences are nonnegotiable and argues that democracy should convert antagonism into agonism and enemies into adversaries through the contestatory, but respectful, engagement across difference (Mouffe 2000). Echoing this approach, proponents of agonistic peace envisage the "peaceful and pluralistic coexistence of multiple, contestatory narratives" (Rumelili and Celik 2017, 284). They argue that this approach is based on a realization that "the dominant narratives, symbols, and rituals of parties in intractable conflicts often prove resistant to conflict transformation" (284). Therefore an agonistic peace requires accepting the complex and open-ended nature of conflicts and challenges the liberal association of peace with consensus (284).

Proponents of agonistic peace suggest that institutions regulated by democratic norms and procedures could operate as outlets to articulate difference and facilitate encounters based on mutual respect to gradually reconstruct conflictual narratives toward accommodation (Peterson 2013). Thus far they do not specify how such agonistic encounters would be facilitated. Therefore, while the concept of agonistic peace may provide safety valves for situations in which people's identities and interests irreconcilably clash, as currently framed it lapses into an ontological view informed by liberal assumptions, as it relies on agonistic encounters taking place within democratic institutions. This poses the challenge of how conflicting parties could negotiate their opposing narratives within institutions if the nature of the institutions themselves are contested, let alone the nature and identities of those parties.

Feeling Peace

The recent "emotion turn" (Hutchison and Bleiker 2014, 492) in international relations can also inform a constructivist analysis of peace, as it emphasizes how emotions can "form part of the sociocultural structure by which . . . agents choose meaning frames and interpretations" (Koschut 2017, 321). In this context, emotions are understood as "the *inner states* that individuals *describe to others as feelings*" (Crawford 2000, 125). Reflecting constructivist thought, emotions are seen as relational and intersubjective (Gergen 2011) and can "serve both regulatory and constitutive functions in the maintenance and cultivation of shared meanings" (Koschut 2017, 328). Emotions can also regulate social behavior by "assigning emotional meaning to norms which members of the group care about, on the one hand, and by retraining undesirable attitudes and behaviour on the other hand" (328). Emotions can also constitute communities, which "provide a 'structure of feeling' through social conventions that constrains and compels members' affective experience in other to facilitate group cohesion" (Williams 1961, 47). Indeed identity "is a feeling"; "it is an emotional connection to a group that gives identity its power" (Mercer 2006, 297).

This highlights that understandings and knowledge about peace are also influenced by agents' emotions, which challenges the dominance of rationalism in much constructivist thought. Constructivism argues that rationality must be based on reason, communication, and persuasion, and contingent on historical, social, and normative contexts (March and Olsen 1998). But constructivism has so far remained skeptical about acknowledging that supposedly irrational, relational, and reflexive emotions and feelings influence knowledge. Yet the emotion turn can articulate well with constructivist analysis, highlighting the role emotions play in "mediat[ing] our receptivity" to beliefs, identities, and norms (Ross 2006, 198), including in relation to peace. However, identifying and studying emotions poses methodological challenges, since emotions are "deeply internal" and difficult to categorize as "genuine" or "instrumental" (Crawford 2000). This means that efforts to understand the role of emotions in the social construction of peace may rely on analyzing how they are represented and communicated, including through "narratives, gestures or other ways of communicating feelings and beliefs" (Bleiker and Hutchison 2008, 129). This will involve challenges of sociocultural interpretation and value judgments by those seeking to interpret and translate these emotional displays.

Practicing Peace

As noted, constructivism holds that identities, interests, and institutions manifest in and through practice; "peace is, first and foremost, itself a *practice*" (Adler 1998, 167). According to Pierre Bourdieu (1977), practice describe people's actions, which not only have physical and material manifestations but also "embody the collective meaning that people give to material reality" (Adler 1998, 167). Reflecting the emphasis in constructivist thought on intersubjectivity, practices are seen as "relational and situational" and as "manifestations of social structures" (Goetze 2017, 26). The constructivist interest in practice has informed the "practice turn" in international relations, which reflects a desire to look beyond discourse to study social action as enacted in and on the world (Neumann 2002; Adler and Pouliot 2011). Indeed social identities are generated by practices, which are "the result of inarticulate, practical knowledge that makes what is to be done appear 'self-evident' or commonsensical" (Pouliot 2008, 258). This highlights that focusing primarily on what "agents think about (reflexive and conscious knowledge)" overlooks what agents "think from (the background know-how that informs practice in an inarticulate fashion)" (260).

Practical knowledge resembles Bourdieu's (2001, 261) concept of "habitus," a "system of durable, transposable dispositions, which integrates past experiences and functions at every moment as a matrix of perception, appreciation, and action, making possible the accomplishment of infinitely differentiated tasks." According to Bourdieu, practical knowledge is "transmitted through practice, in practice, without acceding to the discursive level" (285). Reflecting the insights of constructivism, "habitus is relational; it constitutes the intersection of structure and agency. Thus, what may seem to be a set of individual dispositions is in fact profoundly social" (Pouliot 2008, 274). Habitus has also been characterized as a "structuring structure" as it "expresses, on the one hand, actors' ways of thinking and behaving as the result of their position in a larger social structure, and, on the other hand, actors' reproduction of these structures in the way they behave, act, and think" (Goetze 2017, 28).

Recent studies of the practices of international actors engaged in peacebuilding interventions have revealed the influence of their habitus, or practical knowledge, and have found that interveners are guided by shared "common practices, habits, and narratives" (Autesserre 2017, 120). While the critical peacebuilding literature has seen an increasing number of attempts to analyze the practical knowledge of local actors in conflict-affected societies in order to determine how their practices construct peace, the translation of these studies to international peacebuilding practice remains shallow. This lacuna demonstrates the role of power and politics in the identification of what practices and events influence our understanding of peace. Particular attention needs to be paid to the questions of who has the "authority to define what peace is and how it should be built" (Goetze 2017, 2). Bourdieu's (1997) concept of "capital," that is, the economic, social, and cultural resources actors use to gain social positions, has been used to analyze the role of power in the social construction of peace. It has been argued that "the definition of peace results from knowing and understanding who is speaking, and on which grounds these persons have gained the authority to speak and to define peacebuilding" (Goetze 2017, 16).

Reflecting the dominant influence of the liberal peace on international peacebuilding interventions, in these contexts peace has been defined by the practice of international peacebuilding actors. Consequently many "non-events" or practices that occur at the micro or local level are "ignored, marginalised and unregistered" even though they play a "crucial role" in shaping peace (Visoka 2016, 56). But a more contextual and relational analysis demonstrates that "events are often what we make of them, revealing the social construction of reality, which is based on distinct discursive framings that translate words into social reality" (57). Despite this, the liberal peace narrative attempts to construct peace as an essentially technical concept, built out of "liberal peace events" (or practices) such as elections, state institutions, laws, disarmament, and market-led economic development. These practices are then translated into discourse via the reports of the UN and other international and state agencies involved in peacebuilding (61).

The more reflexive and contextual approach advocated here argues that international scholars and peacebuilders should pay attention to non-events and seek local knowledge. But this will be difficult, for if these new epistemics are to be emancipatory, they will need to be created by the populations affected by them rather than sought out by international scholars and peacebuilders. Indeed postcolonial scholars have questioned the degree to which international scholars and peacebuilders really want to engage with local knowledge given that it challenges their claimed "expertise" (Spivak 1988). There are also questions about whether international scholars and peacebuilders can ever really know local knowledge, given its contextual and relational nature. It also raises ethical considerations, including whether international scholars and peacebuilders can accurately represent local knowledge, whether our attempted translation of that knowledge has consequences for that knowledge, particularly if we are seeking to use local knowledge instrumentally, and whether by writing about knowledge we displace and appropriate local expertise (Briggs and Sharp 2004; Smith 1999; Wallis and Richmond 2017). The attempts of international scholars and peacebuilders to represent and translate local knowledge will also be deeply affected by language, which is a key medium for the construction of intersubjective meanings and social reality (Kratochwil 1989). Yet international scholars and peacebuilders seldom speak the language of populations in conflict-affected societies.

There are no easy answers to these challenges, but they do suggest that international scholars and peacebuilders will need to pay more attention to the way we attempt to identify, interpret, and represent local knowledge (Forster 2007). In conflict-affected societies local knowledge will rarely be written down, and attention needs to be paid to all forms in which local knowledge is recorded or communicated. Attempts at interpretation need to be sensitive to language and "holistic, so that knowledge is interpreted in its context, be that historical, cultural, geographical or otherwise" (Wallis and Richmond 2017, 433). They also need to be reflexive, acknowledging the positionality affects and ontological assumptions of the international scholars and peacebuilders attempting to interpret local knowledge. This underscores how local knowledge as understood by international scholars and practitioners is "by definition generated in an unequal, tension-ridden, and contingent event of social interaction" (Kalb 2006, 582).

One partial solution is for international practitioners and scholars to include local practitioners and scholars in knowledge production about peace. For both pragmatic and principled reasons there has been a growing engagement with local actors during peacebuilding over the past decade. Yet this engagement is often instrumental, such as in the use of anthropological studies to inform counterinsurgency tactics or to facilitate the development of natural resource extraction projects, which echoes the colonial practice of indirect rule. By necessity this engagement also requires local practitioners and scholars to translate their knowledge in the language of (typically) the West and to engage with Western "knowledges, theories and explanations," in the process decentering their own (Briggs and Sharp 2004, 664). Even well-meaning attempts to achieve engagement (and possibly syncretism) between local and international knowledge, as envisaged by advocates of hybrid approaches to peacebuilding (Richmond 2009), can see international practitioners and scholars accept and add on the knowledge of local practitioners and scholars only to the extent that it does not ontologically challenge their own. Attempts at cooperation in knowledge production also faces practical challenges, such as the structural barriers of the financial constraints of universities and other agencies involved in peacebuilding, or the monolingualism of policy and academic publishing (Wallis and Richmond 2017).

Including local practitioners and scholars in knowledge production about peace also poses epistemic challenges to international practitioners and scholars. As noted, local knowledge is rarely written down; it may be nonrepresentational and "embedded" in the form of cosmology, worldviews, memories, and oral and visual histories (Briggs and Sharp 2004, 667). Local practitioners and scholars can play a role in identifying and translating that knowledge, but understanding that knowledge will require international practitioners and scholars to engage with it on ontological terms different from their own. This process of "ontological translation" will likely be both challenging and unsettling, but it could allow local and international practitioners and scholars to intersubjectively generate "ontologically appropriate conceptual constructs and tools for engaging in the multiplicity of worlds" that play a role in the social construction of peace (Trownsell et al. 2019).

Conclusion

We began with the observation that there is no universally valid definition of peace. Instead an analysis informed by constructivism reveals that peace is a social construction, and ideas and practices of peace will be guided by the history, identities, interests, and politics of the agents of peace. This challenges the notion of the normative universality of the liberal peace at the international level (UN General Assembly 1992). It also challenges the ontological assumption that liberal principles provide an accurate description of the nature of peaceful and prosperous social reality (or at least what social reality should look like).

A constructivist analysis allows us to instead take a more reflexive and contextual, less "rational" or "scientific" approach to rethink the tacit ontological assumptions on which our analyses of peace are premised. It suggests that we need to make less certain, more dynamic, contingent, and flexible ontological assumptions regarding how peace is understood and practiced in conflict-affected societies. While acknowledging the ethical issues involved in seeking to know local knowledge, we should expand our analysis to consider the everyday lived experiences, beliefs, feelings, ideas, and understandings of those affected by conflict. Although local or everyday peace is often characterized as second best or "good enough" (World Bank 2011, 106) in the international peacebuilding discourse, it represents the lived experience and understanding of the people it affects. Reorienting our understanding of peace to consider how it is known and practiced from a subaltern positionality will raise questions of historical and distributive justice, alterity, and commonality. Our challenge will then be to consider how these can be mitigated, mediated, and dealt with by entrenched power, hierarchies, structures and institutions, and dominant liberal conceptions of norms, economics, rights, and law in the modern world.

References

Abdelal, R., M. Blyth, and C. Parsons. 2010. *Constructing the International Economy.* Ithaca, NY: Cornell University Press.

Adler, E. 1997. "Seizing the Middle Ground: Constructivism in World Politics." *European Journal of International Relations* 3 (3):319–363.

Adler, E. 1998. "Condition(s) of Peace." *Review of International Studies* 24 (5):165–192.

Adler, E. 2012. "Constructivism in International Relations: Sources, Contributions, and Debates." In *Handbook of International Relations*, edited by T. Risse, W. Carlsnaes, and B. A. Simmons, 112–144. London: Sage.

Adler, E., and M. Barnett. 1998. "Security Communities in Theoretical Perspective." In *Security Communities*, edited by E. Adler and M. Barnett, 3–28. Cambridge: Cambridge University Press.

Adler, E., and A. Pouliot. 2011. "International Practices." *International Theory* 3 (1):1–36.

Antony, L. 2006. "The Socialization of Epistemology." In *The Oxford Handbook of Contextual Political Analysis*, edited by R. E. Goodin and C. Tilly, 58–77. Oxford: Oxford University Press.

Autesserre, S. 2017. "International Peacebuilding and Local Success: Assumptions and Effectiveness." *International Studies Review* 19 (1):114–132.

Ayoob, M. 2002. "Inequality and Theorizing in International Relations: The Case for Subaltern Realism." *International Studies Review* 4 (3):27–48.

Barnett, M., and R. Duvall. 2005. "Power in International Politics." *International Organization* 59 (1):39–75.

Barnett, M., and M. Finnemore. 1999. "The Politics, Power and Pathologies of International Organizations." *International Organization* 53 (4):699–732.

Beier, J. M. 2005. *International Relations in Uncommon Places: Indigeneity, Cosmology, and the Limits of International Theory.* New York: Palgrave Macmillan.

Blaike, N. 1993. *Approaches to Social Enquiry.* Cambridge, UK: Polity.

Bleiker, R., and E. Hutchison. 2008. "Fear No More: Emotions and World Politics." *Review of International Studies* 34 (S1):115–135.

Bourdieu, P. 1977. *Outline of a Theory of Practice.* Cambridge: Cambridge University Press.

Bourdieu, P. 1997. "The Forms of Capital." In *Education: Culture, Economy, Society*, edited by A. H. Halsey, H. Lauder, P. Brown and A.S. Wells, 46–58. Oxford: Oxford University Press.

Bourdieu, P. 2001. *Esquisse d'une théorie de la pratique, précédé de trois études d'ethnologie Kabyle.* Paris: Seuil.

Briggs, J., and J. Sharp. 2004. "Indigenous Knowledges and Development: A Postcolonial Caution." *Third World Quarterly* 25 (4):661–676.

Bubandt, N. 2005. "Vernacular Security: The Politics of Feeling Safe in Global, National and Local Worlds." *Security Dialogue* 36 (3):275–296.

Buzan, B., and O. Waever. 2003. *Regions and Powers: The Structure of International Security.* Cambridge: Cambridge University Press.

Chandler, D. 2010. "The Uncritical Critique of 'liberal peace.' " *Review of International Studies* 36 (1):137–155.

Crawford, N. C. 2000. "The Passion of World Politics." *International Security* 24 (4):116–156.

Croft, S., and N. Vaughan-Williams. 2017. "Fit for Purpose? Fitting Ontological Security Studies 'into' the Discipline of International Relations: Towards a Vernacular Turn." *Cooperation and Conflict* 52 (1):12–30.

de Leon, J. 2018. "Lakota Experiences of (In)Security: Cosmology and Ontological Security." *International Feminist Journal of Politics*, doi.org/10.1080/14616742.2018.1527183.

Doyle, M. 2005. "Three Pillars of the Liberal Peace." *American Political Science Review* 99 (3):463–466.

Finnemore, M., and K. Sikkink. 2001. "Taking Stock: The Constructivist Research Program in International Relations." *Annual Review of Political Science* 4:391–416.

Forster, M. N. 2007. "Hermeneutics." In *The Oxford Handbook of Continental Philosophy*, edited by M. Rosen and B. Leiter, 30–74. Oxford: Oxford University Press.

Gergen, K. J. 2011. "The Social Construction of Self." In *The Oxford Handbook of the Self*, edited by S. Gallagher, 633–653. Oxford: Oxford University Press.

Goetze, C. 2017. *The Distinction of Peace: A Social Analysis of Peacebuilding.* Ann Arbor: University of Michigan Press.

Hameiri, S., and L. Jones. 2017. "Beyond Hybridity to the Politics of Scale: International Intervention and 'Local' Politics." *Development and Change* 48 (1):54–77.

Heathershaw, J. 2013. "Towards Better Theories of Peacebuilding: Beyond the Liberal Peace Debate." *Peacebuilding* 1 (2):275–282.

Hopf, T. 1998. "The Promise of Constructivism in International Relations Theory." *International Security* 23 (1):171–200.

Hutchison, E., and R. Bleiker. 2014. "Theorizing Emotions in World Politics." *International Theory* 6 (3):491–511.

Joseph, J. 2018. "Beyond Relationalism in Peacebuilding." *Journal of Intervention and Statebuilding* 12 (3):425–434.

Kalb, D. 2006. "Uses of Local Knowledge." In *The Oxford Handbook of Contextual Political Analysis*, edited by R. E. Goodin and C. Tilly, 579–596. Oxford: Oxford University Press.
Koschut, S. 2017. "No Sympathy for the Devil: Emotions and the Social Construction of the Democratic Peace." *Cooperation and Conflict* 53 (3):320–338.
Kratochwil, F. 1989. *Rules, Norms, and Decisions*. Cambridge: Cambridge University Press.
Latour, B. 1987. *Science in Action: How to Follow Scientists and Engineers through Society*. Milton Keynes, UK: Open University Press.
Lockie, S. 2004. "Collective Agency, Non-human Causality and Environmental Social Movements: A Case Study of the Australian 'Landcare Movement.'" *Journal of Sociology* 40 (1):41–57.
Lupovici, A. 2013. "Pacification: Toward a Theory of the Social Construction of Peace." *International Studies Review* 15 (2):204–228.
Mac Ginty, R. 2014. "Everyday Peace: Bottom-up and Local Agency in Conflict-Affected Societies." *Security Dialogue* 45 (6):548–564.
March, J., and J. Olsen. 1998. "The Institutional Dynamics of International Political Orders." *International Organization* 52 (4): 943–969.
McCourt, D. M. 2016. "Practice Theory and Relationalism as the New Constructivism." *International Studies Quarterly* 60:475–485.
Mercer, J. 2006. "Human Nature and the First Image: Emotion in International Politics." *Journal of International Relations and Development* 9:288–303.
Mitzen, J. 2006. "Ontological Security in World Politics: State Identity and the Security Dilemma." *European Journal of International Relations* 12 (3):341–370.
Mouffe, C. 2000. *The Deliberative Paradox*. London: Verso.
Neumann, I. B. 2002. "Returning Practice to the Linguistic Turn: The Case of Diplomacy." *Millennium: Journal of International Studies* 31 (3):627–651.
Onuf, N. 1998. "Constructivism: A User's Manual." In *International Relations in a Constructed World*, edited by V. Kubalkova, N. Onuf, and P. Kowert, 58–78. Armonk, NY: M. E. Sharpe.
Paffenholz, T. 2015. "Unpacking the Local Turn in Peacebuilding: A Critical Assessment towards an Agenda for Future Research." *Third World Quarterly* 36 (5):857–874.
Peterson, J. H. 2013. "Creating Space for Emancipatory Human Security: Liberal Obstructions and the Potential of Agonism." *International Studies Quarterly* 57 (2):318–328.
Pingeot, L. 2018. "United Nations Peace Operations as International Practices: Revisiting the UN Mission's Armed Raids against Gangs in Haiti." *European Journal of International Security* 3 (3):364–381.
Pouliot, V. 2008. "The Logic of Practicality: A Theory of Practice of Security Communities." *International Organization* 62 (2):257–288.
Richmond, O. P. 2006. "The Problem of Peace: Understanding the 'Liberal Peace.'" *Conflict, Security & Development* 6 (3):291–314.
Richmond, O. P. 2009. "Becoming Liberal, Unbecoming Liberalism: Liberal-Local Hybridity via the Everyday as a Response to the Paradoxes of Liberal Peacebuilding." *Journal of Intervention and Statebuilding* 3 (3):324–344.
Ross, A. G. 2006. "Coming in from the Cold: Constructivism and Emotions." *European Journal of International Relations* 12 (2):197–222.
Rumelili, B., and A. Betul Celik. 2017. "Ontological Security in Asymmetric Conflicts: Reflections on Agonistic Peace in Turkey's Kurdish Issue." *Security Dialogue* 48 (4):279–296.

Smith, L. T. 1999. *Decolonizing Methodologies: Research and Indigenous Peoples.* London: Zed Books.

Spivak, G. 1988. “Can the Subaltern Speak?” In *Marxism and the Interpretation of Culture*, edited by C. Nelson and L. Grossberg, 271–313. Basingstoke, UK: Macmillan.

Swyngedouw, E. A. 1997. “Neither Global or Local: ‘Globalization’ and the Politics of Scale.” In *Spaces of Globalization: Reasserting the Power of the Local*, edited by K. R. Cox, 137–166. New York: Guilford.

Trownsell, T, A.Q. Escobari, G. Shani, N.C. Behera, J. Reddekop, and A. Tickner. 2019. “Recrafting International Relations through Relationality.” *e-International Relations*, 8 January. https://www.e-ir.info/2019/01/08/recrafting-international-relations-through-relationality/.

UN General Assembly. 1992. *An Agenda for Peace: Preventive Diplomacy and Related Matters.* UN Doc. A/RES/47/120, 18 December.

Visoka, G. 2016. “Peace Is What We Make of It? Peace-Shaping Events and ‘Non-events.’” *Peacebuilding* 4 (1):54–70.

Vygotsky, L. S. 1981. “The Genesis of Higher Mental Functions.” In *The Concept of Activity in Soviet Psychology*, edited and translated by J. V. Wertsch, 147–188. Armonk, NY: Sharpe.

Wallis, J., and O. P. Richmond. 2017. “From Constructivist to Critical Engagements with Peacebuilding: Implications for Hybrid Peace.” *Third World Thematics* 2 (4):422–445.

Watts, V. 2013. “Indigenous Place-Thought and Agency amongst Humans and Non-humans (First Woman and Sky Woman Go on a European World Tour).” *Decolonization: Indigeneity, Education & Society* 2 (1):20–34.

Wendt, A. 1992. “Anarchy Is What States Make of It: The Social Construction of Power Politics.” *International Organization* 46 (2):391–425.

Wendt, A. 1999. *Social Theory of International Politics.* Cambridge: Cambridge University Press.

Williams, R. 1961. *The Long Revolution.* Westport, CT: Greenwood.

World Bank. 2011. *World Development Report 2011: Conflict, Security, and Development.* Washington, DC: World Bank.

CHAPTER 6

CRITICAL THEORY AND THE POLITICS OF PEACE

VIVIENNE JABRI AND OLIVER P. RICHMOND

This chapter outlines the development of the concept of peace, peacebuilding, and statebuilding's relationship with critical theory, and what this means for the development of peace formation as a more contextually sensitized policy-oriented praxis. We argue that critical theory has made very significant contributions since Mitrany ([1943] 1975) first challenged the thinking that gave rise to the League of Nations to face the "new world" of the 1940s, both in holding 1990s-style, policy-driven understandings of peacebuilding and statebuilding praxis to account and in creating a platform for further emancipatory expansion. This might be seen in peace formation from below, cognizant of postcolonial renderings of power (Richmond 2016).

This process has been ambivalent, as the history and evolution of intervention through peacekeeping and peacebuilding, and a widening range of tools, has illustrated. Spanning decolonization, proxy wars, post-Soviet wars, and wars of capital accumulation or identity, the more peace praxis has expanded its emancipatory discursive framing in response to different types of war (e.g., comparing the original peacekeeping agenda of Lester Pearson and colleagues in 1956, Agenda for Peace in 1992, or the Sustaining Peace agenda of 2016; UN Secretary-General 1992, 2018), the less plausible in practice it has been and the more its legitimacy has suffered—especially in the era of statebuilding after 9/11. The more opposition it has faced as a result, not just from authoritarian powers but across the UN Security Council and the donor system, the more problematic international peace frameworks have become. This is not to mention the co-option of peace among General Assembly members who also avowed leadership on its key components, such as human rights (we are thinking here of Sri Lanka's role in the UN Human Rights Council in the 2000s), Even in the context of the "norm cascades" (Finnemore and Sikkink 1998) of the Western international community, or the "normative power" of the EU, there has been a certain shallowness, not to say regression, in the way peace has been connected with security and development, yet disconnected from a peace dividend in economic terms, while building a constitutional and regional order, and the overall international system (Rawls 1993; Manners 2002).

The development of theory and the history of its praxis over the past thirty years or so sheds doubt on peacebuilding and statebuilding's emancipatory intent. So much of its Western, liberal rationality has proven inadequate or has been captured by power interests or ideologies, with little material response, apart from relatively sophisticated discussions of "sustaining peace" (UN Secretary-General 2018) or a "feminist foreign policy" that tend to flout conservative and Eurocentric conventions (see, e.g., Aggestam and Bergman-Rosamond 2016; Robinson 2019). It could be argued that this indicates instead peace's hegemonic capture, especially without a far more radical rethinking of the relationships between peacebuilding as an emancipatory practice in relation to the state and the international system's relationship with subaltern positionalities and peace formation (as opposed to the hegemonic interests and preferences of the West). Discursive responses to global and local political wars and violence have neglected the structural aspects of peace.

Accordingly, we argue that peacebuilding cannot survive as a legitimate practice if it is aimed at preserving systemic hegemony or is based upon counterinsurgency or stabilization measures (UK Stabilization Unit 2018), with the unfortunate consequences of peace thinking being co-opted into a tool for Western-dominated governing practices. To understand the radical emancipation that peacebuilding actually requires, one needs to engage with postcolonial, feminist, intersectional (Crenshaw 1991), and environmental debates at a minimum, in the context of variants of war and newer neoliberal and digital forms of violence across global networks (Zuboff 2019, 9). New forms have to be imagined and designed in this context, though they should expect a life span as short as the 1990s liberal peacebuilding or the 2000s neoliberal statebuilding agenda.

We therefore wish to return to the question of emancipation in the context of peace. Critical theorists, including postcolonial theorists, feminists, constructivists, and poststructuralists, have provided ample modes of critique of what has been an ever- expanding research focus in recent years. These have undoubtedly revealed the workings of power imbricated in practices designed to end violent conflict and deal with the postconflict context in the name of peace. Our aim here is to move the agenda forward, to develop thinking that moves scholarship and practice beyond not just the liberal rights-oriented "agenda for peace" that dominated discourse from the 1990s and continues to do so to the present, on paper at least, but also toward an agenda that both reiterates and expands upon radical emancipatory thought on the politics of peace and the praxis of peace. In a global political context of a resurgent nationalism and exclusionist practices, of the precarity of populations and discursive and institutionalized practices that reinstantiate hierarchies of worth, it is time that peace studies makes itself relevant to these, contemporary challenges . First, this chapter investigates the promise of critical theory, before turning to and assessing the various emancipatory strategies that have emerged as a consequence. These include the local turn and its implications for hybrid peace- and localized peace-formation processes that seek to reform the state and the international system itself (Richmond 2014) if they are to have significance for peace with justice and material dividends beyond co-optation and resilience-therapeutic value. Finally, it addresses the possible characteristics of an elusive radical agenda for peace.

The Promise of Critical Theory

Peacebuilding has undoubtedly acquired hegemonic status in discourses and practices related to responses to conflict. These have a wide remit and can include military interventions and statebuilding or infrastructural operations where institutions and populations in conflict zones are rendered targets for redesign. Research has followed but also enabled and reinforced this hegemonic trajectory, so that critical perspectives on the subject, from feminist concerns relating to agency to poststructural and postcolonial concerns relating to the workings of power and control through peacebuilding, have produced some, albeit limited impact. Incorporated in a reflexive re-visioning of practices, though articulated primarily in declaratory form as in, for example, UN Security Council Resolution 1325 on women, peace, and security, critical perspectives are indeed potentially transformative and emancipatory. Indeed the concept of peace formation was an attempt to open the way for a more radical agenda by foregrounding the subaltern over the legal, institutional, state, and top-down rationalities of peacebuilding and statebuilding, raising issues of justice and sustainability.

The promise of critical theory widely conceived is, in the history of ideas, related to critique as well as emancipation, and the latter is conceived as constitutively premised on critique. From the Frankfurt school of social research, latterly associated with Jürgen Habermas, to poststructural thought, variously and relatedly indebted to Michel Foucault, for scholars interested in the complex relationship between knowledge, power, and subjectivity, to Jacques Derrida, for scholars focusing on deconstruction, the task of the critical theorist is to unearth power relations as manifest in language and institutional practices, exclusions in differential and hierarchical structures, and subjectivizations where discourses come to be so deep-rooted and taken for granted that alternative imaginaries seem no longer possible. Yet it is critical theorists, scholars who mobilize the theoretical and conceptual toolbox we inherit for the purpose of unraveling operations of power in situated locations, who have succeeded in revealing counterhegemonic discourses and practices; these theorists have also revealed that there are alternative voices and hence subjectivities not entirely captured by the weight of global institutions and their civil servants at large. We witness these countervoices from a peace formation perspective in, for example, the women of Bosnia actively resisting the imposition of justice and reparation frameworks that render their agency invisible, defining their own conception of what justice means in a postconflict context of continuing and structurated inequalities of access (see O'Reilly 2018); in the former female Maoist combatants in Nepal, subjected to formulaic demobilization and reintegration programs designed elsewhere, and again articulating a positionality in politics far removed from the stereotypes that inform international agencies' formulae for what Nepalese women are or should be (see Ketola 2020).

Critical perspectives such as these reveal the power of critique and its constitutive potential for suggesting alternative and radically different counterhegemonic voices and their potentialities in conflict zones. At the same time the revelation of such voices brings to the fore just how monumental is the task for critical thought in the area. This task becomes especially salient in a hegemonic context that has the resources and capacity to simply co-opt or, at worst, even instrumentalize the counterhegemonic in the microcosmic practices that constitute peace praxis on the ground. The promise of critical theory is precisely to reveal operations of power and resistance to such power; however, it is also premised on an emancipatory perspective that itself is premised on critique rather than substantive prescription. This suggests that the "emancipatory" is not an imported formula but one the potentiality of which is in the voices and practices of those on the receiving end of violence. At the same time, the promise of critical theory points to the larger global forces against which the voices of resistance in peacebuilding zones must contend, as is evident in the cases highlighted.

We argue that the critical lens must now turn to these global forces, the ways in which these permeate spaces of knowledge and subjectivity. Habermas ([1968] 2015) refers to "knowledge-constitutive interests" as a conceptual means of differentiating, but also relating, what he refers to as instrumental, practical, and emancipatory interests. We have an interest in instrumental knowledge, in the effective and the transformative, for example, in what structures of representation work in democratic transformation. For instance, any effort at conflict resolution and postconflict peacebuilding has to have access to contingent factors, from information relating to the displaced to the emotional and affective aspects of conflict. However, in critical theory, and especially, as Habermas reminds us, in the social sciences and humanities, we at the same time have an interest in understanding, locations of language and communication, as well as the constraints on understanding. Here and in this context, we know from postcolonial theorists that the colonial rationality persists in informing Eurocentric schema as the taken-for-granted and legitimate perspectives in peacebuilding (see, e.g., Jabri 2013a). The paradox for any critical theorist in this context is that while there is acknowledgment and respect for local voices of resistance and counterhegemony, the theorist who descends on conflict zones does so for purposes other than the immediate emergencies that confront populations in difficult and trying circumstances. We are all aware of the differential in access to resources and privilege, and this brings with it ethical challenges in theoretical and praxiological thought. The promise of critical theory is to unravel the Eurocentrism of discourses and practices as a constraint on understanding. Yet as both the instrumental and the hermeneutic aspects of knowledge reveal, global structures of inequality and injustice continue to limit knowledge-constitutive interests, so that we may go so far and no further. However, the promise of critical theory focuses on the emancipatory and therefore on imaginaries that at one and same time focus on the workings of global forces and their transformation. Yet the emancipatory imagination is in itself subject to the constitutive power of purposive thinking, that we want our "critical" ideas to translate to action, what we prefer to call "praxis." In this sense, all knowledge is closely tied up with interests, including critical theorists' primary interest in emancipation, contested though this concept is.

Peacebuilding in Theory and Practice: Towards the Politics of Peace

The concept of "peacebuilding" is constitutively premised on instrumental rationality, how to "get peace done," so to speak. The underpinning assumption is that, given competently drawn up plans for a postconflict future, relations and the institutions that uphold them may be transformed. The design of a peacebuilding framework suggests both a formula premised on what is often referred to as the "liberal peace" (or later, a neoliberal state) and a detailed implementation plan involving concrete interventions on the ground. Both may be a product of collaborations between external and internal agencies, though the peacebuilding blueprint is often, if not always, designed well beyond the reach of those directly involved in conflict. Peacebuilding, as the concept suggests, constitutes a set of practices, and how such practices are analyzed is a matter crucial to any critical perspective on the subject.

The critique of peacebuilding and statebuilding both as concept and as a form of practice in response to conflict is now well integrated into the literature and encompasses perspectives that view peacebuilding as governmentality, drawing on Foucault's analytics of power, postcolonial critiques that focus on the practice as a mode of colonial rationality, and perspectives that seek to analyze agency in the peacebuilding context. These include the agency of the subaltern, the agency of women, and the broader question of how local voices come to impact on and resist control by external forces. This is well-worn ground and is covered elsewhere. What is important to state at the very outset is that responses to conflict, and peacebuilding in particular, are hegemonic practices now so deeply rooted in international institutions that they have, by and large, been adapted to incorporate critical reflection on questions of agency. Liberal governmentality adapts to critique and indeed has the facility to co-opt it in its practices (see Jabri 2010). The aim in this chapter is therefore to explore what is left of any critical reflection that may take us beyond the existing critique. Two moves are suggested that may take us beyond critique and toward an emancipatory understanding. The first suggests that in place of "peacebuilding" we adopt the language of the politics of peace. The second examines the concept of the emancipatory and what this might mean in the current era in the context of conflict and peace. The suggestion is that in the absence of the former, the acknowledgment of the *politics* of peace (i.e., *a nonviolent struggle driven by subaltern claims over scarce resources and for justice and sustainability*), the latter engagement with peace formation becomes an impossibility.

It may seem paradoxical to frame practices of peacebuilding in terms of interests and power, though certainly less so with statebuilding. This is normally the language of realism in international relations, and of strategic considerations in foreign policy decision-making, with a view to winning what is conceived to be the zero-sum game of international politics confined to relations between states. Peacebuilding is conventionally seen to foster a language of peace and the end of conflict, of practical, technocratic considerations motivated not by the selfish desire to win but conceivably motivated by what is considered workable in the face of complexity and societal fragmentation. The outcome may be beneficial to peacebuilders, but the primary good is assumed to accrue to the target populations. The critical theorist is primarily motivated by an interest in peace, particularly peace that realizes a transformation of the violence and inequalities that perpetuate states of conflict. But this realization is always tinged with the recognition of the dialectics of social and political life, that peace cannot be conceived as an end point but is rather a temporal state of being and relations that are always precarious and, in their precarity, requiring work and investment. It is therefore possible to conceive of peacebuilding in critical terms, but this can be achieved, or even imagined, only once a prior move is enacted, namely critique, and specifically of what peacebuilding as a practice now championed by academic disciplines and international organizations does to the populations on the receiving end. This involves a subaltern perspective, engaging with methods and ethics from a completely different level to that of the state and international levels, while also recognizing their mutual constitution.

The very spectrum of what is referred to as "peacebuilding" practices, from those involving military operations to the establishment of an institutional infrastructure in target locations and the assemblage of internationals, NGO-related or otherwise, suggests a field of differential capacities determined by both symbolic and material power. However, before analyzing operations of power in the context of peacebuilding, any critical perspective is interested in how a particular discourse and set of practices come to acquire hegemonic standing, a question that seeks to relate knowledge to interests and power. For it is the case that peacebuilding has acquired such a status, now overwhelmingly institutionalized as a universal, taken-for-granted response to conflict, having primacy over and above diplomatic responses relating to negotiation and conflict resolution. Certainly in the field of knowledge systems, in research-funding regimes, and in the bureaucratization of the practice, the weight of investment is not in diplomacy but in peacebuilding: peace as "government" or peace as governance (see, e.g., Duffield 2007; Jabri 2007: Richmond 2005). Specifically, the imperative is the management of populations through interventionist measures that place primacy on their economic and sociopolitical reshaping—as opposed to engaging with their political claims. The method used is one based mainly on pedagogy, consisting of training programs across a range of subjects, all premised on the creation of liberal, self-governing subjects.

The overwhelming weight of emphasis in the peacebuilding and statebuilding literatures and in international documentation on the subject falls on instrumentality, specifically the outlines of techniques in situated contexts of conflict, such as practices relating to the reintegration of former fighters or the training of local security forces in matters relating to human rights or gender equality. These are also interested in the evaluation and assessment of particular interventions relating variously to success, failure, or even the risks of such failure. Such rationalist, even scientific approaches are premised on the assumed power of transformed institutions, though these are often run by international agencies. The wholesale transformation of the institutional infrastructure of a state, its social, economic, and political spheres are constitutive of what is essentially a project of redesign, of statebuilding in what is historically a postcolonial context, and a reimagined order modeled on the liberal polity. The epistemological and ontological backdrop is what is known in the international relations literature as the democratic peace theory: that "democracies do not fight each other." Peacebuilding is statebuilding by another name, but it is also about the building of a particular kind of state (see Richmond 2014), one based on "free markets," democracy and the rule of law, for it is these that, in this Kantian understanding of peace, guarantee the conditions for peace.

We see from this that peace is not merely an outcome to conflict but is a condition that is subject to governing practices, and these are reliant on techniques of control that in effect seek to define categories of population, the assessment of their economic and social needs as well as potential for conflict. Power in the government of peace operates through subjectivization, through the microcosm of interventions in the lived experience of those on its receiving end. An emancipatory understanding of peace shifts the imaginary from government to politics, as the following outlines.

There is a difference between peace as government, or governmentality, to use Foucault, and the *politics* of peace (Jabri 2007), where peace as such is contested in relation to claims relating to the causes of conflict, the sources of injustice, or institutionalized and structural inequalities. The prevailing orthodoxy is that such claims might be tamed through an introduction of liberal rationalities in economic and social relations that might be adapted to local conditions, producing outcomes that adhere to certain neoliberal standards of measurability or at best those deemed to reflect formulaic measures of diversity and inclusion. Such has become the predominant aspiration of peacebuilding and statebuilding practices and the programs that sustain them. There are multiple instrumental interests invested in these programs, their rationales generating habituated commitment and discourses of legitimization. So powerful has this orthodoxy become that even its blatant militarization and securitization are often framed in terms of the "goods" accrued to the populations on the receiving end. The colonial rationality is never far removed from such discourses and practices on the ground, even where, as in the colonial period itself, those on the ground become the co-opted subalterns mobilized by a public-private industry for peace.

There are, of course, instances of resistance to this industry, and these are articulated in ways that question the orthodoxy, its power and legitimacy. The effect of such moments of resistance is to remind us that the government of peace through technocracy and management is not the politics of peace, the creation of spaces for political claim-making. This paradigm shift, from peace as government to peace as politics, enables us to rethink the concept and the practices carried out in its name so that it becomes inextricably linked to interests that are not merely instrumental but are emancipatory. Such a paradigm shift in both knowledge and practice has to build on a historic moment that is postcoloniality, the moment of asserted liberation from colonial rule and the continuing colonial rationality that, even in the twenty-first century, subjects the populations of the postcolonial world to hierarchies of worth and therefore voice. In a global context of continued impoverishment, to borrow from Walter Rodney ([1972] 2018) and extend his thesis to the present, no external actor can gift "goods" to postcolonial populations, for every such gift has come at a heavy price, and the culpable includes peacebuilding operations. The emancipatory politics of peace must therefore be self-authored, within postcolonial political contexts and the contestations that were inevitably and constitutively aspects of the promise of postcoloniality. Such self-authoring cannot be reduced to the all too convenient realm of tribe and ethnicity, locations of customary authority highlighted and condemned by postcolonial leaders and thinkers such as Kwame Nkrumah and Frantz Fanon as sources of colonial exploitation. Western intervention continues to undercut the public spheres of postcolonial societies, contexts of vibrant deliberation and political contestation. An emancipatory politics of peace has its source in self-generated public spheres, the Tahrir Squares of the postcolonial world (Jabri 2013b).

It is in such public spheres that claims relating to justice, inequality, and rights can be articulated and conferred meaning. More significant still, such claims come to be recognized as first and foremost related to what Hannah Arendt referred to as the right to politics. We see in the making of such claims the intersection of peace with wider global issues relating to sources of injustice and inequality, normative and material. The undercutting of postcolonial public spheres has also serviced postcolonial states that have been complicit in the continued structuration of the colonial rationality and the social, political, and juridical dynamics that sustain it. An emancipatory politics of peace must hence have an expansive understanding of the public sphere aimed at a reauthoring of the postcolonial polity, but also of the regions and wider global coalitions that seek to rewrite the rules of a global game that has long favored the most powerful and exploitative forces, now of proven destructive actuality and potential not just to the postcolonial world but the planet as a whole.

Global entanglements require global coalitions, and an emancipatory politics of peace starts with moments of self-authorship in reclaimed public spheres of deliberation and contestation. Technocracy or formulaic instrumental understandings of peacebuilding may be measured by standards defined and limited by a rationality of colonial governmentality, but these cannot be a substitute for what we are referring to here as an emancipatory politics of peace. Some may argue that it is the former that ultimately enable the latter, that a "liberal peace" can ultimately create the conditions for vibrant and peaceful public spaces. The problem with this argument is that it continues to be premised on the idea that the postcolonial world suffers from a lack, that it will inevitably "catch up" with the Western world, if it has a helping hand. If this latter rationality continues to hold sway, the aspirations of the politics of peace will always be held in check, undercut by a limited imaginary based on pro forma notions of "peace."

The paradigm shift advocated here does not suggest one definition of peace, nor a formula for "success" that can be measured. Rather, it suggests a politics of peace, by definition premised on deliberation and contestation. It recognizes the continued promise of postcoloniality, that the right to politics implies self-authored, self-generated and sustained public spheres that can reenliven the postcolonial moment against the forces that seek its continued undercutting, both local and global. The emancipatory politics of peace suggested here recognizes, as these postcolonial public spheres do, that local manifestations of conflict, injustice, and inequality have their sources in wider global dynamics and hence require global coalitions.

Critiquing the Local Turn

Many of the dilemmas and paradoxes that emerge when understanding the evolution of peace and its relationship with politics through a critical, intersectional, environmental, and ultimately postcolonial lens provided the impetus for the so-called local turn in peacebuilding (Mac Ginty and Richmond 2013). This offered a methodological shift from centering peacebuilding on the state or the international system toward the subaltern, particularly positionalities from the Global South. It produced very different understandings of peace within the context of hybridity (Boege, Brown, and Clements 2009): it often valorized self-determination, raising resource, cultural, historical, and rights issues but also questioning the very epistemological frameworks that such issues had been dealt with in the liberal peacebuilding framework (Richmond 2011). Justice across time and space, issues of equality and sustainability center peace and its governance in a very different framework from those offered by geopolitics, imperialism, the liberal or postcolonial states-system, or global capital, even if there was a sense that the state and multilateral system was still vital. The entanglement of hybrid peace and order have followed the epistemic and methodological shift from the international to the state and then to the local or subaltern.

The local turn—an attempt to engage with these very substantial lacunae through peace formation rather than state formation, statebuilding, or peacebuilding—has taken three paths relevant for our understanding of intervention and its relationship with emancipatory debates. First, its postcolonial path has produced an everyday methodology connected to much deeper historical and distributive questions of global justice. It sees local political data as necessary for self-determination, thus being inherently political and harnessed to bottom-up, transversal forms of peacemaking. In this version, external intervention is necessary to unpack power relations and sometimes expected from populations facing adversity, but political autonomy is also paramount in the context of local political claims. The politics of peace and what they mean for emancipation in concrete terms would be authored by local polities, but this also raises significant questions about how they are constituted and act in the context of global power relations. (This was perhaps the logic behind the nonaligned movement.) It points to a discussion about hybrid peace after the experience of liberal peacebuilding and its many problems.

Second, the immanent liberal critique has tended to see the local turn as useful for policymakers from the Global North (e.g., the UN and international donors) in their attempts to inculcate reform in conflict-affected states, producing a form of methodological liberalism often through ethnographic work. This has often produced a massive amount of bureaucratic, technocratic, and descriptive local data about "everyday peace" (Mac Ginty 2014: Millar 2014). In addition, a further stage has emerged in more prescriptive terms, in which such descriptive data is then processed by northern actors working toward the liberal peace, thus supporting a form of neo "native" trusteeship, related policies of local ownership, and to some degree local autonomy, but within parameters set by the international community.

Third, the more realist-related version of this descriptive approach has used the data for counterinsurgency and stabilization measures related to what has been termed methodological nationalism, mainly of hegemonic regional actors and the Global North. It has been relevant for US statebuilding as well as counterterrorism, or pacification strategies that dampen violence but do not aim at emancipatory processes. This points to the core-periphery nature of many contemporary peace processes, which means that though they may ameliorate conflict, they do little to address underlying power dynamics, inequality, or, in fact, the roots of the conflict.

There are pros and cons to all of these subapproaches within the local turn, and clearly the subaltern is on the receiving end of power relations. But a sense of their relative emancipatory potential is fairly clear in that they foreground local experiences of war and violence in order to build solutions and identify obstacles and blocking power structures. Responses often need to be assisted or enabled from the outside, however, by actors and coalitions with a significant range of capacities. This helps explain the massive expansion of interventionary praxis from methodological perspectives related to nationalism, liberalism, and the everyday context of violence. However, external biases are built into these capacity-based epistemic frameworks that are deployed for intervention, raising problems for the legitimacy of peacebuilding over the longer term. Their hegemonic deflection is also more apparent, particularly in terms of the far greater processing power that dominant actors have access to in versions 2 and 3 of the local turn, whereas local actors have far fewer resources for interpreting their own data even than external actors. The trouble with foregrounding local knowledge and autonomy in peacebuilding thus points to what Chandler (2014) has critiqued as resilience-based- and therapeutic- approaches, which have much in common with neoliberalism. The latter gives free rein to the forces of global capital, and the international political economy is often at the roots of such conflict, creating a paradoxical situation where peacebuilding and statebuilding concur with some of the key conflict dynamics rather than addressing them, and expose conflict-affected societies to ungoverned economic practices. As in the Democratic Republic of Congo, for example, this has fed the conflict over the long term even while multiple UN, EU, and donor missions have been ongoing.

Local understandings of peacebuilding are inevitably be related to understandings of peace connected to geopolitics and the state (generally negative), liberalism (generally Eurocentric), capitalism, and neoliberalism (generally extractive). Room remains for the conceptual revision of peace in terms of justice, especially moving beyond its liberal version and its "epistemic violence" (Millan and Yildirim 2014, 197). More specifically, global justice discussions offer historical and distributive agendas, which are clearly often omitted from peacebuilding and peace processes. It raises obvious and scientifically concrete issues of equality and sustainability, across race, class, and gender, in local to global scales of politics. Peace through peacebuilding and peace agreements, claiming the legitimation of practical emancipation, would have to be governed in more complex and ambitious ways than as previously through hegemons like the US in the Middle East, regional actors such as the EU in the Balkans, or the UN in its many peacebuilding and peacekeeping operations. Shifting from liberal understandings of global justice as framed by the state, liberal international law, and public reason, to postcolonial (Zondi 2017), gendered, and environmental perspectives offers a much more ambitious agenda. It also raises the problem of resistance to the reframing of the politics of peace it would entail: the loss of power, authority, rents, and privilege for elites and hegemons that such a reordering of international relations would entail. This pattern of revolt and reaction has been a prominent part of the liberal international order since postcolonial states began to emerge and has been reflected more recently in the dynamics of statebuilding, peacebuilding, and peace processes (Adebajo 2016).

Peace cannot be divorced from a response to the complex imbrications of war across temporal and geographic scales as it regularly was previously. We can highlight the following cases: secluding the Palestinians from the Middle East peace process in the 1970s and thereafter; relying on a long-term ceasefire without an agreement in Cyprus after 1964; glossing over many, though not all, atrocities and losses across the Balkans after the Dayton agreement in 1995; relying for a peace dividend on fickle global capital, as in Cambodia after 1994; ignoring aspirations for economic development, as in Timor after 2002; withdrawing support and capacity before the peace is secured, as in Afghanistan and Iraq in the 2010s; or simply not engaging at all in peace work, as in Yemen and Syria. In all these cases and beyond, we witness the instrumental abuse of the politics of peace precisely to undermine initiatives geared to the ending of violent conflict, to perpetuate war, and to unleash military technologies against what are deemed to be expendable populations.

Concluding Remarks: The Radical Paradoxes of Peace in the Next Phase

One of the aims of liberal internationalism, liberal peacebuilding, statebuilding, and development, and related practices and institutions over the past century has been to align the agency and political claims of civil society, social movements, and the subaltern with constitutional order, the state, and the international architecture in the context of Enlightenment notions of government, peace, justice, and democracy. This has incidentally aligned it with northern hegemonies. Peace has become intensely and critically political as a consequence, representing far more than stopping violence from a peace formation and subaltern perspective, but addressing emancipation and the resultant expansion of rights, law, and institutions. Indeed, merely stopping the violence results in a victor's peace beneficial mainly for elites, as peacebuilding and statebuilding praxis have demonstrated, rather than a platform for self-determination, social and global justice, and more substantially for peace formation. International, multilateral, and bureaucratic power, in the shape of a global liberal or neoliberal governmentality was used as a platform to allow this architecture to emerge and consolidate everyday and local political claims in peace processes and peacebuilding, as long as they remained within the narrow confines of the liberal or neoliberal peace framework and state-centric or civil notions of war.

This led to an international peace architecture taking shape over the past two centuries in which liberal-internationalism responded to nineteenth-century geopolitics, and structuralism and liberal debates about expanded rights took shape in the postwar and Cold War context of liberal-internationalism and democratic peace theory, leading to the liberal peacebuilding framework after the Cold War. These contributions were appended to regional order and global order, framed by the states-system and its rationalities of hegemony and territorial sovereignty, but they were also given the responsibility of expanding rights, development, international law, and democracy. These contradictory demands were soon probed by critical theory, as we have indicated, and with the shifting understanding of violence led to the collapse of the legitimacy of the architecture by the 2000s, with the war on terror (though much of it remains in place as a moribund reminder).

Much of the critique of critical theory in these areas was borne out in its shifting connection of peace with state, regional, and global stability, then social justice, the transnational justice and public reason within a liberal state, and then most recently subaltern political claims. This implies a future shift toward postcolonial framings of global justice. Yet the response of established powers in the states-system and international political economy was premised on a rejection of such Enlightenment rationalities. Nevertheless the discursive weight of the more sophisticated understandings of peace (ceasefires, regional stabilization, state reconstruction pointing toward building a multilateral system, their equation with public reason, international justice, and now global justice and sustainability) has continued to influence the normative, legal, and institutional debates that have resulted from such critical engagements in the postwar world.

Thus a local-to-global peace architecture has slowly been developing, within which the various peacemaking strategies, tools, and theories can be located over the past century. It has been responding to the variety of critiques and has developed some responsiveness in discursive terms to subaltern political claims, but it is being opposed in practice on old grounds: race, class, positionality, national interest/territorial sovereignty, and capital. Furthermore, while it has aged and become more complex as its responses to changing forms of violence (geopolitical, imperial, ideological, ethnic, resource-based, etc.) have developed, it has become more expensive to uphold and less easy to comprehend.

This lacuna has meant critical versions of peace are always heavily entwined with less advanced practices of power in international relations (as in realism, liberalism, structuralism, governmentality, biopolitics, etc.), pointing to well-disguised dynamics of "counterpeace" that have become adept at blocking the advances that have been made, hindering their progress, and obscuring the expanding rights dimensions they suggest. The opponents to the advancement of the critical agenda to some degree are the very actors, processes, and networks that have supported it for so long (e.g., global political elites, states, even aspects of international law, and multilateral actors), along with their political, economic, and technological networks, some of which directly oppose the critical logic of peacemaking because it undermines their power base, and some of which have become moribund as new strategies of violence have developed and new political claims have been made. Hence any peace process can quickly be turned against itself and, indeed, has been designed to switch if "required," this ironically being an important part of its resilience. (Look at the contemporary non-peace processes in Cyprus, the Middle East, Colombia, or Kosovo, for example.)

It can be argued that the recent "Sustaining Peace" agenda of the UN reflects this type of double movement (Polanyi [1944] 2001), maintaining an impossible paradox of seeing peace both relating to established powers and systems and yet overloading them with ever more rights, duties, and responsibilities in relation to subaltern claims and peace formation dynamics. However, this is a hybrid reflection, one that tries to incorporate gender critiques, critiques of capitalism, critiques of the state, expanded rights, relationality, scientific notions of sustainability, and environmental insights into an expanded and not-quite-liberal but perhaps postliberal system for peace (transnational, transversal, and transscalar, also shifting toward heightened political mobility). It is also one that is still closely connected to the residues of imperial order and the industrial global political economy. This new UN agenda carries forward hints of the debates about equity (distributive and intergenerational), global justice, and decolonization and a broader and longer perspective engendered in the politics of sustainable peace in the contemporary world, even if it is currently only discursive and doctrinal rather than practical or supported by any powerful states at all. Nevertheless its potential is well-prepared and lies in wait, built into the forever expanding (and fragile) international peace architecture, for when the older systems of geopolitics, liberal internationalism, liberal peacebuilding, and neoliberal peacebuilding inevitably are found wanting by entwined political elites and social movements in conflict-affected societies.

The numbers of wars unattended and uncontained are now added up, together with their terrible human cost (Yemen, Syria), along with many more peace processes frozen or in stasis (Bosnia, Kosovo, Colombia). On top of this there are systemic problems in the international political economy and states-system, which are clearly connected to extant wars, and the scientific understanding of violence has enormously expanded as a result. Critical thinking about peace needs to respond; in some ways it has, but it is also anchored in a range of systems that restrain and undermine, or indeed counter the progress it has made.

It is all too challenging to imagine a world devoid of hegemony and the inequalities that perpetuate conflict and violence. Nor is it possible to argue that, in the absence of such global forces as the international political economy or the international marketplace of deadly weaponry, peace as such would prevail. Such utopian thinking is not what is being advocated here. Informed as this chapter is by the underlying assumptions of critical theory—namely critique and emancipation in the context of power-relations—the most promising research agenda is premised variously on feminist and postcolonial thought, where conceptualizations of agency come hand in hand with analytics that reveal the contingent matrices of power that differentially enable and constrain such agency. The emancipatory must lie in the promise of counterhegemonic voices where what becomes apparent are alternative imaginaries of the politics of peace, of peace formation, and thus of the entire peace architecture itself.

Some would argue that international peacebuilding practices do exactly this when they aim to govern the postconflict context, seeking to somehow reshape governance and the social order through a combination of coercion, economic incentives, and pedagogical training in liberal institutions and discourses and neoliberal economics. However, such governing practices are exactly the sites at which counterdiscourses can be seen to emerge, including alternative imaginaries relating to distributive and transitional justice as well as reintegration. A constitutive aspect of this proposed "politics of peace" is to recognize that the postconflict context is not simply a technocratic challenge but is primarily a political context of contestation and deliberation. If peacebuilding and statebuilding represent discourses and a set of practices that have been led by a technocratic rationality, the politics of peace advocated here as an alternative would focus on conflict resolution, in recognition of those immediately affected by conflict as subjects of politics with potentially agentic capacities.

At the same time, we are advocating an emancipatory agenda that recognizes the deep imbrication of conflict with wider global political dynamics that can either help or hinder this agenda. We witness the widespread hindering or even undermining impact of recent interventions that have variously contributed to the escalation of violence or, at best, sought to govern populations in accordance with regimes formulated at a distance. Yet we know that crises are shared and have impact globally, in spaces far removed from the immediate contexts of violent conflict. The agenda advocated here is admittedly largely suggestive; it seeks a new research program in peace and conflict studies, one that emerges from the promise of critical theory, and moves on toward an ethical and political engagement with how the politics of peace might relate to questions of inequality, historical and distributive justice, as well as recognition and sustainability, and how these in turn relate to ongoing struggles. These include struggles of postcoloniality, feminist and intersectional politics, and the environmental and climate emergency. We seek a renewed "agenda for peace" fit for a world in crisis.

References

Adebajo, A. 2016. "The Revolt against the West: Intervention and Sovereignty." *Third World Quarterly* 37 (7):1187–1202.
Aggestam, K., and A. Bergman-Rosamond. 2016. "Swedish Feminist Foreign Policy in the Making: Ethics, Politics and Gender." *Ethics and International Affairs* 30 (3):323–334.
Boege, Volker, M. Anne Brown, and Kevin P. Clements. 2009. "Hybrid Political Orders, Not Fragile States." *Peace Review, February*, 13–21.
Chandler, D. 2014. "Beyond Neoliberalism: Resilience, the New Art of Governing Complexity." *Resilience: International Policies, Practices and Discourses* 2 (1):47–63.
Crenshaw, K. 1991. "Mapping the Margins: Intersectionality, Identity Politics, and Violence against Women of Color." *Stanford Law Review* 43 (6):1241–1299.
Duffield, M. 2007. *Development, Security and Unending War: Governing the World of Peoples.* Cambridge, UK: Polity Press.
Finnemore, M., and K. Sikkink. 1998. "International Norm Dynamics and Political Change." *International Organisation* 52 (4):887–917.
Habermas, J. (1968) 2015. *Knowledge and Human Interests.* Cambridge, UK: Polity Press.
Jabri, V. 2007. *War and the Transformation of Global Politics.* London: Palgrave.
Jabri, V. 2010. "War, Government, Politics: A Critical Response to the Hegemony of the Liberal Peace." In *Palgrave Advances in Peacebuilding,* edited by O. Richmond, 41–57. London: Palgrave.
Jabri, V. 2013a. "Peacebuilding, the Local and the International: A Colonial or a Postcolonial Rationality." *Peacebuilding* 1 (1):3–16.
Jabri, V. 2013b. *The Postcolonial Subject. London*: Routledge.
Ketola, H. 2020. "Withdrawing from Politics? Gender, Agency and Women Ex-fighters in Nepal." *Security Dialogue,* online.
Mac Ginty, R. 2014. "Everyday Peace: Bottom-up and Local Agency in Conflict-Affected Societies." *Security Dialogue* 45 (6):548–564.
Mac Ginty, R, and O. P. Richmond. 2013. "The Local Turn in Peace Building: A Critical Agenda for Peace." *Third World Quarterly* 34 (5):763–783.
Manners, I. 2002. "Normative Power Europe: A Contradiction in Terms?" *Journal of Common Market Studies* 40 (2):235–258.
Millan, Anna, and Ali Can Yildirim. 2014. "Decolonising Theories of Global Justice." In *Decolonising Enlightenment: Transitional Justice, Human Rights, and Democracy in a Postcolonial World,* edited by N. Dhawan, 195–2010. Berlin: Barbara Budrich.
Millar, G. 2014. *An Ethnographic Approach to Peacebuilding: Understanding Local Experiences in Transitional States.* Abingdon, UK: Routledge.
Mitrany, D. (1943) 1975. *The Functional Theory of Politics.* New York: St. Martin's Press.
O'Reilly, M. 2018. *Gendered Agency in War and Peace.* London: Palgrave.
Polanyi, K. (1944) 2001. *The Great Transformation: The Political and Economic Origins of Our Time.* Boston: Beacon Press.
Rawls, John. 1993. *Political Liberalism.* New York: Columbia University Press
Richmond, O.P.2005, The Transformation of Peace, London: Palgrave.
Richmond, O. P. 2011. *A Post-Liberal Peace.* London: Routledge.
Richmond O. P. 2014. *Failed Statebuilding.* New Haven, CT: Yale University Press.
Richmond O. P. 2016. *Peace Formation and Political Order.* Oxford: Oxford University Press.

Robinson, F. 2019. "Feminist Foreign Policy as Ethical Foreign Policy? A Care Ethics Perspective." *Journal of International Political Theory*. doi: 10.1177/1755088219828768.

Rodney, W. (1972) 2018. *How Europe Underdeveloped Africa*. London: Verso.

UK Stabilisation Unit. 2018. "The UK Government's Approach to Stabilisation: A Guide for Policy Makers and Practitioners." 19 December. https://www.gov.uk/government/publications/the-uk-governments-approach-to-stabilisation-a-guide-for-policy-makers-and-practitioners.

UN Secretary-General. 1992. "An Agenda for Peace, Preventive Diplomacy, Peacemaking and Peace-Keeping." *A/47/277-S/24111*, 17 June.

UN Secretary-General. 2018. "Peacebuilding and Sustaining Peace." *A/72/707-S/2018/43*, 18 January.

Zondi, S. 2017. "African Union Approaches to Peacebuilding: Efforts at Shifting the Continent towards Decolonial Peace." *African Journal on Conflict Resolution*. 17 (1):105–131.

Zuboff, S. 2019. *The Age of Surveillance Capitalism*. London: Profile.

CHAPTER 7

PACIFISM IN INTERNATIONAL RELATIONS

RICHARD JACKSON

The aim of this chapter is to explore the contribution of pacifism to international relations (IR), as well as to demonstrate its relevance to discussions around peacebuilding, statebuilding, and peace formation. As a form of theory rooted in a real-world critique of violence and its effects, and the historical practices of peacemaking and nonviolent resistance—that is, as a kind of applied political theory that is practiced by its adherents—pacifism is ideally placed to offer insights and suggestions for both practitioners and theorists of peacebuilding. However, most IR scholars, including many peacebuilding experts, remain unaware of the contributions made by pacifism to the discipline, in large part because it is considered to be unscientific, naïve, and of limited relevance (see Jackson 2018a).

In the following sections, I explain what pacifism is, its main types and approaches, and offer a brief outline of the history and legacy of pacifism in IR, some of the main objections to pacifism, and some answers given by pacifists to these criticisms. The remainder of the chapter explores what pacifism can contribute to IR by way of, first, a critique of violence, power, and just war theory and, second, positive contributions to discussions of power and agency, security, civilian protection, peacebuilding, statebuilding, and peace formation. Broadly speaking, the chapter attempts to argue that pacifism is a credible and insightful approach to IR that should be taken far more seriously than it is. Moreover, the current historical juncture provides an ideal moment for the return of pacifism to IR, although there are a number of serious challenges it will first have to surmount.

What Is Pacifism?

The term "pacifism," derived from the Latin for "peacemaking," is commonly understood to be an approach that expresses opposition to war and violence and, at the same time, a commitment to peacebuilding and the construction of peaceful alternatives to the use of violence within politics and society (Cady 2010). In fact, pacifism, like feminism, environmentalism, liberalism, and other approaches, is a political tradition that encompasses a variety of sometimes conflicting positions and forms. This is because within the broader field of concepts and practices that pacifism is concerned with there are differences in the way concepts such as war, the state, sovereignty, power, and the like are understood. For example, some pacifists approach the state as an inherently violent form of domination, while others see it as essential to providing social order and security.

Similarly, pacifism has emerged from, and bases its ethical judgments and actions upon, different traditions and sources. The oldest forms of pacifism derive from religious traditions such as Christianity in the West, Buddhism, Hinduism, and Jainism in the East, and some indigenous traditions. Other forms of pacifism emerge out of political philosophy, including deontology, virtue ethics, and consequentialism, or out of political theories such as anarchism, feminism, and environmentalism. Additionally, some forms of pacifism have evolved directly out of the experiences and experiments of collective political movements, such as Gandhi's anticolonial struggle and Martin Luther King Jr.'s civil rights campaign.

The result is that different pacifist approaches have been described using a variety of terms indicating their emphasis or the kind of values and traditions they are rooted in. The current literature describes the following types, among others: absolute pacifism, personal or vocational pacifism, collectivist or political pacifism, epistemological pacifism, technological pacifism, nuclear pacifism, environmental or ecopacifism, contingent or pragmatic pacifism, practical pacifism, liberal pacifism, transformational pacifism, and anarchopacifism (Fiala 2014, 2018; Holmes 2013; Cady 2010).

Although somewhat arbitrary, there are some important distinctions between different types of pacifism. First, it is helpful to distinguish between what has been termed "absolute pacifism" and what is variously called "pragmatic" or "contingent pacifism." The former takes an uncompromising position on killing and the use of force, while the latter accepts that there may be restricted or exceptional circumstances or occasions when lethal force can be justifiably used. A second important distinction is between what has been called "personal pacifism," which opposes killing as a personal act and therefore refuses to participate in any violent activities, and "political pacifism," which opposes war as a social practice and works to dismantle the broader war system (Ryan 2015). Third, it is important to note the difference between what has been criticized as "liberal pacifism," which accepts the existing state and economic structures of the world but seeks to reform them to be less directly violent, and "transformational pacifism" or "anarchopacifism," which aims at a revolutionary transformation of the violent capitalist state and all the forms of violence that are indelibly part of it, such as racism, sexism, ableism, and the like.

Following the recent growth in studies of nonviolent resistance movements (Chenoweth and Stephen 2011), pacifism and nonviolence are frequently seen as synonymous, if not as interchangeable terms, because they both oppose the use of organized violence in politics. At the very least, they can be seen as complementary in that pacifism provides a series of ethical and theoretical arguments for opposing violence and war, while nonviolence provides a set of practical strategies that can be used as alternatives to violent tactics in collective political struggles (Atack 2012, 158). Another common approach is to draw a distinction between principled nonviolence and pragmatic or strategic nonviolence. Here pacifism is considered to be a form of principled nonviolence because of its basis in ethical principles and its normative commitment to building peaceful societies.

However, a third approach seeks to go beyond the arbitrary distinction between principled and pragmatic nonviolence, in part by following Vinthagen's (2015, 61; original emphasis) proposal that nonviolence properly refers to intentional forms of organized political action which are simultaneously "*without violence* and *against violence*" (see also Atack 2012). Emphasizing that we are referring specifically to intentional action aimed at opposing or removing some kind of direct, structural, or cultural violence, such as the denial of human rights, for example, and that the means of struggle are also nonviolent draws out the normative basis of all forms of nonviolent political action, regardless of its justification. This conception also recognizes the constitutive nature of political action and suggests that for whatever reason nonviolent action is undertaken, one of its effects will be to constitute and strengthen the practice of nonviolent, peaceful politics (May 2015).

In the final analysis, pacifism and nonviolence are more accurately described as a political tradition, a set of moral ideals or an approach to politics, rather than a theory. Although pacifism contains theoretical elements (such as a theory of power and theories about why nonviolent action works) and has theoretical implications (including implications for agency and the nature of politics), it is a family of different values, arguments, and practices that are nonetheless united in their opposition to organized violence and their commitment to the construction of a more peaceful world. In ethical terms, pacifism can be viewed as a cluster of positions that sit along a moral continuum which has militarism and warism at one end, and absolute pacifism at the other—with more restricted forms of pacifism (such as nuclear pacifism), antiwar internationalism, and just war theory in between (see Cady 2010).

Pacifism in International Relations

There are two notable aspects when considering the place of pacifism in international relations. First, pacifism, pacifists, and nonviolent movements have played a major role in shaping the nature and practices of states and the international system over the past one hundred years, and their influence continues to be evident today (Ryan 2013). Second, and somewhat paradoxically, pacifism has come to occupy a subjugated intellectual position within the field; it is regarded, if at all, as irrelevant and inapplicable to the current theory and practice of international politics.

In the first place, despite the appalling violence of the previous century, as seen in two world wars, numerous genocides, the Cold War, and more than three hundred major wars since 1945, a closer look at this historical period demonstrates that pacifism and nonviolence have played a significant parallel role in the development of IR (see Bartkowski 2013). For example, before and after World War I, the influence of pacifists can be seen in the conscientious objector movement, in a great many disarmament campaigns of the time, and in efforts to establish the League of Nations, the precursor to the United Nations. In these international organizations, the first significant successes in limiting the right of states to go to war was achieved, along with drawing up many international humanitarian laws and conventions and establishing peaceful mechanisms for resolving interstate conflict. In fact, it can be argued that the foundations of much international law today, as well as international institutions like the United Nations and all the many international movements and international conventions related to human rights, disarmament, and civilian protection, were the direct result of a great many pacifists who struggled against the tide of militarism and war before and after World War I.

During World War II, although not widely known or studied, nonviolent resistance was a key feature of the struggle against the Nazi occupation of Europe in countries like Denmark and the Netherlands. After the war, prominent pacifists like Mohandas Gandhi led successful nonviolent anticolonial struggles in India and elsewhere which had a major impact on the global order, as European colonial powers were forced to give up their empires. Similarly, nonviolence played a significant role in struggles for racial and social justice, such as Martin Luther King Jr.'s civil rights struggle in the United States, and in struggles for greater political freedoms in countries like South Africa and Poland and several countries in Latin America.

In the past few decades, a great many nonviolent movements have been instrumental in bringing down dictators or winning greater freedom and human rights for citizens (see Chenoweth and Stephen 2011), including the People Power movement, which brought down the Marcos regime in the Philippines; the Carnation Revolution, which ended the military dictatorship in Portugal; and the peaceful revolutions in Poland, Czechoslovakia, East Germany, Ukraine, Bulgaria, and the Baltic republics which contributed to the collapse of the Iron Curtain (see R. Roberts and Ash 2009; Schock 2005). Following the collapse of the Soviet Union, a series of nonviolent revolutions swept across Africa and other parts of the developing world during the so-called Third Wave of Democracy, which transformed the face of the Global South. More recently, nonviolent movements have taken place across the Middle East and North Africa in the Arab Spring and the color revolutions across Eastern Europe, again transforming the politics of these regions.

However, despite their undoubted impact and influence on international politics over the past one hundred years, pacifism and nonviolence are today largely ignored in IR, and when they are acknowledged, they are often dismissed as naïve, unrealistic, outdated, or even as immoral and dangerous. Recent studies have clearly demonstrated that notwithstanding the burgeoning field of nonviolent resistance movement research, pacifism is rarely acknowledged or engaged with by IR scholars in the field's journals, conferences, or publications on the ethics of armed conflict, nor is it taught to students of IR (Ceadel 1987; Jackson 2018a; Howes 2013; Ryan 2013). This situation also pertains to public political discourse, where the term "pacifist" is not uncommonly used as a term of disapprobation, and to the media, which ignores pacifism as irrelevant. In short, despite their historical and contemporary importance, pacifism and nonviolence today exist as a form of subjugated knowledge within the field.

Objections to Pacifism

The subjugated status of pacifism within IR is maintained in part by a series of commonly expressed objections and criticisms of the approach (Jackson 2017). Among other reasons, it is sometimes argued that pacifism represents a single absolute moral position that, because it rejects any and all force and violence, makes it unsuitable for politics and the hard choices necessary in IR. Pacifists respond by pointing to the heterogeneity of their positions, most of which accept the necessity of force under certain circumstances. Further, they argue that the dominant approach to the use of force in IR, just war theory, is no less absolutist in its insistence that sometimes war is necessary and that some actions are prohibited in war.

Second, it is not unusual to see pacifism described as a form of passivity that entails doing nothing in the face of violent attack. The most common form of this argument involves establishing a stark choice between employing military force and "doing nothing." However, as the real-world examples of Gandhi, King, and countless other nonviolent resistance movements have demonstrated historically, pacifism is far from passive, and "pacifists do not claim that it is wrong to resist violence. On the contrary, they claim that violence should be resisted. They just believe that there are strong moral grounds for preferring to do so nonviolently" (Cady 2010, 103). The choice is not between war and doing nothing, pacifists suggest, but between war and creative, organized, and sustained nonviolence (Fiala 2014). Related to this, pacifists also dispute the argument that pacifism is dangerous because it signals weakness and thereby encourages aggression. Instead they suggest that unarmed states, posing no offensive threat and thereby mitigating the "security dilemma," may in fact be subject to less aggression.

A third analogy sometimes used to discredit pacifism is the so-called individual attacker analogy, in which a scenario involving a violent personal attack becomes the basis for arguing that pacifists are either immoral (because they would do nothing to protect themselves or their loved ones from an individual attacker) or inconsistent (because they would not extend an act of individual self-defense to the level of the society or nation). Pacifists respond by noting that apart from the substantive differences that exist between individuals and large social groups that prevent easy comparison, this analogy misses the obvious point that, unlike a contained incident among a small number of individuals, the use of military force is a form of collectively organized violence that requires extensive preparation, the maintenance of a permanent military force, a supporting economic base, the construction of a violence-supporting culture (including the cultivation of enmity sufficient for mass killing), and in practice, the organized and deliberate killing by and of people who have no direct involvement in the dispute itself. Pacifists also question why, if the analogy is genuine, we tolerate behavior at the state level that we would not tolerate at the individual level (Ryan 2013, 998).

A fourth, and probably the most common objection is that pacifism is ineffective, especially in the face of overwhelming force wielded by an unprincipled foe. Nonviolence, it is argued, worked in the past only because it was employed against democracies. In particular, the historical experience of Hitler and the Nazis proves that nonviolence is hopelessly naïve and unrealistic and that military force is the only way to stop certain kinds of threats. The pacifist answer to this objection is that "[w]hen faced with the objection 'it won't work,' the pacifist response must be, simply, that nonviolent action does work and has a history to document the claim" (Cady 2010, 95). The historical success of nonviolence is by now strongly established in the academic literature based on Chenoweth and Stephen's (2011) well-known study. Importantly, these literatures do not suggest that nonviolence works in every case; neither violence nor nonviolence works in *every* case. But we do know that nonviolence works very well in a great many documented cases, including cases where nonviolent movements faced off against brutal and oppressive opponents (Schock 2005).

The argument about the threat posed by the Nazis is challenging, particularly given its seeming self-evidence. However, one response is that "we should remember that there need be no inconsistency in holding that the war against Nazi Germany was justified but that war today is unjustified" (Holmes 2013, 164), given the unique historical conditions and context of World War II. More specifically, it can be argued that nonviolence might have worked to stop the Nazis' rise to power well before 1939, and that "[i]f the historical fact is that military means stopped Hitler once he began to march, it is also an historical fact that reliance upon such means on the part of the world's nations did not prevent his rise to power in the first place . . . [and] had military action not been taken, say, until 1943 . . . it is unlikely that Hitler could have been stopped this way either" (164). Related to this, it is also important to interrogate how effective the use of violence really was in this instance. While force eventually resulted in the military surrender of Germany, it largely failed to protect civilians, save European Jews, end future military aggression, defeat the forces of fascism, or create the conditions for a more peaceful world.

A final objection is that pacifism is naïve and unrealistic about the perfectibility of human nature and the nature of evil and cannot therefore contribute to serious discussions about how to deal with violence and threats in the real world. Here pacifists respond that "[t]he weight of extensive empirical evidence demonstrates that the practitioners of violence are more often the tragic idealists than are pacifists" (Howes 2013, 438). This is based on the observation that a growing body of empirical research on wars, terrorism, drone killing, torture, repression, and the death penalty demonstrates very low rates of success for the use of violence. Further, looking objectively at the long history of failed military interventions and wars over the past century—which have caused more than 100 million deaths, tens of millions of refugees, and immeasurable human misery—it seems naïve and foolish to argue that military force can be used rationally for security and peace. In this regard, nonviolence appears to have a much stronger record (Chenoweth and Stephen 2011).

Pacifist Critiques of IR: Violence and Just War

Pacifism, as noted, comprises a negative opposition to war and violence and a positive promotion of peaceful alternatives to military force. One of the most important pacifist critiques relates to IR's understanding of the nature of political violence. Pacifists argue that the field of IR has an unrealistic and dangerous view of violence because, first of all, it misunderstands the relationship between violence and coercion and between violence and power. That is, it misunderstands how material violence functions in the real world and instead assumes that the application of overwhelming and targeted force will compel people to submit or comply. In fact, the effectiveness of violence to deter or compel depends entirely on how people respond to the violence, not the violence itself (Wallace 2016). The application of violence can provoke either deterrence or retaliation, intimidation or rage, submission or resistance, and the desired response can never be assured (Howes 2009). This is why proponents of violence so often mistake the reliability of violence as a political instrument, and why so many interventions—such as in Vietnam, Iraq, or Libya, for example—so often produce unexpected and unwelcome outcomes.

Related to this, Vinthagen (2015) explains how power and violence are analytically distinct and how power is something inherent in a relationship between two actors. As a consequence, "the most extreme result of violence—the killing of a human being—is something that ensures that there will never again be subordination within that relationship. Killing results in an absolute absence of power. In fact, violence is a . . . failure of power" (193–194). In a very real sense, the use of violence is not an indication of power and control but a sign that one has lost power over one's opponent.

Second, the proponents of violence misunderstand the conditions and processes that make violence possible in the first place, namely, how it requires an enabling set of beliefs and ideas that make it legitimate and meaningful to its perpetrators and its audience (Wallace 2016). This means that the deliberate use of violence as a political tool constitutes the conditions for its own practice. A more realistic assessment of violence shows that while it can achieve immediate results, like dead bodies, pain, suffering, and material destruction, and while it can sometimes achieve certain short-term goals, like the destruction of an enemy's means to fight, its longer term effects are by virtue of its constitutive and world-shattering nature unpredictable and virtually always ends-destroying. As Arendt (1970, 80) put it, "The practice of violence, like all action, changes the world, but the most probable change is to a more violent world."

Another critique of IR's misunderstanding of violence is that it can never be purely instrumental; it is not, and can never be, merely a tool of politics. Rather, violence is productive and constitutive; it constructs the world in the process of being used. At the very least, we know that military violence cannot be simply a tool because to even have the tool available to use, a society must first have military institutions, a political-economic system to supply the weapons, a legal system to regulate killing in war, a cultural system to make killing and dying in war socially acceptable, and so on. All of these processes leave their mark on society both before and long after it has gone to war. In other words, the institutions and practices of militarism construct a world of violent actors who believe that violence is sometimes necessary and justified. When this is added to the international system as it is currently constituted, it makes what pacifists have long called "the war system."

Finally, supporters of violence in IR misunderstand the relationship between means and ends, which cannot in fact be separated. That is, military force cannot be used simply as the means to a separate end because the outcomes of political actions—actually, of all social action—are prefigured in the means: "[h]owever hard we try to separate means and ends, the results we achieve are extensions of the policies we live. . . . Means and ends are aspects of one and the same event" (Cady 2010, 56). From this perspective, pacifists argue that it is implausible that peaceful ends (such as security, protection, or democracy) can be achieved through violent, harmful means such as military force.

The pacifist critique of violence extends to just war theory (JWT), the dominant ethical approach for justifying and legitimizing war and military force in IR. Here pacifists have a number of key critiques. First, pacifists argue that the primary mode of theorizing about the use of force in JWT entails abstracting about the nature of war and employing hypothetical scenarios rather than using real-world cases (Ryan 2015, 33). Specifically, JWT reduces war to the period of lethal fighting between protagonists and fails to take into account the war preparations, the difficulties of determining the beginning or end of actual wars, and the long-term consequences of war after the fighting has ended. Pacifists wonder whether the moral calculations regarding the just cause versus the casualties of war would be altered by taking into account the excess deaths caused by the diversion of funds from healthcare to military preparations or the postwar suicides of veterans, for example.

Taking into account the connection between means and ends and the impossibility of separating cause and effect, pacifists also wonder whether the use of military violence in any instance can be justified when it leads to the reification of the broader war system and the resulting increased likelihood of future wars. From this perspective, it is disingenuous to claim that a particular war will save lives, when in fact the war will, through the chain of causality, likely cost the lives of those lost in subsequent wars. Related to this, pacifists question whether any actual historical cases of war measure up to the justifying principles of JWT. Ryan (2015, 33), for example, describes that out of around 3,200 recorded wars, JWT seems to agree on World War II as being the one clear example of a just war. And yet, even in this case, the atomic attacks on Japan, the terror bombing of German civilians, and the preventable Bengali famine that killed two million to four million people because food supplies were sent elsewhere for military reasons, are all excluded from judgment.

Second, from an ethical perspective, pacifists question the way JWT instrumentalizes human suffering and treats people as means to an end rather than ends in themselves. They also wonder about the way JWT and the decision to go to war rest on epistemological certainty about the (in reality uncertain) outcomes of using violence. That is, killing is an irreversible action, which therefore requires certainty and leaves no room for mistakes—in contrast to nonviolent action, which is rooted in epistemological humility and allows for mistakes and reversing course.

Third, a number of pacifist and nonpacifist critics have criticized some of the internal inconsistencies in JWT, such as the separation between justice of war (*jus ad bellum*) and justice in war (*jus in bello*), which means that soldiers in an unjust war who nevertheless act justly can have greater legitimacy than rebels who are fighting to protect their community but without the sanction of a functioning government because it has been destroyed by the unjust invasion. They have also criticized the way JWT has been used as political cover by powerful states, how it seems to provide states with justifications for engaging in war, and how it has historically failed to either reduce the number of wars or to noticeably improve conduct within wars. Finally, they have criticized JWT for downplaying the requirement to exhaust all nonviolent alternatives before taking the decision to go to war and for ignoring or not taking seriously alternatives to war, such as unarmed peacekeeping, civilian-based national defense, conflict resolution, and so on.

Pacifist Contributions to IR

The pacifist contribution to IR goes beyond the critique of existing ideas and practices, however. Pacifist theory and practice are based on theories and knowledge that contribute to a great many areas of current relevance to IR. For example, the theorists and practitioners of pacifism and nonviolence have devoted considerable effort to exploring how political power works in the real world, and therefore how power can be resisted and wielded by individuals and movements seeking to overthrow entrenched elites or transform structures of oppression. From Gene Sharp's (1973) consent theory of power to Vinthagen's (2015) Foucault-inspired retheorization of resistant power, nonviolent theorists have focused on the relational and collective aspects of power and how power can be resisted and transformed by collective action. In line with Arendt's (1970) conception of power as a form of normative collective action (Howes 2009, 103–105), pacifists have provided an alternative perspective to the dominant IR approach, which focuses primarily on material and military capabilities. Pacifist approaches to power would appear to better explain how unarmed protestors have been able to overthrow brutal militarized regimes on so many occasions (Atack 2009, 195) or how unarmed peacekeepers have been able to provide protection to vulnerable people in the midst of vicious civil wars.

Directly related to this, pacifist approaches have contributed to a deepening of our understanding of political agency in IR, demonstrating the ways individuals and groups can exert normative power and transform what were once considered immutable social structures, such as colonialism, South Africa's apartheid system, and the Cold War. Interestingly, pacifism is "the only political tradition whose major voices have been inordinately people of colour and women" (Ryan 2013, 978), indicating the way nonviolence functions to give voice and agency to oppressed people. Such understandings of power and agency provide an alternative normative and practical foundation for thinking about how best to reform the worst aspects of international politics, as well as specific challenges such as inequality and climate change, particularly as state and international institution–oriented approaches have proven to be so ineffective thus far.

Importantly, in relation to the themes of this volume, pacifism has real relevance for discussions of peacebuilding, statebuilding, and peace formation in IR. As I have argued elsewhere (Jackson 2018b, 2019), the theory and practice of both "peace" and "peacebuilding" remains trapped within a top-down, state-centric and rational-instrumentalist paradigm that limits their analytical reach and emancipatory potential. For example, recent peacebuilding theory has developed a powerful critique of contemporary practices of peacebuilding in societies after civil war or in transitions from authoritarianism, arguing that they are rooted in a Western-centric, top-down statebuilding approach that seeks to construct liberal democratic states that can be integrated into the global economy (Cooper 2007; Mac Ginty 2011). Critical scholars have argued that "liberal peacebuilding" is both flawed and failing because, among other things, it relies on a standardized blueprint that takes little account of the unique historical and cultural settings in which it is being applied. It is also problematic because it prioritizes state-centric, top-down institution-building, security understood as law and order, and the integration of postconflict societies into global economic structures and processes.

In response to the perceived failures of the liberal peace model, critical scholars have proposed new theories and approaches to peacebuilding, including notions such as "hybrid peace" (Mac Ginty 2010), "everyday peace" (Mac Ginty 2014), "resilient peace" (Johansson 2015), and "post-liberal peace" (R. Richmond 2011), among several others. These approaches are important not only for their critique of the dominant liberal approach but also for the way they attempt to focus on local actors and agency and pay attention to bottom-up processes, emerging forms of hybridity, resilient peace formations, and everyday peace indicators defined by local actors. However, while these theories attempt to push beyond the analytical and normative confines of liberal peacebuilding and rethink some its most obvious limitations, it is notable that they largely fail to seriously consider the role of armed peacekeeping, national militaries, military alliances and assistance, the arms trade, and, more fundamentally, the state's monopoly on the legitimate employment of violence. Instead they appear to assume that a strong, peaceful state, even if it is based on hybrid-local communities, resilience, and forms of everyday peace, nevertheless still necessitates a professional military supported by social institutions and a military-industrial complex in order to ensure the security necessary for political stability and economic enterprise.

More broadly, but perhaps unsurprisingly, an analysis of the main concepts and approaches to peace found in IR and peace studies reveals a similar set of problems. Here, it can be argued that the concept of peace remains relatively undertheorized, highly variegated and contested, and with little consensus on precisely what it signifies, the values it embodies, or how to evaluate its presence or absence empirically. Notwithstanding this broader failure, there have been some important attempts to theorize what peace is, including notions of "positive peace" and "negative peace" (Galtung 1969), "democratic peace," "everyday peace" (Mac Ginty 2014), "indigenous peace" (Mac Ginty 2008), "agonistic peace" (Shinko 2008), "resilient peace" (Johansson 2015), "post-liberal peace" (Richmond 2011), "quality peace" (Wallensteen 2015; Davenport, Melander, and Regan, 2018), and, more rarely, "emancipatory peace" (Richmond 2007), among others. The problem is that almost without exception, and similar to the peacebuilding literature, these conceptions of peace also fail to address the issue of state violence and militarism, and instead implicitly (or sometimes explicitly) assume the modern state and its military forces to be a necessary and natural element of peace, if only for its provision of law and order and security (which is assumed to necessitate the use of violence in some circumstances).

This is deeply problematic, pacifists argue, due to the nature of violence, the inherent violence of the state and its alliance with capitalism, the inseparability of means and ends in social action, and the constitutive nature of political action such as the use of violence. That is, pacifists argue that any normative peace theory must account for the violent historical origins of states, their historical record in mass violence against their own citizens (Rummel 1994), their role in entrenching a global war system, and the state's interdependence with capitalism and the harms caused by structural violence and ecocide (see D. Roberts 2007; Leech 2012). From this perspective, pacifist scholars argue that by failing to seriously consider the state and its violence head on, current theorizations of peace in IR fall short because they do not take seriously pacifist peace or anarchist peace as an essential part of any normative conception of peace. As a consequence, it can be argued that most peace theories are—or at least they function as—theories of pacification and stabilization rather than peace. At the very least, they reify the state-based capitalist system that is responsible for millions of annual deaths from poverty, deprivation, and irreversible harm to the environment—in addition to wars and military campaigns.

The point is that pacifism and nonviolence are highly relevant to discussions about both the nature of peace and the theory and practice of peacebuilding for a number of obvious reasons. In the first instance, pacifism represents a mode of theory and practice that is rooted in bottom-up social activism, the agency of individuals and small groups, collective forms of power, local histories and cultures, everyday forms of peace, and grounded forms of ethics and political life. In contrast to just war theory and statebuilding approaches, for example, which are rooted in the derivation of abstract moral and political principles that can then be universally applied by elites and institutions, pacifism has emerged from the lived historical experiences of local struggles for disarmament, social justice, and political change, and is rooted in the desire of ordinary people to live free from direct and structural violence. In other words, it both emerges from and seeks to build and strengthen locally based peace formations. It can therefore speak directly to attempts at going beyond Westphalian, state-centric, top-down forms of peace and peacebuilding.

Second, directly related to this rootedness in local peace formation, pacifist thought has long questioned the nature of the modern state, its primary mode of power based on violent domination (Bloom 2017) and its role in perpetuating forms of structural and cultural violence. Early pacifists like Gandhi and Tolstoy, for example, considered the state to be an inherently violent institution that could not be reformed but that needed to be replaced with peaceful communities based on a commitment to radical nonviolence, equality, nondomination, and democratic participation. More recently, pacifists have been contributing to important discussions about anarchopacifist communities and forms of political life (Llewellyn 2018), discussions that speak directly to peacebuilding, peace formations, and alternatives to statebuilding. Following recent developments in anarchist theory, as well as analyses of nonwarring societies (see Kaplan 2017), such discussions focus on how communities can build resilient, autonomous peace zones that provide people with the material provisions that states have traditionally provided (health, education, utilities, security) as well as direct democratic decision-making, gender equality, ecological balance, social and cultural development, and more. In other words, they speak to a kind of normative political project that seeks to build new political communities from the bottom up, which can in time make the state irrelevant.

Finally, pacifism has a great deal to offer discussions of alternatives to the use of and reliance on violence, particularly as they relate to statebuilding, humanitarian intervention, and national security. For example, there is a long-standing literature on unarmed civilian-based forms of national defense (Burrowes 1996) which offers alternatives to national security strategies based on state militaries. Similarly, there is a growing literature on how unarmed peacekeeping and nonviolent civilian protection can function effectively and more ethically than violence to protect vulnerable people during civil violence and prevent war crimes (Wallace 2016). Such analyses suggest how we might rethink the nature of political community itself and the roles of states, as well as the aims of statebuilding following civil wars or transitions from authoritarian rule. As a consequence, they open up the range of potential policy responses and local community action to include nonmilitary forms of national defense and international peacekeeping and intervention. In effect, they suggest ways in which the modern state could be refashioned and remade in radically nonviolent ways. In turn, such nonviolent, peaceful states would have a potentially profound impact on international interactions and interstate relations.

There is not the space to explore all the topics in IR to which pacifism and nonviolence have contributed important research and theory in recent years. However, it is worth noting that pacifist and nonviolence researchers have also contributed important ideas and data to discussions about realism and politics (Mantena 2012), freedom and violence (Howes 2016), just war theory and the ethics of violence (Ryan 2013), pacifism and international law (Moses 2017), and nonwarring societies (Kaplan 2017), among many others. Collectively these pacifist-oriented literatures have a great deal to contribute to current debates in IR about humanitarian intervention and peacekeeping, peacebuilding, statebuilding, international law, social justice, and revolutionary action to deal with contemporary challenges such as climate change, inequality, militarism, war, and the rise of populism. Based on a deep understanding of the nature and ethics of violence, power, politics, and agency, engaging seriously with pacifist theory and practice would expand the moral, theoretical, and practical imagination of IR (Jackson 2017) and open up new avenues of theory and policy response.

At the same time, and one of the reasons it remains subjugated, taking pacifism seriously would pose a challenge to many of the foundational myths and beliefs of IR, such as the inherent rationality and legitimacy of employing lethal violence, the purported link between the possession of military capabilities and power, and the state as the primary political actor in IR. However, it can be argued that dealing with some of the more intractable challenges facing IR at the current moment requires rethinking many of these limiting orthodoxies in any case.

Conclusion

In recent years, pacifism has made something of a return to IR. Although few scholars openly identify as pacifists and there remains a level of antagonism or indifference to it within the field, there is no doubt that the number of books and articles being published on pacifism and nonviolence, and the number of conference papers, panels, workshops, doctoral studies, and research programs on it have greatly increased in recent years. This is most likely related to the wider war-weariness that Western publics have about the nearly two-decade-long war on terror, the Vatican's recent decision to abandon support for the doctrine of just war, and the global tide of protest against elite dominance and the interlinked challenges of inequality, climate change, and militarism.

Nevertheless pacifism and nonviolence face a number of important challenges, not least the undeveloped state of pacifist theory (Ryan 2015, 26). While warism has had millennia to develop, pacifism remains a marginalized approach in IR and has yet to attract the same level of intellectual and material resources in developing it. This condition also speaks to the institutional support systems in place for militarism, such as the many thousands of scholars, university departments, and research institutes, the massive funding for military and security-based research, and the contrasting dearth of similar funding for peace scholars and institutes. It also suggests that pacifism faces a huge set of vested interests who would strongly oppose a shift away from militarism toward pacifism.

In addition to the challenges, there are a number of substantive issues facing pacifism at the present historical juncture. One of the key debates among scholars and activists includes the question of violence, namely, what the term should include, and whether some kinds of violence are actually necessary for transforming situations of social justice. Some argue that pacifism and nonviolence ought to focus on removing all forms of violence, including structural violence, cultural violence, and epistemic violence; others argue that direct physical violence in the form of war ought to remain the primary focus. Additionally, some scholars argue that in an era of suffocating neoliberal control, violence encompasses any new kind of meaning making or meaning breaking, and that unarmed insurrection, including rioting and clashing with police or neofascist groups, is a necessary part of constructing revolutionary agency and promoting social justice. In this sense, the very idea of nonviolence is a misnomer and an obstacle to transformative action, as creating the conditions for social justice necessitates the violent tearing down of existing epistemic and material structures (see Meckfessel 2016).

In many respects, this is a debate about the relative merits of liberal pacifism and strategic nonviolence versus transformational or revolutionary nonviolence. A key question for both theorists and activists is whether pacifist approaches based on reforming the current system will be effective or are sufficient, or whether a more radical, revolutionary approach is required to bring down the related evils of warism, racism, capitalism, imperialism, and ecocide. Some theorists have argued that the logical conclusion of pacifism is anarchism, as the state, and its alliance with capitalism, constitutes a form of institutionalized violent domination (Atack 2009, 191–192).

Finally, related to all of these debates, the issue of prefiguration and Gandhi's notion of the constructive project arise. In recent years activists and theorists have debated whether the primary focus of pacifism and nonviolent action should be on opposing militarism and violence in IR, or whether the focus should instead be on building alternative communities and institutions that can model alternatives to the violent capitalist state. Many believe that both are necessary (Vinthagen 2015), and one of the main failures of nonviolent resistance in recent decades has been that the constructive project—modeling alternative peaceful communities and forms of international solidarity—has been neglected in favor of employing nonviolent strategies of resistance to reform the current system (Chabot and Sharifi 2013).

In the end, these questions are also directly relevant to peacebuilding, statebuilding, and peace formation in IR. Should peacebuilders focus on assisting postconflict states in integrating more effectively into the global capitalist system? Should they seek to build and strengthen state-based institutions, structures, and processes? Or is this a pathway to reifying the existing violent structures of world politics that perpetuate militarism and violent conflict? Should peacebuilding instead focus on supporting the emergence of locally based, nonviolent peace formations and political communities that can resist neoliberalism, globalization, and militarism? Pacifism, as an applied theory rooted in real-world political activism and the construction of peaceful societies, is in a strong position to speak directly to all of these important questions.

NOTE

I acknowledge the support of the New Zealand Marsden Fund in the preparation of this chapter, the research for which was conducted under the Marsden Fund proposal, 14-UOO-075, "A new politics of peace? Investigations in contemporary Pacifism and Nonviolence."

References

Arendt, H. 1970. *On Violence.* New York: Harcourt, Brace & World.

Atack, I. 2009. "Pacifism in International Relations." In *The Ashgate Research Companion to Ethics and International Relations*, edited by P. Hayden, 183–200. Aldershot, UK: Ashgate.

Atack, I. 2012. *Nonviolence in Political Theory.* Edinburgh: Edinburgh University Press.

Bartkowski, M., ed. 2013. *Recovering Nonviolent History: Civil Resistance in Liberation Struggles.* Boulder, CO: Lynne Rienner.

Bloom, P. 2017. *Beyond Power and Resistance: Politics at the Radical Limits.* London: Rowman & Littlefield.

Burrowes, R. 1996. *The Strategy of Nonviolent Defense: A Gandhian Approach.* New York: State University of New York Press.

Cady, D. 2010. *From Warism to Pacifism: A Moral Continuum.* 2nd ed. Philadelphia, PA: Temple University Press.

Ceadel, M. 1987. *Thinking about Peace and War.* Oxford: Oxford University Press.

Chabot, S., and M. Sharifi 2013. "The Violence of Nonviolence: Problematizing Nonviolent Resistance in Iran and Egypt", *Societies Without Borders* 8 (2): 205–232.

Chenoweth, E., and M. Stephan. 2011. *Why Civil Resistance Works: The Strategic Logic of Nonviolent Conflict.* New York: Columbia University Press.

Cooper, N. 2007. "Review Article: On the Crisis of the Liberal Peace." *Conflict, Security and Development* 7 (4):605–616.

Davenport, C., E. Melander, and P. Regan. 2018. *The Peace Continuum: What It Is and How to Study It.* Oxford: Oxford University Press.

Fiala, A. 2014. "Pacifism." In *The Stanford Encyclopedia of Philosophy*, edited by Edward N. Zalta. Fall 2018 edition, accessed 8 November 2018. https://plato.stanford.edu/archives/fall2018/entries/pacifism/.

Fiala, A. 2018. *Transformative Pacifism: Critical Theory and Practice.* London: Bloomsbury.

Galtung, J. 1969. "Violence, Peace, and Peace Research." *Journal of Peace Research* 6 (3):167–191.

Holmes, R. 2013. *The Ethics of Nonviolence: Essays by Robert L. Holmes.* Edited by P. Cicovacki. New York: Bloomsbury.

Howes, D. 2009. *Toward a Credible Pacifism: Violence and the Possibilities of Politics.* Albany, NY: SUNY Press.

Howes, D. 2013. "The Failure of Pacifism and the Success of Nonviolence." *Perspectives on Politics* 11 (2):427–466.

Howes, D. 2016. *Freedom without Violence: Resisting the Western Political Tradition.* Oxford: Oxford University Press.

Jackson, R. 2017. "Pacifism and the Ethical Imagination." *International Politics*, online, 12 December. https://doi.org/10.1057/s41311-017-0137-6.

Jackson, R. 2018a. "Pacifism: The Anatomy of a Subjugated Knowledge." *Critical Studies on Security* 6 (2):160–175.

Jackson, R. 2018b. "Post-liberal Peacebuilding and the Pacifist State." *Peacebuilding* 6 (1):1–16.

Jackson, R. 2019. "The Challenge of 'Pacifist Peace' to the Varieties of Peace in IR." Paper presented at the International Studies Association 60th Annual Convention, 27–30 March, Toronto.

Johansson, P. 2015. "Nurturing Adaptive Peace: Resilience Thinking for Peacebuilders." Paper prepared for the 15th Biennial Conference of the IASC, The Commons amidst Complexity and Change, 25–29 May, Edmonton, Canada. https://dlc.dlib.

indiana.edu/dlc/bitstream/handle/10535/9838/Johansson_Patrik_Adaptive_Peace_May15.pdf?sequence=1&isAllowed=y.

Kaplan, O. 2017. *Resisting War: How Communities Protect Themselves.* Cambridge: Cambridge University Press.

Leech, G. 2012. *Capitalism: A Structural Genocide.* London: Zed.

Llewellyn, J. 2018. "Building Emancipatory Peace through Anarcho-pacifism." *Critical Studies on Security* 6 (2):259–272.

Mac Ginty, R. 2008. "Indigenous Peace-Making versus the Liberal Peace." *Cooperation and Conflict* 43 (2):139–163.

Mac Ginty, R. 2010. "Hybrid Peace: The Interaction between Top-Down and Bottom-Up Peace." *Security Dialogue* 41 (4):391–412.

Mac Ginty, R. 2011. *International Peacebuilding and Local Resistance.* Basingstoke, UK: Palgrave.

Mac Ginty, R. 2014. "Everyday Peace: Bottom-Up and Local Agency in Conflict-Affected Societies." *Security Dialogue* 45 (6):548–564.

Mantena, K. 2012. "Another Realism: The Politics of Gandhian Nonviolence." *American Political Science Review* 106 (2):455–470.

May, T. 2015. *Nonviolent Resistance: A Philosophical Introduction.* Cambridge, UK: Polity.

Meckfessel, S. 2016. *Nonviolence Ain't What It Used to Be: Unarmed Insurrection and the Rhetoric of Resistance.* Chico, CA: AK Press.

Moses, J. 2017. "Anarchy, Pacifism and Realism: Building a Path to a Non-violent International Law." *Critical Studies on Security* 6 (2):221–236.

Richmond, O. 2007. *The Transformation of Peace.* New York: Palgrave Macmillan.

Richmond, O. 2011. *A Post-Liberal Peace.* London: Routledge.

Roberts, D. 2007. *Human Insecurity: Global Structures of Violence.* London: Zed.

Roberts, R., and T. Ash, eds. 2009. *Civil Resistance and Power Politics: The Experience of Non-violent Action from Gandhi to the Present.* Oxford: Oxford University Press.

Rummel, R. J. 1994. *Death by Government.* New Brunswick, NJ: Transaction Books.

Ryan, C. 2013. "Pacifism, Just War, and Self-Defense." *Philosophia* 41:977–1005.

Ryan, C. 2015. "Pacifism(s)." *Philosophical Forum* 46 (1):17–39.

Schock, K. 2005. *Unarmed Insurrections: People Power Movements in Nondemocracies.* Minneapolis: University of Minnesota Press.

Sharp, G. 1973. *The Politics of Nonviolent Action.* Boston: P. Sargent.

Shinko, R. 2008. "Agonistic Peace: A Postmodern Reading." *Millennium: Journal of International Studies* 36 (3):473–491.

Vinthagen, S. 2015. *A Theory of Nonviolent Action: How Civil Resistance Works.* London: Zed.

Wallace, M. 2016. *Security without Weapons: Rethinking Violence, Violent Action, and Civilian Protection.* Abingdon, UK: Routledge.

Wallensteen, P. 2015. *Quality Peace: Peacebuilding, Victory, and World Order.* Oxford: Oxford University Press.

CHAPTER 8

INTERNATIONAL POLITICAL SOCIOLOGY OF PEACEBUILDING

CATHERINE GOETZE AND
BERIT BLIESEMANN DE GUEVARA

> Il y a deux terrains sur lesquels nous sommes spontanément mythologues: le monde social et le cas où nous parlons des autres.
>
> (Bourdieu 2015, 419)

The study of peacebuilding does not really belong to any established social science discipline. This is due not only to the complex nature of peacebuilding but also to the very different vantage points from which its practice and politics can be analyzed. To take Cox's ([1981] 2000) distinction between problem-solving and critical theory as a starting point, peacebuilding can first of all be analyzed as an internationally deployed policy with the guiding question of how this policy can be improved and better adapted to internal or external performance criteria of effectiveness, cost-efficiency, or feasibility. Yet peacebuilding can also be analyzed from a critical point of view, by asking what kind of world order is materialized in peacebuilding and which, or whether, any alternatives to current practices might exist. Sociological methods, concepts, and theories have been engaged for both: to devise solutions to policy implementation and to critically examine how social and political power patterns are (re)produced in international peace missions.

The academic, institutionalized discipline of sociology offers a wide variety of interpretations of social phenomena and how they should or could be analyzed and to what effect. Accordingly, international political sociology of peacebuilding reflects an extremely wide range of topics, questions, and findings. As will be explored in this chapter, sociology is used in three different ways: as a general ontological understanding of the research object (peacebuilding) as a "society" in which policymakers can intervene in order to achieve specific policy goals; as a set of observation methods; and as a reference in social theory and philosophy, which, among other, allows criticizing peacebuilding's configurations of power and inequality. While ontologically all three ways assume the existence of "society," their substantially different epistemologies draw a very uneven image of what that society actually is, how it "works," and how it affects or is affected by peacebuilding.

What Is Sociology?

In its widest sense, "society" means simply the circumstance that people live together and that this is, at once, a condition of their existence and its result:

> That people look at each other and that they are jealous of each other, that they write letters to each other or that they lunch together, that they sympathize with each other beyond obvious material interests or that they resent each other, that gratitude for altruistic acts can create long lasting indestructible bonds, that one will ask someone else for directions in the street or that they dress and brush themselves up for each other—all these thousands of random relations between individuals bring us incessantly together no matter if these relations are lived momentarily or durably, wittingly or not, and if they are furtive or of long effect. (Simmel, Rammstedt, and Fitzi 1999, 69)

In the narratives of the academic discipline of sociology, the analysis of such living together is traced back to Enlightenment thinkers, the Industrial Revolution, and the scientific awakening in the nineteenth century that brought specific European forms of conceiving the world as modern and scientifically observable progress. While this understanding of sociology certainly contains a truthful historical account of the academic institution, it minimizes and simplifies the wealth of sociological thought and insights that are circulating in today's discussion of social phenomena. Thinking about how humans live together and understanding the tacit or explicit rules they follow in their interactions have always been at the core of human cosmologies, philosophies, and thought. Sociology constitutes a particular way of formalizing and formatting this thought in the institutional practices of the emerging nineteenth-century European universities. Pierre Bourdieu hence argues that a more accurate rendering of sociology's history needs to conceive it as a space of struggles among classical faculties, most notably between the sciences from which it has borrowed the epistemology of wanting to capture "the larger picture" (Simmel, Rammstedt, and Fitzi 1999) and the humanities from which it has taken its research object: human behavior and action (Bourdieu 2015, 432–433).

Reflexive critique of its own workings and origins has always been part of sociology since "doing science" and "being in academia" are important parts of modern society and reflect precisely this position of sociology in the academic field as hanging in between. As the space of academic and social, political and economic inquiry has widened since the first chair of sociology was created, so has the range of thinking about society that is represented in social studies.

This variety is reflected in the wide range of approaches that have inspired peacebuilding research. National differences emphasize the different pathways of academic institutionalization of both the disciplines of sociology and the disciplines of international relations or political science, where IR is a subfield. Scholars who are oriented toward the Anglosphere narratives of liberal institutionalism have drawn more often on the English school theorizing of international and world society (Wiberg 2006). While these works initially engaged mainly with the relation between international law and international society in cases of humanitarian intervention, of the Responsibility to Protect and postconflict reconstruction (Bain 2001, 2003; Wheeler 2002; Lang, Rengger and Walker 2006; Brommesson and Fernros 2009; Breau 2016; Welsh 2019), the literature on postconflict institutions of justice (Chesterman 2001, 2004; Lang 2008; Roff 2013) and some research that situates itself in the "local turn" (Mac Ginty and Richmond 2013) have taken up many of the English school's concerns with the social diffusion of norms through international institutions, law, and practices (e.g., Bertram 1995; Barten et al. 2002; Corrin 2003; Brinkerhoff and Mayfield 2005; Caplan 2005; Hilhorst and van Leeuwen 2005; Harbom, Högbladh and Wallensteen 2006; Kamatsiko 2015; Whalan 2017).

French scholars, on the other hand, commonly mobilized the entire canon of French post–World War II sociology. In France itself this research could build on an already strong tradition of sociological IR that had notably developed fine-grained analysis of the sociology of the state in non-European areas (Bayart 1996; Badie 2000; Chabal 2009), an analysis of globalization and interventionism that employed sociological concepts of power and attraction (Salamé 1996; Laidi 1998), and ethnographic-sociological analyses of humanitarian and development nongovernmental organizations (Smouts 2003; Siméant 2011) or spaces (Agier and Fernbach 2016).

French sociological traditions were also "translated" into English research and publications through the brokerage of French scholars based in the UK or US and non-French scholars having been trained in France. As English translations of Michel Foucault, Pierre Bourdieu, Claude Lefort, Gilles Deleuze and Félix Guattari, and Luc Boltanski became more and more available in the 1990s and 2000s, a new branch of transnational scholars invested questions of war and peace with concepts like Foucault's biopolitics (Duffield 2007; Jabri 2012) or governmentality (Zanotti 2006, 2011; Neumann and Sending 2010), with Deleuze and Guattari's accounts of the everyday (Smirl 2008, 2012, 2015; Mitchell 2014), or Bourdieu's concepts of capital, habitus, and field (Bliesemann de Guevara and Goetze 2012; Madsen 2011; Goetze and Bliesemann de Guevara 2014; Sending 2015; Goetze 2017).

Recent years have particularly seen the influence of postcolonial, anti-Eurocentric, and feminist or queer alternatives to the European, male, and traditional forms of positivist sociology (e.g., Steinmetz 2014; Go 2016a; Connell 2007; Kowal 2014; Gutierrez Rodriguez, Boatca, and Costa 2010). Given the inherent asymmetry of the peacebuilding field, with the interveners being predominantly from northern, high-income countries and the spaces of intervention being predominantly in the Global South, the interest in and inspiration from decolonial and imperialism-critical social thought seems evident (Sabaratnam 2013).

Across this variety two common threads allow bringing these diverse approaches together as "sociology": first, all sociology is characterized by its interest in the research object of *relations* between humans (and also increasingly between humans and nonhumans); second, all these approaches presume in methodological terms that an analysis of relations among humans (and nonhumans) can accrue knowledge about all of us, our living conditions and our life chances. Sociological approaches distinguish themselves by providing at once a research focus, commonly through a shared vocabulary of "society," "social relations," "social constructivism," "social conditions of possibility," "social power," etc., and a shared methodology, namely of trying to read the meaning of human action and behavior into larger sets of social interactions, exchanges, and relations.

Sociology as Method: Assessing the Sociology of Peacebuilders and Peacebuilding Organizations

The methodological focus is commonly first of all on the assessment of the social phenomena sui generis. Since sociology as discipline positions itself between the sciences and humanities, its main methodological concern has been to provide methods of data collection and exploitation that would show that social phenomena exist, i.e., that individual interactions and exchanges do form patterns that are consistent across time and space, independent of individual or conjunctural influences. Such methods range from intersubjective, hermeneutic, and interpretative techniques of tickling socially constructed meanings out of texts, artifacts, or behavioral observations to objectivist, positivist-empiricist approaches of large-N studies in which correlations are supposed to speak for themselves. Peacebuilding studies have used the entire range of these methods.

Since peacebuilding is a relatively novel area of international policymaking much of this research has been used to chart the field; hence, a lot of the research responds first of all to the question "What is happening when peacebuilding is happening?" For example, using originally surveyed quantitative data and external data sources, Neudorfer (2015) and Karim and Beardsley (2017) investigated the incidence and circumstances of sexual abuse by peacekeepers. A first task of both research projects was to establish that sexual abuse and exploitation is not a random and exceptional occurrence. Hence both studies produce convincing data analyses that show that sexual abuse and exploitation have been endemic in UN peacekeeping missions. Both studies show that organizational structures of the peacekeeping mission and the local circumstances of deployment are crucial in enabling or prohibiting sexual abuse and exploitation. Building on the observation of these studies that there is also a very large variation of incidents of sexual abuse, Moncrief (2017) explores the interconnection between socialization processes internal to the militaries that deploy peacekeeping troops and the circumstances on the ground. He finds that disciplinary breakdown at the lowest level of military command structures is most likely a strong enabling condition of sexual abuse by peacekeepers.

All these quantitative analyses seek to identify causes of what is arguably the most blatant violation of a peacekeeping mission's remit to protect local populations. They situate themselves clearly in the problem-solving space of sociological analyses of peacebuilding. Contrary to many other studies on peacebuilding, their gaze is firmly directed inward, into the organizations that do peacebuilding. In a similar vein, a number of studies have been interested in the social composition and socialization mechanisms inside peacebuilding organizations. These have commonly used not quantitative but qualitative methods, among other reasons because of the inherently difficult access to quantitatively exploitable data from international organizations.

In order to circumvent these difficulties, many studies have referred to publicly available biographical data of UN or other organizational staff. Drawing biographical data from the UN's handbooks, Novosad and Werker (2014), for instance, have analyzed the importance of geographical origin for the attainment of senior positions in the UN. From a traditional organizational sociology perspective, the authors' main finding is that Western Europe and North America have "retained control over a disproportionate share of positions in the (UN) Secretariat" (Novosad and Werker 2014, 31) and that large but poor countries like India, China, and Indonesia are significantly underrepresented. This reflects similar findings of other international organizations such as the World Bank and International Monetary Fund (Woods 2006), the UN High Commissioner for Refugees (Fresia 2009), development aid organizations (Mosse 2011), and humanitarian organizations (Siméant 2011; Tarrow 2005). Goetze's (2017; Goetze and Bliesemann de Guevara 2014) prosopography and network analysis of senior officials complements Novosad and Werker's analysis since it shows that common educational background and shared work experience play a major role in establishing career paths in peacebuilding. This finding reflects the wider experience of career paths in international organizations such as they have been analyzed by Coicaud (2008), who also identifies the international organization "circuit" as a labor market of its own.

With quantitative data becoming even scarcer and more difficult to access, sociological investigations of local populations and the interactions between locals and internationals have relied heavily on qualitative methods like participant observation and interviews. Other than a methodological choice, these qualitative approaches also reflect a greater concern with the fragile and disputable power relations between international organizations and local populations. Watching, observing, and asking questions are considered more apt to capture how international peacebuilding is lived, experienced, and understood on the ground by those people whose lives are shaped by it, as Pouligny (2004, 18) points out. Pouligny's study *Ils nous avaient promis la paix* (translated as "Peace from below") was the first to broach the question of how the imposition of peace by militarized peacekeepers impacts local power relations and sociopolitical configurations. Pouligny employed mainly methods that, from a disciplinary point of view, belong to the repertoire of anthropology but which have been claimed also by sociologists, especially in what she dubs "understanding sociology": participant observation of community meetings and associations and more formal encounters with UN peacekeepers; dialogues, "street side chats," and conversations with local folks, political personalities, community leaders; interviews with focus groups; archival materials and eclectic written documentation of local perceptions of the conflicts, conflict resolution, and the peacekeeping mission; and analysis of artifacts like murals, carnival stunts, musical presentations, jokes and sayings, oral history, poetry and storytelling. Her wide scooping for "text" to analyze allows Pouligny to sharply contrast the different ways in which peacekeepers and local populations conceive of their living conditions and circumstances, time horizons, perceptions of self and other, understandings of peace and conflict, etc. She astutely shows that the misunderstandings and clashes between local populations and peacekeepers are surely political, but in the sense of being deeply rooted in very different conceptions of what "peacefully living together" means.

Autesserre's (2014) in-depth analysis of the daily routines and working habits of aid workers develops a similarly broad approach of collecting data. Autesserre equally draws on auto-ethnography, participant observation, and lengthy structured and unstructured interviews to observe what she calls "practices." She draws on growing scholarship in international relations that focuses less on what actors say in world politics than on the ways they actually do politics. What has often been claimed to represent a "practice turn" in international relations, however, is built on unstable theoretical ground. Although practice (or "praxis" or "act" or "action" or "interaction") is indeed a common concept in sociological analysis, there is no unitary practice theory. As Nicolini (2013, 1) points out, "while practice theories can offer a radical new way of understanding and explaining social and organizational phenomena, they can only be approached as a plurality." Observing practice is to be considered a method rather than a theory. Consequently Autesserre's painstakingly detailed study *Peaceland* reads like an insightful catalogue of all the micro-acts that create and maintain frictions, misunderstandings, inequalities, or inefficiencies of aid on the ground. It says little, however, about what these behaviors mean for the broader picture of power and inequality in the world.

Social Thought and Social Policy in Peacebuilding

Pouligny's and Autesserre's works have drawn on a range of sociological methods in order to provide an organizational sociological analysis of the often conflicted encounter between external peacebuilding forces (whether military peacekeepers or civilian aid workers) and local populations. They have not only collected data to show that these conflicts are rooted in social realities but have also provided data to show how the interaction and communication between internationals and locals actually take place. Underlying these analyses is the (tacit) assumption that a better understanding of this interaction and communication will allow for "fixing" it and the conflicts on the ground since the core of local resistance(s) to peacebuilding is its many inefficacies and contradictions (Bliesemann de Guevara 2012, 1).

Such research reflects the relationship between social work and sociology, namely the expectation that better knowledge of "how society works" will lead to better means of intervening in specific social relations and of improving societies more generally. Social policy and social work are integral parts of the European-origin, industrial and postindustrial welfare states and democracies. The research on the "social work" of peacebuilding is articulated around two poles: on gaining a better understanding of local reactions to peacebuilding and on understanding the sociological and organizational obstacles within the peacebuilding organizations. Studies are often articulated around a variety of buzzwords (cf. Cornwall 2007), whether "whole-of-society," as in the quote that follows, or "resilience" (Cohrs et al. 2013), "hybridity" (Mac Ginty and Richmond 2013), or "local ownership" (Donais 2012). These studies share an understanding of civil wars as the expression of social conflicts that turned violent; they also share the belief that peace can be durably anchored in a society on the basis of social justice. Underlying this research is the presumption that the persistence of international peacebuilding efforts in increasing and continuing adversity (the mission in Bosnia and Herzegovina has now lasted twenty-five years; the UN Mission in Kosovo has still not been dissolved; missions in Afghanistan, Haiti, and Democratic Republic of Congo have been ongoing for almost two decades, etc.) will ultimately be beneficial for those societies, once the appropriate modus of delivery has been found. Or as Martin, Bojicic-Dzelilovic, and Benrais (2018, 172) argue, "We have attempted to articulate 'Whole-of-Society' as a practice-based approach which seeks to enhance the effectiveness of externally led peacebuilding and conflict prevention through recourse to the social contexts within which they are implemented, and with the aim of seeing complexity, within local society and in relations between external peacebuilders and local society, as an opportunity to be grasped, as much as impediment to effective outcomes."

Although there have been attempts to formulate these approaches as a coherent "local turn" (Leonardsson and Rudd 2015; Mac Ginty and Richmond 2013; Donais 2015; Paffenholz 2014), most of the research disagrees substantially on the underlying understandings of "the local" (Hughes, Öjendal, and Schierenbeck 2015; Randazzo 2016), the implications of these understandings for international policymaking, and their analyses of the resistances, challenges, and opportunities of peacebuilding. While some notably argue, similar to Autesserre, that "the local" does represent a field of potentially real peacemaking, others, in line with Pouligny's analysis, dispute that international interference has produced any beneficial effects for the societies in question (Chandler 2013; Sörensen 2014).

These different appreciations of peacebuilding's effects on "the local" have produced two different ways of sociologically analyzing the interaction between the international and the local. For those who wish to advise on peace policies, the challenge is to identify the social cleavages that led to conflict (Hilhorst and van Leeuwen 2005) and the legitimate voices that represent or not the various social forces in a conflict zone (Whalan 2017; Martin, Bojicic-Dzelilovic, and Benrais 2018; Jankowitz 2018). Much of this research draws on earlier research on democratic transitions in the 1990s (Goetze and Guzina 2008) and experiments in one way or the other with arguments about state failure, peace spoilers, warlords, or otherwise vested interests of individual violence entrepreneurs that were promoted in the first decade of the 2000s (Malejacq 2016). Successful peacebuilding will, in this perspective, devise strategies of communication, stakeholder consultation, and redistribution that address grievances and create social harmony (McCandless, Abitbol, and Donais 2015; Saulich and Werthes 2018; von Billerbeck 2017; Bräuchler 2017).

This contrasts with a more critical strand of this literature that questions the benefits of interventions, tacitly or openly assumes that interventions play out in highly asymmetrical international power structures, and seeks to analyze the concrete "pressure points" of interventions on local societies, on the conflict, and, increasingly, on the intervening organizations. This literature commonly seeks to pinpoint the social conditions for peacebuilding's failures and is particularly interested in the way peacebuilding reproduces inequalities, injustices, and asymmetries of power in international relations. Such studies scrutinize how interventions reinforce conflicts on the ground (Hensell and Gerdes 2012); how interventions re-create opportunities for violence, for instance in the form of gendered violence (Pruitt 2016; True 2012, ch. 8); or how interventions create stereotyped, orientalist, or neo-imperial views of "the local" (Sabaratnam 2013). In contradistinction to interventionist research, this strand of research critically acknowledges the social engineering that peace missions represent and is wary of its patronizing, stereotyping, erroneous, and ultimately damaging effects. Although being methodologically committed to sociological empirical work on the ground through interviews, organizational and network analysis, participant observation, and sometimes statistical work, these studies inscribe themselves in the larger project of critically analyzing power structures in global politics.

The question of what actually happens when peacebuilding happens also animates research on the effects of peace missions on the peacebuilding organizations themselves. A range of studies applies tools from organizational sociology to show that the interaction between the "field" and the organizations has effects in both directions (Daho, Duclos and Jouhanneau 2019). Gilmore (2015) argues that military personnel and armies more generally do reinterpret their roles as "cosmopolitan" when sent to contribute to missions mandated by the UN Security Council. Duclos (2016) observed the interaction between former combatants of the Kosovo Liberation Army and international police forces in Kosovo in their attempt to build a Kosovo police force. Rather than being interested in the actual new police force, she analyzes the effect that the demobilization and transformation exercise had on the self-understanding and social positioning of both sides. While she did observe an identity crisis of the Organization for Security and Cooperation in Europe (responsible for security-sector reform in Kosovo), she also argued that the international security-sector reform rather solidified both sides and the gap between them. Daho (2015) similarly identifies international interventions as one crucial factor in the reidentification of French militaries in the post–Cold War era and also as an important element of the increased blurring of lines between internal and external militarization of French society.

Social Theory and Critique

Such critical studies argue that the socially constitutive effects of peacebuilding are mutual, yet they do not affect both sides in the same way since they reflect power differentials in the world. A number of studies therefore use sociological analysis to identify and analyze global power structures in more detail and differentiation. Mitchell (2014), for instance, argues that the West's obsession with humanitarian assistance is intimately linked to the European-origin, metaphysical apprehensions of death which she traces back to the secularization of modern lives. Violence, dying, and death have become so unacceptable in modern Western societies that humanitarianism has gained a more than charitable sense in these: more than being an expression of compassion, humanitarianism expresses the hope and urge to overcome death and the fear of it. Goetze (2017) similarly argues, on the basis of an in-depth sociology of civilian staff deployed in peacebuilding, that the epistemology of peace deployed in peacebuilding is fundamentally built upon a highly selective ethical understanding of what constitutes violence and what does not. While the violence of interventions is downplayed or ignored, the mantra of peacebuilding "ending the killing" is reproduced in various narratives, discourses, and myths around peacebuilding, the United Nations, and iconic personalities like Mandela and Gandhi.

Duffield (2001, 2008) pushes this critique back into a slightly more world-political arena when he argues that humanitarian assistance in conflict situations, as in Sudan, are global efforts of disciplining and managing surplus populations in an increasingly networked and globalized world. In a similar vein, Owens (2015) argues that social thought has been applied to counterinsurgency and colonial and postcolonial interventions to strip these military and violent encounters of their political appearance and contents. Representing a violent form of "householding," interventions destroy societies "there" and conceal the politics of the destruction. The languages of society, social work, and sociology, Owens argues, have legitimated a vocabulary of domestication, modernization, development, or pacification to be used for what would otherwise appear to be openly political conflicts over government, territory, self-determination, and statehood.

Fassin (2012) investigates this argument on a more institutional level when he compares social work at home (France, in his case) and humanitarian assistance abroad.

He notices the same dilemmas of action in both cases where aid seems to be reproducing the conditions of possibility of precisely the same marginalization of populations it is supposed to eliminate. Similar to Duffield and Owens, Fassin argues that social and humanitarian assistance are two sides of the same medal of disciplining power, of concealing and suppressing what should not be: poverty, suffering, and failures of modernity and capitalism to allow acceptable life to everyone, whether or not they are "productive." Sabaratnam (2017) concurs that the current peacebuilding interventions constitute "local" populations as "disposable" and as "blank slates." Going beyond the already formulated critique by Pouligny and Autesserre, Sabaratnam articulates the findings of her study explicitly in the concept of "coloniality of power." The disposability of local populations is, in this perspective, constitutive of the power of interveners; it is only because local populations are rendered dependent, stripped of agency, and seen as surplus populations whose main function is to be the object of Western charity and social engineering that interventions exist.

Conclusion

While Owens wants to be critical of sociology, she actually deploys an array of sociological critical tools, not least the conceptualization of one particular social form, the household (or *oikos*), as a fundamental analytical unit (see Go 2016b). Her critique is particularly directed against the use that is made of sociology and social theory, arguing that the world power structures that condition interventions are fundamentally political and not reducible to the social. She warns against using the canonized sociological language, particularly the language of social work and household, to argue for political emancipation and the political more generally.

Consequently, future sociologically inspired research in peacebuilding needs to pursue both a reflective theorization of peacebuilding and a more detailed empirical investigation of the social forces and interactions involved in these processes. What still needs to be done is vast, but two immediate desiderata stand out. First, peacebuilding analysis as a research field can contribute enormously to the conceptualization of world society. Theoretically and conceptually international political sociology needs to grasp the importance and role of peacebuilding situations for the constitution, hierarchization, and plasticity of world society. Since the asymmetry between "internationals" and "locals" has moved to the center of analysis, the field has been struggling to conceptually reconcile the view that peacebuilding is fundamentally structured by this power binary, with the observation that, in practical politics, multiple scales and temporalities of agency overlap and intersect. These conceptual troubles are not unlike those of various approaches to global, world, or international society where the question remains open whether we can, at all, speak of a global society, and if so, what this social is made up of.

Second, empirically very little continues to be known about the sociology of conflict in the first place and, consequently, the political sociology on the ground and in the postconflict situation. From a policy advising perspective, much of peacebuilding still relies on rough macro-views of the social causes, dynamics, and effects of armed conflict and of negotiated "peace." The dominance of quantitative models of analyzing causes of wars has not helped our understanding of who fights for what, under which circumstances, and with which social effects. Wars shift social balances and reorder hierarchies—after all, that is precisely what most conflicts are about. What concretely happens in the real cases on the ground, however, remains difficult to analyze due to a range of problems, from the practical kind when doing research in war and intervention areas (Bliesemann de Guevara and Bøås 2020), to the conceptual. Yet sociological precedents of analyzing concretely how, on the street, in the home, in school, or at work, on the minuscule local to the more or less open global level, social hierarchies intersect, interact, affect, or disrupt human-human relations need to be more developed in conflict and postconflict analyses. As Go (2016a) points out, for many sociologists doing empirical research on social dynamics is necessarily also a critical examination of power and authority. Bourdieu (1977, 1982, 1989, 2000, 2015) specifically has never tried of pointing out that the struggles over social positions, the capital resources to found them, and the "rules of the game" of who gets to have the authority to decide them, are all profoundly political questions, but also that the struggles over symbols, language, and epistemologies are, sui generis, violent and ultimately political.

A sociological analysis that does not take struggles over power and authority as its starting point will, according to Bourdieu, always remain incomplete and simply reproduce the prevailing structures of dominance. And an important part of this accounting for power is also the reflexive analysis that social scientists need to undertake on their own epistemologies, ontologies, and methodologies when they construct their research object. Sociological analyses of peacebuilding can therefore meaningfully produce critical insights only if they are inserted in larger debates over the power, violence, and life opportunities in world politics, or as Sabaratnam (2017) concludes, when the "listening exercise" of sociological (and ethnographic) methods becomes also a hearing exercise.

References

Agier, M., and D. Fernbach. 2016. *Borderlands: Towards an Anthropology of the Cosmopolitan Condition*. Malden, MA: Polity.

Autesserre, S. 2014. *Peaceland: Conflict Resolution and the Everyday Politics of International Intervention*. Cambridge: Cambridge University Press.

Badie, B. 2000. *The Imported State: The Westernization of Political Order.* Stanford, CA: Stanford University Press.

Bain, W. 2001. "The Tyranny of Benevolence: National Security, Human Security, and the Practice of Statecraft." *Global Society* 15 (3):277–294.

Bain, W. 2003. *Between Anarchy and Society: Trusteeship and the Obligations of Power.* Oxford: Oxford University Press.

Barten, F., R. Perez Montiel, E. Espinoza, and C. Morales. 2002. "Democratic Governance: Fairytale or Real Perspective? Lessons from Central America." *Environment and Urbanization* 14 (1):129–144.

Bayart, J.-F. 1996. *La greffe de l'Etat.* Paris: Karthala.

Bertram, E. 1995. "Reinventing Governments: The Promise and Perils of United-Nations Peace Building." *Journal of Conflict Resolution* 39 (3):387–418.

Bliesemann de Guevara, B. 2012. "Introduction: Statebuilding and State-Formation." In *Statebuilding and State-Formation: The Political Sociology of Intervention*, edited by B. Bliesemann de Guevara, 1–19. London: Routledge.

Bliesemann de Guevara, B., and M. Bøås (eds). 2020. *Doing Fieldwork in Areas of International Intervention: A Guide to Research in Violent and Closed Contexts.* Bristol: Bristol University Press.

Bliesemann de Guevara, B., and C. Goetze. 2012. "The 'Statebuilding Habitus,' UN Staff and the Cultural Dimension of International Intervention." In *Statebuilding and State-formation: The Political Sociology of Intervention*, edited by B. Bliesemann de Guevara, 198–213. London: Routledge.

Bourdieu, P. 1977. "Sur le pouvoir symbolique." *Annales: Economies, sociétés, civilisations* 32 (3):405–411.

Bourdieu, P. 1982. *Ce que parler veut dire: L'économie des échanges linguistiques.* Paris: Fayard.

Bourdieu, P. 1989. "Social Space and Symbolic Power." *Sociological Theory* 7 (1):14–25.

Bourdieu, P. 2000. *Propos sur le champ politique.* Lyon: Presses Universitaires de Lyon.

Bourdieu, P. 2015. *Sociologie Générale: Cours au Collège de France, 1981–1983.* Edited by P. Champagne, J. Duval, F. Poupeau, and M.-C. Rivière. Paris: Raisons d'agir, Seuil.

Bräuchler, B. 2017. "Social Engineering the Local for Peace." *Social Anthropology* 25 (4):437–453.

Breau, S. C. 2016. *The Responsibility to Protect in International Law: An Emerging Paradigm Shift.* Abingdon, UK: Routledge.

Brinkerhoff, D. W., and J. B. Mayfield. 2005. "Democratic Governance in Iraq? Progress and Peril in Reforming State-Society Relations." *Public Administration and Development* 25 (1):59–73.

Brommesson, D., and H. F. Fernros. 2009. "Individualisation and Destabilisation of the International Order: The Case of the Responsibility to Protect." *International Review of Sociology* 19 (2):315–330.

Caplan, R. 2005. *International Governance of War-Torn Territories.* Oxford: Oxford University Press.

Chabal, P. 2009. *Africa: The Politics of Suffering and Smiling.* London: Zed Books.

Chandler, D. 2013. "Peacebuilding and the Politics of Non-linearity: Rethinking 'Hidden' Agency and 'Resistance.'" *Peacebuilding* 1 (1):17–32.

Chesterman, S. 2001. *Just War or Just Peace? Humanitarian Intervention and International Law.* Oxford: Oxford University Press.

Chesterman, S. 2004. *You, the People: The United Nations, Transitional Administration, and State-Building.* Oxford: Oxford University Press.

Cohrs, J. C., D. J. Christie, M. P. White, and C. Das. 2013. "Contributions of Positive Psychology to Peace: Toward Global Well-Being and Resilience." *American Psychologist* 68 (7):590–600.

Coicaud, J.-M. 2008. "International Organizations as a Profession: Professional Mobility and Power Distribution." In *The International Mobility of Talent: Types, Causes and Development Impact*, edited by A. Solimano, 263–297. Oxford: Oxford University Press.

Connell, R. 2007. *Southern Theory: The Global Dynamics of Knowledge in Social Science.* Crows Nest, NSW: Allen & Unwin.

Cornwall, A. 2007. "Buzzwords and Fuzzwords: Deconstructing Development Discourse." *Development in Practice* 17 (4–5):471–484.

Corrin, C. 2003. "Developing Policy on Integration and Re/Construction in Kosovo." *Development in Practice* 13 (2–3):189–207.

Cox, R. W. (1981) 2000. "Social Forces, States and World Orders: Beyond International Relations Theory." In *International Relations: Critical Concepts in Political Science*, edited by A. Linklater, 1537–1571. New York: Routledge.

Daho, G. 2015. "La socialisation entre groupes professionnels de la politique étrangère." *Cultures & Conflits* 98 (Summer):101–131.

Daho, G., N. Duclos and C. Jouhanneau. 2019. "Political Sociology of International Interventions: Peacebuilders and the Ground." *Journal of Intervention and Statebuilding* 13 (3):249–262.

Donais, T. 2012. *Peacebuilding and Local Ownership: Post-conflict Consensus-Building.* New York: Routledge.

Donais, T. 2015. "Bringing the Local Back In: Haiti, Local Governance and the Dynamics of Vertically Integrated Peacebuilding." *Journal of Peacebuilding & Development* 10 (1):40–55.

Duclos, N. 2016. "The DDR in Kosovo: Collision and Collusion among International Administrators and Combatants." *Peacebuilding* 4 (1):41–53.

Duffield, M. 2001. *Global Governance and the New Wars: The Merging of Development and Security.* London: Zed Books.

Duffield, M. 2007. *Development, Security and Unending War: Governing the World of Peoples.* London: Polity.

Duffield, M. 2008. *Development, Security and Unending War.* London: Polity.

Fassin, D. 2012. *Humanitarian Reason: A Moral History of the Present Times.* Berkeley: University of California Press.

Fresia, M. 2009. "Une élite transnationale: La fabrique d'une identité professionnelle chez les fonctionnaires du Haut Commissariat des Nations Unies aux Réfugiés." *Revue Européenne des Migrations Internationales* 25 (3):167–190.

Gilmore, J. 2015. *The Cosmopolitan Military: Armed Forces and Human Security in the 21st Century*. Basingstoke, UK: Palgrave Macmillan.

Go, J. 2016a. *Postcolonial Thought and Social Theory.* New York: Oxford University Press.

Go, J. 2016b. "W(h)ither the Social? On the Imperial Episteme and Economy of Force." *Security Dialogue* 47 (3):201–207.

Goetze, C. 2017. *The Distinction of Peace: A Social Analysis of Peacebuilding.* Ann Arbor: University of Michigan Press.

Goetze, C., and B. Bliesemann de Guevara. 2014. "Cosmopolitanism and the Culture of Peacebuilding." *Review of International Studies* 40 (4):771–802.

Goetze, C., and D. Guzina. 2008. "Peacebuilding, Statebuilding, Nationbuilding—Turtles All the Way Down?" *Civil Wars* 10 (4):319–347.

Gutierrez Rodriguez, E., M. Boatca, and S. Costa. 2010. *Decolonizing European Sociology: Transdisciplinary Approaches*. Farnham, UK: Ashgate.

Harbom, L., S. Högbladh, and P. Wallensteen. 2006. "Armed Conflict and Peace Agreements." *Journal of Peace Research* 43 (5):617–631.

Hensell, S., and F. Gerdes. 2012. "Elites and International Actors in Post-war Societies: The Limits of Intervention." *International Peacekeeping* 19 (2):154–169.

Hilhorst, D., and M. van Leeuwen. 2005. "Grounding Local Peace Organisations: A Case Study of Southern Sudan." *Journal of Modern African Studies* 43 (4):537–563.
Hughes, C., J. Öjendal, and I. Schierenbeck. 2015. "The Struggle versus the Song—The Local Turn in Peacebuilding: An Introduction." *Third World Quarterly* 36 (5):817–824.
Jabri, V. 2012. *The Postcolonial Subject: Claiming Politics/Governing Others in Late Modernity.* London: Routledge.
Jankowitz, S. E. 2018. *The Order of Victimhood.* New York: Springer.
Kamatsiko, V. V. 2015. "Grassroots Peacebuilding and Vertical Integration: A Case Study of the Peace-Rings Approach." *Journal of Peacebuilding & Development* 10 (1):56–71.
Karim, S., and K. Beardsley. 2017. *Equal Opportunity Peacekeeping: Women, Peace, and Security in Post-conflict States.* New York: Oxford University Press.
Kowal, E. 2014. "Sociology and Empire." *Postcolonial Studies* 17 (4):415–417.
Laidi, Z. 1998. *A World without Meaning: The Crisis of Meaning in International Politics.* London: Routledge.
Lang, A. F. 2008. *Punishment, Justice and International Relations: Ethics and Order after the Cold War.* London: Routledge.
Lang, A., N. Rengger, and W. Walker. 2006. "The Role(s) of Rules: Some Conceptual Clarifications." *International Relations* 20 (3):274–294.
Leonardsson, H., and G. Rudd. 2015. "The 'Local Turn' in Peacebuilding: A Literature Review of Effective and Emancipatory Local Peacebuilding." *Third World Quarterly* 36 (5):825–839.
Mac Ginty, R., and O. P. Richmond. 2013. "The Local Turn in Peace Building: A Critical Agenda for Peace." *Third World Quarterly* 34 (5):763–783.
Madsen, M. R. 2011. "Reflexivity and the Construction of the International Object: The Case of Human Rights 1." *International Political Sociology* 5 (3):259–275.
Malejacq, R. 2016. "Warlords, Intervention, and State Consolidation: A Typology of Political Orders in Weak and Failed States." *Security Studies* 25 (1):85–110.
Martin, M., V. Bojicic-Dzelilovic, and L. Benrais. 2018. "Mind the Gaps: A Whole-of-Society Approach to Peacebuilding and Conflict Prevention." *Peacebuilding* 6 (3):171–182.
McCandless, E., E. Abitbol, and T. Donais. 2015. "Vertical Integration: A Dynamic Practice Promoting Transformative Peacebuilding." *Journal of Peacebuilding & Development* 10 (1):1–9.
Mitchell, A. 2014. *International Intervention in a Secular Age: Re-enchanting Humanity?* London: Routledge.
Moncrief, S. 2017. "Military Socialization, Disciplinary Culture, and Sexual Violence in UN Peacekeeping Operations." *Journal of Peace Research* 54 (5):715–730.
Mosse, D. 2011. *Adventures in Aidland: The Anthropology of Professionals in International Development.* New York: Berghahn Books.
Neudorfer, K. 2015. *Sexual Exploitation and Abuse in UN Peacekeeping: An Analysis of Risk and Prevention Factors.* London: Lexington Books.
Neumann, I. B., and O. J. Sending. 2010. *Governing the Global Polity: Practice, Mentality, Rationality.* Ann Arbor: University of Michigan Press.
Nicolini, D. 2013. *Practice Theory, Work, and Organization: An Introduction.* Oxford: Oxford University Press.
Novosad, P., and E. Werker. 2014. *Who Runs the International System? Power and the Staffing of the United Nations Secretariat.* Working Paper 15-018. Boston: Harvard Business School.
Owens, P. 2015. *Economy of Force: Counterinsurgency and the Historical Rise of the Social.* Cambridge: Cambridge University Press.

Paffenholz, T. 2014. "In Theory. Civil Society and Peace Negotiations: Beyond the Inclusion-Exclusion Dichotomy." *Negotiation Journal* (January):69–91.

Pouligny, B. 2004. *Ils nous avaient promis la paix*. Paris: Presses de Sciences Po.

Pruitt, L. J. 2016. *The Women in Blue Helmets: Gender, Policing, and the UN's First All-Female Peacekeeping Unit.* Oakland: University of California Press.

Randazzo, E. 2016. "The Paradoxes of the 'Everyday': Scrutinising the Local Turn in Peace Building." *Third World Quarterly* 37 (8):1351–1370.

Roff, H. M. 2013. *Global Justice, Kant and the Responsibility to Protect: A Provisional Duty.* Hoboken, NJ: Taylor and Francis.

Sabaratnam, M. 2013. "Avatars of Eurocentrism in the Critique of the Liberal Peace." *Security Dialogue* 44 (3):259–278.

Sabaratnam, M. 2017. *Decolonising Intervention: International Statebuilding in Mozambique.* London: Rowman & Littlefield.

Salamé, G. 1996. *Appels d'empire: Ingérences et résistances à l'âge de la mondialisation.* Paris: Fayard.

Saulich, C., and S. Werthes. 2018. "Exploring Local Potentials for Peace: Strategies to Sustain Peace in Times of War." *Peacebuilding,* online. doi:10.1080/21647259.2018.1517965.

Sending, O. J. 2015. *The Politics of Expertise: Competing for Authority in Global Governance.* Ann Arbor: University of Michigan Press.

Siméant, J. 2011. "Entrer, rester en humanitaire: Des fondateurs de MSF aux membres actuels des ONG médicales françaises." *Revue Française de Science Politique* 51 (1):47–72.

Simmel, G., O. Rammstedt, and G. Fitzi. 1999. *Gesamtausgabe/Bd. 16: Der Krieg und die geistige Entscheidungen Grundfragen der Soziologie vom Wesen des historischen Verstehens: Der Konflikt der modernen Kultur Lebensanschauung.* Frankfurt a.M.: Suhrkamp.

Smirl, L. 2008. "Building the Other, Constructing Ourselves: Spatial Dimensions of International Humanitarian Response." *International Political Sociology* 2 (3):236–253.

Smirl, L. 2012. "The State We Are(n't) In: Liminal Subjectivity in Aid Worker Auto-biographies." In *Statebuilding and State-Formation: The Political Sociology of Intervention*, edited by B. Bliesemann de Guevara, 230–244. London: Routledge.

Smirl, L. 2015. *Spaces of Aid: How Cars, Compounds and Hotels Shape Humanitarianism.* London: Zed Books.

Smouts, M.-C. 2003. *Tropical Forests, International Jungle: The Underside of Global Ecopolitics.* New York: Palgrave Macmillan.

Sörensen, J. S. 2014. "The Return of Plural Society: Statebuilding and Social Splintering." *Peacebuilding* 2 (3):270–285.

Steinmetz, G. 2014. "The Sociology of Empires, Colonies, and Postcolonialism." *Annual Review of Sociology* 40 (1):77–103.

Tarrow, S. G. 2005. *The New Transnational Activism.* New York: Cambridge University Press.

True, J. 2012. *The Political Economy of Violence against Women.* New York: Oxford University Press.

von Billerbeck, S. B. K. 2017. *Whose Peace? Local Ownership and United Nations Peacekeeping.* Oxford: Oxford University Press.

Welsh, J. M. 2019. "Norm Robustness and the Responsibility to Protect." *Journal of Global Security Studies* 4 (1):53–72.

Whalan, J. 2017. "The Local Legitimacy of Peacekeepers." *Journal of Intervention and Statebuilding* 11 (3):306–320.

Wheeler, N. J. 2002. *Saving Strangers: Humanitarian Intervention in International Society.* Oxford: Oxford University Press.

Wiberg, H. 2006. "From International to World Society? English School Theory and the Social Structure of Globalisation." *Journal of Peace Research* 43 (2):229–230.

Woods, N. 2006. *The Globalizers: The IMF, the World Bank, and Their Borrowers.* Ithaca, NY: Cornell University Press.

Zanotti, L. 2006. "Taming Chaos: A Foucauldian View of UN Peacekeeping, Democracy and Normalization." *International Peacekeeping* 13 (2):150–167.

Zanotti, L. 2011. *Governing Disorder: UN Peace Operations, International Security, and Democratization in the Post-Cold War Era.* Philadelphia: Pennsylvania State University Press.

CHAPTER 9

SPACES OF PEACE

ANNIKA BJÖRKDAHL AND STEFANIE KAPPLER

While peace has often been considered a primarily symbolic concept (cf. Bar-Tal 2009) it equally has concrete material implications for people's everyday lives (Williams 2015). It plays out in the multiple spaces in which people live and move (Forde 2018) and impacts social relations as they are constituted in the material realities of urban and rural inhabitants. Drawing on previous research on peace and spatial transformation that we have conducted (Björkdahl and Kappler 2017), this chapter advances the spatial turn in peace and conflict studies (cf. Soya 1989; Björkdahl and Buckley-Zistel 2016). It suggests that peace has a spatial component and can thus be read through the prism of space and place. Yet this does not mean that the relationship between peace and space is a linear, causal one. Instead, we will show in this chapter that a focus on the spatial dimensions of peace requires a relational and contextual reading in order to appreciate the nonlinear ways in which peace impacts spatiality. Intrigued by the everyday spaces and places of peace and conflict in postconflict landscapes, we ask: How do places of conflict transform into spaces of peace? Who are the peacebuilding agents driving such transformation? How can we, as researchers, read the spatial narratives of contested spaces and places?

Our thinking is inspired by three strands of research that can be seen to make up the spatial turn in peace and conflict studies. First, critical peace studies helps us rethink what peace is and for whom and where it takes place. Second, critical human geography helps us focus on the mutual constitution of space and place. Third, we draw on the broad field of research that centers on the notion of agency in order to theorize agency as situated in the transformation of space and place. Critical peace studies has explored the essentially contested concept of peace. It has come to question the neat distinction between war and peace (Mac Ginty 2006; Richmond 2008, 2014) as well as the assumed linear development of transitions from war to peace. It has concluded that peace and war often coexist and that the binary of war and peace has become unsustainable. Furthermore, scholars find that peace means different things to different people in different times and places (Visoka 2016; Richmond 2016). Consequently, peace holds multiple understandings, and the evolving critical peace research agenda captures peace not as singular but as plural peace(s). It can be built in different places and spaces, as well as at different scales, such as the individual, family, community, the city, the state, the regional, and the global. As peace has increasingly been seen as embedded in place and existing on different scales, geographers seem well placed to explore "how peace makes place and how a place makes peace" (Williams 2015, 2). Critical peace studies has therefore opened up for a dialogue with critical geography to advance the spatial understandings of peace and to probe where peace takes place.

The analysis of a "bridge that divides" and a "wall that unites" casts light on the benefits that a spatial reading of peace can provide to understand the ways spatial infrastructures are lived by the people who use them. Peace is no longer fully determined by the top-down planning of space, but material places are also transformed by the communities that use them. It is therefore impossible to label a particular space as contentious or cooperative. Instead, we will show that the role and function of a given social space is subject to the agents and communities surrounding it. The process of space-making (the generation of meanings from a material location) will help explain the agency that emerges by the creators, users, and inhabitants of (post)conflict spaces. Investigating peacebuilding agency through space-making requires reflexive research methods. Thus we need to rethink the methodologies of peace and conflict research in order to explore where peace develops and to map peacebuilding agency. To do so, we focus on material "place" as this is where people experience peacebuilding and conflict dynamics in their everyday lives. Second, "space" explains where competing ideas of peace circulate. The peacebuilding process and its agency are reflected and manifested in the dynamics at the chosen sites of this chapter, namely the bridge in the city of Mitrovica in northern Kosovo and the peace walls in Belfast, Northern Ireland.

Spaces and Places of Peace: Transscalar Dynamics

In the field of peace and conflict studies, spatial concerns are central, yet they have so far received limited attention. This chapter shows that war-making and peace-making "take place" and that sometimes the legacy of conflict obscures the visible, symbolic, and tangible manifestations of peacebuilding. The chapter brings to the fore how social spaces and material places shape, and are shaped by, agonistic processes. Here such social spaces are constituted by and constitutive of material places. They relate to spatial practices that project narratives of peace and conflict onto material places (Low 1996; Lefebvre 1991). The physical environment shapes social interactions, governs movement through space, and enables or hampers the likelihood of peace being built, while also molding the manifestations of peace on the ground. In this the potential for peace is tied to many material places but not determined by them. Often, mutually incompatible concepts of peace will remain in and among different places and communities.

There have been a number of calls by geographers to pay attention to peace, and the engagement with peace research has expanded and advanced the research agenda of geography. A new agenda for peace in geography is developing (Megoran 2011), and it has contributed to the conceptualization of peace across time and space. Specifically, Koopman's effort to unpack peace by "taking peace into pieces" helped to rethink peace and the research on geographies of peace. This unpacking defines peace as a contested spatial process composed of practices and actors at multiple scales (McConnell, Megoran, and Williams, 2014) and how the notion of everyday peace (Williams 2015) has helped emplace peace in people's everyday lives. We are particularly inspired by Williams's understanding of peace as a political, spatial, and relational construction. It is both the product of and the context through which the political is assembled and negotiated across scale, articulated through different forms of peace narratives, and informed by uneven geographies of power. Rather than upholding the binary view of war and peace, new research points to how the social construction of postwar spaces can be considered a war legacy and an embodiment of peacebuilding alike, because these spaces both reflect prevailing peace(s) and produce new peace(s) (Megoran and Dalby 2018; Björkdahl and Kappler 2017; Björkdahl and Mannergren Selimovic 2016).

Thinking in terms of space and place also opens up the possibility that peace can move between different scales, such as the local and the global. Inspired by the concepts of *scale, scalar politics* (Swyngedouw 1997; Dannestam 2009; Megoran and Dalby 2018), often applied by geographers and critical scholars, and the notion of *transscalarity* (Scholte 2008), we employ the term "transscalar peace" in order to investigate peacebuilding across various spaces and places (Millar 2020). The type of scale we are interested in claims that social patterns and processes take place at multiple scales at once and any scalar process can intersect with processes at local, national, regional, and global levels. When applying this to peacebuilding, we can rethink peace(s) to grasp how various actors strive to define and possibly change the politics of peace at different scales (Swyngedouw 1997, 137–142). As such it reveals the contestations between multiple actors trying to shape peacebuilding in different spaces and places, which are not ontologically given. Spatial scales do not rest as fixed platforms for social processes that connect up or down but are instead the outcome of social activities. This approach offers a means to probe the empirical site in order to establish exactly where peace occurs and also how different actors rescale peacebuilding. Thus we are able to envision multiple spaces of peace and ways in which both conflictual patterns and actions seeking to resolve them interact at various scales (Bank and Van Heur 2007, 596). If peace is not manifested in one place, such as on the bridge or at the wall, peacebuilding actors may try to utilize a transscalar approach to peacebuilding, in order to empower peace at a different scale. In that, the interplay between local and global peace actors may produce different peace processes and interpretations at different levels. When adapting a scalar approach to a peacebuilding context, we hope to provide a stepping stone for further theory development concerning the local and the global frictional dynamics of peacebuilding. With the notion of transscalar peace we can thus move beyond the binary opposition of the local as a space of dependence and the global as a space of engagement (cf. Cox 1998). Instead this approach allows us to view spaces of peace as situated in transscalar networks of agency. Rather than fixed, the function of any material place can be seen as the product of the interplay between actors situated at different points of social hierarchies.

Peacebuilding Agency

"Agency" is a buzzword of critical peace and conflict research. In that, research has aimed to provide a fine-grained understanding of agency as socially and politically situated—an understanding that attempts to move beyond the view of agency as resistance to power (Richmond 2011; Kappler 2014; Björkdahl and Mannergren Selimovic 2015). Empirically the field has been able to provide a contextualized account of agency and revealed how it is expressed through negotiations at different sites and scales. Of particular interest is how peace is reproduced from the margins by agents working under the radar, in the shadow of uneven power relations, unseen and rarely recognized by the relevant authorities. Such a conceptualization of agency is able to differentiate between various types of situated agents producing peace.

This contribution on space-making and place-making adds to the critical debate about the materialization of peace and the emplacement of peacebuilding agents. By conceptualizing peacebuilding agency in relation to space and place we link expressions of agency to spatial practices and thus contribute to the retheorization of agency. In previous work (Björkdahl and Kappler 2017), we have argued that peacebuilding agency can be read in a process of spatial transformation. Assuming that "place" denotes a material, physical aspect, the physicality of a site determines the ways it is socially used and understood. However, while this may often be the case, this is not an automatic, causal phenomenon. Instead what we sometimes see is that a place with certain preconditions is used in very different ways than what its material infrastructure would suggest. This is where we observe agency: the ways a set of actors is able to distort, transform, and multiply the ways in which place is socially and symbolically used and, therefore, made a "space."

Thus, the approach of this chapter takes place at the intersection of place-making and space-making. The ability to turn space into place, and place into space within a broader process of conflict transformation reveals peacebuilding agency and its role in spurring change. We argue that agency is constituted of and by practices of place-making and space-making. This means that, where imaginary ideas of peace are given a physical presence, agency is involved (place-making), and vice versa: where physical places are given new sets of meanings, this is where we can find agency (space-making). Spatial transformation that has an impact on the ways in which peace is emplaced and imagined is thus a product of peacebuilding agency. Rethinking agency in this sense means that we are able to move toward an emplaced understanding, shaped by interaction with the ideational and material environment. The empirical illustrations we present illustrate the process of space-making and its relationship to transscalar peace processes. They demonstrate the emergence of a transscalar peace by showing how a multitude of actors are able to mobilize new symbolisms to given physical infrastructures. As a result, our given sites have to be read as situated in nonlinear processes of meaning-making and in constant negotiation between different peacebuilding actors operating across scales.

Of course, such processes operate within the wider structural constraints that set the limitations on which repertoires of action are even possible (Megoran and Dalby 2018). Indeed, particularly critical geopolitics has emphasized the power of top-down spatial governance in its ability to frame and govern particular locales (cf. Kappler 2018). Within such constraints, though, it is crucial to acknowledge the extent to which different sets of actors—ranging from global and national political and economic elites to ordinary citizens—are able to co-shape and change the meaning and uses of particular places. It is in such processes that we can observe agency as a tool to interfere and produce "unlikely spatial outcomes."

The Bridge That Divides and the Walls That Unite: Unlikely Peace Spaces

Peace has important reference points. We even often think of it in terms of its spatial symbolisms: a bridge, a crossing, or a meeting room are common spatial markers of peace. The famous bridge in Mostar, which was bombed during the 1990s war in the Balkans and reconstructed, as a bridge of reconciliation, in 2004, is perhaps the most prominent example of such spatial metaphors. In contrast, walls and checkpoints tend to be associated with conflict and war; often, peacebuilding ambitions aim to remove such barriers in order to prevent further spatial division of society. Such have been questions around the future of the buffer zone in Cyprus, South Africa's gated communities, and the inter-entity boundary within Bosnia-Herzegovina. A range of local, national, and international actors are reflecting on the ways in which the spatial outlook of a (post) conflict landscape impacts the quality of peace. Will a bridge help bring communities back together? Will a buffer zone be necessary to keep combatants apart? Will a border settle a complex conflict over land?

We find it to be too simplistic to equate the physical outset of a place with its symbolic and social functions (space). Instead we investigate two case studies to show that the transformation of a place into space, i.e., the creation of a socially significant meaning, is a product of peacebuilding agency.[1] It is the social structures and agents in any given context that establish the meanings and functions of places in a wider context. In that, they are part of an ongoing process of transformation during the course of which spaces of peace are (re)created and (re)purposed.

Dividing the Bridge and Bridging the Divide in Mitrovica

The legacy of the Kosovo war divides Kosovo Albanians from Kosovo Serbs and has created a geography of fear. In this postconflict landscape, the Ibar River has become the demarcation line to the Serbian north side of the city of Mitrovica and the Albanian south side. Since 1999 the Main Bridge over the river and its barricades have been the center of contestation. To prevent Kosovo Albanians from crossing, the ethnonationalists among the Kosovo Serbs in Mitrovica North built a barricade made of stone and sand on the Main Bridge. It is a material manifestation symbolizing the reluctance of the Kosovo Serbs to become part of the emerging state of Kosovo. During the years 1999–2002, bridge watchers guarded the bridge and monitored crossings, sitting on rooftops and in cafés near the riverbank, controlling the space and discouraging people from making interethnic contact (Clark 2014). In recent years, the bridge has been watched around the clock by the men of the Kosovo Serbian Civil Protection Force. As the barricades are removed, new ones are put in place in response to political tensions and insecurities felt by the Kosovo Serbs (Hopkins and Peci 2014; Organization for Security and Cooperation in Europe [OSCE] official, authors' interview, Mitrovica, 2016). For most inhabitants of Mitrovica North, keeping the bridge closed for traffic is key to their sense of security.

For several years, the Kosovo Force (KFOR) reproduced the divisions of the city by putting barbed wire and vehicles on the bridge in an effort to prevent interethnic clashes. A stillborn KFOR project set out to fortify the bridge to keep the two communities apart (Lemay-Hébert 2012, 36). As part of confidence-building measures, KFOR removed the two-meter-high sandbag barricades, barbed wire, and armored vehicles from the bridge in order to return the city to normal life (*New Europe* 2005). When Kosovo police and KFOR troops dismantled barricades, Kosovo Serbs came out to protest and the barricades were quickly rebuilt. As a consequence, the Main Bridge remains guarded by Italian carabinieri accompanied by a Serbian member of the Kosovo police, occasionally KFOR troops, and at times a number of officers of the European Union Rule of Law Mission in Kosovo. UN marked cars can also regularly be seen on the bridge (OSCE official, authors' interview, Mitrovica, 2016).

The barricade made of stone and sand was removed in 2014, after three years, because the mayor of the Kosovo Serb part of Mitrovica, Goran Rakić, intended to have a peace park built in its place. The transformation of the bridge materialized by replacing the barricade with concrete flower boxes (Çollaku and Domanovic 2014). As a consequence, the bridge was open only for pedestrians and remained closed to traffic. Still, the peace park attracted a lot of attention from the diaspora visiting the town, and the park allowed people to come together, the trees and the flowers were tended to, and the benches in the garden were used (OSCE official, authors' interview, Mitrovica, 2016). According to Rakić, who stated, "[T]he aim of the Peace Park is an expression of the desire that the point which separated the two nations over the past 15 years becomes a place where they meet," the eventual removal of the park could be seen "as a threat to peace" (Çollaku and Domanovic 2014). From a Kosovo Albanian perspective, the peace park was a new barricade, and the mayor of South Mitrovica, Agim Bahtiri, stated that the bridge would be cleared "very soon." The decision by Rakić to build the peace park followed consultations with Belgrade (Çollaku and Domanovic 2014), but when the peace park sparked violent objections by Kosovo Albanian protestors, he was instructed by Belgrade to remove it. This unpopular action was carried out in the middle of the night to avoid protest from the Kosovo Serbs. In response Kosovo Serbs erected yet another barricade.

Space-making transformed the Main Bridge and transcended the immediate locality of the place. These efforts demonstrated that peacebuilding takes place at multiple scales at once, involving municipal, national, and international actors. Clearly the Main Bridge has attracted the attention of the EU Special Representative in Pristina, Samuel Žbogar, who has been engaged in the transformation of the bridge. During the Brussels normalization talks, he proposed a resolution of the dispute concerning the bridge (EWB Archives 2015). The aim of the talks was to reopen the bridge in June 2016. At the time of writing this has not yet happened. Kosovo-Serbia relations faced a number of setbacks in 2018, which resulted in Kosovo quitting the normalization talks, and the Main Bridge across the Ibar is still a pending issue (Xharra 2018).

Standing at the bridge that has been equipped with contradictory narratives by the two communities, it becomes more apparent than ever that places and spaces are transformative (field notes, 2016). Frequently those driving the transformations are guided by different ideas and understandings of peace. It is possible to see "peace" in its plural manifestations at different times on the bridge. The eventual removal of the peace park will not be enough to reverse or accommodate these clashing spatial narratives. The bridge is an ideational space containing competing meanings in the narratives of the Kosovo Serbs and the Kosovo Albanians. As such, it is a product of the material manifestation of the ethnic divide in Kosovo, the politicization of ethnic cleavages, and the difficulties of these ethnic groups trying to coexist in one state (Kostovicova 2005; Gusic 2015). Moreover, this contestation transcends the immediate locality of Mitrovica. It demonstrates transscalar, contested peace politics as it is linked to the relationship between Belgrade and Mitrovica—which impacts the tools that the Kosovo Serb leadership can use—and the involvement of international peacebuilders in urban reconstruction. In that sense, the bridge has been a key marker for division and unification, depending on whether it is seen from the perspective of KFOR, the communities next to the bridge, or national elites. It would therefore be a simplification to view a bridge as a place of dialogue only; as our Mitrovica example shows, it is situated in the constant tensions between division and unification, depending on the dynamics of the surrounding peace process.

The Walls That Connect: Belfast's "Draw Down the Walls"

Our second empirical illustration draws attention to a different kind of spatial practice embedded in another urban peace process. We are investigating a public arts project in North Belfast, which also deals with the urban infrastructure of separation and segregation. Belfast, like Mitrovica, can be considered a divided city, shaped by the presence of so-called peace walls in what are assumed to be hotspots of violence. Cunningham and Gregory (2014, 65) consider the peace walls as both liminal and performative spaces, in which social relations are enacted in boundary-making and boundary-breaking practices. In their position and role as interfaces between conflictive parties, the walls are public and visible material markers and reminders of the violence that took place in Belfast's "Troubles," representing the peak of sectarian violence in the city and beyond. The painted and printed murals, which can be found on most of the peace walls as well as on buildings and smaller walls within residential areas, serve to mark territory as belonging to a particular national group and are often rather militaristic in nature.

Belfast has not had the same kind of internationally led peacebuilding intervention that Kosovo has had. As a result, there is not the same level of international peacebuilding presence, and the spatial dimensions of peacebuilding still reflect both local and global characteristics—albeit in different ways—yet revealing the transscalar efforts of space-making. Certainly cities can be considered as hosting global and local actors alike (Stevenson 2013, 135). In Belfast it is indeed remarkable that many of the murals are directed at domestic and international audiences, and the latter are often mobilized to lend some credibility to domestic audiences (cf. Kappler and McKane 2019). And while the murals on the walls have, unsurprisingly, long been acting as markers and key sites for urban violence itself, we want to show that this is not an inherent property of the walls themselves. Instead walls as markers of division have been used as sites of transformation, dialogue, and peacebuilding at the same time. The use of public art techniques is key to understanding such transformational processes.

The project we are referring to is entitled "Draw Down the Walls" and was initiated in 2012 in a context in which the majority of those living near the walls were reluctant to see them torn down, even decades after the cessation of open sectarian hostilities in the city. It was the Golden Thread Gallery in cooperation with the North Belfast Interface Network and the Lower Shankill Community Association, which initiated a project through which the removal of the walls could at least be imagined. To do so, different community actors were involved and a Colombian artist was hired to implement the arts-based dimension of the project. The idea was that a series of arts projects, involving both communities and leading to multiple encounters at the physical sites of separation—the walls—would be able to bring people together and generate a degree of momentum in the peace process. The project included a variety of smaller projects such as film productions, youth work, and exhibitions. There were a number of artists and activists involved in the different artistic and peace-related aspects of the project. Overall the curators placed emphasis on public involvement in the curatorial process and invited the public to a meeting in the run-up to what was a somewhat controversial arts project (Young and Bell 2013).

Perhaps most famously, the Colombian artist Oscar Muñoz created the *Ambulatorio* exhibition, located at the Crumlin Road/Flax Street interface and exploring "universal themes of memory, human loss and impermanence" (Golden Thread Gallery 2012). Here it is important to understand the deliberate decision of the curators to reach out to a global vocabulary of peace and memory in the sense of transporting a message that was relevant beyond the immediate urban context of Belfast. Working with a Colombian artist who brings his own interpretation to such global themes was a key step in being less vulnerable to criticism of bias locally, but also was a way of respecting the sensitivity that public art in conflict contexts requires (cf. Clarke 2012). The "international peacebuilders" in this example are therefore not the usual suspects of UN peacekeepers or internationally subcontracted NGOs but instead a partnership between a gallery, community associations, and local and international artists. Transscalar peacebuilding is therefore, as this example shows, not necessarily a process enshrined in formal elite-driven politics, but can also be facilitated by civil society, artists, and local communities.

On the one hand, this case study allows for using the arts as a way of understanding political discourse, since it is situated in a political time and space (Negash 2004, 188). The project was indeed specifically tied to the physical givens of Belfast's cityscape in terms of using the infrastructure of the walls and interfaces to develop its artistic content. On the other hand, alluding to the parallels to the Colombian conflict and related questions of conflict and loss, the project made a claim to the visuality of global politics and memory (Mookherjee 2018, 208), breaking down the assumed distinction between local and global agency. Instead the project reflects the interplay between local places and global discourses. Public art is one way of engaging with both local and global dimensions of conflict. Processes of place-making can thus be seen as an attempt to make global dynamics relevant to specific local places by emplacing peace in the communities in which it is being negotiated. The arts as a tool for "intervention" is a powerful way of doing so and seems to stand somewhat in contrast to the types of military and diplomatic intervention that we are familiar with in more trusteeship-style peacebuilding scenarios.

A Bridge That Divides and Walls That Connect: The Paradox of Transscalar Peace

We could plausibly argue that the Mitrovica bridge and the Belfast peace walls do three things. First, they show how a spatial analysis helps us study the materiality of peace and conflict, i.e., where places are inscribed with and manifest peace or conflict. These two places reveal how the respective material place constitutes and is constituted by everyday micropolitical acts of peace and conflict. Second, they point to how everyday acts of peace or conflict located in a particular place impact the local, national, regional, and global politics of peace and conflict and travel across scales from a local place to a global space. Therefore emplaced micro-acts of peace are transscalar. A third aspect that we hope the two empirical illustrations might help evoke is to generate accounts of whether and how a given local place is constitutive of national as well as global peacebuilding efforts. The acts, agents, events, and processes of peace or conflict emplaced at the Mitrovica bridge are transscalar in the sense that we can connect such emplaced acts, agents, events, and processes to the Brussels normalization talks between Kosovo and

Serbia and to the land-swap negotiations about redrawing borders to end the political deadlock that has persisted for more than ten years since Kosovo declared its independence. Similarly, the projects aiming to transform Belfast's peace walls cannot be seen in isolation from the wider national and international peacebuilding efforts, involving both formal political elites as well as civil society activists.

It is the legitimate concern of students of peace and conflict to understand how the micropolitics of peace are influenced by abstract, global ideas about peace and international peacebuilding efforts to implement such ideas. Likewise, it must be a legitimate concern to inquire if and how emplaced peace(s) and acts of peacebuilding transcend scale, i.e., are transformative and transscalar and generate changes or stagnation to global processes of peacebuilding. That means that peace is transscalar and able to travel not only from the global to the local but also from the local to the global. Peace and conflict studies is not the only discipline that confronts the theoretical explanation of causal or constitutive links that transcend scale, investigating the traveling of micropolitics situated in local places, in the everyday, to macropolitics, global spaces and processes of peacebuilding. So far, peace and conflict studies, as a discipline, has primarily developed tools to study the transscalarity of peace in one direction to understand how global capitalism, neoliberal peacebuilding, and interstate peace agreements shape and shove everyday places. The research so far is, however, much less successful in grasping how peace politics in global spaces is constituted and generated by the everyday dynamics of emplaced peace politics. At the same time, it is arguably the case that the difference between a locally meaningful but isolated everyday diorama and a globally significant event or act of peace may possibly be the linkage of the local specifics to abstract, ideational global spaces, processes, and dynamics.

Conclusion

The ability to turn a place into a space denotes the process of making a physical place relevant and meaningful to global and local discourses. By emplacing certain spatial practices, narratives, ideas, and material artifacts, places are equipped with a social meaning and become part of the social imaginary. Our analysis of the bridge across the Ibar River in Mitrovica reveals how a particular place does not stay the same but is continually emerging; a place is always becoming, and the becoming involves a transformation of the physical places into symbolic, imagined spaces. Similarly, the arts project in Belfast casts light on the different types of peacebuilding agency that contested sites can activate. Spatial peacebuilding must therefore be considered a dynamic process, concerned with the transformation of places of violence into spaces of encounter. Where this fails, as the Mitrovica example shows, even supposedly unifying spaces such as bridges can become key markers of division, conflict, and insecurity. It is the interplay among a variety of peacebuilding actors, at local, national, and global levels, that shapes the ways in which the materiality of (post)conflict places impact transscalar peace processes.

This jigsaw of agency can produce "unlikely spatial outcomes," hinting at the quality of transscalar peacebuilding. The physical landscape of (post)conflict places must therefore not be viewed as purely deterministic for the outcome of spatial peacebuilding. Instead, while acknowledging that a material infrastructure puts certain structural limitations on the possibilities of spatial transformation, it also opens up possibilities for new social configurations. Thus space-making helps us reveal the tension between place (as a material phenomenon) and space (an immaterial, imaginary phenomenon) as well as that place and space are mutually conditioned and intertwined.

Note

1. The two cases studies are also part of the chapters "Kosovo State Building: Emplacing the State and Peace(s)" and "Northern Ireland: The Maze of Peace" (both in Björkdahl and Kappler 2017), but the cases are analyzed here through a different theoretical lens.

References

Bank, A., and B. Van Heur. 2007. "Transnational Conflict and the Politics of Scalar Networks: Evidence from Northern Africa." *Third World Quarterly* 28 (3):593–612.

Bar-Tal, D. 2009. "Reconciliation as a Foundation of Culture of Peace." In *Handbook on Building Cultures of Peace, edited by* J. de Rivera, 363–377. *Peace Psychology Book Series*. New York: Springer.

Björkdahl, A., and S. Buckley-Zistel. 2016. "Spatializing Peace and Conflict: An Introduction." In *Spatializing Peace and Conflict: Mapping the Production of Places, Sites and Scales of Violence*, edited by A. Björkdahl and S. Buckley-Zistel, 1–22. Basingstoke, UK: Palgrave.

Björkdahl, A., and J. Mannergren Selimovic. 2016. "A Tale of Three Bridges: Agency and Agonism in Peacebuilding." *Third World Quarterly* 37 (2): 321–335.

Björkdahl, A., and S. Kappler. 2017. *Peacebuilding and Spatial Transformation: Peace, Space and Place.* London: Routledge.

Björkdahl, A., and J. Mannergren Selimovic. 2015. "Gendering Agency in Transitional Justice." *Security Dialogue* 46 (2):165–182.

Clark, J. N. 2014. "Kosovo's Gordian Knot: The Contested North and the Search for a Solution." *Nationalities Papers* 42 (3):526–547.

Clarke, B. 2012. "Draw Down the Walls." Transconflict, 29 June. http://www.transconflict.com/2012/06/draw-down-the-walls-296/.

Çollaku, P, and M. Domanovic. 2014. "Kosovo Officials Vow Serb 'Peace Park' Removal." *Balkan Transitional Justice*, 15 December. http://www.balkaninsight.com/en/article/peace-park-in-divided-mitrovica-causes-dispute.

Cox, K. 1998. "Spaces of Dependence, Spaces of Engagement and the Politics of Scale, or: Looking for Local Politics." *Political Geography* 17 (1):1–23.

Cunningham, N., and I. Gregory. 2014. "Hard to Miss, Easy to Blame? Peacelines, Interfaces and Political Deaths in Belfast during the Troubles." *Political Geography* 40:64–78.

Dannestam, T. 2009. *Stadspolitik I Malmö: Politikens meningsskapande och materialitet.* Lund: Lund University Press.

EWB Archives. 2015. "Zbogar Proposes Solution for Mitrovica Bridge." *European Western Balkans*, 18 June. https://europeanwesternbalkans.com/2015/06/18/zbogar-proposes-solution-for-mitrovica-bridge/.

Forde, S. 2018. *Movement as Conflict Transformation. Rescripting Mostar, Bosnia-Herzegovina.* Cham: Palgrave Macmillan.

Golden Thread Gallery. 2012. "Ambulatorio Belfast: Oscar Muñoz." Accessed 15 November 2018. http://www.goldenthreadgallery.co.uk/events/ambulatorio-belfast-oscar-munoz/?v=79cba1185463.

Gusic, I. 2015. "Contested Democrac(ies): Disentangling Understandings of Democratic Governance in Mitrovica." In *Divided Cities: Governing Diversity*, edited by A. Björkdahl and L. Strömbom, 215–234. Lund: Nordic Academic Press.

Hopkins, V, and E. Peci. 2014. "'Park' on Bridge Causes Confusion in Divided Mitrovica." *Balkan Insight*, 20 June. http://www.balkaninsight.com/en/article/new-barricade-causes-uncertainty-in-northern-kosovo.

Kappler, S. 2014. *Local Agency and Peacebuilding: EU and International Engagement in Bosnia-Herzegovina, Cyprus and South Africa.* Basingstoke, UK: Palgrave Macmillan.

Kappler, S. 2018. "The Legitimisation of Post-conflict Intervention: Narrative Frames of Backwardness and Progress." *Political Geography* 66:130–138.

Kappler, S., and A. McKane. 2019. "'Post-conflict Curating': The Arts and Politics of Belfast's Peace Walls." *De Arte* 54 (3): 4–21.

Kostovicova, D. 2005. *Kosovo: The Politics of Identity and Space. London*: Routledge.

Lefebvre, H. 1991. *The Production of Space.* Oxford: Blackwell.

Lemay-Hébert, N. 2012. "Multiethnicité ou ghettoïsation? Statebuilding international et partition du Kosovo à l'aune du projet controversé de mur à Mitrovica." *Études Internationales* 43 (1):27–47.

Low, S. 1996. "Spatializing Culture: The Social Production of and Social Construction of Public Space in Costa Rica." *American Ethnologist* 23 (4):861–879.

Mac Ginty, R. 2006. *No War, No Peace: The Rejuvenation of Stalled Peace Processes and Peace Accords.* Basingstoke, UK: Palgrave.

McConnell, F., N. Megoran, and P. Williams, eds. 2014. *The Geographies of Peace: New Approaches to Boundaries, Diplomacy and Conflict Resolution.* London: I. B. Tauris.

Megoran, N. 2011. "War and Peace? An Agenda for Peace Research and Practice in Geography." *Political Geography* 30:178–189.

Megoran, N., and S. Dalby. 2018. "Geopolitics and Peace: A Century of Change in the Discipline of Geography." *Geopolitics* 23 (2):251–275.

Millar, G. .2020. "Toward a Trans-Scalar Peace System: Challenging Complex Global Conflict Systems." *Peacebuilding* 8 (3):261–278.

Mookherjee, N. 2018. "Memory." In *Visual Global Politics*, edited by R. Bleiker, 201–208. Abingdon, UK: Routledge.

Negash, G. 2004. "Arts Invoked: A Mode of Understanding and Shaping the Political." *International Political Science Review/Revue internationale de science politique* 25 (2):185–201.

New Europe. 2005. "Nato Removes Barricades from Bridge in Mitrovica." 10 April. https://www.neweurope.eu/article/nato-removes-barricades-bridge-mitrovica/.

Richmond, O. 2008. *Peace in International Relations.* London: Routledge.

Richmond, O. 2011. "Critical Agency, Resistance and a Post-colonial Civil Society." *Cooperation and Conflict* 46(4):419–440.

Richmond, O. 2014. "The Dilemmas of Hybrid Peace. Negative or Positive?" *Cooperation and Conflict* 50 (1):50–68.
Richmond, O. 2016. *Peace Formation and Political Order in Conflict Affected Societies.* Oxford: Oxford University Press.
Scholte, J. A. 2008. *Globalization. A Critical Introduction.* London: Palgrave.
Soya, E. S. 1989. *Postmodern Geographies: The Reassertion of Space in Critical Social Theory.* New York: Verso.
Stevenson, D. 2013. *The City. Cambridge,* UK: *Polity Press.*
Swyngedouw, E. 1997. "Neither Global nor Local: Glocalization and the Politics of Scale." In *Spaces of Globalization: Reasserting the Power of the Local,* edited by K. Cox, 137–166. New York: Guilford Press.
Visoka, G. 2016. *Peace Figuration after International Intervention: Intentions, Events and Consequences of Liberal Peacebuilding.* London: Routledge.
Williams, P. 2015. *Everyday Peace: Politics, Citizenship and Muslim Lives in India.* Oxford: Wiley Blackwell.
Xharra, J. 2018. "Kosovo Quits Brussels Talks Until Serbia Implements Deals." *Balkan Insight,* 31 May. http://www.balkaninsight.com/en/article/kosovo-rebuffs-brussels-talks-until-reached-deal-are-implemented-05-31-2018.
Young, J., and J. Bell. 2013. "A Model of Consultation? Transformation and Regeneration at the Interface.". Belfast: Institute of Conflict Resolution.

CHAPTER 10

PEACE METHODS AND METHODOLOGIES

PAMINA FIRCHOW

If there is a rift in the study of peace and conflict, it manifests primarily in the methodological preferences of peace researchers. The roots of the fault lines in the field clearly lie in deeper epistemological and ontological differences or the different ways we view the world and gather knowledge, but the different methods we use, and how we use them, or if we use them at all, distinguish scholars from the disciplines that encompass this multidisciplinary area of study. This means that the methods and methodologies we use determine how seriously our work is taken by different sectors of the peace research community. A scholar's choice of methodology signals the underlying perspective and ontological approach implicit in the project. This occurs on a spectrum from universalist, positivist, and empiricist to contextual, contingent, and interpretivist methodological approaches to understanding peace and conflict.

The rift exists for those studying issues of peace and war from the *social sciences*, where interpretivists and positivists dispute whether there is a tangible, universal, or generalized truth about peace and war to be measured and calculated or whether this is a more contextualized endeavor and dependent on circumstances and experiences.[1] It emanates from the discomfort of certain scholars with the normative orientation of peace and conflict studies and the resulting potential for inherent biases, leading them to eschew any work that is not strictly scientific. Others argue that true objectivity in research is impossible (and possibly not even desirable), since bias starts with the kinds of questions we seek to investigate. And the rift extends to those scholars working on questions of peace and war from fields considered to fall more within the *humanities*, such as philosophy and history or even literature. These scholars often dismiss a fixation on methods by social scientists because they see these as deceptive or myopic and argue there is too much credit given to the results of studies that employ empirical procedures.[2] Methodology can also distinguish those who study conflict and violence from those who study peace, since conflict is understood as more concrete and countable, therefore more accessible to empirical study in comparison to peace, which is seen as more vague and obscure and difficult to measure.

These divisions also extend to the peacebuilding sector, where a clear tension exists between those who use "evaluative thinking" in their peacebuilding work and those who don't see its value. A decade ago, Reina Neufeldt made a distinction between circlers and frameworkers in how different peacebuilders approach their work that still applies today and is reflective of a "paradigm war" in the evaluation community more broadly (Neufeldt, 2011; Bamberger, Rugh, and Mabry 2011). There are those circlers who balk at the thought of having to create specific and measurable indicators for linear and causal thinking and instead see their work as dynamic and interactive, with multiple overlapping stories occurring simultaneously. The frameworkers look to develop programs based on concrete data and a specific logic of intervention. Arguably the peacebuilding field has gone increasingly in the direction of the frameworkers. In the past twenty years, peacebuilding has seen an increasing professionalization and technocratization of peace (Mac Ginty 2012). With this increase has come a demand for evidence and a thirst for methods that can empirically study peace-related phenomena. Yet the rift between those who demand data and those who do the work on the ground remains, and it creates a constant tension between the growing monitoring and evaluation sector vis-à-vis those who implement programming.

The study of peace and conflict is contained by a multidisciplinary, sometimes interdisciplinary field, and therefore a multitude of methods and methodologies are used in its analysis. In order to understand the rifts that emerge from these different approaches and perhaps call for some resolution (we are, after all, in that business), this chapter will give a brief overview of the different methodological approaches to studying peace and explore their strengths and weaknesses. It then provides some discussion and suggestions of how we may overcome some of the divisions just discussed.

A caveat is in order here. Although many of the divisions in the field fall along positivist/interpretivist, deductive/inductive, as well as quantitative/qualitative lines, there is a risk that these are oversimplified since the distinctions are not always clear-cut and often overlap. Therefore, although I have tried to organize this chapter along these fault lines, I try to recognize throughout the exceptions and overlaps inherent in these categories.

Quantitatively Driven Research

Peace scientists are scholars studying issues of peace and conflict from a quantitatively driven and positivist perspective. They identify themselves as engaged in peace research that *systematically* studies issues related to conflict and violence in order to understand *what is really happening*, beyond assumptions or guesses (Wallensteen 2011, 15; Avruch 1998, 34).[3] This approach assumes that we do not create the world as we engage with it, rather that it is static and concrete. Peace scientists believe that there is an independent, tangible truth to be studied about social phenomena to do with war and peace. Therefore they exhibit concerns about eliminating bias and partiality that more qualitative peace researchers might assume exist regardless. For example, a peace science study demonstrated that there was inconsistency in surveys administered to Africans by people of the same ethnic group or tribe versus those of different ethnic groups or tribes, something that would be intuitive to an interpretivist researcher and accounted for in their analysis (Adida et al. 2016). In other words, quantitative researchers try to control away bias. Inherent in this is the assumption that bias is external and can be removed with careful enough preparation. In contrast, a qualitative and interpretivist researcher might try to understand and acknowledge the bias and then tell the reader how they have accounted for that bias in the analysis or, at the very least, allow the reader to make their own decisions about the findings given the researcher's bias (Fujii 2010).

In addition, the concept of peace is used narrowly by peace scientists and, for the most part, is understood as the absence of direct violence. These scholars adhere to scientific realism (Putnam 1982) and hold the view that there is a real (or true) world that humans may come to know via science. Some of the pioneers of the study of peace and conflict—Boulding (1962), Schelling (1966), Gurr (1970), Dahl (1971), and Blalock (1989)—are considered peace scientists as they worked toward causal inquiry and explanation (Moore 2015). Peace scientists are identified as working on issues of peace and war primarily from an international relations perspective,[4] and most peace scientists are trained in political science or economics, although other disciplines, such as sociology, psychology and geography, are also included on occasion (Isard, Smith and Anderton 1988, 6–13).[5] Peace science is considered to be the mainstream approach to conducting peace research and is particularly prominent in the United States (Jackson 2015; Davenport, Melander, and Regan 2018; Lottholz 2018).

Peace scientists see their work as primarily to do with theory testing and validation. They assert that in order for a theory to be successful, it must be confirmed by statistical analysis (Achen 2005, 328). For example, democratic peace theory—which argues that democracies do not go to war with one another—is seen as incomplete to peace scientists if it cannot be proven by empirical discovery that relies on statistical analysis (Jutila, Pehkonen, and Väyrynen 2008, 633). This leads to a lack of methodological diversity, even across disciplines (e.g., psychology), where experimental research designs and statistics are prioritized (Vollhardt and Bilali 2008, 21). For example, using mostly lab experiments embracing Galtung's distinction of negative and positive peace, social psychologists have made an important contribution to the study of peace and conflict, with a focus on themes such as aggression, altruism, conformity, ideology, justice, morality, negative stereotyping, obedience, power, social identity theory, contact theory, dehumanization, racism, social dominance orientation, attitudes, emotions, cognitive dissonance, bias, etc. Thus, within the field of social psychology, there is a body of research that Vollhardt and Bilali call social psychological peace research. This social psychological study of peace may be defined, Vollhardt and Bilali suggest, as "the field of psychological theory and practice aimed at the prevention and mitigation of direct and structural violence between members of different sociopolitical groups, as well as the promotion of cooperation and a prosocial orientation that reduces the occurrence of intergroup and societal violence and furthers positive intergroup relations" (13). Therefore, in addition to large-N analysis of multiple cases, scholars in peace science use case study analysis to investigate theoretical claims. However, for the most part, work that is generalizable and universal is prioritized and produced over individual or context-specific claims, which are not considered of sufficient theoretical value.

The analytical agenda of peace science since the 1950s has included various attempts by peace researchers to develop their own methods of inquiry: quantitative measures and statistical methodologies based on data sets such as the Correlates of War; methodologies inspired by game theory to study relations between nations, especially major powers during the Cold War; case studies and comparative methodologies (Wallensteen 2001). Data sets are a primary approach to quantitative analysis in peace science (Salehyan 2015). Scholars use existing data sets, such as the Correlates of War (Singer and Diehl 1990), the Uppsala Conflict Databases (e.g., Sundberg and Melander 2013), the Armed Conflict Location and Event Data Project (Raleigh et al. 2010), the Nonviolent and Violent Campaign Outcomes database (Chenoweth and Lewis 2013), WomanStats (Caprioli et al. 2009), or smaller data sets created by independent researchers. Although there is some limited use of surveys such as those used for reconciliation barometers (Cole and Firchow 2019) or opinion- and attitude-related data (Pham, Vinck, and Stover 2005), most data sets used by peace scientists are created using *event data* coded and collected from media reports, legal proceedings, historical archives, human rights and peacebuilding NGO archives and reports, and other sources that allow for coding and counting. If the data sets are important enough, they are usually introduced to the general public through an article in the *Journal of Peace Research* and sometimes in the *Journal of Conflict Management and Peace Science*.

Qualitatively Driven Research

Qualitatively driven empirical approaches tend to originate from the softer social sciences such as anthropology, some sectors of sociology, participatory action research approaches, and, more recently, a smaller trend in political science toward ethnographic peace research. These studies tend to focus on one specific country or regional case study to work locally with people toward an in-depth understanding of the dynamics related to peace and conflict in a particular context. Scholars who use qualitative methods are typically concerned with concept formation and adequate representation of marginalized people, perspectives, and understandings to guard against methodological exclusion (Sabaratnam 2013, 263).

Most qualitative social scientists contend that they build theory and then test it using interviews, focus groups, participant observation, archival research, and the careful accompaniment of one or two case studies over time. In most cases, they do not purport to make theoretical claims beyond their case studies, although they may discuss how they relates to other, similar cases. In addition, qualitative scholars are less concerned with causality and attribution in the sense that they provide more of the context and discussion surrounding the variables they are studying since they are not forced to place them into a metaphorical data container in order to concretely measure them (Sartori 1970). Of course qualitative scholars still have to make decisions about concept formation and the development of the concepts they study, but they do so less formally and, for the most part, more inductively since most come from the perspective that concepts and social phenomena are relative and not tractable in a way that they can be universally measured.[6]

However, it is important to note that the use of qualitative methods does not preclude a scholar from working from a positivist perspective. Many scholars using qualitative methods, particularly those located in political science, also conduct research from a positivist perspective, seeking to document and collect evidence of some kind of fixed reality.[7] For example, in her recent book, *War, Women and Power: From Violence to Mobilization in Rwanda and Bosnia-Herzegovina*, Marie Berry (2018) investigates gendered power relations and demonstrates that war can lead to increases in women's political agency. She conducted more than 260 interviews in Rwanda and Bosnia-Herzegovina to document and demonstrate the specific ways war can precipitate women's increased political engagement. She shows how women's strength and boldness amid the horrors of war led to shifts in gendered power relations in both countries, but that these gains were short-lived, as the political settlement, international actors, and patriarchal norms intervened to fracture women's organizing and constrain women's progress. Berry's findings and interpretations represent a fixed reality for women's lives after war and makes recommendations for how we might help to change these realities in the future. In her work, Berry (2017) attempts to qualitatively find patterns across contexts in order to make more generalized conclusions about women's experiences during and after war.

However, for the most part, peace researchers who straddle the humanities and social sciences take an entirely different approach that is more in line with the concerns of critical peace researchers (Jackson 2015). For example, history and anthropology have significant overlaps within the social sciences and humanities. Many form part of the pragmatist school of postmodernity, which holds that scholars create theories that use concepts that are not real and are based on unrealistic assumptions. These scholars are less concerned about whether there is a reality out there that we can measure and observe and more concerned with the actual application of what is learned. In other words, they are concerned with heuristics over truth-seeking in the development and testing of theory. Yet others are more concerned with in-depth or descriptive case study research that reflects the priorities and concerns of the communities or populations of interest (Lottholz 2018). This approach is also taken by pracademics, or those individuals that usually identify more as peacebuilding practitioners than as scholars. These are scholar-practitioners who conduct a lot of practical conflict resolution, reconciliation, peacebuilding, or nonviolent resistance activities and then describe and analyze this work in their publications. These pracademics write almost exclusively about their observations and engagements with the practice they participate in, what different applied peacebuilding practices look like and how they should be improved or engaged (e.g., Lederach 1999; Lempereur and Colson 2010; Rothbart and Allen 2019; Aall 1996; Schirch 2005).

The humanities and applied peace studies use methods that are primarily critical, based on observation and experience or speculation, and have a significant historical perspective. There is not one single methodology employed by the humanities. Indeed each subject area engages with multiple methodologies. History uses the historical method and in-depth archival research; anthropology relies primarily on ethnography; philosophy uses conceptual analysis or phenomenology; religion and languages and literature use methodologies derived from textual criticism. For the most part, scholars situated in the humanities are less concerned with gaining knowledge by means of direct or indirect observation and more concerned with the heuristics of the theories and research they produce.

Mixed-Methods Research

Many scholars, in particular in the social sciences, tout the benefits and strengths of mixed methods when conducting empirical research (Druckman 2005). Yet the reality is that the methodological, epistemological, and ontological rifts in peace and conflict studies are so incommensurable that those attempting to work from a mixed-method perspective are often rendered paper tigers. There is little cross-pollination, not to mention cross-citation, between sides, and usually one methodological tradition is entirely ignorant of the other. For example, in a recent special issue on conflict data collection in the *Journal of Peace Research* there is no mention of ethnographic research as a potentially viable form of conflict data collection, even though the special issue is dedicated to investigating ways to gather data that were "valid, or adequately represent[ed] the underlying concept" (Salehyan 2015, 105; Bräuchler 2018).

The result is that qualitative scholars tend toward ethnographic or participatory action research methods without engaging largely any quantitative scholarship, and vice versa. There is a concern that in mixed-methods research the quantitative might silence the qualitative or that an expertise can properly be developed in only one methodological approach, and therefore the other suffers or the results are not sufficiently rigorous (Dixon, Singleton, and Straits 2015). Of course there are effective scholars who combine both (e.g., Kaplan 2017; Arjona 2016), but these are rare and are for the most part approaching their work from a positivist and deductive perspective. In these cases, qualitative methods often become auxiliary to quantitative findings, helping to explain and augment the facts collected.

Participatory statistics could help to bridge this methodological gap. Participatory statistics support bottom-up approaches to measurement and were developed to allow people to participate in the generation of numbers and statistics. The participation of individuals in the production of participatory statistics can happen in many ways, including data collection itself, but fundamentally changes the power dynamic between the researcher and the researched by allowing those we are saying something about to be involved in decisions surrounding the data collection tools we create. A small movement toward quantitative participatory data collection started in the sustainable development community in the 1990s (Chambers 2007) and continues to be utilized and refined in the development community more recently (Holland 2013; Gaillard et al. 2016). It emerged because development scholars recognized that there was a need to understand the multidimensionality of livelihoods and poverty in quantitative research. The development of participatory statistics–related methods allowed them to attribute value to dimensions of people's everyday lives that were usually only qualitatively understood (Gaillard et al. 2016, 1001). There are many ways of collecting participatory numbers, including everyday indicators, ranking, counting and scoring, enumeration, mapping, Venn diagrams, and measuring (Gaillard et al. 2016). Done correctly, participatory statistics have the potential to bring researchers and the researched together on more equal terms to collaborate to produce more valid tools for measurement than traditional top-down measurement tools (Firchow 2018). Participatory statistics allow for some of the power imbalances in quantitative measurement to be recalibrated so that numbers reflect local priorities and people serve as more than solely data sources in research studies.

More recently participatory statistics have been brought into the peace and conflict studies field as a "critique-as-alternative" solution (Visoka 2019) to address some of the concerns raised by critical peace researchers who have called for more inclusive forms of measurement (Firchow 2018). Participatory statistics have the potential to provide the necessary validity for efforts at measuring peace-related phenomena that have created profound divisions and debates surrounding the direction of peace research. The Everyday Peace Indicators (Firchow and Mac Ginty 2020) provides one such approach to the measurement of peace-related phenomena that provides both qualitative and quantitative information in the form of participatory statistics. Regrettably, the use of participatory statistics is severely underutilized in the field of peace and conflict studies, as well as in the applied peacebuilding field. Only recently have participatory approaches at quantification of impact assessments and evaluations been incorporated into the realm of possibilities for peacebuilding and development evaluation professionals (Firchow and Tilton 2019; Firchow and Urwin 2020; Catley et al. 2008).

Concluding Reflections

Perhaps it is propitious that the field of peace and conflict studies has matured enough to develop its own *Methodenstreit* similar to the methodological divisions that emerged in economics and sociology over a hundred years ago (Maclachlan 2016), or perhaps, these conflicts could continue to develop more ominously into hostile camps (Jackson 2015, 31). But how do we begin to remedy the deep-seated methodological discord in the field? Differences in worldview and opinions about how we should create knowledge will always persist, but there are ways to bring us together that would not only contribute to alleviating the rift but could strengthen the work we do significantly. I will suggest three main ways I believe we can move forward.

First and foremost, peace scientists need to engage with the concerns presented by critical peace researchers about mainstream approaches to studying peace and conflict, including issues related to the ethics of studying fragile and violent conflicts in the field (Cronin-Furman and Lake 2018), as well as the human disconnections that can occur when studying it from afar (Sabaratnam 2013). This debate should be integrated into the curricula taught to students in highly quantitative political science and economics programs, as well as peace and conflict studies programs, and students should be allowed to question the established approaches developed by their advisors and engage with critical peace research and interpretivist approaches (Schwartz-Shea 2003). Students should be made aware of the broader "Perestroika" debates in political science at the turn of the twenty-first century and how the concerns raised about how inaccessible the field was to nonspecialists then may influence today's peace research and peace science (Monroe 2005). The questions during these Perestroika debates of power and hegemony of a certain subsector of primarily male political scientists working from a quantitative perspective brought into question the utility and applicability of the research generated by the field of political science. These questions of positionality, perspective, and hegemony should continue to be of concern to peace researchers, in particular those working on issues laden with power and inequality in the arena of peace and conflict. Mainstream analysis should begin to address, if only in the margins at first, the concerns brought forward by those preoccupied with representing the voices of marginalized communities. For the most part this discussion is absent in peace science fora, although recently some peace scientists have begun to push the field to think about peace as more than just the absence of violence (e.g., Davenport, Melander, and Regan 2018). However, even these scholars insist on using peace as a universal concept that has "operational criteria that can be measured across a large number of observable units" in order to uncover generalizable findings (30). This leaves little breathing room for those who understand the relativity and cultural and contextual nature of the study of peace and conflict and represents very limited engagement with the broader concerns expressed by critical peace researchers. It also does not engage with research that studies a world that is constantly evolving and a social world that is not predetermined in concrete, static, and established units of analysis.

Second, as I've already discussed in this chapter, participatory statistics should be studied more seriously and become an integral part of the peace research field, as they have in development studies (Catley et al. 2008). Arguably the present moment is ripe for broadening the scope of methodologies we use to study peace and conflict since in recent years the traditional support for a liberal interventionalism has been called into question even beyond critical peace research inquiries (Jackson 2015, 34). There is potential in finding some meeting ground in this fledgling area of study, where both positivists and interpretivists could overlap regarding their different epistemological and methodological concerns. While peace researchers range in the extent to which they develop an inductive or deductive approach, as well as the extent to which they capture their data using qualitative or quantitative methodologies, all social science researchers must arrive from an abstract concept to concrete observations. Currently, it is rare for any peace researcher to conceptualize their variables with the participation and collaboration of the researched. This omission leads to problems of power and agency, as well as measurement validity. Participatory statistics allow us to include the marginalized in making decisions about what gets included or, perhaps more important, what gets excluded from the conceptual data container (Duursma, Firchow, and Levy 2020; Levy and Firchow 2020; Firchow 2018; Sartori 1970).

Third, field research in peace and conflict studies could also be conducted by teams of scholars coming from different methodological and disciplinary backgrounds. This echoes Galtung's (2010) call for a transdisciplinary, multilevel approach to draw insights from each of the human and social sciences and at the same time transcend their reductionist and biased approaches to peace in order to integrate them in a coherent theoretical system that provides an integrative understanding of the whole. More concretely, as in public health and other fields in international development, projects could be carried out by multidisciplinary teams that conduct interdisciplinary work that is rooted in each peace researcher's individual area of expertise. This way a peace research topic can be addressed in a much more holistic way, from the methods used to investigate it to the questions asked and the underlying dynamics understood. This should extend to including researchers more concerned with the "problem-solving" aspects of a research question and those more interested in the underlying power dynamics at play. Of course this bridging of epistemologies will bring with it inherent challenges and tensions that may be difficult to reconcile. Therefore, in order to prevent some of the pitfalls of mixed-methods research, ample space for each approach to address a particular research question from its own epistemological perspective is paramount in this kind of scenario.

Notes

1. Michael Brzoska (2014) illustrates this rift well with an amusing short vignette about a quantitative evaluator, a qualitative evaluator, and a normal person being hit by a bus.
2. As Professor Kevin Avruch has said to me, paraphrasing Samuel Johnson, "Methodology is the last refuge of scoundrels."
3. I use the terms "peace research" and "peace and conflict studies" interchangeably for the study of peace and war.
4. See the Peace Science Society website, http://sites.psu.edu/pssi/.
5. See the editorial board of *Conflict Management and Peace Science*: https://us.sagepub.com/en-us/nam/journal/conflict-management-and-peace-science#editorial-board.
6. Inductive approaches are also possible using quantitative methods like machine learning and data mining.
7. Lottholz (2018) makes the argument that this is in fact the primary goal of those working from an ethnographic perspective in peace and conflict studies and political science.

References

Aall, P. 1996. "Nongovernmental Organizations and Peacemaking." In *Managing Global Chaos: Sources of and Responses to International Conflict*, edited by P. Aall, C. A. Crocker, and F. O. Hampson, 433–443, Washington, DC: US Institute of Peace Press.

Achen, C. H. 2005. "Let's Put Garbage-Can Regressions and Garbage-Can Probits Where They Belong." *Conflict Management and Peace Science* 22 (4):327–339.

Adida, C. L., K. E. Ferree, D. N. Posner, and A. L. Robinson. 2016. "Who's Asking? Interviewer Coethnicity Effects in African Survey Data." *Comparative Political Studies* 49 (12):1630–1660.

Arjona, A. 2016. *Rebelocracy*. Cambridge: Cambridge University Press.

Avruch, K. 1998. *Culture and conflict resolution*. Vol. 31. Washington, DC: US Institute of Peace Press.

Bamberger, M., J. Rugh, and L. Mabry. 2011. *Real World Evaluation: Working under Budget, Time, Data, and Political Constraints*. Thousand Oaks, CA: Sage.

Berry, M. E. 2017. "Barriers to Women's Progress after Atrocity: Evidence from Rwanda and Bosnia-Herzegovina." *Gender & Society* 31 (6):830–853.

Berry, M. E. 2018. *War, Women, and Power: From Violence to Mobilization in Rwanda and Bosnia-Herzegovina*. Cambridge: Cambridge University Press.

Blalock, H. M. 1989. *Power and Conflict: Toward a General Theory*. Thousand Oaks, CA: Sage.

Boulding, K. E. 1962. *Conflict and Defense: A General Theory*. New York: Harper.

Bräuchler, B. 2018. "The Cultural Turn in Peace Research: Prospects and Challenges." *Peacebuilding* 6 (1):17–33.

Brzoska, M. 2014. "Evaluation and Funding in Peace Research." In *Challenges of Peace Research*. Vol. 7, edited by L. Goetschel, and S. Pfluger, 49–53. Bern, Switzerland: Swisspeace.

Caprioli, M., V. M. Hudson, R. McDermott, B. Ballif-Spanvill, C. F. Emmett, and S. M. Stearmer. 2009. "The Womanstats Project Database: Advancing an Empirical Research Agenda." *Journal of Peace Research* 46 (6):839–851.

Catley, A., Burns, J., Abebe, D., and Suji, O. 2008. *Participatory impact assessment: A Guide for Practitioners. Feinstein International Center, Tufts University. Medford, USA. https://reliefweb.int/report/world/participatory-impact-assessment-guide-practitioners.*

Chambers, R. 2007. *Who Counts? The Quiet Revolution of Participation and Numbers*. Working paper series, 296. Brighton: IDS.

Chenoweth, E., and O. A. Lewis. 2013. "Unpacking Nonviolent Campaigns: Introducing the NAVCO 2.0 Dataset." *Journal of Peace Research* 50 (3):415–423.

Cole, E. and P. Firchow. 2019. "Reconciliation Barometers: Tools for Post-Conflict Policy Design." *International Journal of Transitional Justice* 13 (3):546–569.

Cronin-Furman, K., and M. Lake. 2018. "Ethics *Abroad: Fieldwork in Fragile and Violent Contexts.*" *Political Science & Politics* 51 (3):1–8.

Dahl, R. 1971. *Polyarchy: Participation and Opposition*. New Haven, CT: Yale University Press.

Davenport, C., E. Melander, and P. M. Regan. 2018. *The Peace Continuum: What It Is and How to Study It*. Oxford: Oxford University Press.

Diehl, P. F., and D. Druckman. 2010. *Evaluating Peace Operations*. Boulder, CO: Lynne Rienner.

Dixon, J. C., R. Singleton, and B. C. Straits. 2015. *The Process of Social Research*. New York: Oxford University Press.

Druckman, D. 2005. *Doing Research: Methods of Inquiry for Conflict Analysis*. Thousand Oaks, CA: Sage.

Duursma, A., P. Firchow, and N. Levy. 2020. *Advancing Peace Measurement: The Scalability of Locally-Sourced Indicators*. Working Paper.

Firchow, P. 2018. *Reclaiming Everyday Peace: Local Voices in Measurement and Evaluation after War*. Cambridge: Cambridge University Press.

Firchow, P., and R. Mac Ginty. 2020. "Including Hard-to-Access Populations Using Mobile Phone Surveys and Participatory Indicators." *Sociological Methods & Research* 49 (1):133–160.

Firchow, P., and Z. Tilton. 2019. "*Everyday Peace Indicators: Renegotiating Rigor for Peacebuilding Evidence.*" In *New Directions in Peacebuilding Evaluation*, edited by Tamra d'Estree, 129–149. Lanham, Maryland: Rowman & Littlefield Press.

Firchow, P., and E. Urwin. 2020. "Not Just at Home or in the Grave: (Mis)Understanding Women's Rights in Afghanistan." *Journal of Intervention and Statebuilding*. DOI: 10.1080/17502977.2020.1812893.

Fujii, L. A. 2010. "Shades of Truth and Lies: Interpreting Testimonies of War and Violence." *Journal of Peace Research* 47 (2):231–241.

Gaillard, J. C., J. R. Cadag, A. Gampell, K. Hore, L. Le Dé, and A. McSherry. 2016. "Participatory Numbers for Integrating Knowledge and Actions in Development." *Development in Practice* 26 (8):998–1012.

Galtung, J. 2010. "Peace Studies and Conflict Resolution: The Need for Transdisciplinarity." *Transcultural Psychiatry* 47 (1):20–32.

Gurr, T. R. 1970. *Why Men Rebel*. Princeton, NJ: Princeton University Press.

Holland, J., ed. 2013. *Who Counts? The Power of Participatory Statistics*. Rugby, UK: Practical Action Publishing.

Isard, W., C. Smith, and C. H. Anderton. 1988. *Arms Races, Arms Control, and Conflict Analysis: Contributions from Peace Science and Peace Economics*. Cambridge: Cambridge University Press.

Jackson, R. 2015. "Towards Critical Peace Research: Lessons from Critical Terrorism Studies." In *Researching Terrorism, Peace and Conflict Studies,* edited by Tellidis, I. and H. Toros, 35–53. Abingdon-on-Thames, England, UK: Routledge.

Jutila, M., S. Pehkonen, and T. Väyrynen. 2008. "Resuscitating a Discipline: An Agenda for Critical Peace Research." *Millennium* 36 (3):623–640.

Kaplan, O. 2017. *Resisting War: How Communities Protect Themselves*. Cambridge: Cambridge University Press.

Lederach, J. P. 1999. *Sustainable Reconciliation in Divided Societies*. Washington, DC: US Institute of Peace Press.

Lempereur, A., and A. Colson. 2010. *The First Move: A Negotiator's Companion*. Hoboken, New Jersey: John Wiley & Sons.

Levy, N., and P. Firchow. 2020. Creating a Bottom-up Barometer with the Pasto Indigenous Group in Nariño, Colombia. Working paper forthcoming in PS Political Science.

Lottholz, P. 2018. "Critiquing Anthropological Imagination in Peace and Conflict Studies: From Empiricist Positivism to a Dialogical Approach in Ethnographic Peace Research." *International Peacekeeping* 25 (5):695–720.

Mac Ginty, R. M. 2012. "Routine Peace: Technocracy and Peacebuilding." *Cooperation and Conflict* 47 (3):287–308.

Maclachlan, F. 2016. "Max Weber within the Methodenstreit." *Cambridge Journal of Economics* 41 (4):1161–1175.

Monroe, K. R. 2005. *Perestroika! The Raucous Rebellion in Political Science*. New Haven, CT: Yale University Press.

Moore, W. H. 2015. "Tilting at a Windmill? The Conceptual Problem in Contemporary Peace Science." *Conflict Management and Peace Science* 32 (4):356–369.

Neufeldt, R. C. 2011. "Frameworkers" and "Circlers"–Exploring Assumptions in Impact Assessment. *Advancing Conflict Transformation: The Berghof Handbook II*, 483–504. Berlin: Berghof Foundation.

Pham, P., P. Vinck, and E. Stover. 2005. "Forgotten Voices: A Population-Based Survey of Attitudes about Peace and Justice in Northern Uganda." SSRN, 13 August. http://dx.doi.org/10.2139/ssrn.1448371.

Putnam, H. 1982. "Three Kinds of Scientific Realism." *Philosophical Quarterly* 32 (128):195–200.

Raleigh, C., A. Linke, H. Hegre, and J. Karlsen. 2010. "Introducing ACLED: An Armed Conflict Location and Event Dataset. Special Data Feature." *Journal of Peace Research* 47 (5):651–660.

Rothbart, D., and S. Allen. 2019. "Building Peace through Systemic Compassion." *Conflict Resolution Quarterly* 36 (4):373–386. doi:10.1002/crq.21249.

Sabaratnam, M. 2013. "Avatars of Eurocentrism in the Critique of the Liberal Peace." *Security Dialogue* 44 (3):259–278.

Salehyan, I. 2015. "Best Practices in the Collection of Conflict Data." *Journal of Peace Research* 52 (1):105–109.

Sartori, G. 1970. "Concept Misformation in Comparative politics." *American Political Science Review* 64 (4):1033–1053.

Schelling, T. C. 1966. *Arms and Influence*. New Haven, CT: Yale University Press.

Schirch, L. 2005. *Ritual and Symbol in Peacebuilding*. Boulder, CO: Kumarian Press.

Schwartz-Shea, P. 2003. "Is This the Curriculum We Want? Doctoral Requirements and Offerings in Methods and Methodology." *PS: Political Science and Politics* 36 (3):379–386.

Singer, J. D., and P. F. Diehl, eds. 1990. *Measuring the Correlates of War*. Ann Arbor: University of Michigan Press.

Sundberg, R., and E. Melander. 2013. "Introducing the UCDP Georeferenced Event Dataset." *Journal of Peace Research* 50 (4):523–532.

Visoka, Gëzim. 2019. "Critique and Alternativity in International Relations." *International Studies Review* 21 (4):678–704.

Vollhardt, J. K., and R. Bilali. 2008. "Social Psychology's Contribution to the Psychological Study of Peace: A Review." *Social Psychology* 39 (1):12–25.

Wallensteen, P. 2001. *The Growing Peace Research Agenda*. Occasional paper no. 21. Notre Dame: Joan B. Kroc Institute for International Peace Studies.

Wallensteen, P. 2011. "The Origins of Contemporary Peace Research." In *Understanding Peace Research: Methods and Challenges*, edited by K. Hoglund and M. Oberg, 14–32. Milton Park, UK: Taylor & Francis.

CHAPTER 11

ETHNOGRAPHIC PEACE RESEARCH

GEAROID MILLAR

MANY scholars seem to avoid writing specifically about their research methodology. Most would prefer to be engaged in developing new theories, conducting empirical research, or writing up and presenting their findings to other scholars. These are, after all, the means by which we think about, learn about, and influence the world (or, at least, a particular field of study). This is true also within the field of peace and conflict studies, in which discussion and debate regarding methodology (a decidedly academic endeavor) can seem rather trivial given the more pronounced orientation toward positively impacting the world outside academia. But *how* we come to know the world around us—how we observe, collect, and analyze information—is central to the knowledge we gain about, the theories we hold regarding, and our means of acting in the world. As a result, reflecting on methodology should be central to our theories, our findings, and our recommendations to stakeholders.

This is even more the case when scholarly fields are in flux, as has clearly been the case recently within the literature on peacebuilding and statebuilding. In recent years this literature has been locked in an apparently endless cycle of "turns" (the local, the spatial, the material, the everyday, etc.), each attempting to find some traction by which to overcome stagnant debates regarding power and privilege in international interventions. Indeed while each of these turns has provided some leverage for more nuanced analysis and more fine-grained understanding of the processes and impacts of peacebuilding and statebuilding interventions, they are each largely dependent on what is fundamentally a methodological issue; the need to refocus on the micro level of analysis (the individual, the community, the subnational). It is for this reason that ethnographic peace research (EPR) can appropriately be forwarded as a key methodological tool for scholars working within each of these turns.

What Is Ethnographic Peace Research?

EPR is best viewed from two perspectives. First, EPR should be seen as a new label for traditions of scholarship—primarily within anthropology—which have examined (a) local (meaning subnational, communal, or community) dynamics of conflict, peace, and transition (Avruch 1998; Wilson 2001; Honwana 2006; Vigh 2006; Bräuchler 2015); (b) local conflict resolution or mediation processes (Dillon 1976; Hamer 1980; Podelefsky 1990; Theidon 2000); and (c) local experiences of violence and recovery (Nordstrom 1997, 2004; Das 2007, 2008; Sluka 2009; Theidon 2013). While we should not retroactively tag the work of these scholars with a label they might not appreciate, EPR follows on from and builds on to this past work. Second, and perhaps more important, EPR should be seen as a new effort to form a forceful research *agenda* feeding from and contributing to such research. The overarching goal of this agenda is to emphasize and centralize local experiences, perceptions, concepts, and understandings within research regarding conflict, violence, and peace. While many studies within anthropology, including but not limited to those listed here, have long been examining such dynamics, they have nonetheless been given less attention in peacebuilding and statebuilding scholarship and have largely been marginalized from policy debates. The "local turn," however, has highlighted the need for such research to play a more robust role in both, and EPR is one response to this need.

However, much like many valuable research approaches, EPR cannot be boiled down to a specific methodology for collecting or analyzing data. This is why the term "approach" is helpful as opposed to "methodology." A research approach can incorporate various methodological tools for collecting and analyzing data, as is the case, for example with qualitative, quantitative, realist, constructivist, or interpretivist "approaches." EPR similarly allows for the use of many different, and potentially an eclectic mix of, methodologies for sampling, collecting, and analyzing data, what Hennings (2018, 632) describes in a recent article as a "method repertoire." This diversity of methods may, of course, expose such research to various critiques. What methods, for example, are inappropriately extractive (see Lottholz 2018, 711) or appropriately "ethnographic" (see Millar 2018a, 264), and who gets to decide what is and what is not labeled as such?

In order to respond to these kinds of critiques this chapter argues that EPR may be deployed in any of three different registers, each serving to accomplish different tasks, of different analytical depth, and for different audiences. In its *evaluative* register, EPR is deployed to assess the outcomes of the projects implemented by peacebuilding or statebuilding organizations largely to inform the future decisions of policymakers, planners, and funders. In its *scholarly* register, it is deployed to test existing or to generate new theories regarding the local experiences of conflict, violence, and peace, largely in order to contribute to academic debate. In its *active* register it is deployed as an action research endeavor, seeking to contribute, via research and in partnership with local actors, to a locally defined end, such as peace, security, dignity, development, or justice.

The Evaluative Register

Each of the three registers can be seen to respond to a specific recent demand within the field. The first of these demands is for increased and more accurate evaluation of peacebuilding and statebuilding projects (Blum 2011). While evaluation has always been recognized as important, this demand has increased in urgency parallel to the ongoing professionalization of the field (Sending 2009, 3) and the increasing budgets for such projects—with spending on peacebuilding alone estimated to be approximately $10 billion per year (Institute for Economics & Peace 2017, 3). Processes of evaluation already exist, of course. Many NGOs publish their own evaluation guidebooks (Catholic Relief Services 2010; Organization for Economic Cooperation and Development [OECD], 2012; United Nations 2012), and academics too have developed appropriate methodologies (Bush 1998; Paffenholz and Reychler 2007). However, these models largely fail to examine the local sociocultural, economic, and political contexts of intervention or the relationship of such dynamics to local expectations and experiences of peacebuilding and statebuilding interventions (Millar 2010). They tend to collect standardized information and fail to challenge the fundamental logics of international interventions (Denskus 2012, 151). There remains a great need, therefore, for grounded empirical evaluations of the local impacts of peace interventions, and the EPR approach in its evaluative register fills that need.

In previous work I introduced an admittedly broad framework for what I then called "an ethnographic approach to peacebuilding" (Millar 2014b). This work describes a four-pillar model for evaluating peacebuilding interventions which I would rephrase today as the four pillars of EPR in its evaluative register. These four pillars are (1) seeing peacebuilding as experiential, (2) ethnographic preparation, (3) local engagement, and (4) appraisal of one's own implicit assumptions. The first, *seeing peacebuilding as experiential,* is necessary for EPR as evaluation because all of the individual mechanisms of intervention deployed by international actors in postconflict states have as their eventual end goal the promotion of some experience: of justice, of empowerment, of opportunity, of dignity, of security, of peace, etc. The international community primarily implements institutional and technocratic reforms, but the final goal of any of these is more substantive in nature. If we want to accurately evaluate what impacts those interventions have, therefore, we must understand how local people *experience* those interventions.

Second, EPR in its evaluative register demands a thorough understanding of existing ethnographic research via *ethnographic preparation* regarding the society under study prior to any research. In most peacebuilding and statebuilding projects practitioners and evaluators enter societies about which they know very little and have little knowledge of the local language or culture (Autesserre 2014), and so they speak largely to accessible English-speaking elites who are often culturally dissociated from the experiences of those they claim to represent (Millar 2010, 2017). Indeed in many cases they are employed by or in some other way associated with the intervention under study. In order to overcome these limitations EPR instructs evaluators to conduct a thorough review of existing ethnographic studies of the region or community and have some sociocultural context in which to place the data they collect.

Third, and not surprisingly, those deploying EPR in its evaluative register must commit to *local engagement*, where "the local" is defined as the identified target community of the peace intervention being evaluated (see Millar 2014b, 82). If the intervention is a nationwide project to create national reconciliation, then the entire nation must be considered "the local." However, if the project being evaluated is designed to provide, for example, livelihood training to young men in a particular region, then only young men in that target setting would constitute "the local." As a result of this definition, an EPR approach to evaluation requires engagement with this local, and engagement would not be accomplished by talking to civil society actors or politicians who claim to represent that local. While defining "the local" as the target beneficiaries of a particular intervention may not be appropriate for EPR in its scholarly register, it is appropriate for EPR in its evaluative register.

Finally, and mirroring the anthropological focus on reflexivity (Davies 2012), an EPR approach to evaluation requires that the researcher be willing and able to *question their own implicit assumptions* about the concepts that underpin experiences of interventions: of justice, dignity, security, empowerment, etc. This self-appraisal is necessary because it forces us to recognize and begin to overcome the limitations of our own internalized biases. Once we start to think of peacebuilding as experiential it becomes clear that assessing the ability of any given intervention to foster the experience it claims to provide will require that we see the intervention not through our own eyes but through another's. Scholars deploying EPR in its evaluative register therefore must learn to interpret events via schemas, frameworks, and perspectives other than those dominant in their own culture, and they must directly question their own assumptions in order to evaluate how peace interventions are experienced by those socialized in a foreign sociocultural setting.

The Scholarly Register

The second demand in the field to which EPR responds is the demand for more locally grounded and contextually specific forms of peacebuilding and statebuilding to be theorized and encouraged (Autesserre 2010; Richmond 2013). While this demand led initially to the use of concepts such as "local ownership" in the literature (Keane, 2005), the substantial challenges to facilitating "ownership" amid the complex interaction between unequally privileged international, national, and local actors quickly became a central focus of the literature (Donais 2009, 2012; Richmond 2012). This is specifically visible in the many publications regarding hybridity (Richmond 2009; Mac Ginty 2011; Petersen 2012; Millar 2014a) and, more recently, Richmond's (2013, 2016) concept of peace formation. This extension of the previously mentioned "local turn" (Mac Ginty and Richmond 2013; Paffenholz 2015) has fundamentally reshaped how scholars theorize about peacebuilding and statebuilding and demands a similar reconsideration of *how* we study that subject. EPR responds to this demand in what I will call its *scholarly* register, which has been described recently as composed of two required and two facilitative features (see Millar 2018a, 259).

The first of these required features is *thick description*, which demands the close examination and interpretation of the complex and multidimensional social reality of the local setting. Geertz (1973, 10), who coined the phrase thick description more than forty years ago, described ethnography as the task of working to decipher "a multiplicity of complex conceptual structures . . . which are at once strange, irregular, and inexplicit" but which the ethnographer works to "contrive somehow first to grasp and then to render." EPR in its scholarly register therefore must include close investigation of the local everyday within the setting of peace interventions and attempt to see the local experience of those interventions not from the perspective of the outsider but to understand "the meaning and significance of social phenomena for people in those settings" (Ragin 1994, 91). Thick description must be the goal of EPR in its scholarly register, although the reader can surely see how consistent this goal is with EPR in its evaluative register.

For work to be considered EPR in its scholarly register it must not only describe what has happened or is happening but must *explain why and understand how* particular events unfold: why conflict occurs, how it is resolved, how peacebuilding is experienced, and why it is experienced as it is. Above and beyond simple description, examinations of how and why events unfold serve to generate nuanced understanding of the causal mechanisms driving phenomena. This is fundamental to EPR in its scholarly register because research that is engaged substantively in exploring these deeper questions can test existing theory, contribute to the evolution of such theory, and generate new theory. These tasks clearly set EPR in its scholarly register apart from EPR in its evaluative register. They demand a deeper analysis and more engagement with theory and explanation. EPR in its scholarly register therefore must go beyond describing what happened and examine the subtle causal relationships inherent in peace and conflict dynamics.

The third feature of EPR in its scholarly register, and the first that is considered facilitative but not strictly required, is the deployment of a *diversity of methodologies*. As noted, EPR is not wedded to any specific methodology of data collection but is open to any that provide the locally grounded insight demanded. Among contributions to a recent edited volume on EPR, for example, are studies using semistructured interviews (Tomforde 2018; Sakti and Reynaud 2018), informal conversations (Cole 2018), documentary analysis (Björkdahl and Mannergren Selimovic 2018), narrative cartographies (Slooter 2018), and participant observation (Bräuchler 2018). This diversity provides for methodological flexibility in response to needs on the ground and allows EPR to better observe aspects of postconflict societies not readily accessible to less flexible approaches.

While not a required feature, methodological flexibility can greatly enhance a study's capacity to achieve thick description and nuanced explanation.

Finally, the fourth feature of EPR in its scholarly register, and the second that should be considered facilitative but not strictly required, is *reflexivity*, or the "critical consideration a researcher gives to their own positionality and role in impacting on or influencing the setting or subjects of their research" (Millar 2018a, 8). Reflexivity is a methodological strength, in that it allows EPR scholars to recognize their positionality and respond accordingly with adjustments to their research practice. But it requires substantial time in the field, which is a rarity for studies of peacebuilding and statebuilding interventions. Greater time, however, allows for more research experiences, which can themselves trigger greater reflexivity and allow the scholar to identify the pros and cons of a particular method and to make adjustments as needed (Millar 2018b). Reflexivity and a diversity of methodology are closely related, and they also make clear the substantive differences between the evaluative register and the scholarly, in that none of these features (extensive time, reflexivity, and the evolution of methodologies) are common features of evaluation practices within the fields of peacebuilding and statebuilding.

The Active Register

As should be clear, the evaluative and the scholarly registers of EPR are related. However, researchers working within each of them will have different goals and distinct audiences. The first seeks to evaluate the impacts of peacebuilding and statebuilding projects at the local level and focuses on specific communities or sectors of a population targeted by that intervention in order to positively influence the design and implementation of future projects. It is not generally engaged in academic questions about theory, and while the audience for such research may sometimes be academics, it will more likely be those with the ability to influence future project implementation: policymakers, funders, and NGOs. The second seeks to understand the nuances of local contexts (cultures, expectations, perceptions, and understandings) in order to theorize about what peacebuilding and statebuilding might look like within those contexts or examine the manner in which interventions influence and affect such contexts. The audience for such scholarly endeavors may sometimes be funders, policymakers, or NGOs but will more likely be those similarly engaged in theorizing about conflict, violence, peace, security, and the state, i.e. other academics. The two registers are clearly related, but they are distinguished also by their goals and their target audience.

There is a third potential register that might be seen partially as a bridge between these two and partially as disrupting both; this is the *active* register. This third register has been discussed only partially in the literature about EPR, but proponents argue that it can provide both a practical engagement with agents and institutions and contribute substantively to academic theory regarding power, agency, emancipation, and the everyday. While EPR in its active register has not yet been fully developed, it certainly borrows from action-oriented approaches reaching back as far as Sol Tax (1975) or the more recent advocacy of participatory action research (Schensul et al. 2004) and critical ethnography (Barab et al. 2004). A few scholars writing explicitly about EPR have argued, in fact, that this should be the end goal of EPR if it is to fulfill the normative agenda of peace research more broadly considered. Close (2018, 187), for example, in her focus on collaborative engagement in research with indigenous communities, argues that EPR can help to "strengthen local capacities for peace," while Collins and Watson (2018, 91) explicitly reach back to Tax in their conception of EPR as "collaborative community research." Consistent with this argument, Mike Klein (2018, 83) builds "action planning" into his approach to EPR, with the primary purpose of having research participants develop "strategies to change structures."

Each of these contributions sees EPR as functioning in an active register, engaging with both practical and theoretical issues. They are not simply evaluative, monitoring or assessing the impact of interventions as per the goals or missions set by the external policymakers or funders. Nor are they examining peacebuilding and statebuilding processes purely with an eye to contributing to or generating theories regarding conflict, violence, peace, or power. The effort is instead to both engage and theorize. In doing so, such approaches also have the potential to overcome some of the criticisms of both of the prior registers, i.e., that EPR in its evaluative register is purely extractive and that EPR in its scholarly register is too removed. In a recent article, for example, Lottholz (2018, 698) critiques EPR and argues in favor of what he calls "dialogical, collaborative and 'activist' research strategies." Such strategies, he claims, can respond to what NGOs, policymakers, and social movements require from research instead of simply contributing to the theoretical reflections that academia demands (711).

I would argue that EPR, even in an active register, should avoid becoming "activist research," as this may move away from being research at all and run the risk of bias in favor of a preferred outcome instead of a truth. But EPR in this third register may appropriately perform the task of balancing the practical and the theoretical if it avoids losing sight of rigor in favor of a political position. It probably is true that EPR in its evaluative register can be extractive and largely conducted for the benefit of those in positions of power (funders, policymakers, etc.) and not of recipient communities. And EPR in its scholarly register can be criticized for the opposite reason, as being overly academic and distanced from any practical contributions, which can certainly be seen as serving a different form of power. But both have their roles to play, and both are perfectly legitimate modes of research as long as the researcher is explicit regarding the purpose and limitations of such research. EPR in its active register therefore can bridge these two (addressing both the practical and the theoretical) or disrupt them entirely, as long as it too is explicit in its purpose and limitations.

Conclusion

EPR has recently been developed in response to two different but related demands within the field, the first regarding the need for improved evaluation of projects (and related goals regarding project design and implementation) and the second regarding the need for more locally grounded and salient theories of peace, justice, security, and associated concepts. While these two demands are related, they nonetheless require quite distinct responses and, as a result, necessitate two separate registers of EPR: the *evaluative* and the *scholarly*. Both of these registers have different goals and target audiences and demand different levels of analysis and reflection. However, both are fundamentally focused on the examination of peacebuilding and statebuilding (whether as practice or theory) within *local contexts*, and most readers will have noticed the overlap between the pillars of the evaluative register and the features of the scholarly register.

There is so much in common between these registers, in fact, that it seems perfectly plausible that a significant EPR research project might incorporate both a shorter-term evaluative component to address the needs of planners, funders, and administrators, as well as a longer-term and more scholarly research project to rigorously test and develop theory for the academic community. Further, while it requires more development, the active register may hold some promise of bridging the gap between the evaluative and the scholarly, being neither the first nor the second but something different altogether. Distinguishing among these three registers therefore does not imply that they do not share much in common or could not even be combined in a single project. What the typology provides, quite simply, is a way to identify the distinct benefits of an EPR approach when applied in response to different demands, conducted for different purposes, and with different audiences in mind. Thus I hope it can also help those deploying it to keep their own purpose and limitations in mind and to avoid certain critiques.

As things stand today, however, there are substantial barriers to the promotion, deployment, and influence of EPR in the fields of peacebuilding and statebuilding. First, while there are critical and constructivist alternatives, the highly ranked journals in these fields today tend to be populated predominantly by large-N quantitative studies, and the process of opening those publications to more qualitative or ethnographic studies (whether single-case or small-N studies) has been slow. For the EPR approach to have the kind of impact it should aspire to, such studies must have at least a chance of publication in agenda-setting publications such as the leading journals. Further, while the fields of peacebuilding and statebuilding are open, at least in principle, to many different methodological traditions, in practice few young scholars working in these fields are provided with substantial training in ethnographic methods. As a result, a much more robust effort to expose PhD students to rigorous long-term ethnographic research methodology is necessary to more fully prepare the next generation of peace scholars and practitioners both to conduct and to understand EPR research. Any efforts to generate a robust EPR agenda will have to overcome such challenges in the not too distant future.

References

Autesserre, S. 2010. *The Trouble with the Congo: Local Violence and the Failure of International Peacebuilding.* New York: Cambridge University Press.

Autesserre, S. 2014. *Peaceland: Conflict Resolution and the Everyday Politics of International Intervention.* New York: Cambridge University Press.

Avruch, K. 1998. *Culture and Conflict Resolution.* Washington, DC: US Institute of Peace.

Barab, S., M. K. Thomas, T. Dodge, K. Squire, and M. Newell. 2004. "Critical Design Ethnography: Designing for Change." *Anthropology and Education Quarterly* 35 (2):254–268.

Björkdahl, A., and J. Mannergren Selimovic. 2018. "Feminist Ethnographic Research: Excavating Narratives of Wartime Rape." In *Ethnographic Peace Research: Approaches and Tensions*, edited by G. Millar, 43–64. London: Palgrave Macmillan.

Blum, A. 2011. *Improving Peacebuilding Evaluation: A Whole-of-Field Approach.* Special Report 280. Washington, DC: US Institute of Peace.

Bräuchler, B. 2015. *The Cultural Dimension of Peace: Decentralization and Reconciliation in Indonesia.* London: Palgrave Macmillan.

Bräuchler, B. 2018. "Contextualizing Ethnographic Peace Research." In *Ethnographic Peace Research: Approaches and Tensions*, edited by G. Millar, 21–42. London: Palgrave Macmillan.

Bush, K. 1998. *A Measure of Peace: Peace and Conflict Impact Assessment (PCIA) of Development Projects in Conflict Zones.* Ottawa: International Development Research Centre.

Catholic Relief Services. 2010. *GAIN Peacebuilding Indicators.* Baltimore, MD: PQSD.

Cole, C. E. 2018. "Beyond 'Being There': Space and Mobility in Ethnographic Peace and Transitional Justice Research." In *Ethnographic Peace Research: Approaches and Tensions*, edited by G. Millar, 231–252. London: Palgrave Macmillan.

Collins, B., and A. Watson. 2018. "The Impetus for Peace Studies to Make a Collaborative Turn: Towards Community Collaborative Research." In *Ethnographic Peace Research: Approaches and Tensions*, edited by G. Millar, 89–114. London: Palgrave Macmillan.

Close, S. 2018. "Researching Peace Peacefully: Using Ethnographic Approaches in Timor-Leste." In *Ethnographic Peace Research: Approaches and Tensions*, edited by G. Millar, 181–206. London: Palgrave Macmillan.

Das, V. 2007. *Life and Words: Violence and the Descent into the Ordinary.* Berkeley: University of California Press.

Das, V. 2008. "Violence, Gender, and Subjectivity." *Annual Review of Anthropology* 37:283–299.

Davies, C. A. 2012. *Reflexive Ethnography: A Guide to Researching Selves and Others.* London: Routledge.

Denskus, T. 2012. "Challenging the International Peacebuilding Evaluation Discourse with Qualitative Methodologies." *Evaluation and Program Planning* 35 (1):148–153.

Dillon, R. G. 1976. "Ritual Resolution in 'Meta' Legal Process." *Ethnology* 15 (3):287–299.

Donais, T. 2009. "Empowerment or Imposition? Dilemmas of Local Ownership in Post-conflict Peacebuilding Processes." *Peace and Change* 34 (1):3–26.

Donais, T. 2012. *Peacebuilding and Local Ownership: Post-conflict Consensus-Building.* New York: Routledge.

Geertz, C. 1973. *The Interpretation of Cultures.* New York: Basic Books.

Hamer, J. H. 1980. "Preference, Principle, and Precedent: Dispute Settlement and Changing Norms in Sidoma Associations." *Ethnology* 19 (1):89–109.

Hennings, A. 2018. "With Soymilk to the Khmer Rouge: Challenges of Researching Ex-Combatants in Post-war Contexts." *International Peacekeeping* 25 (5):630–652.

Honwana, A. 2006. *Child Soldiers in Africa.* Philadelphia: University of Pennsylvania Press.

Institute for Economics & Peace. 2017. *Global Peace Index: Measuring Peace in a Complex World.* Sydney: Institute for Economics & Peace.

Keane, R. 2005. "The Partnership-Conditionality Binary in the Western Balkans: Promoting Local Ownership for Sustainable Democratic Transition." *Cambridge Review of International Affairs* 18 (2):247–257.

Klein, M. 2018. "Institutional Ethnography as Peace Research." In *Ethnographic Peace Research: Approaches and Tensions*, edited by G. Millar, 65–88. London: Palgrave Macmillan.

Lottholz, P. 2018. "Critiquing Anthropological Imagination in Peace and Conflict Studies: From Empiricist Positivism to a Dialogical Approach in Ethnographic Peace Research." *International Peacekeeping* 25 (5):696–720.

Mac Ginty, R. 2011. *International Peacebuilding and Local Resistance: Hybrid Forms of Peace.* New York: Palgrave Macmillan.

Mac Ginty, R., and O. P. Richmond. 2013. "The Local Turn in Peace Building: A Critical Agenda for Peace." *Third World Quarterly* 34 (5):763–783.

Millar, G. 2010. "Assessing Local Experiences of Truth-Telling in Sierra Leone: Getting to 'Why' through a Qualitative Case Study Analysis." *International Journal of Transitional Justice* 4 (9):477–496.

Millar, G. 2014a. "Disaggregating Hybridity: Why Hybrid Institutions Do Not Produce Predictable Experiences of Peace." *Journal of Peace Research* 51 (4):501–514.

Millar, G. 2014b. *An Ethnographic Approach to Peacebuilding: Understanding Local Experiences in Transitional States.* London: Routledge.

Millar, G. 2017. "For Whom do Local Peace Processes Function? Maintaining Control through Conflict Management." *Cooperation and Conflict* 52 (3):259–276.

Millar, G., ed. 2018a. *Ethnographic Peace Research: Approaches and Tensions.* New York: Palgrave Macmillan.

Millar, G. 2018b. "Ethnographic Peace Research: The Underappreciated Benefits of Long-Term Fieldwork." *International Peacekeeping* 25 (5):653–676.

Nordstrom, C. 1997. *A Different Kind of War Story.* Philadelphia: University of Pennsylvania Press.

Nordstrom, C. 2004. *Shadows of War: Violence, Power, and International Profiteering in the Twenty-First Century.* Berkeley: University of California Press.

OECD. 2012. *Evaluating Peacebuilding Activities in Settings of Conflict and Fragility: Improving Learning for Results.* DAC Guidelines and References Series. Paris: OECD.

Paffenholz, T. 2015. "Unpacking the Local Turn in Peacebuilding: A Critical Assessment Towards an Agenda for Future Research." *Third World Quarterly* 36 (5):857–874.

Paffenholz, T., and L. Reychler. 2007. *Aid for Peace: A Guide to Planning and Evaluation for Conflict Zones.* Baden-Baden: Nomos.

Petersen, J. H. 2012. "A Conceptual Unpacking of Hybridity: Accounting for Notions of Power, Politics and Process in Analysis of Aid-Driven Interfaces." *Journal of Peacebuilding and Development* 7 (2):9–22.

Podelefsky, A. 1990. "Mediator Roles in Simbu Conflict Management." *Ethnology* 29 (1):67–81.

Ragin, C. C. 1994. *Constructing Social Research: The Unity and Diversity of Method.* Thousand Oaks, CA: Pine Forge Press.

Richmond, O. P. 2009. "Becoming Liberal, Unbecoming Liberalism: Liberal-Local Hybridity via the Everyday Response to the Paradoxes of Liberal Peacebuilding." *Journal of Intervention and Statebuilding* 3 (3):324–344.

Richmond, O. P. 2012. "Beyond Local Ownership in the Architecture of International Peacebuilding." *Ethnopolitics* 11 (4):354–375.

Richmond, O. P. 2013. "Peace Formation and Local Infrastructures for Peace." *Alternatives* 38 (4):271–287.

Richmond, O. P. 2016. *Peace Formation and Political Order in Conflict Affected Societies.* New York: Oxford University Press.

Sakti, V., and A. Reynaud. 2018. “Understanding Reconciliation through Reflexive Practice: Ethnographic Examples from Canada and Timor-Leste.” In *Ethnographic Peace Research: Approaches and Tensions*, edited by G. Millar, 159–180. London: Palgrave Macmillan.

Schensul, J., M. Berg, D. Schensul, and S. Sydlo. 2004. “Core Elements of Participatory Action Research for Educational Empowerment and Risk Prevention with Urban Youth.” *Practicing Anthropology* 26 (2):5–9.

Sending, O. J. 2009. *Why Peacebuilders Fail to Secure Ownership and Be Sensitive to Context.* NUPI Working Paper 755. Oslo: Norwegian Institute for International Affairs.

Slooter, L. 2018. “Conflicting Boundaries and Roles: Impressions of Ethnographic Fieldwork in the French Banlieues.” In *Ethnographic Peace Research: Approaches and Tensions*, edited by G. Millar, 115–136. London: Palgrave Macmillan.

Sluka, J. 2009. “In the Shadow of the Gun: ‘Not-War-Not-Peace’ and the Future of Conflict in Northern Ireland.” *Critique of Anthropology* 29 (3):279–299.

Tax, S. 1975. “Action Anthropology.” *Current Anthropology* 16 (4):514–517.

Theidon, K. 2000. “‘How We Learned to Kill Our Brother’? Memory, Morality and Reconciliation in Peru.” *Bulletin of the French Institute of Andean Studies* 19 (3):539–554.

Theidon, K. 2013. *Intimate Enemies: Violence and Reconciliation in Peru.* Philadelphia: University of Pennsylvania Press.

Tomforde, M. 2018. “How Much Peace Can the Military Instigate? Anthropological Perspectives on the Role of the Military in Peace Intervention.” In *Ethnographic Peace Research: Approaches and Tensions*, edited by G. Millar, 207–230. London: Palgrave Macmillan.

United Nations. 2012. *Planning Toolkit.* New York: United Nations.

Vigh, H. 2006. *Navigating Terrains of War: Youth and Soldiering in Guinea-Bissau.* New York: Berghahn Books.

Wilson, R. 2001. *The Politics of Truth and Reconciliation in South Africa: Legitimizing the Post-apartheid State.* Cambridge: Cambridge University Press.

CHAPTER 12

VISUALITY OF PEACE AND CONFLICT

EMMA HUTCHISON AND ROLAND BLEIKER

INTRODUCTION

PHOTOGRAPHS, cinema, and television influence how we view and reflect on all aspects of conflict and peace: they shape our understanding of war, humanitarian disasters, protest movements, and peace negotiations. The dynamics of visual politics—and indeed the emerging "visual turn" in society and politics (Bleiker 2018b, 4–8; Callahan 2015)—reach in all directions and go well beyond traditional media outlets. Digital media, such as Twitter, YouTube, Facebook, and Instagram, play an increasingly important role across the political spectrum, from terrorist recruitment drives to social justice campaigns. Fashion and video games are frequently derived from and enact the militarized world we live in. Drones, satellites, and surveillance cameras profile terrorist suspects and identify military targets.

This omnipresence of the visual is political and has fundamentally changed how we understand and approach questions of peace and conflict. There is, obviously, no way we can address these complex and far-reaching issues in a comprehensive manner in this chapter. This is why we highlight and examine a couple of selective issues. We focus on two particular aspects of the visuality of peace and conflict, one conceptual and one empirical. In doing so, we draw on research we have conducted over several years (notably Bleiker, Hutchison, and Campbell 2014; Hutchison and Bleiker 2015; Hutchison 2019; Bleiker 2017, 2018a, 2018b).

Conceptually we stress the need to examine the historical dimension of images and the way they have shaped understandings of suffering and conflict as well as, in turn, ensuing approaches to peace. Images always have a history. This is the case because images make no sense by themselves. They need to be seen and interpreted. They gain meaning in relation to other images and in relation to the personal and societal assumptions and norms that surround us. Our viewing experience is therefore intertwined not only with previous experiences, such as our memory of photographs we have seen in the past, but also with the values and visual traditions that are accepted as common sense by established societal norms. There are thus inevitable power relationships involved in this nexus between visuality and dominant ways of seeing and perceiving of society and politics. Likewise, structures of power also pervade the images typically employed to visualize conflict and the practices of peace promotion, peacebuilding, postconflict development, and statebuilding that emerge in the aftermath of violence.

Empirically we illustrate the issues at stake by focusing on both historical and contemporary photographs of humanitarian crises and their links with peace photography. We define peace photography in a broad sense and as articulated by Frank Möller (2017) as photographs that promote the peaceful and nonviolent transformation of conflict. We consequently focus on humanitarian photography for two reasons. The first is because humanitarian ethics have been increasingly accepted as an important normative element of a peaceful international order. The second, and explicitly related to the visuality of peace and conflict, is that images of human hardship and suffering have been critical to not only the rise of humanitarianism but also the understandings of conflict and peace that eventually ensued.

We begin by outlining the increasingly important role that images play in relation to the politics and ethics of peace and conflict. We then situate the respective insights in the context of the so-called visual and aesthetic turns in international relations and peace studies scholarship.

Having set the context, we embark on examining the role that images played in the emergence of humanitarian values and the push for nonviolent, peaceful resolutions to violence and conflict. To do so, we stress the need to go back to and appreciate the importance of visual traditions that predate the invention of photography. We highlight how political attitudes to pain and suffering changed, at least in part, in response to a range of powerful visual representations, be they sketches of cruel, inhumane practices of slavery in the late eighteenth century, emerging photographs of atrocity, deprivation, and disadvantage in the nineteenth century, or the proliferation of media and aid images following calamities of the twentieth century (Baer 2001). This is significant as it is the first step in "a history of the present" in terms of how conceptualizations of peace and also, crucially here, peace photography have come to be (Foucault 1984, 94–95).

In this light, we then examine how these visual legacies continue to shape current practices of humanitarianism. In particular, we show that the visual frames typical of much abolitionist artwork bear out in present-day images of crisis and conflict. Key to this is a hierarchical framing that depicts victims as helpless and assigns power and agency to privileged Western actors. The political consequences are significant, for such dominant visual frames align with hierarchical power relations that perpetuate the very inequalities that generate the need for humanitarianism—and unfortunately also often promote forms of liberal postconflict transformation and structurally inhibit sustainable peace (Richmond 2005; also 2019)—in the first place.

Finally, we explore the significance of these insights for broader—present-day and emerging—conceptualizations of peace and conflict. Here we present peace not as a perpetual state of harmony but as a process that directly engages the sources of conflict, including the unequal power relationships that are manifest in humanitarian politics and imagery. However, emerging forms of photography that focus on more everyday peace negotiations and contexts may in this way contribute to more enduring peace processes. Indeed new forms of peace photography are referred to as such precisely because they seek to address and potentially transform unequal power relations and redress both structural and overt forms of conflict in a nonviolent manner. We outline how the ensuing insights can be of relevance to the future studies of visuality in peace and conflict.

Visualizing Peace and Conflict

Before exploring the concrete links between this history of humanitarian imagery and visualizing peace, it is necessary to highlight, at least briefly, how the visual realm is inherently political and also a crucial aspect of conceptualizing and promoting the politics of peace.

Images, both still and moving, are political forces in themselves. They often shape politics as much as they depict it. Early modern cartographic techniques played a key role in legitimizing the emergence of territorial states. Hollywood films provide us with well-rehearsed and deeply entrenched models of heroes and villains to the point that they shape societal values. A terrorist suicide bombing is designed to kill people with maximum visual impact: images of the event are intended to circulate around the world and spread fear.

Images, and visual artifacts in general, also tell us something about the world. They are witnesses of our time and of times past. A satellite image provides information about the world's surface. Photographs document wars or diplomatic summits or protest movements. Monuments remind us of past events and their significance for today's political communities.

Images and visual artifacts are neither progressive nor regressive. They are neither for conflict nor for peace. They are simply political in nature. Sometimes images and artifacts entrench conflict and political oppression. Consider how a variety of seemingly mundane visual performances, from hairstyles to body movements, signal and normalize gendered systems of exclusion. Or look at the paradigmatic here: Leni Riefenstahl. Her stunning films of Nazi rallies, such as *Triumph of the Will* and *Olympia*, helped the Nazi regime turn mere propaganda into a broader mythology that was instrumental in gaining popular support for a racist and militaristic state apparatus: "fascinating fascism," as Susan Sontag (1975) called it. But visuals can also challenge domination and bring about peace. Some even associate visual creativity with the potential to fundamentally reorient our political world. A work of art can lead us to see the world in a new light and help us rethink ideas we have taken for granted, including those about politics. Or so believes Alex Danchev (2016, 91), who was convinced that "contrary to popular belief, it is given to artists, not politicians, to create a new world order." Look, as an example, at how Pablo Picasso's famous painting *Guernica* has become an important and influential icon of peace that reminds us of the atrocities of war.

An immediate reaction to a concept of politics as a struggle over visibility is to ask what happens to people, issues, and phenomena we do not see. What happens when we do not see violence, human rights violations, mass rape during war? And what happens to political phenomena that are hard to visualize? How can we "see" peace, for instance (Figure 12.1)? We have numerous visual icons that signify war but hardly any that stand for peace. If peace is seen as the absence of violence, then there is literally an unlimited and meaningless number of images that can depict this (Möller 2018b).

There are countless ways of approaching the complex issues of visualizing peace and conflict. We focus here on one aspect in particular: the historical dimension of images and the manner in which they include current practices of seeing peace and conflict. In this sense, images are not new, nor have they necessarily replaced words as the main means of communication. Images and visual artifacts have been around from the beginning of time. The visual has always been part of life. Images were produced not only to capture key aspects of human existence but also to communicate these aspects to others. Examples range from prehistoric cave paintings that document hunting practices to Renaissance works of art. Some of these images and cultural artifacts we still see today, and they continue to influence our perception and understanding of the world.

Fig. 12.1. "How to visualise peace? United Nations Buffer Zone in Nicosia, Cyprus, July 2014."
Copyright Roland Bleiker.

Aesthetics, Emotions, and the Visual Turn in Global Politics

There are increasing scholarly debates on the significance of visuality in the realms of international politics and peace and conflict studies. Some authors, such as W. J. T. Mitchell (1986), go as far as speaking of a "visual" or "pictorial" turn, stressing that people often perceive and remember key events more through images than through verbal accounts. Mitchell writes of a "new heightened awareness" of the role of visuality, and of how the problem of our time is the "problem of the image." The result is an unprecedented visualization of both our private lives and our political landscapes: a global communication dynamic that is fundamentally new and rooted in various networks and webs of relations (Favero 2014, 66).

The visual turn can be seen as part of a larger shift toward the study of aesthetics in international relations and peace and conflict. Numerous scholars have now revealed how aesthetic sensibilities can help us rethink some of the most serious problems in global politics. An entire new generation of interdisciplinary academics has arisen, pushing the boundaries of how we can understand world politics. They explore different forms of insights, including those that emerge from images, narratives, and sounds, such as literature, visual art, music, cinema, and other aspects of popular culture. These scholars also show that aesthetics is about far more than art: aesthetics is how individuals and collectives see, sense, feel about, and perceive of the social world, and can as such provide new insights into fundamental issues that drive global politics.

The key task for aesthetically oriented scholars consists of legitimizing broader engagements with the political. This approach is about validating the whole register of human perceptions and sensations, not only practices of reason and logos that prevail in conventional scholarship but also a range of other, more creative, and open-ended, yet equally important forms of insights. In this sense aesthetic politics is about cultivating a critical attitude to how we understand and engage the political world around us. An appreciation of aesthetics offers us possibilities to rethink, re-view, re-hear and re-feel the political world we live in (see Rancière 2004). Drawing on these and other insights from aesthetic theory, numerous international relations scholars have thought to challenge the boundaries of what is visible and invisible, thinkable and unthinkable, and thus of what can and cannot be debated in politics (for overviews, see Bleiker 2001, 2009).

Scholars who write in the wake of the aesthetic turn investigate a range of phenomena. They explore the relevance of art, film, television, video, comics, and maps. They examine the politics of monuments, performances, and visual artifacts or the role of drones and satellite images. They also deal with the role of emotions and affect. The aesthetic fields to be investigated are as limitless as the political topics that are linked to them.

A particularly prominent role in the aesthetic turn is occupied by investigations into the role of photography. This is, in part, because photographs are memorable and shape how communities make sense of themselves. At a time when we are saturated with information stemming from multiple media sources, photographs remain influential for their ability to capture social and political issues in succinct and mesmerizing ways. They serve as "visual quotations" (Sontag 2003, 22). There are few realms where the power of photographs is more obvious than with icons, as Robert Hariman and John Louis Luciates (2007) have prominently shown. Icons shape public opinion because they are part of the collective fabric through which people and communities make sense of themselves.

Images of Suffering and the Emergence of Humanitarianism

Writing in the wake of and drawing upon these aesthetic and visual turns, we now embark on our particular historical-political investigation. We examine historical links between images and peace and conflict in one particular case: how representations of pain, suffering, and hardship in early modern times were instrumental to the emergence of humanitarian sensibilities. To do so we reach back to representations that predate the invention of photography. Numerous forms of aesthetic representation reshaped what it meant to experience and witness pain. They include novels, poems, and memoirs, as well as artistic, aesthetic engagements, including sketches, paintings, caricatures, and even silverwork and other jewelry. It is in this manner that suffering groups were brought to light as they had never been before.

In particular, representations mobilized and enabled audiences to enact what was at the time an emerging culture of "spectatorial sympathy," which worked in conjunction with new humanitarian meanings and narratives (Halttunen 1995; Wilson and Brown 2009). Visual representations were especially powerful: images brought to life foreign and unknown hardship and reshaped how audiences perceived and felt for others' suffering. Graphic sketches, paintings, and caricatures drew attention to suffering in ways that textual narratives arguably could not (Fehrenbach and Rodogno 2015).

To illustrate these dynamics, we examine dominant imagery associated with one pivotal historical humanitarian moment: the abolitionist movement and the push to outlaw slavery from the mid-eighteenth century. This period is commonly considered humanitarianism's founding era. Antislavery visual culture has additionally been viewed by commentators as crucial to the movement's progress and success (see Cutter 2017). Yet it can also be seen as the period when attitudes toward cruelty and the suffering began to change. Looking at this period we focus on how dominant imagery and visual frames align with the cultivation of sympathetic sentiments in viewers and, consequently, humanitarian meanings and actions. Yet in doing so we also show how the visual dynamics at play implicate a particular type of humanitarian agency: a form of humanitarianism that emerges from and replicates a hierarchical social relationship between viewer and the enslaved subjects being depicted.

Beginning in the late eighteenth century, visual culture played a central role in generating new understandings of the suffering of slaves. And it was through visuals—caricatures and cartoons, at first—that the sympathy that in part underpinned the anti-slavery movement was cultivated.

Certain recurring visual frames were particularly significant to generating such dramatically different understandings and sympathetic sentiments. Much abolitionist artwork appealed to viewers' sensibilities, foremost by depicting slaves as victims situated within a controlling, domineering white society (Wood 2000). Various stereotypes typically associated with victimhood—prominently of race and gender—were repeatedly employed. The overall effect of such victim-focused frames is clear: slaves are depicted and come to be perceived as powerless, devoid of agency even when presented in obvious positions of resistance. Slaves are predominantly presented as passive, primitive, and silent while enduring unimaginable cruelty and pain. This sense of passivity and helplessness is especially the case with female slaves, whose representation typically hinged on erotic and voyeuristic images that rendered their usually naked bodies "sexualized, submissive, and subservient" (Cutter 2017; see also Lasser 2008). Female slaves are in this way sensualized and objectified, their meanings simplified in a manner that leads toward the ready recognition of a stereotypically feminine powerlessness and vulnerability. Yet it is precisely through this mode of representing slaves' suffering—with an implied powerlessness—that the viewer's sympathy and growing sense of humanitarian need are evoked. One example of such an image is the Isaac Cruikshank caricature *The Abolition of the Slave Trade* that became well-known in the 1790s (Figure 12.2).

Cruikshank's sketch depicts a case in which a captain (Kimber) of a slave ship was accused of murdering a young slave girl who refused to dance naked on his ship's deck. In the sketch, the girl is almost completely naked, strung up to be whipped. She hangs by her foot, her face turned away from the viewer and her body arched slightly backward, her hands raised to her head. The horizon line and focal point of the picture draw the viewer to the girl's naked body, the round of her buttocks, breast, and nipple, the arch in her torso, and her outflung leg. Her torturers, fully clothed in formal attire of the day, look on; one is focused on the task of hoisting her up, and the other stands in a domineering position, leering. The positioning of the young girl's body brings instantly to mind her agony and distress. To add to her humiliation, at least five other people in the picture are witnessing. Three enslaved women in the background appear to be laughing, presumably—as might be indicative of the era in which cruelty was often a voyeuristic pleasure and delight (Dickie 2011; Noble 1997)—at the girl's modesty, humiliation, and pain (a point also made by Cutter 2017; Klarer 2005; Sliwinski 2018).

While this type of imagery was influential in raising awareness and providing a platform to discuss the cruelty and injustices endured by slaves, Cruikshank's sketch replicates a range of stereotypes and meanings common to abolitionist imagery. Key to the image's emotional and ensuing humanitarian meanings are, as mentioned earlier,

Fig. 12.2. *The Abolition of the Slave Trade: or, the inhumanity of dealers in human flesh exemplified in Capt'n Kimber's treatment of a young negro girl of 15 for her virgin modesty.* Attributed to Isaac Cruikshank, 1792.

those associated with gender and race. Imaging the flogging of a young woman's naked body in this manner presents a female slave as an almost pornographic spectacle. Through her torture, the enslaved girl is presented as a primitive yet erotic object; she is passive and helpless—her agency denied completely—in face of the power of the white perpetrator-captain. Scholars have also suggested that such imagery "romanticized" slaves' suffering and played on the voyeuristic and potentially pornographic elements of such scenes. In this imagery, the woman is shown as a graceful sufferer and hence exists as "an object to be enjoyed, as compassion is centred on her desirability" (Lydon 2016, 63; see also Favret 1998).

Stereotypical frames such as those evident in Cruikshank's caricature—most prominently the powerlessness and pitiable nature of slaves' suffering and plight—were common in highlighting a moral duty to abolitionism. These visual frames resonate with audiences' newfound and growing sense of disgust at and dismay toward cruelty and suffering and evoke a concomitant, emotionally laden sympathy and responsibility to do something to help. However, through this type of abolitionist artwork, a new emotional awareness and ensuing humanitarian duties are achieved at a cost. While influential in gaining public attention, depicting slaves as devoid of agency renders them passive and powerless in a manner that places viewers in yet another position of hierarchy and power, albeit a more ethical one. Through the sympathetic and humanitarian duties that are implicated, it is viewers—rather than victims—that have the capacity to feel, act, and respond.

Contemporary Humanitarian Photography and How Visual Legacies Help to Conceptualize Hierarchies of Conflict and Peace

The struggle over the abolition of slavery was just one historical moment, but it played an important role in the early establishment of a humanitarian consciousness. Subsequent periods and political movements cemented the emergence of a humanitarian order. These include imperialist expansion and colonialism in the nineteenth century, which was justified on humanitarian grounds but in fact precipitated cruelty and violence that is now seen as antithetical to the humanitarian project (e.g., Lester and Dussart 2014). Then there is the increasing formal institutionalization of humanitarianism in the twentieth century, manifest in a range of intergovernmental and nongovernmental organizations and also informal yet established norms and customs.

The legacies of these historical struggles continue to shape humanitarian practices today. During the abolition movement sufferers were repeatedly portrayed as voiceless, powerless to the cruelty they were subjected to. Through their abject pain, sufferers appealed to audiences emotionally through a victimhood that increasingly necessitated compassionate feelings.

Just like the images of slaves that circulated during the abolition movement, contemporary photographs tend to bring disaster into focus through the use of frames that disempower sufferers, rendering them helpless and needy victims. Humanitarian photography achieves this by depicting sufferers as vulnerable, "struck down" by tragedy, devoid of agency, and dependent. Employing these frames is powerful as it harks back to time-honored humanitarian tropes that provide audiences with easily recognized points of emotional (sympathetic) identification.

In many instances today, survivors of catastrophic events are often depicted as passive recipients rather than active participants in humanitarian aid processes. Figure 12.3, featured on the front page of the *New York Times* following the 2004 Asian tsunami, utilizes this technique.

In this image the viewer sees a blur of outstretched arms and anxious faces, communicating a sense of desperation. Focusing on the image more carefully, one is gripped by the dramatic reality of the bodies stretching and faces painfully pressed against the wire. So desperate are the men and children that they focus intently on the aid workers, even despite (or possibly ignorant of) the camera. This manifests not only a

Fig. 12.3. "At the airport in Bandah Aceh, in one of the worst battered provinces in Sumatra, Indonesia, refugees reached for relief supplies yesterday." *New York Times*, 2 January 2005.

Source: New York Times/Photograph by Abdullah Azam. (Given that the text discusses and engages this image in great detail, reproduction here occurs under the "fair dealing for criticism and review" stipulated as outlined in Berne Convention for the Protection of Literary and Artistic works, signed by 170+ countries.)

sense of their dire need but also the idea that as a consequence of the misery and desperation of their situation, victims are portrayed as pitiable objects in want of helping.

Critically for viewers, in these images victims are presented as powerless, submissive and dependent humanitarian receivers. They seem reliant upon the distribution of foreign aid rather than, for instance, helping themselves, or even helping Western aid workers to distribute aid. It also seems that while victims are in need, the Western relief workers are "saving the day"; the relief workers are the "white knights," so to speak (Orford 2003). Of course in some instances this may have been the case in relation to victims' day-to-day survival, but as a prominent front-page image it helps to mobilize the idea that victims could not help themselves. The Western relief workers are, in contrast, the active agents; their giving hands indicate assurance and control.

The passive/active dichotomy in this style of imagery is significant in that it functions to convey both the urgent situation and seeming dependence of survivors while simultaneously prompting the viewer to identify with the aid workers—the giving hands. That such imagery suggests a type of paralysis or inability to respond can be seen as instrumental to generating an understanding of victims as disempowered and consequently reliant upon overseas aid. A social relationship that is based in a colonial—as well as a cultural and racial—gaze is in this way perpetuated. Graphic photographs of distressed faces embody a sense of desperation and communicate a sense of victims' powerlessness that functions to affirm wider, historically constituted aid discourses that maintain hierarchical humanitarian relations between the Western and the developing world (see, e.g., I'Anson and Pfeifer 2013; Nair 2013).

It is through such framing that particular emotional dispositions and associated humanitarian meanings begin to emerge. Specifically, this style of humanitarian imagery, which promotes colonial understandings of developing-world others and implicitly delineates their suffering in opposition to influential Western viewers, constructs victims as pitiable objects of sympathy and encourages audiences to engage with the disaster through an emotional lens, much like during times of slavery and colonialism.

From Humanitarian Photography to Visualizing Peace

While we have investigated the specific links between photography and humanitarianism, the ensuing insights are pertinent to broader inquiries into visual politics and peace studies. There are, in particular, compelling links between humanitarian photography and what is increasingly known as peace photography.

Peace photography is a broad and fluid concept, just as is peace itself. If peace is defined as the absence of violence, then almost every photograph—and especially a seemingly "humanitarian" photograph—could be a peace photograph. What, then, is the point of peace photography? Möller (2017; also 2018a, 2018b) draws attention to this dilemma and suggests a way forward by asserting that peace photography is about photography that promotes—or should promote—the nonviolent and peaceful transformation of conflict. By opposing a deeply entrenched visual bias for violence, peace photography offers alternative ways of depicting the political. It might not always focus on grand, spectacular events but can also be a way of highlighting everyday social processes that lead to enduring, nonviolent political transformation (Möller 2017; Möller and Shim 2019). As Möller (2018b, 3) relates, "[I]n order to change the rules of [the] world, the image of the world had to be changed as well." In this sense, photographs that work for peace are about reimaging and in turn revisualizing situations of conflict. Such an approach to peace photography is aligned with more recent scholarly approaches in peace studies, which see peace not just as the absence of violence but also as a broader attempt to address questions of injustice, including injustices linked to the types of structural inequities that often lie at the heart of conflict (for a summary discussion, see, Richmond 2005; see also Barash and Webel 2002).

Humanitarian photography can contribute to these forms of peace photography. But to do so one needs to approach humanitarian photography—and indeed all images of suffering and hardship—with both historical awareness and critical reflection. One needs to see beyond the processes by which photography contributed to the emergence of a seemingly unproblematic unifying global humanitarian consensus, that is, the somewhat settled idea that both states and the international community have a responsibility to people in need. Key here is that the politics of humanitarianism—and also that of humanitarian photography—must be approached with caution. No matter how well intentioned and noble the global humanitarian order may be, it too is linked to power, exclusion, and, inevitably, violence. And, importantly, images can be complicit in these ethical dilemmas. Throughout history and into the present day, humanitarian imagery has not merely represented suffering and violence, and often in horrific ways, but in doing so it has perpetuated the structural inequalities that run counter to the nonviolent ethos crucial for conceptualizing and imaging peace.

Pointing to the problematic legacy and the undersides of humanitarian politics is not to devalue the humanitarian impulse or ensuing nonviolent political ethics and imperatives. Nor is it to dismiss the crucial role that photography and images in general played in raising awareness of violence and suffering and, as such, creating the conditions through which a turn to theorizing and working toward peace has taken place. It is, above all, to stress that if the history of visualizing violence is to contribute more positively to broader peace processes, then one needs to investigate—and transform—the power relationships that are involved in depicting human suffering. It is in this vein that Fehrenbach (2017) similarly contends that peacebuilding requires the establishment of a political order that is "attuned to the truth of past abuses" and also acknowledges them publicly. Doing so is not to eschew the depictions of suffering or conflict or even of violence that contribute to practices of peace. It is instead to suggest that when moving forward, scholars and practitioners need to be aware of the problematic legacies that underpin and have shaped the turn to peace photography.

There are numerous realms in which peace photography and visual art in general have made significant contributions to reconciliation, statebuilding, and peace formation. We do not attempt to offer anything resembling a comprehensive overview here. But consider a few samples. Photographs, for instance, can not only influence how a state represents itself but also form parts of resistance strategies that highlight violence committed in the past and open up new ways of finding more just and peaceful political relations in the future. Manifestations of this visual political dynamic can be seen in the South Africa Pavilion at the Venice Art Biennale (Garnsey 2019) and in Australia's use of indigenous art for cultural diplomacy initiatives (Bleiker and Butler 2016, 65–74). Or consider how media images of traumatic events, such as the 2004 Southeast Asian tsunami, contributed to the emergence significant aid and transnational solidarity, even a form of "affective community" (Hutchison 2014; 2016, 157–82). We can also see how photography and other forms of visual art serve as moral witnessing, documenting past crimes and opening up ways of imagining a better world. Danchev (2009, 2016) has illustrated this process in a range of contexts, from coming to terms with the Holocaust to offering new perspectives on terrorism and war. It is in this sense as well that museum exhibitions play important roles in shaping public opinion on issues that range from reconciliation to peace- and statebuilding (Luke 2002; Sylvester 2009).

Conclusion

The purpose of this chapter has been to examine the role that visuality plays in conceptualizing conflict and in working toward peace. Presenting a comprehensive overview in a short chapter is impossible, which is why we have focused on the historical emergence of images of hardship and suffering and how they have influenced the imaging of conflict and peace.

We have illustrated the key issues at stake by examining how images have shaped what can be seen as one of the first modern peace movements: the emergence of humanitarian ideals and practices. To do so, we highlighted the need to look back in time to appreciate how visual images represented suffering and, in doing so, played a key role in generating a range of compassionate emotions that were central to mobilizing humanitarian sentiments.

We have also consequently sought to show that humanitarian photography can be seen as part of broader and ultimately still emerging practices of peace photography. We have outlined how these practices can influence peace and reconciliation processes in numerous different contexts. So what, then, does our small case study tell us about larger links between visuality, conflict, and peace? Peace photography has to be more than just visualizing peaceful situations. What is equally important is the need to visualize and image conflict and postconflict contexts in critical and transformative ways: in ways that enable scholars, practitioners, and affected communities to learn from past injustices and violence in order to conceptualize and initiate political resistance and action in nonviolent and more just political ways. To do so, one needs to explicitly engage the power dynamics, as well as the colonial legacies, that are implicit and yet often unacknowledged in images of suffering and violence. One needs to engage the politics of depicting pain and suffering in a politically reflective way, with sensitivity and debate, and crucially with dialogue and engagement with those being visualized. Attempts to diffuse the power dynamics implicit in humanitarian photography must also start with greater acknowledgment of its existence—and its many guises throughout the politics of disaster assistance, development, medical humanitarianism, and even human rights. Genuine political transformation and sustainable peace can emerge only out of such critical and self-reflective processes. They would also need to entail developing photographic practices that consciously break with colonial pasts and visualize conflict and suffering in ways that avoid stereotypes and genuinely empower postconflict communities to work toward inclusive and sustainable forms of peace.

References

Baer, U. 2001. *Spectral Evidence: The Photography of Trauma.* Cambridge, MA: MIT Press.

Barash, D. P., and C. P. Webel. 2002. *Peace and Conflict Studies.* Thousand Oaks, CA: Sage.

Bleiker, R. 2001. "The Aesthetic Turn in International Political Theory." *Millennium: Journal of International Studies* 30 (3):509–533.

Bleiker, R. 2009. *Aesthetics and World Politics.* Houndmills, UK: Palgrave Macmillan.

Bleiker, R. 2017. "In Search of Thinking Space: Reflections on the Aesthetic Turn in International Political Theory." *Millennium: Journal of International Studies* 45 (2):258–264.

Bleiker, R. 2018a. "Aesthetic Turn in International Relations." In *Oxford Bibliographies in International Relations*, edited by P. James. New York: Oxford University Press. doi: 10.1093/obo/9780199743292-0236.

Bleiker, R. 2018b. "Mapping Visual Global Politics." In *Visual Global Politics*, edited by R. Bleiker, 1–29. London: Routledge.

Bleiker, R., and S. Butler. 2016. "Radical Dreaming: Indigenous Art and Cultural Diplomacy." *International Political Sociology* 10 (1):56–74.

Bleiker, R., E. Hutchison, and D. Campbell. 2014. "Imaging Catastrophe: The Politics of Representing Humanitarian Crises." In *Negotiating Relief: The Dialectics of Humanitarian Space*, edited by M. Acuto, 47–60. Oxford: Oxford University Press.

Callahan, W. A. 2015. "The Visual Turn in IR: Documentary Filmmaking as a Critical Method." *Millennium* 43 (3):891–910.

Cutter, M. J. 2017. *The Illustrated Slave: Empathy, Graphic Narrative, and the Visual Culture of the Transatlantic Abolition Movement, 1800–1852.* Athens: University of Georgia Press.

Danchev, A. 2009. *On Art and War and Terror.* Edinburgh: Edinburgh University Press.

Danchev, A. 2016. *On Good and Evil and the Grey Zone.* Edinburgh: Edinburgh University Press.

Dickie, S. 2011. *Cruelty and Laughter: Forgotten Comic Literature and the Unsentimental Eighteenth Century.* Chicago: University of Chicago Press.

Favero, P. 2014. "Learning to Look beyond the Frame: Reflections on the Changing Meaning of Images in the Age of Digital Media Practices." *Visual Studies* 29 (2):166–179.

Favret, M. A. 1998. "Flogging: The Anti-Slavery Movement Writes Pornography." In *Romanticism and Gender*, edited by A. Janowitz, 19–43. Cambridge, UK: Brewer.

Fehrenbach, H. 2017. "Photography, Civilians and the Polemics of Peace." Unpublished manuscript.

Fehrenbach, H., and D. Rodogno. 2015. "Introduction: The Morality of Sight." In *Humanitarian Photography: A History*, edited by H. Fehrenbach and D. Rodogno, 1–21. Cambridge: Cambridge University Press.

Foucault, M. 1984. "Nietzsche, Genealogy, History." In *The Foucault Reader*, edited by P. Rabinow, 76–100. New York: Pantheon Books.

Garnsey, E. 2019. "On Representation(s): Art, Violence and the Political Imaginary of South Africa." *Critical Review of International Social and Political Philosophy* 22 (5):598–617.

Halttunen, K. 1995. "Humanitarianism and the Pornography of Pain in Anglo-American Culture." *American Historical Review* 100 (2):303–334.

Hariman, Robert and John Louis Luciates. 2007. *No Caption Needed: Iconic Photographs, Public Culture, and Liberal Democracy.* Chicago: University of Chicago Press.

Hutchison, E. 2014. "A Global Politics of Pity: Disaster Imagery and the Emotional Construction of Solidarity after the 2014 Asian Tsunami." *International Political Sociology* 8 (1):1–19.

Hutchison, E. 2016. *Affective Communities in World Politics: Collective Emotions after Trauma.* Cambridge: Cambridge University Press.

Hutchison, E. 2019. "Humanitarian Emotions through History: Imaging Suffering and Performing Aid." In *Emotional Bodies: Studies on the Historical Performativity of Emotions*, edited by D. Moruno and B. Pichel, 219–241. Champaign: University of Illinois Press.

Hutchison, E., and R. Bleiker. 2015. "Art, Aesthetics and Emotionality." In *Gender Matters in Global Politics*, edited by L. J. Shepherd, 349–360. London: Routledge.

I'Anson, C., and G. Pfeifer. 2013. "A Critique of Humanitarian Reason: Agency, Power, and Privilege." *Journal of Global Ethics* 9 (1):49–63.
Klarer, M. 2005. "Humanitarian Pornography." *New Literary History* 36 (4):559–587.
Lasser, C. 2008. "Voyeuristic Abolitionism: Sex, Gender, and the Transformation of Antislavery Rhetoric." *Journal of the Early Republic* 28 (1):83–111.
Lester, A., and F. Dussart. 2014. *Colonization and the Origins of Humanitarian Governance: Protecting Aborigines across the Nineteenth Century British Empire.* Cambridge: Cambridge University Press.
Lydon, J. 2016. *Photography, Humanitarianism, Empire.* London: Bloomsbury.
Luke, T. W. 2002. *Museum Politics: Power Plays at the Exhibition.* Minneapolis: University of Minnesota Press.
Mitchell, W. J. T. 1986. *Iconology: Image, Text, Ideology.* Chicago: University of Chicago Press.
Möller, F. 2017. "From Aftermath to Peace: Reflections on the Photography of Violence." *Global Society* 31 (3):315–335.
Möller, F. 2018a. "Peace." In *Visual Global Politics*, edited by R. Bleiker, 220–223. London: Routledge.
Möller, F. 2018b. *Peace Photography.* London: Palgrave Macmillan.
Möller, F., and D. Shim. 2019. "Visions of Peace in International Relations." *International Studies Perspectives* 20 (3):246–264
Nair, S. 2013. "Governance, Representation and International Aid." *Third World Quarterly* 34 (4):630–652.
Noble, M. 1997. "The Ecstasies of Sentimental Wounding in *Uncle Tom's Cabin*." *Yale Journal of Criticism* 10 (2):295–320.
Orford, A. 2003. *Reading Humanitarian Intervention: Human Rights and the Use of Force in International Law.* Cambridge: Cambridge University Press.
Rancière, Jacques. 2004. *The Politics of Aesthetics: The Distribution of the Sensible.* Translated by Gabriel Rockhill. London: Continuum.
Richmond, O. 2005. *The Transformation of Peace.* New York: Palgrave.
Richmond, O. 2019. "Human Rights and the Development of a Twenty-First Century Peace Architecture: Unintended Consequences." *Australian Journal of International Affairs* 73 (1):45–63.
Sliwinski, S. 2018. "Human Rights." In *Visual Global Politics*, edited by R. Bleiker, 169–175. London: Routledge.
Sontag, S. 1975. "Fascinating Fascism." *New York Review of Books*, 6 February. http://www.nybooks.com/articles/1975/02/06/fascinating-fascism/.
Sontag, S. 2003 *Regarding the Pain of Others.* New York: Farrar, Straus and Giroux.
Sylvester, Christine. 2009. *Art/Museums: International Relations Where We Least Expect It.* Boulder, CO: Paradigm.
Wilson, R. A., and R. D. Brown. 2009. "Introduction." In *Humanitarianism and Suffering: The Mobilization of Empathy*, edited by R. A. Wilson and R. D. Brown, 1–28. Cambridge: Cambridge University Press.
Wood, M. 2000. *Blind Memory: Visual Representations of Slavery in English and America 1780–1865.* London: Manchester University Press.

CHAPTER 13

PEACE IN NON-WESTERN THEORY

NECATI POLAT

THAT contemporary peace thinking should include also the broadly unfamiliar visions of peace away from the center, indigenous or somehow extraneous, is now a constant of peace research, with the presence, if not the easy reach, of such forms of knowledge or savoir faire out there often taken for granted. There appear to be various reasons for this presumed need in theory. The prevalent conceptualizations of peace, dictating mainstream research and guiding action on terrain, are insular in some sense. To be sure, no longer conducted in limited loci, that is, in Western Europe and North America, and via limited concerns, as it once used to be, current research nevertheless continues to lean on a recognized intellectual heritage and is as such less than fully reflective of the actual diversity in global reality. This in turn may be an issue that is at once moral and epistemological, for its apparent ethnocentrism and for presenting a false sense of universality that research naturally assumes. For more pragmatic reasons, visions of peace not in the hub may be required for greater resourcefulness in peace thinking when it is about or operating in the periphery. Indeed, quite paradoxically, the bulk of the material of peace research is formed by cases situated in the political margins in question. It may therefore make sense, particularly from the established "civilizational" slant that insists on a heavily fragmented world inherited from the past, to deduce that the native sensibilities in such localities may not always tally with those behind the postulations of peace that in effect dictate mainstream research, consequently risking the wisdom and viability of the outputs. Last but by no means least, considering that the established notions of peace are far from efficient in practice, as often admitted, a scholarly desire to remedy their probable flaws in places, or improve them generally, may be part of the expectations. Accordingly, the standard thinking can be enriched and strengthened overall by being more inclusive, not only in its application to what would be the perimeters of the political nexus but also in its altogether functioning, regardless of the context.

This chapter looks critically into this strain in peace thinking, problematizing the premise of a radical distinction between the West and the non-West in cultural terms, and sketches some of the characteristic oversimplifications, or reductionisms, around the theme in peace research.

What Exactly Is the West?

Peace in non-Western theory should perhaps be understood as tropes of peace that are somewhat outlying and relatively obscure, often with a vague promise of empowering the underprivileged who subsist within the global political rim, rather than vistas of peace that are *literally* alien to the governing epistemology in the center, issuing from sources that are conceivably in some radical opposition to it. "The West" as a cultural term may undoubtedly have some validity in strictly defined contexts, referring to the geographies and the peoples that appear to have been in the thick of the onset of modernity in Europe historically. Colonialism formed one of the sources of energy in the process, with overseas possessions of the European powers effectively constituting part of the fuel for the production of cosmopolitanism, for the making of the West, out of the earlier, provincial Europe. In the face of the apparent actors of modernity, that is, there have also been sites and subjects in the process that were in some sense merely *acted on*. It is dubious, however, that the overall set of acts under increasingly global modernity left ultimately intact the traditions on either front, the European or the non-European. More, the domains of respective cultures even before this momentous encounter, as far back as antiquity, were not necessarily mutually exclusive. Later, modernity would spawn wide-reaching and profound effects on cultural formations on the face of the globe that can hardly be explained now through a clear-cut dichotomy of those acted on, the non-West, and the actors, the West.

Therefore, reducing a highly intricate cultivation that would eventually become the West, conflictual in itself through and through, to a fixed set of cultural registers, with virtually no beginning, may be more problematic than helpful. The hallmark intellectual attributes that could possibly be located in modes of thought and practice purportedly endemic to what is deemed today to be the West are in truth relatively recent, not traceable further back than modern times, more than a millennium after Christianity. Modernity as such is none other than a gradually thickening amalgam of a number of decisive factors, chiefly capitalism, and, closely tied to it, the modern state, the Christian Reformation, the Renaissance, modern science, and modern philosophy. In the midst of virtually all of these strands seems to be a sharp differentiation between the subject and the object, the autonomous human person and the world exterior. To be clearly defined in thought not before the seventeenth and the eighteenth centuries, the independent and self-sufficient individual, that is, the subject, appears to run through virtually all of these distinct facets of modernity: the enterprising capital owner and the contract-making laborer; the autonomous individual of liberal political theory; the sovereign state vaguely modeled on the latter; the believer free of ecclesiastical mediation in worship; the creative artist; the scientist detached from the world of facts and thus capable of full objectivity in manufacturing knowledge; and, in philosophy, the Cartesian ego above the rest of existence and experience, encapsulated in pure cognition and cogitation. This process would unfold in history roughly from the thirteenth and fourteenth centuries, with capital accumulation preceding plausibly all else, enabled by novel historical incidents such as hired labor, commodity production for mass consumption, and, later, a system of financial savings transformed into investment and credit.

The West as a cultural term makes sense arguably only with this unique set of configurations in the background. An instant test of this specific denotation for "the West" is to try to use the designation for Europe *before* modernity. The term is likely to cease to signify in its strong cultural sense. Yet somehow the same West is at once assumed to be innately linked to a number of characteristics that have been perennial, stretching further back in time, presumably to ancient Greek thought and to early Christian ethics. Such retrospectively discovered "origins" of the West presumably serve, in a metaphysics that ensures purity and continuity, no more than ideologically to sanction a concrete result that altogether appears to have been the outcome of various accidents that occurred only later, long after the ancient Greeks and the early Christians. A factor behind much under the unraveling modernity, capitalism, for instance, may have had something to do also with the mercantilism and even perhaps "proto-capitalism" of the medieval Muslims, responsible for key trade practices early on and, in particular, for the widening of global trade networks, as suggested by Braudel (1995, 71; 1982, 558–559). The fact that European terms of transterritorial trade such as douane and tariff are loan words from Arabic (the former, "register", being originally Persian, *dīwān*, "office/court," and the latter *taʿrīfa*, "rates") should in itself speak volumes. Long in use in the established trade cultures in Asia and in the Levant, Europeans actually began using bills of exchange that enabled payment in the absence of face-to-face trade, and that prolonged credit when needed, not before the thirteenth century (Braudel 1981, 472). In return, the classical antiquity to which the West is habitually traced is known to have had considerable influence on Muslims (Rosenthal 1975) from as early as the eighth century. Combined with what Muslims borrowed from other sources, especially from India, this process would create a remarkable synthesis and lead in time to some Muslim sway on learning in medieval Europe, from philosophy to medicine, chemistry, geography, astronomy, and mathematics, well into modern times (on this, see Hobson 2009).

In the face of such history, the immaculateness and self-sufficiency attributed to the West, in an uninterrupted flow from the past to the present, carrying in linear continuity some fixed, unadulterated substances in thought and in practice, appear therefore to assume (1) a questionable notion of cultural transmission in time that is highly counterfactual. Practically lost from sight under this assumption are two historical features that are crucial in understanding both the West and the non-West: (2) the actual hybridities, rather than closed structures, that seem to have defined almost all value settings *before* modernity, around the Mediterranean in particular; and, more significant perhaps, (3) the key function of modernity in its *later* global hegemony, recasting much of what has been out there, however named or classified, in its own redoubtable image.

The famed "twain" in Kipling's (1989, 233) ballad "East and West," which seemed, in the late nineteenth century, to have a rather insurmountable gap between them, may in truth have been in rather close contact in the ancient world, as opposed to presenting extremities in outlook, thinking, and conduct. The ancient asceticism from Greece to India had converging strains that would define much of the later monotheistic piety, particularly Hellenic Judaism and Christianity. To come virtually to embody the imagined antipode of the West by the late modern times, Islam at its dawn in seventh-century Arabia was also in marked kinship with the same asceticism and would flourish by sharing common sources and a common milieu with Judaism and Christianity. The early Muslim interest in politics beyond the immediate horizon should perhaps be considered instructive. One of the chapters in the Qurʾān is called Romans (*al-Rūm*), referring to the Byzantines, whom the earliest Muslims around the Prophet ostensibly supported in their extended wars against the Sasanid Persians (Qurʾān 30:2–6). The adherents of the new faith in far-flung Arabia seem to have displayed rather strong passions rooting for the Byzantines (Wāḥidī 1991, 354). Among other highly revealing signs, the Qurʾān appears to have incorporated the Alexandrian romances of antiquity (Tesei 2011) in the Qurʾānic (18:83–98) figure of Dhū al-Qarnayn (literally "the one with two horns"), who, recognized as possibly a prophet of Allah, is none other than Alexander the Great, according to most Muslim exegetes. Aristotle, "the famous philosopher," had been his "vizier," noted the famous Qurʾānic commentator Ibn Kathīr (1999, 189–190), who went on to explain that Alexander may have been called "the one with two horns" for having triumphed over "the two horns of the sun, the east and the west."

As is well known, Muslim philosophers drew heavily on ancient Greek thought, even in theological matters, and some Muslim theologians considered Greek sagacity as part of the same monotheistic tradition as Islam, instructed through direct revelations from Allah, though likely to have later been corrupted in places. Ibn Khaldūn (2004, 250) had no doubt that another Qurʾānic figure, Luqmān, after whom, again, a chapter is named, had been a Greek sage, whose wisdom, he thought, had been transmitted through Socrates (Buqrāṭ al-Dann, a rather common medieval Arabic composite of Socrates, Hippocrates, and Diogenes), Plato, Aristotle, and others, down to the Stoics. The seminal Muslim historian of religion Shahrastānī (1992, 379, 288) identified Empedocles as a pupil of David, a Qurʾānic prophet; Hermes as Idrīs, also a Qurʾānic prophet; and Hermes's teacher Agathodaemon as Shīth, a prophet in the Muslim tradition, although not in the Qurʾān. The exact sense and "accuracy" of these references are obviously immaterial. The fact is that Islam culturally emerged out of a common pool of Mediterranean symbolisms and, as such, hardly formed a typical instance of the alien other to Greek antiquity and Christianity. Yet the same Islam would in time, under modernity, occupy considerable space in everyday conceptualizations of the non-West.

Modernity, which followed this rough framework that refuted clear demarcation lines between the civilizations of the Mediterranean, would remain even less isolated and exclusive. The multiple threads modernity would spawn, virtually in all geographies and cultures, would mix modern cosmopolitan notions with native inputs. Inevitably diverse as a result, these modernities in a wide ambit would nevertheless submit to the one image most readily recognizable in the European modernity. Even avowed gestures of purported antimodernity, of non-Western nativism, as in Islamism from the late nineteenth century, would be cast strictly in basic modern templates. Authenticity by now would be what modernity defined as authenticity. At a more formal level, multiple modernities, with various fine-tunings locally, might still enable subgroupings, highlighting possible chasms, in transnational topologies. It is dubious, however, that radically distinct cultural formations to keep in check forms of modernity, especially European modernity, presenting genuine alternatives, would be part of the big picture. Divergences among forms of modernity globally might at times be confusingly salient, yet such variations would barely obliterate the overall homology under modernity. Let alone diversities of the type, dissimilitudes of a wide variety could be observed also *within* specific filaments of modernity, such as European modernity, otherwise viewed as monolithic. Consider current distinct notions of privacy, of personal space, and modes of communication in interpersonal relations such as physical touch, in European societies, or the practice of tipping following services bought. Such differences under modernity are more ways of *being* perhaps than inert and static modes of *thought*, ontology rather than epistemology, which can barely be accommodated within a radical binary of the West and the non-West.

The assertion that the economic development of several East Asian states—Hong Kong, Singapore, South Korea, and Taiwan—with sustained rapid growth from the 1960s had to do with some "Asian values" centered on Confucianism seems in this regard to have been more jejune scholarship, avoiding the study of much more complicated structures in the region, than about a form of genuine cultural authenticity in Asia in uninterrupted flow, guiding action. Recent ethnographic studies of some of the ostensibly less violent communities (Fabbro 1978; Bonta 1996; Ross 1996; Kemp and Fry 2004) point overall to such intricate structures in a host of otherwise distant cultural settings, which appear to foster procedures of mutual compromise, patience, and tolerance in situations likely to induce conflict, and promote nonviolent settlements of disagreements. Such techniques of nonviolent interaction seem to be practical first and foremost, sustained by routines and conventions that are instinctively acquired by the natives rather than cultivated through what would be forms of cultural authenticity outside modernity.

Does the Non-West Exist?

Ironically, a sharp distinction of the West and what historically seemed to be "outside" of it would in time acquire considerable purchase value also in settings deemed to be the non-West. Produced mostly in modern Orientalist scholarship that ultimately patronized modes of being outside Europe, the dichotomy would nevertheless be welcomed by some intellectuals of the purported non-West who would find it flattering. Relished would be the suggestion that realms of culture somehow outside the hegemonic West possessed unique traits, possibly more scrupulous, given the various forms of discontent under the cultural influence of the West. The following is a description of what Rabindranath Tagore apparently thought of the dichotomy: "The West would subdue Nature. The East would seek unity with Nature. The West sees a break between the world of things and the world of man. The East sees kinship and continuity. The scientific man of the West sees the interaction of natural forces. The Eastern seer finds an eternal will working and manifesting itself in these forces. For the one, the goal is conquest. For the other, it is the realization of the infinite" (Fenn 1929, 318). It is hardly surprising that the ideas attributed to the East in this passage are in truth strictly European, products of the cycles of European romanticism historically. That is, the fundamental binary Tagore is said to have embraced, tightly homogenizing both the West and the East, respectively, is none other than a mixed, vaguely lamenting sentiment in European scholarship itself and nothing to do with the East as such. The East is used in this highly familiar imagination simply to fill the part of the noble savage. The part would later be handed over through popularizations of this imagination to a host of other civilizational misfits, including Native Americans—all assumed to have remained, unlike the Europeans, at a stage of infantile naïveté and innocence in cultural development. A more recent production of cultural authenticity as such seems to have been carried out in relation to the Australian Aboriginals, who, bemused to a point, are said to have nevertheless taken it up (Shilliam 2011, 16). The binary has been received enthusiastically in settings outside Europe and North America not only for what was understood to be its oblique privileging of "deviant" others, who appear to have lost much else to their powerful admirers, as in the case of Tagore perhaps, but also for readily justifying a historical grudge. For instance, Islamists worldwide would find much convenience in a formidable polarity of Islam and the West, two intellectual traditions and life forms presumed to be in perpetual conflict.

Here, then, is the sad oxymoron: since power is consolidated through the production of knowledge, by the same token, what seems to be available as knowledge—the knowledge of purportedly authentic visions or practices of peace, in this case—should be expected to have already been cast in the image of power. Such knowledge made accessible in native settings, and only parading as authenticity, is likely to be equally firmly aligned with power. Retrieving authenticities imagined to be out there, and genuinely in excess of the West, may have to be considered, that is, as no more than an extended play of power, seeking to co-opt the narratives of authenticity—as in the active integration of native cultural sketches to European modernity by native literati from the nineteenth century, practically in all peripheral settings: India, China, Japan, and the Islamicate world. Such apologies by the native literati have often followed a discernible pattern, (1) advocating originality, complexity, and dignity for native routines, yet only by, and paradoxically, (2) treating such routines as no more than mere versions, often earlier or original, of those under the hegemonic model. Illustrations of this gesture, incessantly reading modern notions into the native tradition that show the tradition either in full harmony with the model or, as is not infrequently the case, *ahead* of the model in the same direction, are provided next.

The Indian poet and scholar A. K. Ramanujan (1989) offers a rather ingenious justification of this compulsive move by the natives, bestowing a new form on this all too pervasive native apology. "No Indian text comes without a context, a frame," he observes, "till the 19th century" (48). Modernity, he then argues, could be understood as roughly a historic move in the production of knowledge from the immediate context, which parochializes thought, to a sphere of context independence, which is the domain of rationality and universalism. Left intact and assumed tacitly, as with the rest of the apologies, is the basic continuity between the indigenous wisdom (when abstracted from its context) and the hegemonic, "universal" knowledge. The native co-optation effectively forces into that which is *native* themes from a well-defined canon that is otherwise assumed to be *extrinsic* to it, namely European modernity.

Typical of such obsessive drives by the native literati toward serving power in co-opting the local is the modern Muslim view of Islamic peace. Unlike the concept prevalent in "the West," the Islamic view of peace is said to be "positive." Signifying, that is, more than a mere absence of violence ("negative" peace), the Islamic view is thus reportedly alert to "various forms of aggression including individual, institutional and structural violence" (Kalin 2005, 333.) In one modern apology, Confucianism seems to boast a similar compass, over and above that in "the West," through the "weak" and "strong" senses of the native peace, *p'ing* (Yu 2014, 67). In the weak sense, peace refers to the mere cessation of hostilities, negative peace, as that of "the West," while the strong sense, "true peace," is positive and hints at a requisite structure of mutual toleration, harmony, probity, and equitability in context. More or less the same notion of positive peace is said to be sanctioned in the Confucian-Buddhist term *wa*, which is integral to Japanese peace, *heiwa* (Chiba 2014, 85–86). The notions (negative vs. positive) through which local references to peace seem to be privileged in such examples are obviously already part of a "Western" canon, although established only recently. According to this newly formed canon, the prevalent understanding of peace in "the West," traceable to either Romans (Galtung 1981, 194) or the early Christians (Salem 1993, 362), is a negative concept altogether. It treats peace as the mere absence of violence and in so doing effectively ignores the purely systemic violence that may go on practically unabated in times of formal peace. The canon has in turn rendered current peace thinking highly receptive to views of peace in major non-European traditions as likely in excess of the reigning negative peace, with "positive" connotations. That is, the cue for a search for ideas purportedly outside the canon is provided in the canon itself. Meanwhile, why "Western" peace—as *eirēnē* and *pax* historically, and, say, peace, *paix*, *Frieden* presently—would have to be confined solely to the *negative* sense is of course barely clear. European history, in which peace in this sense is said to have dominated, seems to be the only excuse. It is open to question that the evidence of history is, by contrast, for a *positive* sense of peace concerning the actual Muslim, Hindu, Chinese, and Japanese practices of the past.

As is the wont of current peace thinking on peripheral notions of peace, it appears to be more than possible to secure a firm grounding through etymology for the view of Islamic peace offered in modern Muslim apologies. The root of the very term "Islam" appears to refer specifically to peace (*silm*, *salām*, cognates of the Hebrew *shalom*). At once, the term seems to connote the close and complementing notions of submission, health, welfare, and security (Iṣfahānī 2009, 421–424), besides peace as settlement of violent conflicts. Finally, peace in this sense may have to be considered as clearly and outstandingly valued in the Islamic tradition, judging from the fact that peace has been suggested to be one of Allah's names (*al-salām*) (422). In short, Islam is quite literally a "religion of peace," as devout Muslim literati are fond of stressing.

This widespread Muslim take on Islamic peace, supported by etymology, would have of course made scant sense for much of Muslim piety for at least a millennium after the dawn of Islam, not to mention the early Muslims in the seventh century. Yet contemporary Muslims could perhaps explain away this discrepancy, rather effortlessly too, by merely recalling that Islam is revealed, transcendent knowledge. Timeless, it is to be interpreted—against historical evidence, if necessary. It should be construed under evolving circumstances and is naturally applicable free of, and uncontaminated by, actual past uses, which may appear to be discordant. That is, a set of current, modern values that may have been alien to earlier Muslims may very well function now as effective prompters in attributing meaning to such Islamic notions as peace. Besides interpretation, free of earlier practices, some vague empirical support for this view of Islamic peace could still be found perhaps in the historical Sufi notions of amity, harmony, and tolerance, celebrating rather than negating difference. The fact remains, however, that these Sufi notions may have been reflective more of the wider pre-Islamic ascetic piety, from Greece to India, than Islam.

In Japan local apologists seem to have taken a similar path. The native literati critical of the wholesale and slavishly mimetic modernization locally, especially in the stretch immediately following the Meiji period (1868–1912), would tend to reread local wisdom with European modernity strictly in the background, thereby giving native notions some contemporary relevance. A case in point is the Japanese philosopher Watsuji Tetsurō (1971) on *ningen*, the Japanese term for "man" or "human being," with concerns that largely iterate those of modern Muslim apologists. According to Tetsurō, the concept of human person under Confucianism did not necessarily ignore the European social particularism held in high esteem by latter-day Japanese modernizers and was as such far from endorsing solely order and hierarchy. Rather, the notion hinted at a more balanced synthesis between individualism *and* social integrity, which served a critical function, holding the essentially selfish, and thus perilous, individualism under control—a dynamic that is gapingly absent, Tetsurō argued, in European modernity.

Counterintuitive as it may sound, the purportedly more authentic notions of Islamic peace, closer to historical practice, may have to be viewed as no less modern. According to the fringe Salafi Islamic view, for instance, modern apologetic accounts have taken Islamic peace out of context and sugarcoated it with novel values, effectively distorting it, with the apparent aim of making it more palatable for the modern taste. "Positive" peace, said by apologists to be endorsed in formal sources of Islam, is mostly, if not exclusively, peace in the sense of personal resignation and submission to Allah, confined to the domain of the heart. In other contexts, beyond human repose, peace also meant to early Muslims the order and subjection of nature to the divine will, bringing about a unique, universal harmony. Otherwise, early Muslims believed historically, the Salafi view continues, that they inherited no less than a perpetual conflict on earth between truth (*ḥaqq*) and falsehood (*bāṭil*), which would go on relentlessly, until the unconditional victory of truth over falsehood. Indeed, starting with the formative period of the faith, Muslims discontinued war (*dār al-ḥarb*) with non-Muslims only in the case of the total surrender of the latter, submitting (*dhimma*) to Muslim rule (*dār al-islām*). More precisely, this was a privilege of those non-Muslims who possessed scriptural traditions (*ahl al-kitāb*), in return for the payment of a special tax (*jizya*) and for contentment under Muslim rule with a generally inferior legal status. The rest of the non-Muslims had to choose between conversion to Islam and death. Peace in the sense of coexistence and mutual toleration with non-Muslims (*dār al-ʿahd, dār al-ṣulḥ*) would become the case only in the event of a sort of "power balance" transiently formed in the relations with non-Muslim polities.

It is less than certain, however, that this avowedly "historical" notion of Islamic peace, or lack thereof, held nowadays almost exclusively by Islamists and Islamophobes, is actually more authentic than the more pliant, modern view sketched earlier. A number of concepts read into it somewhat evasively, such as the "unprovoked aggression" practiced by early Muslims, "religious intolerance" sanctioned, and "civic inequality" exercised, are clearly anachronistic, alien to the early Muslims—as is the more blatantly out of place view of Islamic peace in modern Muslim apologies.

To be brutally obvious in the matter, take the earlier wisdom in scholarship about Mayan practices in Mesoamerica, which would turn out to be unfounded. According to the earlier account, unlike the Aztecs, the Mayans did not practice human sacrifice and were less prone to conflict (Raaflaub 2016, 18; Hassig 2007, 312–328). Ignoring later evidence, and assuming that the Mayans had indeed been starkly different from the Aztecs in those regards, it might still be difficult to draw conclusions of a more peaceful Mayan culture. Attributing to the Aztecs the modern traits of cruelty, violence, and aggression, when the Aztecs themselves had no understanding of those, is without doubt problematic. In good acting, they say, an actor who plays the part of a highly questionable character, to be instantly recognized and abhorred as such by viewers, is nevertheless expected to *believe* in the character that barely saw himself or herself in the same light. Cultural commentary would do well perhaps to be heedful of this rule and seek to reflect the object of analysis, such as the Aztec practices, unacceptable as it is in this day and age, with some allowances, with some credence to be lent to what would have been the fuzzy authentic meaning of the object in its now confusingly distant setting.

That which is purported to be *outside* the paradigm—an alien other conceptually, and as such incommensurate with modern sensibilities—is in effect what has already, and inevitably, been rendered paradigmatic (Kuhn 1970). Operative notions of peace back in history, or in some distant setting, such as Islamic peace, were answers not to our present-day questions but to questions posed under highly unfamiliar conditions. The assumption that such questions should be perennial, that people of all ages and places must have asked basically the same questions, seeking and rehearsing answers for all ages and places, is a pervasive fallacy, famously demonstrated by Kuhn in relation to the history of science. Not surprisingly, the tension-ridden character of the enterprise is often betrayed in modern takeovers of remote and obscure notions in the name of unearthing authenticity. One example is the modern appropriation of the historical Buddhist concept of *dhammavijaya*, the overpowering of an adversary through nonviolent, incorporeal force, through *dharma*, as opposed to physical combat, *digvijaya* (Roy 2014, 52). This concept, which was apparently part of handling conflicts in India from the third century BCE, is said to be of particular significance in fully appreciating Mahatma Gandhi's pacifism in the twentieth century. Recast into modern peace thinking, this ancient concept naturally assumes a new context, with attendant elements that were crucially absent in the tradition. The *disparity* of power between the disputants, the British Raj and the natives, for instance, was a major feature in Gandhi's pacifism. This fact in turn had some relevance to the "choice" of resorting to nonviolent force. Again, it is dubious that Indian civil disobedience would have made sense, let alone succeeded, in the absence of a number of novel factors, starting with modern European sensibilities, conceptually open to the natives staging civil disobedience, and a global public sphere that avidly reflected and responded to the whole event. Moreover, removed from the original context, such notions are likely to yield incongruities of an altogether different kind that are to do with modernity. Achieving results through nonviolent pressure has an obvious moral appeal for relinquishing physical coercion. Yet, under different circumstances, the same may be construed as rather a stealthy form of violence, possibly more severe than the merely physical one. The more nuanced present-day assessments of violence may indeed consider noncoercive orderings or administrations of conflict, as in the case of Foucauldian biopower (Foucault 2007, 1–4) or, more simply, parenting through emotional blackmail (Žižek 1999), as *more* sinister forms of subjugation.

Conclusions

None of these notions touted as non-Western are genuinely non-Western of course, forming new and authentic alternatives. Arguably, the reintroduced native terms do little more than use the moral authority of the periphery to speak for visions that are very much out in the open, if not always in the center. Some other native reflexes that, by contrast, have left local heritage aside as "useless" may therefore have done an unwitting service to that heritage. In China, for instance, embarrassed by Confucianism as a "symbol of conservatism and backwardness" (Qin 2010, 38), some of the literati chose simply to ignore it. Against "the old and evil practices" (Fukuzawa quoted in Takenori 2008, xv) that had flourished around Buddhism and Confucianism in Japan, the philosopher Yukichi Fukuzawa advocated strong skepticism of local cultural leanings, setting them squarely against rationality, individual autonomy, freedom, equality, and pluralism, tools that he thought would help Japan on the way to "civilization." Fukuzawa (2008, 17–18) spoke of *the* civilization, the one that was universal and necessary, which occurred in stages: primitive, semideveloped (where Japan was at the time), and full. In his fascinating autobiography, Fukuzawa (1934, 229–231, 259–260) details his crusade as an educator during the early Meiji era, fighting the "degenerate," "retrogressive" Chinese influence. He strove to teach his pupils not only the European "laws of 'number and reason'" but at once the "fool[ishness]," "in this enlightened age," of some of the local customs.

The likes of Fukuzawa may have done the native tradition an unwitting service by leaving it alone, but the cultural reductionism of the sort that they have at once displayed, in this case confining rationality to European thinking from the ancient Greeks onward, is arguably as grave as Tetsurō's reading of the tradition along modern lines. Premodernity or periphery pushed into the domain of esoteric unreason, however attractive this could be in some sense, given the occasional discontent over reason, is ultimately an oversimplification. Rationalism in a broader sense is barely exhausted by a notion of reason as an instrument indifferent to ends. Consider in this regard the cautioning by Lao Tzu (2008, verses 30–31) against the violent subduing of the enemy. A vicious military takeover was not likely to last, and the violence that brought it about might later spring back to harm one, he counseled, not to mention the overall devastation, strictly unnecessary, which such ruthlessness might cause in the first place. If overwhelming the enemy was an imperative, he advised, it had to be done through means other than the use of brute force. In those cases when a violent clash seemed to be inevitable, one had to use only proportionate force. And if ultimately victorious in such confrontation, one had to respect the enemy, feel for their losses, and avoid ostentatious displays of victory.

Oversimplifications of the use of reason, challenged here by Lao Tzu's basic rationalism, may typically ensue also from false universalisms, an apparent contrast, in treating specific categories, such as religion. A habitual instance of this in research is the frequently entertained yet almost certainly less than meaningful question as to whether "religion" facilitates or impedes peace in practice (Haar and Basuttil 2005). The bewildering fact, eschewed in the gesture, is that religion per se, namely religion as a symbolic language that responds at a universal level to human needs for signification and value, and free of the configurations of power in specific contexts where it appears to have this function, may never have existed (Asad 1993, 27–54).

Particularly alarming is the effective reification of culture on the basis of a stable set of parameters, done almost invariably from a distance and usually to the exclusion of much else about the object of research, not to mention the possibly countless gray areas in the actual practice of a culture beyond the formal sources. Such "culturalism" in peace thinking, hunting and lodging authenticities out of the greatly exhausted space of modernity, may lead to various forms of reductionism and end up, rather perversely, affirming rather than remaining alert to the unsightly ethnocentrism that the interest in peripheral notions of peace is meant to moderate, if not entirely banish, in the first place. Take the following view of the Arab-Israeli conflict, held fleetingly by two of the principal figures in peace research with a marked concern to accommodate non-Western perspectives. By this take, the intensity of the contention, "the fierce antagonism between Israel and the Arab countries," is located, at least "partly," in "a common tradition of monotheism and a similar concept of peace as a realization of justice by the divine will" (Galtung 1981, 186, quoting and affirming Ishida 1969, 137). In the account, not only is the conflict reduced to the cultural visions of the immediate parties, but those visions are minimized to a formal, superficial understanding of the respective faiths. More baffling, the complexities of the actual conflict, the myriad factors in its making and development, extending far beyond the region and the immediate parties, are at once reduced, at least "partly," to almost a quirk, to some amusing exoticism, in the assumed common cultural traits of the parties to the conflict.

Such culturalism may also lead to menacing intellectual segregationism, regardless of the initial bona fide intention of greater egalitarianism between the assumed center and the periphery. Take the following statement in a genuine effort to urge the community of scholars in the study of international politics to be open to thematizations beyond the center, in turn strongly advocating the incorporation of the perimetric local wisdom, concerns, and thought processes in the mainstream theory: "[O]ne can ask whether there are cultural differences between the West and the non-West that make the former more generally inclined to approach issues in abstract terms, and the latter less inclined" (Acharya and Buzan 2010, 20). Being "inclined" more to a case-by-case understanding of issues, away from sweeping generalizations that are often deceptive, may no doubt be considered somewhat flattering in the known vein that looks down on grand narratives. That is, not necessarily offensive in itself, the comment may on the contrary retain some goodwill. Let us recall that the same goodwill applied historically also to most of the observations of the Orient within classical Orientalism. As has often been noted, some Orientalists were practically in love with the communities of the Orient they studied, not infrequently expressing their fully convinced laudations. All this would barely prevent the same Orientalists, however, from virtually reducing the Orient to a fantastic shore of infantile innocence, naïveté, and simple cunning, generated largely by setbacks in cultural growth and the attendant benightedness.

Much of the interest in peripheral notions of peace, either historically or across the globe, appears also to be inflicted with a rather facile theory of language, famously refuted by the later Wittgenstein. This reckoning of language understands meaning as a function of the sign, of a term, as opposed to the context wherein the term is part of a performance, namely the intricate and fluctuating, rather than stable, play of life forms and practices. Hence the painstaking effort often put in research into dissecting peace terms from the past or from other cultures, into etymology, and into the formal sources where those terms appear. Meaning on this view is a "shadowy being," as Wittgenstein (1989, 36) put it, accompanying the sign, intrinsic to it, and as such independent of time and place. The result may be a virtual fetishism around peace terms in various languages, such as *shanti*, *p'ing*, *heiwa*, *shalom*/*salām*. Once again, let us remember that historically Orientalist reductions of the Orient relied precisely on such formal evidence, as has been discussed by Said (1978, 52): "[T]he Orient studied was a textual universe by and large." The assumption that formal values may have a life outside and despite the material and largely unfathomable practice is clearly a problem impairing much of such cultural commentary.

References

Acharya, A, and B. Buzan. 2010. "Why Is There No Non-Western International Relations Theory?" In *Non-Western International Relations Theory: Perspectives on and beyond Asia*, edited by Amitav Acharya and Barry Buzan, 1–25. London: Routledge.

Asad, T. 1993. *Genealogies of Religion: Discipline and Reasons of Power in Christianity and Islam*. Baltimore, MD: Johns Hopkins University Press.

Bonta, B D. 1996. "Conflict Resolution among Peaceful Societies: The Culture of Peacefulness." *Journal of Peace Research* 33:403–420.

Braudel, F. 1981. *Civilization and Capitalism*. Vol. 1: *The Structures of Everyday Life, the Limits of the Possible*. Translated by Siân Reynolds. London: William Collins Sons.

Braudel, F. 1982. *Civilization and Capitalism*. Vol. 2: *The Wheels of Commerce*. Translated by Siân Reynolds. London: William Collins Sons.

Braudel, F. 1995. *A History of Civilizations*. Translated by Richard Mayne. London: Penguin.

Chiba, S. 2014. "A Historical Reflection on Peace and Public Philosophy in Japanese Thought: Prince Shotoku, Ito Jinsai and Yokoi Shonan." In *Visions of Peace: Asia and the West*, edited by Takashi Shogimen and Vicki A. Spencer, 85–102. Farnham, UK: Ashgate.

Fabbro, D. 1978. "Peaceful Societies: An Introduction." *Journal of Peace Research* 15:67–83.

Fenn, P. T, Jr. 1929. "An Indian Poet Looks at the West." *International Journal of Ethics* 39:313–323.

Foucault, M. 2007. *Security, Territory, Population: Lectures at the Collège de France, 1977–78*. Translated by Graham Burchell. Basingstoke, UK: Palgrave Macmillan.

Fukuzawa, Y. 1934. *The Autobiography of Fukuzawa Yukichi*. Translated by Eiichi Kiyooka. Tokyo: Hokuseido Press.

Fukuzawa, Y. 2008. *An Outline of a Theory of Civilization*. Translated by David A. Dilworth and G. Cameron Hurst III. New York: Columbia University Press.

Galtung, J. 1981. "Social Cosmology and the Concept of Peace." *Journal of Peace Research* 18:183–199.

Haar, G, and J. J. Basuttil. 2005. *Bridge or Barrier: Religion, Violence and Visions for Peace*. Leiden: Brill.

Hassig, R. 2007. "Peace, Reconciliation, and Alliance in Aztec Mexico." In *War and Peace in the Ancient World*, edited by Kurt A. Raaflaub, 312–328. Oxford: Blackwell.

Hobson, J. M. 2009. "What Have the Muslims Ever Done for Us? Islamic Origins of Western Civilization." In *The Challenge of Eurocentrism: Global Perspectives, Policy, and Prospects*, edited by Rajani Kannepalli Kanth, 217–235. New York: Palgrave Macmillan.

Ibn Kathīr, Abū al-Fidā' Ismā'īl b. 'Umar. 1999. *Tafsīr al-qur'ān al-'aẓīm*. Vol. 5. Edited by Sāmī b. Muḥammad al-Salāma. Riyadh: Dār Ṭayyiba.

Ibn Khaldūn, 'Abd al-Raḥmān b. Muḥammad. 2004. *Muqaddima*. Vol. 2. Edited by 'Abd Allāh Muḥammad al-Darwīsh. Damascus: Dār Ya'rub.

Iṣfahānī, Abū al-Qāsim al-Ḥusain b. Muḥammad al-Rāghib al-. 2009. *Mufradāt alfāẓ al-qur'ān*. Edited by Ṣafwān 'Adnān Dāwūdī. Damascus: Dār al-Qalam and Dār al-Shāmiya.

Ishida, T. 1969. "Beyond the Traditional Concepts of Peace in Different Cultures." *Journal of Peace Research* 6:133–145.

Kalin, I. 2005. "Islam and Peace: A Survey of the Sources of Peace in the Islamic Tradition." *Islamic Studies* 44:327–362.

Kemp, G, and D. P. Fry. 2004. *Keeping the Peace: Conflict Resolution and Peaceful Societies around the World*. London: Routledge.

Kipling, R. 1989. *Complete Verse: Definitive Edition*. New York: Anchor Books.

Kuhn, T. S. 1970. *The Structure of Scientific Revolutions*. Chicago: University of Chicago Press.
Lao Tzu. 2008. *Tao Te Ching*. Translated by Jonathan Star. New York: Penguin Tarcher Cornerstone Edition.
Qin, Y. 2010. "Why Is There No Chinese International Relations Theory?" In *Non-Western International Relations Theory: Perspectives on and beyond Asia*, edited by Amitav Acharya and Barry Buzan, 26–50. London: Routledge.
Raaflaub, K. A. 2016. "Abhorring War, Yearning for Peace: The Quest for Peace in the Ancient World." In *Peace in the Ancient World: Concepts and Theories*, edited by Kurt A. Raaflaub, 18–42. Oxford: Wiley Blackwell.
Ramanujan, A. K. 1989. "Is There an Indian Way of Thinking? An Informal Essay." *Contributions to Indian Sociology* 23:41–58.
Rosenthal, F. 1975. *The Classical Heritage in Islam*. Translated by Emile Marmorstein and Jennie Marmorstein. London: Routledge.
Ross, R. 1996. *Returning to the Teachings: Exploring Aboriginal* Justice. Toronto: Penguin.
Roy, K. 2014. "Concept of Peace in Hinduism: A Historical Analysis." In *Visions of Peace: Asia and the West*, edited by Takashi Shogimen and Vicki A. Spencer, 47–66. Farnham, UK: Ashgate.
Said, E. W. 1978. *Orientalism*. London: Penguin.
Salem, P. E. 1993. "A Critique of Western Conflict Resolution from a Non-Western Perspective." *Negotiation Journal* 4:361–369.
Shahrastānī, Abū al-Fatḥ M. b. ʿAbd al-Karīm al-. 1992. *Al-Milal wa al-niḥal*. Edited by Aḥmad Fahmī Muḥammad. Beirut: Dār al-Kutub al-ʿIlmiya.
Shilliam, R. 2011. "The Perilous but Unavoidable Terrain of the Non-West." In *International Relations and Non-Western Thought: Imperialism, Colonialism and Investigations of Global Modernity*, edited by Robbie Shilliam, 12–26. London: Routledge.
Takenori, I. 2008. "Introduction." Translated by Ry Beville. In Yukichi Fukuzawa, *An Outline of a Theory of Civilization*, xiii–xxvi. New York: Columbia University Press.
Tesei, T. 2011. "The Chronological Problems of the Qurʾān: The Case of the Story of Dhū al-Qarnayn." *Rivista Degli Studi Orientali* 84:457–464.
Tetsurō, W. 1971. "The Significance of Ethics as the Study of Man." Translated by David A. Dilworth. *Monumenta Nipponica* 26:395–413.
Wāḥidī, ʿAlī b. Aḥmad al-. 1991. *Asbāb nuzūl al-qurʾān*. Edited by Kamāl Basyūnī Zaghlūl. Beirut: Dār al-Kutub al-ʿIlmiya.
Wittgenstein, L. 1989. *The Blue and Brown Books*. Oxford: Blackwell.
Yu, K. 2014. "The Confucian Vision of Peace." In *Visions of Peace: Asia and the West*, edited by Takashi Shogimen and Vicki A. Spencer, 67–84. Farnham, UK: Ashgate.
Žižek, S. 1999. "The Superego and the Act." Lecture delivered at European Graduate School, 1 August. Accessed 24 May 2019. https://zizek.uk/the-superego-and-the-act/.

CHAPTER 14

GENDER, SECURITY, AND PEACEBUILDING

SARAH SMITH

It is no small feat to write a handbook chapter on the topic of gender, security, and peacebuilding.[1] Feminist (and nonfeminist) scholars, activists, and policymakers have produced a wealth of resources that demonstrate how gender, understood from different ontological positions, is integral to, produced through and constitutes security and peacebuilding (Meintjes, Pillay, and Turshen, 2001; Schnabel and Tabyshalieva 2012; Sjoberg 2010). Moreover, security and peacebuilding themselves can be understood as broad overarching concepts that capture a significant range of contributions from those working in specific areas and subfields or dealing with particular policy agendas within gender, security, and peacebuilding, such as transitional justice (Scanlon 2008), peace agreements (Bell and O'Rourke 2010), security sector reform and/or disarmament demobilization and reintegration (Bastick 2008; Clarke 2008; MacKenzie 2009), private military security companies (Eichler 2015), peacekeeping missions and postconflict reconstruction (Heathcote and Otto 2014; MacKenzie 2012; Smith 2019), gender, terror, and counterterror (Khalili 2010), and gender policy in security governance (Shepherd 2008, 2017). Both security and peacebuilding praxis hold relevance across contexts too; while conflict and postconflict contexts might immediately come to mind, gendered security concerns are also relevant in humanitarian emergencies and postdisaster crises (Rumbach and Knight 2014; Fisher 2010). Perhaps most important, "gender" is not, and cannot be, singularly understood and is not a singular perspective in peace and security studies.

Given this, it is worth examining how gender is understood and utilized as an analytic category. Essentialist understandings of gender see it as synonymous with sex, rooted in biology, and binary: one is either male or female, which in turn will define lived experience and behavior (Harding 1986, 24–25; Shepherd 2010, 13–14). In security and peacebuilding, from this view, the essentialist qualities of men (battle-ready, warmongering) and women (passive, peacemaking) are emphasized. Constructivist understandings of gender separate sex (biology) from gender (ideas about masculinity and femininity that are socially produced and constructed) (Shepherd 2010, 14). Social expectations of gender as well as systemic gender discrimination will shape experience, meaning men and women will be impacted by and experience (in)security and peacebuilding differently. Contributions from this perspective have emphasized women's experiences of war and peace as an essential site of knowledge on social and political life (Loney 2018) as well as how socially constructed gender shapes laws and policies, such as who is seen as civilian and who as combatant (Carpenter 2003). Thus, despite the conceptual delinking of sex and gender, gender in the constructivist framework remains mapped onto differently categorized bodies and often within a male-female binary. Finally, poststructural perspectives hold an ontology of both gender and sex as produced and performative, challenging binary frameworks (Butler 1990; Shepherd 2010, 14–15). Judith Butler (1990, 7) argues that the splitting of sex/biology and gender/culture has problematically naturalized sex as "prediscursive, prior to culture, a politically neutral surface on which culture acts." From this epistemological position, feminist scholars have demonstrated how security policy is animated by the discursive representation of men and women, which in turn constitutes and governs gendered bodies and identities (Shepherd 2008, 2017).

Therefore it is important to remain attentive to which gender lens is used in examining security and peacebuilding processes. Moreover, gender cannot be understood as a "single-axis issue" and must be situated and decentered in relation to race, ethnicity, religion, age, (dis)ability, class, and sexuality (Crenshaw 1991). An intersectional view of gender highlights that gender is co-constituted alongside race, ethnicity, class, and other identity markers. While intersectionality is a framework that highlights multiple interlocking oppressions on the basis of race/sex/class, Crenshaw's formulation of the concept equally brings into view the institutional structures in which these oppressions are produced, addressing the "larger ideological structures in which subjects, problems, and solutions [are] framed" (Cho, Crenshaw, and McCall 2013, 791). In the context of security and peacebuilding, this exposes how both the violence of war and the processes of peacebuilding target and produce raced and gendered bodies or seek to protect some at the expense of others, delineated by race, gender, class, sexuality, ethnicity, and religion (Butler [2009] 2016; Kinsella 2005).

This chapter begins with a brief overview of how gender plays out in, is relevant to, shapes, and is shaped by security and peacebuilding, before examining gender policy in peacebuilding and providing a reflection on the future of the field based on the significant contributions of feminist work to security and peace theorizing. Given space constraints, the chapter for the most part stays focused on notions of security and practices of peacebuilding in postconflict contexts in the post–Cold War era. However, I acknowledge that the partiality of this approach is challenged by feminist theorizing that broadens our understanding of insecurity to also encapsulate the direct and structural violence of militarism and patriarchal politics in and out of war. Such a framing can also ascribe the issues under examination to certain sites and locations, and this limitation is addressed in the concluding section of the chapter, wherein I argue that an intersectional and nonessential view of gender must come to the forefront of both academic and policy work on security and peacebuilding alike, challenging binary frameworks such as international/local, and delineate the dire necessity of peace and conflict studies taking heed of feminist contributions in their theory-making.

Gender Issues in Security and Peacebuilding

Dominant notions of security and institutional processes of peacebuilding have historically been almost exclusively linked to the experiences and knowledge of (some) men. Activists and scholars have long demonstrated the impacts of the androcentric construction of security and peace by highlighting the direct and structural violence that occurs in the name of security, as well as the insecurity of differently placed groups and individuals in supposed times of peace (Rehn and Sirleaf 2002). By highlighting the gendered violence, direct and structural, that can characterize peacebuilding moments, gender perspectives challenge the notion that these processes are linear progressions that provide security for all equally. It is important to examine experiences of gendered violence in and through peacebuilding and security in order to challenge the production and projection of these in a manner that excises a vast array of experiences and can thus perpetuate insecurity; in short, whose peace and whose security are we talking about?

Activists and scholars have long drawn attention to the gendered violence of war and how gender structures this violence, such as the targeting of men and boys for death and the use of sexual violence (Carpenter 2003; Jones 2000; Rehn and Sirleaf 2002). However, while a peace agreement or ceasefire can bring an end to formal hostilities between warring parties, violence by no means ends with the signing of these deals, and this is especially true for gendered violence. Postconflict and peacebuilding phases remain unstable, and insufficient security provisions combined with lingering impunity for violence means that direct and physical violence can continue unabated, or in some cases exceed occurrence during the conflict period (Pankhurst 2008b). Feminist scholars, for example, have highlighted high rates of sexual and gender-based violence (SGBV) in peacebuilding periods, such as rape and domestic violence, which includes that perpetrated by peacekeepers and humanitarian staff (Smith 2017). They have connected this with gendered understandings of security wherein the cessation of armed aggressions between conflict parties is the focus, while other forms of violence continue unabated, such as violence against lesbian, gay, bisexual, trans, and intersex (LGBTI) individuals, and domestic and family violence. Indeed peacebuilding narratives often evoke nostalgic narratives or "golden-ageism" that seeks a return to romanticized notions of a traditional past that is conservative, gender-discriminatory, and/or heteronormative (Myrttinen, Naujoks, and El-Bushra 2014, 9–10). Legacies of conflict can exacerbate violence through a lack of justice mechanisms as well as the prevalence of weapons in society. In Timor-Leste, for example, researchers found that although the number of firearm incidents decreased once the Indonesian military occupation had come to an end, the "sizeable majority" of cases involving small arms and light weapons involved intimate femicide and domestic and sexual violence (Abdullah and Myrttinen 2009).

Understood narrowly, gender perspectives can be limited to focusing on women and those issues perceived as "women's issues" only; however, it is important to take a broader view of gender to reveal the multiple ways that gendered norms and assumptions can impact security and insecurity. LGBTI people, or simply those who do not present consistent with expected social gendered norms, also face precarious security situations in postconflict and emergency moments. Throughout the statebuilding period in Iraq, for example, Human Rights Watch reported on targeted violence against men and boys who do not present as manly enough or are perceived as effeminate and assumed to be gay (cited in Hagen 2016, 315–316). Forced relocation or migration can expose individuals to greater violence as their usual coping and protection frameworks—such as networks of family and friends—are disrupted by broader conflict and emergency situations (Rumbach and Knight 2014). Moreover, and further discussed later, these types of violence have traditionally been obscured in security narratives that are preoccupied with formal armed conflict and policy responses that understand gender in heteronormative and binary ways.

Failure to recognize the gendered violence of war has meant that the (re)distribution of resources in the aftermath of conflict, as well as transitional justice mechanisms, can provide little if any redress for these violations, which is problematic because direct violence is manifested by and exacerbates broader conditions of structural violence. A limited notion of postconflict security that does not include substantive acknowledgment or understanding of domestic violence, for instance, creates conditions in which this violence can occur with impunity or is normalized. Moreover, much like in times of peace, gendered inequalities function as social, political, and economic hierarchies that can shape the distribution of resources postconflict and in peacebuilding. For example, case study research in Timor-Leste and Libya has demonstrated how male or male-identified individuals are understood as the archetypal veteran subject and thus have been the focus of veteran payment schemes instituted as part of peacebuilding processes (Smith 2019, 88–89; Kent and Kinsella 2015; Langhi 2014, 205). In these case studies, women were more likely to be understood as victims only, which excludes them from veteran assistance schemes. In the Timor-Leste case, victims and veterans have been constructed in political discourse as two mutually exclusive categories of subjects with differing postwar needs, a gendered binary that was further projected onto a hierarchy of needs, with the needs of veterans—a category that captured only armed and male resistance members—prioritized (Smith 2019, 88–89). This is despite significant contributions and sacrifices on the part of female resistance fighters, both armed and not. The understandings of "veteran" here reflects a gendered conceptualization of who or what constitutes a veteran, and women's labor was not understood as the work that makes a veteran subject, entrenching significant gendered inequalities in the redistribution of material benefits postwar. Similarly, in Sierra Leone, Megan MacKenzie (2009) found that demobilization and reintegration policies were framed on the assumption that men were its exclusive targets, ignoring the fact that women had been armed combatants, and, given this lack of gender sensitivity, were found to have limited, if any, benefits for former female combatants, leading them to disengage from programs. Far from being an issue of mere oversight, these cases demonstrate how gender is embedded in the basis of understanding and engaging in postconflict sites, and the deleterious effects this can have. It is deeply problematic that the material and economic benefits of postconflict redistribution schemes entrench gendered inequalities, which in turn underpin the continuance of direct violence.

Therefore the gendered logics that inform how conflict is understood and practiced precipitate into peacebuilding moments. Indeed gender perspectives have demonstrated how a simplified binary between war and peace can be difficult to sustain. It is not simply that gendered violence persists after conflict but that it can be and is exacerbated by masculinist interpretations of what constitutes peace and security, which can undermine the security of groups and individuals who sit outside these frames. Violence perpetrated by peacekeepers and humanitarian staff is perhaps the clearest example of how peacebuilding periods open up new and complex forms of insecurity for some, and the lack of institutional response to these cases is emblematic of how some violence is constructed to fall outside the peacebuilding purview (Smith 2017, 2019). Peacebuilding and statebuilding must therefore account for diverse experiences constituted by gender, race, sexuality, religion, and class. Policy that fails to do so can perpetuate inequalities and hierarchical power structures that contribute to direct and structural violence and thus ultimately an incomplete or unstable peace. The next section examines the development of gender policy in peacebuilding architectures.

Gender Policy in Peacebuilding and Statebuilding

Since attention has been brought to these issues, and more, there has been significant policy development around gender mainstreaming in peacebuilding, which has largely focused on including women, their concerns and experiences. The most notable framework here is the UN Security Council's Women, Peace and Security agenda. This agenda consists of ten Security Council resolutions that emphasize variously the necessity of women's participation in political and security decision-making and peace agreements and the prevention of SGBV as a matter of international peace and security. There is a wealth of literature examining the normative framework, conceptualization, and implementation of the Women, Peace and Security agenda (Davies and True 2019). This agenda has seen rhetorical commitments to women and gender become commonplace in peace and security governance, especially within the UN and its agencies, as well as the EU and NATO (Wright 2016). The agenda has coalesced around the twin pillars of preventing SGBV and increasing women's participation in decision-making and at the peace table; however, as discussed shortly, progress in each of these areas remains limited and the process through which "gender" becomes a mechanism implemented within peacebuilding problematic.

The implementation of Women, Peace and Security in peacebuilding processes has had a mixed record, and the agenda is marginalized in peace operations and security governance more broadly (Coomaraswamy 2015). This is particularly evident given that despite repeated rhetorical commitments to gender mainstreaming in peacebuilding, programs and policies that seek to enact this are routinely underfunded and/or established with circumscribed power and resources (Coomaraswamy 2015). In her analysis of a trove of peacebuilding documents, Laura Shepherd (2017) found that both women and gender are discursively constructed as marginal, either through overt objections, placement at the end of documents (reproducing a "nonessential" position), or simply through absence. Shepherd's work provides significant evidence of the depreciation of different spaces, places, and bodies in peacebuilding. Both "gender" and "women" are routinely linked with national and local spaces, which in turn are positioned as nonexpert and subordinate in the institutional hierarchies of power and knowledge. This represents a process of relegating women and gender to peripheral spaces invested with less power in peacebuilding processes and the corollary of the arguable feminization of "local" spaces through their consistent rhetorical linking with women's organizations and civil society, which is valued less than political and security bodies (Shepherd 2017; see also Smith 2019, 138–164).

The Women, Peace and Security agenda is also subject to politicization, which likewise circumscribes its normative potential. A clear example of how Women, Peace and Security is circumscribed in the face of political will is the adoption of the ninth Security Council resolution in the agenda, Resolution 2467. The draft resolution contained stipulations on the necessity of sexual and reproductive health services for survivors of sexual violence in conflict, language that reflects earlier resolutions on the matter. In debates leading up to the resolution's adoption, however, conservative actors, primarily the US, ensured provisions for reproductive health were dropped because of fears it could connote abortion (Leimbach 2019). This represented a reversal of previous advancement of the agenda because of the evangelical Christian beliefs of some in the then Trump administration, most especially former Vice President Mike Pence (Leimbach 2019).

In examining gender policy in peacebuilding, then, it is important to remain cognizant of the political environments in which it operates as well as the extent to which it is positioned as nonessential in security governance. In addition to, or more likely because of, these issues, though, is the essentialist understanding of gender that is often enacted via policy, in particular the synonymization of gender and women. Given the complexity of postconflict gender issues, understanding "gender" as synonymous with women and then seeking to add women to existing practices to ameliorate the gendered causes and consequences of violence, without other substantive change, is limited (Myrttinen, Naujoks, and El-Bushra 2014). This also produces a flattened approach to women's experiences and needs. According to Hilary Charlesworth (2008), an institutional orthodoxy has developed on women and peacebuilding, one that assumes women are better than men at sustaining peace, chiefly characterizes women as vulnerable, and uses "gender" as synonymous with "women." This orthodoxy and the representations of women it produces hold productive power, relying on reifying a certain "woman subject" (Shepherd 2008, 2017) and marginalizing experiences and needs that fall outside this frame.

This limited conceptualization of women and gender and its productive consequences are well demonstrated by the narratives of SGBV prevention in security governance and peacebuilding. SGBV has become a cornerstone of the Women, Peace and Security agenda, a move that has been critiqued by many who see it as circumscribing women's agency and framing them as victims only. As a UN gender advisor told me during my fieldwork, "I actually think that in places where this is debated, in donor organizations . . . they're much more comfortable about giving money to the woman as victim rather than the empowerment of women. . . . [T]here is a comfort zone around women as victims. There is no comfort zone around women who are empowered."[2] This is not to deny that SGBV in and after war is a serious concern or to obscure this form of suffering. Instead feminist perspectives seek to acknowledge SGBV on a broader continuum of violence and see prevention as integrally linked to the alleviation of other socioeconomic inequalities. Similarly, Hanna Muehlenhoff (2017), in her discourse analysis of EU security policy, found that women were represented as victims of conflict who needed to be empowered to become resources for peace, which is underpinned by neoliberal rationalities. Moreover, it is not only women who are subject to SGBV in and after war.

Nadje Al-Ali and Nicola Pratt's (2009) work on statebuilding in Iraq demonstrates the interlocking and intersectional ways that gender works to structure statebuilding processes. They focus in particular on the difficulties subsequently faced by women in accessing political and security decision-making. This case study is particularly illuminating because the US-led invasion was enacted partly under a banner of liberating women specifically from the oppressive regime of Saddam Hussein (see, for example, Shepherd 2006). The chaos and humanitarian crisis that ensued from this intervention, though, has done little to secure Iraqi women's (or men's) security and stability. Indeed Al-Ali and Pratt (2009, 167) found that because of this rhetoric of an intervention to liberate women there was actually tremendous backlash against women's rights: "On one level, a strong knee-jerk reaction [took] place across many sectors of Iraqi society . . . opposing and resisting military occupation and Western cultural encroachment. . . . In this attempt to create and harden an authentic Iraqi or Muslim culture different from the culture of 'the West,' culture is reconfigured in far more exclusive ways than historically was the case."

These issues also speak to how security is conceptualized and pursued in peacebuilding processes. Peacebuilding and statebuilding norms are based on Weberian notions of the state, with the monopolization of the use of force in the hands of the state a chief concern and that militarized approaches to security therefore ensue. However, as feminist scholars have highlighted, "more guns do not mean more security" (Bhagwan Rolls 2014, 122). Simply adding more female peacekeepers or security personnel, for example, has not equated to the kinds of substantive changes demanded by feminist theorizations of peace, ones that account for the alleviation or removal of structural inequalities based on gender, class, race, ethnicity, sexuality, and religion. Feminist scholars view violence, and gendered violence specifically, on a broader continuum than is allowed for in simplified war/peace binaries: "war and conflict are merely the explicit expressions of deeply gendered, as well as ethnicized and classed, long-term dynamics that precede the outbreak of conflict, escalate dramatically, and persist long after 'peace' has been officially declared and the transition from overt warfare is taking place" (Mama and Okazawa-Rey 2008, 3). SGBV is widespread globally and is exacerbated in times of conflict, meaning that in times of peace women and women-identified and LGBTI individuals "do not enjoy peace and security in their homes, workplaces or on the streets" (3).

The linking of gender with women has also reproduced a binary "M/F" understanding of gender (Shepherd 2010, 13–14). This has had a dual effect of obscuring gender identities not captured in this framework as well as relegating men's experiences and identities outside of gendered concerns or policy. The binary understanding of gender that permeates the implementation of the Women, Peace and Security agenda can erase violence targeted toward nonbinary individuals, LGBTI people, and men and male-identified individuals (Hagen 2016, 315). Myrttinen, Naujoks, and El-Bushra (2014) found that there is substantial activism from local actors and civil society organizations around the inclusion of LGBTI persons in peacebuilding, more so than has been done under the banner of Women, Peace and Security instituted by international institutions and interventions. The provision of health services for women who have suffered SGBV in and after conflict has been a notable success of the Women, Peace and Security agenda, but by focusing on women exclusively these same programs have operated to exclude others who are also subject to rape and sexual violence (Myrttinen, Naujoks, and El-Bushra 2014). This is not an unfortunate oversight but constitutes a form of structural violence, reproducing a norm of gender that marginalizes those who do not "fit" the framework. More recent research is demonstrating that men and boys are also victims of SGBV during conflict and in forced migration yet struggle to access health and psychosocial supports in peacebuilding and statebuilding programs (Chynoweth 2017). These cases demonstrate the productive consequences of limited conceptualizations of gender in peacebuilding and statebuilding activities.

This discussion exemplifies the central concerns in relation to gender and peacebuilding policy to date: who is included in gender policy frameworks and how, and how gender policy is framed and implemented. Beyond this, though, an essentialist approach fails to untangle how gendered logics inform peacebuilding practices, that is, what is prioritized and valorized against what is feminized and made peripheral.

Conclusion: Where Now for Gender, Peacebuilding, and Security?

This discussion highlights the negotiation around what kinds of stability and security count, and what or whose actions are legitimate and why in peacebuilding and statebuilding, questions with which feminist scholars have long grappled (Meintjes, Pillay, and Turshen 2001; Pankhurst 2008a). Recently in peace and conflict studies broadly, the attribution of legitimacy to different actors and actions in peacebuilding has been in focus, manifested in theorizing that shifts attention from institutions and sites of elite power to "local" and everyday practices of peace (O'Reilly and McLeod 2019, 137–138). Feminist work has long challenged reliance on simplistic binaries that obscure lived experience and perpetuate hierarchical social relations, while at the same time foregrounding similar or related methodological and epistemological commitments, although without losing sight of the intersectionalities of power that constitute and are constituted by gender (Kappler and Lemay-Hébert 2019). Looking ahead to future directions for research on gender, security, and peacebuilding, what is most notable is the insistence and demonstration that theory cannot be developed without attention to gender as an analytic category or without reference to the existing wealth of feminist work and the methodological and epistemological richness therein.

The previous section delineated some limitations in gender policy in peacebuilding; this is a site requiring imperative change for peacebuilding and statebuilding processes to, at the very least, not perpetuate gender-unequal relations of power. But academic work in peace and conflict studies has also continued to generate theory without attention to gender or indeed to the operation of the Women, Peace and Security agenda, and the pitfalls of doing so are evident (Martin de Almagro 2018; O'Reilly and McLeod 2019; Smith 2019, 138–157). Utilization of an international/local binary to theorize peace as a contested terrain has failed to take account of the productive consequences of this binary (Hameiri and Jones 2018; Nadarajah and Rampton 2015). Through the lens of feminist knowledge production, such a binary also fails to account for the alliances and antagonisms that emerge in peacebuilding around gender policy that do not reflect international or local categorizations (Smith 2019).

This demonstrates the need for a reflective practice in peace and conflict studies, one that is cognizant of the role of theory-makers in producing the contexts, signs, and value of which they speak (see Visoka and Musliu 2019). This additionally links with the need for decolonizing knowledge production in peace and conflict studies (Sabaratnam 2013). Decolonizing knowledge requires "re-engagement with that which Eurocentric thinking suppresses or discounts [and] finding different political sites from which to think about the world and constructing different problematiques for analysing it" (270–271). Gender, race, ethnicity, violence, and sexuality all mark some bodies for violence and others for protection; thus taking gender as a category of analysis is done with attention to its co-constitution with other identity markers. This includes recognition of colonial histories and relations of power and their reproduction in contemporary relations, as gender and race have been utilized to claim difference and perpetuate processes of othering that can foster violence.

This chapter has highlighted two significant contributions of feminist and gender theory in security and peacebuilding: making visible previously marginalized experiences and knowledge and exposing the gendered logics that both inform this exclusion and are fundamentally entwined with and productive of the priorities and practices of security and peacebuilding. This means that rectifying gender-discriminatory understandings of peace and security is not simply a matter of adding previously marginalized experiences or bodies but requires a fundamental rethink of what constitutes security and peace and the institutions and processes that pursue these goals. Gender perspectives in security and peacebuilding provide normative and theoretical imperatives to rethink approaches to peacebuilding and statebuilding, take a broader view of what constitutes security and peace, and reflect critically on the work that "gender" does in peacebuilding policy. These imperatives are directed at both peace and security governance as well as peace and conflict studies as a field of academic study.

Notes

1. Gender, security, and peacebuilding is such a broad canvas that a number of handbooks and special journal issues are devoted to the topic, demonstrating the contributions and essential theorizations this field makes, demonstrating its depth and richness as well as the difficulty of capturing it in a single chapter. See, for example, Sharoni, Welland, and Steiner (2016); Gentry, Shepherd, and Sjoberg (2019); Davies and True (2019); Olonisakin and Okech (2011); "Building Peace: Feminist Perspectives" (2016); Singh (2017); O'Reilly and McLeod (2019).
2. UN gender advisor, interview with author, Melbourne, 27 February 2014.

References

Abdullah, S., and H. Myrttinen. 2009. "'Now They Have Guns, Now They Feel Powerful': Gender Perspectives on Small-Arms Violence in Timor-Leste." In *Sexed Pistols: The Gendered Impacts of Small Arms and Light Weapons*, edited by V. Farr, H. Myrttinen, and A. Schnabel, 177–208. Tokyo: United Nations University Press.

Al-Ali, N., and N. Pratt. 2009. *What Kind of Liberation? Women and the Occupation of Iraq*. Berkeley: University of California Press.

Bastick, M. 2008. *Integrating Gender in Post-Conflict Security Sector Reform*. Geneva: Geneva Centre for the Democratic Control of Armed Forces.

Bell, C., and C. O'Rourke. 2010. "Peace Agreements or Pieces of Paper? The Impact of UNSC Resolution 1325 on Peace Processes and Their Agreements." *International and Comparative Law Quarterly* 59 (4):941–980.

Bhagwan Rolls, S. 2014. "Thinking Globally and Acting Locally: Linking Women, Peace and Security in the Pacific." In *Rethinking Peacekeeping, Gender Equality and Collective Security*, edited by G. Heathcote and D. Otto, 118–130. Basingstoke, UK: Palgrave Macmillan.

Butler, J. 1990. *Gender Trouble.* New York: Routledge.

Butler, J. (2009) 2016. *Frames of War: When Is Life Grievable?* London: Verso.

Carpenter, R. C. 2003. "'Women and Children First': Gender, Norms, and Humanitarian Evacuation in the Balkans 1991–95." *International Organization* 57 (4):661–694.

Charlesworth, H. 2008. "Are Women Peaceful? Reflections on the Role of Women in Peace-Building". *Feminist Legal Studies* 16 (3):347–361.

Cho, S., K. Crenshaw, and L. McCall. 2013. "Toward a Field of Intersectionality Studies: Theory, Applications, and Praxis." *Signs* 38 (4):785–810.

Chynoweth, S. 2017. *"We Keep It in Our Heart": Sexual Violence against Men and Boys in the Syria Crisis.* Geneva: UNHCR.

Clarke, Y. 2008. "Security Sector Reform in Africa: A Lost Opportunity to Deconstruct Militarised Masculinities?" *Feminist Africa* 10:49–66.

Coomaraswamy, R. 2015. *Preventing Conflict, Transforming Justice, Securing the Peace: A Global Study on the Implementation of United Nations Security Council Resolution 1325.* New York: UN Women.

Crenshaw, K. 1991. "Mapping the Margins: Intersectionality, Identity Politics, and Violence against Women." *Stanford Law Review* 43 (6):1241–1299.

Davies, S. E., and J. True, eds. 2019. *The Oxford Handbook on Women, Peace and Security.* Oxford: Oxford University Press.

Eichler, M., ed. 2015. *Gender and Private Security in Global Politics.* Oxford: Oxford University Press.

Fisher, S. 2010. "Violence against Women and Natural Disasters: Findings from Post-Tsunami Sri Lanka." *Violence against Women* 16 (8):902–918.

Gentry, C., L. J. Shepherd, and L. Sjoberg, eds. 2019. *Routledge Handbook on Gender and Security.* London: Routledge.

Hagen, J. J. 2016. "Queering Women, Peace and Security." *International Affairs* 92 (2):313–332.

Hameiri, S., and L. Jones. 2018. "Against Hybridity in the Study of Peacebuilding and Statebuilding." In *Hybridity on the Ground in Peacebuilding and Development,* edited by J. Wallis, S. Dinnen, L. Kend, M. Forsyth and S. Bose., 99–112. Canberra: ANU Press.

Harding, S. 1986. *The Science Question in Feminism.* Ithaca, NY: Cornell University Press.

Heathcote, G., and D. Otto, eds. 2014. *Rethinking Peacekeeping, Gender Equality and Collective Security.* Basingstoke, UK: Palgrave Macmillan.

Jones, A. 2000. "Gendercide and Genocide." *Journal of Genocide Research* 2 (2):185–211.

Kappler, S., and N. Lemay-Hébert. 2019. "From Power-Blind Binaries to the Intersectionality of Peace: Connecting Feminism and Critical Peace and Conflict Studies." *Peacebuilding* 7 (2):160–177.

Kent, L., and N. Kinsella. 2015. "A Luta Kontinua (The Struggle Continues)." *International Feminist Journal of Politics* 17 (3):473–494.

Khalili, L. 2010. "Gendered Practices of Counterinsurgency." *Review of International Studies* 37 (4):1471–1491.

Kinsella, H. 2005. "Securing the Civilian: Sex and Gender in the Laws of War." In *Power in Global Governance,* edited by M. Barnett and R. Duvall, 249–272. Cambridge: Cambridge University Press.

Langhi, Z. 2014. "Gender and State-Building in Libya: Towards a Politics of Inclusion." *Journal of North African Studies* 19 (2):200–210.

Leimbach, D. 2019. "At the UN, the US Darkens Women's Right to Abortion." *Pass Blue,* 23 April. https://www.passblue.com/2019/04/23/at-the-un-the-us-darkens-womens-right-to-abortion/.

Loney, H. 2018. *In Women's Words: Violence and Everyday Life during the Indonesian Occupation of East Timor, 1975–1999*. Brighton, UK: Sussex Academic Press.
Mama, A., and M. Okazawa-Rey. 2008. "Editorial: Militarism, Conflict and Women's Activism." *Feminist Africa* 10(1): 1–8.
Martin de Almagro, M. 2018. "Hybrid Clubs: A Feminist Approach to Peacebuilding in the Democratic Republic of Congo." *Journal of Intervention and Statebuilding* 12 (3):319–334.
MacKenzie, M. 2009. "Securitization and Desecuritization: Female Soldiers and the Reconstruction of Women in Post-Conflict Sierra Leone." *Security Studies* 18 (2):241–261.
MacKenzie, M. 2012. *Female Soldiers in Sierra Leone: Sex, Security, and Post-Conflict Development*. New York: NYU Press.
Meintjes, S., A. Pillay, and M. Turshen, eds. 2001. *The Aftermath: Women in Post-Conflict Transformation*. London: Zed Books.
Muehlenhoff, H. 2017. "Victims, Soldiers, Peacemakers and Caretakers: The Neoliberal Subject Constitution of Women in the EU's Security Policy." *International Feminist Journal of Politics* 19 (2):153–167.
Myrttinen, H., J. Naujoks, and J. El-Bushra. 2014. *Re-Thinking Gender in Peacebuilding*. London: International Alert.
Nadarajah, S., and D. Rampton. 2015. "The Limits of Hybridity and the Crisis of the Liberal Peace." *Review of International Studies* 41 (1):49–72.
Olonisakin, F., and A. Okech, eds. 2011. *Women and Security Governance in Africa*. Cape Town: Pambazuka Press.
O'Reilly, M., and L. McLeod. 2019. "Critical Peace and Conflict Studies: Feminist Interventions." *Peacebuilding* 7 (2):127–145.
Pankhurst, D., ed. 2008a. *Gendered Peace*. London: Routledge.
Pankhurst, D. 2008b. "Introduction." In *Gendered Peace*, edited by D. Pankhurst, 1–30. London: Routledge.
Rehn, E., and E. Sirleaf. 2002. *Women, War and Peace: The Independent Experts' Assessment on the Impact of Armed Conflict on Women and Women's Role in Peacebuilding*. New York: UNIFEM.
Rumbach, J., and K. Knight. 2014. "Sexual and Gender Minorities in Humanitarian Emergencies." In *Issues of Gender and Sexual Orientation in Humanitarian Emergencies*, edited by L. W. Roeder, 33–74. London: Springer International.
Sabaratnam, M. 2013. "Avatars of Eurocentrism in the Critique of the Liberal Peace." *Security Dialogue* 44 (3):259–278.
Scanlon, H. 2008. "Militarization, Gender and Transitional Justice in Africa." *Feminist Africa* 10:31–48.
Schnabel, A., and A. Tabyshalieva, eds. 2012. *Defying Victimhood: Women and Post-Conflict Peacebuilding*. Tokyo: United Nations University Press.
Sharoni, S., J. Welland, and L. Steiner, eds. 2016. *Handbook on Gender and War*. Cheltenham, UK: Edward Elgar.
Shepherd, L. J. 2006. "Veiled References: Constructions of Gender in the Bush Administration Discourse on the Attacks on Afghanistan Post-9/11." *International Feminist Journal of Politics* 8 (1):19–41.
Shepherd, L. J. 2008. *Gender, Violence, and Security: Discourse as Practice*. London: Zed Books.
Shepherd, L. J. 2010. "Feminist Security Studies." In *Critical Approaches to Security*, edited by L. J. Shepherd, 11–23. London: Routledge.
Shepherd, L. J. 2017. *Gender, UN Peacebuilding, and the Politics of Space: Locating Legitimacy*. Oxford: Oxford University Press.

Singh, S. 2017. "Gender, Conflict and Security: Perspectives from South Asia." *Journal of Asian Security and International Affairs* 4 (2):149–157.

Sjoberg, L., ed. 2010. *Gender and International Security: Feminist Perspectives.* London: Routledge.

Smith, S. 2017. "Accountability and Sexual Exploitation and Abuse in Peace Operations." *Australian Journal of International Affairs* 71 (4):405–422.

Smith, S. 2019. *Gendering Peace: UN Peacebuilding in Timor-Leste.* London: Routledge.

Visoka, G., and V. Musliu, eds. 2019. *Unravelling Liberal Interventionism: Local Critiques of Statebuilding in Kosovo.* London: Routledge.

Wright, K. A. M. 2016. "NATO's Adoption of UNSCR 1325 on Women, Peace and Security: Making the Agenda a Reality." *International Political Science Review* 37 (3):350–361.

CHAPTER 15

PEACE PSYCHOLOGY

DANIEL J. CHRISTIE

At a basic level, peace psychologists engage in research and practice aimed at the promotion of harmony and equity in human relations. However, a closer look reveals that the primary focus of peace psychologists varies with geohistorical context. Viewed through a psychological lens, the first wave of peace psychologists focused on the Cold War, which featured a bipolar power rivalry between East and West (Frank 1982). Central to the rivalry was a nuclear arms race between the United States and the Soviet Union and the threat of nuclear war. In this geohistorical context, the primary goal of peace psychologists and peace psychology organizations was the prevention of nuclear war through the application of psychological concepts and perspectives (Wessells 1996). Accordingly, peacekeeping, which separates would-be combatants but does not resolve the underlying issues, along with peacemaking though conflict resolution were among the focal interventions proposed by Western scholars in peace psychology (Christie et al. 2008). Conceptually the goals of peace psychology centered around the prevention of violent episodes and the preservation of status quo power arrangements in the international system.

After the Cold War, peace psychologists turned their attention away from the nuclear threat and toward the problem of intrastate conflict and violence, which was raging in many parts of the world. This discursive turn in scholarship recognized that the usual psychological analyses employed during the Cold War were ill suited to intrastate conflicts that were intractable and divided along ethnopolitical lines (Gurr 1994). Hence the focal concerns in Western peace psychology were enlarged from the prevention of interstate violence to interventions tailored to intrastate cycles of violence and the underlying structures that fueled and sustained the violence (Christie, Wagner, and Winter 2001; Coleman et al. 2007).

The growing appreciation for the problem of structural and cultural violence ushered in a second wave of peace psychology scholarship led by scholars who were not aligned with the East or the West during the Cold War. Nonaligned countries have the majority of the world population and remain characterized by low levels of human development and the concentration of wealth. Drawing on the distinction between negative and positive peace (Galtung and Höivik 1971), the primary goal of nonaligned countries was not so much the pursuit of negative peace (i.e., absence of direct violence) as positive peace, which is aimed at disrupting the status quo and pursuing more equitable political and economic arrangements locally and globally (Montiel 2001, 2003). In addition to mitigating structural violence, the second wave addressed cultural violence through a critical psychology lens that foregrounded the voices of oppressed people who were challenging the neoliberal order and colonization of the mind. Going by various names, these liberation and critical psychologists are leading the way in a movement that puts an emancipatory agenda at its center (Seedat, Suffla, and Christie 2017a, 2017b). In this chapter, the two waves of scholarship and activism in global peace psychology are placed in geohistorical context and the primary concepts and themes of each are elaborated.

First Wave of Peace Psychology: The Prevention and Mitigation of Violence

The Cold War and Nuclear Arms Race

While the discipline of psychology is primarily based on the individual or, in the case of social psychology, the individual within a social context, the advent of peace psychology substantially enlarged the scope of psychology by looking at peace not only within individuals, families, and communities but also within and between nations (Christie and Montiel 2013). The impetus for peace psychologists to examine larger units of analysis was animated by the Cold War, a bipolar power struggle between the United States and the Soviet Union that threatened to escalate to a hot war and the use of nuclear weapons. The history of psychology features a number of psychologists who were interested in peace; however, it was not until the 1980s that the growing tensions of the Cold War gave rise to peace psychology organizations and psychologists who began to identify themselves as peace psychologists, especially in the United States, Germany, and Australia (cf. Christie 2011).

These early peace psychologists could draw on the interdisciplinary field of conflict resolution (Kelman 1981) and its flagship publication, *Journal of Conflict Resolution*, which was established in 1957 with a mission of addressing the "most important practical problem facing the human race today . . . the prevention of global war." At the same time, in the field of psychology, Morton Deutsch (1973) was conducting empirically based work on cooperation and effectively made the case that conflict could be viewed as an opportunity to build constructive relationships. Other resources were available from the Harvard Negotiation Project. Instituted in 1979, the project offered training in negotiation and published the popular cross-over book by Fisher, Ury, and Patton (2011), *Getting to Yes*. Fisher and colleagues worked out the details of a principled approach to negotiations. In contrast to positional bargaining (such as haggling over the price of something), principled negotiation elevates the importance of the relationship between the parties and frames the conflict as an opportunity to engage in mutual problem-solving; thus the parties work together to arrive at a mutually beneficial outcome. Key elements of the approach are (1) separating the people from the problem, which allows the parties to focus on the issues rather than damaging the relationship; (2) focusing on underlying common interests rather than positions, thereby encouraging a positive-sum rather than a zero-sum outcome; (3) generating a wide range of options before settling on an option that is mutually beneficial; and (4) insisting the agreement is based on objective, measurable outcomes.

In addition to the literature on conflict resolution, a watershed of publications in psychology began to delineate the contours of the intellectual space that would later be occupied by the emerging area of peace psychology. Many of these concepts and themes appeared in special issues of the *Journal of Social Issues* (Russell 1961; Wagner 1988). Themes from these publications included the meanings and recommended actions to develop trust in the US-Soviet relationship, "mirror images" of each other in the perceptions of US and Soviet citizens, the concept of a diabolical enemy image when viewing the Other, the role of misperceptions in escalating tensions in the relationship, and a critique of deterrence as the centerpiece of US and Soviet foreign policy. One concept that received a great deal of attention as a remedy for superpower tensions was GRIT: graduated and reciprocal initiatives in tension reduction (Osgood 1962). GRIT is a method for de-escalating tensions and conflicts between nations through the actions of national leaders who engage in a series of small, tension-reducing initiatives in a turn-taking fashion. GRIT was thought to be operative (Etzioni 1967), though not explicitly acknowledged by the principals, when President Kennedy and Premier Khrushchev drew down tensions in a step-by-step fashion, culminating in the Limited Test Ban Treaty. The Treaty banned nuclear weapons tests on land, in the sea, and in the atmosphere, allowing only underground testing.

The notion of groupthink was also elaborated upon and applied to a range of foreign policy fiascos (Janis 1972). Groupthink occurs when a cohesive decision-making group engages in concurrence seeking rather than generating multiple viewpoints and dispassionately analyzing policy options. The concept of groupthink resurfaced more recently when policymakers and pundits alike used the concept to characterize the run-up to the US-led Iraq War in 2003.

Another readily available stream of scholarship for peace psychology in the 1980s was political psychology, which was increasingly addressing the topics of conflict and international relations (Jost and Sidinius 2004). Most notably, Ralph K. White's (1984) book *Fearful Warriors: A Psychological Profile of US-Soviet Relations* and White's (1986) edited volume *Psychology and the Prevention of Nuclear War* offered a rich set of concepts, themes, and perspectives on the superpower rivalry and approaches to tension reduction.

Clearly some of the most durable concepts in peace psychology were established during the Cold War. However, in a sense, the domain of peace psychology was rather narrow, focusing mainly on concepts and themes that centered around the prevention of nuclear war. As Smith (2001) noted, the prevention of nuclear war seemed to be the preeminent threat at the time, and therefore little attention was paid to the pursuit of peace in relation to intractable identity-based conflicts and other forms of violence, such as structural and cultural violence.

Intractable Conflicts and Cycles of Episodic Violence

As the Cold War wound down and the bipolar geostrategic struggle between the United States and the Soviet Union receded, Western peace psychologists turned their attention to a complex pattern of intergroup conflicts. The fault lines for intergroup conflicts did not neatly trace the contours of national borders; instead divisions were primarily intercommunal and defined by oppositional social identities (Tajfel 1974; McKeown, Haji, Ferguson 2016). Divisions emerged along ethnic, linguistic, socioeconomic, and religious lines. These identity-based conflicts were characterized by deeply divisive ideologies and violent episodes that were personal—exacting a heavy toll on the lives and well-being of family members, friends, and neighbors (Eriksson, Wallensteen, and Sollenberg 2003).

All but five of the twenty-three wars being fought during the first few years following the Cold War were based on communal rivalries or ethnic challenges to states (Gurr 1994). Ethnic identities played a major role in the Balkans and the Caucasus, Africa and South Asia. Some of these conflicts were a result of the unwinding of centralized economies and other Soviet influences. For instance, the collapse of communism and the Yugoslav wars left the Balkan region (roughly Southeast Europe) with a host of conflicts and difficult transitions: from centralized to market-driven economies, multiculturalism to ethnonational identities, rapidly changing migration patterns, and the difficulty of coping with a traumatic past (Simić and Volčič 2012).

The post–Cold War period also saw the attention of scholars and activists migrate toward the conflicts in Northern Ireland, South Africa, and the Middle East, especially the Israeli-Palestinian conflict. Many of these conflicts were marked by repeated cycles of violent episodes and could be classified as "intractable conflicts" because they were intergenerational and viewed by the participants as existential and zero-sum in nature (Bar-Tal 2007; Coleman 2003). Not surprisingly, peace psychological research and practice began to move toward an analysis of these intractable conflicts and potential interventions designed to prevent and mitigate these repeated cycles of violence.

Each cycle of violence in an intractable conflict can be divided into three phases: conflict, violence, and postviolence. Interventions that interrupt cycles of violence can be tailored to fit each particular phase (Christie and Louis 2012). In the following sections, we examine each phase of a cycle and some corresponding interventions.

Conflict Phase of a Cycle

The conflict phase can be treated as an opportunity to engage in creative problem-solving and arrive at mutually beneficial outcomes (Kriesberg 2007). The opportunity to prevent violence during the conflict phase arises because the field of psychology rather sharply distinguishes thoughts (conflict) from actions (violence). During the conflict phase, the opposing groups have differences in interests, views, or goals, which may be real or imagined. Typically each side in an intractable conflict believes that the aspirations of both parties cannot be achieved simultaneously (Pruitt and Kim 2004); thus each party perceives its interests as incompatible with the interests of the other party. Unlike violence, in the conflict phase the parties may or may not have violent feelings toward each other, but importantly, if they do have violent feelings they do not act on them, which makes it possible to intervene and attempt to resolve issues that are fueling the conflict before the outbreak of violence.

Many problem-solving approaches have been proposed in relation to the conflict phase of a cycle: A partial list includes interactive conflict resolution, preventive diplomacy, negotiation strategies, constructive controversy, a cooperative orientation, intergroup empathy, creating an emotional climate for peace, adopting nonviolent values, dialogue methods, appreciative inquiry, a common ingroup identity model, deprovincialization, and the use of communication and listening skills (cf. Christie 2011).

Violence Phase of a Cycle

If interventions during the conflict phase do not result in a peaceful resolution, the conflict phase can become an antecedent to organized violence. Typically in the violence phase, the parties become even more polarized and options for interventions narrow considerably. When violence is raging, a host of cognitive, emotional, and behavioral processes militate against movement toward peace between the parties (Christie and Louis 2012). The result can be a "mutually hurting stalemate" in which neither party can win or defeat the other, yet both parties experience unacceptable costs. Mutually hurting stalemates can become ripe and lead to the de-escalation of violence if several conditions are met: both parties perceive a way out of the stalemate, believe a negotiated solution is possible, and are willing to engage in negotiations (Zartman 2000). Pruitt (2011) has recast ripeness theory in psychological terms by explaining that "readiness" to negotiate is a function of the degree to which parties are motivated to escape the situation and are optimistic about the possibility of reaching a mutually beneficial outcome.

Because political leaders are often under pressure to show toughness and resolve, it can be difficult for either side to conciliate. However, in a number of cases, readiness has been helped along through back-channel talks or unofficial diplomacy and other interventions that fall within the general rubric of ICR, or interactive conflict resolution (Fisher 2005; d'Estrée 2012). ICR is a collection of methodologies that brings together influential members of the parties in conflict. Although agreements between the parties are not binding, because of their status in society, they may catalyze changes in the larger political community and ultimately influence official diplomatic negotiations. ICR has been applied in Cyprus, Malaysia/Indonesia, Northern Ireland, Sri Lanka, the Horn of Africa, and the Israeli-Palestinian conflict (Christie and Louis 2012).

Postviolence Phase of a Cycle

Violence eventually subsides, with or without interventions. UN documents often refer to the postviolence phase as "postconflict." However, given the sharp distinction psychologists make between thought (conflict) and action (violence), the phase following violence is more accurately captured by the concept of postviolence. In 1991 United Nations Secretary-General Boutros Boutros-Ghali introduced an Agenda for Peace, which recognized the importance of "postconflict" interventions as a means of preventing additional cycles of violence. At about the same time, there was increasing awareness among psychologists and members of the medical community that trauma was widespread in postviolent contexts and required effective interventions (Agger 2001; de Jong 2006). After all, the relief of suffering and the promotion of well-functioning people are important prerequisites for rebuilding political institutions, physical infrastructures, and public services.

However, as trauma practitioners proliferated around the world, critiques arose about the imposition of Western conceptions of mental disorders and treatment approaches that might be applicable only in a particular cultural context (Kienzler 2008), and therefore trauma work might not comport with the maxim "Do no harm." Accordingly some practitioners advocated a "blended approach" that honored indigenous forms of healing and resilience combined with culturally sensitive psychosocial interventions to promote intergroup dialogue, community mobilization, and cooperation toward common goals on issues such as the reintegration of former soldiers and displaced people into their communities (Wessells and Montiero 2001). Taking a similar tack, some scholars and practitioners argued that a trauma-focused approach to intervention overemphasizes the impact of direct exposure to violence on mental health and fails to give full consideration to the daily stresses associated with disrupted social and material conditions (Miller and Rasmussen 2010). From such a perspective, the problem of daily stressors could be addressed first, and if distress was not reduced, then specialized treatments would seem to be indicated.

The growing appreciation for the postviolence stage has also given rise to scholarship and practice around the issues of forgiveness and reconciliation (Iyer and Blatz 2012; Kalayjian and Paloutzian 2009; Lederach 1999) as a means of preventing additional cycles of violence and, more generally, promoting peace. Peace psychological analyses have also benefited from a number of national and international initiatives on reconciliation that have taken place since 1990. The United Nations has vigorously promoted reconciliation through UNICEF and UNESCO and has drafted declarations, perhaps most notably the Declaration on the Rights of Indigenous Peoples. State-level initiatives have also been undertaken. Australia and Canada have issued formal apologies to indigenous people for forcibly removing children from their families, and Japan has apologized for its involvement in World War II as well as its exploitation of girls and women who were forced into sexual slavery by the Imperial Japanese Army during its occupation of territories both before and during World War II (Claggett-Borne 2013).

Most notable was South Africa's Truth and Reconciliation Commission (TRC), which sought to promote forgiveness and reconciliation nationwide as apartheid was being dismantled. Guided by the leadership of Archbishop Desmond Tutu, the TRC has served as a model for more than twenty other truth commissions around the world. While formats vary, generally speaking these commissions provide a forum for perpetrators of violence to publicly describe their transgressions, seek forgiveness from their victims, and work toward envisioning a common future. The effectiveness of truth commissions has been mixed. A review of recent research emphasizes several components of reconciliation worthy of further research: acknowledgment of the perpetrator's group having engaged in deliberate and undeserved material or cultural harm to victims; accepting responsibility for causing this harm; and having both parties agree on the feasibility of making restitution for doing harm (Iyer and Blatz 2012). While the growth of research and practice on reconciliation processes is encouraging, intergroup peace can be elusive if the underlying political and economic structures that privilege some segments of the population over others are not fully addressed.

In short, there is a great deal to learn about the conditions within a postviolent context that foster trust, cooperation, social cohesion, and other psychosocial processes that are necessary for the construction of a well-functioning society. Among the important considerations that have been proposed when intervening in the postviolence contexts are whether the intervention is culturally informed, supports healing processes, relieves suffering, and promotes reconciliation throughout society. And although the tools for peaceful interventions in cycles of violence are becoming more differentiated and precisely targeted, especially with greater attention given to the postviolence stage, the protracted nature of intractable conflicts suggests the problem is rooted in deep-seated structural and cultural violence.

Second Wave of Peace Psychology: Structural and Cultural Transformation

Mitigating Structural Violence

Just as intrastate conflicts were overshadowed and ignored during the Cold War, the enormity of structural violence that was taking place outside of the East-West conflict also failed to register in peace psychology. Many of the psychologists in the "periphery" (from the West's perspective), or so-called Third World peace scholars, regarded "peace" as suspect and associated with pacifism in the face of oppression and exploitation. For these psychologists, the focal concern was structural violence, not episodic violence (Montiel and Noor 2009). In the first text with "peace psychology" in it title (Christie, Wagner, and Winter 2001), structural violence was described as chronic and normalized rather than an acute and dramatic insult to well-being, which characterizes direct forms of violence. As an insidious form of violence, structural violence is built into the political and economic institutions of a society; it kills more people than episodic violence worldwide and does so slowly and indirectly through the deprivation of human need satisfaction.

Most of the population in the world is coping with structural violence; hence the term "Majority World" is often applied to people living under these conditions. Deprived of voice and representation in matters that affect their well-being, the preeminent task for people in the Majority World is what Galtung refers to as "positive peace," or the pursuit of social justice. Because peace and social justice are indivisible, peace psychology in the Majority World has often focused on nonviolent social activism as a means of transforming political and economic structures into more socially equitable and inclusive arrangements that satisfy basic human needs (Montiel and Noor 2009). In contrast to the tension-reduction approach that characterized peacemaking efforts during the Cold War, structural peacebuilding often involves tension-inducing actions through people power in order to transform vertical power structures into horizontal structures (Montiel 2001; Montiel and Noor 2009).

The tension that accompanies the creative destruction of status quo power arrangements is a prerequisite for the kind of social, political, and cultural transformations that enlarge the scope of human well-being (Christie 2018). Efforts to promote destructive and constructive processes can be found in the work of Latin American scholars of liberation psychology, such as Ignacio Martín-Baró (1994), whose notions of liberation psychology spawned emancipatory agendas all over the world.

The spread of liberation psychology has generally followed, but is not limited to, the geographic contours of the Global South. After all, liberation psychology takes the perspective of the oppressed wherever they may be situated geohistorically. The liberation process seeks to demystify or expose the cultural violence of ideologies that obscure the workings of power structures in a society. These ideologies normalize structural violence and justify the inferior status of marginalized people who experience material privations and a lack of voice and representation in matters that affect their well-being. Demystification occurs through conscientization, a collective bottom-up process in which individuals develop a critical capacity to contest power structures and develop a sense of personal and political empowerment (Montero and Sonn 2009). A host of nonviolent activist movements can be enacted to pursue social change (Louis and Montiel 2018). The outcome of structural and cultural transformation is a society in which the satisfaction of "life-extending human needs" is enlarged to include people who occupy subaltern strata of a society (Christie 2018).

Mitigating Cultural Violence

In 1995 the fourth Biennial Symposium on the Contributions of Psychology to Peace took place in Cape Town, South Africa. For most South Africans, this was a tremendously exhilarating, heady, and hopeful time. The structural violence of apartheid and its widespread enforcement by direct violence was finally being dismantled. Twenty years later the fourteenth Biennial Symposium returned to South Africa. In many ways, much had changed as the democratization process and the constitution that enshrined the principle of human rights had fully enfranchised the Black-majority population. But at the same time, little had changed materially for Blacks, as evidenced in widespread poverty and economic inequality (cf. Gini coefficient of 0.65), both brutal reminders that political democracy does not ensure a higher standard of living that satisfies everyone's basic life-extending needs.

Unlike in 1995, in 2015 little attention was given to South African psychology's complicity with the oppressive ideological discourses and practices of apartheid (Foster 2008) or the moral bankruptcy of a clinical psychology that failed to take a critical position on the harmful effects of apartheid (Dawes 1985). Instead what took center stage was South Africans' desire to contribute to international psychology scholarship with a distinctive African perspective. To date, the voice and representation of African scholars remains marginalized in the knowledge-creation process (Seedat and MacKenzie 2008). Redressing this injustice was precisely why biennial symposia were instituted, given their stated mission to promote cultural peace by bringing forward the voices of people who are typically not represented in peace-related discourses.

In the published proceedings of the symposium (Seedat, Suffla, and Christie 2017b), we are reminded that systems of knowledge-making (epistemologies) and the way we construct reality (ontologies) are not only culturally bound in time and space; they are also embedded in power relations (Law and Bretherton 2015). After all, those who occupy the apex of power in a society control social discourses centering on which group is superior, whose history matters, and what knowledge is regarded as valid (Foucault 1980). A number of presentations sought to redress cultural violence by eliciting the stories of marginalized people from the townships in South Africa. In some instances, the narratives had an emancipatory quality and were accompanied by social activism (Lau and Seedat 2017; Lazarus et al. 2017; Stevens, Duncan, and Canham 2017). Using a critical lens and elicitive methodologies, these researchers provided a safe space for marginalized people to express their thoughts and feelings, thereby contributing to the emancipatory and knowledge-generation process in peace psychology.

Hence a second wave of research in peace psychology is seeking to hear the voices of people from the Global South who share a history of colonialism and neo-imperialism. The cultural heritage of many of these countries has been decimated and repurposed for a consumption-oriented neoliberal economic order that values individualism, acquisitiveness, and competition. The contestation of ideas and values puts cultural violence at the center of the conflict between the Global North and the Global South (Christie 2017).

The enactment of a critical peace psychology was recently on display at the sixth International Conference on Community Psychology, held in Durban, South Africa, in 2016. The methodology of choice for the critical peace psychology movement is participatory knowledge co-creation, critical reflexivity, and emancipatory action (Seedat, Suffla, and Christie 2017a). Community members take the lead in the problem-identification process and involve themselves in all phases of the research process, including the development of interventions, analysis of the findings, implementation of findings, and follow-up action. While this kind of research is admittedly messy and situated, in a visceral way it evokes a methodology of the heart (Seedat 2012).

Conclusion

Peace psychology grew out of a collective effort to apply psychology to the problems of episodic, structural, and cultural violence, and therefore peace psychologists find a congenial intellectual home in the transdisciplinary field of peace and conflict studies (Galtung 1969). From this perspective, peace psychology is not so much a subfield of psychology; instead research and practice in psychology is positioned at the center of a contextual analysis that draws heavily on the perspectives of other disciplines, including political science, economics, sociology, and cultural studies (Christie 2006). The most recent iteration of peace psychology, which foregrounds cultural violence and emancipatory praxis, substantially enlarges the scope of peace psychology beyond a more narrow focus on individual and collective actions aimed at the prevention and mitigation of episodic and structural violence. It remains to be seen whether peace psychology in the future will further enlarge its scope to include the role of individual and collective emotions in manifestations of violence and peace. Moreover peace psychology has yet to realize the promise of substantially reducing some of the most consequential threats to human survival and well-being, including the continuing prospect of nuclear war, polarized political identities, social inequalities, and climate change.

References

Agger, I. 2001. "Reducing Trauma during Ethno-political Conflict: A Personal Account of Psychosocial Work under War Conditions in Bosnia." In *Peace, Conflict, and Violence: Peace Psychology for the 21st Century*, edited by D. J. Christie, R. V. Wagner, and D. D. Winter, 240–250. Upper Saddle River, NJ: Prentice-Hall.

Bar-Tal, D. 2007. "Sociopsychological Foundations of Intractable Conflicts." *American Behavioral Scientist* 50 (11):1430–1453.

Christie, D. J. 2006. "What Is Peace Psychology the Psychology Of?" *Journal of Social Issues* 62:1–17.

Christie, D. J., ed. 2011. *Encyclopedia of Peace Psychology*. Malden, MA: Wiley-Blackwell.

Christie, D. J. 2017. "Emancipatory and Participatory Methodologies in Peace, Critical, and Community Psychology." In *Enlarging the Scope of Peace Psychology: African and World-Regional Contributions*, edited by M. Seedat, S. Suffla, and D. J. Christie, 19–38. New York: Springer.

Christie, D. J. 2018. "The Meaning and Metrics of Social and Political Transformation." *Peace and Conflict: Journal of Peace Psychology* 24:77–84.

Christie, D. J. and W. R. Louis. 2012. "Peace Interventions Tailored to Phases within a Cycle of Intergroup Violence." In *The Oxford Handbook of Intergroup Conflict*, edited by L. Tropp, 252–272. New York: Oxford University Press.

Christie, D. J., and C. J. Montiel. 2013. "Contributions of Psychology to War and Peace." *American Psychologist* 68:502–513.

Christie, D. J., B. S. Tint, R. V. Wagner, and D. D. Winter. 2008. "Peace Psychology for a Peaceful World." *American Psychologist* 63(6):540–552.

Christie, D. J., R. V. Wagner, and D. D. Winter, eds. 2001. *Peace, Conflict, and Violence: Peace Psychology for the 21st Century*. Upper Saddle River, NJ: Prentice-Hall.

Claggett-Borne, E. 2013. "Definitions of Peace and Reconciliation." In *International Handbook of Peace and Reconciliation*, edited by K. Malley-Morrison, A. Mercurio, and G. Twose, 11–21. New York: Springer.

Coleman, P. T. 2003. "Characteristics of Protracted, Intractable Conflict: Toward the Development of a Metaframework-I." *Peace and Conflict: Journal of Peace Psychology* 9(1):1–37.

Coleman, P. T., R. R. Vallacher, A. Nowak, and L. Bui-Wrzosinska. 2007. "Intractable Conflict as an Attractor: A Dynamical Systems Approach to Conflict Escalation and Intractability." *American Behavioral Scientist* 50(11):1454–1475.

Dawes, A. 1985. "Politics and Mental Health: The Position of Clinical Psychology in South Africa." *South African Journal of Psychology* 15 (2):55–61.

de Jong, J., ed. 2006. *Trauma, War, and Violence: Public Mental Health in Socio-cultural Context*. New York: Springer Science & Business Media.

d'Estrée, T. P. 2012. "Addressing Intractable Conflict through Interactive Problem-Solving." In *The Oxford Handbook of Intergroup Conflict*, edited by L. R. Tropp, 229–251. New York: Oxford University Press.

Deutsch, M. 1973. *The Resolution of Conflict: Constructive and Destructive Processes*. New Haven, CT: Yale University Press.

Eriksson, M., P. Wallensteen, and M. Sollenberg. 2003. "Armed Conflict, 1989–2002." *Journal of Peace Research* 40 (5):593–607.

Etzioni, A. 1967. "The Kennedy Experiment." *Western Political Quarterly* 20:361–380.

Fisher, R. J., ed. 2005. *Paving the Way: Contributions of Interactive Conflict Resolution to Peacemaking*. Minneapolis, MN: Lexington Books.

Fisher, R., W. L. Ury, and B. Patton. 2011. *Getting to Yes: Negotiating Agreement without Giving In*. London: Penguin.

Foster, D. 2008. "Critical Psychology: A Historical Overview." In *Interiors: A History of Psychology in South Africa*, edited by D. Painter, 92–124. Pretoria: University of South Africa Press.

Foucault, M. 1980. *Power/Knowledge: Selected Interviews and Other Writings, 1972–1977*. New York: Pantheon.

Frank, J. D. 1982. *Sanity and Survival in the Nuclear Age: Psychological Aspects of War and Peace*. New York: Random House.

Galtung, J. 1969. "Violence, Peace, and Peace Research." *Journal of Peace Research* 6:167–191.

Galtung, J., and T. Höivik. 1971. "Structural and Direct Violence: A Note on Operationalization." *Journal of Peace Research* 8:73–76.

Gurr, T. R. 1994. "Peoples against States: Ethnopolitical Conflict and the Changing World System." *International Studies Quarterly* 38 (3):347–377.

Iyer, A., and C. Blatz. 2012. "Apology and Reparation." In *The Oxford Handbook of Intergroup Conflict*, edited by L. R. Tropp, 309–327. Oxford: Oxford University Press.

Janis, I. L. 1972. *Victims of Groupthink: A Psychological Study of Foreign-Policy Decisions and Fiascos*. Oxford: Houghton Mifflin.

Jost, J. T., and J. Sidanius. 2004. "Political Psychology: An Introduction." In *Political Psychology: Key Readings*, edited by J.T. Jost, and J. Sidanius, 1–17. New York: Psychology Press.

Kalayjian, A., and R. F. Paloutzian, eds. 2009. *Forgiveness and Reconciliation: Psychological Pathways to Conflict Transformation and Peace Building*. New York: Springer.

Kelman H. C. 1981. "Reflections on the History and Status of Peace Research." *Conflict Management and Peace Science* 5 (2):95–110.

Kienzler, H. 2008. "Debating War-Trauma and Post-traumatic Stress Disorder (PTSD) in an Interdisciplinary Arena." *Social Science & Medicine* 67 (2):218–227.

Kriesberg, L. 2007. *Constructive Conflicts: From Escalation to Resolution*. Lanham, MD: Rowman & Littlefield.

Lau, U., and M. Seedat. 2017. "Structural Violence and the Struggle for Recognition: Examining Community Narratives in a Post-Apartheid Democracy." In *Emancipatory and Participatory Methodologies in Peace, Critical, and Community Psychology*, edited by M. Seedat, S. Suffla, and D. J. Christie, 203–220. New York: Springer.

Law, S. F., and D. Bretherton. 2015. "The Imbalance between Knowledge Paradigms of North and South: Implications for Peace Psychology." In *Emancipatory and Participatory Methodologies in Peace, Critical, and Community Psychology*, edited by M. Seedat, S. Suffla, and D. J. Christie, 19–38. New York: Springer.

Lazarus, S, J. R. Cochrane, N. Taliep, C. Simmons, and M. Seedat. 2017. Identifying and Promoting Factors that Promote Community Peace. In *Enlarging the Scope of Peace Psychology: African and World-Regional Contributions*, edited by M. Seedat, Suffla, S. and D. J. Christie, 161–184. New York: Springer.

Lederach, J. P. 1999. *Sustainable Reconciliation in Divided Societies*. Washington, DC: US Institute of Peace Press.

Louis, W. R., and C. J. Montiel. 2018. "Social Movements and Social Transformation: Steps towards Understanding the Challenges and Breakthroughs of Social Change." *Peace and Conflict: Journal of Peace Psychology* 24:3–9.

Martín-Baró, I. 1994. *Writings for a Liberation Psychology*. Cambridge, MA: Harvard University Press.

McKeown, S., R. Haji, and N. Ferguson. 2016. *Understanding Peace and Conflict through Social Identity Theory: Contemporary Global Perspectives*. New York: Springer.

Miller, K. E., and A. Rasmussen. 2010. "War Exposure, Daily Stressors, and Mental Health in Conflict and Post-conflict Settings: Bridging the Divide between Trauma-Focused and Psychosocial Frameworks. *Social Science & Medicine* 70 (1):7–16.

Montero, M., and C. C. Sonn, eds. 2009. *Psychology of Liberation: Theory and Applications*. New York: Springer.

Montiel, C. J. 2001. "Toward a Psychology of Structural Peacebuilding." In *Peace, Conflict, and Violence: Peace Psychology for the 21st Century*, edited by D. J. Christie, R. V. Wagner, and D. D. N. Winter, 282–294. Upper Saddle River, NJ: Prentice-Hall.

Montiel, C. J. 2003. "Peace Psychology in Asia." *Peace and Conflict* 9 (3):195–218.

Montiel, C. J., and N. M. Noor, eds. 2009. *Peace Psychology in Asia*. New York: Springer.

Osgood, C. E. 1962. *An Alternative to War or Surrender*. Urbana: University of Illinois Press.

Pruitt, D. G. 2011. "Ripeness Theory." In *The Encyclopedia of Peace Psychology*, edited by D. Christie, 968–972. Malden, MA: Wiley-Blackwell.

Pruitt, D. G., and S. H. Kim. 2004. *Social Conflict: Escalation, Stalemate, and Settlement*. New York: McGraw-Hill.

Russell, R. W., ed. 1961. "Psychology and Policy in a Nuclear Age." Special issue, *Journal of Social Issues* 17 (3): 1–87.

Seedat, M. 2012. "Community Engagement as Liberal Performance, as Critical Intellectualism and as Praxis." *Journal of Psychology in Africa* 22 (4):489–498.

Seedat, M., and S. MacKenzie. 2008. "The Triangulated Development of South African Psychology: Race, Scientific Racism and Professionalisation." In *Interiors: A History of Psychology in South Africa*, edited by C. Van Ommen and D. Painter, 63–91. Pretoria: Unisa Press.

Seedat, M., S. Suffla, and D. J. Christie, eds. 2017a. *Emancipatory and Participatory Methodologies in Peace, Critical, and Community Psychology*. New York: Springer.

Seedat, M., S. Suffla, and D. J. Christie, eds. 2017b. *Enlarging the Scope of Peace Psychology: African and World-Regional Contributions*. New York: Springer.

Simić, O., and Z. Volčič, eds. 2012. *Transitional Justice and Civil Society in the Balkans*. New York: Springer.

Smith, M. B. 2001. "Foreword." In *Peace, Conflict, and Violence: Peace Psychology for the 21st Century*, edited by D. J. Christie, R. V. Wagner, and D. D. Winter, vii–viii. Upper Saddle River, NJ: Prentice-Hall.

Stevens, G., N. Duncan, and H. Canham. 2017. "Interrogatory Destabilization and an Insurgent Politics for Peacebuilding: The Case of the Apartheid Archive Project." In *Enlarging the Scope of Peace Psychology: African and World-Regional Contributions*, edited by M. Seedat, S. Suffla, and D. J. Christie, 185–202. New York: Springer.

Tajfel, H. 1974. "Social Identity and Intergroup Behaviour." *International Social Science Council* 13 (2):65–93.

Wagner, R. V. 1988. "Distinguishing between Positive and Negative Approaches to Peace." *Journal of Social Issues* 44 (2):1–15.

Wessells, M. G. 1996. "A History of Division 48 (Peace Psychology)." In *Unification through Division: Histories of the Divisions of the American Psychological Association*, vol. 1, edited by D. A. Dewsbury, 265–298. Washington, DC: APA Press.

Wessells, M., and C. Monteiro. 2001. "Psychosocial Interventions and Post-war Reconstruction in Angola: Interweaving Western and Traditional Approaches." In *Peace, Conflict, and Violence: Peace Psychology for the 21st Century*, edited by D. J. Christie, R. V. Wagner, and D. D. Winter, 262–275. Upper Saddle River, NJ: Prentice-Hall.

White, R. K. 1984. *Fearful Warriors: A Psychological Profile of U.S.-Soviet Relations*. New York: Free Press.

White, R. K. 1986. *Psychology and the Prevention of Nuclear War*. New York: New York University Press.

Zartman, I. W. 2000. "Ripeness: The Hurting Stalemate and Beyond." In *International Conflict Resolution after the Cold War*, edited by P. Stern and D. Druckman, 225–250. Washington, D.C.: National Academic Press.

PART II

PEACEBUILDING AND STATEBUILDING IN GLOBAL POLITICS

CHAPTER 16

INTERNATIONAL INTERVENTIONS

AIDAN HEHIR

SINCE the fabled Treaty of Westphalia in 1648, the state has inexorably become the unit of currency in international politics; today the ubiquity of states, and their exalted status as *the* dominant political unit, marks them out as uniquely powerful entities. Indeed, according to Danilo Archibugi (2003, 32), "Never in the history of the human race has there been such a successful structure, one which has, de facto, become of crucial importance to all the inhabitants of the planet. No single religion—not even all the religions put together—has ever held as much power as the world's states possess today." According to the UN Charter, all states are equal; in practice, of course, while all states are formally equal, *some states are more equal than others*. This inequality arguably manifests most obviously with respect to intervention.

The issue of intervention, Martin Wight (1979, 191) wrote, "raises questions of the utmost moral complexity: adherents of every political belief will regard intervention as justified under certain circumstances." Complexity and subjectivity have certainly been perennial features of the debate around intervention, and they continue to influence contemporary debates. One need think only about the nature of the contestation surrounding the crises in Syria, Myanmar, Palestine, and Yemen to see that the issue of intervention—and indeed *nonintervention*—continues to stir both controversy and emotion.

In this chapter I will outline the status of the sovereign state and how its position of preeminence frames the debate about intervention. I then compare the laws governing intervention—in its various forms—with the reality of actual state practice. International law has undeniably expanded both in remit and depth in the past hundred years, yet while these advancements have clarified what constitutes lawful ground for intervention, the lack of concomitant evolution in the means by which international law is *enforced* has meant that these rules remain prey to the deleterious influence of power.

Sovereignty, Power, and Intervention

Though sovereignty is greatly valued by states, there remains significant confusion as to what sovereignty actually is. As Arthur Watts (2001, 5) notes, "States through their rulers or governments, think of themselves as sovereign. They do not, of course, always know what sovereignty means, but it is clearly worth having and keeping." Given the pronounced sensitivities surrounding sovereignty—however states understand it—there is naturally significant unease at the very idea of intervention. Intervention, by definition, challenges sovereignty; the act of one state intervening forcibly in another state—whatever the rationale—constitutes the leveraging of external influence over the internal affairs of another state.

In the pre–UN Charter era, intervention was an overtly subjective practice; states justified their interventions on moral grounds or claimed self-defense, but in the absence of any meaningful consensus on either, in practice intervention was purely a function of realpolitik (Bellamy 2006, 93). International law began its tentative evolution only toward the end of the nineteenth century; prior to this, intervention was, essentially, constrained only by issues relating to capacity, will, and power. Some legal and moral norms existed; some—such as the Treaty of Westphalia—were even codified; but all were inherently weak and based on a very obviously hierarchical view of states and indeed humanity itself. In particular, while the basic tenets of sovereign inviolability began to emerge amongst European states from the mid-seventeenth century onward, these norms were clearly not extended to the non-Western world (Cohen 2004, 12). Thus outside Europe there existed a "terra nullis" populated by peoples who ostensibly lacked the "standards of civilisation" necessary to enjoy the right to nonintervention (Bain 2003, 66).

Thus, while by the mid-seventeenth century the treaties of Augsburg (1555) and Westphalia (1648) had begun the evolution of the state system, and specifically the related principles of internal and external sovereignty, these developments did not usher in consistent state practice or anything approximating a global legal order based on the principle of sovereign equality. Indeed, by 1815 the number of states in Europe had decreased since the Treaty of Westphalia as larger states simply absorbed smaller ones, while by 1914 the entire continent of Africa—with the exception of Ethiopia—was colonized, as were large parts of South America and Asia (Bull 2002, 17).

In the wake of the Second World War, the UN emerged as an attempt to "save succeeding generations from the scourge of war," and to this end the UN Charter outlined specific rules on sovereignty and intervention. The Charter's rules did not, of course, emerge in a vacuum; they were built, rather, on the foundations laid by a number of prior junctures in the evolution of international law, such as the League of Nations, the 1928 Kellogg-Briand Pact, and the 1933 Montevideo Convention. By virtue of consolidating these previous developments, the Charter established a set of principles which have come to frame the debate about the legality of intervention ever since.

After the establishment of the UN, it no longer made sense to talk about states possessing unilateral rights to intervene; the legal framework established by the Charter outlined the rules governing intervention and the use of force binding on all. While signatories to the Charter could of course disagree with a particular characterization of their intervention, they could no longer argue—plausibly at least—that it was *their* view that this practice was legal, whatever the Charter may say. Thus, at the very least, the UN framework created a normatively bounded space in which states could engage in a debate about an intervention—actual or imminent—but there were at least limits to the scope of their disagreements (Armstrong and Farrell 2005, 7). Of course, while the UN Charter, and supplementary international law, did much to clarify the legal basis for intervention, the practice of intervention remains both controversial and subject to the influence of power; put bluntly, powerful states have often demonstrated a willingness to subvert international law.

The influence of power on both the decision to intervene and the responses by the international community to acts of intervention has, naturally, been a key focus of debate for International Relations theorists seeking to explain the motivation driving intervention by states. On the one hand, the realist view—in both the "classical" and the "neo" manifestations (though for different reasons)—holds that intervention is inevitable; classical realists largely attribute this to intrinsic human avarice and an inherent drive to acquire more resources (Niebuhr 2013), while neorealists argue that in the absence of global governance, states exist in a condition of international anarchy; as a result, they face a perennial security dilemma and are impelled to behave at all times with their national interests to the fore and operate on the basis that they may be attacked at any time (Mearsheimer 1994–1995). States are not, so the theory goes, moral actors, and thus their interventions are always motivated by their particular reading of the prevailing distribution of power rather than by legal or moral concerns. The only feasible barrier to intervention—and means by which it can be in any way regulated—is the achievement of a balance whereby the cost of intervention would be deemed prohibitively high. By contrast, liberal theorists argue that states do in fact act on the basis of values; rather than being single-minded interest-maximizing entities, states—at least certain states—understand their security is dependent on the proliferation of certain ideas, values, and norms (Keohane and Nye 1977). When states share similar values the prospects of war between them diminishes; indeed the democratic peace theory holds that democracies do not go to war with democracies precisely because they share a common ideology (Levy 1989, 88). The liberal view attributes far more importance to international institutions as a means by which intervention is limited than do realists, as they see these bodies as capable of socializing states into a particular way of behaving through the creation of common rules and norms (Smith 2002, 217). Interventions occur, this view holds, in those cases where there is a dramatic disjuncture between the values adhered to by states and a triggering event that impels intervention. There is, however, a significant difference between the classical liberal position, which urges caution and restraint (Walzer 2011), and the modern liberal internationalist view, which advocates a much more expansive justificatory framework for military intervention (Chopra and Weiss

1992). English school theorists are divided as to the motivations behind, and legitimacy of, intervention; solidarists believe that groups, or societies, of states emerge within which shared norms and laws ensure order. States within these societies, however, are occasionally compelled to take action against states that do not share their particular set of norms (Wheeler 2002, 6). The pluralist view, by contrast, holds that states are only ever bound together by minimalist shared standards of behavior and that the influence exercised by law and norms is always limited, especially when in competition with issues of national interest (Jackson 2000). Whereas solidarists believe that the society of states can grow both in numbers and in the depth of the values they share, the pluralist view holds that the most that can be achieved is a world order that enables diversity and independence; ill-conceived attempts to impose or spread values through intervention will cause instability and lead to more instances of interstate war (Bull 1984, 13).

Intervention: The Laws and the Practice

It is beyond the scope of this chapter to provide a detailed analysis of the Charter's laws on all types of intervention, and of course since 1945 international law has evolved in tandem with judicial rulings and customary practice, all of which is again impossible to cover here. Rather, my intention is to summarize the law on five types of intervention—though debate continues to rage about each—and then compare these rules with actual state practice.

UN-Authorized Intervention

Article 2.7 of the UN Charter states, "Nothing contained in the present Charter shall authorize the United Nations to intervene in matters which are essentially within the domestic jurisdiction of any state or shall require the Members to submit such matters to settlement under the present Charter." While this appears to consolidate sovereign inviolability—thereby outlawing intervention—this Charter provision is superseded by the powers afforded to the Security Council under Chapter VII; indeed Article 2.7 goes on to state, "[T]his principle shall not prejudice the application of enforcement measures under Chapter VII." Chapter VII—collectively, Articles 39–51—outlines the Security Council's powers to enforce international law and, when deemed necessary, authorize military intervention. The key wording appears in Article 42, which states that the Security Council "may take such action by air, sea, or land forces as may be necessary to maintain or restore international peace and security." Thus, from its inception, the Charter included an exception to the nonintervention rule. The practical application of Chapter VII, however, has always been controversial.

The obvious problem with Chapter VII relates to the fact that it can be activated only if the permanent five members of the Security Council (P5) all agree to, or at the very least don't oppose, a proposal to intervene. Given the divisions wrought by the superpower standoff, this was, unsurprisingly, effectively impossible to achieve during the Cold War. With the dissolution of the Soviet Union came the increased invocation of Chapter VII–mandated action by the Security Council; yet throughout the 1990s, while resolutions citing Chapter VII were passed more frequently, there was manifestly little consistency. Chapter VII action always cohered with the national interests of the P5, meaning that when these states had no national interests involved there was no action, while in cases where their interests collided, collective action under Chapter VII was impossible (Chesterman 2002, 5).

Another problem relates to the ambiguity surrounding the wording of Article 42; specifically, what exactly constitutes a threat to "international peace and security," and who decides this? Throughout the Cold War this question was largely of limited academic interest, but once the P5 began to authorize more Chapter VII resolutions it naturally became a more pressing issue. Did it include, many wondered, cases where governments were persecuting their own people? While such cases were clearly a violation of the human rights of the victims, did they actually constitute a threat to *international peace and security*? Furthermore, if Chapter VII could be interpreted to enable the P5 to authorize an intervention to protect human rights, would this be applied in all cases or just those of interest to the P5? The record clearly suggested the latter, and as such Chapter VII interventions—though legal—were a source of controversy precisely because they were by definition inherently political (Chesterman 2011, 280).

The application of Chapter VII has been further undermined by the fact that the UN has no standing army; as a result, intervention authorized under Chapter VII requires that states are willing to supply the troops and logistics required to actually carry out the intervention (Hehir 2012, 236). This means in practice that a UN-authorized intervention manifests as a state-led military intervention, with all the attendant problems associated with national interests determining both the motives and means involved. This was illustrated in 2011, when the Security Council authorized an intervention in Libya; under the terms of Resolution 1973, authorization was granted to impose a no-fly zone over Libya. In practice, however, the intervening coalition engaged in far more than just establishing a no-fly zone; the goal of the intervention soon became "regime change," and those involved became, in effect, active parties in the struggle against Muammar Gaddafi's forces. While many of course welcomed this, others—such as Russia, China, and South Africa—declared that this constituted a violation of the terms of Resolution 1973 (Hehir 2018a, 10).

Thus while intervention by the UN Security Council is permissible under international law, it remains highly controversial. The P5 *can* authorize intervention, but it certainly does not *have to* (Berman 2007). Any intervention will in effect be carried out by states with their respective interests to the fore, and owing to the inherently political nature of P5 deliberations, there has yet to be, nor is there likely to be, anything approximating consistent practice with respect to the authorization of Chapter VII intervention.

Aggression

At the Nuremberg trials aggression was described as "not only an international crime" but "the supreme international crime differing only from other war crimes in that it contains within it the accumulated evil of the whole" (Thomas 2003, 193). Given the destruction caused by the First and Second World Wars, this view was clearly not unfounded. There was broad consensus that aggression should be proscribed, and this was reflected in the Charter, specifically Article 1.1.

Yet there were two problems with this proscription. First, if a state is determined to use force against another state, is it conceivable that it will be dissuaded from doing so because the UN Charter proscribes it? The UN was afforded very limited powers for actually enforcing international law; as detailed earlier, without the Security Council's consent, punitive action cannot be initiated. Thus, absent realistic prospects of censure, why *wouldn't* a state initiate an aggressive intervention if it felt it was in its interests to do so? Of course, the military capabilities of the other state—and its allies—may be sufficiently prohibitive, but again, in a situation where there is a power asymmetry and Security Council division, the system essentially relies upon states honoring their promises. Second, while we may understand aggression simply as one state using force against another for material gain, many instances of alleged aggression are not presented by the aggressors in these terms; states rarely justify their interventions in terms that clearly denote aggression. They invariably describe their interventions instead as humanitarian and/or some form of self-defense. In the absence of an impartial body within the UN's architecture empowered to both issue binding judgements on whether an intervention was aggression and subsequently punish those found guilty of so doing, the laws proscribing intervention are inherently weak. Additionally, the mechanisms for punishing aggression are so inherently political that censure—at least through the UN—has been all but impossible to invoke. Thus, while the US-led intervention in Iraq in 2003 and the Russian intervention in Georgia in 2008 appeared to constitute acts of aggression—and indeed were widely deemed to be illegal—both the US and Russia avoided any censure by virtue of their veto power at the UN Security Council.

Nonetheless the International Criminal Court (ICC) has taken steps toward activating its jurisdiction over the crime of aggression. While aggression was listed as one of the ICC's four core crimes in its 1998 Rome Statute, the term was not initially defined. In June 2010 the first Review Conference of the Rome Statute of the ICC finally established a definition, namely, "the planning, preparation, initiation or execution, by a person in a position effectively to exercise control over or to direct the political or military action of a State, of an act of aggression which, by its character, gravity and scale, constitutes a manifest violation of the Charter of the United Nations."

In December 2017 a resolution was formally passed granting the ICC the right to pursue alleged perpetrators of aggression after 17 July 2018. While this is certainly a step toward both further clarifying aggression and potentially deterring acts of aggression, the ICC is itself hamstrung by power politics; the US, Russia, and China are not signatories, and their capacity to refuse to support the delegation of cases to the ICC impairs the Court's ability to work impartially (Kersten 2016, 7).

External Support

Perhaps the polar opposite of aggression is intervention undertaken with the consent of the host state. This occurs when the government of a state appeals to another state, or group of states, to intervene to support them. Though not mentioned in the UN Charter, this practice also includes peacekeeping; by definition, UN peacekeeping can be undertaken only with the host state's consent. If a state *invites* another state to intervene militarily or accedes to the UN's desire to establish a peacekeeping force inside its territory, there is no issue with international law.

Yet the issue is in practice more complicated. First, establishing that consent has actually been given can be difficult; it is dependent on the consent of the host state's government, though this raises the question, who is the legitimate government? This was an issue in the case of Haiti in 1994; the de jure government of the president supported US intervention, though the de facto government—a cabal that came to power after a military coup in 1992—opposed the intervention (Hehir 2013b, 145). Also, there is the possibility that consent can be coerced (Johnstone 2011, 170; Cassese 2005, 345). For example, Indonesia technically consented to an intervention by an Australian-led peacekeeping mission into East Timor in 1999, though this consent was the product of overt threats made against the government by a number of powerful states. More recently, Russia's intervention in Syria has been undertaken with the consent of President Bashar al-Assad; whether Assad's government has any legitimacy within Syria is a moot point, and many states have claimed that—by virtue of his widespread oppression—he no longer has legitimate authority to call upon Russian support.

Self-Defense

The Charter's rules on self-defense are, on one level, clear; Article 51 states, "Nothing in the present Charter shall impair the inherent right of individual or collective self-defence if an armed attack occurs against a Member of the United Nations, until the Security Council has taken measures necessary to maintain international peace and security." However, while this provision recognizes the long-established right of states to defend themselves once attacked, it has proved ambiguous and malleable. In particular, the phrase "after an armed attack has occurred" has led to much criticism (Gow 2011). No state, many have claimed, should have to *wait* to be attacked before it can defend itself; if a state has issued a threat and amassed troops for an imminent attack, surely, so the logic goes, the threatened state has the right to act in self-defense prior to actually incurring the assault. This argument has led to long-running debates about the legality of *preemptive* self-defense, namely military action taken against a state or group of states *about* to launch an attack. While there is clearly some logic behind the legitimacy of preemption, in practice this has proved difficult to regulate as *preemptive* self-defense has often become conflated with *preventative* self-defense. In the case of the latter, action is taken against a foe on the basis that an assault in the future is *likely*, not that it is clearly *imminent*. This distinction between preemptive and preventative self-defense became especially controversial in the wake of the 2002 US National Security Strategy; while the US outlined its view that preemption was both prudent and legal, the actual action described in the document cohered more with preventative action, which has long been deemed illegal. The US intervention against Iraq was justified as preemptive self-defense, for example, but the actual alleged threat the intervention was designed to nullify—namely that posed by Iraq's possession of weapons of mass destruction—was in reality *not* imminent.

An additional issue with the law governing self-defense is proportionality; while a state may have the right to defend itself against attack—actual or imminent—in practice the scale of the response has proved highly controversial. This has manifested a number of times with respect to Israel's disproportionate response to attacks perpetrated by Hamas and Hezbollah. Israel has a long history of interpreting its right to self-defense in an expansive way; in 1967 it launched attacks against Syria on the basis of its inherent right to self-defense, culminating in the occupation of Syrian territory in the Golan Heights, and in 1981 it destroyed a reactor plant in Iraq, claiming it was to be used to make nuclear weapons. Likewise, in the wake of the September 11, 2001 terrorist attacks, the US launched a military intervention into Afghanistan, asserting it was acting in self-defense. The US had secured some degree of Security Council backing for its actions; in September 2001 the UN Security Council passed Resolution 1373, which constituted a significant reinterpretation of self-defense. Though the intervention in Afghanistan was not explicitly authorized by the Security Council, Resolution 1373 did constitute a new conception of self-defense, one that significantly stretched the preexisting understanding of the concept and appeared to many to legitimize military intervention in Afghanistan (Gow 2011). It is at the very least debatable, however, whether a full-scale military invasion, regime change, and the creation of an international administration within a previously sovereign state was ever envisaged by the UN Security Council as permissible under Resolution 1373.

Humanitarian Intervention

Humanitarian intervention has long existed in the "shadowy periphery" of international law (Farer 2003, 143). From a strictly positivist reading of the Charter it is simply illegal; no provisions within the Charter explicitly legitimize military intervention in the event that human rights violations—or even a mass-atrocity crime—are taking place inside a sovereign state. While the UN Charter's preamble speaks of upholding human rights as being a core goal of the organization, there is in practice nothing in the Charter that permits humanitarian intervention, and much that in fact appears to explicitly proscribe it (Hehir 2013a, 102). Likewise the generally agreed view is that supplementary international law also proscribes the measure. In practice, however, the lack of any legal basis for humanitarian intervention has ended neither the practice itself nor the debate surrounding its permissibility.

During the Cold War instances of putative humanitarian intervention were rare; India's intervention in East Pakistan in 1971, Vietnam's 1978 intervention in Cambodia, and Tanzania's intervention in Uganda in 1979 are generally seen as the cases that most closely cohere with the generally understood meaning of the term. Even these cases, however, proved highly controversial and not just because of their legality. These cases each involved the states intervening offering a justification for their actions that included, to varying degrees, appeals to self-defense alongside a determination to end suffering, thus whether these actually constituted cases of humanitarian intervention was in some doubt. The international community's response to each failed to clarify the permissibility or otherwise of humanitarian intervention, given the highly politicized nature of the response by states to each (Hilpold 2001, 445).

When the Cold War ended, many hoped that the UN, ostensibly freed from the paralysis imposed by the superpower stand-off, would be able to take a more proactive approach to humanitarian intervention (Berdal 2003). Initially this optimism appeared to come to fruition; the UN's response to the plight of the Kurds in Iraq in 1992, coupled with the intervention in Somalia were heralded as evidence of a new disposition. The response to the 1994 Rwandan genocide, the 1995 massacre at Srebrenica, and the retreat from Somalia, however, soon highlighted that the UN's ability to generate—and sustain—international action in response to intrastate crises was limited; while the Cold War had ended, there was little evidence that states had cultivated a new collective determination to engage in humanitarian intervention even in situations of dire need.

NATO's intervention in Kosovo in 1999 led to a renewal of optimism, at least among some (Human Rights Watch 2000). The intervention was legitimized as an effort to protect Kosovo's Albanian population from President Slobodan Milošević's state-sponsored aggression, but the military action was taken without Security Council authorization and was therefore technically illegal. The majority of Western states supported the intervention, however, the rationale being that it was "illegal but legitimate" (Independent International Commission on Kosovo 2000, 4). This, of course, led to much discussion at the UN, and among academics, as to the legal status of humanitarian intervention. It was unconscionable, many argued, that P5 states like Russia and China, with poor human rights records, should have the power to stop action that would prevent or halt the slaughter of innocent people (Bellamy 2002, 212). Critics argued, however, that humanitarian intervention—whatever the rhetoric of those engaged—was an act of power, determined by national interests, which would only ever be reserved for a small number of powerful states (Simpson 2004, 70). It would, they argued, degrade international law, facilitate neo-imperialism, and ultimately jeopardize international peace and stability.

The legacy of the intervention in Kosovo is a matter of some debate; for some, it led directly to the ill-fated interventions in Afghanistan (2001) and Iraq (2003), and ultimately to the Russian interventions in Georgia (2008) and Ukraine (2014). Kosovo, so this argument goes, undermined international law and ushered in a new era of unilateral interventions cynically portrayed as humanitarian. For others, however, Kosovo led directly to the creation of the Responsibility to Protect (R2P), which, so its proponents claim, has come to "change the world" (Bellamy 2015, 111). R2P certainly emerged from the debates about humanitarian intervention which increased after Kosovo, though its efficacy remains a matter of some dispute. R2P has undeniably become "part of the world's diplomatic language," having been recognized by all the world's states in 2005 and increasingly affirmed by the UN General Assembly and Security Council (Welsh 2013, 378). Yet since it was recognized, atrocity crimes and human rights violations have *increased*, and the international community's response to the crises in Darfur (2005–2008), Sri Lanka (2009), Syria (2011–2018), and Myanmar (2015–2018) was by any standards derisory (Hehir 2018a, 3). The efficacy of R2P has also suffered as a result of the intervention in Libya in 2011; while many cautioned that the intervention did not in fact signal a new willingness on the part of the international community to act in response to looming or actual human rights violations, others adopted a far more celebratory stance, heralding the intervention as evidence that R2P "works" (Hehir 2018b, 40). As Libya has degenerated into civil war since the intervention, however, this has naturally cast a shadow over the effectiveness of humanitarian intervention and R2P more generally.

The international response to the situation in Syria has highlighted the ongoing contestation and dilemmas surrounding humanitarian intervention. While the scale of human suffering in Syria has been enormous, the international response has been widely criticized for being inept, if not in fact counterproductive. External powers—most particularly Russia and Iran—have been derided for their support of Assad, while many have criticized the West for encouraging the rebel groups but at the same time refusing to actually intervene to remove Assad. In 2017 and 2018 the US and its allies *did* launch air strikes against Assad's forces, and President Donald Trump claimed the motivation was his own revulsion at the images of "beautiful babies" killed by Assad's chemical weapons. Aside from questioning the effectiveness of these limited strikes, many wondered about their legality. In neither case were the airstrikes authorized by the UN, and the UK's legal defense of the 2018 airstrikes—that they were permitted under the "doctrine of humanitarian intervention"—led many to reject this reading of the law and lament that once again "humanitarian intervention" proved to be little more than an ill-defined term employed by big powers to obscure the illegality of their military interventions.

Postintervention?

One of the more prominent issues to have emerged in the post–Cold war era related to the debates on intervention is the issue of what happens *postintervention*. Whether an intervention has taken the form of peacekeeping, such as in East Timor in 1999, preemptive self-defense, such as in Iraq in 2003, or humanitarian intervention, such as in Libya in 2011, the question "What next?" has proved controversial, and the actual impact of interventions on peace and stability in the target state has been mixed.

During the Cold War, state-initiated interventions were largely undertaken for overtly geopolitical reasons, while UN peacekeeping was essentially designed only to keep warring parties apart; as such, the postintervention phase was essentially oriented toward maintaining a narrow understanding of order. Since the Cold War, however, more emphasis has been placed on achieving ambitious goals related to "postconflict reconstruction," "statebuilding," and "peacebuilding." The more expansive aims of these endeavors have tended to include the development of democratic government, fostering reconciliation, and implementing transitional justice as long-term aims of the intervention (Visoka 2016). Hence the interventions in Kosovo, East Timor, Iraq, Afghanistan, and Libya in particular were legitimized not just on the basis of the need to provide immediate solutions to a pressing problem but as a means to achieve transformative long-terms goals in the postintervention phases.

The record of these ambitious postintervention reconstruction projects, however, has been decidedly mixed; Afghanistan, Iraq, and Libya stand out as obvious failures, and while the situations in Kosovo and East Timor are less clearly calamitous, neither can reasonably be held up as a template to emulate. This record of postintervention reconstruction has had a profound effect on the debates around, and the practice of, intervention; when facing contemporary intrastate crises and calls for "something" to be done, many have cautioned that intervention also requires postintervention reconstruction and noted that there is limited evidence to suggest that this can be achieved. This indeed has been one of the arguments made against intervention in Syria; while removing Assad may be possible, many warn that holding Syria together after his departure will be impossible and in fact likely result in another Iraq/Libya scenario. Linked to this in the current era when Western power is declining, when Western states have adopted a more insular foreign policy, and when the prospects for agreement at the Security Council for expansive peacekeeping operations have diminished as old rivalries reemerge, the prospects of interventions designed to achieve ambitious ends have arguably decreased markedly (Hehir 2017).

Conclusion: Intervention Today

A review of the practice of intervention in its various guises shows that it is determined much more by politics and power than by law. Taken together, the UN Charter and supplementary international law comprise detailed rules as to the legality of intervention. While some issues, such as preemptive self-defense and humanitarian intervention, are by definition difficult to tightly regulate due to their inherent subjectivity, international law on intervention has been violated not because it is unclear but because it is possible for certain states to ignore or subvert it. Rules on intervention are, in effect, little more than words on paper, given that they cannot actually be enforced consistently. Nonetheless it is worth noting that in the past hundred years international law has made dramatic progress, and its evolutionary trajectory—when considered holistically—points toward the strengthening of the legal order.

While the end of the Cold War constituted a tumultuous change in the global distribution of power, and the post-2008 global financial crisis has seen the power of the West steadily diminish, these changes have not fundamentally altered the underlying factors that govern intervention; until such time as international law becomes more robust—if it ever does—we are left with national interests and military capacity as the key determinants of when and where intervention occurs. There is therefore a pressing need for more innovative thinking on how best to reform international law so as to better regulate intervention; the current international legal order is, as Hans Kelsen (1945, 338) wrote, "primitive," and yet surprisingly little attention has been paid to examining how to move beyond this stage. For many, the idea of UN reform and/or substantive legal change is little more than a utopian fantasy; there is, however, as E. H. Carr (2001, 10–12) notes, a significant difference between utopianism and idealism. While the former is divorced from reality, the latter takes the current state of affairs as a starting point and articulates prescriptions that may not be feasible now but do point toward an evolutionary path that is at least conceivable, while also being ambitious (Hehir 2017). Unless more effort is devoted to thinking in these terms about the regulation of intervention, it can hardly come as a surprise if we continue to live in a world where the practice of intervention is essentially a function of narrow state interests and, as a result, a profound threat to international peace and security.

References

Archibugi, D. 2003. "Cosmopolitical Democracy." In *Debating Cosmopolitics*, edited by D. Archibugi, 1–15. London: Verso.

Armstrong, D., and T. Farrell. 2005. "Introduction: Force and Legitimacy in World Politics." In *Force and Legitimacy in World Politics*, edited by D. Armstrong and T. Farrell, 3–14. Cambridge: Cambridge University Press.

Bain, W. 2003. *Between Anarchy and Society*. Oxford: Oxford University Press.

Bellamy, A. 2002. *Kosovo and International Society*. Basingstoke, UK: Palgrave Macmillan.

Bellamy, A. 2006. *Just Wars: From Cicero to Iraq*. London: Polity.

Bellamy, Alex 2015. *The Responsibility to Protect: A Defence*. Oxford: Oxford University Press.

Berdal, M. 2003. "The UN Security Council: Ineffective but Indispensable." *Survival* 45 (2):7–30.

Berman, F. 2007. "Moral versus Legal Legitimacy." In *The Price of Peace*, edited by C. Reed and D. Ryall, 157–176 Cambridge: Cambridge University Press.

Bull, H. 1984. *Justice in International Relations: Hagey Lectures*. Ontario: University Publications Distribution Service.

Bull, H. 2002. *The Anarchical Society*. Basingstoke, UK: Palgrave Macmillan, Columbia University Press.

Carr, E. H. 2001. *The Twenty Years' Crisis*. Hampshire, UK: Palgrave.

Cassese, A. 2005. *International Law.* Oxford: Oxford University Press.

Chesterman, S. 2002. *Just War or Just Peace?* Oxford: Oxford University Press.

Chesterman, S. 2011. "'Leading from Behind': The Responsibility to Protect, the Obama Doctrine, and Humanitarian Intervention after Libya." *Ethics and International Affairs* 25 (3):279–285.

Chopra, J., and T. G. Weiss. 1992. "Sovereignty Is No Longer Sacrosanct: Codifying Humanitarian Intervention." *Ethics & International Affairs* 6 (1):95–117.

Cohen, J. 2004. "Whose Sovereignty? Empire versus International Law." *Ethics and International Affairs* 18 (3):1–24.

Farer, T. 2003. "The Ethics of Intervention in Self Determination Struggles." In *Ethics and Foreign Intervention*, edited by D. Chatterjee and D. Scheid, 143–167. Cambridge: Cambridge University Press.

Gow, J. 2011. "Principles of Pre-emption." In *International Law, Security and Ethics*, edited by A. Hehir, N. Kurht, and A. Mumford, 111–128. London: Routledge.

Hehir, A. 2012. *The Responsibility to Protect: Rhetoric, Reality and the Future of Humanitarian Intervention.* Hampshire, UK: Palgrave Macmillan.

Hehir, A. 2013a. *Humanitarian Intervention: An Introduction.* Hampshire, UK: Palgrave Macmillan.

Hehir, A. 2013b. "The Permanence of Inconsistency." *International Security* 38 (1):137–159.

Hehir, A. 2017. "'Utopian in the right sense': The Responsibility to Protect and the Logical Necessity of Reform." *Ethics and International Affairs* 31 (3):335–355

Hehir, A. 2018a. "Denial, Fatalism, and the Protection of Human Rights." In *Protecting Human Rights in the 21st Century*, edited by A. Hehir and R. W. Murray, 1–16. London: Routledge.

Hehir, A. 2018b. *Hollow Norms and the Responsibility to Protect.* Hampshire, UK: Palgrave Macmillan.

Hilpold, P. 2001. "Humanitarian Intervention: Is There a Need for a Legal Reappraisal?" *European Journal of International Law* 12 (3):437–467.

Human Rights Watch. 2000. "Introduction." In *World Report 2000,* accessed 10 October 2018. https://www.hrw.org/legacy/wr2k/Front.htm#TopOfPage.

Independent International Commission on Kosovo. 2000. *Kosovo Report.* Oxford: Oxford University Press.

Jackson, R. 2000. *The Global Covenant: Human Conduct in a World of States.* Oxford: Oxford University Press.

Johnstone, I. 2011. "Managing Consent in Contemporary Peacekeeping Operations." *International Peacekeeping* 18 (2):168–182.

Kelsen, H. 1945. *General Theory of Law and State.* Cambridge MA: Harvard University Press.

Keohane, R., and J. Nye. 1977. *Power and Interdependence: World Politics in Transition.* Boston: Little, Brown.

Kersten, M. 2016. *Justice in Conflict.* Oxford: Oxford University Press.

Levy, J. 1989. "Domestic Politics and War." In *The Origin and Prevention of Major Wars*, edited by R. Rotberg and T. Rabb, 79–100. Cambridge: Cambridge University Press.

Mearsheimer, J. J. 1994–1995. "The False Promise of International Institutions." *International Security* 19 (3):5–49.

Niebuhr, R. 2013. *Moral Man and Immoral Society.* Louisville, KY: John Knox Press.

Simpson, G. 2004. *Great Powers and Outlaw States.* Cambridge: Cambridge University Press.

Smith, M. 2002. "Liberalism and International Reform." In *Traditions of International Ethics*, edited by T. Nardin and D. R. Mapel, 201–224. Cambridge: Cambridge University Press.

Thomas, R. 2003. "Wars, Humanitarian Intervention and International Law." In *Yugoslavia Unraveled*, edited by R. Thomas, 3–40. Oxford: Lexington.

Visoka, G. 2016. *Peace Figuration after International Intervention.* London: Routledge.

Walzer, M. 2011. "The Case against Our Attack on Libya." *New Republic*, 20 March 20. http://www.tnr.com/article/world/85509/the-case-against-our-attack-libya.

Watts, A. 2001. "The Importance of International Law." In *The Role of International Law in International Politics*, edited by M. Byers, 5–16. Oxford: Oxford University Press.

Welsh, J. 2013. "Norm Contestation and the Responsibility to Protect." *Global Responsibility to Protect* 5 (4):365–396.

Wheeler, N. 2002. *Saving Strangers: Humanitarian Intervention in International Society.* Oxford: Oxford University Press.

Wight, M. 1979. *Power Politics.* Harmondsworth, UK: Penguin.

CHAPTER 17

PEACEKEEPING

MICHAEL PUGH

WHEN Canadian Foreign Minister Lester Pearson and UN Secretary-General Dag Hammarskjöld invented the peacekeeping concept, they were building on activities conducted by the International Committee of the Red Cross, such as prisoner rescue convoys, at the end of the Second World War. These were patently humanitarian and nonoffensive. Indeed the image conjured up by the word "peacekeeping" is of uniformed soldiers in blue berets or helmets and white-painted vehicles on patrol, keeping hostile forces apart and supervising a prearranged armistice between combatants. Operations had a modest beginning, with UN military observers monitoring the borders between Greece and its northern neighbors, defusing flashpoints between Israeli and Arab forces in the Middle East, and monitoring borders in Kashmir.

The main contention here is that peacekeeping represented an integral part of power asymmetry in global politics for disciplining unruly states. Peacekeeping interacted with other instruments, such as preventive diplomacy and peacebuilding. It heralded the Atlanticist and Anglophone concept of liberal peace (peacebuilding ideas having been incubated in North America). In practice the system operated as a double asymmetry; peripheral and former colonial states were objects of intervention, and former colonies lately did most of the peacekeeping. Moreover, since the 1940s and 1950s peacekeeping has become more robust and expansive, more ambitious in attempting to change conflict environments to prepare for and overlap with peacebuilding.

This chapter considers where peacekeeping occurs, who does it, and why. It examines the institutional growth that became an industry, briefly discusses evolutions in the Cold War and after, interrogates the "robust turn" in peacekeeping and its consequence for peacebuilding, and assesses the 2018 recommitment to peacekeeping and its cohesive relationship with peacebuilding.

Where Peacekeeping Occurs, Who Contributes, and Why

Peacekeeping continues to be involved in the Middle East and Asia, operating as a multilateral legacy of imperialism in a dialectic of power between former colonial powers (metropolitans) and former colonies. Of the thirteen peacekeeping operations at the end of 2020, all were in territories that had been European "possessions" or mandates, apart from Kosovo, until the Second World War or thereafter. Seven were in Africa, four in the Middle East (including Cyprus) and one in Asia. Moreover, from 1945 to the end of 1996, 90% of all wars occurred in parts of the world with high poverty rates and high vertical and horizontal inequalities (Jakobeit 1966–).

Neither peacekeeping nor peacebuilding could intrude on metropolitan homelands. Two of the five permanent UN Security Council members (the UK and France) are former colonizers, and a third (the US) an ally and locomotive of economic diffusion and control. As Philip Cunliffe (2013, 261) astutely observes, peacekeeping in the Cold War period was contingent on a unipolar world and could be "conceived as an adumbration of US power, even if one mediated through international institutions." Other close allies, such as Israel, were also exempt from hosting peacekeepers. In addition, the metropolitans also took advantage of de facto, as well as UN Security Council de jure, dispensations to undertake interventionist policies.

Contributors and Motivations

A further asymmetry involves postcolonial states sending peacekeepers into other postcolonial states. For Palestine, in 1948 the UN Truce Supervision Organization of military observers had been drawn from Belgium, France, and the US, and until the 1970s prominent peacekeeping contributors were from the Nordic states, former British "Dominions" and Ireland, and European countries such as Poland and neutral Austria. But of the top twenty troop-contributing countries (TCCs) in November 2020 (each volunteering more than a thousand uniformed personnel to the UN), thirteen were from Africa, four from the Indian subcontinent, two from East Asia/Pacific, and one from Latin America. China was the only exception in having avoided full European colonization. Those in the top ten (i.e., providing more than two thousand each) are shown in Box 17.1.

The dominating presence of low- and low- to middle-income TCCs suggests more than one paradox. They are not all among leading "freedom" ranked democracies, and yet since the 1980s peacekeeping was anticipated to clear the way for freedom and democracy. They themselves do not lack supposed measures of "fragility": provincial unrest, vulnerability to corruption, gross inequalities, and extreme poverty. Yet leading TCCs participate in an enterprise that generally aims to establish stable security in other

Box 17.1 Top 10 Military and Police Contributions to Peacekeeping and Peacebuilding as of October 2020

Bangladesh	6,726
Ethiopia	6,725
Rwanda	6,363
Nepal	5,714
India	5,424
Pakistan	4,478
Egypt	3,160
Indonesia	2,828
China	2,548
Ghana	2,513

Source: UN Peacekeeping Open Data Master Sets.

former colonies to allow peacebuilding to implant "good governance." Peacekeeping could be described as South-South intervention were it not for the weak influence of the TCCs in its modeling.

Motivations encompass a mélange of factors that can vary over time. Several states do not volunteer (e.g., Vanuatu, United Arab Emirates, Turkmenistan); decisions to contribute range widely. A spare military capacity and the lure of operational experience are significant for some contributors. Traditions of refugee protection and multilateralism, as in Ghana's constitutional provision to support peacekeeping (including the Kofi Annan Training Centre), can contribute to a high participation rate. Other incentives include bureaucratic rationales in state security bodies: bidding for a temporary Security Council seat or a quest for international standing; promoting regional, precolonial, or geopolitical interests; and cultivation of economic connections. China is a particularly interesting contributor in light of its traditional sovereignty principles. It maintains peacekeepers as symbols of its international status as a major diaspora nation and fosters "good business" rather than liberal "good governance." Commentators often assume that states commit to peacekeeping for financial rewards and hard currency, but this is unconvincing. Reimbursing the costs of deployment is sluggish, and the UN is usually in arrears. Although peacekeeping earnings for Rwanda in 2011 represented 46% of its military expenditure, for India and Pakistan it was less than 1% (Cunliffe 2013, 172; Bellamy and Williams 2013).

Significantly, India was the first to deploy a female unit (for noncoercive policing). However, in spite of strong social justice rationales (women usually bear heavy economic burdens during and after conflict), and also for enhancing the durability of peace (see Olsson and Gizelis 2014; Hudson 2012; Willett 2010), campaigns since the early 2000s to increase female recruitment have had limited results. Although scholars claim that women assist in making peacekeeping acceptable to conflict-affected societies, it also means co-opting women into an enterprise forged largely by men for treating postcolonial conflicts. Peacekeeping has remained an overwhelmingly masculine enterprise, and scandals arising from gender abuse and prostitution have hardly abated in the twenty-first century. In 2015 a new series of scandals broke concerning sexual and other human rights abuses, committed not only by militias and government forces but also by peacekeepers in the Central African Republic (CAR), who had a mandate to protect civilians.[1] When it comes to finance, the peacekeeping budget (US$6.5 billion for FY2019–2020) is based on a formula that includes relative economic wealth; the US, China, Japan, Germany, the UK, and France contribute nearly 70% of the budget. A common contention is that the rich provide the money and the poor shed the blood, as if personnel from poor TCCs were mercenaries. While compelling, this fails to account for the complexity of relationships. Peacekeeping and policing are disciplinary formulations constructed and circulated globally (Hönke and Müller 2016). TCCs exert agency in various ways. Beginning in 2009 they have secured consultation about mandates, which can be extended and changed during deployments. Several TCCs, such as Rwanda and the Democratic Republic of the Congo (DRC) have been on the receiving end of peacekeeping, and individuals locally employed by intervention missions sometimes seek employment opportunities in other interventions. A redistribution of US policing techniques by Colombia enabled it to exert influence in Latin America, a limited example of privileged knowledge shaping relations in the Global South (Tickner 2016). Whether this circulation of experience should be considered evidence of postcolonial autonomy is doubtful. The exchange is asymmetric. High-income states control technical prowess, logistical support, global intelligence capacity, peacekeeping frameworks, and mandate determination in the Security Council. These mandates grow in scope, along with institutions to support them.

Institutional Growth

The number of UN personnel involved in early deployments was tiny relative to subsequent missions, except in the Congo. Nor did the UN have a recognizable "peacebuilding" horizon to scan until Secretary-General Boutros Boutros-Ghali's *An Agenda for Peace* of 1992. But peacekeeping became institutionalized and its principles established in a code of 1973 (see Box 17.2). It was operated from the Department of Political Affairs until the creation of the UN Department of Peacekeeping Operations (DPKO) at the start of 1992. In a manner of speaking, it had became a cog in a peace-formation industry (see Woodward 2017). Furthermore, peacekeeping and peacebuilding required the functional support of international police; a Police Division was added to DPKO in 2007. Independent but linked, a Peacebuilding Commission, also added in 2007, primed the Secretary-General about threats to peace and postconflict reconstruction. Logistical matters grew exponentially, and in 2010 the DPKO handed

Box 17.2 Peacekeeping Code (1973)

- Only deployed by consent of parties to a dispute.
- Strict impartiality in deployment and actions.
- Use of force only in self-defense as a last resort.
- Mandated and supported by the UN Security Council.
- Reliance on voluntary contributions by UN member states of personnel, equipment, logistics.

over the administration of field personnel, finances, and some logistical matters to a new Department of Field Support. Further institutional overlapping followed with the UN's creation of a Counterterrorism Centre and Task Force (2012) and a Counterterrorism Office in the Department of Political Affairs (2017).

The common use of UN terminology to characterize the building of peace produced a metaphorical conceit about designing peace and constructing something that resembled the work of architects. The components—diplomacy, peacekeeping, peacebuilding, and development—had various approaches to peace, but the UN Secretariat pointed out they were all part of the edifice and were working to plans for a more peaceable world. Peacekeeping had a crucial part to play in organizing the stability necessary for diplomacy to survive and for peacebuilding to flourish. Although the rationale for silo behavior by bureaucracies was weak, the linkage did not necessarily work well. Former UN personnel drew public attention to what they regarded as an unfit bureaucracy, mismanagement of staffing and finance, and the longevity of missions that failed to achieve goals (Banbury 2016; Karlsrud 2015).

UN Peacekeeping Reforms

Peacekeepers had first worn blue berets in supervising the separation of Egyptian forces and the British and Israeli invaders in Suez in 1956. Subsequently they worked in Lebanon and policed the transfer of West Irian from the Netherlands to Indonesian administration. The first major shift in peacekeeping rationale took it toward more ambitious roles than observation by impartial military personnel.

Cold War

Hammarskjöld famously argued that "peacekeeping is not soldiers' work but only soldiers can do it," in the belief that such intercessions between forces entailed risks and required discipline, organization, and symbolic power in uniform. As the murder by

Zionists of the UN diplomat Count Folke Bernadotte in 1948 had exposed, even mediation could cost lives. Hammarskjöld himself was killed in the Congo in 1961 , his plane allegedly attacked by pro-colonial forces. An operation to facilitate Belgium's withdrawal and assist the newly independent Congolese government to restore order was the UN's most ambitious deployment yet. It involved twenty thousand troops and international civilians during 1960 to 1964. Committed to interference in Congo's internal politics, mandated under Chapter VII of the UN Charter to forcibly prevent secession by Katanga province and maintain the Congo's territorial integrity, the intervention was also a forerunner of peacebuilding. It had a civilian component to advise the government. In Cyprus in 1964 peacekeepers also had humanitarian tasks as well as quelling Greek and Turkish clashes and eventually maintaining a buffer zone between the two communities.

After the Cold War

Following the chastening Congo experience and a retreat from peacekeeping, a revival and a second acceleration occurred beginning in the 1980s. The end of the Cold War made it possible for the Security Council to increasingly address conflicts within states as threats to international peace and security that qualified for peacekeeping and peacebuilding. The shift embraced functions and doctrines that by the late 2010s amplified traditional peacekeeping. Reminiscent of the Congo operation in the 1960s, situations involving violent contest for authority and control of a state or a collapse of state functions (the "failed state" notion) accounted for the largest UN deployments. In other words, peacekeepers intruded where there was no peace to keep or peace was at best fragile. UN civilian policing functions expanded in the 1990s to substitute for absent or depleted local forces in dealing with public order, civilian protection, and crime fighting. Coordination between civilian and military elements became increasingly important, as activated in Eastern Slavonia (1996), for example. However, distinctions between soldiers and civilian police grew blurred through the use of heavily armed, gendarmerie-type forces under military orders. They were deemed particularly effective for public order in Bosnia and Herzegovina (Fearon and Laitin 2004, 23; Holm and Eide 2000).

The UN not only had civil "partners"; it opened the door to collaboration and coordination with military enforcement organizations. Other forces drawn, for example, from the Africa Union, the EU, and NATO, also claimed to be keeping the peace in cooperation with the UN. They had their own agendas, often connected to postcolonial interests, as in the case of *L'Organisation internationale de la francophonie* (Charbonneau and Chafer 2014). The African Union invasion of Ajouan (Comoros) in 2008 was backed by France, which, along with Libya, provided logistical support. The EU supported French interests in Chad in 2007 with a force that was replaced after a year by a UN peacekeeping mission which also operated in the CAR, until the elected dictator of Chad refused to renew its presence in 2019. Collaboration also worked in reverse in the Balkans, where the UN assigned a role, or gave way, to NATO troops and an EU force in Bosnia and Herzegovina. Either way, UN peacekeepers could no longer anticipate a neutral or peaceable role or count on lasting host government consent. In reality, peacekeeping attended to Western goals of controlling disorder. To put it simply, the traditional symbolism of universality was increasingly marginalized. Peacekeepers had traditionally relied on their symbolism as neutral and impartial observers to engage in tactical negotiation in the field—a UN presence signifying international concern about war that commanded respect.

The Robust Turn

Consent remained significant for access to conflicts through negotiated Status of Forces Agreements and Memoranda of Understanding with governments. But conflict participants did not necessarily respect the salvation promised by peacekeeping. During its war in Lebanon in 2006, Israel repeatedly bombed a clearly marked UN Truce Supervision Organization post, killing four peacekeepers (Fisk 2016). Deaths of UN personnel increased throughout the 1990s, notably in Rwanda in 1994 and Sierra Leone in 2000. This prompted the Security Council to strengthen mandates.

Following inflated and disappointed expectations about peacekeeping's ability to subdue attacks on civilians—in Sierra Leone, Rwanda, and Bosnia and Herzegovina—and prompted by the High-Level Panel report of experts (the Brahimi Report of 2000), the Security Council extended mandates to include civilian protection and robust, but flexible, escalation of force under Chapter VII of the UN Charter. Although the DPKO's Guidelines and Principles (UN 2008) for peacekeeping emphasized force as a last resort, Western doctrine writers and military commentators favored doctrine revision on the grounds that it had already proved the UN meant business. (For rules of engagement in UN missions, see Findlay 2002, appendix 2.) Western militaries were already engaged in doctrine revision, affecting NATO and US intervention doctrines, the latter categorizing peacekeeping as Operations Other Than War (see Mackinlay and Chopra 1992; Mackinlay 2004; Findlay 2002; Pugh 2016).

Correlations between robust mandates, large personnel deployments, and budgetary provision on the one hand, and low recurrence of relapses into conflict on the other, appeared to justify the shift (Doyle and Sambanis 2000). From a political perspective, however, correlations hide complexity. In Timor Leste a threat by the US that Indonesia would lose International Monetary Fund (IMF) and World Bank access (thus flagging where control of those bodies lay) kept Indonesian militia resistance to a vicious but sporadic minimum. In Bosnia and Herzegovina the cessation of war owed more to the US-backed Croatian army's advances than to NATO's UN-mandated bombing (Berdal 2001, 67). El Salvador's peace aftermath brought oppressive authoritarianism and the labeling of resistance to it as "criminal" violence. Muscularity in Somalia also proved extremely counterproductive. Moreover, metric correlations underestimate the significance of geopolitical interests among proponents of force. The UK in Sierra Leone, France in Côte d'Ivoire and Rwanda, and the US in Liberia had postcolonial diplomatic and economic linkages to protect. In Kosovo, NATO had its reputation to safeguard.

What "robust" meant was hard to pin down and controversial. Not all military personnel favored a peacekeeping spectrum of force, for they understood the significant problems of flexible switching between muscularity and traditional symbolism. Some doubted the credibility of UN force and the ability and willingness of TCCs to undertake escalation. Although host governments, as in Chad and the CAR, regarded peacekeepers as allies against rebels, failure to pursue them or, alternatively, appearing to undermine state sovereignty could fray consent; it could raise civilian expectations that were unlikely to be met. In the CAR, refugees in sanctuaries near UN camps were slaughtered. The UN did not necessarily assume political responsibility for the consequences; escalation of force would require additional protection measures for the force itself and security backup for UN-employed civilians in peacebuilding (who had to satisfy insurance companies about safety). Aid agencies, in Somalia for example, were torn between urging forceful intervention to protect aid and concern that it would militarize aid provision and increase risks to workers. Indeed 329 were killed, injured, or kidnapped in 2014 (UN 2016). According to Banbury (2016, 5), over 80% of the budget for sending troops, who had no counterterrorism experience, to Mali in 2013 was spent on logistics and self-protection. Preoccupation with interventionists' security disrupts interaction with populations (Duffield 2010).

The absence of parallel conflict resolution diplomacy and the potential for widening conflict were bound to affect the modes of follow-on peacebuilding. It sometimes fostered assumptions that peacekeeping could take sides while remaining neutral, sometimes aligning with governments (as in Chad and the DRC) and at other times supporting irredentists (Kosovo), a partiality that would skew peacebuilding.

Nevertheless, robust peacekeeping persisted, reinvigorated as a response to insurgency and terrorism. The UN launched a Counterterrorism Strategy in 2006, and for the first time the Security Council (Resolution 2036) specified an "enemy" (al-Shabab) for the 2012 African Union counterinsurgency mission in Somalia. Subsequently the Security Council inscribed Islamic groups with enemy status in the CAR and DRC. However, counterinsurgency peacekeeping in Central Africa courted disaster. In Mali and other Chapter VII operations in Africa, "malicious deaths" of UN personnel (i.e., excluding the more numerous accidents and fatal illness) totaled 274 in the period 2008–2017 with a further 67 from January 2018 to the end of October 2020. As John Karlsrud (2015) explains, the consequence of instructions to neutralize and disarm rebels resulted in the UN becoming a party to intrastate wars. It compromised the purposes of the Peacebuilding Commission to institute peace processes when peacekeeping was mandated to adopt any necessary means to defeat an enemy (see also *Guardian* 2018). Nor did it avoid potentially harmful consequences of a victor's peace and the embedding of grievances. As in Somalia, coercive demobilization of militias adversely affected the willingness of fighters to respect agreements (UN and World Bank 2018, 199). It must be noted also that regime-change wars conducted by independent military coalitions had not produced particularly satisfactory results for peacebuilding in Iraq, Afghanistan, and Libya.

A more sophisticated, but also problematic, approach infused UN Secretary-General Ban Ki-Moon's 2015 "Action Plan to Prevent Violent Extremism." He judged that offensive military operations failed to deal with violent extremism, partly because it fed on grievances that arose from political and economic oppression, and partly because attacks were designed to provoke strong reactions. Therefore adherence to the Sustainable Development 2030 goals and creation of "open, equitable, inclusive and pluralist societies, based on the full respect of human rights and with economic opportunities for all, represent[ed] . . . the most promising strategy for rendering it unattractive" (UN 2016). Nevertheless, so Richard Atwood (2016) argues, the terminology of "violent extremism" ignored the enormously varied politics of radicalism and offered catch-all answers to it. "Which is more extreme," Atwood asks, "the Yemen 'rebels' or the Saudi bombardment"?

Action for Peacekeeping and Peacekeeping for Peacebuilding

In response to mission crises, partly accounted for by robustness, Secretary-General António Guterres urged the General Assembly of 25 September 2018 to improve the safety and security of peacekeepers. (Attacks had killed fifty-nine peacekeepers in 2017.) He called for better assessment of mandate viability and greater emphasis on political processes to end conflicts. Mandate inflation had to be curbed since a peacekeeping mission was "not an army, or a counter-terrorist force, or a humanitarian agency." Seemingly in contradiction, Guterres complained, "Too often in the past, our troops have been reduced to waiting in a defensive posture, giving hostile forces time and space to plan attacks" (UN Secretary-General 2018). Could mandates be curtailed if peacekeepers had to be more active? If they were not an army or counterterrorist force, why would they need to upgrade their own security like a regular army? Peacekeepers provide not only security for themselves but also security backup for UN-employed civilians in peacebuilding. If peacekeepers were more secure, would civilian inhabitants also be safer?

According to the weighty UN and World Bank (2018, 201) report of 2018, *Pathways for Peace*, "experience demonstrates that if the internal-security challenge is not handled early, 'old' habits and structures will prevail and undermine other efforts to enhance post-conflict peace building." In fact uniformed personnel were directly as well as indirectly involved in peacebuilding. Directly, peacekeepers and follow-on forces helped to track down war criminals, oversee the disarmament of combatants, retrain and reform militaries, help establish security institutions, assist in prisoner exchanges and refugee returns, and even survey signs of economic development (Marley 2016).

International police undertook training and reconstruction of forces of law and order, justice and prison systems, and undertook "executive policing" in Timor Leste and Kosovo (conducting investigations and arresting suspects).

UN policymakers formalized the role of peacekeepers in preparing the way for peacebuilding and transitional administrations through a series of reforms: Integrated Missions (2005), New Horizon (2009), Sustaining Peace (2017), and Action for Peacekeeping (A4P; 2018) (see Box 17.3). For instance, the UN's 2017 "Sustaining Peace Agenda" explicitly located peacekeeping on a "peace continuum" that traveled from conflict prevention and peacekeeping to peacebuilding and development (*UN News* 2017). All instruments and agencies should be mobilized, recommended the *Pathways* report. Because the plethora of international agencies had different and often contradictory goals, reformers constantly pleaded for greater coordination, even as the institutional expansion created more agent involvement and a need for greater management. In

Box 17.3 Action for Peacekeeping Declaration of Shared Commitments, 16 August 2018

The Declaration committed signatories to:

- Renewed political support for peacekeeping to confront higher risks, lack of political process, unfocused mandates, civilian protection expectations, resource problems, and incapacities for contributing to sustainable peace.
- The primacy of politics and comprehensive solutions embracing security, reconciliation, rule of law, and development in parallel.
- Reaffirm basic principles of impartiality, consent, nonuse of force except for self-defense and mandate defense.
- Clear, prioritized, and sequenced mandates and strengthened consultation with all agencies and governments.
- Implementation of the Women, Peace and Security agenda.
- Inclusive and participatory action with recipient governments, greater attention to youth inclusion, engagement with civil society and all segments of local populations.
- Tailored, context-specific approaches to protect civilians.
- Greater effort to deal with war crimes and those who attack peacekeepers.
- Avoid caveats in contributions that are detrimental to implementation and performance.
- Better training, preparation, and environmental awareness among peacekeepers and adoption of a UN Secretariat policy performance framework for standards of effectiveness and conduct (with stronger accountability processes).
- Support UN country teams in building peace during transitions.
- Foster better collaboration and planning with other organizations, including enhancement of African Union capacity, and to implement human rights policy for UN support to non-UN security forces.

Source: Author's summary, based on UN Peacekeeping n.d.-a and links.

overcoming the divergences, linkages were justified as improving coordination between agencies, preventing duplication of effort, for example (Jennings and Kaspersen 2008). Duffield (2001) shows that the continuum also served as a laboratory for advancing a "security-development" nexus, by which ideas of security were merged with development. Third World development could be geared up to underpin international security, and thus the status quo.

In A4P consultations, member states raised the need for coherence between peacekeeping and peacebuilding, and coherence among key bodies such as the Peacebuilding Commission, host countries, and international financial institutions. States also acknowledged the need for continuity of peacebuilding tasks during peacekeeping transitions and drawdown (UN Peacekeeping 2018). The key question is "Coherence about what?" Fulfillment of the peacekeeping mandate, although extremely difficult, could at least have relatively clear goals compared to transforming social relations and the international financial institutions–determined criteria of "good governance," which, as if to deny (or prove) its vacuity, involved thousands of benchmarks (Woodward 2017, 57). Not only does good governance mean that violence-affected societies need to absorb structural adjustment and payment of toxic debts in return for development loans; it can also mean increased inequalities and continued precarity for the population. When a surge in the number of conflicts and deaths and length of conflict duration occurred after 2005, a scholarly study of twelve cases in Africa concluded that international political and economic governance "was directly or indirectly involved in perpetuating a poverty-conflict trap" (Salih 2009; see also UN Conference on Trade and Development 2004; Flassbeck et al. 2013). Given the emphasis in peacebuilding on the mobility of capital and foreign direct investment, peace and political economy based on competition was not seen as contradictory. As *Pathways to Peace* indicated (UN and World Bank 2018, 276), achieving sustainable peace required the phasing in of good governance "within political and investment cycles." Peacekeeping thus cleared the path for a stressful monetization of social relations.

Indirectly, then, by striving for security and stability, peacekeeping underpins particular forms of governance. Clearing a pathway for peace had given way to ambitious peacebuilding enterprises that were largely cultivated not by TCCs but by the designers of liberal and neoliberal governance. Bretton Woods institutions (the World Bank Group and IMF) and other international financial institutions, donors, and potential investors in providing financial aid "with strings" played a key role in reformed governance. Furthermore, the UN and World Bank (2018) report suggested that local leaders who sought peace formation and polices of inclusion or power sharing should merit international backing. But mutually constitutive processes with local leaders and business interests (often the same) were unlikely to protect populations from political disempowerment and economic exclusion. The ground prepared by peacekeepers who, together with international police, escort financial peacebuilders into conflict economies served to engage with local oligarchs to co-constitute rentier economies (see Belloni and Strazzari 2014; Lemay-Hébert and Murshed 2016). Such collusion derived equally from a common interest in legitimizing primitive accumulation of war-time assets and improving the environment for capital acquisition and maintaining precarity for labor forces. Reformed states would thereby engage in global competition—though, from a critical point of view, with comparative disadvantages in technological and human capacity.

Conclusion

To develop peacekeeping for conflicts within states, UN policymaking had to consider that the superiority of its power depended less on symbolism and consent and more on resources of technology, funding, and competent military escalation. It is important to question who or what was served by peacekeeping. It purported to serve "the people," the "victims," and the governments affected by conflict; there should be no doubt that it saved countless lives. It also served, by UN proclamation, to link up interventions by blue berets to peacebuilding. But of what kind? In shorthand, the term used by the World Bank was "good governance." In longhand it meant structural adjustment to engender global integration through such measures as privatizing the state, decentralization, macroeconomic orthodoxy, and buttressing capital accumulation. The role of the Bretton Woods institutions in peacebuilding (the World Bank has a reserved seat on the Peacebuilding Commission and organizes budgetary interventions) as bulwarks of corporate capital promotion cannot be underestimated. The Bank inscribed poverty rather than capital's restructuring of social relations as the source of Third World violence. As Slavoj Žižek (2002, 36) remarks about corporate finance, "[T]he link between these structural decisions and the painful reality . . . is broken; the 'specialists' taking the decisions are unable to imagine the consequences, since they measure the effects of these decisions in abstract terms (a country can be 'financially sane,' even if millions are starving.)"

As power balances in the international system transform, former imperial metropolitan powers can hardly pretend to be in control of peace formation. The British and French veto power in the Security Council looks continually inequitable, as is the Council's overall composition, and at odds with their influence and their flexible interpretation of international norms of conduct. Additionally, US leadership of the supposed "free world" and its authoritarian client states was not merely undermined by system multipolarity and its own client status in relation to "small" but assertive illiberal states such as Saudi Arabia exerting economic or political leverage. Dissatisfied with multilateral economic regulation to advance capital accumulation, its own international policies contradicted both neoliberal political economy and liberal internationalism. Ironically, the US was one of the 151 UN member states that in November 2018 endorsed the A4P Shared Declaration of Commitment to bolster peacekeeping—as did NATO, the EU, the African Union, and *L'Organisation internationale de la francophonie*. However, in December the ultranationalist government in the United States announced a parsimonious approach to peacekeeping and a go-it-alone strategy to reorder Africa in its own interests. A4P may have been less akin to reconfirmation of wedding vows in a Christian church by the elderly and more a desperate visit to a marriage counselor.

Peacekeeping and peacebuilding models will need to change, though one cannot be overly sanguine about this. The liberal concept of good governance and the potential for universal reasoning about peace, though seemingly independent of culture or depth of understanding, had stood the Atlanticist states in good stead for ordering a world in transition from colonialism to Cold War and to globalizing capitalist development. Structural foundations have shifted again, and liberal intentions for peace are in retreat. Perhaps peace formation as an imaginary construction could not withstand the contradictory dynamics that maintains gross inequalities and unstable tensions within states and in the international system as a whole.

Note

1. At various times, troops from France, Canada, Gabon, Burundi, and Democratic Republic of the Congo have been accused of serious abuses. Anthony Banbury (2016) had long warned about it and resigned in protest as UN Assistant Secretary-General for Field Support Operations. After (often) flawed and drawn-out investigations, the UN is powerless to do more than insist that the TCCs take action against their own personnel and order the unit's withdrawal.

References

Atwood, R. 2016. "The Dangers Lurking in the U.N.'s New Plan to Prevent Violent Extremism." Reuters, 8 February. http://blogs.reuters.com/great-debate/2016/02/07/why-is-the-wolf-so-big-and-bad/.

Banbury, Anthony. 2016. "I Love the UN but It Is Failing." *New York Times Sunday Review*, 20 March, 5.

Bellamy, Alex J., and Paul D. Williams, eds. 2013. *Providing Peacekeepers: The Politics, Challenges and Future of United Nations Peacekeeping Contributions.* Oxford: Oxford University Press.

Belloni, R., and F. Strazzari. 2014. "Corruption in Post-Conflict Bosnia and Herzegovina and Kosovo." *Third World Quarterly* 35 (5):855–871.

Berdal, M. 2001. "Lessons Not Learned: The Use of Force in Peace Operations in the 1990s." In *Managing Armed Conflicts in the 21st Century*, edited by Adekeye Adebajo and Chandra Lekha Sriram, 55–74 London: Cass.

Charbonneau, B, and T. Chafer, eds. 2014. *Peace Operations in the Francophone Spaces.* London: Routledge.

Cunliffe, P. 2013. *Legions of Peace: UN Peacekeepers from the Global South.* London: Hurst.

Doyle, M W., and N. Sambanis. 2000. "International Peacebuilding: A Theoretical and Qualitative Analysis." *American Political Science Review* 94 (4):779–801.

Duffield, M. 2001. *Global Governance and the New Wars: The Merging of Development and Security.* London: Zed Books.

Duffield, M. 2010. "Risk Management as the Fortified Aid Compound: Every-day Life in Post-Interventionary Society." *Journal of Intervention and Statebuilding* 4:453–474.

Fearon, J, and D. Laitin. 2004. "Neotrusteeship and the Problem of Failed States." *International Security* 28 (4):5–43.

Findlay, Trevor. 2002. *The Use of Force in UN Peace Operations.* Oxford: Oxford University Press.

Fisk, R. 2016. "It Is Ten Years since UN Peacekeepers Were Killed." *Independent* (London), 25 July.

Flassbeck, H, P. Davidson, J. K. Galbraith, R. Koo, and J. Ghosh. 2013. *Economic Reform Now: A Global Manifesto to Rescue Our Sinking Economies.* Basingstoke, UK: Palgrave Macmillan.

Guardian. 2018. "'Killing, Abuse, Sexual Violence beyond Belief': Fears Grow of All-out War in CAR." 16 November.

Holm, T.T, and E.B. Eide, eds. 2000. *Peacebuilding and Police Reform.* London: Cass.

Hönke, Jana, and Markus–Michael Müller, eds. 2016. *The Global Making of Policing: Postcolonial Perspectives.* Abingdon, UK: Routledge.

Hudson, H. 2012. "A Double-Edged Sword of Peace? Reflections on the Tension between Representation and Protection in Gendering Liberal Peacebuilding." *International Peacekeeping* 19 (4):443–460.

Jakobeit, C. 1966–. War Data Archive [in German]. Hamburg University. www.wiso.uni-hamburg.de/fachbereich-sowi/professuren/jakobeit/forschung/akuf/kriegearchiv.html.

Jennings, K. M., and A. T. Kaspersen, eds. 2008. "Integrated Missions Revisited: Policy and Practitioner Perspectives." Special issue, *International Peacekeeping* 15 (4).

Karlsrud, J. 2015. "The UN at War: Examining the Consequences of Peace-Enforcement Mandates for the UN Peacekeeping Operations in the CAR, the DRC and Mali." *Third World Quarterly* 38 (6):40–54.

Lemay-Hébert, N, and S. M. Murshed. 2016. "Rentier Statebuilding in a Post-conflict Economy: The Case of Kosovo." *Development and Change* 47 (3):517–541.

Mackinlay, J. 2004. "Opposing Insurgents during and beyond Peace Operations." In *Peace Operations after 11 September 2001*, edited by Thierry Tardy, 160–178. London: Cass.

Mackinlay, J, and J. Chopra. 1992. "Second Generation Multinational Peacekeeping." *Washington Quarterly* 15 (3):113–131.

Marley, J M. 2016. "Whose Line Is It Anyway? Understanding the Military Role in Delivering Rights-Based Policies in Post-conflict Territories." PhD thesis, London School of Economics and Political Science.

Olsson, L, and T. Gizelis. 2014. "Advancing Gender and Peacekeeping Research." *International Peacekeeping* 21 (4):520–528.

Pugh, M. 2016. "Lineages of Aggressive Peace." In *The Politics of Intervention: The Tyranny of Peace*, edited by Mandy Turner and Florian Kühn, 77–93. Abingdon, UK: Routledge.

Salih, M. A. M. 2009. "A Critique of the Political Economy of the Liberal Peace: Elements of an African Experience." In *New Perspective on Liberal Peacebuilding*, edited by Edward Newman, Roland Paris, and Oliver P. Richmond, 133–158. Tokyo: United Nations University Press.

Tickner, A B. 2016. "Associate Dependent Security Cooperation: Columbia and the United States." In *The Global Making of Policing: Postcolonial Perspectives*, edited by Jana Hönke and Markus–Michael Müller, 96–113. Abingdon, UK: Routledge.

UN. 2008. *United Nations Peacekeeping Operations: Principles and Guidelines.* https://www.un.org/ruleoflaw/files/Capstone_Doctrine_ENG.pdf

UN. 2015. "Report to the General Assembly on a Plan of Action to Prevent Violent Extremism, A/70/674." 24 December. https://www.undocs.org/A/70/674.

UN Conference on Trade and Development. 2004. *Least Developed Countries Report 2004: Linking International Trade and Poverty Reduction*. E.04 11.0.0.27. Geneva: UN.

UN News. 2017. "'Sustaining Peace' Strategy Must Cover Entire Peace Continuum—UN Deputy Chief." 29 August. https://news.un.org/en/story/2017/08/564082-sustaining-peace-strategy-must-cover-entire-peace-continuum-un-deputy-chief.

UN Peacekeeping. 2018. "High Level Meeting on Action for Peacekeeping." 25 September. https://peacekeeping.un.org/en/ga73-high-level-meeting-action-peacekeeping.

UN Peacekeeping. n.d.-a. "Action for Peacekeeping (A4P)." Accessed 13 October 2020. https://peacekeeping.un.org/en/action-for-peacekeeping-a4p.

UN Peacekeeping. n.d.-b. Peacekeeping Master Open Datasets. Accessed 13 October 2020. https://peacekeeping.un.org/en/peacekeeping-master-open-datasets.

UN Secretary-General. 2018. "Secretary-General's Remarks to Security Council High-Level Debate on Collective Action to Improve UN Peacekeeping Operations." 28 March. www.un.org/sg/en/content/sg/statement/2018-03-28/secretary-generals-remarks-security-council-high-level-debate.

UN and World Bank. 2018. *Pathways for Peace: Inclusive Approaches for Preventing Violent Conflict*. Washington, DC: World Bank. doi:10.1596/978-1-4648-1162-2.

Willett, S, ed. 2010. "Women, Peace and Conflict: A Decade after Resolution 1325." Special issue, *International Peacekeeping* 17 (2).

Woodward, S. L. 2017. *The Ideology of Failed States: Why Interventions Fail*. New York: Cambridge University Press.

Žižek, S. 2002. *Welcome to the Desert of the Real*. London: Verso.

CHAPTER 18

PROTECTION OF CIVILIANS

WALT KILROY

THE protection of civilians (PoC) by UN missions is one of the most challenging tasks given to peacekeeping operations and has become increasingly relevant. The obligation is now included in most UN-mandated operations, which are often situated in areas of ongoing conflict. PoC refers to preventing direct violent attacks on civilians rather than structural violence, and it goes to the core of what is expected of peacekeepers. However, the ways of promoting protection go far beyond the immediate tasks of patrolling, presence by peacekeepers, or direct military action. UN policy on protection is more holistic than the traditional image of "peacekeeping by presence" or the use of force, and includes support for a peace process, political inclusion, and human rights monitoring. The challenges raised by the task include sovereignty, legitimacy, maintaining host state consent, and the creation and management of expectations. There are important questions around the training, gender representation, and risk aversion of personnel deployed and the troop-contributing countries.

The original concept of peacekeeping was very limited and aimed to build confidence in a ceasefire or initial agreement, limiting the use of force to protecting the mission's own personnel—and only as a last resort. The conflict parties might be facilitated in developing a sustainable settlement by the security and confidence created by the peacekeepers in their monitoring role. Missions have changed radically since the first experiments with peacekeeping in the 1950s, and they are now deployed in complex environments of ongoing conflict with little peace to keep and multiple armed actors and threats to the civilian population. The international community's understanding of security has also evolved to include human security and now takes account of forced displacement, war crimes, gender-based violence, destruction of cultural heritage, and genocide in its various forms. UN missions have become multidimensional, extending to human rights and support for political processes. In line with these changes, PoC has become one of the many responsibilities included in the mandate. The expanded role has brought to the surface tensions between peacekeeping and state sovereignty. The tricky question of maintaining consent for a mission from all parties to the conflict also comes into play, especially that of the host state, which may even be a player in attacking civilians. A UN mission is not an occupation force; it has neither the authority to use force nor the military capacity to impose its presence if a host state does not want it in the country. The strains created by expanding peacekeepers' responsibilities come as concerns emerge about a possible overlap between peacekeeping on one hand and the "stabilization" agenda and countering and preventing radicalization on the other (Karlsrud 2017; Gelot 2017; Williams 2013; Willmot et al. 2015).

Protection of civilians is defined by the United Nations (2015a, 5, paragraph 13) as "all necessary means, up to and including the use of deadly force, aimed at preventing or responding to threats of physical violence against civilians, within capabilities and areas of operations, and without prejudice to the responsibility of the host government." It is a narrower concept than that used by humanitarian or development actors, which relates to a broader notion of protecting or realizing human rights in general, and not just physical safety.

This chapter looks at the historical context in which PoC moved up the agenda of international organizations and how peacekeeping mandates started to include this task. The way these mandates evolve and have become more robust is also described. How this has worked in practice, and the issues that have arisen, is then considered, using examples from the Democratic Republic of the Congo and South Sudan. The way PoC relates to liberal peacebuilding and peace formation is also described, along with how the effectiveness of initiatives for PoC might be measured.

The Evolution of UN Peacekeeping Mandates

The genocides in Srebrenica (1995) and Rwanda (1994) had a significant impact on thinking about the international community's obligations to act in the face of atrocities unfolding under the gaze of the world's media. What shocked many was the fact that these mass killings took place despite the presence of relatively small UN operations with very restrictive mandates and limited military capacity. This was still early in the post–Cold War era, when there had been optimism about a "peace dividend" following the supposed end of the East-West military rivalry. The optimism was further undermined as the breakup of the former Yugoslavia continued, with the ominous signs of yet more violence and ethnic cleansing in Kosovo (1999). These wake-up calls prompted several debates, one of which centered on the limits of state sovereignty and ultimately the development of the separate concept of Responsibility to Protect (R2P), which was adopted at a summit of world leaders in 2005.[1] There was also a change in emphasis from state or regime security to human security, and the recognition that the vast majority of casualties in war are civilians. Alongside these debates, the limitations on traditional peacekeeping operations were impossible to ignore. The *Report of the Secretary-General to the Security Council on the Protection of Civilians in Armed Conflict* (UN Secretary-General 1999,

22, paragraph 68) stated, "The plight of civilians is no longer something which can be neglected, or made secondary because it complicates political negotiations or interests. It is fundamental to the central mandate of the Organization. The responsibility for the protection of civilians cannot be transferred to others." The willingness to act more decisively was expressed shortly afterward in a Security Council resolution that seems remarkably timid by today's standards. Resolution 1265 says the Council "expresses its willingness to consider how peacekeeping mandates might better address the negative impact of armed conflict on civilians" (UN Security Council 1999a). Later in the same year, peacekeepers in Sierra Leone became the first mission to have a PoC mandate, albeit one riddled with caveats (UN Security Council 1999b). This stated that acting under Chapter VII of the UN Charter, the mission "may take the necessary action . . . within its capabilities and areas of deployment, to afford protection to civilians under imminent threat of physical violence, taking into account the responsibilities of the Government of Sierra Leone and ECOMOG [the military force of the Economic Community of West African States]."

The rapid incorporation of PoC into the mandates for new and existing UN operations has been traced by Hultman (2013) and is now a standard task. In all, 98% of military and police personnel in UN operations now have a mandate to protect civilians (United Nations 2015b). The wording in mandates has also become tougher as the caveats have become weaker or disappeared. The mandate for the joint UN–African Union mission in the Darfur region of Sudan faced considerable hostility from the government of Sudan, which strongly opposed any UN role in what was initially an African Union operation. One of the qualifications was that they could deal with civilian protection issues if they came across them in the course of their normal operations (the implication being that they should not go looking for them). The restriction on acting within the mission's capabilities is a very real one, given the vast areas a limited number of troops were supposed to cover and the difficulties of moving safely at night, or at all during the rainy season.

Another restriction that has fallen away is reference only to threats of "imminent" violence. This was dropped in the 2015 renewal of the mandate for Mali, and has also disappeared in the case of the Central African Republic, the Democratic Republic of Congo (MONUSCO), and the UN Mission in South Sudan (UNMISS). Other revisions to the Mali mandate that year called on the mission to "move to a more proactive and robust posture to carry out its mandate" and to "take robust and active steps to protect civilians, including through active and effective patrolling" (UN Security Council 2015, paragraph 18).

A more recent discussion within the UN about protection and the questions it raises can be seen in the *Report of the High Level Independent Panel on Peace Operations* (United Nations 2015b, 11), which states that PoC "is a core obligation of the United Nations, but expectations and capability must converge." Protection mandates must be realistic, it advises, and ultimately linked to a wider political approach. Like R2P, the primary responsibility to protect civilians lies with national governments. The report says that a UN presence does not diminish the responsibilities of a host government to protect civilians, but also that the government's primary responsibility does not take away from the UN's obligation to act when the government is unwilling or unable to protect its citizens.

Protection of Civilians in Practice

It is important to stress that although the use of force has significant implications, PoC is a great deal more than a military solution to a human security problem. There are parallels with the related concept of R2P, in that both may be discussed in one-dimensional terms, as force, despite the wide range of actions that are not only possible but necessary. The nonmilitary part of the agenda is not simply some kind of preamble to the "main event" that serves the purpose of making military action appear more acceptable. A holistic, integrated approach is needed when dealing with conflict dynamics that often resemble complex adaptive systems (Brusset, de Coning, and Hughes 2016; de Coning 2016), and a similarly nuanced analysis is required when assessing the reality of PoC in practice. The repeated reference to the primary responsibility of the state for protecting its own citizens (seen in both R2P and PoC) is not just a diplomatic nicety that aims to soothe sensitivities about sovereignty. It is also more than an aspiration. The reality is that even an enormous peacekeeping force would be spread very thinly across the territory where it is deployed and would be unable to police an entire country (even if this was desirable). Ownership and the buy-in of state institutions are also essential for any sustainable solution. The social contract, social capital, and ability of a society to handle conflict by nonviolent means are essential elements to any lasting solution to human insecurity, along with the opportunity to address the causes of conflict. The recognition in peacekeeping of the "primacy of politics" is important: UN or African Union operations cannot impose peace, but they can help to provide a conducive environment and facilitate the development of solutions by the main actors. In the end, protection (or threats to it) happens within communities (Gorur and Carstensen 2016).

The UN PoC policy therefore consists of a three-tiered approach (United Nations 2015a; Johnson 2018). The first tier is essentially political and involves dialogue, mediation, and negotiation with conflict parties and potential perpetrators, as well as public information. The second tier involves a police and military role to deter or respond to attacks and includes patrolling, presence, and a show of force—and possibly its use. The third is a broader programmatic tier that helps to create a better environment, often with medium- and long-term peacebuilding objectives, such as human rights monitoring, accountability, and security-sector reform. These three fields require an integrated approach from different teams in a mission, working ultimately with the state, conflict parties, and civil society.

In planning for PoC and assessing how it works in practice, it is essential that a gender-blind approach is not taken. This would overlook the important differences in which women and men experience war, personal violence, and opportunities for peacemaking

(Ní Aoláin, Haynes, and Cahn 2011). The agenda for Women, Peace, and Security is now well-established. Without intending to replicate the stereotype of women as helpless victims with no agency, one threat they experience in particular ways is that of sexual and gender-based violence, not least because of the stigma associated with it. Very high rates of sexual violence have been reported during the war in South Sudan, one of the areas where a UN operation has a specific PoC mandate (UN Mission in South Sudan and Office of the High Commissioner for Human Rights 2017).

Every context is specific, with its own dynamics, so the experience of major peace support operations with significant PoC mandates varies a great deal. Each provides important opportunities to understand the dilemmas better. It is also clear that policy has been developed and improved through field-led innovations, sometimes in response to particular challenges. Two cases are worth looking at in particular: South Sudan and the Democratic Republic of the Congo (DRC). The DRC is in fact host to one of the largest and longest-running operations, which is called MONUSCO in its latest incarnation and has had close to twenty thousand troops in the country for more than a decade. There are many armed actors in the eastern part of this vast country, some of which regularly target civilians. This includes the armed forces of the state, which MONUSCO operates alongside in some situations. The problems this creates for the UN are considerable. One response is the organization's Human Rights Due Diligence Policy, which requires it to vet individuals in the military for possible involvement in human rights abuses, before working with them (Boutellis 2013).

While the use of force by peacekeepers has always been controversial (Hunt 2017; Slim 2001; Piiparinen 2016; Bellamy and Hunt 2015; Karlsrud 2015; Howard and Dayal 2017), the DRC is unique in the creation of a specific unit to tackle and disable armed groups targeting civilians. The Force Intervention Brigade was an innovation set up by the Security Council in 2013 (UN Security Council 2013) in response to the actions of the M23 group, which had taken over the city of Goma. The brigade was a real shift in approach and was effectively a combat force of about three thousand troops (mainly from Africa) with attack helicopters and artillery (Murphy 2017; Berdal and Ucko 2015; Müller 2015). It defeated the M23, thus meeting its immediate objectives and no doubt sending a message to others, but this left dozens of other armed groups operating in the area. Scaling up an operation like this would completely change the nature of the entire mission, even if agreement were possible. It demonstrates how force—even if effective—is still secondary to politics when it comes to long-term solutions.

The other case worth considering is South Sudan, which quickly progressed from independence in 2011 to civil war just over two years later. UNMISS had been present since before the country gained independence from Sudan, but quickly found itself dealing with a massive civilian protection crisis. By 2018 an estimated four million people had been displaced by the conflict, amounting to a third of the population. Having started as a power struggle within a power-sharing government, the ethnic aspect acted as an accelerant. The Dinka-dominated government forces have been active players in the abuses. The conflict has included gender-based violence, killings of civilians, and ethnic cleansing (UN Mission in South Sudan and Office of the High Commissioner for

Human Rights 2017), and by 2018 the death toll had been estimated at close to 400,000 (Checchi et al. 2018).

The fact that the host state is a player, and often shows hostility to the UN, is a significant aspect. UNMISS was found wanting on a number of occasions, including during the outbreak of renewed violence in the capital, Juba, in July 2016 (Center for Civilians in Conflict 2016b). During this time, UNMISS itself came under attack from government forces, and not just with small arms fire. Some contingents did act to protect civilians, and two Chinese peacekeepers lost their lives. However, other units' lack of response to direct orders to intervene to stop violence against civilians nearby and other failures resulted in the force commander being dismissed. The UN's own report is scathing, citing failure to plan for foreseeable scenarios. The criticisms include lack of leadership, a "chaotic and ineffective response," and reporting and acting in silos, which "inhibited effective action [when] swift, joint action was essential." The force "did not operate under a unified command, resulting in multiple and sometimes conflicting orders" (United Nations 2016). Clearly the mission was put to the test, while coming under attack itself, and was widely criticized for not responding adequately. It later faced obstruction by the government when the Security Council authorized a specific Regional Protection Force to join the UN mission as a response to the violence (Spink 2016).

UNMISS is unique in the way it has hosted civilians who sought refuge at UN bases. There is in fact a long history of this, including in Lebanon, and more infamously in Srebrenica. But the numbers involved, and what ultimately became a formalized if somewhat reluctant response was unlike anything seen before. It should be said that this was not something sought or encouraged by the operation, but was initially a case of people voting with their feet. The ethnic component of the violence was clear when fighting erupted in December 2013 and more than twenty thousand people moved into the UNMISS base at Malakal almost immediately in search of protection. The mission was obliged to come up with a response, which later was formalized as "PoC sites" (Kilroy 2018; Arensen 2016; Lilly 2014). Tens of thousands of people are sheltered at each of the temporary camps beside UN bases in Bentiu, Juba, and Malakal. While the numbers fluctuate, about one-tenth of the two million internally displaced people (IDPs) in South Sudan are located in these PoC sites.[2] While PoC sites are not a solution in themselves, they have become part of the emergency response. Unfortunately civilians living there are still subject to ethnically motivated attacks, as seen in Malakal in February 2016 (Center for Civilians in Conflict 2016a). Dozens of people died, and UNMISS again was criticized for failing to respond effectively to attacks happening where they were located.

Many issues have emerged already in the twenty years since PoC was put on the agenda. There are considerable risks and consequences linked to using force (and indeed failing to use it at times). Relations with and consent of all the conflict parties—but especially the host state—are highly sensitive, as these groups can prevent freedom of movement or indeed deployment at all. This is especially difficult when the host government is one of the conflict parties targeting civilians. Other issues include the capacity of missions in logistical and physical terms such as transport. Being able to move effectively, safely, and without delay in insecure or difficult terrain can deter violence through presence and the possibility of abuses being documented and investigated. However, buy-in and capacity at the human level is also fundamental—and sometimes found wanting. This underlines the importance of training (Curran 2017), especially given the regular rotation of troops. Gender sensitivity and human rights training is important, as is female participation in military, police, and civilian roles. Troop- and police-contributing countries can also display considerable risk aversion. There can be an unofficial "double line of command," whereby UN orders are not acted on until a contingent's national headquarters has approved the action. This question of buy-in and internalization of protection norms applies also to members of the Security Council, whose mandates need to be coherent with operational realities on the ground. But in the end, what matters is how these complex mandates with competing demands are prioritized, operationalized into specific plans, and implemented.

Protection of Civilians and Liberal Internationalism

Situating PoC in relation to liberal internationalism helps to shed light on both this theory and the concept of protection itself. One of the key questions here is what are presumed to be the aims of liberal internationalism. If its objectives (even if these are undeclared or implicit) is the promotion of a liberal world order, it is hard to fit it under this heading. While sovereignty of weaker states may be at issue, the primary responsibility of the state to protect its citizens is central to PoC, and this is reiterated in most if not all UN Security Council resolutions dealing with the issue. If liberal internationalism is seen as firmly rooted in human rights and human security—without being intended as a Trojan horse for neocolonial adventures—then we can see PoC as clearly linked to this worldview and set of norms. Its referent is civilians rather than regime or state security. The concept and practice of PoC is closely linked to the importance both liberalism and liberal institutionalism give to international bodies such as the UN or the African Union, as a means to promote the well-being of citizens and to regulate the behavior of states that threaten peace. With the emergence of both PoC and R2P as legal and political concepts, the UN Security Council in particular has become more significant, as the primary body that can legitimize the use of force in defense of these principles.

In terms of aims, the PoC agenda is not to promote a neoliberal world order but rather to protect civilians from attacks by nonstate actors or indeed governments. It most closely relates to the way R2P may confront state sovereignty. Should the norm of civilian protection be seen as a Western liberal imposition, especially when it has been endorsed by the United Nations, and not just by the UN Security Council?

There may be a temptation to put PoC into the category of Western military intervention. However, although the funding for peacekeeping operations comes mainly from richer countries, troops and military leadership are overwhelmingly from developing countries. Furthermore, the mandate for the use of force comes directly from the UN Security Council, unlike the invasion of Iraq in 2003 or indeed NATO's action in Kosovo in 1999. The veto-wielding permanent five members of the Council do indeed have the power to block decisions. But they cannot push through action that is unpopular, as was seen in the lead-up to the Iraq War in 2003. A total of nine votes in favor out of fifteen members is still needed to pass a resolution. The African Union, whose membership is of course limited to that continent, faces similar dilemmas in its operations when civilians come under attack. This was seen in the African Union missions that preceded UN or "double-hatted" peacekeeping in Darfur, Mali, and the Central African Republic.

Perhaps PoC can be seen as a conceptual tool that clarifies and brings to the forefront immediate human security needs and dilemmas. These may raise difficult questions, and many of the actions (or inaction) may create new problems. However, the dilemmas and contradictions are not the result of the conceptual framing; they exist whether or not we choose to see them. The problem of attacks on civilians exists regardless of one's attitude to whether any kind international response is ever advisable. Deciding that PoC initiatives are part of a Western interventionist agenda does not somehow make the killing of civilians disappear as a problem. It may take gross human rights abuses off our radar, but that is hardly a solution without negative consequences.

One argument may be that PoC concerns are part of a spectrum that includes actions such as the unjustified invasion of Iraq in 2003 or other examples of disastrous military intervention by Western powers. There may be a suggestion of "guilt by association" rather than an explicit argument based on evidence of such a spectrum. Another reason for suspicion is that well-meaning concepts like PoC or R2P will be used as cover for military adventurism. Again, making this argument explicit helps to clarify our thinking: just because a concept can be misrepresented or applied in contravention of its underlying principles does not invalidate the concept itself. Ideas about the safety of minority groups or human rights have been invoked to justify offensive action, but that does not mean that minorities or rights are not legitimate concerns.

PoC is framed broadly and does not need to involve international actors or initiatives, given that the fundamental responsibility for protecting civilians is repeatedly asserted as lying with the state. When the state fails to act (possibly because it is the aggressor against its own citizens), then other actors come into the picture. Protection can also be framed in terms of international versus local ownership or control, and awareness of this dynamic is in fact essential if space is to be allowed for sustainable solutions to emerge. However, this measure should not be applied in a simplistic fashion. Chambers (1974, 109, quoted in Cornwall 2008, 276), who has done so much to highlight genuine ownership and participation, also warned of the reification of "the people." There will usually be many "local" voices and interests, some of which directly contradict each other, and we are left with the choice of which of them to prioritize. Some of them may be marginalized or victimized, while others may even be those carrying out gross human rights abuses. Peace support operations can and must create the space for political processes with local ownership leading to a sustainable peace process—opportunities the parties may or may not seize. A process where there is genuine agency by local actors sees the realization of peace formation (Richmond 2016), using the insights and capacities of those most affected, even where those capacities are not understood or noticed by external actors.

In terms of state formation (the creation of new states from within rather than externally driven statebuilding), the agency of the state is recognized in the way its primary responsibility for PoC is made explicit. This agency—and sovereignty—is of course challenged if abuses continue and the international community then responds. But even then, the state's acquiescence at the very least is required. All of this underlines the tensions that exist regarding sovereignty, agency, legitimacy, the local, and power.

The chaos and suffering following the overthrow of the Gaddafi regime in Libya in 2011 (albeit on an R2P rather than a PoC agenda) are a painful illustration of how actions can come with great risks. (It could also be argued that further atrocities were averted in Kosovo and Timor Leste in 1999, but that is a debate for another place.) On the other hand, there can also be enormous risks associated with failing to act, as seen in the "hard cases" that provoked some of the thinking about civilian casualties, such as Srebrenica and Rwanda. Moral purity may be claimed by those highlighting the very real problems of civilian protection, but if inaction means gross human rights violations, it begins to look much less pure. The vast majority of casualties in armed conflict since the end of World War II are civilian.

Whether or not we see PoC as being part of a statebuilding agenda—possibly within a liberal peace paradigm—the threats faced by civilians are at least compounded by state fragility, failure, or capture, or indeed the collapse of the social contract. In some cases of armed conflict, these are in fact significant causal factors. Whoever is doing the "building" of the state, its capacities, or institutions, the task remains to be done, whether the process is driven locally, by outsiders, through elites, or with genuine participation of those most affected. The question of how it is done, and in whose image, is just as significant as who does it, and has profound implications for the kind of peace that emerges and its sustainability. But deciding that these issues are too entangled with unintended consequences and hidden agendas does not make a laissez-faire approach any more likely to bring an inclusive peace, especially if this leaves society in the grip of powerful or armed elites. One of the many questions for international actors is their ability to recognize and work in harmony with existing or potential capacities within society for protection and conflict resolution—dynamics that may not be immediately apparent to outsiders whose stay in each context is of a limited duration.

Conclusions

Has protection been a success? It is important not to approach this question as a binary of "success/failure." The context is one of causal complexity (Brusset, de Coning, and Hughes 2016), with many interacting variables besides a peacekeeping mission. Even the timeframes over which impact may appear is a factor. How do we measure the absence of violence against civilians? It may be that the greatest impact of (for example) active patrolling or peacekeeper presence has been to deter or reduce further attacks. There are also competing long-term and short-term goals, such as the need to prevent immediate harm and the more open-ended task of facilitating a sustainable and inclusive political settlement, for which the parties to the conflict may or may not be ready. The reality of the context is that many (and perhaps most) solutions offered generate additional problems to be solved or perhaps accepted. Facilitating local ownership of a peace process may involve dealing with actors with poor human rights records. From the perspective of an academic commentator, a facile analysis is always possible—and perhaps even pleasing—in which every initiative is declared to be a failure, even when they were better than all of the alternatives on offer. It does not mean that we should not also think about creating new alternatives. But if everything is inevitably declared a failure, how useful is the analysis? What simple solutions, entirely without disadvantages, are available for situations faced by Rwanda in 1994, Kosovo in 1999, or Myanmar in 2017? So an assessment of success or failure must deal with the complexities of the situation. Failings can be pointed to in South Sudan, the DRC, and Darfur. Some of these imply blame, such as the lack of an adequate response to the attacks in Juba in July 2016. Others can be argued to be the result of decisions taken to avoid creating further problems, such as civilian casualties arising from use of force, or managing relations with a host government that restricts peacekeepers' movements and effectiveness. In the end, it is impossible to say that peacekeepers have had no effect in terms of civilian protection, and local populations have been quick to express anxiety about the downsizing of UN operations.

The success of PoC as a concept can also be considered by looking at cases where it has not been on the agenda: the expulsion of the Rohingya people from Myanmar in 2017, and associated violence, and mass casualties and displacement in Syria since 2011 are what can happen when no action is possible due to the stranglehold on the UN Security Council held by veto-wielding powers. The range of tasks now given to peace support operations in "Christmas tree" mandates on which many additional responsibilities are hung creates contradictions and sometimes difficult choices. How can inclusive political dialogue be supported, human rights promoted, and civilians protected when key players in a political solution are also responsible for some of those abuses and attacks? Is it for the international community to decide which actors should be excluded from future government (and how can it do so without being accused of neocolonial interference in national affairs)? Assessment of how far the agenda for PoC has advanced must take into account this wider context. The development of this agenda, clarification of concepts and responsibilities, and lessons learned are necessary for understanding the dynamics at play and identifying better policies and practices.

PoC goes to the very heart of human security and peacekeeping and resonates with the ideal of what is often expected of the UN. And yet it challenges peacekeeping by raising difficult choices about sovereignty, legitimacy, neutrality, and the use of force. The UN is frequently criticized for failing to act robustly or effectively, while also being damned if it does. Many of these fundamental questions and contradictions are not a function of the organization, however, but of the nature of conflict and the possible responses, few of which are without risk or cost. The dilemmas are still there in some form, no matter how ambitious or cautious the responses to them. A clearer set of policies and practices, together with an understanding of the consequences of acting and failing to act, has been emerging gradually since PoC was added to peacekeeping mandates in 1999. This at least expands the range of options available for dealing with armed conflict and mass atrocities.

Notes

1. Protection of course builds from the same concerns as R2P, which is related but yet entirely distinct from it. One of the fundamental differences is that the ultimate (but not inevitable) level of action for R2P is the use of force in the territory of another state, whereas peacekeeping can be done only with the consent of the parties. These are entirely different types of operations, with different levels of risk and possibilities for exiting the situation. PoC in which force is contemplated is not simply another point on a spectrum shared with robust measures to ensure R2P; the question of operating with the consent of a host nation (among other actors) puts them in different categories altogether.
2. As of May 2019, 178,000 people were living in PoC sites located beside UNMISS bases, out of 1.78 million IDPs in South Sudan (UN Office for the Coordination of Humanitarian Affairs 2019).

References

Arensen, M. J. 2016. *If We Leave We Are Killed: Lessons Learned from South Sudan Protection of Civilian Sites 2013–2016.* International Organization for Migration, South Sudan. https://publications.iom.int/system/files/pdf/if_we_leave_0.pdf.

Bellamy, A. J., and C. T Hunt. 2015. "Twenty-First Century UN Peace Operations: Protection, Force and the Changing Security Environment." *International Affairs* 91 (6):1277–1298.

Berdal, M., and D. H. Ucko. 2015. "The Use of Force in UN Peacekeeping Operations." *RUSI Journal* 160 (1):6–12.

Boutellis, J. A. 2013. "From Crisis to Reform: Peacekeeping Strategies for the Protection of Civilians in the Democratic Republic of the Congo." *Stability: International Journal of Security & Development* 2 (3):1–11.

Brusset, E, C. de Coning, and B. Hughes. 2016. *Complexity Thinking for Peacebuilding Practice and Evaluation.* Basingstoke, UK: Palgrave Macmillan.

Center for Civilians in Conflict. 2016a. *A Refuge in Flames: The February 17–18 Violence in Malakal POC.* Washington, DC: Center for Civilians in Conflict. https://civiliansinconflict.org/publications/research/refuge-flames-february-17-18-violence-malakal-poc/.

Center for Civilians in Conflict. 2016b. *Under Fire: The July 2016 Violence in Juba and UN Response.* Washington, DC: Center for Civilians in Conflict. https://civiliansinconflict.org/wp-content/uploads/2017/09/civic-juba-violence-report-october-2016.pdf.

Chambers, R. 1974. *Managing Rural Development: Ideas and Experience from East Asia.* Uppsala: Scandinavian Institute of African Studies.

Checchi, F., A. Testa, A. Warsame, L. Quach, and R. Burns. 2018. *Estimates of Crisis-Attributable Mortality in South Sudan, December 2013–April 2018: A Statistical Analysis.* London: London School of Hygiene and Tropical Medicine.

Cornwall, A. 2008. "Unpacking 'Participation': Models, Meanings and Practices." *Community Development Journal* 43 (3):269–283.
Curran, D. 2017. "Muddling On Through? Cosmopolitan Peacekeeping and the Protection of Civilians." *International Peacekeeping* 24 (1):63–85.
de Coning, D. 2016. "From Peacebuilding to Sustaining Peace: Implications of Complexity for Resilience and Sustainability." *Resilience* 4 (3):166–181.
Gelot, L. 2017. "Civilian Protection in Africa: How the Protection of Civilians Is Being Militarized by African Policymakers and Diplomats." *Contemporary Security Policy* 38 (1):161–173.
Gorur, A., and N. Carstensen. 2016. "Community Self-Protection." In *Protection of Civilians*, edited by Haidi Willmot, Ralph Mamiya, Scott Sheeran, and Marc Weller, 409–427. Oxford: Oxford University Press.
Howard, L Morjé, and A. Kaushlesh Dayal. 2017. "The Use of Force in UN Peacekeeping." *International Organization* 72 (1): 71–103.
Hultman, L. 2013. "UN Peace Operations and Protection of Civilians: Cheap Talk or Norm Implementation?" *Journal of Peace Research* 50 (59):59–73.
Hunt, C. T. 2017. "All Necessary Means to What Ends? The Unintended Consequences of the 'Robust Turn' in UN Peace Operations." *International Peacekeeping* 24 (1):108–131.
Johnson, H. F. 2018. "Protection of Civilians in the United Nations: A Peacekeeping Illusion?" In *United Nations Peace Operations in a Changing Global Order*, edited by Mateja Peter and Cedric de Coning, 133–152. Cham: Palgrave Macmillan.
Karlsrud, J. 2015. "The UN at War: Examining the Consequences of Peace-Enforcement Mandates for the UN Peacekeeping Operations in the CAR, the DRC and Mali." *Third World Quarterly* 36 (1):40–54.
Karlsrud, J. 2017. "Towards UN Counter-Terrorism Operations?" *Third World Quarterly* 38 (6):1–17.
Kilroy, W. 2018. "The Evolution of Civilian Protection in Peacekeeping Mandates: The Reality of UNMISS Operations in South Sudan." *Irish Studies in International Affairs* 29:133–143.
Lilly, D. 2014. "Protection of Civilians Sites: A New Type of Displacement Settlement?" *Humanitarian Exchange* 62 (September):31–33.
Murphy, R. 2017. "The United Nations Mission in South Sudan and the Protection of Civilians." *Journal of Conflict and Security Law* 22 (3): 367–394.
Müller, L. 2015. "The Force Intervention Brigade—United Nations Forces beyond the Fine Line between Peacekeeping and Peace Enforcement." *Journal of Conflict and Security Law* 20 (3):359–380.
Ní Aoláin, F., D. Francesca Haynes, and N. Cahn. 2011. *On the Frontlines: Gender, War, and the Post-Conflict Process*. Oxford: Oxford University Press.
Piiparinen, T. 2016. "Intervening to Strengthen Sovereignty: The Lessons of the UN Intervention Brigade for Global Peacekeeping." *International Relations* 30 (2):154–175.
Richmond, O. P. 2016. *Peace Formation and Political Order in Conflict Affected Societies*. Oxford: Oxford University Press.
Slim, H. 2001. "Violence and Humanitarianism: Moral Paradox and the Protection of Civilians." *Security Dialogue* 32 (3):325–339.
Spink, L. 2016. *Challenges and Conditions for Deploying an Effective Regional Protection Force to South Sudan*. Washington, DC: Center for Civilians in Conflict.https://civiliansinconflict.org/publications/research/challenges-conditions-for-deploying-effective-regional-protection-force-south-sudan/.

United Nations. 2015a. *The Protection of Civilians in United Nations Peacekeeping.* New York: UN.

United Nations. 2015b. *Report of the High Level Independent Panel on Peace Operations on Uniting Our Strengths for Peace: Politics, Partnership and People.* 17 June. A/70/95-S/2015/446.

United Nations. 2016. *Executive Summary of the Independent Special Investigation into the Violence Which Occurred in Juba in 2016 and UNMISS Response.* https://reliefweb.int/sites/reliefweb.int/files/resources/N1635591.pdf.

UN Mission in South Sudan and Office of the High Commissioner for Human Rights. 2017. *A Report on Violations and Abuses of International Human Rights Law and Violations of International Humanitarian Law in the Context of the Fighting in Juba, South Sudan, in July 2016.* January. http://reliefweb.int/sites/reliefweb.int/files/resources/ReportJuba16Jan2017.pdf.

UN Office for the Coordination of Humanitarian Affairs. 2019. *South Sudan Humanitarian Snapshot.* May. https://reliefweb.int/sites/reliefweb.int/files/resources/ss_20190614_humanitarian_snapshot_may.pdf

UN Secretary-General. 1999. *Report of the Secretary-General to the Security Council on the Protection of Civilians in Armed Conflict.* S/1999/957. https://undocs.org/S/1999/957

UN Security Council. 1999a. Security Council Resolution 1265. 17 September. S/RES/1265 (1999). https://undocs.org/S/RES/1265(1999)

UN Security Council. 1999b. Security Council Resolution 1270. 22 October. S/RES/1270 (1999). https://undocs.org/S/RES/1270(1999).

UN Security Council. 2013. Security Council Resolution 2098. 28 March. UN Doc S/RES/2098.

UN Security Council. 2015. Security Council Resolution 2227. 29 June.

Williams, P. D. 2013. "Protection, Resilience and Empowerment: United Nations Peacekeeping and Violence against Civilians in Contemporary War Zones." *Politics* 33 (4):287–298.

Willmot, H., R. Mamiya, S. Sheeran, and M. Weller, eds. 2015. *The Protection of Civilians in International Law.* Oxford: Oxford University Press.

CHAPTER 19

THE UNITED NATIONS AND THE RESPONSIBILITY TO REBUILD

ALEX J. BELLAMY

ARISING out of the collective failure of the UN to prevent or respond effectively to genocides in Rwanda and Bosnia, the Responsibility to Protect (R2P) is an international principle unanimously agreed by heads of state and government at the 2005 World Summit. It insists that states have a responsibility to protect their populations from genocide, war crimes, ethnic cleansing, and crimes against humanity (hitherto referred to as "atrocity crimes"); that the international community has a responsibility to encourage and assist states to fulfill their responsibility to protect; that all states should use diplomatic, humanitarian, and other peaceful measures to protect populations; and that when states are "manifestly failing" to protect their populations the international community should take "timely and decisive" action through the UN. The term "R2P" was first coined in 2001 by the International Commission on Intervention and State Sovereignty (ICISS), established by the Canadian government to figure out ways of preventing future Rwandas (where the international community fails to act to prevent or stop atrocity crimes) and future Kosovos (where a group of states acts outside the bounds of international law to protect populations from these crimes). Since 2005, R2P has become an established international norm. It has been reaffirmed by the UN General Assembly and incorporated into more than sixty UN Security Council resolutions and twenty-five Human Rights Council resolutions. More than a quarter of the UN's member states have appointed a senior official to serve as their national focal point for R2P. Over that time the UN Security Council has grown more likely to respond to atrocity crimes, governments and international organizations have begun (albeit hesitantly) to incorporate atrocity prevention into their foreign policies and development partnerships, and the protection of populations from atrocity crimes has moved from the periphery to the core of the UN's agenda. According to Secretary-General António Guterres, atrocity prevention is a core element of his broader strategy for prevention, which he intends to make the hallmark of his secretary-generalship.

Peacebuilding and statebuilding were integral parts of R2P when the principle was first articulated in 2001. ICISS recognized that states and societies are at their most vulnerable as they emerge out of violence. As such, it maintained that the international community had a moral responsibility to help rebuild societies after violence and argued that effective peacebuilding was a crucial part of the R2P agenda. But since 2005, R2P and peacebuilding have developed quite separately, creating a gap between the theories and practices of protection and peacebuilding. The effects of this gap are not just theoretical but practical too. The UN's failure to properly follow through with rebuilding support in Libya contributed to that country's descent into chaos and civil war, especially after 2013. Likewise a failure to incorporate atrocity-prevention concerns into ongoing peacebuilding efforts in places like Sri Lanka and the Central African Republic (CAR) meant that the UN's field presences did not do all they could to prevent atrocities or protect vulnerable populations.

This chapter examines the relationship between peacebuilding and R2P in the UN context. It shows how the two were conceived as being mutually supporting activities but were separated during the UN's wider deliberations on reform. It describes the effects of this gap between peacebuilding and protection before arguing that the two agendas are closely aligned and should be integrated. In the final section, it points to practical work to ensure that atrocity prevention is mainstreamed into peacebuilding efforts, and vice versa.

The Responsibility to Rebuild

The ICISS, whose 2001 report, *Responsibility to Protect,* helped set the R2P agenda, maintained that the international community's responsibility for protecting populations from atrocity crimes comprises three elements: responsibilities to prevent, react, and rebuild. It understood the "responsibility to rebuild" in fairly limited terms—as the duty of interveners to protect those in their care and to remain engaged after any international intervention undertaken under the auspices of the "responsibility to respond." Thus, as the ICISS saw it, the moral duty to support rebuilding was established by a prior act of intervention. The ICISS did not address questions about the relationship between prevention and peacebuilding or about duties to support rebuilding in the event of nonintervention. Nor did it consider the question of states simply avoiding their moral duty to support rebuilding by not intervening to stop atrocity crimes in the first place. As the report put it, "[I]f military intervention action is taken . . . there should be a genuine commitment to helping to build a durable peace, and promoting good governance and sustainable development" (International Commission on Intervention and State Sovereignty 2001, 39). This third component of R2P drew heavily from earlier efforts to develop the conceptual, operational, and institutional dimensions of peacebuilding following Boutros Boutros-Ghali's (1992) introduction of this term in his *Agenda for Peace* report. The ICISS quoted extensively from Kofi Annan's 1998 report on conflict in Africa. As such, it maintained that the responsibility to rebuild demanded action in three specific areas: "security, justice and reconciliation, and economic development." The repertoire of rebuilding measures it recommended were drawn from this earlier work. Interveners should work toward disarming and demobilizing former combatants and establishing effective and legitimate national armed forces. They should support justice and reconciliation by helping establish local judicial systems, fostering local opportunities for reconciliation, and guaranteeing the legal rights of returnees. They should use all possible means to foster economic growth and in all this be mindful of the principle of local ownership. But neither Boutros-Ghali nor Annan had addressed the question of how societies could best recover from atrocity crimes, as distinct from other forms of armed conflict, and neither did the ICISS report (Bellamy and Luck 2018, 26).

The ICISS's discussion of rebuilding exhibited some of the same strengths and weaknesses as its treatment of prevention (Bellamy 2009, 62–64). As the report acknowledged, peacebuilding could well encounter sovereignty concerns, as would intrusive preventive measures by the international community. In both cases, obtaining and maintaining a sustainable balance between local ownership and global engagement could be a demanding and delicate task. In a postconflict situation, the ICISS advised, international actors should strive "to do themselves out of a job" (International Commission on Intervention and State Sovereignty 2001, 45). It might well be even more difficult to achieve this balance in countries that had been engulfed in horrific communal violence, as had been the case in the Balkans, but this point was not explored by the report. Peacebuilding received thoughtful attention in the text of the report but was not included among the ICISS's principal recommendations. Like prevention, rebuilding was accorded secondary or tertiary coverage compared to its principal focus, which was placed squarely on military intervention (Bellamy and Luck 2018, 26). More substantively, there were issues with the chronological sequencing of the three responsibilities, since in real-life crises there is not always such a clear differentiation among prevention, reaction, and rebuilding, nor did events or policy choices uniformly follow such an orderly sequencing and progression. The idea that interveners bore a moral duty to rebuild was innovative (Paris 2016, 512)—and in keeping with the emerging idea of a *jus post bellum*—but it was a limited duty, acquired only by interveners. (It was not a general responsibility borne by all UN member states or the organization itself.) What is more, there was no clear sense of how rebuilding related to the other dimensions of R2P.

In preparation for the 2005 World Summit, Secretary-General Annan assembled the High-Level Panel on Threats, Challenges and Change to spur fresh thinking about UN reform. The Panel's December 2004 report, *A More Secure World: Our Shared Responsibility*, reinforced the prioritization of the "reaction" component of R2P over both prevention and rebuilding. Indeed the whole discussion on R2P came not in the report's section on human rights but in a section titled "Collective Security and the Use of Force." The High-Level Panel endorsed "the emerging norm that there is a collective international responsibility to protect, exercisable by the Security Council authorizing military intervention as a last resort," but separated this out from prevention. The responsibility to rebuild was sidelined, the Panel instead dealing with the UN's peacebuilding work separately (UN General Assembly 2004, 66). This separation of peacebuilding and R2P was then carried over into the 2005 World Summit itself. The Summit redesigned R2P, eschewing the responsibilities to prevent, react, and rebuild, relegating the use of force, and emphasizing prevention. In that context, paragraphs 138–139 of the Summit outcome put considerable emphasis on helping or assisting states—notably including those "under stress"—as a preventive measure, not a rebuilding one. Here the treatment of collective military measures as just one of many options for protecting populations (as opposed to the principal concern, as it had been in 2001) made the need for rebuilding afterward a much lower priority in the overall conception of R2P in 2005 (Bellamy and Luck 2018, 27).

This is not to say that peacebuilding itself was marginalized in 2005. Among the most important achievements of the 2005 World Summit was the establishment of new peacebuilding architecture for the UN: the Peacebuilding Commission, Support Office, and Fund. The problem was that from this point on, R2P and peacebuilding came to be seen—and practiced—as separate and distinct agendas despite the strong normative and function overlap between the two.

Since member states chose not to support the ICISS's prevent-react-rebuild paradigm, a new conceptual architecture was needed. The Summit outcome did not articulate a clear framework, leading to some significant confusion among states as to what had actually been agreed. In that context, Secretary-General Ban Ki-moon proposed a three-pillar implementation strategy:

- Pillar 1, the protection responsibilities of the state, as affirmed in paragraph 138 and derived from well-developed legal principles.
- Pillar 2, a concomitant responsibility of the international community to encourage and assist states to meet their R2P responsibilities.
- Pillar 3, an international commitment to a timely and decisive response, employing the full range of tools under Chapters VI, VII, and VIII, as appropriate and under the Charter, to protect populations from atrocity crimes and their incitement.

The secretary-general repeatedly stressed that the three pillars should be thought of as of equal length and strength and should not be approached sequentially or chronologically—overcoming one of the problems associated with the earlier tripartite paradigm. Recognizing that no two crises were alike, the secretary-general's strategy called for an "early and flexible response tailored to the specific circumstances of each case" (Ban 2009, 2). It is this approach that states have subsequently committed themselves to.

This new architecture was set out in the secretary-general's January 2009 report on implementing the responsibility to protect (Ban 2009). Peacebuilding was not explicitly included as part of his strategy for R2P, but there were striking functional similarities between what was judged necessary support for the implementation of R2P and the UN's post-2005 peacebuilding agenda. In particular, as Roland Paris (2016) shows, much of the language and substance of the secretary-general's proposals on R2P—particularly its first two pillars—echoed that of his recommendations for peacebuilding. Thus assistance to help states fulfill R2P should focus on "effective, legitimate, and inclusive governance." It should support "participatory and accountable political institutions," the rule of law, equal access to justice, and the transparent and legitimate allocation of economic resources and assets. There should be work to reinforce the rule of law, support security-sector reform, improve judicial institutions, and build local capacities for reconciliation (see Paris 2016, 514). All this bears a close resemblance to the secretary-general's recommendations on peacebuilding. What is more, conceptually the approach that evolved from the World Summit had two advantages over its predecessor. First, the responsibility to support societies under stress became a general responsibility owed by all UN member states (as well as the UN itself and regional organizations), not a specific responsibility that encumbered only interveners. Second, the approach embraced the fluidity of prevention, protection, and rebuilding. What mattered, by the secretary-general's reckoning, was whether steps were taken to address sources of risk and vulnerability, not the temporal sequencing or conceptual ordering of those steps.

Yet the obvious connections between R2P and peacebuilding were not recognized, let alone fashioned into a coherent and coordinated agenda for action. This created significant gaps in R2P theory and practice, especially a gap in the moral responsibility to rebuild and a gap in understanding of how postatrocity rebuilding ought to proceed. Precisely *why* this gap emerged remains unclear, but Paris (2016, 515–516) suggests two possible explanations. The first is straightforward path dependency. When the World Summit decided to separate rebuilding from R2P, this established peacebuilding and R2P as separate agenda items with their own distinct institutional frameworks and offices (the Peacebuilding Commission and the UN Office on Genocide Prevention and R2P). This encouraged the separate and distinct development of both activities. In this, it might be argued that the R2P-peacebuilding gap is an unremarkable product of the UN's bureaucratic politics, no different from divides between peacebuilding and peacekeeping, or the protection of civilians and human rights, for example. Second, Paris suggests that actors involved in implementing both agendas had additional incentives to separate them. On the one hand, because R2P was seen as politically controversial, those involved with peacebuilding saw it as potentially counterproductive to draw links between it and their work. Outi Donovan (2018, 394) also suggests that the "responsibility to rebuild's" connection to military intervention made it controversial by association. On the other hand, determined to set out the distinctiveness of atrocity prevention, R2P advocates were cautious about linking their agenda to others'. Whatever the reason, the key point is that as R2P evolved from idea to practice, the "responsibility to rebuild" was hived off, creating a gap between R2P and peacebuilding.

António Guterres came to the secretary-generalship in 2017 promising to reform the UN's peace and security architecture and overcome the problem of policy silos. Although progress has been slower than some activists had hoped for, there are clear signs of moves that might help reduce the gap between R2P and peacebuilding. In this the secretary-general was helped by two developments inside the General Assembly: the 2015 adoption of Sustainable Development Goal 16, calling for the promotion of peaceful and inclusive societies, and the Assembly's 2017 resolution and focus on "sustaining peace." The secretary-general's own blueprint for sustaining peace included a call for strengthening partnerships with regional organizations, civil society, and the private sector; improving internal coordination through joint assessments, planning, and financing; and integrating the UN's development agencies (most notably the UN Development Program) more fully into its conflict prevention (Guterres 2018a). By trying to elevate civilian action in peacebuilding to the same level as the UN's more established agendas for peacemaking and peacekeeping, the sustaining peace agenda opens avenues for thinking more creatively about developing and harnessing civilian expertise found in peacebuilding for the purpose of atrocity prevention. Specifically, in his report on R2P in the same year the secretary-general referred explicitly to the value of civilian action for atrocity prevention and committed to reviewing and examining ways of strengthening capacity in that area. To that end, he promised to articulate "a comprehensive plan on the basis of extensive consultation to strengthen civilian action for atrocity prevention" (Guterres 2018b, 49(f)). Ideas under consideration include the creation of a global fund for atrocity prevention to support community-based action and the development of a five-hundred-strong civilian response capacity. Although the process is at an early stage and not all member states are enthusiastic about these proposals, it is easy to see how the development of civilian capacities to respond to support conflict prevention, response, or rebuilding might help give practical sense to closing the gap between R2P and peacebuilding. It is also possible to see how such innovations might contribute to a shift toward greater support for local peacebuilders to mitigate some of the well-known problems with top-down state-based approaches to peacebuilding and harness the latent potential of everyday peacebuilders (Autesserre 2020).

On institutional reform, Guterres moved the peacebuilding portfolio into the new Department of Political and Peacebuilding Affairs (DPPA) to reinforce the idea that approaches to peacebuilding be driven by political considerations and strategies. Although seemingly modest, this reform has the potential to overcome the gap between R2P and peacebuilding since it brings the two mandate holders, the Office of the Special Advisers on Genocide Prevention and R2P and the Peacebuilding Support Office, into the same department, both formally sitting within the DPPA. Whether that translates into practical and programmatic cooperation remains to be seen, however.

It also remains to be seen whether the reforms introduced by Guterres will help to close the gap between R2P and peacebuilding. This is an important question for, as the next section demonstrates, this gap is not just a conceptual problem; it has had practical consequences too.

The R2P-Peacebuilding Gap

The R2P-peacebuilding gap is troubling for several reasons (Bellamy and Luck 2018, 93–94). One is because of the long-observed tendency for atrocity crimes to occur in places where they have previously occurred. A second is because preventing future occurrences in places where atrocities have already been committed is at least as urgent as, if not more urgent than upstream preventive work in advance of atrocity crimes. In this context, the temporal sequencing of prevention-reaction-peacebuilding can create false and counterproductive distinctions. A third reason is that rebuilding strategies may well need to be cast differently in places that have experienced atrocity crimes than in those recovering from more conventional conflict. In the aftermath of atrocity crimes, it may be particularly critical to take into account the intergroup effects of economic, political, and institutional rebuilding so as to avoid perpetuating or exacerbating ethnic, ideological, or sectarian differences. Fourth, because of the separation of R2P and peacebuilding, we do not have a keen understanding of what kind of rebuilding measures would be most appropriate and productive under distinct sets of circumstances, as there has been remarkably little study or analysis of rebuilding after atrocity crimes under any conditions.

One effect of the R2P-peacebuilding gap is that the UN's efforts to refine its peacebuilding strategies have not incorporated R2P and atrocity prevention. This is despite the fact that while experience since 2005 has not yet provided compelling answers to the question of how best to rebuild societies and prevent recurrence after atrocities, it has underscored the importance of incorporating concerns about rebuilding into calculations about the utility of different models for the use of force to protect populations and of building atrocity-prevention considerations into peacebuilding efforts.

Clearly Libya provides the best example of the first kind of problem—where insufficient consideration, will, and resources were dedicated to rebuilding after armed intervention. There international support for major investment in long-term rebuilding was tepid, not only on the part of the West but also across the wider UN membership. As such, when the immediate atrocity-prevention goals were achieved by the NATO-led air strikes, there was an impulse to declare "Mission accomplished" and downscale investment. The requisite political and material support for postintervention peacebuilding in Libya was not sustained, and the longer-term outcomes remain uncertain at best. Partly as a result, the country fragmented into two halves, governed by rival entities in Tripoli and Benghazi. Armed conflict and human rights abuses have persisted. A more comprehensive understanding of R2P, which included a responsibility to rebuild that might have given rise to a more far-reaching approach to supporting the transitional government that emerged after Muammar Gaddafi's ouster, could have helped consolidate the mission's early successes and prevent the country's subsequent slide into chaos. This is precisely what the Chinese scholar Ruan Zongze (2012) had in mind when he argued that "responsible protection" meant that the "protectors" should assume responsibility for the reconstruction of the state concerned.

But Libya is not the only kind of problem that arises as a result of the R2P-peacebuilding gap. Problems also arise when peacebuilding activities pay insufficient attention to understanding and addressing the risks of atrocity crimes and vulnerabilities of certain populations. This problem—the failure to include atrocity-prevention considerations in peacebuilding—is also far more common than the first type. There are numerous examples, but I will limit myself to three.

During the closing stages of Sri Lanka's civil war in 2009 the international community was slow and confused in its response to civilian destruction, despite advance warning from the UN about the potential for atrocity crimes that accompanied the resumption of conflict. The UN maintained a country team in Sri Lanka focused on development and humanitarian issues but which also included a focus on peacebuilding in Tamil communities. Not only did these programs themselves not address sources of atrocity crime risk (and there are suggestions that some development projects might have exacerbated the dangers by rewarding the Liberation Tigers of Tamil Eelam), but as the violence escalated, none of this work was reconfigured with a focus on preventing atrocities or protecting vulnerable populations. In announcing his "Human Rights Up Front" action plan, which aimed to ensure that protection was mainstreamed throughout the UN's field missions, Secretary-General Ban Ki-moon acknowledged this "systematic failure" to protect in Sri Lanka.

In the CAR, the presence of a UN peacebuilding support office did not prevent the organization being criticized for responding too slowly and timidly to the onset of atrocities in 2013. Although the Integrated Peacebuilding Office contained conflict prevention and human rights components in the CAR, which identified the risk of atrocity crimes associated with the escalation of conflict between the anti-Balaka militias and those loyal to the government as early as late 2012 (prompting a decision to withdraw nonessential staff), it was not until September–October 2013 that the UN's Special Adviser on Genocide Prevention and High Commissioner for Refugees began to publicly discuss the commission of atrocity crimes and risk of genocide. By that stage, the anti-Balaka militia had already embarked on a systematic campaign of ethnic cleansing and crimes against humanity that resulted in the forced migration of approximately 80% of the country's Christian population (Cinq-Mars 2016).

In Myanmar in 2017 the UN and other actors, including the EU, had extensive programs designed to support the country's democratic transition, humanitarian needs, and economic development. As in Sri Lanka, none of these programs sought to address and reduce the very obvious risks of violence and atrocity crimes against the Rohingya population in Rakhine state. Nor did the UN (or anybody else, for that matter) place human rights–based conditionality on the delivery of assistance or foreign investment. Thus financial support for the government poured into the country despite the fact that it took few tangible steps to ease systematic discrimination and systemic violence against the Rohingya. The widespread and systematic atrocity crimes, which may constitute genocide, perpetrated by Myanmar's military were widely anticipated, and the risks widely understood, and yet these were not factored into the international community's peacebuilding engagement with Myanmar.

An independent investigation into the UN's handling of the crisis in Myanmar, commissioned by the secretary-general and led by the seasoned Guatemalan diplomat Gert Rosenthal (2019), found that there had been a "systematic failure" comprising five main elements. First, there was insufficient intergovernmental support for addressing the situation—something beyond the immediate control of the UN but to which the UN may have contributed by not sensitizing states to the atrocity risks. Second, the UN's response lacked a coherent and unifying strategy to protect civilians—an obvious consequence of the siloing of peacebuilding, development assistance, humanitarian affairs, and atrocity prevention. Third, the UN's efforts lacked a clear coordinating point. Fourth, perhaps because of all this, the UN's country team—led by a development expert with no training or experience operating in conflict or atrocity zones—responded to the crisis in a dysfunctional and incoherent fashion. Fifth, the UN lacked a unified and coherent assessment of the situation. In particular, the UN's engagement at both headquarters and in the country team was not guided by a unified assessment, and the assessments that were done did not include an atrocity crimes perspective (Rosenthal 2019).

Putting It Back Together

One of the principal sources of risk of future atrocities is a recent past of atrocity crimes. It is therefore important to pay close attention to those countries that have recently experienced atrocity crimes and to make additional efforts to prevent the recurrence of these crimes. Two interrelated issues are especially important in this regard: (1) the need to integrate an atrocity-prevention perspective into peacebuilding, which includes measures aimed at addressing the sources of past atrocity crimes, and (2) the need to address atrocity-specific issues of truth, justice, and reparation relating to crimes from the past. The first points to the need for a closer relationship between atrocity prevention and peacebuilding. The second points to the need for more attention to be paid to the UN Human Rights Council's thematic agenda on the promotion of truth, justice, reparation, and guarantees of nonrecurrence. This section addresses each in turn.

Integrating Atrocity Prevention into Peacebuilding

As conceptions of R2P have evolved, so should notions of related peacebuilding. The ICISS approach to rebuilding was based on two premises: that coercive military intervention would be a common response to atrocity crimes and that it would cause wide societal damage that should be addressed by those undertaking such interventions or others in the wider international community. Rebuilding plans, it was asserted, should be in place before the decision to use force is taken. In practice, of course, most engagements to prevent atrocities or protect populations have not involved the coercive use of force. Either force has not been employed, or it has been used with the consent of the local government, so the question arises of how rebuilding plans and strategies might differ depending on whether consent had been obtained and/or other tools for prevention and protection had been utilized. R2P's first two pillars provide the basis for thinking of a more comprehensive approach to rebuilding, grounded in the international community's general responsibility to support the prevention of atrocity crimes. In practice, this might mean incorporating an "atrocity prevention lens" into the UN's peacebuilding activities and ensuring closer collaboration between the organization's peacebuilding and atrocity-prevention entities.

Promoting Nonrecurrence

A second, and more focused, dimension relates to the prevention of recurrence. Few questions are more sensitive—or more important—than that of addressing historical cases of mass atrocity. The question goes to the very heart of sovereignty, national and political identities, and contested understandings of the past. Opening up past issues is painful and potentially dangerous, yet one of the principal sources of risk of future atrocities is a recent past of atrocity crimes not properly dealt with.

Within these rubrics, there are a number of discrete areas relevant to the prevention of atrocity crimes. First, in postatrocity settings, requirements for disarmament, demobilization, and reintegration and security-sector reform are especially pressing. Second, in order to break past cycles of impunity that could encourage past perpetrators to reoffend or past victims to commit reprisal atrocity crimes, it is imperative that perpetrators of past atrocities be held legally accountable for their crimes. There are a number of ways in which this might be achieved, including referrals to the International Criminal Court, special international tribunals, hybrid courts comprising national and international elements, national processes, and processes that combine formal prosecutions for senior leaders with more traditional forms of restitution for lesser offences. Third, attention needs to be paid to the promotion and protection of human rights, and especially to combating forms of discrimination that may have given rise to atrocities in the first place.

According to Pablo de Grieff (2015), the UN Human Rights Council's special rapporteur on the promotion of truth, justice, reparation, and guarantees of nonrecurrence, guaranteeing nonrecurrence requires a comprehensive strategy, which should be adopted by states in partnership with the international community, in the aftermath of atrocity crimes. De Grieff (2015) argued that promoting nonrecurrence requires a comprehensive strategy, which should be adopted by states in partnership with the international community, in the aftermath of atrocity crimes. Core elements of such a strategy would include the following:

- The ensuring of security for all and ending of violations, especially against the most vulnerable.
- Recognition of the legal right to identity for all, so that members of all groups can be legal persons before the law and access their rights through relevant institutions.
- The ratification of relevant instruments of international human rights and humanitarian law and the passage of enabling legislation.

- Legal reform aimed at de-incentivizing the perpetration of atrocity crimes.
- Judicial reform to ensure judicial competence and independence.
- Constitutional reform to remove discriminatory provisions, incorporate international human rights standards, regulate the security sector, ensure separation of powers, and facilitate judicial oversight of constitutional law.
- Enabling civil society (understood as including trade unions, religious organizations, etc.) to contribute to prevention by limiting legal restrictions and refraining from harassing civil society organizations.
- Enabling civil society by removing barriers to their constructive participation.
- Establishing programs designed to promote the legal empowerment of marginalized groups, including women.
- Ensuring that education promotes critical thought and peacefulness by emphasizing different perspectives, international standards of human rights, and the resolution of disputes. The proper teaching of history is also an important dimension.
- Utilizing cultural initiatives—museums, exhibitions, monuments, and theater—to memorialize past crimes, recognize victims, and build empathy and understanding.
- Ensure that survivors are provided psychosocial support and trauma counseling.

These considerations address some of the risk factors associated with atrocity crimes and ought to inform comprehensive strategies for peacebuilding in societies shattered by past atrocity crimes. In such settings, the structural prevention of atrocity crimes is an important aspect of peacebuilding.

When it comes to promoting nonrecurrence, the record thus far is mixed, not least because of the R2P-peacebuilding gap. On the one hand, significant strides have been taken toward achieving the nonrecurrence of atrocity crimes in West Africa, especially in Sierra Leone and Liberia. In both of those countries, a combination of state-led prosecutions and civil society–driven reconciliation processes have supported the embedding of peace. On the other hand, in both CAR and Myanmar, societies with past experience of atrocity crimes were plunged into violence once again while the international community was slow to respond.

One emerging—and concerning—practice is a move toward transitioning from violence to peace by avoiding questions of historical memory. In Indonesia and Myanmar, for example, peacebuilding has been prefaced upon the idea that past crimes should not be confronted until peace has been well embedded. In both cases, the key to negotiating political transition rested upon the capacity of reformers to persuade the military to accept reform in return for implicit guarantees that individual officers would retain their social and economic status. Reform in Indonesia has proven very successful, transforming the state from a serial perpetrator of atrocity crimes into a functioning democracy while steadfastly avoiding many of the historical memory issues described earlier. But in Myanmar, impunity for past crimes helped create the conditions in which the military felt able and willing to perpetrate atrocities on a massive scale against the Rohingya. The relationship between peace and justice is, of course, an enduring problem in peacebuilding, but it is especially significant—and difficult to resolve—in the context of rebuilding after atrocity crimes and preventing their recurrence.

Beyond this, there are at least three deeper sets of questions that will need to be addressed. The first is that the normative contours of the responsibility to rebuild require further elaboration (see Keranen 2016). The second is the need to think more carefully about how peacebuilding itself is conceptualized and practiced. There are well-known limits to the sorts of approaches described here (as explained in other chapters in this volume) and many good ideas about how these might be addressed (ranging from hybrid and post-hybrid forms of peacebuilding to everyday peacebuilding and much else), but as yet little consensus on how it all hangs together or, rather, how it ought to hang together. The third is that we remain unsure about how to measure peace and the effectiveness of peacebuilding (Caplan 2019).

Conclusion

Initially an integral part of R2P, albeit in limited form, the responsibility to rebuild has been sidelined from discussions about the prevention of atrocity crimes and protection of vulnerable populations. This was largely due to the UN General Assembly's decision to treat R2P and peacebuilding as separate domains, but this gap has had significant practical implications. On the one hand, it has allowed interveners—and indeed the wider UN membership—to avoid assuming responsibility for reconstruction after military operations to protect populations. On the other hand, it has excluded atrocity-prevention concerns from peacebuilding, leaving the UN's country teams and field missions inadequately prepared to help mitigate the underlying risks of atrocities, stem escalation, or protect vulnerable populations. The R2P-peacebuilding gap, however, is being increasingly recognized. It can be closed by action focused on integrating an atrocity-prevention lens into peacebuilding programs and strengthening the UN's existing work on promoting nonrecurrence.

References

Annan, K. 1998. *The Causes of Conflict and the Promotion of Durable Peace and Sustainable Development in Africa: Report of the Secretary-General.* A/52/871-S/1998/318, 13 April.

Autesserre, S. 2020. *The Frontlines of Peace: An Insider's Guide to Changing the World.* Cambridge: Cambridge University Press.

Ban Ki-moon. 2009. *Implementing the Responsibility to Protect: Report of the Secretary-General.* A/63/677, 12 January.

Bellamy, A. J. 2009. *The Responsibility to Protect: The Global Effort to End Mass Atrocities.* Cambridge, UK: Polity.

Bellamy, A. J., and Edward C. Luck. 2018. *The Responsibility to Protect: From Promise to Practice.* Cambridge, UK: Polity.

Boutros-Ghali, B. 1992. *An Agenda for Peace: Report of the Secretary-General.* A/47/277, 17 June
Caplan, R. 2019. *Measuring Peace.* Oxford: Oxford University Press.
Cinq-Mars, E. 2016. *Too Little, Too Late: Failing to Prevent Atrocities in the Central African Republic.* New York: Global Centre for the Responsibility to Protect.
de Grieff, P. 2015. *Report of the Special Rapporteur on the Promotion of Truth, Justice, Reparation and Guarantees of Non-recurrence.* A/HRC/30/42, 7 September.
Donovan, O. 2018. "Changing Ideas, Changing Norms: The Case of the 'Responsibility to Rebuild.' " *Cooperation and Conflict* 53 (3):392–410.
Guterres, A. 2018a. *Peacebuilding and Sustaining Peace: Report of the Secretary-General.* A/72/707-S/2018-43, 18 January.
Guterres, A. 2018b. *Responsibility to Protect: From Early Warning to Early Action. Report of the Secretary-General.* A/72/884-S/2018/525, 1 June.
International Commission on Intervention and State Sovereignty. 2001. *The Responsibility to Protect.* Ottawa: IDRC.
Keranen, O. 2016. "What Happened to the Responsibility to Rebuild?" *Global Governance* 22 (2):331–348.
Paris, R. 2016. "The Blurry Boundary between Peacebuilding and R2P." In *The Oxford Handbook of the Responsibility to Protect*, edited by Alex J. Bellamy and Tim Dunne, 509–523. Oxford: Oxford University Press.
Rosenthal, G. 2019. *A Brief and Independent Inquiry into the Involvement of the United Nations in Myanmar from 2010 to 2018.* 29 May. https://www.un.org/sg/sites/www.un.org.sg/files/atoms/files/Myanmar%20Report%20-%20May%202019.pdf.
UN General Assembly. 2004. *A More Secure World: Our Shared Responsibility.* Report of the Secretary-General's High Level Panel on Threats, Challenges and Change. 2 December. http://undocs.org/A/59/565.
Zongze, R. 2012. Responsible Protection: Building a Safer World. Beijing: China Institute of International Affairs.

CHAPTER 20

THE EUROPEAN UNION AND PEACEBUILDING

NATHALIE TOCCI

In 2003 the European Union was full of confidence and hope. Back then Europeans lived in an international hyperliberal order (Gray 2018) in which the belief in an imminent *end of history* was widespread (Fukuyama 1992): human rights, democracy, liberal market economies were slowly and surely spreading and within everyone's reach. Liberal peacebuilding (Richmond 2005) was the international, Western, and European flavor of the day, to which the European Security Strategy (EU High Representative 2003) firmly adhered.

The EU thought it knew what was the best and in fact only avenue to peace, notably for the myriad open and seemingly frozen conflicts in its eastern and southern surrounding regions. Federal, consociational, and on vary rare occasions secessionist—Palestine, Kosovo, or Montenegro—solutions were advocated (Tocci 2007; Diez, Stetter, and Albert 2006), all of which included provisions for good governance, human and minority rights protection, rule of law, democracy, and market liberalization The question was thus not *what kind* of peace the EU should pursue from Cyprus to Abkhazia, from Transnistria to Palestine. The answer to that question was self-evident to all back then. The unanswered question was rather *how* the EU should act to bring about its desired peace outcomes.

The optimism, confidence, idealism, and, yes, hubris of the early 2000s could simply not be replicated in the EU Global Strategy in view of the more connected, contested, and complex world Europeans inhabited over a decade later (EU HR/VP 2015, 2016). At the same time, the Union did not want to convey a political message of closure, defensiveness, defeatism, or crude realpolitik to Europeans and to the wider world. To the contrary, the Global Strategy wanted to send an opposite message, of openness and self-confidence. The EU would have to stand united in engaging the wider world, and it would do so responsibly, promoting peace in an integrated manner both internally and in partnership with others. But it could not pretend that a ring of well-governed countries beyond its borders was easily within reach, nor that it had a magic wand to fix all failed states, regional conflicts, and human rights abuses within and beyond its borders.

The pendulum had to move away from the outward-looking hyperliberal idealism of the early 2000s, without swinging all the way back to the opposite end of defensive realpolitik. "Principled pragmatism" was the notion that sought to square the circle. Priorities such as "resilience" and the "integrated approach" to conflicts and crises in the Global Strategy suggest an EU questioning not simply the means it should pursue in peacebuilding but the actual goals it should support. With the Global Strategy, the EU conceptualized a 2.0 foreign policy approach, which aimed at revising the 1.0 hyperliberal peacebuilding mechanisms of the 1990s and early 2000s. To trace this journey, this chapter will briefly recount what the EU's approach to peacebuilding looked like in the 1990s and 2000s and what it is now transforming into, largely as a result of the changed strategic context both in Europe and in the wider world.

EU Liberal Peacebuilding 1.0

Promoting peace has long been identified by the EU as one of its top foreign policy priorities. Particularly when it comes to conflict resolution in areas straddling or bordering the Union, the EU in the post–Cold War period used integration incentives and conditionalities embedded in its contractual relationships with third countries to pursue variants of the liberal peace that it was uncritically wed to (Tocci 2007; Diez, Stetter, and Albert 2006). These contractual ties ranged from the accession process to looser forms of integration, such as the association agreements for southern Mediterranean countries, the partnership and cooperation agreements for the former Soviet countries, the stabilization and association agreements for the western Balkan states, and the European Neighbourhood Policy (ENP) for both the eastern and the southern EU neighbors. The declared aims of these contractual relations were to achieve varying degrees of cooperation and integration in the EU, as well as to foster long-running structural change, such as peace, within and between third countries. Liberal peace within conflict-ridden countries in the EU's surrounding regions was pursued within the framework of these contractual relations through three principal means and mechanisms: conditionality, learning, and passive enforcement.

Conditionality

The first and most well-known mechanism through which the EU sought to spur liberal peace beyond its borders was conditionality (Grabbe 2001; Smith 1998). This would typically take the form of "positive conditionality"—the promise of a benefit in return for the fulfillment of a predetermined condition—such as economic assistance or progress in the enlargement process. The delivery of EU benefits was at times made directly conditional on peace efforts, such as the case of the 1995 Stability Pact promoted by Prime Minister Édouard Balladur of France to diffuse minority and border tensions in Eastern Europe. On other occasions the link was more diffuse but nonetheless well appreciated by the conflict parties and thus no less effective. The agreement between Serbia and Montenegro in 2003 was achieved when High Representative Javier Solana's skillfully leveraged the prospects for western Balkan enlargement, which was first formally recognized in the Thessaloniki European Council that year to broker an agreement between Belgrade and Podgorica (Coppieters et al. 2004). Another case of positive conditionality concerned Turkey and the Kurds. By making respect for the Copenhagen political criteria—including human rights, minority rights, and rule of law—a precondition for the opening of accession negotiations, the EU successfully persuaded Ankara in the early 2000s to embark on a set of reforms that positively impacted the Kurdish population, such as a degree of liberalization of language, expression, and association rights, as well as the abolition of the death penalty (Kurban 2003). Negative conditionality, notably sanctions, was far more sparingly used (Portela 2005) but occasionally applied in the 1990s and early 2000s in sub-Saharan Africa, Belarus, and a few other often faraway places.

Learning

Beyond conditionality, a second mechanism through which the Union promoted liberal peace was social learning and persuasion, which took place through the institutional, political, economic, and wider societal contact between the EU and conflict parties. As opposed to conditionality, which alters decision-makers' cost-benefit calculus, domestic change through learning occurs with a transformation of perceived interests, as conflict parties voluntarily internalize the liberal peace norms promoted and personified by the EU (Diez 2013). Through participation in or close contact with the EU institutional framework, conflict parties would thus alter their substantive beliefs, visions, and purposes, as well as their preferred negotiation strategies in a manner conducive to peace. In northern Cyprus, for instance, beyond the reality that membership in the EU was conditional upon reunification, the degree of Turkish Cypriot dissatisfaction with the leadership of Rauf Denktaş at the time and the appeal of entering a liberal Union through a carefully construed federal compromise goes far in explaining their overwhelming support for the Annan Plan in 2004. The Turkish Cypriots remain outside the Union to this day, the persistence of the frozen conflict largely due to the unwillingness of their Greek Cypriot compatriots, comfortably within the EU, to accept meaningful compromise.

Passive Enforcement

The final and perhaps most subtle mechanism the EU used in the 2000s to promote its vision of a liberal peace was the passive enforcement of its own rules, norms, and regulations. Rather than highlighting the logic of punishment, which sets in when rules are violated, this mode of foreign policymaking hinged on a system of rule-bound cooperation (Tocci 2007). Unlike conditionality, passive enforcement does not seek to alter the incentives by altering the cost-benefit calculus of a conflict party. The EU's delivery of benefits does not come as a reward for a party's compliance with a given condition. Obligations constitute the necessary rules that make mutually beneficial cooperation with the EU possible. For passive enforcement to work, there must be a clear set of legally defined and definable rules embedded in EU contracts. Passive enforcement cannot be easily used for conditions that the EU simply considers politically desirable but that have no legal bearing. Furthermore, this system of rules must be considered a necessary price to pay for EU engagement. The most telling example of a successful case of passive enforcement regards the Israeli-Palestinian conflict and in particular EU-Israeli free trade and Israel's participation in the EU's Research Programme Horizon 2020. In view of Israel's—and the US's—leverage on the EU, the use of conditionality was never seriously contemplated by the EU and its member states. At the same time, EU persuasion on the Palestinian file has never been particularly effective vis-à-vis Israelis. Yet the EU could not accept, as a matter of law, that the benefits it was willing to grant internationally recognized Israel—i.e., Israel within the pre-1967 borders—would accrue also to settlements and settlers beyond the 1967 lines. The EU thus resorted to passive enforcement. It did not try to persuade, condition, or bribe Israelis. It simply stood firm on the EU's rules of the game. If Israel wanted to benefit from inclusion in a free-trade zone with the EU or participate in the European research area, the rules of engagement were clear. Israeli officials kicked and screamed. But ultimately and particularly in the case of Horizon 2020, Israel ultimately complied and Israeli entities and individuals beyond the 1967 borders were excluded from receiving EU research funds.

EU Liberal Peacebuilding 2.0

The EU Global Strategy and the Interests versus Values Debate

This was the Union that was: a Union that all recognized as the most successful peace project in history; a Union that enlarged its internal peace by reuniting the continent after the scourge of the Cold War; a Union that, after acknowledging its utter failure to put an end to war in the Balkans, rolled up its sleeves and extended its integration promise to the western Balkans and Turkey too; a Union that, well aware of the impossibility of enlarging forever, sought to export its liberal values beyond the confines of enlargement by devising a neighborhood policy and embedding human rights conditionalities in its trade and development policies with all third countries and regions. In short, a Union that, while pursuing its interests, did so always couching such interests in its liberal peace values. Back then, through its self-image as a normative power (Manners 2002), the EU and most member states felt uncomfortable in acknowledging they had any interests at all.

Fast-forward to 2015, when Europe and the European project looked fundamentally different, and not for the better. Externally an assertive Russia had lifted its gaze, bringing geopolitics forcefully back into the eastern neighborhood and leading the EU to respond with sanctions to Moscow's annexation of Crimea and its subsequent destabilization of eastern Ukraine (Youngs 2017). In the south, after the short-lived promise of an Arab Spring, the Middle East and North Africa plunged into an unprecedented cycle of violence and repression (Del Sarto 2015). And within the Union itself, shattered by the Eurozone crisis and a deep crisis of solidarity triggered by migration flows, political unity and above all unity around liberal values was no longer a given.

Stemming from this context, the EU Global Strategy worked on in 2015–2016 sought to chart a different route to the EU's approach to fragility and conflict within and beyond its borders (Tocci 2017). To do so it started by acknowledging openly its interests rather than hypocritically pretending these did not exist. The Global Strategy stated that the EU has four basic interests: security, prosperity, democracy, and a rules-based global order. With the exception of a rules-based global order, which should be seen as the external condition for the fulfillment of the first three interests, these are first defined by their internal connotations. The EU has an interest in promoting the security and prosperity of its citizens and the vibrancy of its democracies. These internal interests, however, can be fulfilled only if accompanying external interests are pursued too.

European citizens will not be safe so long as there is violence, repression, and instability in the EU's surrounding regions. Terrorism is the most obvious case in point. Likewise in the twenty-first century in view of the greater connectivity and complexity of the international system, Europeans will not be prosperous without sustainable development and an open economic system worldwide. With most global growth expected to take place outside the EU in the coming decades, the EU's own development will increasingly hinge on its relationship with the outside world, be it through trade, investment, the digital economy, or the movement of people. As Europe (and the United States) undergo a wave of populist-inspired closure, this point is far from uncontested within Europe and the West more broadly.

Finally, and perhaps most interestingly, the state of health of European democracies cannot be detached from the way Europeans engage with the outside world. When walls and fences are built to keep away refugees in search of protection in the EU, the mortal blow to European democratic systems is incalculable. Hence, before preaching human rights to others, a first fundamental interest lies in securing the liberal laws and values that underpin the EU's own democratic systems. The way the EU conducts its external relations plays an important role in that respect.

In this regard, the EU Global Strategy can still be viewed as a "liberal" rather than as a "postliberal" strategy (Juncos 2016). But its liberal connotations are internal more than external: the Strategy is devoid of the hyperliberal hubris that devoured the West in the early post–Cold War era. In other words, the Global Strategy stands firm on the affirmation of the EU's internal liberal values, which lie at the core of the European project. Its firmness on this point is key given that those values are being questioned within, as evident with the rise of extreme right-wing populism, nationalism, and nativism across the continent. But this does not mean that the EU expects its internal liberal values to be adopted externally too, be it through conditionality or to a lesser extent persuasion. What it does mean is that the EU's internal democratic rules and values should pragmatically guide its external policy. In this respect, of the three methods pursued in EU peacebuilding 1.0, passive enforcement, which is intrinsically connected to the EU's internal system of norms, is the one that continues to resonate most in the EU's peacebuilding 2.0 in the making. Elaborating on this conceptual approach, the Global Strategy presents two new approaches: resilience and the integrated approach to conflicts and crises.

The Resilience of States and Societies in Surrounding Regions

The security of the Union hinges on peace and stability beyond its borders. This observation is not new. Already in the 2003 European Security Strategy, the EU set out to promote a ring of well-governed countries to the east and south as an integral element of its own security. The recipe back then was that of Europeanization. By radiating its norms and values outward, the EU would promote peaceful and well-governed countries beyond its borders. It would do so through the enlargement policy and the ENP. Even when it came to countries such as Russia, particularly under Dmitry Medvedev's rule, the name of the game was Europeanization. The four common spaces for cooperation between the EU and Russia, inscribed into the framework of the partnership and cooperation agreement between the two, were informed by the general understanding that Russia would eventually approximate EU laws and standards across different policy areas.

Today that world is gone. True, there are still countries that have opted for reconciliation, democratization, and modernization through Europeanization. The countries of the western Balkans and a handful of countries in the ENP, would still like to move closer to EU norms and standards. The EU Global Strategy acknowledges this and recommits to these countries, which often feel abandoned by a Union that has no wind left in its enlargement sails (EU HR/VP 2016). But beyond this limited number of countries, most neighbors to the east and south are not pleading to become more like the EU. In truth, this was also the case in the early 2000s. But the Union, buoyed by the enlargement success at the time, failed to see this. Today no one can hide the fact that for all these countries the challenge is to develop a real foreign policy and not just a surrogate for enlargement—a foreign policy that addresses state fragility, listens more than it preaches, supports more than it dictates.

Resilience reflects the EU's newfound humility in foreign policy without giving up on its principles. The world had changed, and the EU had to become more pragmatic as a result. It had to remove its rose-tinted glasses that depicted a world that simply wanted to look like the EU. Many countries to the EU's east and south have no such intention. At the same time, the EU couldn't simply abandon its transformational agenda in favor of a crude transactional one, in which even the most egregious violations of rights and laws by states beyond its borders would be ignored nonchalantly by the Union. Resilience sought to capture that middle way. As put by Wagner and Anholt (2016, 4), resilience provides "a middle ground between over-ambitious liberal peacebuilding and under-ambitious stability." The EU Global Strategy saw resilience as the ability to absorb, react to, and respond to shocks and crises. A resilient state is thus one that is able to survive change by changing itself, just like a resilient metal that bends but does not break. Hence the EU Global Strategy claimed that authoritarian states are not resilient in the long term. They may appear to be extremely stable, at times immobile. They may indeed remain so for years and even decades. One only needs to think about regimes such as North Korea to realize that the long term may be very long indeed. But when a shock or crisis does occur, the brittleness of authoritarian states emerges in full force. Faced with shock, they tend to break altogether. The 2011 Arab uprisings are a testimony of that and should not be forgotten at times in which escalating violence and instability in the Middle East lure many in Europe to back unconditionally apparently stable yet deeply fragile authoritarian regimes in the region.

Resilient states and societies therefore need to be secure, but they must also be inclusive, well-governed, developed, cohesive, and sustainable. This does not mean that there is a single recipe to achieve such resilience, namely, the implementation of the EU's *acquis communautaire* and more generally the ready-made, off-the-shelf liberal peace agenda. The EU's *acquis* served its members very well, including those that entered the Union in 2004. In this respect, it is worth recalling that Poland and Ukraine shared similar economic standards when the Soviet Union collapsed. Clearly Poland today, despite its dangerous dip back into authoritarianism, is far more resilient than Ukraine, whose weakness invited Russian intervention. But while Europeanization "works" for some, the EU Global Strategy recognizes that it is not the way for a bright future for all. There is no one-size-fits-all resilience: "there are many ways to build inclusive, prosperous and secure societies" (EU HR/VP 2016, 25).

The EU would therefore support different paths to build resilient states and societies in its surrounding regions. The rather self-referential term "neighborhood" was deliberately dropped. In fact, while the Global Strategy makes references to the ENP and on occasions refers factually to the EU's "neighbors," nowhere does it define its surrounding geography as the "neighborhood." Instead it speaks more broadly of "surrounding *regions*" (emphasis added). On the one hand, the EU Global Strategy wanted to signal that it viewed the resilience of states and societies as important not only in the countries falling within the scope of the enlargement policy and the ENP, but also beyond, stretching east into Central Asia and south into Central Africa. On the other hand, the Strategy wanted to abandon the term "neighborhood," which conveys a Eurocentric vision of a homogeneous space beyond the EU's borders, a vision blatantly detached from realities on the ground. Not only are there huge differences between east and south, but the predicament and aspiration of each "neighbor" are to be accounted for in their own right.

An Integrated Approach to Conflicts and Crises

Integral to the EU's 2.0 understanding of peacebuilding is the "integrated approach" to conflicts and crises in surrounding regions to the east and south. One of the offspring of the 2003 European Security Strategy was what is commonly known as the "comprehensive approach." The most common interpretation of the comprehensive approach is the notion that the EU ought to approach conflicts and crises by blending all its policies, most importantly its security and development instruments. This traditional interpretation of the comprehensive approach remains relevant to this day. For years, policymakers working on security and development have had an uneasy relationship, the latter staunchly resisting the former's attempt to use development funds for security purposes.

But the EU Global Strategy did not want to stop at instilling a new lease of life in the old comprehensive approach. It therefore used a different term, "integrated approach," to signal what had to be added to it. Beyond the multidimensional approach blending security and development as well as many other policy instruments, the EU Global Strategy called for a multiphase approach along the whole conflict cycle; a multilevel approach focusing on the local, national, regional, and global levels of conflict; and a multilateral approach of partnering with all the relevant regional and global actors in a given conflict configuration.

When highlighting the need to act at all stages of the conflict cycle, the Global Strategy—and particularly the follow-up work on civilian crisis management in 2018—emphasized the prevention and stabilization phases of conflict. Conflict prevention is an area where the EU has already had more than occasional successes, in Montenegro, Macedonia, Kosovo, Indonesia, and elsewhere. Recognizing this fact, the Strategy highlighted the need to further invest in prevention, notwithstanding the multiple crises that have already broken out around the Union. The Strategy also emphasized the stabilization phase of conflict, after the crisis has exited its most acute violent phase but before it enters into long-term peacebuilding. As always, the choice was influenced heavily by context, particularly the need to pay attention to those areas of Iraq, Syria, and Libya that had been liberated from the Islamic State but could easily fall back under the Islamic State's control if abandoned to their own devices.

Also worth highlighting is the emphasis on the local and regional levels of conflicts when discussing a multilevel approach to conflicts and crisis. The EU Global Strategy recognized that most conflicts, particularly those to the EU's south, are marked by local, national, regional, and global elements. It is the interaction of these four elements that makes such conflicts so intractable and protracted. It is only by acting on all four that the EU can hope to make a meaningful difference on the ground. So, whereas the European Security Strategy, informed by European passivity in the Balkans, mentioned the need for rapid and robust intervention at the international level, the EU Global Strategy, influenced by the failure of such interventions in Afghanistan, Iraq, and Libya, does not. This is not to say that the Security Strategy believed in international military quick fixes and the Global Strategy does not. It means that in view of the different experiences that preceded these two strategic exercises, the Global Strategy took a rather sober view. It discussed what the EU can do at the international level—including through military means—to facilitate, mediate, and support locally owned and regionally embedded agreements. Rather than a reduced level of military ambition compared to the European Security Strategy, this reflected a recognition in 2016 that the robust interventions in 2003 and beyond did not produce sustainable peace. Hence different approaches, including to the way military instruments are deployed, had to be considered. It may well be that many of these conflicts will remain unresolved for many years. Cases such as Syria, Libya, and Yemen most certainly fall in this category. Yet even in these cases, the EU must act patiently on the ground, aware of the intricacies of everyday local politics and regional dynamics, doing what it can to support locally owned peace (Richmond and Mitchell 2011). Rather than preparing for the crisis of yesterday, the EU Global Strategy implicitly argued that it is worth learning from the past and preparing for the crises of today and of tomorrow.

In view of the multiple levels of conflicts, the integrated approach also highlights the imperative of developing a multilateral approach to their resolution. The EU cannot solve conflicts alone. It must develop agile and responsive ways of partnering with all relevant actors in a given conflict configuration. This partnering does not necessarily mean that all such actors will be partners of the EU, but rather that the Union recognizes that a particular conflict cannot be solved without working with them. Hence the EU must and does partner with regional players such as Iran and Saudi Arabia and international ones like Russia when addressing the conflict in Syria. But this does not mean that such players will be EU partners in every international theater. To the contrary, it factors in that in other situations, these players may have different interests and goals, or may simply not be relevant at all.

Squaring the Circle through Principled Pragmatism

The EU's transition from a 1.0 to a 2.0 approach to peacebuilding can be captured by its new guiding philosophy, "principled pragmatism," which aims to square the circle of promoting liberal peace in an increasingly illiberal geopolitical world. On this point the EU Global Strategy reads, "We will be guided by clear principles. These stem as much from a realistic assessment of the strategic environment as from an idealistic aspiration to advance a better world. In charting the way between the Scylla of isolationism and the Charybdis of rash interventionism, the EU will engage the world manifesting responsibility towards others and sensitivity to contingency" (EU HR/VP 2016).

But what does "principled pragmatism" mean? The most commonly heard definition is that the EU should "act in accordance with universal values (liberal ones in this case), but then follow a pragmatic approach which denies the moral imperatives of those universal categories" (Juncos 2016, 2). However, as correctly noted by Ana Juncos herself, such an interpretation entails a contradiction in terms: "The EU needs to be either pragmatic or principled; it cannot have it both ways" (2).

But this is not how principled pragmatism should be defined. The correct interpretation is not that the EU should compromise on its principles as a result of pragmatic interest-based considerations. The point is rather that the EU should remove its rose-tinted glasses and pragmatically look at the world as it is, not as it would like to see it. The world is no longer (if it ever was) on an irreversible march toward a (hyper)liberal peace. The pragmatism comes in the diagnosis of the geopolitical predicament the EU finds itself in, and such a predicament both within and outside Europe is part of a growing illiberal challenge to what many believed were ironclad liberal principles underpinning the EU and international society as a whole.

The EU's embrace of pragmatism echoes a broader rediscovery of pragmatist philosophy that entails a rejection of universal truths, an emphasis on the practical consequences of acts, and a focus on local practices and dynamics (Joseph 2016, 379). This means accepting different recipes to build resilient states and societies and supporting locally owned pathways to peace which may look quite different from the standard EU textbook cases. Nowhere was this clearer in 2018 when the EU sought to usher in the final push for peace between Serbia and Kosovo. The outlines of the agreement included a territorial swap, which certainly clashed with much of what the EU had been preaching in the region for the past two decades. But after so many mistakes made by Europeans (and Americans) in their 1.0 promotion of (hyper)liberal peace, many EU officials pragmatically and perhaps humbly asked themselves "Who are we to say no?" if the parties themselves reached such an agreement.

The Union must be aware not to fall into the trap of cultural relativism: EU pragmatism should be and remain firmly anchored to principles. While different pathways, recipes, and models are to be embraced, international law and its underlying norms, first and foremost human rights, should be the benchmark of what is acceptable for the EU and what is not. Principled pragmatism is no panacea. Policymakers will often find themselves at a crossroads, having to make difficult choices without the luxury of knowing the chain of events their decisions will unleash. Seeking to be pragmatic while at the same time principled is not easy. But neither is it a contradiction in terms. Rather it is a complex, time-consuming, and at times convoluted approach to navigate the dilemmas of a more humble but perhaps more effective 2.0 version of EU peacebuilding.

References

Coppieters, B., M. Emerson, G. Noutcheva, and N. Tocci, eds. 2004. *Europeanization and Conflict Resolution*. Gent: Academia Press.

Del Sarto, R. 2015. "Normative Empire Europe: The European Union, Its Borderlands, and the 'Arab Spring.'" *Journal of Common Market Studies* 54 (2):1–18.

Diez, T. 2013. "Normative Power as Hegemony." *Cooperation and Conflict* 48 (2):194–210.

Diez, T., S. Stetter, and M. Albert. 2006. "The European Union and Border Conflicts: The Transformative Power of Integration." *International Organization* 60 (3):563–593.

EU High Representative. 2003. *A Secure Europe in a Better World: European Security Strategy*. December. https://www.consilium.europa.eu/media/30823/qc7809568enc.pdf.

EU HR/VP. 2015. *The European Union in a Changing Global Environment: A More Connected, Contested and Complex World*. June.https://eeas.europa.eu/archives/docs/docs/strategic_review/eu-strategic-review_executive_summary_en.pdf.

EU HR/VP. 2016. *Shared Vision, Common Action: A Stronger Europe. A Global Strategy for the EU's Foreign and Security Policy*. June. https://eeas.europa.eu/archives/docs/top_stories/pdf/eugs_review_web.pdf

Fukuyama, F. 1992. *The End of History and the Last Man*. New York: Free Press.

Grabbe, H. 2001. "How Does Europeanization Affect CEE Governance? Conditionality, Diffusion and Diversity." *Journal of European Public Policy* 8 (6):1013–1031.

Gray, J. 2018. "The Problem of Hyper-Liberalism." *Times Literary Supplement*, 27 March. https://www.the-tls.co.uk/articles/public/john-gray-hyper-liberalism-liberty/.

Joseph, J. 2016. "Governing through Failure and Denial: The New Resilience Agenda." *Millennium Journal of International Studies* 44 (3):370–390.

Juncos, A. E. 2016. "Resilience as the New EU Foreign Policy Paradigm: A Pragmatist Turn?" *European Security*, October. http://www.tandfonline.com/doi/pdf/10.1080/09662839.2016.1247809?needAccess=true.

Kurban, D. 2003. "Confronting Equality: The Need for Constitutional Protection of Minorities in Turkey's Path to the European Union." *Columbia Human Rights Law Review* 35:151–214.

Manners, I. 2002. "Normative Power Europe: A Contradiction in Terms?" *Journal of Common Market Studies* 20 (2):235–258.

Portela, C. 2005. "Where and Why Does the EU Impose Sanctions?" *Politique Européene* 17:83–111.

Richmond, O.P. 2005. *The Transformation of Peace*. London: Palgrave Macmillan.

Richmond, O.P, and A. Mitchell. 2011. *Hybrid Forms of Peace: From Everyday Agency to Post-Liberalism*. London: Palgrave Macmillan.

Smith, K. 1998. "The Use of Political Conditionality in the EU's Relations with Third Countries: How Effective?" *European Foreign Affairs Review* 3 (1):253–274.
Tocci, N. 2007. *The EU and Conflict Resolution.* London, Routledge.
Tocci, N. 2017. *Framing the EU's Global Strategy*. London, Springer–Palgrave Macmillan.
Wagner, W., and R. Anholt. 2016. "The EU Global Strategy." *Contemporary Security Policy* 37 (3):414–430.
Youngs, R. 2017. *Europe's Eastern Crisis: The Return of Geopolitics*. Washington, DC: Carnegie.

CHAPTER 21

RISING POWERS AND PEACEBUILDING

KAI MICHAEL KENKEL

The past two decades have seen the rise of a set of states from the Global South to positions of greater influence in the discourse and practice of international politics; peacebuilding and statebuilding have become one of the major arenas where this heterogeneous group has mounted a challenge to the Northern-dominated liberal peacebuilding paradigm. These states show both obvious divergences and significant commonalities in their position toward the liberal peace; while they have been able to bring a wealth of experience into the practice of peacebuilding over the past two decades, they have done so largely from within the paradigm set by major Northern players. Specific domestic experiences have been harnessed to ameliorate the working of the liberal paradigm from within, particularly where institutions and the economy are at stake; however, in the absence of well-developed independent cultures of peace, these countries' contributions have remained state-driven and have made quite limited progress toward broader notions of peace formation as conceived of by Oliver Richmond (2014). This chapter provides a definition of this class of states apt for analyzing their peace policies and crystallizes elements of a Southern-based rising-power contribution to global peace. Examples are given of how these states have sought to multilateralize their cooperation and act consistently within the United Nations. The analysis then takes a closer look at the practices of Brazil and India, two major rising powers from the Global South, focusing on their specificities in both the international and domestic contexts.

Defining Rising Powers

It is important to arrive at the nomenclature most accurate for both the role these countries play in building peace and the role contributing to peace plays in their own perception of their role in the international system. In this sense, primarily market-based labels such as "emerging powers" capture neither the breadth of their engagement across areas of international policy nor the broad set of factors that underlies the change in their position in the international system. Given the states in question, the idea of emergence, or of "new powers," fails to account for past status as great powers, as in the case of China and Russia.

For those actors who have recently taken on increased roles in peace processes throughout the globe, the most apt characterization for the present topic is "rising powers." Besides connoting change in these states' position in the global hierarchy (Call and de Coning 2017a, 244) and a positive dynamism, the term contains two key definitional elements: these states have indeed experienced a change in their perceived status sufficient to influence responses from other actors, and they self-define and behave as powers in the midst of a positive trajectory (Carvalho and de Coning 2013, 2). In other words, they are actively contesting current hierarchies both in form and in content, challenging the normative underpinnings of the rule-based international order in which they have historically played a subordinate role (Peter 2014, 3).

At the same time, with greater influence upon the system comes the expectation of greater contributions to its maintenance, conceived of as a global good (Narlikar 2011, 1608). Peacebuilding is a crucial activity in this respect: it allows rising powers to demonstrate responsibility in this way (Visoka and Doyle 2014) while challenging a global paradigm that favors Western experience based on successful domestic experiences. States that fit these criteria and have chosen to emphasize peacebuilding as a contribution to security in their quest for increased status include Brazil, India, South Africa, Turkey, and Qatar. Russia and China largely do not self-identify as newly emergent actors and have a broader international profile in security issues (Carvalho and de Coning 2013, 3–4). Other rising actors who have received less attention are Turkey and Indonesia (Sazak and Woods 2017a, 2017b; Satana 2016; Alexandra 2017; Capie 2016).

Oliver Richmond and Ioannis Tellidis (2014, 564–566) have identified three archetypal positions rising powers take in their policy toward peacebuilding: they can seek to uphold the status quo ante, take a critical position toward the extant paradigm, or adopt a revolutionary posture. Status quo powers largely support the system and adopt the dominant Western-led paradigm; critical states, while still working within the system, seek to improve it and to challenge it from within (Stephen 2014, 914). The third option of seeking to overthrow the system in its entirety has not materialized.

Critical states, while accepting the basic premises of the Western-led liberal peace, have sought to bring their own subaltern historical experiences to bear within it and have adapted their approaches to reflect and further their own agendas within this context (Richmond and Tellidis 2014, 565). Additionally, alternative proposals from these powers often involve successful domestic development and peace-creation experiences (565–566).

Central to rising powers' critical engagement with peacekeeping is their normative challenge to both the hierarchies in international peacebuilding practices and their liberal, and particularly interventionist, content (Richmond and Tellidis 2014, 565–566;

Adebajo 2016). Here peacebuilding and other aspects of intervention such as the "responsibility to protect" have become a preferred locus for the normative contestation of inequalities and hierarchy in the international system as a whole (Weinlich 2014, 1836–1837). Rising powers are intent on countering a one-way flow of ideas and rules from North to South, resisting co-optation without influence (Pu 2012; Ghimire 2018, 43). Tied to this is the idea of bringing domestic experiences to bear in ways distinct from the liberal model, with a clear emphasis on the exceptionalism of rising powers' experience (Abdenur and Call 2017; Nyuykonge and Zondi 2017).

As rising powers' ascent is often tied to leadership within their region, there can be significant resistance to the imposition of values considered tied to the West in their own perceived spheres of influence (Ghimire 2018, 38; Adhikari 2018, 168–169). With regard to the liberal peacebuilding model, the two main areas of resistance from rising powers are both rooted in their colonial past: aversion to the use of force and emphasis on local ownership and respect for host state sovereignty (Carvalho and de Coning 2013, 5); taken together, these factors have given rise to a distinct approach to peace in the Global South.

Southern Definitions of Peace and Peacebuilding

The first step toward fully grasping Southern rising powers' approach is to adopt a wider approach than that usually applied to and by Northern actors. Where Northern practice enforces the separation between development assistance, peacebuilding, and security-related spending, states such as Brazil, India, and South Africa operate along a much broader and interconnected spectrum in their contributions, taking tentative steps toward peace formation (Call and de Coning 2017b, 7–8). Two main factors explain emerging powers' differentiated practice, with a third recently attaining more relevance as well. First, donors from developing countries are not bound by the Organization for Economic Cooperation and Development's Development Assistance Committee guidelines obliging them to separate development, peacebuilding, and security activities in both budgetary and normative terms. This allows for a greater degree of integration across these states' presence in postconflict and peace formation contexts. Second, differently from many Northern actors, rising powers can base the policies they implement abroad on successful policies at home (Call and de Coning 2017a, 247–250). These include policies designed to combat local underdevelopment through program delivery across a broad range of agencies, as in the case of Brazil (Kenkel 2013, 285) as well as programs modeled after nationwide policies designed to address structural inequalities, such as South Africa's Reconstruction and Development Programme (Nyuykonge and Zondi 2017), as well as approaches that have proven their mettle in violent domestic postconflict situations, as in the case of India (Adhikari 2018; Riddle 2017; Singh 2017).

However, a very significant fraction of peacebuilding from these states and the Global South continues to be implemented—and heavily influenced—by the state. Innovative concepts are taken from specific state agencies, as well as some from civil society organizations, but efforts to break out of state-centric approaches, and indeed the liberal paradigm on the whole, are hampered by financial constraints and, in the case of Brazil, Turkey, and some other emerging powers, state engagement based on repression and exclusion rather than independent cultures of peace. South Africa's efforts in the Democratic Republic of the Congo belie a similar adherence to the precepts of institution- and project-led peacebuilding (Nyuykonge and Zondi 2017, 116–118).

Finally, Southern actors' approaches have been shaped by—and been instrumental in shaping—the "sustaining peace" model currently being discussed in the UN development architecture; indeed this broader approach may afford more success in integrating the latter's various pillars. The common characteristics in Southern rising powers' approach to peace formation can largely be traced to their postcolonial heritage. Where the Western model is seen as top-down, state-centric, and focused on security-related institution-building (particularly of the kind conducive to enhancing security in the North itself), major Southern states describe their burgeoning paradigm in contrast to this approach.

Based on the Brazilian example, South-South cooperation, the centerpiece of rising powers' engagement with other developing states, is described as demand-driven, respectful of sovereignty through nonconditionality and an emphasis on local ownership, based on equal partnership rather than hierarchy, and generally involving a broad gamut of state agencies across a variety of issue areas (Siqueira 2018). These include "health projects, student exchanges, education support such as building schools, food security, infrastructure development of any sort, as well as political/security cooperation like security advisers, mediation support, dialogue facilitation, and elections support" (Call and de Coning 2017a, 248). Rising powers, clearly driven by an aversion to the use of military force, have sought to compensate for not using hard power through the soft power generated by their domestic successes. Similarly, this emphasis on technical cooperation and the exchange of expertise honed domestically can be seen as compensatory for a general lack of material resources as compared to Northern peacebuilders (Sinha 2017, 130).

As Anita Mathur (2014, 9–10) highlights:

> The two sets of guiding principles of South-South cooperation as confirmed in the Nairobi outcome document adopted in 2009 by the [High-Level United Nations Conference on South-South Cooperation] are as follows:
>
> (a) Normative principles, including respect for sovereignty and national ownership, non-interference in domestic affairs, partnership among equals, demand-driven engagement for mutual benefit, non-conditionality.
> (b) Operational principles, including mutual accountability, development effectiveness (transfer of knowledge with a view to strengthening local capacity and

developing national resources), coordination of evidence- and results-based initiatives, and a multi-stakeholder approach.

Mathur (2014, 17) goes on to highlight that South-South cooperation, while the nucleus of a differentiated Southern approach, represents only part of peacebuilding activities as a whole; it is, however, designed for maximum complementarity in that role. Several authors have focused on emerging actors' coordinated activities in Guinea-Bissau as an example of policy integration both within and among Southern actors, of a nature clearly adapted to the African context (Abdenur and Marcondes 2014; Roque 2009; Olayode and Ukeje 2017).

As a result, whereas Northern peacebuilding actors possess consolidated agencies that coordinate state efforts in the capital, rising powers from the South possess neither state-level peacebuilding agencies to channel their efforts nor a common multilateral organization providing guidelines and priorities, although the issue has recently been placed on the BRICS agenda. In this sense, the "rising powers" or "Southern" peacebuilding approach remains an amalgam of the efforts undertaken by its individual representatives. This ranges from the numerous but smaller-scale concentrated efforts like those undertaken by Brazil to an Indian understanding of peacebuilding as almost any development in postconflict contexts, to South Africa's formula of "good governance, dialogue and reconciliation, human resource and infrastructure development, policy implementation, economic development and trade, information sharing and exchange visits among South African dignitaries, as well as humanitarian assistance" (Call and de Coning 2017a, 248).

As a result, non-Western approaches to peacebuilding have achieved notable success within the confines of specific projects and even national contexts. Due to their nature as a mosaic of small projects across a number of areas, and often without higher-level coordination, these projects have yet to coalesce into an approach that can be said to have had an imprint on peacebuilding practice as a whole. With the exception of the US$3 million IBSA Trust Fund for development assistance, no major multilateral peacebuilding initiatives from emerging powers have come to fruition since these states have taken on a larger role. However, it is possible to say that improvements on individual practices based on domestic experiences has led to increases in effectiveness in concrete cases, as well as tentative steps in the direction of a more hybrid peace formation paradigm.

In this sense, one element that unites rising powers' approaches is the perceived increase, through the addition of experiences from the Global South, in local ownership of the ensuing initiatives. Among academic critics of the liberal peace this has raised hope that rising powers' approach might serve to return not only ownership but primary agency in peacebuilding endeavors to local actors. Increasingly this is coupled with a desire to harness the emancipatory potential of moving peacebuilding beyond the confines of state-centered approaches, in what Richmond (2002, 2010) has termed a fourth-generation peacebuilding approach (see also Jabri 2013, 4). While there is some consensus that the liberal paradigm has disappointed in this regard, rising powers are an unlikely hero here; due to their focus on respecting sovereignty and vision of peacebuilding as a foreign policy exercise between governments, Southern actors' approaches are not likely to lead down this avenue of emancipation. Indeed domestic experiences have often placed the loci of emerging cultures of peace directly in the firing line of state policies based on the criminalization of poverty and focused on repression.

While it may return agency to local actors, the cohesiveness among rising powers' approaches remains insufficient to identify a distinct and unified theory of change in the commonalities underpinning rising powers' approaches to transformation and peace (Call and de Coning 2017b, 4). Domestic differences continue to result in a less coordinated overall presence from rising donors, but it is clear that overall, major Southern actors such as Brazil, India, South Africa, and Turkey, as well as even established non-Western powers such as China and Russia, are acting within the dominant liberal paradigm. This includes, in large part, the transformational logic that undergirds that approach, albeit modified by domestic learning processes. Critical analysts claim that Southern states are no less beholden to clientelistic logics in their interactions with recipient states than their Northern counterparts (Ghimire 2018).

This underlying ambiguity—both critical of and acting within the liberal paradigm—is reflected in the way the term "peacebuilding" itself is used by Southern powers. The term or its direct translation is rarely used in domestic parlance; it is, however, used regularly—often with reference to distinct sets of practices—in or in reference to global debates and paradigms centered on the United Nations. In countries such as Brazil, related terms such as pacification are fraught with the legacy of foreign intervention and armed forces' repression, during the Cold War, of putative internal enemies. Often it is mentioned in the context of a foreign policy project focused on exporting the know-how of a broad set of government instances (Call and de Coning 2017b, 246, 251). Indeed, rising powers from the Global South have been considerably more successful in using peacebuilding to advance their quests for global influence individually than through multilateral efforts.

Rising Powers and Peacebuilding Multilateralism

The two main multilateral bodies through which rising powers have sought to challenge Northern hegemony are the BRICS and IBSA arrangements;[1] of these the BRICS has enjoyed considerably more impact and continuity. Both are focused on economic issues and have been able to mobilize a concerted counterpoint to Western interests. The same cannot be said of security issues, both broadly writ and in the area of peace formation. Russia and China are distinguished from the IBSA states by their possession of a UN Security Council veto, and all five rising powers are bound to the calculus of the regional security complexes of which they are a part (Buzan and Waever 2003, 31–40).

Three main elements unite efforts to build a common BRICS position: attempts to coordinate positions on specific security issues, namely armed conflicts and related normative stances; efforts to coordinate policies; and institution-building initiatives (Abdenur 2017). Abdenur concludes that what progress has been made has come in the first area. Rising powers' agency through the BRICS mechanism shows a clear dichotomy: while there is a recurring effort to formulate a common position on international peace and security,[2] divergences over responses to specific conflict situations have to date prevented effective collective action (Allouche and Lind 2014, 4).

These divergences have led to tensions in the relationship with the existing global peacebuilding architecture: the core of a common BRICS position is the importance of respecting host-state sovereignty through nonconditionality and adherence to principles of nonintervention, particularly regarding military force. At the same time, however, BRICS states in particular seek to integrate into the extant peacebuilding framework, albeit without the threat posed to their own recipient status by mandatory contributions (Richmond and Tellidis 2013). Poor coordination at the foreign policy level reflects mechanisms for concordance at the internal level that "appear to have developed by happenstance" rather than through joint planning (Rowlands 2012, 641).

Brazil, India, and South Africa in particular have worked largely within the liberal paradigm when such collaboration has been propitious in terms of the deployment of soft power. As such no contrasting model of peacebuilding has been created; rather these states have improved the liberal model's topical effectiveness with experiences from their domestic contexts (Richmond and Tellidis 2013; 2014, 567–568). Russia and China, in contrast, tend to favor bilateral approaches more directly geared toward economic advantage (Richmond and Tellidis 2013; Allouche and Lind 2014; Rowlands 2012, 642–643).

Whereas collaboration between rising powers and Northern states, as well as between rising powers themselves, has been rare, collaborative contributions to the United Nations peacebuilding and development architectures are common (Ghimire 2018, 43). Today's rising powers have been key players in peacebuilding and development debates at the UN since the Organization's inception (Abdenur 2016; Freeman, 2017). Brazil is particularly proud of what it sees as its seminal role in the creation of the Peacebuilding Commission (Neves 2010). South Africa has similarly been a driver of efforts by the African Union, particularly within the framework of its own New Partnership for Africa's Development, launched in 2001.

Two current movements within the UN architecture are set to have divergent effects on rising powers' peacebuilding efforts in the near future. In positive terms, there has recently been a discernible shift in logic underlying the Commission's efforts from the military-heavy Security Council view to the more inclusive and holistic outlook practiced at the General Assembly (Kmec 2017). This is partially the result of the resurgence of military force, to the detriment of long-term liberal peacebuilding, in Security Council practice as peacekeeping operations increasingly adhere to a Northern-dominated logic of stabilization (Karlsrud 2018). This latter development likely spells the beginning of a trend toward sharply reduced emerging power influence within UN peace operations as a whole, as the role of peacebuilding dwindles. A look at individual rising states' peacebuilding practices illustrates their commonalities and divergences, as well as the increasing obstacles to maintaining current levels of contribution for some key players.

Brazil

Brazil's ascendance as a provider of peace—which had a leadership role in MINUSTAH (2004–2017), the UN's stabilization mission in Haiti, at its core—is closely tied to the foreign policy platform of the Workers' Party. It reached its heyday under Lula, began to wane under his successor, Dilma Rousseff, and was definitively entered upon her impeachment and replacement with the center-right and Western-aligned Michel Temer in 2016. Under the extreme rightwing government of Jair Bolsonaro there has been a renunciation of principles that previously guided Brazilian policy, such as human rights, gender equality, multilateralism, and environmental protection. The historical Brazilian case, however, retains analytical value and presumably represents the default position for less illiberal leaders in the future.

Indeed recent political developments in Brazil's case demonstrate the close relationship between domestic politics, peacebuilding initiatives, and their mobilization in foreign policy. By their inherent nature, peacebuilding initiatives link development security, identifying the link between inequality, underdevelopment, and the potential for conflict. As such, when coupled with the export of domestic policies and experiences, there is a greater likelihood that this type of initiative will be highlighted in the foreign policies of progressive governments. This is even more true where foreign policy rhetoric is already rooted in a challenge to Western powers' dominant role in the international system and in the extant peacebuilding paradigm. Ironically, in many developing countries where governments have adopted neoliberal policies at the domestic level, they have reduced their involvement in supporting the liberal peace paradigm abroad. This is perhaps due to the statist nature of successfully exported programs, noted earlier.

Under Lula, Brazil's engagement in MINUSTAH and peacebuilding in Africa was clearly designed to demonstrate the country's capacity to take on responsibility in the security sphere, and to do so in a way not subordinate to the West. The goal was to highlight successful domestic policies and the exceptionalism of a Brazilian way of peacebuilding/peacekeeping (Kenkel 2010, 2013; Abdenur and Call 2017). Boasting an overall aid budget that at its height was about US$1 billion a year (Overseas Development Institute 2010), Brazil equaled rising rivals China and India:

> The crux of the incipient Brazilian model is the coordination of numerous state agencies with domestic experience in combating underdevelopment such as the Brazilian Cooperation Agency (ABC), the Brazilian Agricultural Research Corporation (EMBRAPA), [the Oswaldo Cruz Foundation on health policy], the National Service for Industrial Apprenticeship (SENAI), Ministries such as Health,

> Agriculture and Education, and, importantly, civil society organizations such as the NGO Viva Rio that are active in conflict reduction in the favelas of Rio de Janeiro. . . . The emphasis lies in agricultural innovation, health policies, and the "social protection" provided by hunger prevention and poverty reduction. (Kenkel 2013, 287; for details and numbers, see Abdenur and Call 2017; Call and de Coning 2017a, 244ff.; Inoue and Costa Vaz 2012; Mattheis, Stolte, and Cisneros 2016)

This breadth of expertise never encountered, however, the necessary degree of coordination or consistent funding to constitute a peacebuilding paradigm in its own right. Despite the emergence of the Brazilian Cooperation Agency, this body still operates largely within the confines of the Foreign Ministry; efforts across other areas and ministries are not bundled effectively by any other agency with overall responsibility (Abdenur and Call 2017, 17). While many projects have been impactful, this has doubtless reduced the overall political and discursive impact of Brazil's efforts; the country's peacebuilding should, in the view of most experts, be considered an extension, rather than a concerted contestation, of the liberal peace (Richmond and Tellidis 2014, 570), The country's stance on the liberal peace is nonetheless a profoundly ambivalent one (Fernández and Gama 2016). One important element of difference is that due to its own possession of extensive natural resources, Brazil's engagement with Africa in particular, while no less driven by commercial interests, does not possess the same extractive character as many other states (White 2013; Mattheis, Stolte, and Cisneros 2016, 18–19). The country remains a significant donor whose future as a peacebuilding contributor should be followed closely.

India

Of the major rising powers, India has the broadest definition of peace formation; virtually all development activity in postconflict settings is viewed under the rubric of peacebuilding (Call and de Coning 2017a, 248). Alongside this, the country has been one of the largest contributors to UN peacekeeping operations for over a decade, including the most robust missions authorized under Chapter VII of the UN Charter (Choedon 2017), as well as a major contributor to regional interventions (Adhikari 2018, 172). Indeed India stands out among Southern rising powers for its experience with violent domestic uprisings and greater degree of reconciliation with the use of military force (Vieira and Alden 2011). Nevertheless field research has shown that state-led peacebuilding in the Indian domestic context has encountered similar difficulties in its exportability as have such efforts in Brazil and South Africa (Pogodda and Huber 2014).

While it is a status-quo power on peacekeeping (Richmond and Tellidis 2014; see Cunliffe 2013a), it has avoided the more liberal elements of the dominant paradigm, such as democracy promotion (Adhikari 2018, 169–172). Its core peacebuilding principles favor the Southern approach, being demand-driven, without conditionalities, decentralized, and respectful of sovereignty (Singh 2017), and it is likely that Indian participation particularly in regional peacebuilding processes will increase (Adhikari 2018, 173).

India's legacy builds on its history of leadership of the Non-Aligned Movement; though it is a major regional player it has not yet organized its peacebuilding efforts around one consolidated agency. Similarly, despite extensive investment in both peacekeeping and peacebuilding, the main initiatives around which its peacebuilding presence is organized, such as the Indian Development Partnership and the Indian Technical and Economic Cooperation Programme, have yet to reach their full potential.

India's current main area of peacebuilding investment is its own region, particularly Afghanistan. This commitment stands alongside a decades-long presence in Africa, which combines postcolonial legacies with commercial interests (Fair 2011; Kragelund 2010). India's Afghanistan National Development Strategy "lists eight pillars for their national development: security; good governance; infrastructure and proper utilization of natural resources; education and culture; health and nutrition; agriculture and rural development; social protection; and economic governance and private sector development" (Singh 2017, 81).

The country has been the fifth-largest donor in Afghanistan over the past fourteen years (Sinha 2017, 130, 137), totaling over US$2 billion (Singh 2017, 81), and takes pains to differentiate its efforts from the Western model. Indian projects are implemented based on demands directly from the Afghan government and operate according to a model significantly different from that employed by the various Western Provincial Reconstruction Teams in the country (Sinha 2017, 144–145). India will continue to grow as a peacebuilding actor, also beyond its region; as it does so, in the short term it looks likely to continue its role as a status-quo country deeply involved in UN peacekeeping operations, while employing a peacebuilding approach more aligned to the other Southern rising powers.

Conclusion

Rising powers are here to stay as influential actors in the peacebuilding arena. Their challenge to problematic underlying characteristics of the liberal peace is crucial to both the legitimacy and the efficacy of future peacebuilding efforts. To date, however, there is significant room for increased coordination at national and multilateral levels, as well as for improvements in planning capacity and continuous funding. Without a concerted effort in this regard their efforts are destined to remain ultimately supportive of the status quo and their national interests rather than a larger postcolonial agenda. Ultimately the fact that emerging powers' efforts at peacebuilding have not effectively left the confines of the liberal peace paradigm—whether due to financial/organizational constraints or lack of political will—has prevented the initiatives exported from their domestic contexts from unfurling their full emancipatory potential. Indeed in some cases—as has been argued was the case with Brazilian initiatives in Haiti—rather than effecting a lasting impact on the dominant paradigm, they have provided it with a mantle of putatively local legitimacy not reflected in actual reforms.

This does not detract, however, from the current tendency for their role to increase in the decades to come. There are nevertheless two factors that may prove problematic for rising powers' participation in the future. First, UN peacekeeping operations are, as a general trend, moving away from development-heavy liberal peacebuilding—where rising powers can bring domestic experiences to bear—and toward NATO-inspired stabilization missions that raise qualms in Southern states about their very robust use of force (Karlsrud 2018). Second, evidence is mounting that "diversionary peace theory"—the notion that involving armed forces in peace operations makes domestic political interference less likely (Sotomayor 2013)—may not hold true in all cases (Cunliffe 2013b; Ghimire 2019, 4–5). Brazil is a clear case of Chapter VII peacebuilding participation leading to increased armed forces' internal deployment for the provision of public security and to deal with an increase in refugees (Müller 2016); armed forces' influence is sure to rise under the extreme right-wing government that took power in 2019. Indeed many rising powers—notably each of the BRICS countries—is experiencing varying degrees of crisis in its democratic system. It remains to be seen whether these turbulences will lead to a turn away from multilateral and bilateral peacebuilding.

Notes

1. The BRICS consist of Brazil, Russia, India, China, and South Africa; IBSA joins the same members without Russia and China.
2. Clear evidence of these efforts can be seen in the text of the Johannesburg Declaration, issued at the 10th BRICS Summit in July 2018, and in the regular meetings of BRICS countries' security advisors, now in their eighth iteration (Indian Ministry of External Affairs 2018).

References

Abdenur, A. E. 2016. "Emerging Powers and the Creation of the UN: Three Ships of Theseus." *Third World Quarterly* 37 (7):1171–1186.

Abdenur, A. E. 2017. "Can the BRICS Cooperate in International Security?" *International Organisations Research Journal* 12 (3):73–93.

Abdenur, A. E., and C. T. Call. 2017. "A 'Brazilian Way'? Brazil's Approach to Peacebuilding." In *Rising Powers and Peacebuilding: Breaking the Mold?*, edited by C. T. Call and C. de Coning, 15–38. Cham, Switzerland: Palgrave Macmillan.

Abdenur, A. E., and D. Marcondes de S. N. 2014. "Rising Powers and the Security-Development Nexus: Brazil's Engagement with Guinea-Bissau." *Journal of Peacebuilding and Development* 9 (2):1–16.

Adebajo, A. 2016. "The Revolt against the West: Intervention and Sovereignty." *Third World Quarterly* 37 (7):1187–1202.

Adhikari, M. 2018. "India in South Asia: Interaction with Liberal Peacebuilding Projects." *India Quarterly* 74 (2):160–178.

Alexandra, L. A. 2017. "Offering Support and Sharing Experiences: Indonesia's Approach to Peacebuilding." In *Rising Powers and Peacebuilding: Breaking the Mold?*, edited by C. T. Call and C. de Coning, 39–68. Cham, Switzerland: Palgrave Macmillan.

Allouche, J., and J. Lind. 2014. *Beyond the New Deal: Global Collaboration and Peacebuilding with BRICS Countries*. IDS Policy Briefing 64. Brighton, UK: Institute of Development Studies, University of Sussex.

Buzan, B., and O. Waever. 2003. *Regions and Powers: The Structure of International Security*. Cambridge: Cambridge University Press.

Call, C. T., and C. de Coning. 2017a. "Conclusion: Are Rising Powers Breaking the Peacebuilding Mold?" In *Rising Powers and Peacebuilding: Breaking the Mold?*, edited by C. T. Call and C. de Coning, 243–272. Cham, Switzerland: Palgrave Macmillan.

Call, C. T., and C. de Coning. 2017b. "Introduction: Why Examine Rising Powers' Role in Peacebuilding?" In *Rising Powers and Peacebuilding: Breaking the Mold?*, edited by C. T. Call and C. de Coning, 1–12. Cham, Switzerland: Palgrave Macmillan.

Capie, D. 2016. "Indonesia as an Emerging Peacekeeping Power: Norm Revisionist or Pragmatic Provider?" *Contemporary Southeast Asia* 38 (1):1–27.

Carvalho, B., and C. de Coning. 2013. *Rising Powers and the Future of Peacekeeping and Peacebuilding*. NOREF Report, November. Oslo: Norwegian Peacebuilding Resource Centre.

Choedon, Y. 2017. "India's UN Peacekeeping Operations Involvement in Africa: Change in Nature of Participation and Driving Factors." *International Studies* (Delhi) 51 (1–4):16–34.

Cunliffe, P. 2013a. "From Peacekeepers to Praetorians—How Participating in Peacekeeping Operations May Subvert Democracy." *International Relations* 32 (2):218–239.

Cunliffe, P. 2013b. *Legions of Peace: UN Peacekeepers from the Global South*. London: Hurst.

Fair, C. C. 2011. "Under the Shrinking U.S. Security Umbrella: India's End Game in Afghanistan?" *Washington Quarterly* 34 (2):179–192.

Fernández, M., and C. F. P. S. Gama. 2016. "Modernization In-between: The Ambivalent Role of Brazil in Contemporary Peacebuilding Efforts in Africa." In *Brazil as a Rising Power: Intervention Norms and the Contestation of Global Order*, edited by K. M. Kenkel and P. Cunliffe, 65–80. Abingdon, UK: Routledge.

Freeman, D. 2017. "The Global South at the UN: Using International Politics to Re-vision the Global." *Global South* 11 (2):71–91.

Ghimire, S. 2018. "Rising Powers and Security: A False Dawn of the Pro-South World Order?" *Global Change, Peace & Security* 30 (1):37–55.

Ghimire, S. 2019. *The Politics of Peacebuilding: Emerging Actors and Security Sector Reform in Conflict-Affected States*. Abingdon, UK: Routledge.

Indian Ministry of External Affairs. 2018. "10th BRICS Summit Johannesburg Declaration." 26 July. https://www.mea.gov.in/bilateral-documents.htm?dtl/30190/10thBRICSSummitJohannesburgDeclaration.

Inoue, C. Y. A., and A. Costa Vaz. 2012. "Brazil as 'Southern Donor': Beyond Hierarchy and National Interests in Development Cooperation?" *Cambridge Review of International Affairs* 25 (4):507–534.

Jabri, V. 2013. "Peacebuilding, the Local and the International: A Colonial or a Postcolonial Rationality?" *Peacebuilding* 1 (1):3–16.

Karlsrud, J. 2018. "From Liberal Peacebuilding to Stabilization and Counterterrorism." *International Peacekeeping*. Early release article. https://www.tandfonline.com/doi/pdf/10.1080/13533312.2018.1502040?needAccess=true.

Kenkel, K. M. 2010. "South America's Emerging Power: Brazil as Peacekeeper." *International Peacekeeping* 17 (5):644–661.

Kenkel, K. M. 2013. "Brazil's Peacekeeping and Peacebuilding Policies in Africa." *Journal of International Peacekeeping* 17:272–292.

Kmec, V. 2017. "The Establishment of the Peacebuilding Commission: Reflecting Power Shifts in the United Nations." *International Peacekeeping* 24 (2):304–325.

Kragelund, P. 2010. *The Potential Role of Mon-Traditional Donors' Aid in Africa*. Issue Paper 11. Geneva: International Centre for Trade and Sustainable Development.

Mathur, D. 2014. *Role of South-South Cooperation and Emerging Powers in Peacemaking and Peacebuilding*. NUPI Policy Brief 10/2014. Oslo: Norwegian Institute of International Affairs.

Mattheis, F., C. Stolte, and E. J. Cisneros. 2016. "Los desafíos de un donante emergente: La cooperación Sur-Sur de Brasil con Ghana / The Challenges of Being an Emerging Donor: Brazil's South-South Cooperation with Ghana." *Revista CIDOB d'Afers Internacionals* 114:157–181.

Müller, M.-M. 2016. "Entangled Pacifications: Peacekeeping, Counterinsurgency and Policing in Port au-Prince and Rio de Janeiro." In *The Global Making of Policing: Postcolonial Perspectives*, edited by J. Hönke and M.-M. Müller, 77–95. Abingdon, UK: Routledge.

Narlikar, A. 2011. "Is India a Responsible Great Power?" *Third World Quarterly* 32 (9):1607–1621.

Neves, G. M. S. 2010. "Comissão das Nações Unidas para Consolidação da Paz--Perspectiva Brasileira." PhD thesis, Instituto Rio Branco, Brasília.

Nyuykonge, C., and S. Zondi. 2017. "South African Peacebuilding Approaches: Evolution and Lessons." In *Rising Powers and Peacebuilding: Breaking the Mold?*, edited by C. T. Call and C. de Coning, 107–126. Cham, Switzerland: Palgrave Macmillan.

Olayode, K., and C. Ukeje. 2017. "South-South and Regional Cooperation for Peace Building in West Africa." *Modern Africa: Politics, History and Society* 5 (1):33–54.

Overseas Development Institute. 2010. *Brazil: An Emerging Aid Player*. ODI Briefing Paper 64, October. London: ODI.

Peter, M. 2014. *Emerging Powers and Peace Operations: An Agenda for Research*. NUPI Policy Brief 10/2014. Oslo: Norwegian Institute of International Affairs.

Pogodda, S., and D. Huber. 2014. "India's Peacebuilding between Rights and Needs: Transformation of Local Conflict Spheres in Bihar, North-East India and Jammu and Kashmir?" *International Peacekeeping* 21 (4):443–463.

Pu, X. 2012. "Socialisation as a Two-way Process: Emerging Powers and the Diffusion of International Norms." *Chinese Journal of International Politics* 5:341–367.

Richmond, O. P. 2002. *Maintaining Order, Making Peace*. London: Palgrave.

Richmond, O. P. 2010. "Resistance and the Post-liberal Peace." *Millennium* 38 (3):665–692.

Richmond, O. P. 2014. *Failed Statebuilding: Intervention, the State, and the Dynamics of Peace Formation*. New Haven, CT: Yale University Press.

Richmond, O. P., and I. Tellidis. 2013. *The BRICS and International Peacebuilding and Statebuilding*. NOREF Report, January. Oslo: Norwegian Peacebuilding Resource Centre.

Richmond, O. P., and I. Tellidis. 2014. "Emerging Actors in International Peacebuilding and Statebuilding: Status Quo or Critical States?" *Global Governance* 20:563–584.

Riddle, K. C. 2017. "Structural Violence, Intersectionality, and Justpeace: Evaluating Women's Peacebuilding Agency in Manipur, India." *Hypatia* 32 (3):574–592.

Roque, S. 2009. *Peacebuilding in Guinea-Bissau: A Critical Approach*. NOREF Report, May. Oslo: Norwegian Peacebuilding Resource Centre.

Rowlands, D. 2012. "Individual BRICS or a Collective Bloc? Convergence and Divergence amongst 'Emerging Donor' Nations." *Cambridge Review of International Affairs* 25 (4):629–649.

Satana, N. S. 2016. "Multilateral Interventions as a Power-Enhancing Instrument: Rising Powers' Path from the Periphery to the Center." In *Brazil as a Rising Power: Intervention Norms and the Contestation of Global Order*, edited by K. M. Kenkel and P. Cunliffe, 127–146. Abingdon, UK: Routledge.

Sazak, O., and A. E. Woods. 2017a. "Breaking with Convention: Turkey's New Approach to Peacebuilding." In *Rising Powers and Peacebuilding: Breaking the Mold?*, edited by C. T. Call and C. de Coning, 93–106. Cham, Switzerland: Palgrave Macmillan.

Sazak, O., and A. E. Woods. 2017b. "Thinking outside the Compound: Turkey's Approach to Peacebuilding in Somalia." In *Rising Powers and Peacebuilding: Breaking the Mold?*, edited by C. T. Call and C. de Coning, 163–190. Cham, Switzerland: Palgrave Macmillan.

Singh, P. K. 2017. "Peacebuilding through Development Partnership: An Indian Perspective." In *Rising Powers and Peacebuilding: Breaking the Mold?*, edited by C. T. Call and C. de Coning, 69–92. Cham, Switzerland: Palgrave Macmillan.

Sinha, S. 2017. "Rising Powers and Peacebuilding: India's Role in Afghanistan." In *Rising Powers and Peacebuilding: Breaking the Mold?*, edited by C. T. Call and C. de Coning, 129–166. Cham, Switzerland: Palgrave Macmillan.

Siqueira, I. R. 2018. *The Case for South-South Cooperation on Peace and Development*. Rio de Janeiro: BRICS Policy Center.

Sotomayor, A. C. 2013. *The Myth of the Democratic Peacekeeper: Civil-Military Relations and the United Nations*. Baltimore, MD: Johns Hopkins University Press.

Stephen, M. D. 2014. "Rising Powers, Global Capitalism and Liberal Global Governance: A Historical Materialist Account of the BRICS Challenge." *European Journal of International Relations* 20 (4):912–938.

Vieira, M. A., and C. Alden. 2011. "India, Brazil, and South Africa (IBSA): South-South Cooperation and the Paradox of Regional Leadership." *Global Governance* 17 (4):507–528.

Visoka, G., and J. Doyle. 2014. "Peacebuilding and International Responsibility." *International Peacekeeping* 21 (5):673–692.

Weinlich, S. 2014. "Emerging Powers at the UN: Ducking for Cover?" *Third World Quarterly* 35 (10):1829–1844.

White, L. 2013. "Emerging Powers in Africa: Is Brazil Any Different?" *South African Journal of International Affairs* 20 (1):117–136.

CHAPTER 22

GLOBALIZATION OF PEACE

JACKIE SMITH

THE logic of hegemonic globalization is inherently violent. Globalized capitalism institutionalizes competitive, territorial states within an economic order that incentivizes greed, exploitation, extraction, individualism, and competition. Over several hundred years this logic has expanded to encompass a growing array of social relations across the globe, and it confounds efforts to promote peaceful and sustainable societies. But if we shift our gaze from global to local, we see another logic and vision of how to organize the world. Globalization from below centers people and their livelihoods and reveals popular experimentation in the work of peace formation. According to Oliver Richmond (2013, 383), peace formation "implies a reconstruction of political community, the state and international organisations from the ground up, if they are to be representative, democratic and responsive to the situations of their subjects in local, state, regional and global contexts." As a *political project*, peace formation involves work at multiple scales from local to global simultaneously to create possibilities for the emergence of alternatives while working to envision and construct those alternatives and the strategies needed to realize them. Importantly, this work involves attention to culture and requires the activation of political and legal imagination to support the emergence of new identities and political possibilities that lie outside existing political and conceptual frameworks. It also evades classification into simplistic binary conceptions of state/nonstate or grassroots/global dualisms, and the work of activists engaged in peace formation frequently crosses such divides.

By looking at the work of subaltern groups challenging dominant patterns of social relations and promoting cooperative and life-supporting alternatives, we can uncover a counterhegemonic project to articulate a globalized model for peace and peace formation that will be far more successful than the top-down model centered on states and state-led "peacebuilding." In this chapter I summarize key critiques of the hegemonic peacebuilding project before presenting some evidence of an emergent global project of peacebuilding based in human rights frameworks.

Globalization and Hegemonic Peacebuilding

Interstate peacebuilding initiatives are inseparable from the global project of capitalist development, a project based in what David Harvey (2004) calls "accumulation by dispossession." They prioritize the reduction of violent conflicts that disrupt the operation of global capitalism over work to address the structural violence underlying social conflicts. They seek to maintain status quo power relations rather than transform such relations in ways that advance equity, justice, and lasting peace. To do this, the state-led peacebuilding project must preempt the question "What is peace?" and confine participation in peacebuilding to those willing and able to support the neoliberal capitalist arrangements.

A growing body of research provides evidence of the persistent and widespread failures of these mainstream approaches to peacebuilding (Paris 2003; Pugh, Cooper, and Turner 2008; O. Richmond and Franks 2009b; O. P. Richmond, Pogodda, and Ramović 2016; Smith 2010; Smith and Verdeja 2013). Analysts have critiqued the state-centered, top-down nature of international peacebuilding interventions for their timing, their priorities, and the range of actors they involve. In regard to timing, peacebuilding interventions tend to come only after conflicts have escalated into militarized violence, when underlying positions and tensions have become more entrenched. The priority then is necessarily to end armed conflicts and demobilize militants, and these tasks distract resources and attention from the underlying sources of the conflict. Interventions aimed at preventing escalation would instead focus on root causes of conflicts and would engage more stakeholders in constructive interchanges than is possible in the aftermaths of armed struggles. Yet in the absence of a crisis, it is unlikely that powerful states would agree to the redistribution of power and resources such interventions would require.

A second critique is of the priorities that have characterized most peacebuilding interventions. The state-centered nature of these enterprises and the timing of their interventions so late in the evolution of conflicts has meant that the work of statebuilding has been prioritized. In part this is a functional necessity, since war disrupts the provision of basic services and internal security for residents, threatens mass dislocation, and affects regional peace and security. But it also results from the privileging of national governments and their priorities in their design and implementation. O. Richmond and Franks (2009a, 182, emphasis added) write, "We have found major gaps in an empathetic connection between statebuilding and the everyday aspects of peace. . . . We are all too aware of the failings of the global statebuilding and peacebuilding industry and the hidden ideological camouflage of its technical expertise. This is not to say that all is lost, but clearly that *the local, the everyday, the customary, human needs, human security and the social contract need to be positioned far higher up the practical agenda than they currently are.*"

In addition to prioritizing statebuilding, mainstream international peacebuilding initiatives forcibly implement the project of neoliberal globalization. They require that postwar societies quickly reintegrate themselves into the global economy and enact policies favoring trade liberalization and foreign investment—regardless of local context. However, such policies are often linked to and help reproduce institutionalized discrimination and structural violence, perpetuating and often exacerbating the underlying causes of intergroup conflict (Smith 2010). Not only do these policies threaten to reignite conflicts by intensifying underlying inequities and exclusions, but they also evade the resource redistribution necessary for creating conditions for peace in postwar contexts. For this would require the channeling of resources into local initiatives alongside policies that would remedy past inequities and injustices.

A third major critique of conventional peacebuilding initiatives is that their origins in the interstate system and their centering of states and state sovereignty (and in particular the most politically influential states) produce systematic neglect of grassroots and civil society actors while leaving unquestioned the ways militarized state sovereignty exacerbates systemic violence.[1] Despite the fact that most analyses of peacebuilding efforts show civil society actors to be key players in postwar peacebuilding, the literature has been largely silent in regard to offering concrete programs or strategies for integrating civil society actors in the design and implementation of peace initiatives (Leonardsson and Rudd 2015; Smith and Verdeja 2013; Paffenholz 2010). Instead, where these approaches do engage civil society it is to instrumentalize grassroots actors and co-opt them into state-led peacebuilding projects.

In response to this lacuna, I orient this chapter around the question of what a "people-centered," counterhegemonic peace process might look like. Rather than starting from the top down and assuming that any postwar peace must sustain the existing state-led capitalist order, I begin with the discourses and mobilizations of civil society working from the bottom up to articulate visions of more peaceful social relations that support grassroots peace formation projects. Such work is grounded in the social movements operating at grassroots levels in places around the world, where actors are working to build and sustain community well-being—often against the impacts of capitalist globalization. In recent decades, such grassroots struggles have become increasingly globalized through "translocal" connections supported by proliferating horizontal networks and formal coalitions and convergences like the World Social Forums as well as more informal exchanges of ideas and resources (Desai 2015; Escobar 2008). Addressing problems of poverty and social exclusion, racism, patriarchy, environmental degradation, and violence, these movements have been generating increasingly coherent analyses of the links between neoliberal globalization and a variety of conflicts and intersecting oppressions in places all over the world (Desai 2005; Falcón 2016; Alvarez 2000). Yet because the state and leading market actors are the chief beneficiaries of the status quo and the extreme inequalities it has produced, the analyses and prescriptions emerging from these movements remain outside the realm of what is considered possible in mainstream peacebuilding initiatives. Moreover, powerful actors see their interests threatened by social movements' attention to inequities in resource distribution and institutional privilege and thus use their power to exclude or marginalize grassroots and other movement actors.

I believe the world-systems perspective can provide a helpful corrective to the existing literature on peace and peacebuilding processes. This approach situates states and other actors in a world-historical framework. That is, it considers how the interstate system and particular social relationships among and within states are related to long-term processes of capitalist development. Since capitalism is necessarily expansionist and ultimately global in nature, even the most localized conflicts reflect broader tensions or contradictions in the larger world-system (Wallerstein [1974] 2011). Furthermore, the world-system structures social relationships in ways that perpetuate inequalities and obstruct any fundamental reordering of social hierarchies. Fundamentally competitive logics of colonialism, patriarchy, and racism enable the exploitation needed for capital accumulation. Yet effective work to reconstruct political community for peace requires efforts to confront fundamental conflicts over resources and identities while promoting tools for cooperative social relations. Such work is not supported by hegemonic logics.

Hegemonic Logics and Conflict

Pointing to how the global capitalist system furthers "accumulation by dispossession" and thus fuels structural violence and conflict, Saskia Sassen (2014b) uses the term "expulsions" to describe the experiences of growing numbers of people today. Rising inequalities and the concentration of wealth among elites has reduced the economic space available for growing populations to participate in local economies. At the same time, the globalized economy has commodified more and more of the commons, limiting access to basic material resources such as land, water, and healthy habitats to those who can pay market rates. This produces a "brutal sorting of winners and losers" in the global economy (Sassen 2014a, 199). In this sense, capitalism's "logic of expulsion" is contributing to polarization and putting growing numbers of people into situations where they must struggle for their basic survival. Illustrating this logic in action, Edelman (2018) uses the term "sacrifice zones" to characterize the state of rural and nonmetropolitan communities around the United States (see also Lerner 2010). The term refers to the ways market ideologies normalize the denial of basic rights and justify exclusion of people from social life and health. Because neoliberal globalization has generated few living-wage jobs or other economic opportunities outside major urban centers, it has forced growing numbers of people into conditions of precarity and marginalization—even in the world's richest countries. At the same time, neoliberal prescriptions have gutted state social support systems and limited public revenues available for the kinds of public projects and social economies that analysts such as Brendan Murtagh (2016, 120) have deemed essential to peacebuilding.

Significantly, although it is absent from most formal peacebuilding work, attention to colonialism and its legacies is seen as indispensable to any effort to understand and shape contemporary social relations and the conflicts they engender. Specifically, peace requires the decolonization of our modes of thought as well as institutions (Jabri 2016; Santos 2006).[2] World-systems analysts, in short, see the basic—and highly unequal—structures of the world-system as fundamental obstacles to justice and peace. A foundation for peace would thus be a fundamentally different world-system, one organized around principles of equity, justice, and human emancipation rather than around the exclusionary and competitive logics of capitalism. Interestingly, this conclusion is emergent in critical perspectives on peacebuilding:

> Recent and critical . . . theorizing points to the fact that peace probably requires bottom-up, social ontologies developing an empathetic account of everyday forms of emancipation. These should shape institutions and law so that states, economies, laws and the international community are able to respond to socially mediated claims. It should provide social, economic and political resources sufficient to meet the demands made upon it by its local constituencies and an international community of which it should be a stakeholder. Any viable concept of peace that conforms to the above conditions must not displace localized forms of legitimacy with preponderant institutions that are inflexible and actually obscure the local. (O. P. Richmond 2016, 65)

Although lacking explicit reference to globalized capitalism and its role in defining states and the inequities of the global economy, one could conclude from Richmond's assessment that effective peacebuilding requires a fundamental shift in power away from states and toward local communities. Attention must shift from the security interests of states to the everyday needs and concerns of people and their livelihoods, and this requires a fundamental redistribution of resources and power at local and global scales. Envisioning a path toward such large-scale transformation is essential to the work of peace formation—work that will necessarily involve grassroots social movements in projects that center the needs and experiences of local communities rather than states.

In summary, conventional state-oriented approaches are forms of "hegemonic" peacebuilding that seek to preserve a fundamentally inequitable and unjust social order, leaving unaltered the underlying inequities and structures—such as militarized states and the state-centered global political and economic order—that fuel the conflicts they presume to resolve. If our aim is to secure peace and to build peaceful societies, then we need to dig down to the world-systemic roots of conflicts and consider how to transform the social relations and institutions that perpetuate conflict and fuel polarization and violence. Flipping our attention from the commanding heights of the state and leading market players to the places where people struggle to survive and thrive, we can better understand the forces driving conflicts and observe efforts at what we might call "counterhegemonic peacebuilding." This refers to work by people and communities to develop their own visions of peace and to shape projects aimed at advancing the difficult tasks of fundamentally transforming global power structures in ways that support lasting peace (Smith and Verdeja 2013). Fetherston (2000, 210) employs Gramsci to capture this idea: "In societies where civil space can be characterized as complex networks of social relations and institutions which support a prevailing hegemonic view of the meaning and direction of social life, and which are resistant to direct, immediate influence, successful social transformation must take the form described by Gramsci as a 'war of position.' Such a *counterhegemonic project* aims to build a consensus among non-dominant groups which articulates an alternative direction for social life. If consensus can be reached and maintained among enough groups, a new hegemony can emerge."

Thus a defining feature of counterhegemonic peace formation is that it explicitly challenges the organization of the global economy, treating this system as a root cause of local conflicts around resource distribution and environmental well-being. Often this work is invisible to those whose attention is on states and state-led projects such as peacebuilding. Without an effort to understand and address the links between localized conflicts and global processes of wealth accumulation, and without efforts to fundamentally restructure social relations, the sources of violent conflict will persist. Emerging scholarship reflecting a "local turn" in peacebuilding research acknowledges this, stressing the need to reduce the chasms between top-down and bottom-up understandings of peace and peacebuilding processes (Leonardsson and Rudd 2015; Schierenbeck 2015).

Human Rights as Counterhegemonic Peacebuilding

In his analysis of peacebuilding literatures, Trimikliniotis (2016, 106) calls for rethinking that "locate[s] 'peace' within the processes of transformation struggles which generate new socialities." This gets at what Donnelly (2016, 278) calls the "infrapolitics of peacebuilding," which involves "how the 'local' and 'everyday' are being constituted, and by whom." Attention to the local level helps situate people and social relations in particular places and contexts. Environmental conditions and the relative ease of access to the basic resources needed for survival become harder to ignore at this level of analysis, and the factors that facilitate or obstruct livelihoods more apparent. Also evident are the ways people work in communities to mobilize collective identities and practices to secure basic coexistence and survival, regardless of how such efforts articulate with state-led projects (Mac Ginty 2014; Mac Ginty and Firchow 2016).

In the remainder of this chapter I provide examples of work by social movement actors who are usually left out of mainstream accounts of peacebuilding because they are local and less formally organized and because the ways they frame their work don't fit within the conventional peacebuilding frameworks and discourses defined by states. I argue that these increasingly "translocal" actors are part of a global counterhegemonic movement engaged in transformative peacebuilding. The proliferation of international

NGOs over recent decades, coupled with the expansion of digital communications technology, has enabled much more extensive and direct communication and coordination across locales. As a result, analysts are seeing more horizontal linkages between local communities in different countries and more organizing that links organizers working at different scales from the local through the global levels (Desai 2015). These networks and the communication they entail are creating spaces that are generating more coherent, globalized analyses and projects for social transformation and bottom-up peace formation. Some of these projects reflect new framings of struggles over basic resources such as land and water that are helping build broad, transversal alliances of actors around shared visions and strategies for global transformation (Smith 2014). I show how increasingly globalized networks of social movement actors are working to build an alternative to neoliberal peacebuilding that centers people and human rights rather than capitalist markets and territorial states.

Most transnational activism addresses human rights claims, and indeed the development of the international system of human rights norms and laws is directly linked to the work of civil society actors (Rajagopal 2003; Sikkink 2011; Smith 2008). The centrality of rights to liberalism makes it vulnerable to pressures from groups pointing to the increasingly evident contradictions between the operation of capitalism and the realization of human rights. Illustrating how a focus on human rights can advance peace, Suárez (2013) surveyed accounts from a variety of rural organizers working in different parts of the Global South. For many, the decision to organize around a human rights framework was pragmatic, as violent opposition would only generate a brutal response from the state. Yet there is evidence that movements for human rights have been successful at achieving some of what is deemed essential for durable peace formation processes: "The impacts of using the human rights framework greatly vary and can range from contributing to the empowerment of oppressed groups to stand up for their rights, decreasing violence in land conflicts, changing the way conflicts over resources are framed, opening up space for policy dialogue centered on people's lives, fighting against agrarian legislation biased in favor of corporate interests and formulating alternative legal frameworks" (Suárez 2013, 251).

The UN Human Rights System has been extremely important in supporting the legitimacy of the claims raised by marginalized groups that have long struggled to overcome structural inequalities and violence. While huge gaps remain between global norms and the actual practices of states and other powerful groups, human rights have provided an important focal point for transnational movements, and as Rajagopal (2006, 768) observes, "the radical democratic potential in human rights . . . can be appreciated only by paying attention to the pluriverse of human rights, enacted in many counter-hegemonic cognitive frames. In this new approach, official and sanctioned human rights discourse becomes one of many languages of resistance, enacting a cultural politics at many scalar levels." Four examples stand out as international legal resources for addressing basic sources of inequality and both structural and other violence: the Convention on the Elimination of All Forms of Racial Discrimination (CERD, ratified 1969), the Convention on the Elimination of All Forms of Discrimination against

Women (CEDAW, 1979); the UN Permanent Forum on Indigenous Issues (2002) and Declaration on the Rights of Indigenous Peoples (2007); and the recently adopted UN Declaration on the Rights of Peasants and Other People Working in Rural Areas (2018). While these mechanisms have limited enforcement capacity, they provide legitimacy for the claims of local advocates whose rights are threatened, and they serve as an international court of appeal that can be activated long before a conflict escalates into armed struggle.

Although states have often resisted their passage and ignored their operation, over time the work of transnational networks of activists working at multiple scales have contributed to a gradual strengthening of international human rights machinery. The proliferation of translocal activist networks over recent decades has enhanced international human rights monitoring and reporting by all states—including the most powerful ones.[3] This also amplifies the potential for less powerful groups to utilize international mechanisms to address local injustices and develop practices and social relations conducive to peace (see, e.g., Keck and Sikkink 1998; Kaldor 2003; Smith et al. 2018). Despite limited enforcement capacity, these mechanisms subjecting powerful elites to international scrutiny can prevent abuses and even force concessions from authorities. Moreover, as activists and activist lawyers make moral and formal appeals to international law, they advance local agency vis-à-vis more powerful extralocal actors and strengthen precedents connecting national and local practices with international norms and legal obligations.[4] They also help constitute new global actors and transversal identities that supersede nationalist and other parochial identities and support global norms that contribute to justice and peace (e.g., Mouly 2008; Tsutsui 2018).

What do these international legal mechanisms do for marginalized groups? The CERD provides a regular review mechanism requiring states to file reports on their compliance with the Convention. Civil society groups are given an opportunity to provide their own accounting of local-level compliance, and their assessments figure in the final report issued by the committee of independent experts that monitors CERD. Activist groups like the US Human Rights Network (n.d.) have used the treaty-monitoring process to mobilize grassroots constituencies and raise consciousness about international processes organized to reduce racial discrimination and violence. The review process has impacted important struggles, such as that of Black activists resisting discriminatory police violence in the United States.[5]

CEDAW has provided the basis of legitimacy for groups working to end gender-based violence, including structural violence linked to patriarchal norms related to land tenure and employment (see, e.g., M. K. Meyer and Prügl 1999). A more recent and promising development that has been enabled by the strengthened translocal connections among social movements is an initiative known as Cities for CEDAW in the United States. Because the US is one of the only remaining countries that have not ratified CEDAW, activists in communities across that country are demanding that local authorities adopt the treaty provisions without waiting for national leaders to do so (Frug and Barron 2006; MacNaughton and McGill 2012; A. Meyer 2009). These initiatives involve public education that enhances local knowledge of international human rights institutions while also building skills at local monitoring of gendered inequities in local policies and practices. Activists are sharing lessons across cities to help strengthen local ordinances, and transnational connections across cities have inspired activists around the world to mobilize local efforts to monitor and press for greater local compliance with CEDAW and other treaties (Oomen 2016; Rajagopal 2006; Shawki 2011). Working internationally, translocal networks of activists use international review processes to push states to improve their compliance with CEDAW and highlight specific laws and practices that violate its principles.[6]

Advances in the United Nations by indigenous peoples have enabled indigenous leaders to participate in interstate dialogues and defend the rights of indigenous peoples to be treated as equals among states (Morgan 2007; Sargent 2012). These institutional developments have encouraged and supported both global and local initiatives to advance the rights of indigenous peoples, and they have even empowered groups to articulate and advance long-suppressed cultural identities (Tsutsui 2018). Finally, the Declaration on the Rights of Peasants is critical both for strengthening other international human rights precedents and for advancing recognition of rights that are at stake in most intractable conflicts, including land rights, collective rights, food sovereignty, and rights involving ecosystem health (Edelman and James 2011; Smith and Schroering 2018). The Declaration is the direct result of efforts by a network of translocal activists, Via Campesina, to defend and celebrate peasants and their rights to land and culture in the face of modernization's aim to obliterate them (McMichael 2008).

What is noteworthy in these four areas of international human rights law is that they are a source of legitimacy for groups that have been most subjected to dispossession and denial of basic humanity and dignity via globalized capitalism. They involve mobilization of diverse alliances that share a critique of capitalism's expulsions as well as concrete visions of alternatives to capitalism (e.g., Carroll 2016). And significantly these ideas are linked to an increasingly dense network of organizations and movement-led processes that are less focused on interstate politics than was true in earlier decades (Smith and Wiest 2012). That contemporary translocal activism is expressing the interests and voices of these particular groups demonstrates its role in helping " 'decolonize' relationships and institutional arrangements," something that critical analysts view as essential to effective peace formation (Jabri 2016, 156).

Conclusion

The globalization of neoliberal peace has failed to address the fundamental inequities of the world economic and political order that marginalizes and excludes growing segments of humanity. A focus on states and official accounts that see peacebuilding as a top-down and often militarized process has limited critical inquiry into the structural sources of conflict and violence as well as the pathways to peace. Yet, invisible to most elites and mainstream analysts is a long history of counterhegemonic peacebuilding by advocates working for human rights, dignity, and basic survival. Attention to these actors can provide lessons in what constitutes peace formation as a process that must be understood as distinct from state-led peacebuilding projects.

The expansion of global networks that strengthen communication and collaboration across locales has supported a global counterhegemonic project and globalized constituencies who have articulated demands for fundamental transformations of unequal social relationships as a necessary step to ensuring dignity and survival for all. Through movement-centered processes like the World Social Forums, growing translocal networks of social movements like Via Campesina are helping to articulate and amplify demands of historically marginalized and excluded groups, illuminating structural violence, and creating conditions where the root causes of conflicts can be addressed. As Mouly (2008, 309) observes, "[t]he only way to build peace is to foster counter-hegemonic discourses that will gradually replace dominant ones. Peace constituencies should intend to reform the state as well as to change the norms that dominate civil society."

By advancing human rights frames, grassroots organizers focus attention on transforming social relations while developing nonviolent paths toward their realization. By engaging global human rights machinery to, for instance, condemn racism and promote the rights of indigenous peoples, women, and peasants, human rights advocates are working to transform global hierarchies and exclusions and bring new actors onto the global stage. In doing so, they challenge statebuilding as a social project and draw attention to its critical flaws. By making visible the basic structural conflicts that pervade the world-system and supporting the creation of formal institutions centered on human rights and equity, they are attacking the world-system's pillars of competitive markets and militarized nationalism, while helping define a new foundation based on principles of equity, nonviolence, and cooperation (Scoones et al. 2018). The growing visibility of translocal networks of activists is helping generate a broader base of support for these networks' projects of peace formation. While my focus here has largely been on the global arena and activist work to shift global norms and discourses, peace formation must be understood as a multiscalar project that extends from the grassroots to global levels and that involves both very localized, community-based practices as well as practices that contest state practices and seek to transform states through the work of peace formation. This work requires far more attention from scholars of peacebuilding as well as more substantial resources and support from proponents of peace.

Notes

1. Drawing on her research on the Israel/Palestine conflict, Dalsheim (2019, 16–17) highlights this fundamental contradiction: "Much of the scholarly critique of nationalism has focused on those it excludes, such as stateless people, refugees, and internal Others within a polity. Certainly, national sovereignty seems preferable to such precariousness . . . And so, we continue to both criticize the dangers of popular sovereignty in the form of ethnonationalism, while simultaneously recommending it as a means of achieving liberation."

2. For an overview of world-systems analysis, see Shannon (1992); Wallerstein (2004).
3. In addition to the review mechanisms provided in the growing numbers of human rights treaties, the UN Human Rights Council's Universal Periodic Review process requires all states to undergo human rights reviews that include reports from civil society and assessments by independent experts.
4. Kaldor (2003) refers to this as the "double boomerang," referencing Keck and Sikkink's (1998) "boomerang model" of transnational advocacy networks that target national authorities by appeals to international mechanisms and allies. Sikkink and her collaborators have described this process of strengthening international institutions as a "norms spiral" (Risse, Ropp, and Sikkink 1999).
5. The United States ratified CERD in 1994, and activists in that country could engage with the treaty monitoring process only following ratification.
6. For instance, in the most recent Universal Periodic Review of the United States, a number of member states called for the US government to ratify (among other specific human rights treaties) CEDAW and to implement its protections for women's equity and reproductive justice in its domestic and foreign policies (https://www.upr-info.org/sites/default/files/document/united_states/session_22_-_may_2015/a_hrc_30_12_en.pdf).

References

Alvarez, S. E. 2000. "Translating the Global: Local Effects of Transnational Organizing on Local Feminist Discourses and Practices in Latin America." *Meridians: Feminism, Race, Transnationalism* 1 (1):29–67.

Carroll, W. 2016. *Expose, Oppose, Propose: Alternative Policy Groups and the Struggle for Global Justice*. New York: Zed.

Dalsheim, J. 2019. *Israel Has a Jewish Problem: Self-Determination as Self-Elimination*. New York: Oxford University Press.

Desai, M. 2005. "Transnationalism: The Face of Feminist Politics Post-Beijing." *International Social Science Journal* 57 (184):319–330.

Desai, M. 2015. *Subaltern Movements in India: Gendered Geographies of Struggle against Neoliberal Development*. New York: Routledge.

Donnelly, F. 2016. "Critical Security Studies and Alternative Dialogues for Peace: Reconstructing 'Language Barriers' and 'Talking Points.'" In *The Palgrave Handbook of Disciplinary and Regional Approaches to Peace, edited by* O. P. Richmond, S. Pogodda, and J. Ramović, 272–285. London: Palgrave Macmillan.

Edelman, M. 2018. "Sacrifice Zones in Rural and Non-metro USA: Fertile Soil for Authoritarian Populism." *Open Democracy*. https://www.opendemocracy.net/marc-edelman/sacrifice-zones-in-rural-and-non-metro-usa-fertile-soil-for-authoritarian-populism

Edelman, M., and C. James. 2011. "Peasants' Rights and the UN System: Quixotic Struggle? or Emancipatory Idea Whose Time Has Come?" *Journal of Peasant Studies* 38 (1):81–108.

Escobar, A. 2008. *Territories of Difference: Place, Movements, Life, Redes*. Durham, NC: Duke University Press.

Falcón, S. 2016. *Power Interrupted: Antiracist and Feminist Activists inside the United Nations*. Seattle: University of Washington Press.

Fetherston, A. B. 2000. "Peacekeeping, Conflict Resolution and Peacebuilding: A Reconsideration of Theoretical Frameworks." *International Peacekeeping* 7 (1):190–218.

Frug, G. E., and David J. Barron. 2006. "International Local Government Law." *Urban Lawyer* 38 (1):1–62.

Harvey, D. 2004. "The 'New' Imperialism: Accumulation by Dispossession." *Socialist Register* 40:63–87.

Jabri, V. 2016. "Post-Colonialism: A Post-Colonial Perspective on Peacebuilding." In *The Palgrave Handbook of Disciplinary and Regional Approaches to Peace*, edited by O. P. Richmond, S. Pogodda, and J. Ramović, 154–169. London: Palgrave Macmillan.

Kaldor, M. 2003. *Global Civil Society: An Answer to War*. Cambridge, UK: Polity.

Keck, M., and K. Sikkink. 1998. *Activists beyond Borders*. Ithaca, NY: Cornell University Press.

Leonardsson, H., and G. Rudd. 2015. "The 'Local Turn' in Peacebuilding: A Literature Review of Effective and Emancipatory Local Peacebuilding." *Third World Quarterly* 36 (5):825–839. http://dx.doi.org/10.1080/01436597.2015.1029905.

Lerner, S. 2010. *Sacrifice Zones: The Front Lines of Toxic Chemical Exposure in the United States*. Cambridge, MA: MIT Press.

Mac Ginty, R. 2014. "Everyday Peace: Bottom-Up and Local Agency in Conflict-Affected Societies." *Security Dialogue* 45 (6):548–564.

Mac Ginty, R., and P. Firchow. 2016. "Top-Down and Bottom-Up Narratives of Peace and Conflict." *Politics* 36 (3):308–323. https://journals.sagepub.com/doi/abs/10.1177/0263395715622967.

MacNaughton, G., and M. McGill. 2012. "Economic and Social Rights in the United States: Implementation without Ratification." *Northeastern University Law Journal* 4 (2):365–406.

McMichael, P. 2008. "Peasants Make Their Own History, but Not Just as They Please." In *Transnational Agrarian Movements: Confronting Globalization*, edited by S. M. Borras Jr., M. Edelman, and C. Kay, 37–60. Malden, MA: Wiley-Blackwell.

Meyer, A. 2009. "Local Governments and Human Rights Implementation: Taking Stock and a Closer Strategic Look." *Pace diritti umani* (3):7–25.

Meyer, M. K., and E. Prügl, eds. 1999. *Gender Politics in Global Governance*. Boulder, CO: Rowman & Littlefield.

Morgan, R. 2007. "On Political Institutions and Social Movement Dynamics: The Case of the United Nations and the Global Indigenous Movement." *International Political Science Review* 28 (3):273–292.

Mouly, C. 2008. "Peace Constituencies in Peacebuilding: The *Mesas de Concertación* in Guatemala." In *Whose Peace? Critical Perspectives on the Political Economy of Peacebuilding*, edited by N. Cooper, M. Pugh, and M. Turner, 302–317). New York: Palgrave Macmillan.

Murtagh, B. 2016. "Economics: Neoliberal Peace and the Politics of Social Economics." In *The Palgrave Handbook of Disciplinary and Regional Approaches to Peace*, edited by O. P. Richmond, S. Pogodda, and J. Ramović, 110–123. London: Palgrave Macmillan.

Oomen, B. 2016. "Introduction." In *Global Urban Justice: The Rise of Human Rights Cities*, edited by B. Oomen, M. F. Davis, and M. Grigolo, 103–120. New York: Cambridge University Press.

Paffenholz, T. 2010. "Civil Society and Peacebuilding." In *Civil Society and Peacebuilding: A Critical Assessment*, edited by T. Paffenholz, 43–64. Boulder, CO: Lynne Rienner.

Paris, R. 2003. "Peacekeeping and the Constraints of Global Culture." *European Journal of International Relations* 9 (3):441–473.

Pugh, M., N. Cooper, and M. Turner, eds. 2008. *Whose Peace? Critical Perspectives on the Political Economy of Peacebuilding*. New York: Palgrave Macmillan.

Rajagopal, B. 2003. *International Law from Below: Development, Social Movements, and Third World Resistance*. New York: Cambridge University Press.

Rajagopal, B. 2006. "Counter-Hegemonic International Law: Rethinking Human Rights and Development as a Third World Strategy." *Third World Quarterly* 27 (5):767–783.
Richmond, O. P. 2013. "Failed Statebuilding Versus Peace Formation." *Cooperation & Conflict* 48 (3): 378–400.
Richmond, O., and J. Franks. 2009a. "Conclusion: Evaluating the Achievements of the Liberal Peace and Revitalising a Virtual Peace." In *Liberal Peace Transitions: Between Statebuilding and Peacebuilding*, edited by O. Richmond and J. Franks, 181–215. Edinburgh: Edinburgh University Press.
Richmond, O., and J. Franks, eds. 2009b. *Liberal Peace Transitions: Between Statebuilding and Peacebuilding*. Edinburgh: Edinburgh University Press.
Richmond, O. P. 2016. "Peace in International Relations Theory." In *The Palgrave Handbook of Disciplinary and Regional Approaches to Peace*, edited by O. P. Richmond, S. Pogodda, and J. Ramović, 57–68. London: Palgrave Macmillan.
Richmond, O. P., S. Pogodda, and J. Ramović, eds. 2016. *The Palgrave Handbook of Disciplinary and Regional Approaches to Peace* London: Palgrave Macmillan.
Risse, T., S. C. Ropp, and K. Sikkink, eds. 1999. *The Power of Human Rights: International Norms and Domestic Change*. New York: Cambridge University Press.
Santos, B. d. S. 2006. "Globalizations." *Theory, Culture & Society* 23:393–400.
Sargent, S. 2012. "Transnational Networks and United Nations Human Rights Structural Change: The Future of Indigenous and Minority Rights." *International Journal of Human Rights* 12 (1):123–151.
Sassen, S. 2014a. "Expelled: Humans in Capitalism's Deepening Crisis." *Journal of World-Systems Research* 19 (2):198–201. http://dx.doi.org/10.5195/jwsr.2013.495.
Sassen, S. 2014b. *Expulsions: Brutality and Complexity in the Global Economy*. Cambridge, MA: Belknap/Harvard University Press.
Schierenbeck, I. 2015. "Beyond the Local Turn Divide: Lessons Learnt, Relearnt and Unlearnt." *Third World Quarterly* 36 (5):1023–1032. https://www.tandfonline.com/doi/full/10.1080/01436597.2015.1043991.
Scoones, I., M. Edelman, M. Saturnino, M. Borras Jr., R. Hall, W. Wolford, and B. White. 2018. "Emancipatory Rural Politics: Confronting Authoritarian Populism." *Journal of Peasant Studies* 45 (1):1–20.
Shannon, T. R. 1992. *An Introduction to the World-System Perspective*. Boulder, CO: Westview Press.
Shawki, N. 2011. "Global Norms, Local Implementation—How Are Global Norms Translated into Local Practice?" *Globality Studies Journal* (26):1–13.
Sikkink, K. 2011. *The Justice Cascade: How Human Rights Prosecutions Are Changing World Politics*. New York: Norton.
Smith, J. 2008. *Social Movements for Global Democracy*. Baltimore, MD: Johns Hopkins University Press.
Smith, J. 2010. "Globalization and Strategic Peacebuilding." In *Strategies of Peace: Transforming Conflict in a Violent World*, edited by D. Philpott and G. F. Powers, 247–270. New York: Oxford University Press.
Smith, J. 2014. "Counter-Hegemonic Networks and the Transformation of Global Climate Politics: Rethinking Movement-State Relations." *Global Discourse: An Interdisciplinary Journal of Current Affairs and Applied Contemporary Thought* 4 (1):1–19. https://www.tandfonline.com/doi/abs/10.1080/23269995.2013.874111.

Smith, J., B. Gemici, S. Plummer, and M. M. Hughes. 2018. "Transnational Social Movement Organizations and Counter-Hegemonic Struggles Today." *Journal of World-Systems Research* 24 (2):372–403. https://doi.org/10.5195/jwsr.2018.850.

Smith, J., and C. Schroering. 2018. "Global Peasant Declaration Represents Huge Advance for Human Rights." *Common Dreams*, 18 December. https://www.commondreams.org/views/2018/12/18/global-peasant-declaration-represents-huge-advance-human-rights.

Smith, J., and E. Verdeja, eds. 2013. *Globalization, Social Movements and Peacebuilding*. Syracuse, NY: Syracuse University Press.

Smith, J., and D. Wiest. 2012. *Social Movements in the World-System: The Politics of Crisis and Transformation*. New York: Russell Sage Foundation.

Suárez, S. M. 2013. "Grassroots Voices: The Human Rights Framework in Contemporary Agrarian Struggles." *Journal of Peasant Studies* 40 (1):239–290.

Trimikliniotis, N. 2016. "Sociology: A Sociological Critique of Liberal Peace." In *The Palgrave Handbook of Disciplinary and Regional Approaches to Peace*, edited by O. P. Richmond, S. Pogodda, and J. Ramović, 95–109. London: Palgrave Macmillan.

Tsutsui, K. 2018. *Rights Make Might: Global Human Rights and Minority Social Movements in Japan*. New York: Oxford University Press.

US Human Rights Network. n.d. "International Mechanisms." Accessed 15 October 2020. https://ushrnetwork.org/our-work/international-mechanisms.

Wallerstein, I. (1974) 2011. *The Modern World-System I: Capitalist Agriculture and the Origins of the European World-Economy in the Sixteenth Century*. New York: Academic Press.

Wallerstein, I. 2004. *World-Systems Analysis: An Introduction*. Durham, NC: Duke University Press.

CHAPTER 23

GLOBAL CIVIL SOCIETY, PEACEBUILDING, AND STATEBUILDING

MARY KALDOR AND DENISA KOSTOVICOVA

PEACE-BUILDING, it can be argued, is based on an old-fashioned concept of peace, wherein war is understood as a deep-seated political contest between two sides that are assumed to be elites. In these circumstances, peace can be achieved only through compromise between the sides or by the victory of one side over the other. Contemporary wars are very different; they are better understood as a sort of persistent social condition, in which a multitude of armed groups benefit from violence itself rather than winning or losing. They benefit economically through predatory behavior and politically through the construction of exclusive ideologies based on fear. Of course all wars represent a profound social reordering. But whereas wars in the past involved centralization and strengthening of the state that provided a basis for top-down peace agreements, contemporary wars could be described as an extreme form of neoliberalism involving disassembly of the state (Sassen 2008) and a combination of fragmentation and globalization.

Peacebuilding and statebuilding start from the premise that, once a peace agreement is signed, the war is over and that what is required is a technical exercise of building institutional capacity destroyed during the war. But if we understand contemporary wars as a social condition, then peacebuilding is a much more complex undertaking that aims to reverse the persistent social condition by contributing to new forms of political legitimacy at different levels (local, national, and international). The role of global civil society is critical to this endeavor not as technical implementers but as a form of emancipatory civic agency.

This chapter addresses the civil society aspect of peacebuilding, which spans global and local connections below and beyond the state. It discusses how civil society can both contribute to peace- and statebuilding, but also undermine them, by analyzing different types of civic agency in a new war context. We first discuss the meaning of global civil society and reflect on the implications of different definitions that reduce the complexity of civic agency. We then analyze a technocratic approach to peace- and statebuilding that reduces the multitude of global and local civil society actors to one type: NGOs and their limited ability to address the social condition created by war. Last, we provide two contrasting cases of civil society's involvement in Bosnia and Herzegovina and in Syria and their implications for peacebuilding.

What Is Global Civil Society?

The terms "civil society" and "globalization" both entered the political lexicon after the end of the Cold War. Unlike the term "globalization," "civil society" has a long history originating in the Aristotelian concept of political community. But it was rarely used from the mid-nineteenth to the mid-twentieth century except in Marxist circles. It was activists resisting totalitarianism in Eastern Europe and military dictatorship in Latin America who revived the concept. For them it arose from the realization that they could never overthrow their regimes through violence because the regimes were so much stronger than they in military terms. Adam Michnik's (1985) essay "The New Evolutionism" explains that it was through civil society understood as an autonomous political space that self-organized groups (free-trade unions like Solidarnosc in Poland or independent intellectuals) could challenge the regime in political terms. A similar idea was simultaneously developed in Latin America, especially in Brazil under the military dictatorship (Stepan 2001).

From the beginning their understanding of civil society was global. Actually, it was a key member of the Hungarian Democratic Opposition, Gyorgy Konrad (1984), who was one of the first to coin the term "globalization," in his book published in English as *Anti-Politics.* For Eastern Europeans, the Helsinki Agreement of 1975, which their governments had signed and which contained far-reaching commitments to human rights, provided a key tool for legitimizing their activities. Moreover, support from Western European peace and human rights groups was critical in expanding the space in which they operated. Something similar was happening in Latin America. Keck and Sikkink (1998) in their book *Activists beyond Borders* talk about the "boomerang effect" whereby links with activists in other countries and international law provided an alternative way to pressure regimes. In other words, "global civil society" means not just civil society that crosses borders but the way civil society operates in these global times.

In the aftermath of the 1989 revolutions, "civil society" became an international buzzword, but it was understood differently by different groups of people. It is possible to delineate two broad interpretations (Pearce and Howell 2001). For activists, "civil society" was a term that legitimized their role as active citizens; it provided a public platform and as such spread rapidly in India, for example, and South Africa. It was associated with the notion of a public sphere. According to Jürgen Habermas (quoted in Kaldor 2003, 21–22), "What is meant by 'civil society' today, in contrast to its usage in the Marxist tradition, no longer includes the economy. . . . Rather, its institutional core comprises those non-governmental and non-economic connections and voluntary associations that anchor the communication structures of the public sphere in the society component of the life-world. Civil society is composed of those more or less spontaneously emergent associations, organizations, and movements that, attuned to how societal problems resonate in private life spheres, distil and transmit such reactions to the public sphere."

This could be said to be the activist version of civil society. But Western governments and donors interpreted the term in a more conservative way as what the West already has. They identified civil society with associationalism, and drawing on Tocqueville's ([1835] 1945) argument as well as Robert Putnam's (1994) understanding of "social capital," it was argued that a high level of associationalism contributes to democracy and development. Western governments and donors understood civil society as a substitute for the state rather than pressure on the state, a political "invisible hand" equivalent to the market in the economic sphere. This was a key component of the so-called Washington Consensus that markets and civil society are the standard recipe for countries "in transition." Increasingly development aid was subcontracted to NGOs, leading to a dramatic growth of NGOs and to the identification of NGOs with civil society. During the decade of the 1990s the number of international NGOs grew from 22,200 to 56,000 (see Kaldor, Moore and Selchow, 2012). We call this the neoliberal version of civil society.

Both these meanings can be found in conflict zones. On the one hand, NGOs are the main implementers of peacebuilding programs. On the other hand, antiwar and countersectarian activists tend to use the term "civil society" to describe themselves. (This is in contrast to activists in less violent situations, such as Tahrir Square, where the term was eschewed because of its association with contracts and NGOs.) Many contemporary conflicts are preceded by democracy protests. This was the case in Bosnia, for example, and in Ukraine or Syria. When the protestors are attacked, some turn to violence in response, but the majority tend to echo Michnik's argument that the use of violence will always lead to defeat. As the violence spreads, those who oppose the turn to violence organize themselves as civil society groups, providing immediate humanitarian assistance, including healthcare and schools, mediating ceasefires, documenting war crimes, or protecting the cultural heritage. When asked how they define civil society, they tend to say that civil society is composed of people who act in the public interest (as opposed to the private or sectarian interest; Theros 2019) or that members of civil society are those who care about human rights and values (Theros and Turkmani 2018). Their understanding of civil society as an alternative to violence is both directly influenced by the 1989 revolutions and shares the assumption that protest, activism, debate, and deliberation are the main mechanisms for change.

As was the case for Eastern European and Latin American activists, the existence of international law and the links with international networks—either international NGOs or support groups in other countries—are crucial in enabling these civil society groups to operate, hence the term "global civil society." In other words, global civil society does not refer to groups operating at the global as opposed to a local level. It refers to civil society in these global times, when international law, foreign governments, and international institutions provide a necessary condition for the operation of civil society and are in turn shaped by civil society pressures. To be sure, activist groups in the Global North may provide support for civil society groups in war zones and act as a bridge to foreign governments and international agencies, but both constitute global civil society. What is poorly understood in international circles is the extent to which these groups potentially provide the basis for legitimate and inclusive governance that might be the key to effective peacebuilding and statebuilding. Instead a technocratic type of statebuilding has been paralleled by assigning a technocratic role to civil society embodied exclusively by NGOs.

Peacebuilding and Statebuilding

The end of the Cold War ushered in a new kind of engagement between external actors and volatile postconflict states. In contemporary wars, peace agreements are not the moment wars end because most wars involve a persistent social condition; rather, they are the moment that international peacebuilding missions are deployed. Foreign states and multilateral institutions (such as the UN, EU, NATO, and Organization for Security and Cooperation in Europe) have taken on the task of rebuilding states in the aftermath of peace agreements (Caplan 2005, 16–44). Liberal internationalism was defined by exporting democracy and the free market as the model of governance into diverse states and societies worldwide. It was underpinned by a premise that to build states equates to building peace (Call and Wieth 2008). As spelled out in the *Supplement to an Agenda for Peace* by the UN secretary-general (1995), a response to the insecurity emerging from state failure was "the re-establishment of effective government."

The global context in which "new internationalism" emerged was consequential to this logic. Whether we are talking about the invasions that were part of the war on terror, so-called humanitarian interventions like Kosovo or Libya, or multilateral peacekeeping deployments after peace agreements, it was assumed that the only way to stabilize the violence was capacity-building and statebuilding. The task of building state capacity was defined by a Weberian approach, which is embodied by its focus on institution-building. It draws on the German sociologist Max Weber's conceptualization of the state as a set of institutions that wields the monopoly of authoritative rule-making, alongside the monopoly of legitimate violence. Accordingly, the state's legitimacy is determined by its ability to administer impartial rule, which ultimately safeguards the state from a recurrence of violence (Lottholz and Lemay-Hébert 2016). Consequently, driven by the "ideology of failed states," international statebuilding was guided not by the needs of the countries where the intervention was undertaken but by a priority to create capacity to manage outsiders' aid (Woodward 2017, 250). Mounting evidence of the effects of external statebuilding since the end of the Cold War reveals a tautology whereby the policies that aimed to address the challenges of the fragile states actually end up causing the state failure and fragility.

Nearly three decades of external involvement in rebuilding local political authority has, in some instances, kept the recurrence of violence at bay, as is the case in the Balkans.

But the goal of building sustainable and legitimate states has remained largely elusive (Kostovicova and Bojicic-Dzelilovic, 2009). The idea of liberal peace through capacity-building, until recently a dominant framework for theorizing external statebuilding interventions, has come under critical scrutiny. Demonstrations of local resistance to external actors and their statebuilding projects, epitomized by the Taliban resurgence in Afghanistan or continuing violence in the Democratic Republic of the Congo, point to a crisis of statebuilding in both theory and practice. While criticism comes from a variety of normative and political standpoints, arguments centered on the technocratic approach to statebuilding as peacebuilding capture the misalignment between international statebuilding as public policy and the complexity of restoring legitimate public authority in new war contexts.

The technocratic approach to statebuilding rests on the conception of the state "as ideally divorced from politics, economics, and society" (Wesley 2008, 380). However, contemporary wars are a form of social transformation. The extent of this transformation is particularly consequential in wars fought in the global age, when cross-border dynamics—such as arms and commodities trafficking facilitated by transnational networks—shape local politics. In effect, this technocratic approach exacerbates and entrenches the social condition that characterizes contemporary conflicts. The armed groups that have a vested interest in violence for both political and economic reasons are generally part of global networks involving a combination of organized crime and exclusivist ideology (Kaldor, 1999). They are the ones who are chosen as the participants in peace talks, and their predatory activities are legitimized and entrenched through peace agreements. The agreements might reduce overt violence, though these groups rarely fight each other since most violence is directed against civilians. But the statebuilding efforts that follow are continually subverted because these groups benefit from disorder, and at the same time, capacity-building merely builds the capacities of these neophyte elites.

Therefore the institutionalist approach to statebuilding overlooks not only how the social base—in which the state is embedded—is restructured by violence, but also how these social relations have penetrated the realm that is formally designated as being part of the state. Scholars have pointed to the limits of the institutionalist approach to peacebuilding through statebuilding. Whether this is because the international actors fail to understand contemporary conflicts and still see them as a political context among elites, or whether there are external interests in continuing disorder is, of course, open to inquiry.

A relational perspective provides an alternative to the neo-Weberian conceptualization of state capacity. From this perspective, "capacity is not something that resides necessarily within certain institutions or with certain individuals, but [is] an attribute that relates to broader social and political structures, such as those affecting class and ethnicity, within which institutions develop" (Hameiri 2007, 124). In line with this, administrative capacity should be conceived "as products of social and political relationships both inside and outside the state" (Jayasuriya 2005, 382). A relational view of the state draws attention to different types of relations and challenges of building peace. In contrast to building institutions in a neo-Weberian sense, this perspective puts emphasis on dismantling informal relationships as critical to statebuilding as peacebuilding.

The technocratic approaches to statebuilding have shaped critically how the role of civil society in postconflict contexts was envisaged. The building of civil society from outside was envisaged as being directly in support of statebuilding. This has been carried out either by injecting financial support into existing or newly created local NGOs or by creating the local outposts of international NGOs. The activities supported by external funding typically mapped the areas of state capacity-building, earmarked in accordance with this technocratic approach to governance. Typically they included areas of democratization, human rights, development, security-sector reform, reconciliation, and most recently counterterrorism.

As a result of the practice pursued since the end of the Cold War, the premise—once taken for granted—that civil society building will support state- and peacebuilding has become problematic. The NGO-ization of civil society has been recognized as contributing to the failure to rebuild legitimate war-torn states (McMahon 2017). Just as it has proved challenging, and even counterproductive, to transplant a Weberian model of a legitimate state underpinned by functioning institutions into postconflict environments, the problem of creating civil societies on the Western model has proved equally misplaced. NGOs, and even the very concept of civil society, are being resisted in peacebuilding contexts for being divorced from the citizens and their interests. Far from embodying a democratic ideal of citizens' voice and grassroots action, the neoliberal form of civil society is being castigated as a new form of predation. How civil society should be supported in order to underpin its legitimacy and increase its efficiency has been a subject of considerable debate in the scholarship. This debate points to a need for greater local ownership in determining spending priorities, expanding the recognition of a variety of organizational forms eligible for support and recognizing the significance of local civic capital that has survived the conflict, rather than approaching the task as creating civil society anew. Critical theorists of peacebuilding have pointed to a need to rethink emancipatory agency in conflict contexts. This in turn has led to propositions to consider hybrid forms of governance, where civic agency as such will not exist necessarily as a separate sphere from the conventional formal sources of authority.

What the debate misses, however, is the central point that if we reconceptualize conflict as a social condition, then activist civil society needs to be regarded not merely as a recipient of donor funds, although that might be desirable, but as a partner in transforming the social condition by countering sectarian and fundamentalist narratives and by increasing the accountability of corrupt elites. On this reading, civil society linked up with international actors can provide the basis for a strategy for constructing legitimate institutions at different levels. By focusing on what are considered to be elites associated with the armed groups as participants in the peace talks, the international community has at least implicitly endorsed the marginalization of activist civil society; reversing this relationship opens up new possibilities for reducing violence. The theorizing about civil society in relation to state- and peacebuilding is at a juncture. The critique of the neoliberal model of civil society building by supporting local NGOs has resulted in a backlash against local NGOs. At the same time, there are no clear emerging alternatives in practice, although recent theoretical approaches point to new possibilities. However, the emerging dichotomy between the "old" and "new" approaches should be treated with caution. Between them there lies a whole array of roles that civil society undertakes with various effects on statebuilding, both positive and negative. These often counter our assumptions on their effectiveness.

Dominant research paradigms have overshadowed investigations that demonstrate the complexity of civil society's contribution to state- and peacebuilding. Scholars have shown that the externally supported NGOs with global connections can enjoy local legitimacy as well as meaningfully impact state policymaking. Similarly, the focus on global connections often prioritizes the links with progressive civil society in the Global North to the detriment of equally consequential cross-border connections between illiberal civil society groups that impact both local states and security beyond them.

Identifying the two broad interpretations of global civil society could suggest directions in which the link between international actors and local civil groups could be reconceptualized. The neoliberal version envisages global civil society as an implementer of a technical exercise of institution-building. By contrast the activist understanding of global civil society suggests civil society rather as a possible partner in the political construction of legitimacy. In the two case studies that follow we suggest different ways these two interpretations can be observed, and in the conclusion we offer a way forward.

Civil Society and Statebuilding in Bosnia and Herzegovina

The war in Bosnia and Herzegovina began in 1992 when shots were fired at a civil society demonstration protesting the descent to war and calling for an international protectorate. As in other places, independent voices were the first to be targeted during the war; many were killed or were forced to leave the country. Those who remained organized themselves as civil society. They ran independent radio stations, kept secular culture alive in places like Sarajevo and Tuzla, documented war crimes, were the first providers of help to the victims of war, and transmitted proposals to the international community, including for a no-fly zone, for safe havens and humanitarian corridors, for local protectorates, and for a war crimes tribunal. They used the term "civic" to mean antisectarian, and they talked about "multi-multi-multi" values, meaning "multinational, multireligious, and multicultural."

But despite their role in peacebuilding during the war, they were entirely excluded from the political talks, which followed the logic of ethnic partition and contradicted much that civil society was trying to achieve on the ground. The Dayton Accords of 1995 froze the social transformation that had taken place during the war and entrenched the ethnic warlords, providing little space for actually existing civil society. Yet paradoxically, the international community placed the goal of strengthening civil society at the core of the international statebuilding policy in Bosnia and Herzegovina after the agreement was signed. The policy rested on the premise that all civil society had to be created anew, overlooking any forms of civic agency from before or during the war. The building of civil society primarily entailed support for service provision (Fagan 2005). Consequently the NGO scene was fashioned in a mirror image of the neoliberal state that was being built at the same time. Nearly three decades later, the failure of external statebuilding is largely reflected in the failure of externally sponsored civil society building in the country. This common claim in the scholarship overlooks a more complex dynamics at the local level—both the forms of civil agency and action that arise as a result of Western interventionism, as well as those that exist independent of it.

Scholars have commonly focused on the failure of civil society building that is part of neoliberal statebuilding. From this perspective, local NGOs are disempowered, alienated, ethnically divided, and even corrupt segments of the Bosnian polity, which have delegitimated the statebuilding project itself (Belloni 2001; Chandler 1998). These conclusions are backed up by criticism of statebuilding, of which civil society building is an integral part, as a Western neocolonial practice, disconnected from the local political and cultural contexts. As such it stifles rather than creates civil agency, especially in its political role in constructing state legitimacy. As a result, civic agency can emerge as a form of resistance to liberal peacebuilding and statebuilding (Kappler and Richmond 2011).

A wave of protests in 2013 and 2014 that swept Bosnia and Herzegovina demonstrated a renewed form of activist civic agency (Fagan and Sircar 2017). People rose against a high level of unemployment, corrupt privatization, and dysfunctional governance, which accompanied neoliberal peace and statebuilding through the introduction of a market economy. This led to the creation of "plenums" or "citizens' assemblies," which as a form of bottom-up discursive agency were open to all to express their views. The assemblies resulted in a set of demands that closely reflected ordinary people's needs in a postconflict transitional economy, while overcoming ethnic polarization in the postconflict state (Murtagh 2016). Presenting a left agenda, without international support, and distancing themselves from state structures, this brief experiment with direct participatory bottom-up democracy as a route to a more responsive peacebuilding to local needs fizzled out. In the aftermath of the protests, there were devastating floods, which revealed that government coffers were empty and the civil protection fund had no money—raided by the various patronage networks. The floods "vividly illustrated . . . the cumulative dangers of chronically poor regulatory oversight, bad governance and political infighting" (International Crisis Group 2014). The humanitarian response was entirely provided by local municipalities across the region and across borders and by individual volunteers (interview with activists, Tuzla, June 2014). As in the war, the protestors were transformed into humanitarian workers and used their newly found organization to manage assistance.

However, criticism of NGOs that paints all NGOs with the same brush is dismissive of their emancipatory agency in postconflict statebuilding. Puzzled by externally funded NGOs that enjoy legitimacy with the donors and with the citizens in Bosnia and Herzegovina, Puljek-Shank (2018) points out that they serve as active intermediaries between the donors and local actors, which also earns them legitimacy with local political actors. Not all externally funded NGOs are artificial. Some have emerged out of the consequence of self-organization by activists in the past. They are effective because of their ability to bring in donor financial and symbolic resources, while at the same time facilitating collaboration with other local NGOs that are accepted by constituents, such as the transformation in governance in youth affairs. Several of these NGOs cross ethnic divides and are able to keep alive a Bosnian identity. Effectiveness in shaping state policy therefore can to some extent be captured by the transactions between civil society and state actors.

Civil Society and Peacebuilding in Syria

As in the case of Bosnia and Herzegovina, the war in Syria began with violence against peaceful protests in 2011. While some protestors did join armed groups and argued that the revolution could be saved only through armed resistance, the majority of the protestors opposed the turn to violence. Again like in Bosnia, the activists were transformed into organized civil society, and, also like in Bosnia, over the course of war civil society has been battered: activists have been killed, imprisoned, tortured, divided by checkpoints and partitions, and forced to leave. Nevertheless, even after nine years of war, civil society is almost the only source of social cohesion and undertakes most of the activities needed for survival. The range of issues, roles, campaigns, and services include help for victims of war and immediate humanitarian aid, such as the White Helmets, Nobel Prize nominees who are the first responders after attacks and are probably the best known among the representatives of civil society. Other activities include economic empowerment and education. Civil society's role in relation to conflict resolution is particularly prominent and includes peace negotiations and mediation. For example, one group in a Damascus suburb organizes a global café and interactive theater where people from different viewpoints can engage with each other. The work laying the ground for postconflict justice includes documentation and evidence collection on prisoners; human rights violations and justice monitoring; campaigns against sexual violence, including male rape in Assad's prisons; as well as work on increasing awareness of citizenship rights, political equality, and the role of the International Criminal Court.

Particularly interesting is the emphasis among many civil society groups on gender equality and women's empowerment. Indeed the feminist movement is considered an important reason for the emergence of the democratic movement. One religious group, for example, promotes an understanding of Islam in which the core meaning is a fair life for both men and women and in which "crimes of honour" are completely excluded (Kaldor 2017). Likewise, civil society groups emphasize the importance of values and tend to be very critical of ethnic and religious divisions (Theros and Turkmani 2018).

In contrast to the Bosnian war, the international community has been very divided over Syria, and many countries, including Russia, Turkey, Iran, the Coalition against ISIS, and Israel, have intervened militarily, greatly increasing the level of violence and the rate of casualties. Whereas in Bosnia, UN agencies were deployed in the country during the war and were able to undertake humanitarian aid directly, in Syria external players are dependent on both local armed groups and local NGOs for the delivery of aid, which has led to the NGO-ization of civil society alongside its more activist role. Large communities of Syrian think tanks and NGOs are based in Turkey and Lebanon and go back and forth into the country. On the one hand, these externally funded NGOs find their agenda and activities tend to be shaped by donor priorities rather than activities on the ground. On the other hand, such organizations are dependent on local civil society groups, and this helps to provide some sort of protection and platform.

A further significant difference is the role of civil society in the UN-sponsored political talks. Under pressure from civil society, the United Nations special envoy, Staffan de Mistura, agreed to establish women's advisory groups and a civil society room. A survey undertaken of those who participated shows that skepticism about the formal talks has increased as a consequence of the civil society room. On the other hand, participants argue that the civil society room has enabled them to meet activists from different parts of Syria and to understand different points of view. It has also helped in providing links between civil society and the international community, contributing to better understanding of what is going on and enabling practical steps on issues like humanitarian aid. And importantly, participation in the civil society room has legitimized civil society groups at the local level and thereby provided a degree of empowerment (Theros and Turkmani 2018).

Conclusion

Scholars recognize multiple forms and manifestations of civic agency in conflict contexts, which are both strengthened and weakened by global links (Kostovicova and Bojicic-Dzelilovic 2013). While the critique of the NGO-ization of global civil society has become a mainstream theoretical position, the emancipatory role of alternative forms of civil society in constructing political legitimacy is much less understood. Paradoxically, the social cohesion and capacity that is built by civil society during the war, as is the case in Bosnia and Syria, tend to be stifled by an NGO approach to civil society building in which global political and civil society actors collude. Civil society building as NGO building embodies a neoliberal version of civil society and is shaped by the technical approach to peace- and statebuilding. By contrast, the social condition that is inherited after a conflict, which is characterized by a fuzzy distinction between a formal and an informal political sphere, as war-time networks and actors penetrate the "state," requires a different response. If civil society is understood in the Habermasian sense, it then becomes an essential political partner in peacebuilding—something that may be beginning to happen in the Syrian context because of the failure and absence of traditional approaches. Instead of focusing on the change in statebuilding policies by domestic and international actors as a benchmark, the emancipatory agency of civil society should recognize the discursive political contribution of civil actors to the construction of legitimate political authority.

This requires a different dominant understanding of the nature of contemporary conflict. Another way of describing the social condition that characterizes contemporary conflict is as an attack on civility. Public protest, as in Sudan at the time of writing, threatens regimes that typically combine powerful vested economic and political interests. The only way to sustain their rule is through the construction of disorder and the abandonment of the rule of law and the civic values that necessarily underpin the rule of law. Not only are civilians the target of attack, but the various institutions, including secular culture and education, expertise, security, and civil behavior, are all under attack. This is why the promotion of civility is not an afterthought or a welcome extra but central to any strategy for ending violence.

Treating activist civil society as a partner rather than a recipient of money could provide the basis for such a strategy. Any strategy would be the outcome of debate and discussion among members of civil society, who by no means agree about what should be done, and with counterparts at international levels. It might involve a range of possible approaches, including helping to preserve civic spaces in conflict areas, supporting explicitly countersectarian and counterfundamentalist narratives, contributing to legitimate economic activities, documenting war crimes, and strengthening both international and national justice mechanisms.

After nearly three decades of peacebuilding, there are evidently those within the liberal peace community who are beginning to adopt a different interpretation of contemporary conflicts. The continuing myopia of the dominant international elites suggests that there are vested interests in a top-town view of conflict. Ways of comprehending what is happening are deeply embedded in ways of doing that incentivize certain perceptions of the world; far-reaching change is required in both. This is why any alternative approach to addressing conflict has to involve change at global levels as well, whether we are talking about the way neoliberal assumptions have contributed to the disassembly of the state, the various ways in which practices in the Global North feed various criminal activities, or the cultural relativism that tolerates the construction of ethnic and religious polarization.

References

Belloni, R. 2001. "Civil Society and Peacebuilding in Bosnia and Herzegovina." *Journal of Peace Research* 38 (2):163–180.

Call, C. and V. Wieth (eds). 2008. *Building States to Build Peace*. Boulder, CO: Lynne Rienner Publishers.

Caplan, R. 2005. *International Governance of War-Torn Territories: Rule and Reconstruction.* Oxford: Oxford University Press.

Chandler, D. 1998. "Democratization in Bosnia: The Limits of Civil Society Building Strategies." Democratization 5 (4):78–102.

Fagan, A. 2005. "Civil Society in Bosnia Ten Years after Dayton." *International Peacekeeping* 12 (3):406–419.

Fagan, A, and I. Sircar. 2017. "Activist Citizenship in Southeast Europe." *Europe-Asia Studies* 69 (9):1337–1345.

Hameiri, S. 2007. "Failed States or a Failed Paradigm? State Capacity and the Limits of Institutionalism." *Journal of International Relations and Development* 10 (2):122–149.

International Crisis Group. 2014. *Bosnia's Future.* Europe Report No. 232. 10 July.

Jayasuriya, K. 2005. "Beyond Institutional Fetishism: From the Developmental to the Regulatory State." *New Political Economy* 10 (3):381–387.

Kaldor, M. 1999. *New and Old Wars: Organized Violence in a Global Era.* Cambridge: Polity Press.

Kaldor, M. 2003. *Global Civil Society: An Answer to War.* Cambridge: Polity Press.

Kaldor, M. 2017. "Syrian Civil Society." *Open Democracy* 4 (4 March).

Kaldor, M, H. L. Moore, and S. Selchow (eds). 2012. *Global Civil Society 2012.* Basingstoke: Palgrave, Macmillan.

Kappler, S., and O. Richmond. 2011. "Peacebuilding and Culture in Bosnia and Herzegovina: Resistance or Emancipation?" *Security Dialogue* 42 (3):261–278.

Keck, M, and K. Sikkink, 1998. *Activists beyond Borders.* Ithaca, NY: Cornell University Press.

Konrad, G. 1984. *Antipolitics: An Essay.* New York, NY: Harcourt, Brace and Jovanovich.

Kostovicova, D., and V. Bojicic-Dzelilovic. 2009. "Introduction: State Weakening and Globalization." In *Persistent State Weakness in the Global Age*, edited by Denisa Kostovicova and Vesna Bojicic-Dzelilovic, 1–16. Aldershot: Ashgate.

Kostovicova, D., and V. Bojicic-Dzelilovic. 2013. "Introduction: Civil Society and Multiple Transitions—Meanings, Actors and Effects." In Civil Society and Transitions in the Western Balkans, edited by V. Bojcic-Dzelilovic, J. Ker-Lindsay, and D. Kostovicova, 1–25. Basingstoke: Palgrave Macmillan.

Lottholz, P., and N. Lemay-Hébert. 2016. "Re-reading Weber, Re-conceptualizing State-Building: From Neo-Weberian to Post-Weberian Approaches to State, Legitimacy and State-Building." *Cambridge Review of International Affairs* 29 (4):1467–1485.

McMahon, P. C. 2017. *The NGO Game: Post-Conflict Peacebuilding in the Balkans and Beyond.* Ithaca, NY: Cornell University Press.

Michnik, A. 1985. "The New Evolutionism." In A. Michnik, *Letters from Prison and Other Essays*, 135–148. Berkeley, CA: California University Press.

Murtagh, C. 2016. "Civic Mobilization in Divided Societies and the Perils of Political Engagement: Bosnia and Herzegovina's Protest and Plenum Movement." *Nationalism and Ethnic Politics* 22 (2):149–171.

Pearce, J. and J. Howell. 2001. *Civil Society and Development: A Critical Exploration.* Boulder, CO: Lynne Rienner.

Puljek-Shank, R. 2018. "Civil Agency in Governance: The Role of Legitimacy with Citizens vs. Donors." *Journal of Voluntary and Nonprofit Organization* 29 (4):870–883.

Putnam, R. 1994. *Making Democracy Work: Civic Traditions in Modern Italy.* Princeton, NJ: Princeton University Press.

Sassen, S. 2008. *Territory, Authority and Rights: From Medieval to Global Assemblages.* Princeton, NJ: Princeton University Press.

Stepan, A. 2001. *Arguing Comparative Politics.* Oxford: Oxford University Press.
Theros, M. 2019. "Re-imagining Civil Society in Conflict: Findings from Post-2001 Afghanistan." *Journal of Civil Society* 15 (2):143–161.
Theros, M, and R. Turkmani. 2018. "Civil Society Support Room: Preliminary Research Findings." Paper presented at Political Economy and Governance in Syria Conference, London School of Economics, 3–4 December.
Tocqueville, A. de. (1835) 1945. *Democracy in America.* New York, NY: Vintage Books.
UN Secretary-General. 1995. *Supplement to an Agenda for Peace: Position Paper of the Secretary-General on the Occasion of the Fiftieth Anniversary of the United Nations.* A/50/60-S/1995/1, 3 January. https://digitallibrary.un.org/record/168325?ln=en.
Wesley, M. 2008. "The State of the Art on the Art of State Building." *Global Governance* 14 (3):369–385.
Woodward, S. L. 2017. *The Ideology of Failed States: Why Intervention Fails.* Cambridge: Cambridge University Press.

CHAPTER 24

NETWORKS OF PEACE

NAJI BSISU AND AMANDA MURDIE

In addition to the dozens of different countries contributing military forces, several hundred international nongovernmental organizations (INGOs) descended on Afghanistan in the few years after the start of the US Operation Enduring Freedom in late 2001 (Olson 2006; Mitchell 2017). Although most of these organizations shared concerns about humanitarian assistance, peacebuilding, and statebuilding for Afghanis, they operated across many different sectors and brought with them individuals, structures, experiences, and funding from over a hundred different countries. With so many organizations operating on interrelated issues, it was imperative that groups coordinate their efforts and learn from their shared experiences. However, several factors made coordination difficult, including the revolving door of workers in and out of the country as aid to Afghanistan increased (Higham, Schulberg, and Rich 2014).

Further, NGOs were not the only actors involved in humanitarian relief. As military actors began to rely more heavily on a "hearts and minds" counterinsurgency strategy, both civilian and military agents often carried out tasks previously thought of as largely in the purview of NGOs, including constructing wells and health clinics. For many NGOs, these "blurred lines" between military operations and humanitarian work were seen as problematic, undermining the neutrality of aid work and possibly endangering the lives of their employees and volunteers (Mitchell 2015). Additionally, "differences in cultures, organizational structures, beliefs, and priorities" made coordination between NGOs and militaries especially difficult (N. C. Roberts, 2010, 213).

Given the variety of actors involved in peacebuilding, humanitarian aid, and statebuilding in conflict- and disaster-affected societies, existing accounts from many areas of the world show us how difficult coordination and cooperation can be. The payoffs of such coordination, however, can be substantial, greatly improving the likelihood that human security objectives are met. By working as a collective network instead of isolated organizations, NGOs and military interveners can often improve overall humanitarian outcomes and encourage peace. Using network analysis, Dorussen and Ward (2008) have argued that individual intergovernmental organizations are limited in their ability to enact change due to their institutional weakness; however, membership in intergovernmental organizations creates networks between states, making them more effective as they work collectively. We believe that a similar process can occur between state and nonstate actors after conflicts or disasters, enhancing the cultivation of stable peace. In the remainder of this chapter, we outline how a "network of peace" can form; this network comprises coordinated efforts between NGOs, including interactions between international and domestic NGOs, between NGOs and their government or military counterparts, in addition to interactions with local individuals and civil society organizations (hereafter CSOs). We then examine the effects of these coordinated efforts, both on the achievement of human security and on the organizations themselves. Although we see many benefits from coordination, we also stress the negative externalities that networked efforts can bring. We conclude the chapter with some attention to the questions remaining in this growing area of research. Further theoretical development and empirical examinations of the causes and consequences of the network of peace can help inform policymakers and practitioners, ultimately improving security outcomes on the ground.

The Nature of the Network

During and after a conflict or disaster, many actors are often involved. As mentioned, international and domestic NGOs and military interveners can be present. Additionally, civilian agencies from donor countries and intergovernmental organizations can be present, as well as local government officials, local CSOs, and representatives from businesses. In situations of civil conflict and insurgency, representatives of rebel groups and ethnic communities may carry out their own relief and service delivery efforts (Berman and Laitin 2008; Flanigan 2008).

At the onset, these actors can have drastically divergent goals and tools at their disposal. Although they might all want peace and better conditions for those affected, they may focus on different sectors or strategies. Some may have a religious message or represent a specific ideological framework. Some have a long history of involvement with the United Nations or other intergovernmental organizations. Some may have very limited regional expertise and no local partners or in-country experience. Many will be in competition for the same donor grants and projects (Cooley and Ron 2002). As a result, even if we were to focus on only INGOs, there are many impediments to coordination. Further, given the stress of the job and the environment, even well-meaning individuals may have little time or desire to reach beyond their organizational boundaries (see Visoka 2015).

Looking beyond NGOs, however, we can see even more disincentives for coordination. Norms and training may make NGOs think military interveners are the problem; military involvement in "hearts and minds" counterinsurgency could politicize humanitarian relief (Voluntary Organizations in Cooperation in Emergencies 2008). Military actors may be hesitant to consult or trust local NGOs, who they may see as aiding the insurgents. Although military leaders may want a "whole-of-government" or "hearts and minds" military approach, force structures are often designed with war, not peace, in mind (Miklaucic 2011). Local actors of all sorts may be leery about working with international actors, many of whom they see as opportunistic and profiting from the misery of others. These general impressions or stereotypes may be reinforced by standard operating procedures that limit discussion of strategies or operations outside of specific chains of command. Even the divergent terminology and complex acronyms can limit any incentive for joint planning or cooperation.

Because of these impediments, a lack of coordination between actors in conflict- or disaster-affected communities is the norm. In a recent reflection on the limited success of international efforts in Afghanistan, an Oxfam America background report noted that a lack of coordination, both among NGOs themselves and between NGOs and military actors, undermined the achievement of "development objectives" and led to "ineffective aid and increased conflict" (Blankenship 2014, 12). One individual interviewed in the report noted, "[T]here are too many people doing their own thing. And they are able to do this because there is too much money and too many agendas without the necessary accountability to deliver on a cohesive strategy" (25).

Both within and outside of Afghanistan, however, there are also many anecdotes of coordination between all types of actors in complex emergencies, including coordination between NGOs and military forces. When coordination does occur, the humanitarian outcomes are often improved. In Afghanistan, for example, coordination between local and international NGOs, universities, and "agricultural development teams" from the US military successfully helped improve agriculture output more than what could have been achieved by individual organizations (Peck and Kemmet 2011). In Mali in 2014, UN forces helped with security for humanitarian flights at local airfields (Solomon 2014). Coordination in Mali has been multifaceted since the establishment of the UN peace enforcement mission in 2013; coordination also means "coexistence," with minimal interaction between NGOs and military forces when engagement is not deemed necessary (Solomon 2014).

The UN Office for the Coordination of Humanitarian Affairs (OCHA; 2017) has been instrumental in facilitating best practices when multiple actors are involved in conflict- and disaster-affected societies, particularly when the coordination involves civilian and military actors (often referred to as CMCoord or CIMIC). In addition to formal guidelines for interaction, OCHA actors help in drafting country-specific policies, perform public and private diplomacy with various stakeholders, share information, and help fund joint humanitarian efforts. OCHA's policies reflect a belief that coordinated efforts outperform individualized actions. Coordinated efforts, especially those that actively engage with local populations to build country-specific policies, are key. Networked actors that empower local actors give the local community a sense of ownership in the policies that are derived from their interactions. These policies toward peacebuilding tend to have more localized legitimacy and may ultimately be more successful (Richmond 2007; D. Roberts 2011). Kumar and De la Haye (2012, 13) detail how fourteen African states recently came together to denote an "infrastructure of peace" that explicitly focuses on the dynamic nature of peacebuilding and conflict prevention. This process highlights the continued dialogue and evolving nature of peacebuilding as networked actors include local communities in their peacebuilding initiatives.

In many regards, academic work examining the coordination between the various actors involved in humanitarian space is still relatively underdeveloped. Coordinated efforts and interactive effects are mentioned almost in passing (Wood and Sullivan 2015). This is especially the case in examining coordination between NGOs and military interveners. Until recently much of the literature has simply outlined the same impediments to coordination as found in practitioner work. There is a richer cross-disciplinary literature on NGO-to-NGO coordination. Some of the most well-known work in this area focuses not on humanitarian organizations but on the creation of "transnational advocacy network" between NGOs and other advocates in the areas of human rights and the environment (Keck and Sikkink 1998). There is, however, a robust debate centered on the nature of liberal peacebuilding itself, as evidenced in the work of critical theorists (D. Roberts 2011).[1]

Drawing on this literature and the broader network analysis approach provides us with several tools to examine coordination in conflict- and disaster-affected countries. As Elkins (2009) points out, the study of networks has two broad approaches: "network-as-actor" and "network-as-structure." A network-as-actor approach could prove useful in examining how a somewhat new, collective actor emerges in situations where collaboration occurs; this was the approach used in the canonical transnational advocacy network literature. The network-as-structure approach may be more influential, however, in helping us understand the factors that lead to coordinated efforts and the effects of this coordination on any one organization or outcome. This can be seen in the discussions of hybrid peace in the peacebuilding literature, which will be discussed in further detail throughout the chapter (Mac Ginty 2010; Richmond 2010). With these ideas in mind, we now turn to examine that factors that help create coordinated efforts and the effects coordination has on security outcomes.

What Creates Networks for Peace?

Before delving into the factors that help explain the formation of networks of peace, we must account for the factors that help in determining where and when interveners decide to intervene. A growing consensus seems to indicate that UN peacekeeping missions (PKMs) are sent where they are needed most (Fortna 2004). Essentially, UN PKMs tend to go to weaker states and where civilian casualties are highest (Fortna 2004). Beyond UN PKMs, scholars have also looked at when and where NGOs are more inclined to go. NGOs are found to value both places where they can operate safely as well as places that have international issue salience (Murdie and Webeck 2015). Areas with high levels of international issue salience can assist NGOs with funding and recruitment

(Murdie and Webeck 2015). While NGOs may not always be able to provide their own organizational safety when operating in areas afflicted by conflict or a humanitarian crisis, they may be able to take advantage of the presence of UN PKMs or other military interveners, even if those forces do not have an expressed mandate to protect the NGOs.

Despite the hurdles associated with collaboration, individual humanitarian actors do network with one another and work in tandem to accomplish their goals. Several factors that are posited to assist in this process are shared ideas, trust, and opportunity; organizational capacity; complex tasks requiring cooperation; and the presence of other actors that either actively or passively encourage collaboration. The following paragraphs expound upon each of these factors.

Sharing a belief regarding the most pressing humanitarian crisis and how to best approach it is foundational for actors to consider working together. Consensus in approaching disaster is imperative for efficiency and quality of humanitarian assistance (Spitka 2018). In looking at the effectiveness of multilateral interventions rather than looking at biased versus neutral interventions, it is more important to look at whether the actors participating in the multilateral intervention are united or divided; united multilateral interventions are more successful in achieving their goals (Spitka 2018). A shared vision of how to best approach a humanitarian crisis increases the probability that groups will want to work alongside one another. Not only that, but their collective efforts often end up being more efficient with regard to the quality of the humanitarian assistance they provide. Not only is a shared vision among the various international actors important to the success of the assistance they provide, but incorporating the vision of the local population is also paramount. Doing so generates local legitimacy and enhances the chances for success (Richmond 2007; D. Roberts 2011, 2012). This lends further credence to the idea that collaboration among NGOs, governments, militaries, CSOs, and others benefit all those involved, especially the impacted populations requiring assistance in the first place.

The conditions of the locale that humanitarian actors find themselves in has a large impact on the prospects of collaboration. Dependable and secure governing structures can allow a humanitarian actor to feel that they can trust that any relationships they develop with other humanitarian actors will not fall victim to local corruption or that the local government bureaucracy will not intervene, negatively impacting the new partnership (Murdie 2014). Beyond the conditions necessary to allow for groups to trust that the time and effort placed in developing networks will not be infringed upon, humanitarian actors also need to feel that they have the opportunity to develop these networks safely. Humanitarian state actors can provide the opportunity for NGOs to collaborate by providing strong leadership and coordination resources. Furthermore, the presence of intergovernmental military interveners can help ensure a safer environment for humanitarian actors to operate and collaborate with one another (Murdie 2013). NGOs can also benefit from their logistical support technologies and expertise (Bell et al. 2019). Assistance in networking among NGOs can come from direct actions by military interveners, for example when they set up community meetings bringing together various NGOs and local communities. Collaboration efforts by NGOs can also be indirectly assisted by military interveners through their work paving roads or setting up phone lines, among other infrastructure projects (Bell et al. 2019). Beyond collaborating among themselves, it is imperative to "connect the local to the global through the national and subnational" by enhancing the prospect that international humanitarian and military actors can dialogue with locals, ultimately fostering a more receptive climate in which to operate (D. Roberts 2012, 372). This can occur, for example, with the assistance of CSOs and the distribution of communication devices (laptops and mobile phones) and the appropriate networks to transmit information by working with internet service providers and enhancing 3G network technologies (D. Roberts 2012).

Collaborating with others is often costly; thus organizational capacity is a key factor in the development of cooperative relationships between humanitarian actors. It is for this reason that more powerful actors have dominated interactions and cross-level connections with other groups during humanitarian crises (Gallemore et al. 2015). High levels of organizational capacity can relate to robust finances and access to donors, expertise and experience dealing with humanitarian issues, and numbers of available personnel and volunteers. These factors also assist with levels of success of the humanitarian projects they undertake. Organizations with high organizational capacity are more likely to network and, by extension, more likely to be networked with, resulting in a situation where the majority of NGO collaborative relationships are dominated by a few large NGOs (Murdie and Davis 2012). Due to this, firms with enhanced levels of organizational capacity are more likely to have an outsized presence on the ground dealing with humanitarian issues as well as benefit from a positive reputation in the industry. Having a positive reputation enhances the chances that other firms will trust your ability to perform and make them more likely to be receptive to working with you.

The prerequisite organizational capacity seems to be fundamental in allowing NGOs to network, and the literature does suggest that "the capacity to network transnationally can be an important determinant of NGOs' capacity to develop and implement projects" (Gallemore and Jespersen 2016, 89). In a sense, the organizational capacity needed for effective collaboration is also indicative of an ability to accomplish humanitarian goals. Beyond merely networking with one another, certain actors can even assist in coordinating efforts between other actors. However, it is often only the most powerful of actors that have the capability and reputation to coordinate the efforts of others (Gallemore et al. 2015). One caveat is that the majority of these large NGOs are located in the Global North. Connections between powerful and well-apportioned Northern NGOs and Southern NGOs are often unbalanced, with Southern NGOs often becoming dependent on the larger firms with more organizational capacity (Murdie and Davis 2012). This can result in mission creep and gatekeeping, where powerful organizations from the Global North end up having a disproportionate voice in project implementation and advocacy agendas (Carpenter 2007). Further, given that funding may come from powerful Northern states, this can create a situation where powerful state foreign policy interests are driving programmatic decisions.

Experts believe that population growth and climate change have the potential to increase the number of natural disasters. Combine that with vulnerable populations in low-income states, and the potential devastation of natural disasters is increasing as well (Moshtari and Gonçalves 2017; Van Wassenhove 2006). These same factors can also be responsible for increased devastation when dealing with human-made disasters like conflict. Beyond that, the complexity of how to best respond is ever-increasing as well. Collaboration among different humanitarian actors, despite the costs associated with it, can assist with overcoming some of these challenges. As crises requiring humanitarian aid and involvement increase in number and complexity across the world, the incentives to cooperate often outweigh the challenges of collaboration (Moshtari and Conçalves 2017). And the necessity of collaboration is becoming paramount.

Scholars and practitioners have begun to tackle these issues by categorizing factors that either facilitate or impede cooperation. One attempt at a categorization of factors influencing collaboration among NGOs in the face of disaster relief settled on three categories: contextual factors, interorganizational factors, and inner-organizational factors (Moshtari and Conçalves 2017). Contextual factors look at the characteristics of the setting of the humanitarian crisis that can either facilitate or impede collaboration. Often these are out of the NGOs' control. Interorganizational factors are those that take into account other humanitarian actors and their ability and capacity to coordinate with others. Finally, inner-organizational factors are issues inherent in an individual organization that impact its ability to collaborate with others, such as their available resources or individual capabilities (Moshtari and Conçalves 2017). While this framework is geared toward collaboration in the context of natural disasters, the underlying logic of these conceptualizations can prove useful toward creating networks of peace that can assist in the response to any type of humanitarian issue. Having looked at some factors that may facilitate the collaboration between humanitarian actors, we now turn our attention to what these networks may be able to accomplish.

What Can Networks for Peace Accomplish?

There is an abundance of scholarship that explores the role individual actors have on the prospects for achieving sustainable peace and building stable states. Despite being sent to the most difficult cases, PKMs have been shown to have a positive impact on durable peace (Fortna 2004). However, despite increased operational latitude and material capabilities, the peaceful outcome in war zones following peace operations seems to be more in tune with negative peace (Richmond 2004). Relating to violence against civilians, PKMs may not have much of an impact on government violence toward civilians, and they may even increase rebel violence toward civilians (Hultman 2010). However, UN PKMs, with a mandate to protect civilians, do reduce rebel violence towards civilian (Hultman 2010). UN PKMs, with their multilateral nature in combination with a mandate to protect civilian populations, have been shown to reduce overall levels of violence against civilians (Doyle and Sambanis 2000; Hultman, Kathman, and Shannon 2013). They have also been shown to contribute to institutional and political reform and overall democratization (Doyle and Sambanis 2000).

Despite the wealth of knowledge relating to the effects of individual actors, there remains limited scholarship that examines the joint impact of actors working together on humanitarian issues. Previous studies have identified several potential mechanisms that could lead to improved performance for groups of humanitarian actors working in tandem. First, multiple actors working alongside each other can act as "force multipliers," enhancing the positive impact of their partners. Second, actors can share information and learn from each other's previous experiences. Third, military interveners and larger governmental organizations like the UN have the expertise and organizational capacity to act as central coordinators for joint action. Fourth, UN PKMs and military interveners can, directly or indirectly, provide security for other NGOs to operate safely when administering humanitarian aid. Finally, elements of a "hybrid peace" can enhance long-term success of peace operations (D. Roberts 2011).

Collaborative networks with various humanitarian actors working alongside one another can lead to each partner acting as a force multiplier. In essence, the impact of each individual is amplified when working with others in a collaborative network. Often civilian and military interveners will bring different but complimentary skill sets to humanitarian crises. Military interveners can provide stability and security to areas, allowing NGOs greater access to vulnerable populations. Furthermore, military interveners are often well-funded, and some of that funding can go toward humanitarian projects. Military interveners are also adept at setting up communication networks, such as roads and phone lines, which NGOs can sometimes use to their advantage. Finally, military interventions often result in increased media attention, which NGOs may be able to harness (Bell et al. 2013). On the other hand, NGOs have their own set of unique skills with which to enhance this cooperative relationship. Especially those that have been working on local service provision will have more localized experience and expertise. They are less beholden to bureaucratic structures and may be more responsive with humanitarian aid in the face of rapid changes on the ground. And their general policy of neutrality can aid with their ability to reach actors that would be inaccessible to military interveners (Bell et al. 2013). This does not mean that combined networks of civilian and military actors will always be more successful in dealing with humanitarian crises. The nature of the crisis itself matters, as joint collaborations have been shown to be more effective when dealing with more complex humanitarian disasters relative to less complex ones (Bell et al. 2013).

Working together in collaborative networks also gives different organizations the opportunity to learn techniques and strategies from others. Essentially they could learn from each other's mistakes and previous experiences. Collaborative relationships in commercial industries have benefited all parties involved as they are able to assess each other's performance and provide valuable feedback (Fujimoto 2001). We believe that these same collaborative relationships can exist in the context of humanitarian networks as well. These networks may also be able to assist the state itself to provide humanitarian relief to its vulnerable populations. When operating in democracies, development NGOs have been shown to have a long-term positive impact on the host state's bureaucratic capacity (Campbell, DiGiuseppe, and Murdie 2019).

Collaborative networks are capable of reducing levels of violence. This includes violence originating from the state (Wilson, Davis, and Murdie 2016) and the potential for violence from rebel groups (Wood and Sullivan 2015). Conflict resolution networks are capable of fostering peace through the promotion of norms that encourage peaceful action, while also conveying information to relevant sources to discourage violence (Wilson, Davis, and Murdie 2016). One potential negative externality with humanitarian interventions is that aid itself may end up causing distress. Increased levels of aid in a conflict area may in fact encourage rebel violence as it provides an opportunity for looting and predation (Wood and Sullivan 2015). Additionally, humanitarian aid projects can be seen as delegitimizing the rebel group if they end up providing services and resources that the population may normally be dependent on the rebel group for, opening up those projects to attack by rebel forces (Wood and Sullivan 2015). However, when those dispensing the aid work collaboratively with military interveners, these risks may be mitigated (Wood and Sullivan 2015).

The confluence of various networked actors working together toward a common goal has led to a diverse nomenclature relating to the types of peace they generate. Whether it is called hybrid peace, popular peace, or other variation, they all tend to denote a working relationship between international and domestic actors (Richmond 2010; D. Roberts 2012). This process tends to denote a fusion of bottom-up (local) and top-down (international) approaches toward peacebuilding (D. Roberts 2011). Successful international interventions are to be moderated by local preferences and needs to assist in the creation of both formal and informal institutions geared toward establishing and maintaining long-term peace (D. Roberts 2011, 2012). A successful fusion of local and international actors, often designated as a form of hybrid peace, often necessitates a dynamic working relationship between local and international actors over the process, institutions, and goals of organizations working toward a sustainable peace (Richmond 2010).[2]

Despite the benefits provided by creating networks focused on generating sustainable peace, there are some potential drawbacks to collaboration between various organizations. As briefly mentioned earlier, NGOs often maintain neutrality, especially when in conflict zones. Collaborating with military interveners, while it does provide some benefits, may blur the lines of neutrality, potentially making the NGO a target of violence. Additionally, the actions of one NGO could potentially make other NGOs in the area targets of violence as well. For example, an NGO that engages in humanitarian work focused on policy changes that may threaten or challenge an armed group may find themselves and other NGOs at risk of attack (Murdie and Stapley 2014). Humanitarian crises that attract networks often also attract large sums of funds. Increased levels of donations may lead to the creation of predatory organizations with the goal of capturing that aid to enrich themselves at the expense of those that need it (Higham, Schulberg, and Rich 2014). Finally, it has been argued that certain states may use NGOs and humanitarian causes as tool to "whitewash" their own military objectives (Forte 2015).

One-sided collaborations, in which larger INGOs dominate the peacebuilding process, also have their drawbacks. For example, they may bring successful strategies from a previous mission with the belief that these can be successful again, but this has the potential to provide "the kind of peace it believes people should have, rather than the kind of peace people might seek themselves" (Roberts 2012, 367). There is risk involved in using a predetermined template rather than incorporating the needs and desires of the local population when designing a strategy to achieve sustainable long-term peace (Cubitt 2013).

Conclusion: What Questions Remain?

Given the relative infancy of academic scholarship relating to the role of humanitarian actors and human rights NGOs working collaboratively to encourage and assist with achieving sustainable peace and the development of stable states, it is not surprising that we are left with myriad questions that still need to be explored. The role of individual actors within these networks still needs to be examined. Humanitarian actors tend to operate all over the world, and the individuals working for them develop interpersonal relationships with others as they often find themselves interacting across multiple humanitarian crises. While useful for promoting networks of peace and collaboration, they sometimes lead to an isolated community of peacemakers that find themselves distant from the local population (Autesserre 2014). These boundaries sometimes prevent local input being taken into account when developing strategies to combat specific crises, thus resulting in less effective strategies (Autesserre 2014). This reinforces the argument that additional training and local knowledge would be beneficial (Autesserre 2014), in addition to arguments promoting training for proper and effective collaborative work (Balboa and Deloffre 2015). We should also look at the potential impact that the nationality and background of various actors have working in networks and on humanitarian crises (Stroup 2008). Finally, further work needs to be done assessing the role of gender on networks of peace and their approach to humanitarian issues. Studies have looked at where female peacekeepers get sent (Karim and Beardsley 2013), as well as the impact that peacekeepers in general may have on female populations (Beber et al. 2017).

While the situations are often difficult, the work done by humanitarian actors in alleviating the suffering of affected populations is paramount to improving the lives of people all over the world. Working toward making these groups more effective is certainly important, as unfortunately the need for networks of peace is unlikely to subside anytime soon. Hopefully this work can help with further exploration of the factors that create networks of peace and how they can best positively impact those in need.

Notes

1. More discussion can be found in the following sources: Cubitt (2013); Paris (2010); Richmond (2010); and Roberts (2012).

2. For more information relating to forms of hybrid peace and its benefits, please see Kumar and De la Haye (2012) and Mac Ginty (2010).

References

Autesserre, S. 2014. *Peaceland: Conflict Resolution and the Everyday Politics of International Intervention*. New York: Cambridge University Press.

Balboa, C. M., and M. Z. Deloffre. 2015. "Policymaking in the Global Context: Training Students to Build Effective Strategic Partnerships with Nongovernmental Organizations." *Journal of Public Affairs Education* 21 (3):417–434.

Beber, B., M. J. Gilligan, J. Guardado, and S. Karim. 2017. "Peacekeeping, Compliance with International Norms, and Transactional Sex in Monrovia, Liberia." *International Organization* 71 (1):1–30.

Bell, S. R., P. Blocksome, K. P. Brown, and A. Murdie. 2019. "Help or Hindrance? The Role of Humanitarian Interventions in Human Security NGO Operations." *International Political Science Review* 40 (2):263–278.

Bell, S. R., A. Murdie, P. Blocksome, and K. Brown. 2013. "'Force Multipliers': Conditional Effectiveness of Military and INGO Human Security Interventions." *Journal of Human Rights* 12 (4):397–422.

Berman, E., and D. D. Laitin. 2008. "Religion, Terrorism and Public Goods: Testing the Club Model." *Journal of Public Economics* 92 (10–11):1942–1967.

Blankenship, E. 2014. *Delivering Aid in Contested Spaces: Afghanistan*. Oxfam America Research Backgrounder. https://www.oxfamamerica.org/static/media/files/Aid_in_Contested_Spaces_Afghanistan_1.pdf.

Campbell, S., M. DiGiuseppe, and A. Murdie. 2019. "International Development NGOs and Bureaucratic Capacity: Facilitator or Destroyer?" *Political Research Quarterly* 72 (1):3–18.

Carpenter, R. C. 2007. "Setting the Advocacy Agenda: Theorizing Issue Emergence and Nonemergence in Transnational Advocacy Networks." *International Studies Quarterly* 51 (1):99–120.

Cooley, A., and J. Ron. 2002. "The NGO Scramble: Organizational Insecurity and the Political Economy of Transnational Action." *International Security* 27 (1):5–39.

Cubitt, C. 2013. "Responsible Reconstruction after War: Meeting Local Needs for Building Peace." *Review of International Studies* 39:91–112.

Dorussen, H., and H. Ward. 2008. "Intergovernmental Organizations and the Kantian Peace." *Journal of Conflict Resolution* 52 (2):189–212.

Doyle, M. W., and N. Sambanis. 2000. "International Peacebuilding: A Theoretical and Quantitative Analysis." *American Political Science Review* 94 (4):779–801.

Elkins, Z. 2009. "Constitutional Networks." In *Networked Politics: Agency, Power, and, Governance*, edited by Miles Kahler, 43–66. Ithaca, NY: Cornell University Press.

Flanigan, S. T. 2008. "Nonprofit Service Provision by Insurgent Organizations: The Cases of Hizballah and the Tamil Tigers." *Studies in Conflict & Terrorism* 31 (6):499–519.

Forte, M. C. 2015. "Force Multipliers: Imperial Instrumentalism in Theory and in Practice." *Force Multipliers: The Instrumentalities of Imperialism* 5:1–87.

Fortna, V. P. 2004. "Does Peacekeeping Keep Peace? International Intervention and the Duration of Peace after Civil War." *International Studies Quarterly* 48:269–292.

Fujimoto, T. 2001. "The Japanese Automobile Parts Supplier System: The Triplet of Effective Inter-firm Routines." *International Journal of Automotive Technology and Management* 1 (1):1–34.

Gallemore, C., M. Di Gregorio, M. Moeliono, H. Brockhaus, and R. D. Prasti. 2015. "Transaction Costs, Power, and Multi-level Forest Governance in Indonesia." *Ecological Economics* 114:168–179.

Gallemore, C., and K. Jespersen. 2016. "Transnational Markets for Sustainable Development Governance: The Case of REDD+." *World Development* 86:79–94.

Higham, S., J. Schulberg, J., and S. Rich. 2014. "Doing Well by Doing Good: The High Price of Working in War Zones." *Washington Post,* 4 May. https://www.washingtonpost.com/investigations/doing-well-by-doing-good-the-high-price-of-working-in-war-zones/2014/05/04/2d5f7ca8-c715-11e3-9f37-7ce307c56815_story.html.

Hultman, L. 2010. "Keeping Peace or Spurring Violence? Unintended Effects of Peace Operations on Violence against Civilians." *Civil Wars* 12 (1):29–46.

Hultman, L., J. Kathman, and M. Shannon. 2013. "United Nations Peacekeeping and Civilian Protection in Civil War." *American Journal of Political Science* 54 (7):875–891.

Karim, S., and K. Beardsley. 2013. "Female Peacekeepers and Gender Balancing: Token Gestures of Informed Policymaking?" *International Interactions* 39 (4):461–488.

Keck, M. E., and K. Sikkink. 1998. *Activists beyond Borders: Advocacy Networks in International Politics.* Ithaca, NY: Cornell University Press.

Kumar. C., and J. De la Haye. 2012. "Hybrid Peacemaking: Building National Infrastructures for Peace." *Global Governance* 18:13–20.

Mac Ginty, R. 2010. "Hybrid Peace: The Interaction between Top-Down and Bottom-Up Peace." *Security Dialogue* 41 (4):391–412.

Miklaucic, M. 2011. "Learning the Hard Way: Lessons from Complex Operations." *Joint Center for Operational Analysis Journal* 12 (3):32–39.

Mitchell, D. 2015. "Blurred Lines? Provincial Reconstruction Teams and NGO Insecurity in Afghanistan, 2010–2011." *Stability: International Journal of Security and Development* 4 (1):1–18.

Mitchell, D. F. 2017. "NGO Presence and Activity in Afghanistan, 2000–2014: A Provincial-Level Dataset." *Stability: International Journal of Security and Development* 6 (1):1–18.

Moshtari, M., and P. Gonçalves. 2017. "Factors Influencing Interorganizational Collaboration within a Disaster Relief Context." *International Society for Third-Sector Research* 28:1673–1694.

Murdie, A. 2014. "Scrambling for Contact: The Determinants of Inter-NGO Cooperation in non-Western Countries." *Review of International Organizations* 9:309–331.

Murdie, A., and D. Davis. 2012. "Looking in the Mirror: Comparing INGO Networks across Issue Areas." *Review of International Organizations* 7 (2):177–202.

Murdie, A., and C. S. Stapley. 2014. "Why Target the 'Good Guys'? The Determinants of Terrorism against NGOs." *International Interactions* 40:79–102.

Murdie, A., and S. Webeck. 2015. "Responding to the Call: Human Security INGOs and Countries with a History of Civil War." *International Political Science Review* 36 (1):3–19.

Olson, L. 2006. "Fighting for Humanitarian Space: NGOs in Afghanistan." *Journal of Military and Strategic Studies* 9 (1):1–28.

Paris, R. 2010. "Saving Liberal Peacebuilding." *Review of International Studies* 36:337–365.

Peck, E., and L. Kemmet. 2011. "Kansas National Guard Agribusiness Development in Afghanistan." *Interagency Journal* 2 (1):38–44.

Richmond, O. P. 2004. "UN Peace Operations and the Dilemmas of the Peacebuilding Consensus." *International Peacekeeping* 11 (1):83–101.

Richmond, O. P. 2007. "Emancipatory Forms of Human Security and Liberal Peacebuilding." *International Journal,* Summer: 459–477.

Richmond, O. P. (2010). "Resistance and the Post-Liberal Peace." *Journal of International Studies* 38 (3):665–692.

Roberts, D. 2011. "Beyond the Metropolis? Popular Peace and Post-Conflict Peacebuilding." *Review of International Studies* 37:2535–2556.

Roberts, D. 2012. "Saving Liberal Peacebuilding from Itself." *Peace Review: A Journal of Social Justice* 24:366–373.

Roberts, N. C. 2010. "Spanning 'Bleeding' Boundaries: Humanitarianism, NGOs, and the Civilian-Military Nexus in the Post–Cold War Era." *Public Administration Review* 70 (2):212–222.

Solomon, S. 2014. "Civil-Military Coordination in Complex Emergencies (Mali)" International Peace & Security Symposium 2014. https://www.mod.go.jp/js/jsc/jpc/english/event/proceedings/pro2014/sympo_4_e_1.pdf.

Spitka, T. 2018. "Mediating among Mediators: Building a Consensus in Multilateral Interventions." *International Negotiation* 23:125–154.

Stroup, S. 2008. *Varieties of Activism: The National Origins of International NGOs.* Berkeley: University of California Press.

United Nations Office for the Coordination of Humanitarian Affairs, Civil-Military Coordination Section. 2017. *Guide for the Military 2.0.* http://www.unocha.org/sites/unocha/files/Guide%20for%20the%20Military%20v2.pdf.

Van Wassenhove, L. N. 2006. "Humanitarian Aid Logistics: Supply Chain Management in High Gear." *Journal of the Operational Research Society* 57:475–489.

Visoka, G. 2015. "National NGOs." In *The Routledge Companion to Humanitarian Action*, edited by R. Mac Ginty and J. H. Peterson, 267–278. London: Routledge.

Voluntary Organizations in Cooperation in Emergencies. 2008. *NGO Seminar on Civil Military Relations*. December. https://ngovoice.org/publications/civmil-seminar-report-dec-07.pdf.

Wilson, M., D. R. Davis, and A. Murdie. 2016. "The View from the Bottom: Networks of Conflict Resolution Organizations and International Peace." *Journal of Peace Research* 53 (3):442–458.

Wood, R. M., and C. Sullivan. 2015. "Doing Harm by Doing Good? The Negative Externalities of Humanitarian Aid Provision during Civil Conflict." *Journal of Politics* 77 (3):736–748.

CHAPTER 25

PEACE, INTERVENTION, AND STATE FRAGILITY

NICOLAS LEMAY-HÉBERT

STATE fragility is intractably linked to statebuilding interventions, both in theory and in practice. State weakness is understood to be only "corrected" or addressed through statebuilding—even if both concepts are open to interpretation and their meanings can be debate and contested. How one defines the state will have ramifications on how one conceptualizes state fragility, statebuilding, and the peace that is supposed to follow. Through the modern—and quite mainstream—episteme, the state is understood as a provider of public goods, and the key public good in a Weberian fashion is security provision. Weber's (1948, 78) definition of the state "as a human community that successfully claims the monopoly of the legitimate use of physical force within a given territory" gained traction across various disciplines and has contributed to cement the peace-and-conflict disciplinary focus on security provision and state institutions. What came to be known as the institutional approach to statebuilding (Lemay-Hébert 2013) has also been challenged by other approaches, bundled under the wide umbrella of "social legitimacy" (Lemay-Hébert 2014b). However, the institutional approach has contributed to cement, and to a certain extent continues to drive, both the practice of statebuilding and the literature on state fragility.

This chapter explores the relationship between state fragility and statebuilding and peacebuilding, looking at the ramifications of this approach for the practice of international interventions and peace figuration (Visoka 2016). More precisely, it looks at how discourses of fragility shape international responses to conflicts and the policy solutions thereafter, understanding the specific epistemic community behind state fragility and statebuilding "as the sites of a constant struggle over how to define what qualifies as valid knowledge" (Lemay-Hébert and Mathieu 2014, 233). In this chapter, two different—yet compatible—discourses of state fragility and intervention are identified. The first one is based on a dichotomy between strong and weak states through state rankings and fragile states listings which legitimizes intervention in these weak states. The second one is an emerging discourse on risk and resilience, with more fluid understandings of fragility that go beyond the dichotomy of strong versus weak states, but also opens up new interventionary possibilities for international actors. The chapter will be taking stock of these discourses, emphasizing commonalities and divergences between them, as well as the repercussions for the peacebuilding and statebuilding practice and academic literature.

The Modern Moment: Failed States and Top-Down Interventions

The modern way of listing failed states gained traction in policymaking circles in the mid-1990s, in the midst of the collapse of Yugoslavia and Somalia, and especially after Gerald Helman and Steven Ratner's (1993, 5) seminal article, "Saving Failed States," where they define state failure as "a situation where governmental structures are overwhelmed by circumstances."[1] The authors make a loose distinction between degrees of collapse (failed, failing, and newly independent states) and advocate new conservatorships to deal with failed states—thus exemplifying the interconnectedness of the concept of "failed state" with the interventionist agenda at that time. The agenda was picked up in the following years by a group of "institutionalist," or "neo-Weberian," scholars (Lottholz and Lemay-Hébert 2016), including Robert Rotberg and Francis Fukuyama. Both equate "failed" states with deficit of state capacity and identify top-down international statebuilding interventions as a solution to this deficit of capacity. For Fukuyama (2006, 2), "the underlying problems caused by failed states or weak governance could only be solved through long-term efforts by outside powers to rebuild indigenous state institutions," while for Rotberg (2004, 31), "there is a great need for conscientious, well-crafted nation building—for a systematic refurbishing of the political, economic, and social fabric of countries that have crumbled."

The "statebuilding as capacity building" approach is also operationalized through the "more is better" approach, where "the more intrusive the intervention, the more successful the outcome" (Zuercher 2006, 2). State failure then becomes a convenient legitimizing tool for all forms of international intervention. The "more is better" approach is encompassed in the work of multiple noted academics. For Michael Doyle and Nicholas Sambanis (2006, 4), "the deeper the hostility, the more the destruction of local capacities, the more one needs international assistance to succeed in establishing a stable peace." The "more is better" approach is nowhere clearer than in the report produced by the Rand Corporation on US-led nation-building. For Dobbins, Gordon, and Martini (2003, 160), the lessons of the successful cases of Japan and Germany, and the less successful recent cases of Somalia, Haiti, Bosnia, Kosovo, and Afghanistan, can be boiled down to a simple equation: "the higher level of input accounts in significant measure for the higher level of output measured in the development of democratic institutions and economic growth."

The conflation of "statebuilding as capacity building" and "more is better" approaches also provides a useful legitimization platform for interventions taking the form of international administrations (Lemay-Hébert 2011, 2017). For Roland Paris (2003, 456), for instance, "the most effective means of establishing new governmental institutions in war-shattered states is to rebuild these institutions from scratch, to staff them with international personnel, and then to gradually replace these officials with adequately trained and politically non-partisan locals." The preference for top-down statebuilding approaches—of which international administrations are probably the most extreme examples—has been an integral part of the institutionalist agenda. Emblematizing the "more is better" approach, Helman and Ratner (1993, 7) plead the case for a "more systematic and intrusive approach" to statebuilding. The usual methods of strengthening governance in the developing world, such as providing aid and foreign support, will meet with scant success in failing states and will prove wholly inadequate in those that have collapsed. Thus the conceptual basis for the international effort should lie in the idea of conservatorship. Using a medical analogy, the implications of which have been analyzed in details in recent critical works (Hughes and Pupavac 2005; Lemay-Hébert 2013; Manjikian 2008; Pia 2013), Helman and Ratner (1993) portray the failed state either as an entity affected by a serious illness or as a helpless person who suffered a bad turn of fate. For them, "forms of guardianship or trusteeship are a common response to broken families, serious mental or physical illness, or economic destitution. . . . It is time that the United Nations consider such a response to the plight of failed states" (12). The UN's guardianship role will differ depending on the degree of collapse of governmental institutions. Where the state retains a governmental structure, the UN should act under "governmental assistance" schemes, designed as expansive technical assistance programs.

For the states that have already failed, a second, more intrusive form of conservatorship would be appropriate, in which the state could actually delegate certain governmental functions to the UN. Finally, Helman and Ratner (1993) envision the more radical option for full-fledged failed states: direct UN trusteeship. The intention of the authors is clear: "such a plan would resurrect the old trusteeship system and apply it to failed states" (16). Contrary to Helman and Ratner's recommendation, the "old trusteeship system" has not been reactivated, but ad hoc international administrations saw the light of day in the 1990s and early 2000s, such as in Bosnia and Herzegovina, Kosovo and Timor Leste. Furthermore, recurring calls for the establishment of international administrations have been made following conflicts or disasters, such as in the context of postearthquake Haiti (Lemay-Hébert 2014a), South Sudan (Crocker 2016, 199), and Syria (Dobbins, Gordon, and Martini 2015).

While there is no single consensual definition of state fragility, the UK Department for International Development's (DFID 2005) institutionalist definition of fragile states as "countries where the government cannot or will not deliver core state functions, such as providing security and justice across its territory and basic services to the majority of its people," is considered relatively consensual in the field. Different scholars and organizations have come up with different ways to compile data on fragility and list fragile states over the years; however, despite the differences, there is a consensus on a number of fragile states. I have compiled the latest information available on state fragility indexes (see Table 25.1), bringing together a mix of the main bilateral and multilateral organizations working on state fragility (DFID, the Organization for Economic Cooperation and Development [OECD], and the World Bank) as well as influential scholars (University of Carleton and David Carment's *Country Indicator for Foreign Policy*; and Monty Marshall's Center for Systematic Peace) and think tanks (Fund for Peace). The "consensual" fragile states include Afghanistan, Burundi, Central African Republic, Chad, the Democratic Republic of the Congo, Somalia, South Sudan, and Sudan. Other nearly consensual states—included in all but one of the fragile states lists compiled for this exercise—are Eritrea, Iraq, Mali, Syria, and Yemen. These states are all "hosting" or the prime candidates for hosting international interventions in one form or another—from interventionary coalitions to multilateral peacebuilding.

As mentioned, the concept of "failed" or "fragile" state sits at the center of the top-down, institutionalist logic of intervention. It captures what needs transformation through intervention, while projecting what needs to be achieved: the final goal of statebuilding efforts or the template that needs to be reproduced for peace to take hold. This is the mythical "road to Denmark" for Fukuyama. The linear logic of sociopolitical evolution, with countries arranged on a spectrum of weakness, was (and still is to a certain extent) the hallmark of the modernistic logic of statebuilding. If the institutionalist logic of intervention has informed a large number of interventions—from Kosovo and Timor-Leste in 1999 to Iraq in 2003—the critique of this approach has increasingly been recognized, accepted, and internalized by policy actors, henceforth moving the debates to altogether different grounds, as I will discuss in the "States of Fragility" section.

Peacebuilding, State Fragility, and Interventions

As with state fragility, there is no universally accepted definition of peacebuilding. The general ambition of peacebuilding is "to identify and support structures that will tend to strengthen and solidify peace in order to avoid a relapse into conflict" (UN 1992, para. 46). It connotes activities that go beyond crisis intervention, such as longer-term development and building of governance structures and institutions. In the 2000 Brahimi Report, the United Nations provided a similar definition, describing "activities undertaken on the far side of conflict to reassemble the foundations of peace and provide the tools for building on those foundations something that is more than just the absence of war" (UN 2000, para. 13). Hence peacebuilding aims "to address the sources of current hostility and build local capacities for conflict resolution. Stronger state institutions, broader political participation, land reform, a deepening of civil society, and respect for ethnic identities are all seen as ways to improve the prospects for peaceful governance" (Doyle and Sambanis 2000, 779).

Table 25.1 Listing Fragile States

Countries	World Bank (2017)	State Fragility Index (Marshall and Elzinga-Marshall 2017)	Fund for Peace (2018)	DFID (2017)	OECD (2018)	Country Indicator for Foreign Policy (2017)
Afghanistan	X	X	X	X	X	X
Angola		X				
Bangladesh			X			
Burkina Faso		X				X
Burundi	X	X	X	X	X	X
Cameroon		X	X			X
Central African Republic	X	X	X	X	X	X
Chad	X	X	X	X	X	X
Comoros	X					
Congo, Democratic Republic	X	X	X	X	X	X
Congo, Rep.	X	X				
Côte d'Ivoire	X	X	X			
Djibouti	X					
Eritrea	X		X	X	X	X
Ethiopia		X	X		X	X
Gambia	X					X
Guinea		X	X			X
Guinea Bissau	X	X	X			X
Haiti	X		X		X	X
Iraq	X	X	X	X	X	
Kenya			X			
Kiribati	X					
Kosovo	X					
Lebanon	X					
Liberia	X		X			X
Libya	X		X	X		
Mali	X	X	X		X	X
Marshall Is.	X					
Mauritania		X	X			
Micronesia	X					

Table 25.1 Continued

Countries	World Bank (2017)	State Fragility Index (Marshall and Elzinga-Marshall 2017)	Fund for Peace (2018)	DFID (2017)	OECD (2018)	Country Indicator for Foreign Policy (2017)
Mozambique	X					X
Myanmar	X	X	X	X		
Niger		X	X			X
Nigeria		X	X	X		X
North Korea			X	X		
Pakistan		X	X	X		X
Papua New Guinea	X					
Rwanda		X				
Sierra Leone	X					
Solomon Islands	X					
Somalia	X	X	X	X	X	X
South Sudan	X	X	X	X	X	X
Sudan	X	X	X	X	X	X
Syria	X		X	X	X	X
Togo	X					
Tuvalu	X					
Uganda		X	X			X
Ukraine				X		
Venezuela				X		
West Bank and Gaza	X					
Yemen	X	X	X		X	X
Zimbabwe	X	X	X	X		

Not unlike statebuilding, which can be understood as wider processes of state formation or simply as interventions to build or reinforce state institutions, peacebuilding can be understood either as grassroots initiatives focusing on local peace capabilities—circumventing traditional top-down interventionary processes—or, quite literally, as international interventions to restore or build peace. The first understanding does not connect directly with programmatic policymaking as it is considered a more diffuse societal process, similar to the broader understanding of statebuilding or state formation. The latter understanding of peacebuilding is more in line with institutional statebuilding, focusing on peace interventions (or interventions in the name of peace) as a way to (re)structure societies deemed turbulent or abnormal (Lemay-Hébert and Visoka 2017).

With these two understandings of peacebuilding in mind, one could argue that top-down peace interventions can contribute to undermine peacebuilding efforts on the ground. As Ekaterina Stepanova (2008, 71) asserted in the *SIPRI Yearbook* more than ten years ago, state weakness is among the most critical factors stimulating armed violence, with outside intervention, paradoxically. The same assessment can be made today, with major interventions claiming to restore peace actually reinforcing state fragility, which brings the Stockholm International Peace Research Institute (SIPRI) to claim that the relationship between "the statebuilding paradigm and the peacebuilding paradigm is likely to be the focus of attention for the international community" (Anthony 2015, 4). One way out of this conundrum, and one that is advocated by SIPRI, is to challenge the concept of state fragility itself, moving the conversation closer to the drivers of peace and conflict on the ground.

States of Fragility: New Interventionism in the Age of Resilience and Adaptation

The global discourse on state fragility is now moving away from identifying and listing "fragile states" to focusing on "situations of fragility," "fragile contexts" (UNDP et al. 2016, 2), "fragile situations" (World Bank 2017), or "states of fragility" (OECD 2016b). Helen Clark, administrator of UNDP, notes that "building resilience is now central to the way in which the UN is responding to fragility," which intends to broaden the scope of analysis in addressing fragility to include subnational and global levels at the same time (UNDP et al. 2016, 2; Pospisil and Kuhn 2016). Behind this change of approach, there is recognition that past approaches to state failure and state fragility simply backfired. As the executive boards of various UN agencies and programs state, "the emerging concept [fragile contexts] addresses the stigma associated with 'fragile states' " (UNDP et al. 2016, 3). This is echoed by the OECD's (2016a) noting that "there is an increasing awareness that a binary list does not do justice to the multi-faceted nature of fragility" and committing to moving toward a universal framework of fragility, phasing out its "fragile states" list, dropping the "fragile states" label (as used in its *Fragile States Reports* from 2005 until 2014), and including a multidimensional framework around "states of fragility" (*States of Fragility Reports* since 2014). In this new approach, fragility is defined as the "combination of exposure to risk and insufficient coping capacity of the state, system and/or communities to manage, absorb or mitigate those risks" (OECD 2016a, 22). Similarly, the World Bank moved from a classification of Low Income Countries under Stress List (2006–2009), to the Fragile States List (2010) to the now harmonized List of Fragile Situations (2011–2015). This semantic change "reflect[s] consultations with clients and partners" (World Bank 2016, 3).

This shift from the conventional understanding of fragile states to "states of fragility" and other affiliated concepts has been driven by five distinct debates in policy circles (Lemay-Hébert 2019) which have shaken the foundations of the modern episteme of state fragility. The first debate is whether state fragility should be equated with underdevelopment. State fragility has most of the time been considered in a direct correlation with poverty in the modernist framework, with poor state performance and low levels of national income becoming proxies for state fragility. This relies on the idea that fragility is a measure of the extent to which the actual practices and capacities of states differ from their idealized image. This idealized image is most often not reflecting life within the OECD area. Most typologies in the literature and data sets on fragile states (listed in Table 25.1) reveal a normative orientation toward the Western model. Even certain scholars who try to come up with alternative frameworks of state fragility end up reinforcing the same bias, with a dichotomy following roughly the lines of the inclusion/exclusion of the OECD area (see, for instance, Krasner and Risse 2014, 548–550). The OECD in its 2016 draft working paper clearly reacted to this issue, considering that the future OECD model "should not take income levels as an input, nor make a distinction between 'developing' or other countries. It should be universal. It pits so-called fragile states against non-fragile ones." The World Bank (2016, 4) also noted that "definitions built on the CPIA [Country Policy and Institutional Assessment] and peacekeeping missions can poorly account [for] contexts such as fragilities in middle-income countries, and spatial dynamics. The current direction of analysis underpinning the fragility list aims to include more multi-dimensional approaches to identifying fragile situations."

The second debate is whether state fragility can be equated with a lack of capacity. Understanding state fragility or failure through state capacity (Hameiri 2007; Lemay-Hébert 2013) is increasingly considered outdated and unhelpful, notably because it excludes crucial nonstate actors in service provision and how they can perform functions associated with the state in the Western world (Hagmann and Hoehne 2009; Batley and Mcloughlin 2010). The new discussions on "fragile contexts" tend to recognize that fact, starting from the premise that state strength is more than a simple addition of available infrastructure or a list of services provided by the state.

The third debate is whether one can forecast state failure. To a certain extent this debate is linked to the definition of state fragility one uses. Conceiving state fragility and state failure as a breakdown of government institutions allows one to identify failed or failing states according to "institutional" strength. As the OECD (2016b) indicates in a recent draft working paper, there are no established theories of change that would be the same across different contexts—that's the very essence of "starting with the context" (for a similar discussion, see also African Development Bank 2015, 7). In this context, the OECD describes its new approach as being "primarily descriptive, while having some predictive value." Hence, as Stepputat and Engberg-Pedersen (2008, 28) write, "there is no shortcut circumventing a detailed analysis of the historical evolution and specific characteristics of individual situations," a discussion that reflects the current "culture turn" in peacebuilding and resilience thinking (see Bargués-Pedreny 2015). Following this new discussion, intervening countries should be "realistic about their limited ability to understand, much less to shape, local patterns of legitimacy" (OECD 2010, 13). This is also a discussion that is informed by the failure to predict the Arab Spring.[2]

The fourth debate concerns the modernistic tendency to equate state fragility with the agreed international borders of states. Most organizations now recognize that state fragility affects specific subregions, and when it affects a whole country, state fragility will most of the time have ramifications for neighboring countries. To take this discussion into account, DFID has developed a new category of "neighbouring 'high fragility' states" to include countries at risk of the effects of spillover from fragility in neighboring states.[3] Similarly, the G7+, the African Development Bank, and the OECD are now avoiding the term "fragile states" in favor of "fragile situations" or "fragile contexts" to account for the fact that "fragility does not respect state boundaries," in the words of the African Development Bank (2015, 7). According to the OECD, the concept of fragility should endeavor to connect with analyses that reach beyond the national level and look at various layers of society (from the state to provinces, cities, communities, and even households).

The final debate regards the methodology used in most of the state fragility indicators. Most methodologies are obscure and rely on experts' guesswork (Nay 2013); the most blatant cases are probably the World Bank's CPIA or the Fund for Peace's Indicator of State Fragility. The World Bank's CPIA measuring involves professional calibrating rates in terms of how they should "probably" compare to numbers in similar countries (Rocha de Siqueira 2017 174). In this context, the OECD's new framework for fragility aims to use a transparent methodology and be replicable, and all underlying data will be published. Recognizing the subjective nature of the data analysis on state fragility, the G7+ has pioneered a Fragility Assessment, with the aim of using its own definition of the phenomenon and assessing it itself (Rocha de Siqueira 2014).

Conclusion

These waves of criticisms, debating the value of the premises behind the modernistic state fragility agenda, have led many organizations to reframe the concept of fragility around risks and vulnerabilities. Starting in 2008, the OECD (2008; see also G7+ 2013, 1) conceptualized resilient states as the polar opposite of fragile states—thus playing the role of the "strong" state category in conventional analyses of state fragility. As I indicated, the OECD (2015) now vies to move away from the traditional binary categorization of "fragile states," focusing instead on the diversity of risks and vulnerabilities that lead to fragility. What has been achieved by this reframing of the state fragility agenda? I argue that this new framing has enabled the epistemic community to restructure the liberal internationalism agenda around what is arguably its core feature in the specific context of fragile states: security and securitization. Just as pruning the weaker branches of a tree allows it to grow higher, this has enabled the fragile states agenda to solidify through the integration and co-option of its critique. Whereas the modernistic approach to intervention in the fragile states context was permeated with language about normalizing and transforming societies into "liberal" ones based on the image of intervenors (Lemay-Hébert and Visoka 2017), the new set of practices has emancipated itself from its contentious, normative components, focusing mainly on the security dimension of state fragility. This discursive move represents a retreat from the past position made of unequivocal certainties, but at the same time represents willingness by international actors to regroup and re-engage the terrain in a stronger position.

As I have underlined in this chapter, there is a lot of continuity in the fragile states agenda before and after the "resilience turn." However, the new interventionary framework for responding to fragile situations does raise a number of pressing issues for the research and policy community. In this new framework, resilience, fragility, and crisis are considered moving points along a spectrum of situations including all states (OECD 2016a), thus leaving no state behind and no area of life untouched. Hence I agree with Brad Evans and Julian Reid (2014, 21) that "when everything is the source of endangerment, everyone is endangered." The retreat from one governmentality mechanism—imposing order on societies deemed abnormal or "failed" (Lemay-Hébert and Visoka 2017)—enables the deployment of another form of governmentality, which potentially goes deeper and wider than the previous one.

The repercussions of this discussion for peacebuilding and peace formation are clear. Echoing the distinction between peacebuilding as peace interventions or as local peace capabilities, peacebuilding interventions can be seen as either disruptive, reinforcing governmentality mechanisms of the "other" deemed unruly and turbulent, or as an untapped potential for sustainable solutions on the ground, following in that sense the resilience turn in the state fragility agenda. Even in the first sense of the concept—peacebuilding as peace interventions—there is clearly recognition of the rising complexity and nonlinearity of peace processes, something that is captured well in the peace figuration discussion (Visoka 2016). In a sense, the reframing of the fragility agenda around resilience reinforces the urgency to find new ways to analyze local peace capabilities that tend to be obscured by more high-profile, top-down peace interventions.

Notes

1. While there has been a conceptual proliferation to describe the same basic phenomenon (Grimm, Lemay-Hébert, and Nay 2014), two main concepts emerged. The concept of fragile states was generally preferred by development experts, while the concept of failed states was traditionally favored by security experts and diplomats. The semantics of fragility has appeared to be more consensual in the past few years, mainly because of the negative connotation behind the "failed state" concept.
2. Economic indicators have failed to predict the Arab uprisings (see World Bank 2015), and the Fund for Peace had only one Arab country (Yemen, ranked thirteenth) as failed in 2010. Egypt and Syria barely fell into the top fifty at that time, and their scores were similar to rather more stable, semidemocratic countries like Colombia and the Philippines. Three

Arab countries that experienced a great deal of conflict and violence in 2011—Tunisia, Libya, and Bahrain—actually fell into the bottom half of the index, that is, among the allegedly stronger and more successful states (Goodwin 2011, 452).

3. This category includes ten countries: Algeria, Armenia, Jordan, Laos, Rwanda, Tanzania, Thailand, Tunisia, Turkey, and Zambia (see UK Parliament n.d.).

References

African Development Bank. 2015. *Addressing Fragility and Building Resilience in Africa*. AfDB Group Strategy 2014–2019. https://www.afdb.org/fileadmin/uploads/afdb/Documents/Policy-Documents/Addressing_Fragility_and_Building_Resilience_in_Africa-_The_AfDB_Group_Strategy_2014%E2%80%932019.pdf

Anthony, I. 2015. "Introduction: International Security, Armaments and Disarmament." In *SIPRI Yearbook 2015*. Edited by SIPRI, 3–13. Oxford: Oxford University Press.

Bargués-Pedreny, P. 2015. "Realising the Post-Modern Dream: Strengthening Post-Conflict Resilience and the Promise of Peace." *Resilience: International Policies, Practices and Discourses* 3 (2):113–132.

Batley, R., and C. Mcloughlin. 2010. "Engagement with Non-State Service Providers in Fragile States: Reconciling State-Building and Service Delivery." *Development Policy Review* 28 (2):131–154.

Carment, S., P. Tikuisis, Y. Samy, and J. Floch. 2017. *The CIFP Fragility Index: New Trends and Categorizations*. Carleton University, December. https://carleton.ca/cifp/wp-content/uploads/CIFP-2017-Fragility-Report.pdf.

Crocker, C. A. 2016. "A Transitional International Administration for South Sudan." *American Interest*, 9 September. http://www.the-american-interest.com/2016/09/09/a-transitional-international-administration-for-south-sudan/.

DFID. 2005. *Reducing Poverty by Tackling Social Exclusion: A DFID Policy Paper*. Department for International Development, September. http://www2.ohchr.org/english/issues/development/docs/socialexclusion.pdf.

DFID. 2017. *Methodology Note – FACS*. https://assets.publishing.service.gov.uk/government/uploads/system/uploads/attachment_data/file/624860/Methodology-Note-FACS.pdf

Dobbins, J., P. Gordon, and J. Martini. 2015. "A Peace Plan for Syria." *RAND Perspective*, Doc. PE-182-RC.

Doyle, M., and N. Sambanis. 2000. "International Peacebuilding: A Theoretical and Quantitative Analysis." *American Political Science Review* 94 (4):779–801.

Doyle, M., and N. Sambanis. 2006. *Making War and Building Peace: United Nations Peace Operations*. Princeton, NJ: Princeton University Press.

Evans, B. and J. Reid. 2014. *Resilient Life: The Art of Living Dangerously*. London: Polity.

Fukuyama, F. 2006. "Nation-Building and the Failure of Institutional Memory", in *Nation-Building: Beyond Afghanistan and Iraq*, edited by F. Fukuyama, 1-16. Washington: Johns Hopkins University Press.

Fund for Peace. 2017. *Fragile States Index 2018—Annual Report*. 24 April. http://fundforpeace.org/fsi/2018/04/24/fragile-states-index-2018-annual-report/.

G7+. 2013. Note on the G7+Fragility Spectrum. https://www.pbsbdialogue.org/media/filer_public/17/43/17434d29-eb70-425b-8367-7ccef13bfb0b/g7fragility_spectrum_2013.pdf

Goodwin, J. 2011. "Why We Were Surprised (Again) by the Arab Spring." *Swiss Political Science Review* 17 (4):452–456.

Grimm, S., N. Lemay-Hébert, and O. Nay. 2014. "Fragile States: Introducing a Political Concept." *Third World Quarterly* 35 (2):197–209.

Hagmann, T., and M. Hoehne. 2009. "Failures of the State Failure Debate: Evidence from the Somali Territories." *Journal of International Development* 21 (1):42–57.

Hameiri, S. 2007. "Failed States or a Failed Paradigm? State Capacity and the Limits of Institutionalism." *Journal of International Relations and Development* 10 (2):122–149.

Helman, G., and S. Ratner. 1993. "Saving Failed States." *Foreign Policy* 89 (4):3–20.

Hughes, C., and V. Pupavac. 2005. "Framing Post-conflict Societies: International Pathologisation of Cambodia and the Post-Yugoslav States." *Third World Quarterly* 26 (6):873–889.

Krasner, S., and T. Risse. 2014. "External Actors, State-Building, and Service Provision in Areas of Limited Statehood: Introduction." *Governance* 27 (4):545–567.

Lemay-Hébert, N. 2011. "The 'Empty Shell' Approach: The Setup Process of International Administrations in Timor-Leste and Kosovo, Its Consequences and Lessons." *International Studies Perspectives* 12 (2):188–209.

Lemay-Hébert, N. 2013. "Rethinking Weberian Approaches to Statebuilding." In *Routledge Handbook of International Statebuilding*, edited by D. Chandler and T. Sisk, 3–14. London: Routledge.

Lemay-Hébert, N. 2014a. "Resistance in the Time of Cholera: The Limits of Stabilization through Securitization in Haiti." *International Peacekeeping* 21 (2):198–213.

Lemay-Hébert, N. 2014b. "The Semantics of Statebuilding and Nationbuilding: Looking beyond Neo-Weberian Approaches." In *Semantics of Statebuilding: Language, Meanings and Sovereignty*, edited by N. Lemay-Hébert, N. Onuf, V. Rakic, and P. Bojanic, 89–105. London: Routledge.

Lemay-Hébert, N. 2017. "Exploring the Effective Authority of International Administrations from the League of Nations to the United Nations." *Journal of Intervention and Statebuilding* 11 (4):468–489.

Lemay-Hébert, N. 2019. "From Saving Failed States to Managing Risks: Reinterpreting Fragility as Resilience." In *Governance and Political Adaptation in Fragile States*, edited by J. Lahai, H. Brasted, K. von Strokirch, and H. Ware, 75–101. Basingstoke, UK: Palgrave.

Lemay-Hébert, N., and X. Mathieu. 2014. "The OECD's Discourse on Fragile States: Expertise and the Normalisation of Knowledge Production." *Third World Quarterly* 35 (2):232–251.

Lemay-Hébert, N., and G. Visoka. 2017. "Normal Peace: A New Strategic Narrative of Intervention." *Politics and Governance* 5 (3):146–156.

Lottholz, P., and N. Lemay-Hébert. 2016. "Re-reading Weber, Redefining State-Building? From Neo-Weberian to Post-Weberian Approaches to State, Legitimacy, and State-Building." *Cambridge Review of International Affairs* 29 (4):1467–1485.

Manjikian, M. 2008. "Diagnosis, Intervention, and Cure: The Illness Narrative in the Discourse of the Failed State." *Alternatives* 33:335–357.

Marshall, M., and G. Elzinga-Marshall. 2017. *Global Report 2017: Conflict, Governance and State Fragility*. Center for Systemic Peace. http://www.systemicpeace.org/vlibrary/GlobalReport2017.pdf.

Nay, O. 2013. "Fragile and Failed States: Critical Perspectives on Conceptual Hybrids." *International Political Science Review* 34 (3):326–341.

OECD. 2008. *Concepts and Dilemmas of State Building in Fragile Situations: From Fragility to Resilience*. https://www.oecd-ilibrary.org/docserver/journal_dev-v9-art27-en.pdf?expires=1605505570&id=id&accname=ocid194734&checksum=84263FADE91A66373A4BDA8B2A8A5FF2

OECD. 2010. *The State's Legitimacy in Fragile Situations: Unpacking Complexity*. https://www.oecd-ilibrary.org/docserver/9789264083882-en.pdf?expires=1605505595&id=id&accname=ocid194734&checksum=AD4F8D4993902C57E5C7525D35AA1455

OECD. 2015. *States of Fragility 2015: Meeting Post-2015 Ambitions*. https://www.oecd-ilibrary.org/docserver/9789264227699-en.pdf?expires=1605505620&id=id&accname=guest&checksum=D4C9A51FA99939699FD47D277457A8DE

OECD. 2016a. *States of Fragility 2016: Understanding Violence*. https://www.oecd-ilibrary.org/docserver/9789264267213-en.pdf?expires=1605505654&id=id&accname=ocid194734&checksum=ED218E226ECA7D10433FECDAA92B5AC7

OECD. 2016b. "Towards a Multidimensional Concept of Fragility." Draft document. Paris: OECD.

OECD. 2018. *States of Fragility 2018*. Paris: OECD. https://read.oecd-ilibrary.org/development/states-of-fragility-2018_9789264302075-en.

Paris, R. 2003. "Peacekeeping and the Constraints of Global Culture." *European Journal of International Relations* 9 (3):441–473.

Pia, E. 2013. "Narrative Therapy and Peacebuilding." *Journal of Intervention and Statebuilding* 7 (4):476–491.

Pospisil, J., and F. Kühn. 2016. "The Resilient State: New Regulatory Modes in International Approaches to State Building?" *Third World Quarterly* 37 (1):1–16.

Rocha de Siqueira, I. 2014. "Measuring and Managing 'State Fragility': The Production of Statistics by the World Bank, Timor-Leste and the G7+." *Third World Quarterly* 35 (2):268–283.

Rocha de Siqueira, I. 2017. "Development by Trial and Error: The Authority of Good Enough Numbers." *International Political Sociology* 11 (2):166–184.

Rotberg, R. 2004. "The Failure and Collapse of Nation-States: Breakdown, Prevention, and Repair." In *When States Fail: Causes and Consequences*, edited by R. Rotberg, 1–49 Princeton, NJ: Princeton University Press.

Stepanova, E. 2008. "Trends in Armed Conflicts." In *SIPRI Yearbook 2008: Armaments, Disarmament and International Security*, edited by SIPRI. 43–71. Oxford: Oxford University Press.

Stepputat, F. and L. Engberg-Pedersen. 2008. "Fragile States: Definitions, Measurements and Processes" in *Fragile Situations: Background Papers*, edited by L. Engberg-Pedersen, L. Andersen, F. Stepputat and D. Jung, 21-31. Copenhagen: DIIS.

UK Parliament. n.d. "International Development Committee Inquiry: Allocation of Resources." http://data.parliament.uk/writtenevidence/committeeevidence.svc/evidencedocument/international-development-committee/dfids-allocation-of-resources/written/28276.pdf.

UN. 1992. *An Agenda for Peace: Preventive Diplomacy, Peacemaking and Peace-keeping*. UN Doc. A/47/277-S/24/111, 17 June.

UN. 2000. *Report of the Panel on United Nations Peace Operations*. UN Doc. A/55/305-S/2000/809, 21 August.

UNDP, UNFPA, UNOPS, UNICEF, UN-Women and WFP . 2016. "Working in Fragile Contexts, Including in Middle-Income Countries." Joint Meeting of the Executive Boards of

UNDP/UNFPA/UNOPS, UNICEF, UN-Women and WFP, 3 June. https://sites.unicef.org/about/execboard/files/Topic_2_-_background_paper.pdf
Visoka, G. 2016. *Peace Figuration after International Intervention: Intentions, Events and Consequences of Liberal Peacebuilding*. London: Routledge.
Weber, M. 1948. "Politics as a Vocation." In From Max Weber: Essays in Sociology, edited and translated by H. H. Gerth and C. W. Mills, 77–128. New York: Oxford University Press.
World Bank. 2015. "Economic Indicators Failed to Predict Arab Uprisings." 21 October. http://www.worldbank.org/en/news/press-release/2015/10/21/economic-indicators-failed-to-predict-arab-uprisings.
World Bank. 2016. "Information Note: The World Bank Group's Harmonized List of Fragile Situations." http://pubdocs.worldbank.org/en/586581437416356109/FCS-List-FY16-Information-Note.pdf.
World Bank. 2017. "Harmonised List of Fragile Situations FY18." http://pubdocs.worldbank.org/en/189701503418416651/FY18FCSLIST-Final-July-2017.pdf.
Zuercher, C. 2006. *Is More Better? Evaluating External-Led State Building after 1989*. CDDRL Working Paper 54. Stanford, CA: Stanford University.

CHAPTER 26

TERRORISM AND PEACEBUILDING

IOANNIS TELLIDIS

At first glance, the concepts of terrorism and peace are bound in a dyadic relationship: whoever fights terrorism (wherever it emanates from and however defined) does so in order to achieve peace. Particularly in the post-9/11 setting, this dichotomy has been reinforced in narrative terms. The liberal democratic states that are principle-, freedoms-, and rights-guided are fighting the evil, sadism, and disregard for human life of the terrorists. In other words, those fighting terrorism are the good guys, whereas the terrorists are always the bad guys. This narrative is so potent that non- and antidemocratic regimes find it easier to legitimize via internationalization (de Jonge Oudraat and Marret 2010) their (usually excessive) violence against groups that question the authority of the state by claiming that said violence was necessary in their fight against terrorism.

Although this picture is a simple one, the simple truth is that there is nothing simple in the relationship between the concepts—their definitions, their understandings, and, crucially, the deployment of those understandings. Many attempts to secure peace, for example, were driven not by a desire for peace but for complete annihilation of the opponent, thus fueling the conflict further (Tellidis 2020). In other instances, the dichotomization between good and evil has led to the exclusion and even criminalization of social groups living in the same space as the terrorist group (Lederach 2011). Further still, genuine efforts to engage armed groups in dialogue and create the conditions necessary for the cessation of conflict were equally condemned because, by continuing their efforts and activities, peacebuilders and NGOs working in such regions would have meant facing serious legal repercussions (Cortright et al. 2012; Haspeslagh 2013).

Examples like these make evident the complexity of the relationship between terrorism (or counterterrorism, more precisely) and peacebuilding. This chapter aims to take a closer look at the ways in which the research and practice(s) of terrorism may interconnect but also clash with those of peace, and peacebuilding in particular.

I will do so, first, by providing a brief chronology of how the academic and policy conceptualizations of terrorism and peace developed over the years. I will then evaluate the political implications generated by these conceptualizations. This will be key in assessing whether current ontological, epistemological, and methodological foundations are adequate for an accurate understanding of how to build peace in terrorist conflicts, or whether more nuanced approaches are needed. Recognizing the need for "dissident" research (Stump and Dixit 2013, 4) on both terrorism and peace, I propose that an elimination of the relevant terminology ("terrorism," "terrorists") may be what is needed for peacebuilding efforts to engage in and promote inclusionary human security.

(Mis)Understanding Terrorism, (Mis) Reading Peace

As a subject of scientific research, terrorism suffered from even more dichotomizations than the ones just described. Over forty years ago, as Gill and Corner (2017, 231) explain, early psychological studies on terrorism offered psychopathy as a cause of terrorism involvement, extrapolating that an individual was either a terrorist or not. This conclusion was reached despite the writings of individuals who saw themselves as ideologues of the use of terrorist tactics, like Carlos Marighela (1971, 20), Che Guevara (in Wilkinson 1986, 100), or even Russian anarchoterrorists (Nechayev, in Rubenstein 1987, 98) active in the early 1880s. All of them evaluated terrorism as a tactic that would increase their gains in a conflictive political relationship. Some, like Guevara and Debray, even advocated a limited use of terrorist tactics for fear of endangering the greater objective that was the revolution (Wilkinson 1986, 100).

The assumption that terrorists are simply psychopaths continues to this day, although it is now justified as religion-based. Barbarities committed by groups such as the Islamic State (IS), as well as other groups that are or want to be affiliated with it, cannot be easily comprehended in terms of political strategy. For example, video recording the most horrendous torture and the most appalling ways of execution of groups and/or individuals imprisoned by the IS should lead one to consider whether the ultimate objective was the horror and the torture itself, or whether it was a strategy to provoke a reaction by those whom the IS wants to fight (Barr and Herfroy-Mischler 2017). It is this type of extreme and graphic violence that leads people to presume that "today's terrorists don't want a seat at the table, they want to destroy the table and everyone sitting at it" (National Commission on Terrorism 2000, 2). Nevertheless most scholars agree that, even without a unanimously accepted definition, terrorism is political in nature (Schmid 2011, 99–157) and its objectives serve a strategic function (Walter 1969; Leach 1977; Groom 1978; Laqueur 1986; Wilkinson and Stewart 1987; Schmid and Jongman 1988; Weinberg, Pedahzur, and Perliger 1992; Hoffmann 1998; Malik 2001; Wilkinson

2001; Bjørgo 2005; Goodin 2006; English 2009; Stampnitzky 2013; Richards 2014). As Wilkinson (2001, 61) put it, "What appears at first sight to be a purely religious phenomenon is in fact in large part about political control and socioeconomic demands." This is an evaluation also shared by the US Army, according to whom "terrorist goals are always political, as extremists driven by religious or ideological beliefs usually seek political power to compel society to conform to their views" (in Whittaker 2001, 17).

Attempting to mask or entirely remove the political dimensions of terrorism from one's examination of the phenomenon highlights two very significant implications. The first is that, by virtue of being "evil" and "incomprehensible," terrorism is a more dangerous type of political violence (Mac Ginty 2014) that merits the severity of counterterrorist measures adopted so far, as well as the specialized judicial treatment provided (Finn 2010). As such, any criticism towards those tasked with countering terrorism and/or any attempt to systematically and accurately study terrorism is received with suspicion and tainted as "sympathising with" (Stampnitzky 2013, 191) or "defending" (Tellidis 2015, 3) terrorism. The second implication is related to the first and concerns the exaggeration of an otherwise real threat. Magnifying the risk that terrorism poses allows the state to reinforce its position of control by claiming urgency in implementing extraordinary measures. This affects the ontological and epistemological aspects of research on terrorism, a great deal of which rests on the sensationalist dimensions of the phenomenon (Sageman 2015), ascribing it a statist focus and rendering it "profoundly unscientific" (Durodié 2007, 432).

Although terrorism is not new (Chaliand and Blin 2007), nor more lethal (Piazza 2009), nor indeed more brutal (Jäckle and Baumann 2017) than in the past with regard to some Western countries (Institute for Economics and Peace 2016; Jenkins 2010, 8–9), the occurrence of 9/11 led to the exoticization of the phenomenon (Breen-Smyth 2008), which, in turn, resulted in its acute securitization (Wolfendale 2016). This meant the deployment of counterterrorist measures and policies, both at home and abroad, that rested on a "culture of fear" (Brzezinski 2007; Mavelli 2016) in order to prevent threats that were not only unknown and unsubstantiated (Aradau and Van Munster 2007) but also imagined (de Goede 2008). This is confirmed by the authorities' eagerness to publicize success in foiling a plot but not divulging any information about it (Priest, in Mueller and Stewart 2011, 93)—as opposed to instances of successful counterterrorist operations, where the same agencies "remain anything but tight-lipped about their achievement" (Mueller and Stewart 2011, 94). Even more scandalous, this culture of fear is confirmed by cases where state agencies have manufactured terrorist plots to lure and trap individuals who otherwise would not have been involved in terrorist acts (Kundnani 2014, 188–193).

This indicates the kind of peace that is to be built when an otherwise real threat is exaggerated (Lustick 2006): the purpose is not so much the protection of the population as the strengthening and consolidation of the state's control through the suppression of "bodies deemed representative of a threatening political challenge" (Heath-Kelly 2015, 47). Although it has been argued that such approaches favor state security at the expense of human security (Richmond 2003; Richmond and Franks 2009; Richmond and Tellidis 2012; Mueller 2012), other scholars purport that such counterterrorist frameworks may, in fact, undermine state security, appearing as they do to contribute to the anticipation and creation of terror as much as terrorism itself (Frank 2015; Zulaika 2009), even contributing to the terrorist organization's legitimacy and recruiting strategies (Walker 2000; Blackbourn 2008). This type of top-down peace is recurrent in terrorism studies as well as peacebuilding. As the next section discusses, it is also one of the main obstacles to the resolution of conflicts and their transformation to more emancipatory versions of peace.

Liberal approaches to peacebuilding have long been at the receiving end of criticism regarding their aims, agendas, and approaches. Chief among those is the argument that liberal peacebuilding seeks to "impose norms, rules, practices and standards in distinct areas of governance, institutions, social relations and culture" (Lemay-Hébert and Visoka 2017, 147) that are foreign to the region where intervention is being undertaken. This is because the international system—with particular focus on its interventionary mechanisms—still operates in a colonial fashion (Richmond 2017, 3), seeing how a dominant group of states (globally but also regionally) dictates the practices and discourses of peace, sovereignty, interest, power, and other concepts traditionally linked with realist frameworks of state security. As a result, many such interventions have failed or have produced unintended outcomes and consequences—primarily for the populations they sought to assist (Paris 2004; Richmond 2005).

To a certain extent, it should not come as a surprise that peace is practiced or indeed sought to be imposed according to the interests of the interveners rather than the needs of those who claim it or need it. By extrapolation, it should be equally unsurprising that peace has for so long been researched according to metrics of violence (and its absence) rather than the transformation of power roles, unequal relations, lack of justice, and distribution of and access to resources. This is because preoccupations with peace are usually synonymous with concerns about security, defined as power preservation and/or maximization. Despite scholarly calls to engage with "the power and competence of the allegedly powerless" (Reid and Yanarella 1976, 326), a cursory glance at the titles and content of articles published in the *Journal of Peace Research* or *Conflict Management and Peace Science*, for example, shows that such calls have largely gone unheeded, with researchers and students alike choosing to perpetuate discourses and practices that were more traditionally (and narrowly) conceptualized.

The recent formulation of the conceptual category of "everyday peace" questions this type of thinking and praxis by providing alternative understandings, discourses, and frameworks and a healthy dose of self-reflexivity in the field. Although it may be unevenly matched against the "everyday war" mindset (exhibited by students and scholars alike), its gaining in traction means that, at the very least, the balance of power between war- and peace-studying has tilted in the latter's favor. The multitude of actors, their interests, agendas, assumptions, and presumptions, their goals and standard operating procedures, have often been indicators as to why peacebuilding efforts have failed to deliver what was expected of them. In terms of research, too, this messiness is perceived as having no or very little analytical value. As a result, it rarely finds its way into the publication of findings. However, rather than being disregarded for being analytically useless, this messiness should be embraced (Perera 2017) because not only is the "everyday" not abstract (Richmond 2009; Guillaume 2011; Jarvis and Lister 2013; Sylvester 2013), but its agency is a clear response to and resistance against the established and dominant international agenda(s) (Solomon and Steele 2017). Thus it is in this cacophony of narratives, identities, interests, and incentives that the most valuable information is to be had. This is evident in terrorist conflicts where, more often than not, the conflict transforms into a war of attrition between the state forces and the terrorist network. Yet little to no attention is paid to the needs and expressions of non- or antiviolent groups on the one hand, and to those civil society actors trying to transform the conflict behaviors, attitudes, and dynamics on the other hand.

Despite the wealth of knowledge accumulated in the past few decades about where to look for peace and how to build it, dominant discourses of security continue to render "others" invisible by making their visualization central to the analysis, or, when "others" are visible, to be so "within and through dominant representations" (Dixit 2013, 338). In their effort to minimize any loss of power and control and to guarantee the prevalence of the state's discourse(s), goals, and identity, counterterrorist frameworks continue to exhibit counterproductive mentalities and knee-jerk reactions, leading to exclusion, marginalization, and/or the construction of dangerous Others. This is supported by Barkawi's (2016) research on the decolonization of war. As he argues, a periodization of war and peace as it exists and is understood in the Global North is not to be found in the Global South since their encounter from 1492 until the war on terror. Said encounter "has engendered near continual, if geographically dispersed, warfare and violent repression. Instead of war and peace, we have permanent war" (205). This implies that the dichotomy between war (or, in our case, terrorism) and peace is a false one. Rather, what becomes evident is that someone's peace is experienced as war or terrorism by others. Viewed in this light, it appears pertinent that any effort to build peace in terrorist conflicts will be destined to fail unless the very conceptualization of peace is also decolonized. In that regard, the following section aims to challenge some of the long-held assumptions and "truths" about terrorism and peace and suggest alternative ways of doing and thinking counterterrorism and peacebuilding.

Building Peace by Eliminating "Terrorism"

Viewed from the prism of postcolonialism, it becomes apparent that both fields adopt a particularly narrow view of the phenomena they seek to examine. First and foremost, the narratives and practices of both fields rest on a foundation of otherness from which they operate: counterterrorism does that by refusing to see the political aspects of the conflict and aiming at the annihilation of the terrorists. Meanwhile peacebuilding does that by placing the interests of the interveners above those of the intervened—whether by imposing or usurping a peacebuilding agenda, as was the case in Mali with the French intervention (Charbonneau 2017); by conditionalizing aid and donations, as was the case in Mozambique (Manning and Malbrough 2010); or by seeking to benefit and/or profit from the reconstruction of the state, as was the case in Afghanistan (Mac Ginty 2010).

Second, this dichotomization is distilled into the very concepts with which both fields are deeply concerned: legitimacy, sovereignty, and security. The former two are important because they delineate the contours of authority that gets to decide what kind of violence is permissible and for what aims (Charbonneau 2017). Traditional theories of both terrorism studies and peacebuilding prime the existence of a sovereign state that monopolizes the administration of violence and, usually unquestionably, is bestowed with legitimacy while everyone and everything else is not. Proscription of terrorist organizations is a perfect example of this logic (if it can even be called that, when the aim is to build peace) because any individual or organization attempting to open or maintain negotiations with terrorist actors is deemed criminal (Cortright et al. 2011; Jarvis and Legrand 2016; Jarvis and Lister 2014), thus debilitating (Richmond and Tellidis 2012) or, worse still, outlawing peacebuilding efforts (Haspeslagh 2013).

Security is also understood accordingly and, thus, applied in a mutually exclusive fashion, particularly in the case of terrorism: one can be secure when and only when terrorism and terrorists are eliminated. But as the proscription example made evident, these understandings undermine and sabotage the very aims of peacebuilding missions (or at least those that are genuinely concerned with building peace rather than maximizing interests) because they always presume that terrorists are enemies. Yet, as Charbonneau (2017, 419) points out, the moral argument in peacebuilding is that "the use of force is limited by the normative objectives of keeping or building peace, protecting civilians and/or providing humanitarian aid. Peacekeepers have no enemy to kill and destroy."

Herein, however, lies the problem. As a method of political violence that aims to secure political gains (as explored in the beginning of this chapter), terrorism will never be eliminated. More than that, any policy that is geared toward that goal of annihilating the enemy is bound to be counterproductive because it will be perceived as terrorism by the other side (as is the case with drone killings and other abuses in the name of the war on terror), thus perpetuating a vicious circle. This is particularly true in cases where terrorism is part of a greater social movement or of grassroots mobilization, and therefore the legitimacy ascribed to it as a method of political violence may be much larger than under other circumstances (Bosi and della Porta 2015).

Taking all of this into consideration, it seems to me that there is a way to get around the problem of how to build peace in terrorist conflicts, that is by ignoring or, better still, eliminating the very terminology ("terrorism," "terrorists," etc.). This may sound counterintuitive, but that is only because of the monopolization of the legitimate meaning of counterterrorist security after 9/11 (Butler 2004). Norway is one example where political authorities decided to oppose international norms of counterterrorist behavior in order to commit to and foment peacebuilding initiatives and efforts in conflicts marked by terrorist violence: following Hamas's electoral victory in 2006, Norway was the only Western country to initiate dialogue with the group, and a year later it opted out of adopting the EU's list of terrorist organizations to continue talks with the Liberation Tigers of Tamil Eelam (Lindahl 2017, 536). Furthermore, having experienced terrorism itself, its response was unlike anything seen so far in other Western countries: the country opted to fight terrorism with "more democracy, more openness and greater political participation" (Pidd and Meikle 2011; Tellidis 2019). This indicates not only that alternative understandings of legitimacy and sovereignty can be implemented, but that such an implementation leads to formulations of more inclusive human security.

While such alternative understandings may be harder to implement in policy terms (not least because of how 9/11 and the war on terror have molded all three concepts), it is perhaps easier to do so in terms of research. Indeed both fields have begun to critically question previous and long-held ontological, methodological, and epistemological assumptions. As explored elsewhere (Tellidis 2015; 2016), critical terrorism studies (CTS) and critical peace research now hold great potential for the interconnection between the two fields, whereby each can stand to benefit from the insights of the other. For instance, if, as I advocated earlier, terrorism studies abandon the somewhat automatic vilification of persons or groups that is prevalent among policy circles (Jackson 2007, 2015), then it stands to gain a much broader and much deeper understanding of what motivates groups and individuals to adopt this particular method of political violence; the local, national, and international structures and hierarchies in which these groups emerge; their ideologies, and what role they play in recruitment. Furthermore, besides self-imposed censorship and a refusal to see the forest for the trees, there is nothing that should prevent terrorism studies from exploring the ways in which third-party intervention may resolve or transform terrorist conflicts. These themes were explored both before (Zartman 1990) as well as after (Lederach 2001a, 2001b) 9/11, while more recent research has shown that such talks do not lead to more violence—as policy myths prescribe (Goerzig 2010; Toros 2012; see also Zartman and Faure 2011; Hampson and Zartman 2011; Lederach 2011). But all these came from scholars who were more closely linked to peace research than to terrorism studies. This points to the fact that the point of departure for traditional strands of terrorism studies is the same as that of the policy world: terrorism is evil and needs to be eliminated, and "you're either with us or against us" (Brassett 2007). Yet despite their prevalence and dominance, such frameworks have been shown to be either counterproductive (Pantazis and Pemberton 2009) or wholly inefficient (Amoore and de Goede 2008).

A similar argument can be made about peace studies and how it stands to gain from insights from terrorism studies. The wealth of research on the psychological dimensions of terrorists and terrorist group dynamics is one such area. The ways individuals radicalize and become recruited and the bonds they form inside a terrorist group (Horgan 2005, 2009) can be of extreme value when researching ways of intervening in a terrorist conflict in order to resolve or transform it. The same is true about the internal dynamics of a terrorist group, its legitimacy, possible competition between its fractions, or any potential splintering that may act as a spoiler to peacebuilding efforts. Furthermore, in cases where terrorist groups enjoy relatively high levels of support, social movement and anthropological research on terrorism could shed particularly useful insights on how peace is conceptualized by the networks of people that surround terrorist groups (Bosi and della Porta 2015).

Conclusion

Both terrorism studies and peace research have seen the emergence of more critical strands that aim to question traditional, state-centric, and hard-security-oriented understandings of both terrorism and peace (see Mac Ginty 2014). Both fields appear to have been somewhat successful in turning the tables insofar as academic research and inquiry is concerned, but obstacles and limitations persist, particularly with regard to the implementation of said research by policymakers. Despite the emergence of CTS more than a decade ago, for example, the field contributes to the perpetuation of the narrative that utilizes 9/11 as "a moment of temporal rupture [to] legitimise violent counterterrorism policies, delegitimise non-violent responses such as negotiation and dialogue . . . silence dissenting voices [and] extend US hegemony over world time" (Toros 2017, 207). Similarly CTS has been criticized for its inability to distance itself from the normative role of the state and, like the more recent research on peacebuilding, shed more light on everyday/local levels of analysis (Richmond and Tellidis 2012; Jarvis and Lister 2013).

The role of the state and the extent to which peacebuilding research can go beyond it by identifying poststate alternatives is also a criticism that concerns peace studies (Boege 2014). Cosmopolitan approaches to counterterrorism, for example, are unable to suggest frameworks that look beyond the state—albeit as the guarantor of the rule of law and the protection of freedoms and liberties rather than the prosecutor of war (Held 2010, 133–134)—or that do not adopt the "either with us or against us" mentality that won ground after 9/11 (Brassett 2007). Furthermore, the political economies of research in the Global North often contradict and subvert the aims and objectives of research (Richmond and Mac Ginty 2015, 184), leading to the reproduction of "an essentially neocolonial relationship." Although peacebuilders are now better-informed about the importance of local contexts and the significance of bottom-up approaches (Perera 2017, 45), the narrative that makes Western intervention necessary because of "local backwardness" (45; Kappler 2018) still persists.

This makes evident the fact that, despite the common epistemological, ontological, and methodological advancements in the conceptualization of terrorism, peace, and peacebuilding, both fields face similar limitations and obstacles in their efforts to affect policymaking in ways that will lead to more emancipatory praxes of peace. The deviation of terrorism from the broader study of conflict (Richmond 2003) and the establishment of an autonomous field of terrorism studies (despite the appearance of CTS) may have impeded the advancement and development of frameworks that would have favored alternative understandings of terrorist conflicts, first, before configuring ways in which to become involved in their transformation, resolution, and the building of peace afterward. This is certainly the case with some counterterrorist strategies, such as, for example, proscription. "Far from fostering moderation and shifts to non-violent strategies, proscription can fuel radicalism and create impediments for humanitarian or political negotiations" (Dudouet 2011, 5). Becoming a bully to stop other bullies (Shapiro 2011, 167) not only eliminates any legitimacy claims about security, but it also blocks opportunities for nonviolent transformation that will lead to more inclusionary peace.

References

Amoore, L., and M. de Goede. 2008. *Risk and the War on Terror*. London: Routledge.

Aradau, C., and R. Van Munster. 2007. "Governing Terrorism through Risk: Taking Precautions, (Un)Knowing the Future." *European Journal of International Relations* 13 (1):89–115.

Barkawi, T. 2016. "Decolonising War." *European Journal of International Security* 1 (2):199–214.

Barr, A., and A. Herfroy-Mischler. 2017. "ISIL's Execution Videos: Audience Segmentation and Terrorist Communication in the Digital Age." *Studies in Conflict and Terrorism*, 11 September. doi:10.1080/1057610X.2017.1361282.

Bjørgo, T., ed. 2005. *Root Causes of Terrorism: Myths, Reality and Ways Forward*. Abingdon, UK: Routledge.

Blackbourn, J. 2008. "Counter Terrorism and Civil Liberties: The United Kingdom Experience, 1968–2008." *Journal of the Institute of Justice and International Studies* 8:63–74.

Boege, V. 2014. "Roundtable: Building a Peaceful State—Desirable? Why? Possible? How?" 3rd Annual Conference, International Association of Peace and Conflict Studies, University of Manchester, 12 September.

Bosi, L., and D. della Porta. 2015. "Processes of Disengagement from Political Violence: A Multi-Level Relational Approach." In *Researching Terrorism, Peace and Conflict Studies: Interaction, Synthesis and Opposition*, edited by I. Tellidis and H. Toros, 000-00081–99. London: Routledge.

Brassett, J. 2007. "Cosmopolitanism vs. Terrorism? Discourses of Ethical Possibility before and after 7/7." *Millennium: Journal of International Studies* 36 (2):311–337.

Breen-Smyth, M. 2008. "Geographies of Conflict: Meanings and Effects of Political Violence." Paper presented at International Studies Association annual convention, San Francisco, 28 March.

Brzezinski, Z. 2007. "Terrorized by 'War on Terror.'" *Washington Post*, 25 March. http://www.washingtonpost.com/wp-dyn/content/article/2007/03/23/AR2007032301613.html.

Butler, J. 2004. *Precarious Life: The Powers of Mourning and Violence*. London: Verso.

Chaliand, G., and A. Blin, eds. 2007. *The History of Terrorism: From Antiquity to Al Qaeda*. Los Angeles: University of California Press.

Charbonneau, B. 2017. "Intervention in Mali: Building Peace between Peacekeeping and Counterterrorism." *Journal of Contemporary African Studies* 35 (4):415–431.

Cortright, D., A. Millar, L. Gerber-Stellingwerf, G. A Lopez, K. Wall, E. Fackler, and J Weaver. 2012. *Friend Not Foe: Opening Spaces for Civil Society Engagement to Prevent Violent Extremism*. Notre Dame Journal of International and Comparative Law, 2 (2): 238–256.

de Goede, M. 2008. "Beyond Risk: Premeditation and the Post-9/11 Security Imagination." *Security Dialogue* 39 (2–3):155–176.

de Jonge Oudraat, C., and J. L. Marret. 2010. "The Uses and Abuses of Terrorist Designation Lists." In *The Consequences of Counterterrorism*, edited by M. Crenshaw, 93–129. New York: Russell Sage Foundation.

Dixit, P. 2013. "Decolonizing Visuality in Security Studies: Reflections on the Death of Osama Bin Laden." *Critical Studies on Security* 2 (3):337–351.

Dudouet, V. 2011. *Anti-terrorism Legislation: Impediments to Conflict Transformation*. Berghof Policy Brief No. 2, November. Berlin: Berghof Conflict Research and Berghof Peace Support.

Durodié, B. 2007. "Fear and Terror in a Post-Political Age." *Government and Opposition* 42 (3):427–450.

English, R. 2009. *Terrorism: How to Respond*. Oxford: Oxford University Press.

Finn, J. E. 2010. "Counterterrorism Regimes and the Rule of Law: The Effects of Emergency Legislation on Separation of Powers, Civil Liberties, and Other Fundamental Constitutional Norms." In *The Consequences of Counterterrorism*, edited by M. Crenshaw, 33–93. New York: Russell Sage Foundation.

Frank, M. 2015. "Conjuring Up the Next Attack: The Future-Orientedness of Terror and the Counterterrorist Imagination." *Critical Studies on Terrorism* 8 (1):90–109.

Gill, P., and E. Corner. 2017. "There and Back Again: The Study of Mental Disorder and Terrorist Involvement." *American Psychologist* 72 (3):231–241.

Goerzig, C. 2010. *Talking to Terrorists: Concessions and the Renunciation of Violence*. Abingdon, UK: Routledge.

Goodin, R. 2006. *What's Wrong with Terrorism?* Cambridge, UK: Polity.

Groom, A. J. R. 1978. "Coming to Terms with Terrorism." *British Journal of International Studies* 4:62–77.

Guillaume, X. 2011. "The International as an Everyday Practice." *International Political Sociology* 5 (4):446.

Hampson, F. O., and I. W. Zartman. 2011. *The Global Power of Talk: Negotiating America's Interests*. Boulder, CO: Paradigm.

Haspeslagh, S. 2013. "'Listing Terrorists': The Impact of Proscription on Third-Party Efforts to Engage Armed Groups in Peace Processes—A Practitioner's Perspective." *Critical Studies on Terrorism* 6 (1):189–208.

Heath-Kelly, C. 2015. "Counter-terrorism: The Ends of a Secular Ministry." In *Critical Perspectives on Counter-terrorism*, edited by L. Jarvis and M. Lister, 41–55. Abingdon, UK: Routledge.

Held, D. 2010. *Cosmopolitanism, Ideals and Realities*. London: Polity Press.

Hoffman, B. 1998. *Inside Terrorism*. London: Victor Gollancz.

Horgan, J. 2005. *The Psychology of Terrorism*. London: Routledge.

Horgan, J. 2009. *Walking Away from Terrorism: Accounts of Disengagement from Radical and Extremist Movements*. London: Routledge.

Institute for Economics and Peace. 2016. *Global Terrorism Index 2016: Measuring and Understanding the Impact of Terrorism*. http://economicsandpeace.org/wp-content/uploads/2016/11/Global-Terrorism-Index-2016.2.pdf.

Jäckle, S., and M. Baumann. 2017. "New Terrorism = Higher Brutality? An Empirical Test of the 'Brutalization Thesis.'" *Terrorism and Political Violence* 29 (5):875–901.

Jackson, R. 2007. "The Core Commitments of Critical Terrorism Studies." *European Political Science* 6 (3):244–251.

Jackson, R. 2015. "Towards Critical Peace Research: Lessons from Critical Terrorism Studies." In *Researching Terrorism, Peace and Conflict Studies: Interaction, Synthesis and Opposition*, edited by I. Tellidis and H. Toros, 19–37. London: Routledge.

Jarvis, L., and T. Legrand. 2016. "Legislating for Otherness: Proscription Powers and Parliamentary Discourse." *Review of International Studies* 42 (3):558–574.

Jarvis, L., and M. Lister. 2013. "Vernacular Securities and Their Study: A Qualitative Analysis and Research Agenda." *International Relations* 27 (2):158–179.

Jarvis, L., and M. Lister, eds. 2014. *Critical Perspectives on Counter-Terrorism*. Abingdon, UK: Routledge.

Jenkins, B. 2010. *Would-Be Warriors: Incidents of Jihadist Terrorist Radicalization in the United States since September 11, 2001*. Santa Monica, CA: RAND.

Kappler, S. 2018. "The Legitimisation of Post-Conflict Intervention: Narrative Frames of Backwardness and Progress." *Political Geography* 66:130–138.

Kundnani, A. 2014. *The Muslims Are Coming! Islamophobia, Extremism and the Domestic War on Terror*. London: Verso.

Laqueur, W. 1986. "Reflections on Terrorism." *Foreign Affairs* 65:86–100.

Leach, E. 1977. *Custom, Law and Terrorist Violence*. Edinburgh: Edinburgh University Press.

Lederach, J. P. 2001a. "The Challenge of Terror: A Traveling Essay." https://emu.edu/now/news/2001/challenge-terror-traveling-essay/.

Lederach, J. P. 2001b. "Quo Vadis? Reframing Terror from the Perspective of Conflict Resolution." *Mediate*. http://www.mediate.com/articles/lederach2.cfm.

Lederach, J. P. 2011. "Addressing Terrorism: A Theory of Change Approach." In J. P. Lederach, R. Clos, D. Ansel, A. Johnson, L. Weis, J. Brandwein and S. Lee, Somalia: Creating Space for Fresh Approaches to Peacebuilding, 7–19. Uppsala: Life and Peace Institute.

Lemay-Hébert, N., and G. Visoka. 2017. "Normal Peace: A New Strategic Narrative of Intervention." *Politics and Governance* 5 (3):146–156.

Lindahl, S. 2017. "A CTS Model for Counterterrorism." *Critical Studies on Terrorism* 10 (3):523–541.

Lustick, I. 2006. *Trapped in the War on Terror*. Philadelphia: University of Pennsylvania Press.

Mac Ginty, R. 2010. "Warlords and the Liberal Peace: State-Building in Afghanistan." *Conflict, Security and Development* 10 (4):577–598.

Mac Ginty, R. 2014. "Against Critical Terrorism Studies: Don't Give It the Oxygen of Publicity." Unpublished manuscript.

Malik, O. 2001. *Enough of the Definition of Terrorism!* London: Royal Institute of International Affairs.

Manning, C., and M. Malbrough. 2010. "Bilateral Donors and Aid-Conditionality in Post-conflict Peacebuilding: The Case of Mozambique." *Journal of Modern African Studies* 48 (1):143–169.

Marighela, C. 1982 [1969]. "Minimanual of the Revolution." In *Terror and Urban Guerrillas: A Study of Tactics and Documents*, edited by J. Mallin, 67-115. Coral Gables: University of Miami Press.

Mavelli, L. 2016. "The Governmentality of Terrorism: Uncertainty, Risk Management and Surveillance." In *Routledge Handbook of Critical Terrorism Studies*, edited by R. Jackson, 237–247. Abingdon, UK: Routledge.

Mueller, J. 2012. "The Terrorism Delusion: America's Overwrought Response to September 11." *International Security* 37 (1):81–110.

Mueller, J., and M. G. Stewart. 2011 *Terror, Security and Money: Balancing the Risks, Benefits and Costs of Homeland Security*. New York: Oxford University Press.

National Commission on Terrorism. 2000. *Countering the Changing Threat of International Terrorism: Report of the National Commission on Terrorism*. Washington, DC: Government Printing Office.

Pantazis, C., and S. Pemberton. 2009. "From the 'Old' to the 'New' Suspect Community: Examining the Impacts of Recent UK Counter-terrorist Legislation." *British Journal of Criminology* 49 (5):646–666.

Paris, R. 2004. *At War's End: Building Peace after Civil Conflict*. New York: Cambridge University Press.

Perera, S. 2017. "Bermuda Triangulation: Embracing the Messiness of Researching in Conflict." *Journal of Intervention and Statebuilding* 11 (1):42–57.

Piazza, J. 2009. "Is Islamist Terrorism More Dangerous? An Empirical Study of Group Ideology, Organization and Group Structure." *Terrorism and Political Violence* 21 (1):62–88.

Pidd, H., and J. Meikle. 2011. "Norway Will Not Be Intimidated by Terror Attacks, Vows Prime Minister." *Guardian*, 27 July. http://www.theguardian.com/world/2011/jul/27/norway-terror-attacks-prime-minister.

Reid, H., and E. Yanarella. 1976. "Toward a Theory of Critical Peace Research in the United States: The Search for an 'Intelligible Core' ", *Journal of Peace Research*, 13 (4): 315–341.

Richards, A. 2014. "Conceptualizing Terrorism." *Studies in Conflict and Terrorism* 37 (3):213–236.

Richmond, O. P. 2003. "Realizing Hegemony? Symbolic Terrorism and the Roots of Conflict." *Studies in Conflict and Terrorism* 26 (4):289–309.

Richmond, O. P. 2005. *The Transformation of Peace*. London: Palgrave.

Richmond O. P. 2009. "A Post-liberal Peace: Eirenism and the Everyday." *Review of International Studies* 35 (3):557–580.

Richmond, O. P. 2017. "Peace in the Twenty-First Century: States, Capital, and Multilateral Institutions versus Positionality Arbitrage, Everyday Mobility, Networks, and Multiverticality." *Globalizations* 14 (6):1014–1028.

Richmond, O. P., and J. Franks. 2009. "The Impact of Orthodox Terrorism Discourses on the Liberal Peace: Internalisation, Resistance or Hybridisation?" *Critical Studies on Terrorism* 2 (2):201–218.

Richmond, O. P., and R. Mac Ginty. 2015. "Where Now for the Critique of the Liberal Peace?" *Cooperation and Conflict* 50 (2):171–189.

Richmond, O. P., and I. Tellidis. 2012. "The Complex Relationship between Peacebuilding and Terrorism Approaches: Towards Post-terrorism and a Post-liberal Peace?" *Terrorism and Political Violence* 24 (1):120–143.

Rubenstein, R. 1987 *Alchemists of the Revolution: Terrorism in the Modern World*. London: I. B. Tauris.

Sageman, M. 2015. "The Stagnation in Terrorism Research." *Terrorism and Political Violence* 26 (4):565–580.

Schmid, A. 2011. *The Routledge Handbook of Terrorism Research*. Abingdon, UK: Routledge.

Schmid, A., and A. Jongman. 1988. *Political Terrorism: A New Guide to Actors, Authors, Concepts, Data Bases, Theories and Literature*. New Brunswick, NJ: Transaction.

Shapiro, I. 2011. *The Real World of Democratic Theory*. Princeton, NJ: Princeton University Press.

Solomon, T., and B. J. Steele. 2017. "Micro-moves in International Relations Theory." *European Journal of International Relations* 23 (2):267–291.

Stampnitzky, L. 2013. *Disciplining Terror: How Experts Invented "Terrorism."* Cambridge: Cambridge University Press.

Stump, J. L., and P. Dixit. 2013. "Studying Terrorism and Practicing Criticism." In: *Critical Methods in Terrorism Studies*, edited by J. L. Stump and P. Dixit, 3–16 (Abingdon: Routledge).

Sylvester, C. 2013. "Experiencing the End and Afterlives of International Relations Theory." *European Journal of International Relations* 19:609–626.

Tellidis, I. 2015. "Researching Terrorism, Peace and Conflict: An Introduction." In *Researching Terrorism, Peace and Conflict Studies: Interaction, Synthesis and Opposition*, edited by I. Tellidis and H. Toros, 1–15. Abingdon, UK: Routledge.

Tellidis, I. 2016. "Terrorism and Peace Studies." In *Routledge Handbook of Critical Terrorism Studies*, edited by R. Jackson, 298–308. Abingdon, UK: Routledge.

Tellidis, I. 2019. "Street Art as Everyday Counterterrorism? The Norwegian Art Community's Reaction to the 22 July 2011 Attacks." *Cooperation and Conflict* 54 (2):191–210.

Tellidis, I. 2020. "Peacebuilding beyond Terrorism? Revisiting the Narratives of the Basque Conflict." *Studies in Conflict and Terrorism* 43 (6):516–534.

Toros, H. 2012. *Terrorism, Talking and Transformation: A Critical Approach*. London: Routledge.

Toros, H. 2017. "'9/11 Is Alive and Well': Or How Critical Terrorism Studies Has Sustained the 9/11 Narrative." *Critical Studies on Terrorism* 10 (2):203–219.

Walker, C. 2000. "Briefing on the Terrorism Act 2000." *Terrorism and Political Violence* 12 (2):1–36.

Walter, E. V. 1969. *Terror and Resistance: A Study of Political Violence with Case Studies of Some Primitive African Communities*. New York: Oxford University Press.

Weinberg, L., A. Pedahzur, and A. Perliger. 1992. *Political Parties and Terrorist Groups*. London: Frank Cass.

Wilkinson, P. 1986. *Terrorism and the Liberal State*. Basingstoke, UK: Macmillan.

Wilkinson, P. 2001. *Terrorism versus Democracy: The Liberal State Response*. London: Frank Cass.

Wilkinson, P., and A. Stewart. 1987. *Contemporary Research on Terrorism*. Aberdeen: Aberdeen University Press.

Whittaker, D. 2001. *The Terrorism Reader*. Abingdon, UK: Routledge.

Wolfendale, J. 2016. "The Narrative of Terrorism as an Existential Threat." In *Routledge Handbook of Critical Terrorism Studies*, edited by R. Jackson, 114–123. Abingdon, UK: Routledge.

Zartman, I. W. 1990. "Negotiating Effectively with Terrorists." In *The Politics of Counterterrorism: The Ordeal of Democratic States*, edited by B. Rubin, 163–188. Washington, DC: Foreign Policy Institute.

Zartman, I. W., and G. O. Faure, eds. 2011. *Engaging Extremists: Trade-offs, Timing and Diplomacy*. Washington, DC: US Institute of Peace.

Zulaika, J. 2009. *Terrorism: The Self-Fulfilling Prophecy*. Chicago: University of Chicago Press.

CHAPTER 27

PEACE AFTER REVOLUTIONS

SANDRA POGODDA

HISTORICALLY revolutions have presented a complex set of political dilemmas for neighboring countries as well as international powers. Given the potential of antisystemic ideologies to spark revolutionary waves in other parts of the world (Halliday 1999, 94–132), neighboring countries and powerful global players have often tried to contain, sanction, or militarily defeat revolutions. Yet, as emancipatory projects, revolutionary processes are hostile grounds for intervention. In pursuit of the overarching goal of freedom (Arendt [1963] 1990), revolutionary movements are wary not to replace the toppled regime with subordination to external control. From France to Haiti, Cuba, and Iran, revolutionary movements have therefore used their capacity to mobilize the masses in order to defend their revolution against external counterrevolutionary forces. Coercive forms of counterrevolutionary intervention have thus remained notoriously unsuccessful in stemming the "irresistible flow" of revolutions (Arendt [1963] 1990, 265; Halliday 1999, 224–226).

In 1989, however, a Western-idealist perspective started to portray contemporary revolutions in a different light: not as a challenge to Western interests but as their vindication (Hobson 2010). Seeing the revolutions in Eastern Europe as the ultimate victory of market-based democracy, Western intervention aimed to support rather than undermine the revolutionary process. Many revolutionaries of 1989, however, consider this support a form of colonialism, which suffocated the emancipatory impulses of the anticommunist uprisings (Anheier and Seibel 1998). Given the enduring political consequences of widespread frustration over the transformation processes of the 1990s in Eastern Europe (Krastev and Holmes 2019; Hockenos 2019), this chapter critically examines the notion of "supportive" interventions in contemporary revolutionary contexts.

In the twenty-first century, revolutionary struggles as well as their international and domestic contexts tend to diverge from historically established patterns. In contrast to previous waves of revolutions, the Arab uprisings emerged in an ideological vacuum. Communism had failed during the Cold War, and liberalism became tainted by rising social inequality and pervasive corruption in the two decades thereafter (Krastev and

Holmes 2019). In the absence of homegrown revolutionary liberation ideologies, the unexpectedly rapid toppling of their dictators left Tunisian, Egyptian, Libyan, and Yemeni revolutionaries without any preconceived plans for political and economic transformation (Bayat 2017). In terms of their organization and strategies, the Arab uprisings were equally difficult to read. In contrast to the takeover of state power by armed revolutionary movements and vanguard parties, contemporary revolutionaries in the Arab region tended to be organized in loose horizontal networks and showed more modest ambitions: overcoming oppressive power structures at home rather than taking power, and reshaping the state without owning it. Hence the Arab uprisings have been criticized for not adding up to more than fragmented struggles whose organizational weakness accounts for their eventual ineffectiveness (Abdelrahman 2015).

However, across history, a revolution's success or failure has been determined by a set of factors beyond internal characteristics. Among them is the role played by dominant outside powers (Foran 2005, 24), which will be the focus of this study. International responses to the Arab uprisings had been shrouded in inconsistencies from the onset (Morey et al. 2012). Patterns emerged and blurred again in accordance with the fast pace of events on the ground and the fluidity of interest constellations. In order to connect to the theme of this book, this chapter is limited to analyzing liberal peacebuilding interventions and their compatibility with revolutionary agency. Similar to the revolutions of 1989, Western governments and international organizations (such as international financial institutions [IFIs], UN, and EU) read the Arab uprisings as democratization movements and responded with liberal peacebuilding to promote the importation of Western governance models. The revolutionary moment provided a window for rapid transformation (Skocpol [1979] 2015), which peacebuilders could exploit for the rapid liberalization inherent in the liberal peace (Paris 2004).

But were these interventions compatible with the emancipatory drive of the Arab uprisings? Without taking power, Arab revolutionaries cannot disrupt the field of intervention that has engulfed their countries. In Tunisia, Egypt, Libya, and Syria, revolutionary networks have been marginalized, while domestic governments and their foreign counterparts negotiate the forms, extent, and costs of international intervention that their countries have to bear. By 2011, when the Arab uprisings broke out, liberal peacebuilding and statebuilding had come under severe criticism (e.g., Richmond 2006; Chandler 2010; Paris 2010). Neoliberal development policies as part of the statebuilding agenda were even explicitly identified as fueling the political upheavals across the Arab region (Azmeh 2014; Hanieh 2013; Achcar 2013; Joya 2011; Adly 2011). Yet, translating revolutionary calls for regime change into demands for democratization, an inclusive constitution, rule of law, a vibrant civil society, job growth, security-sector reform, and human rights might have raised interveners' hopes that liberal peacebuilding—despite its shortcomings—could find fertile ground on the southern shore of the Mediterranean.

Since international conflict management, peacebuilding, statebuilding, and development were designed with different problems and contexts in mind (e.g., for the aftermaths of civil wars, socialism, decolonization, or ethnic conflict), the revolutionary states of the Arab region, from Tunisia to Yemen, pose fresh questions. Where civil society can connect its claims against powerful actors and the state with the international community or with donors, revolutionary networks are far more isolated even where they make similar emancipatory claims. Paradoxically both depend on the state and oppose it: there is no equivalent of a peace process to support them.

This chapter focuses on the tensions between the liberal peace's orthodox and emancipatory strands (Richmond 2006) by asking: Does liberal peacebuilding support or hinder revolutionary agency in the Arab region? The "constitutional peace" that domestic reforms aim at and the "institutional peace" that liberal peacebuilding promotes follow what Richmond describes as the orthodox model of the liberal peace. Emancipatory approaches in the Arab uprisings, by contrast, are more difficult to identify. Revolutions are complex phenomena insofar as the veneer of unity among the revolting masses crumbles as soon as the previous regime gives way to the revolutionary challenge. Among the revolutionaries, this marks the beginning of the political struggle over state and class transformation (Skocpol [1979] 2015). In this process, hierarchical power diffuses in a way that disadvantages the revolutionaries more than state power (Vaneigem [1967] 2017): The latter broadens, while revolutionary networks fragment and their struggles shift to the micropolitics of everyday life. This chapter argues that in the aftermath of the Arab uprisings patterns of *everyday state formation* emerged, here understood as micropolitical agency to transform the state, based on a fundamental reconsideration of state-society relations. The overall objective of this open-ended struggle is to refocus the state away from the interests of elites and toward broader notions of justice, dignity, and participation. This might imply constraining the coercive instruments of the state apparatus, reinforcing participatory forms of governance, challenging state property as well as the state's defense of property rights, or cleansing state institutions of their networks of corruption and patronage. The institutional and political strategies of everyday state formation as well as its sustainability differ depending on its political context. Transcending their differences is the quality, which makes everyday state formation relevant for international relations and peace and conflict studies—its capacity to reenvision the state and help rebalance state-society relations, even if only gradually and sometimes limited to the micro level of state organization.

Hence this chapter juxtaposes revolutionary agency with liberal peace interventions. First, it contrasts international statebuilding with regional state formation dynamics prior to and after the Arab uprisings. Then it explores resistance to democratization and development. While a desire for transformation is shared by liberal peacebuilders, neoliberal statebuilders, and Arab revolutionaries, this chapter aims to explore how different, in parts even incompatible, their strategies and objectives are.

The research presented here is based on fieldwork conducted since 2014 through semi-structured interviews with revolutionaries in Tunisia and Egypt,[1] combined with analysis of secondary literature and policy documents. Rather than focusing on one country or the policies of one peacebuilding actor, this chapter analyzes a range of connections between orthodox and emancipatory strands of the liberal peace. By elaborating everyday state formation and its relationship with liberal peacebuilding and domestic reforms, this chapter hopes to advance our understanding of contemporary revolutionary struggles and their implications for peace and conflict studies. Thereby, it seeks to close a gap in that field's local turn (Mac Ginty and Richmond 2013) and international relations' interest in micropolitics (Solomon and Steele 2016). While ethnographic methods in peace and conflict studies have recently helped to ascertain political claims of subaltern actors as well as different variations of local agency and peace formation (Richmond 2016), political struggles in revolutionary societies have remained the discipline's blind spot.

State Formation versus Statebuilding

Marx lamented in the *Eighteenth Brumaire of Louis Bonaparte* (1869) that bourgeois revolutions have led to the reinforcement of the state apparatus through the constant expansion of its bureaucracy and army. Since this entrenched the bourgeoisie's rule over the proletariat, Lenin ([1932] 2011) envisaged that the proletarian revolution would seize the state only to let it "wither away." In reality, however, the Russian and Chinese revolutions gave rise to a highly centralized state bureaucracy (Skocpol [1979] 2015). Indeed this state formation impulse has been so strong that theorists regard the institutionalization of a new administration, army, and government as the minimum requirement of a social revolution (Skocpol [1979] 2015). Revolutionary objectives such as the pursuit of freedom and emancipation, by contrast, were often abandoned once revolutionary movements had seized the state. Growing suspicious of the masses whose mobilization brought them to power, revolutionary governments increased the surveillance and repressive capacities of the state in order to control social movements (Trotsky [1937] 2004, 120–128). Revolutionary state formation was thus often marked by oppression and exclusion, amounting in extreme cases to totalitarianism (Skocpol 1994, 280).

Similarly, Arab state formation has been a history of exclusion and oppression, from the early despotic state formation period to colonial control, nationalist governments, and previous revolutionary periods. When the Arab republics emerged from colonialism in a conflict-prone environment with arbitrarily set borders, radicalized social forces, and other forms of in-built instabilities, "defensive state formation" was accompanied by wealth distribution to appeal to the masses (Hinnebusch 2015). In the state consolidation phase of the 1970s and 1980s, domestic elites established a plethora of security institutions to negate any internal "threat" as well as bureaucratic structures to penetrate society (Bellin 2004). Hence the Arab republics have developed into "fierce" rather than strong states (Ayubi 2009, 447–450): They deal with society mainly through coercion, while their infrastructural capacity to reach into society for the purpose of regulation and law enforcement has remained weak.

Yet postcolonial state formation processes were also marked by massive public investment in education and state employment, aiming to cultivate citizens' loyalty to the state. The populist nation-building strategies of 1950s to 1980s thus instilled societal expectations of state services and redistribution (Hinnebusch 2015). Since the 1980s, however, external pressure to roll back the state has become prevalent in the Arab region. When Soviet patronage dried up in the 1980s and many Arab republics focused their foreign policies westward, they became subject to structural adjustment measures (Hinnebusch 2015, 99–103). Conditionality resulted in a diversion of funds from much needed public services to regime security and patronage, shifting the balance of forces toward the neopatrimonial state (Fukuyama 2005, 22–23).

The pressure to shrink the state caused two problems that revolutionaries are now struggling with. First, privatization as a source of personal enrichment increased inequality and created an entrenched state-dependent elite (Marfleet 2009, 22; Naggar 2009, 44–47). In the aftermath of the Arab uprisings, connections between senior state officials and businessmen resurfaced in attempts to restore the "deep state". Second, corruption became all pervasive through the impoverishment[2] of the civil service. Pressured by the IFIs to reduce their payroll, Arab states inflicted wage compression on the public sector to avoid mass layoffs. Hence conditionality to enforce neoliberal statebuilding had arguably transformed the state into a more exclusive, less caring, and more corrupt entity.

It is thus not surprising that among the Arab uprisings' different driving forces, rebellion against the nexus of state oppression and economic deprivation loomed large in 2011. Mohamed Bouazizi's self-sacrifice in the face of corrupt policing, unaccountable governance, and a lack of socioeconomic perspectives resonated sufficiently with Tunisians to prompt mass mobilization across the interior. The Egyptian uprising took off on National Police Day (25 January) as a protest against the brutality of the security state, but joined workers, pro-accountability and democratization protesters. Indeed demands for "bread/work, freedom and dignity" merged political and economic regime change in both uprisings (Beinin 2016).

In such contexts, international statebuilding dissects socioeconomic from security-related demands, supporting the latter but ignoring the former. Statebuilding's mixture of demobilization disarmament reintegration (DDR)/security sector reform (SSR), structural adjustment, democratization, rule of law programming, marketization, and support for civil society (Paris 2004) prioritizes security provision and political reforms at the expense of a social contract that states historically provide (Richmond 2014). In the light of the Arab uprisings' demands for bread and social justice, the neoliberal nature of statebuilding constitutes a counterrevolutionary response to antiregime mobilization.

Within its security portfolio, however, statebuilding interventions could help to meet the revolutionaries' demands for accountability, rule of law, and human rights. Indeed statebuilding seeks to export the model of a strong state that provides security, ensures the rule of law and accountability and establishes a legal framework for development through good (enough) governance (Fukuyama 2011, 420). Hence the objective to transform the authoritarian architecture of states in order to rectify the excesses of unmitigated state formation processes represents a shared endeavor between statebuilders (Richmond 2014, 240) and revolutionaries. In the aftermath of the Arab uprisings, however, statebuilding's accountability-related agenda was neglected in favor of its capacity-building functions (Hanau Santini and Cimini 2019).

Designed as an intervention to help weak or failing states to contain threats to regional or international stability, statebuilding not only aims to reform security sectors but strives to build up their coercive capacity (Chandler 2006). A deteriorating security context in the aftermath of the uprisings shifted popular support and international assistance from security sector reform with an emphasis on accountability to security assistance, which prioritized effectiveness (Hanau Santini and Cimini 2019). After the attacks on foreign tourists in Sousse and Tunis's Bardo Museum in 2015, even the EU lifted its pressure for accountability-related reforms in the Tunisian security sector and settled for an effectiveness-focused approach (Hanau Santini and Cimini 2019). While the G7+ mechanism achieved a better coordination of security-related aid with domestic reform efforts,[3] it also shifted reform toward external priorities, such as counterterrorism measures and the security of tourism, borders, and airports. Hence the focus on regime security during the dictatorial era was replaced by a focus on stability rather than human security. Indeed a security analyst recently remarked that in Tunisia's security sector, none of the revolutionary goals (e.g., achieving accountability, anchoring human rights standards and the rule of law in policing, obtaining civilian oversight) had been achieved.[4] Outside of revolutionary networks, however, the stabilization agenda is remarkably popular across the Arab region. Due to widely shared perceptions of growing crime, terrorism, and the looming specter of state collapse and civil war in Libya, Syria, and Yemen, stability is again in stronger demand than human rights and accountability (Sayigh 2016, 9–10).

This widespread desire for stability has enabled remnants of the old regimes (*feloul*) to restore their power through popular support for their law-and-order and counterterrorism platforms since 2013. In Egypt, President Abdel Fattah al-Sisi's upgraded authoritarianism might become a test case for this reversal to traditional Tillyan forms of state formation (i.e., the expansion of coercive capacity to eliminate internal rivals and insurgents; Tilly 1985). By neglecting the demands of the 2011 uprising—"bread, freedom, social justice"—he has pegged his political fate to delivering stability through pacification. This strategy not only led to massacres in July and August 2013 but unleashed unprecedented levels of violence in the competition between different police and security agencies (Marfleet 2016, 201). Moreover, by simultaneously eroding democracy through the constitutional changes of 2019 (Hawthorne and Miller 2019) and perpetuating the model of a fierce but weak state with an impoverished population, he is currently re-creating the same conditions that led to the toppling of his predecessor. In the eyes of many revolutionaries, financial support for al-Sisi's oppressive regime implicates his donors (especially the Gulf Cooperation Council countries, but also the IMF, the US, and the EU) in the regime's crushing of the democratic opposition and the restoration of dictatorship in Egypt.[5]

In stark contrast to external statebuilding and domestic pacification approaches stand revolutionary forms of state formation, which aim to create participatory political orders at the grassroots. The weakening of central control in 2011 allowed Egyptians,

Syrians, and Libyans to rediscover the "lost treasure" of revolutionary history: the spontaneous organization of people into councils (Arendt [1963] 1990, 244–281). As projects of self-government with a focus on local needs, local councils are examples of *everyday state formation* and represent a rupture in the historical trajectories of Arab state-society relations (El-Meehy 2017). While some local councils perpetuated social hierarchies in their composition or failed to become truly participatory (El-Meehy 2017), the councils shifted the volatile demarcation between state and society (Mitchell 1999) and inscribed revolutionary principles such as participatory governance in the local level of state organization.

The Libyan municipal councils as the successors of local councils have recently attracted the attention of international statebuilders. In contrast to statebuilding's tendency to replace, co-opt, or deconstruct customary systems of order (Boege et al. 2008, 8), the municipal councils integrate traditional authorities into vital governance mechanisms. Shura councils, tribal councils, religious leaders, councils of wise men, and city notables are participating in newly formed Libyan institutions for dispute settlement (Detzner and Perroux 2017, 3–4). Crucially the more successful councils have developed extraconstitutional tax systems to address local needs, while also asserting control over local militias (Detzner and Perroux 2017, 3–5). Due to the effectiveness of local governance, international statebuilders such the UN Development Program (UNDP) and the EU have shifted their financial support from internationally coordinated mechanisms to the municipal councils (Megerisi 2018, 7–9). They are the only examples of *everyday state formation* that are financially supported by liberal peacebuilding. Despite their various shortcomings, some of Libya's municipal councils have maintained a progressive antisystemic stance insofar as they have established alternative forms of political order to both: the statelessness of the Gaddafi era as well as the dysfunctional and contested national state institutions that succeeded it.

Democratization versus Participation

According to Arendt ([1963] 1990), revolutions strive for freedom but often stop at liberation. Essential to distinguish different types of revolutions is an exploration of revolutionaries' concept of freedom. In defense of bourgeois revolutions, Tocqueville ([1856] 1983, 169) suggested that "The man who asks of freedom anything other than itself is born to be a slave." The communist revolutions, by contrast, were driven by the goal of achieving freedom from poverty, and thus prioritized the transformation of class structures at the expense of political freedoms and civil liberties. For Lenin and Trotsky democracy implied the rule by and for the bourgeoisie, which was characterized by "social inequality and lack of freedom hidden under the sweet shell of formal equality and freedom" (Luxemburg [1937] 2006, 220). Crucially, where democratization did occur in the revolutionary process, it opened up political space for counterrevolutionary forces to block political transformation through authoritarian populism (Israel 2014; Foran 2005).

While the demands for freedom in the Arab uprisings pointed in the direction of democratization, this was never a foregone conclusion. In Syria, Yemen, and Libya, emancipatory movements have been buried under civil war or militia rule. Only in Tunisia and Egypt have substantial democratization processes been prompted by the uprisings of 2011. Yet revolutionaries in both countries were acutely aware of the potential of rapid democratization to undermine the emancipatory potential of their revolutions (e.g., early elections, leading to Islamization of parliaments and biased constitutional assemblies).[6] Rather than cherishing elections a hard-won gain of their revolution, many revolutionary factions have boycotted ballots. Political developments in Egypt validated their opposition and confirmed Marx's suspicion towards bourgeois revolutions: parliamentary democracy has allowed authoritarian forces and vested interests to consolidate their grip on power behind a democratic façade (Kandil 2016, 340–343). The pinnacle of this development was the election of the former general Abdel Fattah al-Sisi in 2014, who first transformed the parliament into a rubber stamp of the executive (Fahmy 2017, 50–53; Kandil 2016) then exempted himself from presidential term limits and expanded the military's influence in politics (Hawthorne and Miller 2019). In Tunisia the failure of democratization to reduce socioeconomic inequalities severely undermines its potential to achieve positive peace. Hence the difficulties in Tunisia's and Egypt's political liberalization warrant analysis of the suitability of the standard model of majority democracy for revolutionary societies in the Arab region.

Since 2012 Egypt's and Tunisia's reform processes have been contested by revolutionary networks and counterrevolutionary forces. The latter strive to reinstate a leadership vanguard to provide security and safeguard economic privileges at the expense of human rights, socioeconomic justice, and democracy. By contrast, many Tunisian and Egyptian revolutionaries regard leadership as a means of oppression and seek rights and social justice through more direct forms of participation (Filiu 2011). Many of the revolutionary networks that emerged from the uprisings of 2011 did not aspire to become political parties, since their horizontal organizational structures did not square with the hierarchical order of traditional party politics.[7] Instead they hoped for a more fluid system of political participation, in which they could continue to influence the transformation process through mass street mobilization. Initially this strategy seemed successful. Several interim governments and underperforming elected governments had to resign under the pressure of protests between 2011 and 2013. Moreover the Tunisian constitution-makers established a process of consultation with civil society as well as national dialogues, which allegedly resulted in changes in the new constitution (Jasmine Foundation 2014).

Revolutionaries who considered establishing their own parties, by contrast, realized quickly that the newly reconfigured political system was an uneven playing field, biased against their participation. Without the financial capacity to campaign across the country, revolutionary networks were unable to compete for power at the ballot box.[8] Instead activists had to join parties that revolved around well-known personalities and found themselves systematically marginalized within those structures (Abdalla 2016; Honwana 2013). Worse still, the legalization of new parties and new private media outlets allowed the restoration of pre-revolt patronage networks to power. By bankrolling new parties and buying media influence to bias the electorate in their favor, former regime cronies managed to rekindle their intimate relationship with executive power through the democratization process (Roll 2013).

National elections thus created a conflict between revolutionary and democratic legitimacy. In Egypt until 2013 and still in Tunisia, the marginalized revolutionaries initiated protest campaigns as an informal system of checks and balances. Initially, street protests allowed revolutionaries to continue their activism in the margins of their everyday lives, while the distance to executive and legislative power prevented a corruption of revolutionary agency. Yet being excluded from formal decision-making left their organizations vulnerable to attacks from the resurgent counterrevolution. Indeed the arsenal of traditional state formation strategies came to be deployed against revolutionary movements in the form of restored security apparatuses and biased justice systems.

Democratization arguably became a strategy of the counterrevolution: newly elected governments in Egypt and Tunisia used their lawmaking capacity to criminalize protests and justified crackdowns on peaceful demonstrators with the institutionalization of parliamentary politics (Ansari and Zaree 2012).[9] Thousands of activists found themselves in courts charged with offenses against laws that enacted authoritarian policies through parliamentary means. Revolution fatigue spread throughout Arab societies, permitting the state to crack down on peaceful protests (Abdalla 2016, 49).

Contesting state power in the streets became more dangerous still when the embattled Ennahda and Muslim Brotherhood governments organized massive counterprotests. Demonstrations and counterdemonstrations turned into "competitive street mobilisation," which further polarized society, heightened tensions, and resulted in periods of intense political instability (Boubekeur 2015). The summer of 2013 constituted a watershed moment in the Egyptian democratization process. The military responded to mass protests by overthrowing President Mohamed Morsi and committing massacres against Muslim Brothers before a court banned them as a terrorist organization. The subsequently passed protest law in combination with mass military trials of peaceful protesters crushed the revolutionary networks in Egypt, quelling any potential for an emancipatory peace. To escape this fate, the Ennahda-led government in Tunisia adopted "strategic minimalism," engaged in negotiations with its political opponents, and eventually resigned amid mass protests (Marks 2017).

Democratization as a strategy of liberal peacebuilding to instill a political culture of compromise and consensus-building proved unable to reconcile Islamists and secular liberals in Egypt. The latter regard Islamist movements as a threat to democracy, religious minorities, and women's rights. Islamists, by contrast, perceive these allegations as a continuation of their prosecution under authoritarian governments (e.g., Hallaq 2013; Hamid 2011; Roy 2013). In the practice of postrevolt politics, both sides showed antidemocratic reflexes. The Muslim Brotherhood in Egypt displayed its illiberal tendencies by hijacking the constitution-making process and in President Morsi's power grab in November 2012 (Rashidi 2013). In fear of Islamist domination, Egyptian liberals first supported the military coup against an elected president, then accepted the massacres of 2013 as measures of "necessity" and concurred with the abortion of the democratization process under President al-Sisi (Fahmy and Faruqi 2017). Hence democratization ultimately set liberals and Islamists on a collision course that ended with a reversal of the democratization processes in Egypt.

At the heart of this clash between Islamists and liberals lies the lack of a historically evolved societal consensus on core values, secured in the form of constitutionally non-negotiable rights (Hamid 2016, 250–257). In its absence, distrust between the two sides escalated into the perception that group interests cannot be realized through democratization and compromise. Moreover, by expanding their political energy on fighting one another rather than working to mitigate the dire socioeconomic and political issues the countries faced, both liberals and Islamists ensured that democratization became irrelevant to poverty-stricken segments of society.

In this context, international peacebuilding actors—including the UN, IFIs, donors, and international NGOs—insist on elections to create legitimate authorities (Reilly 2019) but are struggling to connect with demands for alternative forms of representation. In particular, they are not equipped to deal with leaderless networks or those marginalized youth who have turned to Salafism as a form of solidarity (Lamloum 2016). Instead liberal peacebuilding focuses largely on NGO funding as a strategy to overcome the various exclusions that pervade postdictatorial contexts. Yet this support fundamentally misjudges the underlying asymmetry of power. A fragmented civil society is no match for the alliances between counterrevolutionary forces: an unreformed judiciary, military, and police as well as illiberal media and vested business interests. As the 2014 presidential elections in Egypt showed, the collaboration between those forces might be able to brutally crush all opposition and still see their victory legitimized domestically by an electoral mandate, externally validated by EU election observers (Asseburg 2014, 2).

Many Tunisian revolutionaries had rejected peacebuilding's civil society funding as a form of postcolonial control even before those alliances surfaced.[10] They recognized that support for NGOs (most of which are based in metropolitan centers) would further entrench the divisions between highly educated political activists in coastal cities and marginalized youth in the interior. Faced with the shortcomings of domestic reform processes and liberal peacebuilding, new practices of *everyday state formation* evolved. Bridging class divisions and countering geographical differentiation requires horizontally organized protest movements with chapters across the country. Democratization, by contrast, focuses political agency on national parliaments as protest campaigns have to lobby parties in order to achieve their goals.[11] This centripetal effect on activism erodes movements' capacity to join disparate pockets of discontent across the country as the success of campaigning ultimately depends on finding allies in the capital, not on their mobilizing capacity in the countryside.

After almost a decade of trying to foster state transformation from the margins of the political spectrum, activists sought closer connections to power in order to achieve progress on the goals of the revolution. Alienated by a democratization process that has been dominated by conservative parties (Nidaa Tounes and Ennahda), many Tunisian activists threw their weight behind an independent candidate: Kais Saied. Through canvassing and campaigning on social media,[12] they helped to mobilize the youth vote, which won Saied the presidency. What made him appealing to activists is his radical platform for transforming the state through decentralization, popular assemblies, anticorruption drives, electoral reforms, social economy, and direct democracy (*Arab Weekly* 2019). Here everyday state formation actors connected their agenda more closely than before to high office in recognition of the fact that democratization had so far marginalized revolutionary agendas. Given the president's weak constitutional position though, the effectiveness of this new approach might be limited.

Development, Precarity, and Resistance

What makes revolutions attractive to radicals and reviled by conservatives is their potential for fundamental transformation—not only of the state but also of the economic system and the class structure that underpins it. For the poor, political equality may appear to be mere deception if not bolstered by economic rights (Moore 1966). Hence social revolutions across the world have overturned class structures and upended traditional social hierarchies (Skocpol [1979] 2015, Foran 2005). Yet the failure of socialism in Eastern Europe suggests that freedom from poverty might lose mass support if achieved at the expense of all other freedoms. Tyranny, oppression, surveillance and corruption were answered by resistance at all levels (Chirot 1999; Verdery 1999), revealing that the dystopian reality of socialist rule had discredited its utopian ambitions. Hence, only twenty years after the collapse of socialism the Arab uprisings burst onto the scene in search of both: bread and freedom.

At the heart of the Arab uprisings lies another struggle for an alternative to unfettered capitalism. Economic grievances have played a large role among the different driving forces of antiregime mobilization that erupted in 2011 (Achcar 2013). Since the 1980s, most governments in the region have implemented neoliberal policies, here understood as "political economic practices that propose that human well-being can best be advanced by liberating individual entrepreneurial freedoms and skills within an institutional framework characterized by strong private property rights, free markets, and free trade" (Harvey 2005, 2). For the masses, the outcome was a sustained erosion of basic economic and social rights, condemning an entire generation of youth to "a future of unemployment, low-wage work and social exclusion" (Hanieh 2013, 71). On the eve of the uprising, 40 to 50% of all employment in Egypt and Tunisia had moved to the informal economy and thus left workers devoid of any rights (UNDP 2009, 111). Even formal employment promised no certain escape from misery as working conditions and wages had been eroded. In Egypt real wages had declined significantly since moderate GDP growth levels (4.3% from 1983 to 2007) had been swallowed up by record levels of inflation (12% during the same period; Naggar 2009, 44). The conditions of many industrial jobs in the Arab region are reminiscent of the conditions of the working class described by Marx and Engels. Neither host communities of industrial plants nor their workers enjoy even the most basic protections against chemical contamination, life-threatening hazards at the workplace, or pollution (see Shenker 2016).

Precarity is central to the neoliberal development model as institutionalized insecurity cements the domination of the state and capital over labor (Bourdieu 2003, 29). While workers were systematically impoverished in the private and public sectors, networks of patronage made businessmen and the state officials who struck deals with them rich. Mafia-like networks of military and state bourgeoisies developed with the decay of the ruling regimes (Achcar 2013, 91/92). Endemic corruption was the result: at the top end of the state it served to shore up privileges; at the lowest end, it ensured survival.

However, precarity also enables new forms of cross-class solidarity such as the emergence of social movements like April 6 in Egypt or revolutionary student networks in Tunisia.[13] Since the threat of economic insecurity affects not only traditionally marginalized groups (unemployed, unskilled workers, women, youth, small peasantry) but also the middle class (Joya 2011; Adly 2011), cross-class mobilization around shared economic interests is much more likely. Moreover neoliberal deregulation frustrates popular expectations for a social contract, which had been instilled by the expansion of the state since the 1950s. Worse still, neoliberal development has failed to deliver: despite exploitative working conditions, low wages, and at least partial liberalization of Arab labor markets, the region is lagging behind in development (Achcar 2013).

The big questions after the uprisings were thus how to create jobs that offer decent conditions and appropriate wages while remaining competitive, how to expand state services and infrastructure without bankrupting the state, and how to cleanse the state and the economy of pervasive corruption without large-scale institutional destruction. Instead of developing policies suitable for the revolutionary contexts in the Arab region, the IFI that had imposed the neoliberal development model on many resource-poor countries through a vicious cycle of aid and debt (Hanieh 2013) continued to peddle its old policy prescriptions of deregulation after the uprisings (Hanieh 2015). Even the EU continues to push for deeper market penetration through a Deep and Comprehensive Free Trade Area, which is so unpopular in Tunisia that no government has yet dared to sign on to it.

Due to entrenched economic interests or a lack of vision, none of the elected governments in postrevolt Tunisia and Egypt has attempted to transform its economy either (e.g., Alexander and Bassiouny 2014; Achcar 2016). Mass grievances aired throughout decades of growing strike waves and the uprisings (Beinin 2016) have been neglected by domestic governments and external interveners alike.

While the uprisings did not produce a counterhegemony to the prevailing neoliberal orthodoxy, they generated ample revolutionary practices of contestation. In Egypt, workers occupied their workplaces, cleansed their management of representatives of the old regime, and democratized their organizations (Alexander and Bassiouny 2014;

Shenker 2016). In Tunisia's southern interior, a takeover of the oasis of Jemna by the local community inspired young revolutionaries throughout the country. Jemna's community managed to reverse colonial dispossession, challenge state property based on communal claims, and develop the oasis's infrastructure through self-management (Hamouchene 2017). Meanwhile protest campaigns, targeting extractivism and stark regional wealth disparities were able to achieve increased investment in poor regions and reveal the weakness of the army in asserting state control (Cherif 2017; Grewal 2019).

Yet the disjointed character of these struggles undermined their transformational potential (Alexander and Bassiouny 2014). Even where different initiatives of *everyday state formation* are demonstrating mutual support (e.g., by deploying each other's slogans and symbols; see Mahmoud 2018), disparate pockets of dissent have only rarely and briefly joined forces since 2011. Given the asymmetry of power between scattered resistance on the one hand and the state's arsenal of oppression combined with the external pressure of IFI conditionality on the other, resistance often had to back down. The IFIs as well as liberal peacebuilders (e.g., EU, UN) failed to support economic transformation, even in the one aspect where the revolutionary and international development agendas are aligned: in their belief that corruption needs to be rooted out in order to foster development. While most international organizations have financed projects on anticorruption, they have often invested in technical fixes and transparency procedures rather than engaging with the structural issues driving corruption (see, e.g., UN Development Program, Independent Evaluation Office 2019).

Despite their limited effectiveness, the struggles to cleanse the private and public sectors of regime holdovers and to reverse the economic marginalization of rural communities continue to be relevant objectives in the pursuit of a civil peace. These manifestations of *everyday state formation* aim to alter state-society relations in pursuit of grassroots democratization, disruption of political control, and the leveling of illegitimate hierarchies (Alexander and Bassiouny 2014). Moreover this resistance fundamentally reconsiders the role of the state. It refuses to accept the neoliberal state's prioritization of capitalist interests and its defense of exploitative property rights over the needs of the poor. Revolutionary transformation of the economy after the Arab uprisings thus implies a reordering of the state's core functions by prioritizing the formulation of a new social contract. Continued waves of economic protests and strikes, not only in Tunisia but also in heavily controlled Egypt, demonstrate that the resistance against dysfunctional development models continues.

Conclusion

This analysis has elaborated the divergence between the orthodox and emancipatory strands of the liberal peace in the aftermath of the Arab uprisings. Liberal reforms and peacebuilding interventions as the two orthodox strands of the liberal peace have often reinforced one another in practice since the uprisings of 2011. By contrast, emancipatory approaches to peace as pursued by revolutionary activism in the Arab region have tried to steer domestic reforms toward state transformation, while remaining largely disconnected from liberal peacebuilding.

In response to the shortcomings and exclusions of domestic reform processes, networks of political activists have been striving to transform the state in pursuit of revolutionary goals. Certain patterns of *everyday state formation* emerged. In the initial stages of the Arab uprisings, the revolutionaries in large parts of the region[14] steered deliberately clear of power. In contrast to established patterns of revolutionary processes such as social revolutions (Skocpol [1979] 2015), multiple sovereignty (Tilly 1973) and negotiated revolutions (Lawson 2005), contemporary Arab revolutions may rather fall into the category of Garton-Ash's "refolutions" (Garton-Ash 1989, Bayat 2013). Yet in contrast to the refolutions in Eastern Europe, Arab revolutionaries were more suspicious of copying Western blueprints and developed their own emancipatory practices for continuous state transformation outside of the formal channels of decision-making as discussed in this paper. With revolutionary legitimacy on the wane, some (and in other cases all) avenues of everyday state formation became blocked by the rise of counterrevolutionary alliances. Over time, revolutionary agency had to seek closer connections to power in pursuit of revolutionary objectives. In the absence of an overarching ideology as well as internal cohesion and due to its micro-level outcomes, this type of continued revolution in the margins has been largely ignored.

Within this field of contentious politics, liberal peacebuilding has rarely supported revolutionary agency in the postuprising period. In the spheres of democracy promotion, statebuilding, and development, liberal peacebuilding interventions have at times accentuated the flawed nature of the domestic reform processes. Most clearly counterrevolutionary among these interventions is the continuous pressure for neoliberal development, regardless of the exclusions, inequalities, and resistance that it produces. Peacebuilders' initial support for revolutionary objectives such as cleansing the state of patronage networks or establishing accountability in the security sector was over time abandoned due to competing security goals. While domestic counterrevolutionary elites forged alliances with international supporters under the stabilization paradigm, alliances among revolutionary networks and between them and other progressive forces seem tenuous. Due to its isolation and internal fragmentation, everyday state formation is even more transient and limited than Richmond's peace formation. In this field of political forces, liberal peacebuilding's tendency to side with counterrevolutionary alliances creates a new type of hostile environment for revolutionary agency.

Notes

This research has been supported by the British Academy project "The International Peacebuilding Architecture and State Formation in Revolutionary Societies" (award no. SG140803) and by the European Union's Horizon 2020 research and innovation program (EUNPACK- A conflict sensitive unpacking of the EU comprehensive approach to conflict and crises mechanism), under grant agreement no.: 693337).

1. Initially spokespeople for revolutionary networks or campaigns were selected for interviews, but over time snowballing expanded the pool of interviewees that participated in this research.
2. According to Naggar (2009, 49), 95% of Egypt's 5.8 million state employees and their families were considered poor or extremely poor by World Bank standards (living on $2/$1 per day) on the eve of the uprising.
3. The Tunisian government was included in the G7+ mechanism in order to tailor external statebuilding interventions to the domestic reform agenda.
4. Participant in Global Challenges Research Fund workshop in Tunis, 1 November 2019.
5. It should be noted that these donors have very different aid strategies and reviewed their funding in different ways after the army's massacres of Muslim Brotherhood supporters in July 2013.
6. In conversations with the author, Tunisian revolutionaries questioned the notion that their uprising had a clear postdictatorial agenda which could be characterized as a pro-democracy movement. For Egypt, see Tohamy (2016, 201–217).
7. Author's interview with leader of the 6 April movement in Egypt (15 February 2014) and revolutionaries in Tunisia (7 April 2014).
8. Author's interview with a revolutionary in Tunis, 17 March 2014.
9. On Tunisia, see the reporting of Nawaat on the topic: http://nawaat.org/portail/tag/criminalisation-des-mouvements-sociaux/.
10. Author's interview with a revolutionary in Tunisia, 12 July 2018.
11. The campaign Manich Msamah (I Don't Forgive) in Tunisia exemplifies this: after two years of campaigning against the government's efforts to protect perpetrators of economic crimes from the transitional justice process, it was not the growth of the protest movement but its lobbying of lawmakers that led to partial success.
12. Author's interview with a political analyst, 1 November 2019.
13. Student revolutionaries explained in interviews in April 2014 that their political struggle did not start in 2010 but during the uprising in the mining region of Gafsa in 2008.
14. In the first wave of the Arab uprisings (which included Tunisia, Egypt, Libya, Syria, Yemen and Bahrain), Libya and Yemen constituted an exception to this rule. In the second wave (Lebanon, Algeria, Iraq, Sudan), Sudan may fall into the recognized pattern of negotiated revolutions.

References

Abdalla, N. 2016. "Youth Movements in the Egyptian Transformation: Strategies and Repertoires of Political Participation." *Mediterranean Politics* 21 (1):44–63.

Abdelrahman, M. 2015. *Egypt's Long Revolution: Protest Movements and Uprisings*. London: Routledge.

Achcar, G. 2013. *The People Want: A Radical Exploration of the Arab Uprisings*. London: Saqi Books.

Achcar, G. 2016. *Morbid Symptoms: Relapse in the Arab Uprisings*. London: Saqi Books.

Adly, A. I. A. 2011. "When Cheap Is Costly: Rent Decline, Regime Survival and State Reform in Mubarak's Egypt (1990–2009)." *Middle Eastern Studies* 47 (2):295–313.

Alexander, A., and M. Bassiouny. 2014. *Bread, Freedom Social Justice: Workers and the Egyptian Revolution*. London: Zed Books.

Anheier, H., and W. Seibel. 1998. "The Non-Profit Sector and the Transformation of Societies: A Comparative Analysis of East Germany, Poland and Hungary." In *Private Action and the Public Good*, edited by W. W. Powell and E. S. Clemens, 177–192. New Haven, CT: Yale University Press.

Ansari, M. E., and M. A. Zaree. 2012. *Criminalizing the Egyptian Revolution: Commentary on Draft Laws Restricting Freedom of Expression and Assembly in Egypt*. Cairo: Cairo Institute of Human Rights Studies.

Arab Weekly. 2019. "The Many Faces of Kais Saied." 19 September.

Arendt, H. (1963) 1990. *On Revolution*. London: Penguin Books.

Asseburg, M. 2014. "The EU in the Middle East and North Africa." *SWP Comments* 50/2014. https://www.swp-berlin.org/fileadmin/contents/products/comments/2014C50_ass.pdf.

Ayubi, N. 2009. *Over-Stating the Arab State: Politics and Society in the Middle East*. London: I. B. Tauris.

Azmeh, S. 2014. "The Uprising of the Marginalised: A Socio-Economic Perspective on the Syrian Uprising." LSE Middle East Centre, Paper Series No. 6, November. http://eprints.lse.ac.uk/60243/1/__lse.ac.uk_storage_LIBRARY_Secondary_libfile_shared_repository_Content_LSE%20Middle%20East%20Centre%20Papers_Uprising%20of%20marginalised_2014.pdf.

Bayat, A. 2013. Revolution in Bad Times. New Left Review No. 80 (Mar./Apr.). https://newleftreview.org/issues/ii80/articles/asef-bayat-revolution-in-bad-times

Bayat, A. 2017. *Revolution without Revolutionaries: Making Sense of the Arab Spring*. Stanford, CA: Stanford University Press.

Beinin, J. 2016. *Workers and Thieves: Labour Movements and Popular Uprisings in Tunisia and Egypt*. Stanford, CA: Stanford University Press.

Bellin, E. 2004. "The Robustness of Authoritarianism in the Middle East: Exceptionalism in Comparative Perspective." *Comparative Politics* 36 (2):139–157.

Boubekeur, A. 2015. "The Politics of Protest in Tunisia." *SWP Comments* 13/2015. https://www.swp-berlin.org/en/publication/tunisia-politics-of-protest/.

Boege V, Brown A, Clements K and Nolan A. 2008. On Hybrid Political Orders and Emerging States: State Formation in the Context of 'Fragility'. *Berghof Handbook Dialogue* No. 8. Available at https://www.berghof- foundation.org/fileadmin/redaktion/Publications/Handbook/Articles/boege_e tal_handbook.pdf

Bourdieu, P. 2003. *Firing Back: Against the Tyranny of the Market 2*. London: Verso.

Chandler, D. 2010. *International Statebuilding: The Rise of Post-liberal Governance*. Abingdon, UK: Routledge.

Chandler, D. 2006. *Empire in Denial: The Politics of State-Building*. London: Pluto.

Cherif, Y. 2017. "*The Kamour Movement and Civic Protests in Tunisia*." Carnegie Endowment for International Peace, 8 August. https://carnegieendowment.org/2017/08/08/kamour-movement-and-civic-protests-in-tunisia-pub-72774.

Chirot, D. 1999. "What Happened in Eastern Europe in 1989?". In *The Revolutions of 1989*, edited by V. Tismaneanu, 19–50. New York: Routledge.

Detzner, S., and J.-L. R. Perroux. 2017. "Libyan Local Governance Case Studies." EU Delegation to Libya, July. https://www.docdroid.net/ce1aKnu/00-libya-local-government-report-sept-2017.pdf.

El-Meehy, A. 2017. "Governance from Below: Comparing Local Experiments in Egypt and Syria after the Uprisings." *Century Foundation Report*, 7 February. https://tcf.org/content/report/governance-from-below/.

Fahmy, D. F. 2017. "Egypt's Structural Illiberalism." In *Egypt and the Contradictions of Liberalism*, edited by D. F. Fahmy and D. Faruqi, 31–54. London: Oneworld.

Fahmy, D. F., and D. Faruqi, eds. 2017. *Egypt and the Contradictions of Liberalism*, London: Oneworld.

Filiu, J.-P. 2011. *The Arab Revolution: Ten Lessons from the Democratic Uprisings*. Oxford: Oxford University Press.

Foran, J. 2005. *Taking Power: On the Origins of Third World Revolutions*. Cambridge: Cambridge University Press.

Fukuyama, F. 2005. *State Building: Governance and World Order in the Twenty-First Century*, London: Profile Books.

Fukuyama, F. 2011. *Origins of Political Order: From Prehuman Times to the French Revolution*. New York: Farrar, Straus and Giroux.

Garton-Ash, T. 1989. "Revolution: The Springtime of Two Nations." *New York Review of Books*, 15 June.

Grewal, S. 2019 "Military Defection during Localized Protests: The Case of Tataouine." *International Studies Quarterly* 63 (2):259–269.

Hallaq, W. 2013. *The Impossible State: Islam, Politics, and Modernity's Moral Predicament*. New York: Columbia University Press.

Halliday, F. 1999. *Revolution and World Politics*. London: Palgrave.

Hamid, S. 2011. "The Rise of the Islamists. How Islamists Will Change Politics, and Vice Versa." *Foreign Affairs* 90 (3):40–47.

Hamid, S. 2016. *Islamic Exceptionalism: How the Struggle over Islam Is Reshaping the World*. New York: St. Martin's Press.

Hamouchene, H. 2017. "Jemna in Tunisia: An Inspiring Land Struggle in North Africa." *Open Democracy*, 13 April. https://www.opendemocracy.net/north-africa-west-asia/hamza-hamouchene/jemna-in-tunisia-inspiring-land-struggle-in-north-africa.

Hanau Santini, R., and G. Cimini. 2019. "The Politics of Security Reform in post-2011 Tunisia: Assessing the Role of Exogenous Shocks, Domestic Policy Entrepreneurs and External Actors." *Middle Eastern Studies* 55 (2):225–241.

Hanieh, A. 2013. *Lineages of Revolt: Issues of Contemporary Capitalism in the Middle East*. Chicago: Haymarket Books.

Hanieh, A. 2015. "Shifting Priorities or Business as Usual? Continuity and Change in the Post-2011 IMF and World Bank Engagement with Tunisia, Morocco and Egypt." *British Journal of Middle Eastern Studies* 42 (1):119–143.

Harvey, D. 2005. *A Brief History of Neoliberalism*. Oxford: Oxford University Press.

Hawthorne, A., and A. Miller. 2019. "Worse than Mubarak." *Foreign Policy*, 27 February. https://foreignpolicy.com/2019/02/27/worse-than-mubarak/.

Hinnebusch, R. 2015. *The International Politics of the Middle East*. Manchester: Manchester University Press.

Hobson, J. M. 2010. "Back to the Future of Nineteenth-Century Western International Thought." In *The Global 1989: Continuity and Change in World Politics*, edited by G. Lawson, C. Armbruster and M. Cox, 23–50. Cambridge: Cambridge University Press.

Hockenos, P. 2019. "Germany's Far-Right Freedom Fighters." *Foreign Policy*, 29 August. https://foreignpolicy.com/2019/08/29/germanys-far-right-freedom-fighters-afd-merkel-saxony-brandenburg/.

Honwana, A. 2013. *Youth and Revolution in Tunisia*. London: Zed Books.

Israel, J. 2014. *Evolutionary Ideas: An Intellectual History of the French Revolution from The Rights of Man to Robespierre*. Princeton, NJ: Princeton University Press.

Jasmine Foundation. 2014. "An Interview with Badreddine Abdelkefi: The Role of Civil Society in Drafting the New Constitution." http://www.jasminefoundation.org/en/?p=903.

Joya, A. 2011. "The Egyptian Revolution: Crisis of Neoliberalism and the Potential for Democratic Politics." *Review of African Political Economy* 38 (129):367–386.

Kandil, H. 2016. *The Power Triangle: Military, Security and Politics of Regime Change*. Oxford: Oxford University Press.

Krastev I., and S. Holmes. 2019. *The Light That Failed: A Reckoning*. London: Allen Lane.

Lamloum, O. 2016. "Politics on the Margins in Tunisia: Vulnerable Young People in Douar Hicher and Ettadhamen." International Alert, March. https://www.international-alert.org/sites/default/files/Tunisia_PoliticsOnTheMargins_EN_2016.pdf.

Lawson, G. 2005. *Negotiated Revolutions: The Czech Republic, South Africa and Chile*. New York: Routledge.

Lenin, V. I. (1932) 2011. *State and Revolution*. New York: International Publishers.

Luxemburg, R. (1937) 2006. *Reform or Revolution and Other Writings*. New York: Dover.

Mac Ginty, R., and O. Richmond. 2013. "The Local Turn in Peace Building: A Critical Agenda for Peace.". *Third World Quarterly* 34 (5):763–783.

Mahmoud, I. 2018. "Tunisia's Next Revolution." *Jacobin*, 17 February. https://www.jacobinmag.com/2018/02/tunisias-next-revolution.

Marfleet, P. 2009. "State and Society." In *Egypt: Moment of Change*, edited by R. El Madhi and P. Marfleet, 14–33. London: Zed Books.

Marfleet, P. 2016. *Egypt: Contested Revolution*. London: Pluto Books.

Marks, M. 2017. "Tunisia's Unwritten Story: The Complicated Lessons of a Peaceful Transition." Century Foundation, 14 March. https://tcf.org/content/report/tunisias-unwritten-story/.

Marx, K. 1869. "The Eighteenth Brumaire of Louis Bonaparte." Translated by Saul K. Padover. https://www.marxists.org/archive/marx/works/download/pdf/18th-Brumaire.pdf.

Megerisi, T. 2018. "Order from Chaos: Stabilising Libya the Local Way." European Council on Foreign Relations, Policy Brief No. 265. https://ecfr.eu/publication/order_from_chaos_stabilising_libya_the_local_way/.

Mitchell, T. 1999. "Society, Economy, and the State Effect." In *State / Culture: State Formation after the Cultural Turn*, edited by G. Steinmetz, 76–97. Ithaca, NY: Cornell University Press.

Moore, B., Jr. 1966. *Social Origins of Dictatorship and Democracy*. London: Penguin.

Morey, D. S., C. L. Thyne, S. L. Hayden, and M. B. Senters. 2012. "Leader, Follower, or Spectator? The Role of President Obama in the Arab Spring Uprisings". *Social Science Quarterly* 93 (5): 1185–1201.

Naggar, A. E.-S. E. 2009. "Economic Policy: From State Control to Decay and Corruption." In *Egypt: Moment of Change*, edited by R. El Madhi and P. Marfleet, 34–50. London: Zed Books.

Paris, R. 2004. *At War's End*. Cambridge: Cambridge University Press.

Paris, R. 2010. "Saving Liberal Peacebuilding." *Review of International Studies* 36 (2):337–365.

Rashidi, Y. 2013. "Egypt: The Rule of the Brotherhood." *New York Review of Books*, 7 February 7.

Reilly, B. 2019 "Post-Conflict Elections." In *Palgrave Encyclopedia of Peace and Conflict Studies*, edited by G. Visoka and O. Richmond. London: Palgrave. https://link.springer.com/referenceworkentry/10.1007/978-3-030-11795-5_73-1.

Richmond, O. P. 2006. "The Problem of Peace: Understanding the 'Liberal Peace.'" *Conflict, Security and Development* 6 (3):291–314.

Richmond, O. 2014. *Failed Statebuilding: Intervention, the State and the Dynamics of Peace Formation*. New Haven, CT: Yale University Press.

Richmond, O. 2016. *Peace Formation and Political Order in Conflict Affected Societies*. Oxford: Oxford University Press.

Roll, S. 2013. "Egypt's Business Elite after Mubarak: A Powerful Player between Generals and Brotherhood." Stiftung Wissenschaft und Politik Research Paper, August. https://www.swp-berlin.org/en/publication/egypts-business-elite-after-mubarak/.

Roy, O. 2013. "There Will Be No Islamist Revolution." *Journal of Democracy* 24 (1):14–19.

Sayigh, Y. 2016. "Dilemmas of Reform: Policing in Arab Transitions." Carnegie Middle East Centre, March. https://carnegieendowment.org/files/CEIP_CMEC61_Sayigh_Final.pdf.

Shenker, J. 2016. *The Egyptians: A Radical Story*. Milton Keynes, UK: Penguin Random House.

Skocpol, T. (1979) 2015. *States and Social Revolutions*. Cambridge: Cambridge University Press.

Skocpol, T. 1994. *Social Revolutions in the Modern World*. Cambridge: Cambridge University Press.

Solomon, Tye, and B. Steele. 2016. "Micro-moves in International Relations Theory." *European Journal of International Relations* 23 (2):267–291.

Tilly, C. 1973. "Revolutions and Collective Violence". CRSO Working Paper No. 83. https://deepblue.lib.umich.edu/bitstream/handle/2027.42/50863/83.pdf

Tilly, C. 1985. "War Making and State Making as Organized Crime." In *Bringing the State Back In*, edited by P. Evans, D. Rueschemeyer, and T. Skocpol, 169–191. Cambridge: Cambridge University Press.

Tocqueville, A. de. (1856) 1983. *The Old Regime and the French Revolution*. New York: Anchor Books.

Tohamy, A. 2016. *Youth Activism in Egypt: Islamism, Political Protest and Revolution*. London: I. B. Taurus.

Trotsky, L. (1937) 2004. *The Revolution Betrayed*. Mineola, NY: Dover.

UN Development Program, 2009. *Arab Human Development Report: The Challenge to Human Security*. New York: UNDP.

UN Development Program, Independent Evaluation Office. 2019. "Independent Country Programme Evaluation: Tunisia Evaluation of UNDP Contribution." *Issuu*, 12 May. https://issuu.com/undp-evaluation/docs/icpe_tunisia_2018_bc79b97329dfb7.

Vaneigem, R. (1967) 2017. *The Revolution of Everyday Life*. London: Rebel Press.

Verdery, K. 1999. "What Was Socialism and Why Did It Fall?". In *The Revolutions of 1989*, edited by V. Tismaneanu, 63–88. New York: Routledge.

CHAPTER 28

PEACE AND SECURITY IN THE AGE OF HYBRID WARS

MARIA RAQUEL FREIRE AND LICÍNIA SIMÃO

Hybrid wars have become a central concept in the security jargon of the past decade, reflecting new combinations of actors and means as well as decision-making, strategic, and operational procedures that threaten international peace and security. Narcotraffic, transnational terrorism, and state capture, among others, serve as examples of the range of new threats bearing on national, regional, and international security (Hoffman 2010; Wither 2016). Hybrid warfare has been increasingly used to refer to the combination of regular and irregular tactics by states and nonstate actors in their attempts at gaining advantage over direct adversaries, including through recourse to other agents and means of subversion, such as disinformation and propaganda, and the use of cyber technologies, among others. Russia's actions in Ukraine since 2014, and Russia's broader opposition with the West under President Vladimir Putin's leadership, have been particularly associated with the concept (Monaghan 2016; Charap 2016; Renz and Smith 2016; Galeotti 2016), leading analysts to consider that much of the current usefulness of the term "hybrid warfare" is political, as the term has become part of the propaganda war itself (Fridman 2018; Schroefl and Kaufman 2014; Gray 2012). Following on this line, this chapter argues that the concept of hybrid wars, although useful for understanding the complexity of contemporary security dynamics, is not totally new. It is essentially a politically appropriated concept with implications at the local, national, and international levels. This chapter addresses the issue of conceptual vagueness and novelty of hybrid wars, linking it to other concepts used to analyze (in)security since the end of the Cold War, including the so-called new wars, state collapse, and the more recent focus on resilience. What issues have been raised with the increased focus on hybridity? How useful has the concept been in helping to better understand dynamics of insecurity and instability worldwide? How does hybrid warfare impact international peace and security? And what are the prospects for peacebuilding and statebuilding in societies that experience hybrid wars or are exposed to hybrid threats?

In order to address these issues, the chapter discusses hybrid wars both conceptually and as an integral part of political strategies by key actors of global security. It starts by mapping the different understandings of the concept and its relation with other key concepts in peace and security studies. It then looks at the institutionalization of the concept in the context of the policies of key contemporary security actors, namely the North Atlantic Treaty Organization, the European Union, and the United Nations. It investigates how the concept has been included in strategic documents and operational decisions, and which is the setting these provide for its enactment. From the analysis, it becomes evident that the post-9/11 context provided an important impulse for the development of greater awareness and preparedness for fighting asymmetric threats, a trend that has been accelerated by Russia's intervention in Ukraine since 2014, visible particularly in the practice of Western security institutions. Finally, the chapter looks at the future of peace and security policies in the age of hybrid wars and what might be expected with regard to current and future challenges.

Mapping the Concepts: Hybrid Wars, Security, and Peace

The concept of hybrid wars is not new, as the term "hybridity" has been used to mean a combination of civilian and military means, of global and local levels of analysis, or even of conventional or traditional means with postmodern tools. Hybrid threats include transnational terrorism, the use of governmental and nongovernmental agencies for subversive purposes, and the combination of public and private spheres in security operations, where the processes of undermining an adversary may seek to impact governments, transnational actors, and civilian populations alike (Deep 2015; Wither 2016; Giegerich 2016; Freire 2017). The term was first used by two officers in the US military, James Mattis and Frank Hoffman, in 2005, who referred to the use of "irregular methods" to gain tactical advantage, such as "terrorism, insurgency and narco-crime" (IISS 2014; Hoffman 2010). The concept includes therefore a combination of "different modes and means of war" (IISS 2014). According to the authors, these irregular moves imply the exploitation of tactical advantage in places and times that are not within traditional warfare rules. By playing according to their own rules, hybrid threats become more difficult to grasp, as the aim is to build up from small tactical effects to gain a wider presence, particularly through the use of media and information warfare. In this way, the authors argue, hybrid threats and hybrid wars challenge the efficacy of state response (IISS 2014).

Thus hybridity points to a complex mix of means across a wide spectrum, from decision-making processes to implementation dynamics, and differentiated actors, such as governments or irregular groups. Some suggest the branding of hybrid warfare "take[s] place beneath the threshold that western observers would consider armed conflict, much less war" (Palmer in Giegerich 2016, 67). Others claim it is a new label that points to a "war-winning formula" conceived as a "means to an end" and thus more of an operational than strategic approach (Renz and Smith 2016, 3). Others still claim it is essentially centered on the Western failure to deal with asymmetric threats, which, until the case of Ukraine, were thought to be mainly the work of nonstate actors (Veljovski, Taneski, and Dojchinovski 2017). Thus the term "hybrid" to qualify war and threats is new, whereas the concept itself is not. In fact, it might be argued that not being a new concept, as it was already in place, for example, in the Napoleonic Wars, and that it gained "prominence as a tool in international relations" in direct proportion to the increasing costs of conventional conflict (Stratfor 2017). In his critical assessment of the politicization of the concept, Colin Gray (2012) argues that the nature of warfare has not fundamentally changed, but the concept allows an expansion to new dimensions.

Thus the novelty of this concept lies not so much in the possible combinations it implies as in the new contexts where it has been applied. The systematic and refined way in which hybrid warfare is backed by official state discourse, simultaneously supporting and denying it, makes it difficult for the international community to respond (Polese, Kevlihan, and Ó Beacháin 2016, 365). In this line of reasoning, the central issue is "understanding conflict in the spaces in-between, in the grey areas" (cited in Giegerich 2016, 68). According to Mälksoo (2018, 375), hybrid warfare "capsizes an embedded cognitive structure about what war is, thus defying attempts of organizing life and social relations in a particular way, with fundamental consequences for the ontological security of the European Union (EU) and NATO." If the phenomenon is not new, as propaganda, disinformation, subversion have been tactics applied for a long time, the contexts have changed. In fact, a highly technological environment, more advanced instruments in the digital and conventional dimensions, such as social media, and the spaces for intervention, both territorially and transnationally defined, confer on these threats the need for innovative thinking. By embracing the concept of hybrid warfare, Western security institutions have managed to establish a framework for political and operational responses that allows them to overcome the uncertainties associated with the complex security environment they perform in. Most interesting in this context is the way security issues labeled as hybrid wars and hybrid threats are introduced to a military and strategic language. The terminology involves regular/irregular and conventional/unconventional tactics, strategic operations and procedures, and belligerent groups, and becomes visible in the literature as well as in institutional practice. This move is clearly aligned with a more militarist disposition by Western actors, evidenced by NATO's refocusing on territorial security and the EU's development of its Common Security and Defense Policy.

New Wars, State Fragility, and Resilience

The concept of hybrid wars emerges in the framework of a broader adjustment of the understandings and practices of international security, possible partly due to the openings of the post–Cold War context. The current trend for a reterritorialization and militarized focus on security in response to the perceived threats to liberal peace (Schilde 2017; Māris and Viljar 2017) creates pressure on the more cosmopolitan views of security that gained ground in the early 1990s (Buzan, Waever, and de Wilde 1997; Buzan and Hansen 2009; Booth 1991). This section addresses how the emergence of a new narrative on hybrid wars interacts with existing security conceptual frameworks and practices and these new contemporary trends toward territorialization and militarization. Three central concepts assist in analyzing the evolving security context in the past decade and a half, namely Mary Kaldor's (2007) "new wars," Helman and Ratner's (1993) work on "failed states," and the concept of "resilience" (OECD 2014; European Union 2016). These concepts sought to provide a clearer understanding of threats and responses to international (in)security, in the context of civil wars, genocide, the proliferation of private and economic actors in conflicts, transnational flows, privatization of security, and threats to individuals from state and nonstate actors, visible in the 1990s and early 2000s. The connections between these concepts soon became evident, at both theoretical and practical levels, raising issues that are still relevant to understanding how hybrid threats and hybrid wars interact with these lasting insecurity trends.

Kaldor (2007) introduced "new wars" to refer to a different approach to war from conventional approaches, where there is a mix of war, organized crime, massive violations of human rights, involving global and local actors, public and private actors, informal criminalized economic networks, constituting thus both a military and a political challenge, with the centrality of the state being often questioned. Kaldor (2013, 6) highlights how the new wars end up becoming more of a political challenge about breaches of legitimacy and the need for distinct political responses: "War does imply organised violence in the service of political ends. This is the way it legitimises criminal activity. Suicide bombers in their farewell videos describe themselves as soldiers not as murderers." By incorporating " 'disruptive social behavior' or criminality," the connections between criminal and terrorist networks become clearer in warfare, with all the associated illicit practices undermining the legitimacy of local or national government (Hoffman 2010, 443). The answer, Kaldor (2007, 2013) argues, needs to include cosmopolitan policies that are inclusive, opening the political space in a context of legitimacy. A response to hybrid warfare needs to acknowledge that it challenges governance practices and statebuilding (Schroefl and Kaufman 2014; Kaldor 2013).

This reading of new wars as multifaceted challenges to the state is clearly relevant to our understanding of hybrid wars. It also contributes to our understanding of the issue of state fragility, which in the early 2000s became a label Western countries used to legitimize security interventions in countries facing violence and poverty (Menocal 2011; Heathershaw 2008). The absence of strong state structures capable of exercising sovereignty both domestically and externally was perceived as a threat to international security and thus in need of external intervention. The promotion of liberal peace, understood as the need to assist these countries to develop liberal state structures, was fully incorporated in Western institutions' assistance and security interventions (Richmond 2014). Nevertheless the "receiving end" in these fragile states seems to have also "incorporated and reinterpreted the concept to fit their own political agendas" (Grimm, Lemay-Hébert, and Nay 2014, 197). State fragility seems perfectly in line with the diagnosis made in the context of hybrid threats (where lack of state capacity can increase vulnerability and pose direct and indirect challenges to others), as well as in the context of the approach required to address these threats, namely by improving state resilience. The term "fragile state" is a "policy narrative" reflecting the strategic visions and political goals of its main advocates (205); thus it is also in line with the political use of concepts for certain purposes.

The political dimensions of these concepts therefore become clear and must be taken into account when discussing the reach and implications of hybrid warfare. This is challenging in contexts where human security issues "are difficult to redress [because] warfare takes hybrid forms and major actors are non-recognized entities that are not members of international organizations" (Asmussen 2014, 287). This dynamic affects the response by traditional actors, namely states and international organizations, in the face of hybrid threats. Along this same line, traditional methods related to statebuilding, including in contexts of fragility, seem not to be able to encompass the whole spectrum of challenges that are implied in hybrid threats. The starting point is that the social and political context is complex and the state is weak (Schroefl and Kaufman 2014, 863), meaning that the "counterinsurgent—whether the incumbent government itself or an external ally—therefore finds itself in the position of trying to stabilize a state that is corrupt, dysfunctional and to a large extent absent" (871). Schroefl and Kaufman understand that "[t]hese conflicts transcend national boundaries, social and economic classes, and political ideologies. In these types of warfare there are fewer clearly defined enemies and allies. In this current (dis)order, hybridized actors have the means to surprise and spread fear throughout the traditional nation-state community, but not to establish a viable alternative order. As a result, the world will continue to face hybrid conflicts: weak states are hybridized states, governed by corrupt, predatory, and sometimes criminal elites or warlords who exercise power through patron-client ties instead of through the institutions of government" (877).

The linkages to the new concept of resilience as a possible response to hybrid threats and hybrid wars in the context of fragile states, as much as inside Western liberal countries, demonstrate how the concern is now focused on the capacity to respond locally, besides external intervention and support. Resilience emerges as a response to hybrid threats in the way it seeks to reduce vulnerabilities, be it in Ukraine, Afghanistan, or the EU. It contributes to "deterrence in a hybrid context by reducing the potential gains any attacker might hope to reap" (Giegerich 2016, 69). The concept of societal resilience has been put forward, underlining the capacity to adapt and cope with disturbances while essentially retaining functions and feedback by the people (Van Metre and Calder, 2016). This response-related reading points to a formula whereby "(1) 'fragile states' gradually turn into 'fragile situations,' and later into de-territorialised 'fragility'; (2) 'resilience,' which was just briefly present at the beginning of the fragile states debate, returns to become the main catalyst. . . ; and (3) along with resilience, several conceptual figures enter the central stage of analysis: state-society relations, which should be constructive and mutually reinforcing, political settlements, which ought to be inclusive and the adaptive capacity of (state and social) institutions to cope with shocks and crises, the latter highlighted in particular in the attempt to substitute the still popular stabilisation paradigm" (Pospisil and Kühn 2016, 7). The criticism that emerged to the concept of resilience for its ambiguities translates well in the understanding that it is politicized by both intervening and recipient actors in the way it fosters "continued practices which can be tailored to the institutional interests of implementing agencies, to the (geo-)strategic visions of intervening states, and providing a back door for recipients' attempts to steer practices in favour of their individual or group interests" (2016, 9). If the promise of the concept was emancipatory through empowering local and national agents, the politicization of the concept and associated practices seems to undermine this potential. These linkages between conceptual tools, empirical implications, and institutional practice are analyzed in the next section.

Hybrid War in the Discourse and Practice of Major Security Actors

Concerns with countering hybrid threats have become prominent in the agenda and operational procedures of major security actors, including NATO, the EU, and the UN, since the early 2010s. The context for this renewed focus on hybrid threats is closely related to the lessons learned by Western states from fighting terrorism after the 9/11 attacks in the United States, including in Afghanistan. Moreover Hezbollah's actions against Israel in 2006 showed how a nonstate actor could deploy a series of conventional and nonconventional means to pose a threat to a major Western partner country. The unfolding Arab Spring and NATO's intervention in Libya further evidenced the complex scenarios from which threats could emerge. Failing states, civil unrest, proliferation of sophisticated weaponry among radical and extremist groups, the combination of terrorism with cyber-attacks, and massive population shifts toward Europe (Bachmann and Gunneriusson 2014, 3) constitute examples of this complexity. Adding to this scenario, Russia's 2008 and 2014 interventions in Georgia and Ukraine, respectively, pushed Western states toward a greater focus on the challenge Moscow may pose to their collective security, on many levels. The range of conventional and unconventional means deployed by Russia in its attempt to destabilize the US and European states further pushed these countries, and particularly security institutions, such as the Atlantic Alliance, to focus on hybrid threats and the development of operational means to better understand, predict, and counter such threats. NATO, the EU, and the UN have addressed hybrid war in their strategic documents and operational measures.

NATO

According to NATO (2010b), hybrid threats "are those posed by adversaries, with the ability to simultaneously employ conventional and non-conventional means adaptively in pursuit of their objectives." The 2010 NATO strategic concept does not refer to hybrid war or hybrid threats explicitly. It does, however, list a series of potential threats to NATO members, ranging from a renewed focus on conventional threats and collective security to issues dealing with cyber-attacks, attacks on critical infrastructure fundamental for prosperity in the NATO area, terrorism, and energy security, among others—all areas associated with hybrid war (NATO 2010a). The concept does not address the scope, nature, or likelihood of the "emerging security challenges," nor the possibility of having to address a convergence of these different elements (Aaronson, Diessen, De Kermabon, Long, and Miklaucic 2011). It is in the September 2014 Wales Summit declaration that hybrid war is fully recognized as a major threat facing the Alliance. There it is referred to as a situation where "a wide range of overt and covert military, paramilitary, and civilian measures are employed in a highly integrated design," requiring enhanced strategic communications, developing exercise scenarios, and strengthening coordination between NATO and other organizations (NATO 2014).

NATO's response to hybrid threats has focused on the development of a comprehensive approach and a greater focus on resilience (Hartmann 2017). However, the military alliance has found that important hurdles remain in the incorporation of the military dimension in broader mandates, including statebuilding functions, the promotion of political and economic reforms, and the important work of humanitarian assistance and social regeneration (Aaronson, Diessen, De Kermabon, Long, and Miklaucic 2011). Moreover, as a response to Russia's (re)emergence as a significant challenger of the Alliance's security and thus a central target of NATO's operational measures, the Alliance adopted in 2015 a Readiness Action Plan, including assurance and adaptation measures (NATO 2015). These, however, remain largely focused on enhancing the ability of the Alliance to project its military means, suggesting that some difficulty exists in developing a more integrated response to complex challenges. Seeking to overcome this, the ministers of foreign affairs also adopted, in their meeting in Antalya, Turkey, in 2015, a new strategy on hybrid warfare focused on "improving situational awareness through enhanced intelligence and reconnaissance, and better information sharing between allies as well as other international organisations." Issues related to preempting imminent attacks and improve timely and correct decision-making were also addressed (Oğuz 2016, 174). NATO also established an accredited Strategic Communications Centre of Excellence in Latvia, tasked with dealing with hybrid warfare alongside the implementation of the Readiness Action Plan. A European Centre of Excellence for Countering Hybrid Threats, articulated between NATO and the EU, further contributes to the goal of understanding hybrid threats and finding effective responses, building on the existing capacities of their member states (Standish 2018).

Similar initiatives include the Strategic Communications Centre of Excellence in Riga, Latvia; the Cooperative Cyber Defense Centre of Excellence in Tallinn, Estonia; and the Energy Security Centre of Excellence in Vilnius, Lithuania. Contributing to NATO's level of preparedness, within the Joint Intelligence and Security Division at NATO Headquarters a hybrid analysis branch has been set up, designed to provide the Alliance and its members with the means to analyze potential threats. A focus on sharing of expertise and training has also been developed. Nevertheless the way these decisions will be empowered is still to be seen: interinstitutional coordination has long been on the agendas, and the instruments that have been developed specifically targeting the analysis of hybrid threats still have to prove effective in threat management and in the policies of response and intervention that follow suit.

European Union

European Union initiatives aimed at dealing with hybrid threats have been closely articulated with NATO's, considering the large overlap in the membership of both organizations. The EU-NATO joint declaration, adopted in July 2016 at the margins of the Warsaw NATO Summit, is the cornerstone of the EU's approach to hybrid warfare, identifying new areas for practical cooperation, in particular with regard to hybrid threats, building resilience in cybersecurity and strategic communications (Pawlak 2017). The Council conclusions of 6 December 2016 stressed the strategic importance of this partnership and endorsed a common set of proposals focused on better coordination, situational awareness, strategic communication, crisis response, and bolstering resilience. The European Commission has also advanced a series of proposals both in its 2015 European Agenda on Security and in its 2016 communication "Joint Framework on Countering Hybrid Threats: A European Union Response," including an attempt at defining the phenomena: "[T]he concept aims to capture the mixture of coercive and subversive activity, conventional and unconventional methods (i.e. diplomatic, military, economic, technological), which can be used in a coordinated manner by state or non-state actors to achieve specific objectives while remaining below the threshold of formally declared warfare. There is usually an emphasis on exploiting the vulnerabilities of the target and on generating ambiguity to hinder decision-making processes. Massive disinformation campaigns, using social media to control the political narrative or to radicalise, recruit and direct proxy actors can be vehicles for hybrid threats" (European Commission 2016, 2).

The European Union's approach focuses clearly on the contributions it can make to help its member states improve its resilience to hybrid threats, combining EU strategies and sectoral policies, namely the European Agenda on Security, the European Union Global Strategy for Foreign and Security Policy, the European Defense Action Plan, the EU Cybersecurity Strategy, the Energy Security Strategy, and the European Union Maritime Security Strategy, among others.

The EU has fully incorporated hybrid threats into its broader steps aimed at establishing an "effective and genuine Security Union," strongly rooted in the nexus between internal and external security and reflecting a perceived security context of fast changing challenges, including issues of terrorism inside the EU, radicalization, migration and border management, and cyber warfare and disinformation (European Commission 2018). Operational measures to address these challenges range across sectors of EU policymaking, including legislation proposed by the European Commission, the interoperability of surveillance and data processing systems, namely those controlled under the Area of Liberty, Security and Justice, the deployment of specific missions under the Common Security and Defense Policy, and policies of regional stabilization for strategic neighboring regions, such as the western Balkans, Ukraine, and North Africa. The EU's approach is closely interlinked with the concept of resilience as moving "from crisis containment to a more structural, long-term, non-linear approach to vulnerabilities, with an emphasis on anticipation, prevention and preparedness" (European Commission and High Representative of the Union for Foreign Affairs and Security Policy 2017), thus providing for an integrated approach to hybrid threats. However, the implementation of a resilient approach is not yet assessed in the context of the EU peacebuilding actions and "would require significant institutional reforms and a much more adaptive and responsive foreign policy" (Juncos 2018, 565). Juncos adds that experience has shown that the way peacebuilding strategies are designed has been leading more to the development of a comprehensive approach to peacebuilding than to the empowerment of local actors (570). Regarding hybrid threats, this means the flexibility to adapt to complex contexts remains limited.

United Nations

The UN's main objective is to "maintain international peace and security" and deploying the necessary means to achieve that purpose. Considering that the organization is composed of all fully recognized states in the world, a consensual definition of what constitutes a threat to peace and security lies on the Security Council, particularly its five permanent members. Over its history, the UN has become increasingly committed to broader understandings of (in)security and to more comprehensive and multifaceted approaches to promoting peace. Concepts such as peacebuilding bring together the military means to keep peace, but also civilian means in the field of economic and political reforms (UN Secretary-General 1992). Naturally debates in the UN have also been influenced by ongoing dynamics in the international setting. Hybrid threats have been put on the UN agenda in an informal way, mainly in the context of Russia's actions in Ukraine following the annexation of Crimea. A major discussion took place on 31 March 2017 during an Arria-formula UN Security Council meeting chaired by Ukraine and dedicated to the topic of "hybrid war as a threat to international peace and security." Overall, however, it can be argued that the UN has shown little interest in the conceptual possibilities of hybrid wars, partly due to the politically charged understanding of the concept.

This is not to say that the UN has failed to take steps in setting up important capabilities in conventional and nonconventional areas of security, directly contributing to dealing with hybrid threats to its members. One of the areas is transnational terrorism, which since 9/11 has been a major issue in the security agenda of many UN member states. In 2006 the UN General Assembly adopted the Global Counter-Terrorism Strategy, which is revised by the General Assembly every two years and addresses issues relating to the sources of radical mobilization, specifically the challenges posed by the use of conventional and nonconventional means by terrorist groups, among others. The strategy endorsed the work of the UN Counter-Terrorism Implementation Task Force, established by the secretary-general in 2005, and subsequent efforts led to the establishment of a UN Centre for Counter-Terrorism in 2011 to promote international counterterrorism cooperation and support member states in the implementation of the Global Counter-Terrorism Strategy. Finally, in 2017, the General Assembly established the United Nations Office of Counter-Terrorism, providing greater autonomy and visibility to this area of action. The UN doctrine has thus been evolving, trying to keep up with the challenges hybrid threats bring to peacebuilding. In the 2015 revision of the UN peacebuilding architecture, the concept of "sustaining peace" was introduced, calling for the need to focus beyond a technical peace deal or elections (UN General Assembly and Security Council 2015, 48), as these are only moments in a long-term and deeper process. Sustaining peace was defined "as a goal and a process to build a common vision of a society . . . which encompasses activities aimed at preventing the outbreak, escalation, continuation and recurrence of conflict," thus pressing for a different approach from the technical liberal peace formula. "At the UN level, the adoption of the sustaining peace concept is a manifestation of the pragmatic turn in peacebuilding. It reflects a shift away from the preoccupation (associated with the liberal peace) with identifying and addressing conflict drivers to prevent imminent relapse into violent conflict. Instead, the focus is now on identifying and supporting the political and social capacities that sustain peace" (de Coning 2018, 304; see also de Coning 2016).

The understanding that current challenges are complex and require flexible and adjustable approaches, such as hybrid threats, seems to be gaining ground. Cedric de Coning (2018) advances the concept of "adaptive peacebuilding" to materialize the new approach, seeking to assist societies in building the resilience needed to address challenges, cope with change, and adapt to new circumstances. According to the author, this allows for a differentiated approach to the linear, cause-effect liberal peace approach, in that it foresees "uncertainty, focuses on processes not end-states, and opts to invest in the resilience of local and national institutions and thereby their ability to promote change" (317). Building resilience from below, rooted in local processes, is part of the new peacebuilding readings that might impact the UN response to hybrid threats. Nevertheless, considering these efforts and the politically sensitive nature of the concept of hybrid threats in the current context, the UN is likely to continue to improve its coordination mechanisms to assist countries in highly volatile regions to address complex threats, including those posed by nonstate actors using hybrid means, but still falling short of significant conceptual clarification.

Conclusion: The Future of Peace and Security Policies in the Age of Hybrid Threats

Current discourse on hybrid threats and hybrid warfare remains largely rooted in a view by Western actors of the many vulnerabilities that open political systems present when directly challenged. Hybridity has thus been used to encapsulate the many ways in which conventional and nonconventional challenges can create security challenges to Western democracies, particularly in a context of economic crisis and political contestation, both at home and abroad. Part of the response to this perceived context has rested on increased investment in military spending and the development of means to address the challenges posed by the nexus between internal and external security. Issues like transnational terrorism, cyberwarfare, and the links between hostile powers and nonstate entities and individuals are good illustrations of these dynamics. Given the fact that the central problem of hybrid war is political, it seems likely that "the answer must also be at least as much political as military . . . to seek multilateral cooperation in undermining the transnational links that keep hybrid conflicts going" (Schroefl and Kaufman 2014, 878).

Providing answers to international conflict that is perceived as caused by hybrid threats requires the political alignment of views among major actors and the ability to streamline existing tools for effective answers on the ground. If hybridity is a political concept more than a radically new way of driving conflict, then international organizations and their members will act to address these threats when there is political consensus to do so. Pulling efforts to address asymmetrical threats may be perceived as a relevant step, but trust is also eroded due to the fact that hybrid conflict uses the vulnerabilities presented by open societies to create unrest. Reinforcing resilience of systems and societies both at home and abroad will most likely remain the central approach to promote peace and security, either through concrete multilateral efforts (as in the case of the EU and NATO) or unilaterally, adjusting national defense and security strategies, in an age of growing global competition.

Hybrid threats are nevertheless a reality to many members of the international community, and their vulnerability to the potentially negative effects of globalized transnational dynamics beyond their control requires concerted efforts from the international community. The evolution in doctrine away from the Western neoliberal model of intervention as a response to the challenges hybrid threats pose needs to be underlined as a step forward in the building of more adaptive, flexible, and tailor-made responses. These seek to reinforce resilience, building stronger local dynamics and putting in motion processes of peacebuilding that are multifaceted and multi-actor-oriented. This predisposition to change the locus from ends to processes seems to open new avenues to think about how to counter hybrid threats and wars. Peacebuilding interventions planned in a process-oriented manner gain room for maneuvering in terms of continuously adjusting goals on the basis of local needs, adopting various instruments to reply to the complexity of the contexts, involving distinct actors to better reflect multiple visions, and combining various strategies at the same time. This makes of this new peacebuilding approach a process better suited to respond to the complex challenges hybrid threats and wars imply. Moreover a clarification of the concept and of the added-value it presents vis-à-vis other existing concepts of international security, given the politicization dynamics analyzed, is an important step toward more transparent and effective decisions for operational measures.

This approach also presents an important challenge, which is the need to measure success and to be able to exit long-term interventions in contexts of instability and conflict. The focus on the process rather than on the results may raise issues regarding the political sustainability of peace operations when the political will changes. It is also harder to monitor whether the identified needs are in line with the deployed means. For societies undergoing hybrid conflict, having the international community invest in their own capabilities to respond to these challenges will provide the tools for resilience along the road. However, in the short term there are significant obstacles to having these societies responding adequately to these threats, considering that this requires transfer of complex knowledge and technology, as well as their integration into networks of support and decisions that would upgrade their immediate resilience. NATO and EU membership may be an option for some countries in Europe, but no such equivalent exists in other parts of the world. Having the African Union, the Association of Southeast Asian Nations, or other regional organizations creating the capacity to assist their members in developing capable responses may be an important step.

References

Aaronson, M., S. Diessen, Y. De Kermabon, M. B. Long, and M. Miklaucic. 2011. NATO Countering the Hybrid Threat,. NATO ACTPrism 2(4), 23 September. Available at https://apps.dtic.mil/dtic/tr/fulltext/u2/1042838.pdf https://www.act.nato.int/nato-countering-the-hybrid-threat, accessed 30 October 2018.

Asmussen, J. 2014. "International Crisis Management and Human Security in the Framework of 'Hybrid Wars' and Unrecognized States: Lessons Learned from Ukraine." *Security and Human Rights* 25:287–297.

Bachmann, S, and H. Gunneriusson. 2014. "Terrorism and Cyber Attacks as Hybrid Threats: Defining a Comprehensive 21st Century Approach to Global Security. *Journal on Terrorism and Security Analysis* 9 (Spring):1–13.

Booth, Ken. 1991. "Security and Emancipation." *Review of International Studies* 17 (4):313–326.

Buzan, B, and L. Hansen. 2009. *The Evolution of International Security Studies*. Cambridge: Cambridge University Press.

Buzan, B, O. Waever, and J de Wilde. 1997. *Security: A New Framework for Analysis*. London: Lynne Rienner.

Charap, S. 2016. "The Ghost of Hybrid War." *Survival* 57 (6):51–58.

de Coning, C. 2016. "From Peacebuilding to Sustaining Peace: Implications of Complexity for Resilience and Sustainability." *Resilience* 4 (3):166–181.

de Coning, C. 2018. "Adaptive Peacebuilding." *International Affairs* 94 (2):301–317.

Deep, A. 2015. "Hybrid War: Old Concept, New Techniques." *Small Wars Journal*, 2 March. http://smallwarsjournal.com/jrnl/art/hybrid-war-old-concept-new-techniques.

European Commission. 2015. "The European Agenda on Security." Communication from the Commission to the European Parliament, the Council, the European Economic and Social Committee, and the Committee of the Regions, COM/2015/0185 final. Strasbourg, 28 April.

European Commission. 2016. "Joint Framework on Countering Hybrid Threats: A European Union Response." Joint Communication to the European Parliament and the Council, JOIN/2016/018 final. Brussels, 6 April.

European Commission. 2018. "Fourteenth Progress Report towards an Effective and Genuine Security Union." Communication from the Commission to the European Parliament, the European Council, and the Council, COM(2018) 211 final. Strasbourg, 17 April.

European Commission and High Representative of the Union for Foreign Affairs and Security Policy. 2017. "A Strategic Approach to Resilience in the EU's External Action." Joint Communication to the European Parliament and the Council, JOIN(2017) 21 final. Brussels, 7 June.

European Union. 2016. *Shared Vision, Common Action: A Stronger Europe. A Global Strategy for the European Union's Foreign and Security Policy*. June. Brussels: EEAS.

Freire, M. R. 2017. "War in Ukraine and European Security: Reset, Reverse or Revoke?", In Sadik, Giray, ed. Europe's Hybrid Threats: What Kinds of Power Does the EU Need in the 21st Century? Newcastle upon Tyne: Cambridge Scholars.

Fridman, Ofer. 2018. *Russian "Hybrid Warfare": Resurgence and Politicization*. New York: Oxford University Press.

Galeotti, M. 2016. "Hybrid, Ambiguous, and Non-linear? How New Is Russia's 'New Way of War'?" *Small Wars & Insurgencies* 27 (2):282–301.

Giegerich, B. 2016. "Hybrid Warfare and the Changing Character of Conflict." *Connections* 15 (2):65–72.

Gray, C. S. 2012. *Another Bloody Century: Future Warfare*. London: Weidenfeld & Nicolson.

Grimm, S., N. Lemay-Hébert, and O. Nay. 2014. "'Fragile States': Introducing a Political Concept." *Third World Quarterly* 35 (2):197–209.

Hartmann, U. 2017. *The Evolution of the Hybrid Threat, and Resilience as Countermeasure. NATO Research Paper*, no. 139, September. Rome: NATO Defense College.

Heathershaw, J. 2008. "Unpacking the Liberal Peace: The Dividing and Merging of Peacebuilding Discourses." *Millennium* 36 (3):597–621.

Helman, G B., and S. R. Ratner. 1993. "Saving Failed States." *Foreign Policy* 89 (4):3–20.

Hoffman, F. G. 2007. *Conflict in the 21st Century: The Rise of Hybrid Wars*. Arlington, VA: Potomac Institute for Policy Studies.

Hoffman, F. G. 2010. "'Hybrid Threats': Neither Omnipotent nor Unbeatable." *Orbis* 54 (3):441–455.

IISS. 2014. "Countering Hybrid Threats: Challenges for the West." *Strategic Comments* 20 (8):x–xii.

Juncos, A. E. 2018. "Resilience in Peacebuilding: Contesting Uncertainty, Ambiguity, and Complexity." *Contemporary Security Policy* 39 (4):559–574.

Kaldor, M. 2007. *New and Old Wars: Organized Violence in a Global Era*. 2nd ed. Stanford, CA: Stanford University Press.

Kaldor, M. 2013. "In Defence of New Wars." *Stability* 2 (1):1–16.

Mälksoo, M. 2018. "Countering Hybrid Warfare as Ontological Security Management: The Emerging Practices of the EU and NATO." *European Security* 27 (3):374–392.

Māris, A, and V. Viljar. 2017. "Deterrence Dilemma in Latvia and Estonia: Finding the Balance between External Military Solidarity and Territorial Defence." *Journal on Baltic Security* 3 (2):29–41.

Menocal, A. R. 2011. "State Building for Peace: A New Paradigm for International Engagement in Post-conflict Fragile States." *Third World Quarterly* 32 (10):1715–1736.

Monaghan, A. 2016. "The 'War' in Russia's 'Hybrid Warfare.'" *Parameters* 45 (4):65–74.

NATO. 2010a. *Active Engagement, Modern Defence*. Strategic Concept for the Defence and Security of the Members of the North Atlantic Treaty Organisation, adopted by Heads of State and Government in Lisbon, 19 November.

NATO. 2010b. *BI-SC Input for a New NATO Capstone Concept for the Military Contribution to Countering Hybrid Threats*. 25 August. http://www.act.nato.int/images/stories/events/2010/20100826_bi-sc_cht.pdf.

NATO. 2014. *Wales Summit Declaration*. Issued by the Heads of State and Government Participating in the Meeting of the North Atlantic Council in Wales, 5 September.

NATO. 2015. *Fact Sheet: NATO's Readiness Action Plan*. October. https://www.nato.int/nato_static_fl2014/assets/pdf/pdf_2015_12/20151130_1512-factsheet_rap_en.pdf.

OECD. 2014. *Guidelines for Resilience Systems Analysis*. OECD. http://www.oecd.org/dac/Guidelines%20for%20Resilience%20Systems%20Analysis%20-%20Landscape%20Facilitation%20Guide.pdf.

Oğuz, S. 2016. "The New NATO: Prepared for Russian Hybrid Warfare?" *Insight Turkey* 18 (4):165–180.

Pawlak, P. 2017. *Countering Hybrid Threats: EU-NATO Cooperation*. European Parliamentary Research Service Briefing, PE 599.315, March.

Polese, A., R. Kevlihan, and D. Ó Beacháin. 2016. "Introduction: Hybrid Warfare in Post-Soviet Spaces. Is There a Logic Behind?" *Small Wars and Insurgencies* 27 (3):361–366.

Pospisil, J, and F. P. Kühn. 2016. "The Resilient State: New Regulatory Modes in International Approaches to State Building?" *Third World Quarterly* 37 (1):1–16.

Renz, B., and H. Smith. 2016. *Russia and Hybrid Warfare: Going beyond the Label*. Aleksanteri Papers, no. 1/2016. Helsinki: Kikimora.

Richmond, O. 2014. *Failed Statebuilding: Intervention and the Dynamics of Peace Formation*. New Haven, CT: Yale University Press.

Schilde, K. 2017. "European Military Capabilities: Enablers and Constraints on EU Power?" *Journal of Common Market Studies* 55 (1):37–53.

Schroefl, J., and S. J. Kaufman. 2014. "Hybrid Actors, Tactical Variety: Rethinking Asymmetric and Hybrid War." *Studies in Conflict & Terrorism* 37:862–880.

Standish, R. 2018. "Inside a European Center to Combat Russia's Hybrid Warfare." *Business Insider*, 18 January. https://foreignpolicy.com/2018/01/18/inside-a-european-center-to-combat-russias-hybrid-warfare/.

Stratfor. 2017. "This Is Why Russia Is Increasingly Turning to Hybrid Warfare." *Business Insider*, 8 August. https://www.businessinsider.com/this-is-why-russia-is-increasingly-turning-hybrid-warfare-2017-8.

UN General Assembly. 2006. *The United Nations Global Counter-Terrorism Strategy*. A/RES/60/288, Resolution Adopted by the General Assembly. 8 September.

UN General Assembly and Security Council. 2015. *Report of the High-Level Independent Panel on Peace Operations on Uniting Our Strengths for Peace: Politics, Partnership and People*. A/70/95–S/2015/446, 17 June.

UN Secretary-General. 1992. *An Agenda for Peace: Preventive Diplomacy, Peacemaking and Peacekeeping*. Report by the Secretary-General pursuant to the Statement Adopted by the Summit Meeting of the Security Council on 31 January 1992.

Van Metre, L and Calder, J. 2016. "Peacebuilding and Resilience: How Society Responds to Violence", *PeaceWorks Report*. Washington D.C.: United States Institute of Peace.

Veljovski, G., N. Taneski, and M. Dojchinovski. 2017. "The Danger of 'Hybrid Warfare' from a Sophisticated Adversary: The Russian 'Hybridity' in the Ukrainian Conflict." *Defense & Security Analysis* 33 (4):292–307.

Wither, J. K. 2016. "Making Sense of Hybrid Warfare." *Connections* 15 (2):73–87.

CHAPTER 29

TECHNOLOGIES OF PEACE

ALLARD DUURSMA AND JOHN KARLSRUD

THE mass production and widespread use of the computer, smartphones, and the internet have dramatically changed our world. The digital revolution has particularly picked up pace in the twenty-first century. Hilbert and López (2011) estimate that in 2002, the worldwide digital storage capacity overtook the total analog information capacity for the first time. As of 2007 around 94% of the total amount of information was stored in digital form. Not only has the amount of available data grown, but the computing power to analyze these data has also grown exponentially: computing capacity has grown 58% per year between 1986 and 2007. More smartphones than non-smartphones were sold in developing countries in 2015 (Internet Society 2015, 17). It is estimated that half of the world's households (53.6%) in 2017 had access to the internet at home, compared with fewer than 20% in 2005 and just over 30% in 2010 (International Telecommunication Union 2017), and this figure is rising steadily. In short, we are living in the information age. This chapter examines the effects of the digital revolution and new information and communication technologies on peacekeeping and peacebuilding, focusing on how technology applications can prevent armed fighting, as well as foster inclusive societies and the peaceful resolution of conflicts.

We define peace technologies as mostly digital and web-based information and communication technologies used to prevent and manage armed violence and build a durable and high-quality peace (Sandvik et al. 2014, 220).[1] While the United Nations is arguably still lagging behind commercial industries and governments in terms of its adaption of new technologies, it is no longer a fair assessment to refer to the UN as a Remington typewriter in a smartphone world (Banbury 2016). In fact, the UN is increasingly adopting new technologies, particularly within the field of peacekeeping (Karlsrud 2014, 2017b; Dorn 2016). New technologies are also increasingly adapted within peacebuilding projects. New technological tools help people to foster collaboration, transform attitudes, and give a stronger voice to local communities (Larrauri and Kahl 2013).

This chapter outlines the current use of technologies aimed at peacekeeping and peacebuilding and reflects on the *possible use* of existing technologies that currently have not yet been implemented to help peacekeeping and peacebuilding efforts. Since one cannot simply assume that technology will change everything for the better, we also reflect upon several ethical implications of the use of new peace technologies. We conclude that peace technologies significantly influence the prospects for peacekeeping and peacebuilding, particularly when the use of technology helps peacekeepers and peacebuilders to be more people-centric. Yet the use of these technological advances are not without risk. Third parties using peace technologies should be vigilant when it comes to mitigating the possible abuse and negative side effects of the use of peace technologies.

Current Peace Technologies

Around the turn of the century, the UN began to adopt new technologies to support its peacekeeping missions. This trend was in response to a greater need to produce more efficient field intelligence to identify risks to peacekeepers and risks to achieving the mandate of peace missions. When he assumed office in 2011, the under-secretary-general for UN peacekeeping, Hervé Ladsous, embarked on a program of bringing "the UN into the 21st century" (Karlsrud 2017a, 51), adding controversial elements such as surveillance and an intelligence policy to the tools of UN peacekeeping operations. In 2014 Secretary-General Ban Ki-moon launched an expert panel on technology and innovation in peacekeeping, which advanced the idea in its report that besides being troop-contributing countries, UN members must also be technology-contributing countries (Dorn 2016; United Nations, 2014). The report of the High-Level Independent Panel on Peace Operations published in 2015 also stressed the role of technology, concluding that the UN "must embrace innovation and the responsible use of technology to bridge the considerable gap between what is readily available to and appropriate for United Nations peace operations and what is actually in use in the field today" (UN General Assembly and Security Council 2015). In spite of this gap, the UN is increasingly improving its peacekeeping efforts through technological innovation.

A first major improvement of peacekeeping through technological innovation is the use of new communication technologies to get information from locals. Several UN peacekeeping missions have experimented with giving locals a place to which they can send observations and insights. This type of "participatory peacekeeping" by crowdsourcing information dramatically increases the situational awareness of the information analysts within a peacekeeping mission. The High-Level Panel reports that "[t]he best information [for peacekeepers] often comes from communities themselves. To use that information, missions must build relationships of trust with local people, leading to more effective delivery of protection of civilians mandates and better protection for peacekeepers" (UN General Assembly and Security Council 2015, paragraph 98). Dorn (2016, 1) succinctly summarizes this conclusion by pointing out that the UN's ability to protect depends on its ability to connect with locals. An analysis based on UN data on Darfur conducted by Duursma (2017) seems to support this point, showing that local information can be used to predict where armed violence is going to occur (see also Duursma and Karlsrud 2019). In short, the crowdsourcing of information from locals may increase the situational awareness of peacekeepers.

Another area of technological innovation within the UN is the use of satellite imagery. To look at population movements many peacekeeping missions use Google Earth images with time intervals of one to three months, but some missions have adopted satellite imagery with much smaller time intervals (Dorn 2016). A telling example of how the use of near real-time satellite imagery can be useful is how the United Nations Mission in South Sudan (UNMISS) was able to use satellite imagery to issue an early warning about an impending attack. After the start of the civil war in South Sudan in 2013, geospatial analysts at UNMISS started to monitor a two-hundred-kilometer stretch of road between Juba and Bor. On 10 January 2014, a satellite image was taken that showed a convoy of twelve armored vehicles and three hundred soldiers heading toward Bor. Convergne and Snyder (2015, 571–572) note that given "the direction of the convoy, it was identified as potentially belonging to pro-government forces. Six hours elapsed between the time the image was taken and when its analysis was presented to decision makers at UN headquarters and in the mission." This information subsequently allowed the UNMISS peacekeepers to take preventive action by taking up strategic positions and protecting pockets of civilians.

The use of drones is another form of remote surveillance that has been introduced in UN peacekeeping missions. The UN Security Council granted the UN Department of Peacekeeping Operations permission to use drones for the first time in 2013. This permission was granted in the context of the UN peacekeeping mission in the Democratic Republic of the Congo (DRC) acquiring surveillance drones (Karlsrud and Rosén 2013; Dorn 2016). While this was the first time the Security Council endorsed the use of drones, the UN already had some experience with drones at this point. In 2006, for instance, the UN peacekeeping mission in the DRC was supported by a European force, in which a Belgian contingent brought drones with them. Another case involved the UN's inheriting drone capacity when a European force in eastern Chad was rehatted into a UN peacekeeping mission in 2009. Karlsrud and Rosén (2013, 2) explain that the use of drones in Chad "proved very useful to the mission, as UN forces could closely monitor the movement of the opposition forces and enhance the protection of refugees, IDPs [internally displaced persons], and humanitarian aid workers accordingly, thereby living up to the mandate of the mission." The situational awareness of peacekeepers has thus improved in those missions where these technologies have been implemented.

A lot of attention has been paid in the scholarly literature to how technology can help improve peacekeeping missions, but technological advances can also support peacebuilding processes. One of the most well-known examples in this regard is the use of the *Ushahidi* platform (https://www.ushahidi.com). *Ushahidi*, which means "testimony" in Swahili, was a website set up to map incidents of violence in Kenya after the postelection fallout at the beginning of 2008. Though specifically developed for the Kenyan context, *Ushahidi* now acts as an open-source platform used to produce visual map information of a crisis based on data crowd-seeded in real time. The data can be provided via text messages, email, Twitter, and web forms. The *Ushahidi* platform has been widely used to monitor elections. The platform makes it possible to easily collect reports from thousands of volunteers on the ground across a country in which elections are held. This greatly helps in the detection of possible voter fraud, thus holding political parties accountable (Breuer and Welp 2014).

Mapping is also increasingly used to document human rights abuses (Mancini 2013). For example, the Voix des Kivus used SMS input from trusted reporters to document abuses in the eastern DRC (*Ushahidi* 2011). Similarly the Syria Tracker documents human rights violations via reports from the public (Humanitarian Tracker 2018). Indeed documenting human rights violations and the use of armed violence via local input is now commonplace. A major advantage of the early detection of armed violence is that it can trigger a timely response, for example in the form of community mediation (Bailey and Ngwenyama 2016; Tellidis and Kappler 2016).

The collection of data on human rights abuses is important because it can deter potential human rights violators and because it can inform policy and help peacekeepers and peacebuilders develop an effective response; it is also important in building a knowledge base for when the armed conflict has ended. When a war is over, documentation on human rights abuses and the experiences of victims can be helpful to truth and reconciliation committees (Miklian and Hoelscher 2018; Tellidis and Kappler 2016). Data on human rights are collected not only from local sources; geospatial technologies have also been used to for this purpose in countries like Sudan and Myanmar. "Geospatial technology" refers to a range of tools that help to conduct a spatial analysis of the Earth and human geography (Bromley 2009). In the case of Sudan, Amnesty International used satellite imagery to document the destruction of villages. The ultimate goal of this project, called Eyes on Darfur, was to deter future attacks in this region (Bromley 2009; Levinger 2009).

While the majority of the data collection and analysis initiatives pertain to data on armed violence, some initiatives have also turned to the measurement of stability and development. For instance, UN Global Pulse has analyzed perceptions about food and fuel prices. Global Pulse is a flagship innovation initiative of the UN secretary-general to harness big data for humanitarian action. In a study conducted on Indonesia, Global Pulse (2014) showed that it is possible to mine tweets in order to measure Indonesians' perceptions of food prices in order to anticipate social unrest in relation to spikes in food prices. Indeed the analysis of big data can help peacebuilders understand changes in human well-being, which, in turn, helps to formulate a timely policy response.

Crucially the digital revolution has also made it easier to communicate across different countries and communities. This has strong implications for peacebuilding. Conflict often uses a certain discourse to legitimize violence (Lederach 2015; Galtung 1990). Peacebuilding is therefore about bringing opposing groups with divergent discourses together and creating a discourse conducive to peace (Larrauri and Kahl 2013, 15). The internet has greatly facilitated the ease with which different sides in a conflict can interact, as well as the ability for peacebuilders to reach a certain audience. Blogs and platforms such as Facebook and WhatsApp make it possible for people in sites of armed conflict to organize themselves, but also to engage with the views of the other side (Gallagher 2018;

De Ville, Buckley, and Butler 2015; Salazar 2008). An initiative called the Peace Factory was launched in March 2012, when tensions between Israel and Iran were high, asking people to share positive messages from Israel to Iran and vice versa. The website soon also began posting messages from other pairs of countries in conflict. Some research is beginning to emerge that provides evidence of the effectiveness of initiatives like these. For instance, Martin-Shields and Stones (2014) find that the use of smartphones helped create social bonds between different ethnic groups in Kenya, contributing to more peaceful relations. More generally, they conclude that new communication technologies can facilitate "inter-group bridging" (6). However, the use of technology to improve peacebuilding efforts is still lagging behind the use of technology in the context of stabilization, statebuilding, crisis management, and humanitarian assistance. The next section therefore considers how technology can advance and complement existing peacebuilding efforts, as well as how it can improve peacekeeping practices.

The Best Has Yet to Come? Implementing New Peace Technologies

While peacekeepers and peacebuilders have started to implement an impressive array of new technological tools to support their work, there is still a lot of room for improvement. At least three major opportunities for how peace technologies can support peacebuilding can be recognized. First of all, peacebuilding organizations can use new technologies to better facilitate exchange of information, while taking into account that some other parties might be interested in spreading false information to undermine the peacebuilding process or access information flows in order to target those who provide information. Peacebuilding has traditionally been a communication process largely based on face-to-face interaction between different sides of a conflict. New information and communication technologies can help close information gaps between parties by providing a lot of information to many people. For instance, in a Humanitarian Dialogue publication solely devoted to peacemaking and new technologies, Jenny et al. (2018, 14) note that "a shuttle diplomacy approach focused on passing factual information becomes less pertinent, as groups are less likely to be isolated and can easily access information online." However, this also means that peacebuilders can no longer control the information flow—and some information that appears online might not be true. Indeed the misinformation and disinformation crisis that is accompanying the rise of social media greatly enhances the value of a third party that establishes confidence in the veracity of information (14). Peacebuilding organizations can act as independent organizations that verify information, providing all sides with confidence in the accuracy of certain pieces of information.

Second, peacebuilding organizations could tap into new technologies to allow for greater problem-solving to resolve conflicts, especially for society-wide problem-solving rather than elite-level problem-solving. Jenny et al. (2018, 15) write, "The archetypal model of a mediation as two political leaders from opposing groups sitting across a table with the mediator in the middle has never quite matched the complicated reality of multilevel and multi-track dialogue and consensus-building processes. It is likely to be even less reflective of reality in a future where power is diffused and ICT [information and communications technology] offers new platforms for engagement." Online platforms could be utilized for engaging citizens in brainstorming and constructive consensus-building. This type of peacebuilding process has arguably much more potential to lead to an emancipatory peace than the elite-based problem-solving processes typically conducted as part of peacebuilding.

Third, and relatedly, new technologies make it possible for peacebuilders to inform the attitudes of not only elites but also entire societies. The analysis of big data has opened up the possibility of analyzing the psychological needs and attitudes of groups. Companies like Google and Facebook optimize algorithms based on big data for marketing purposes, but peacebuilding organizations could potentially alter algorithms to privilege conciliatory outcomes in social media exchanges. It would even be possible to develop "peace bots" that put the spotlight on conciliatory messaging while diminishing polarizing messaging (Build Up and MISTI 2018). In other words, peacebuilders now have possibilities for operating at a societal level to address emotions, shift conflict narratives, and perhaps even change hostile attitudes into more reconciliatory attitudes (Jenny et al. 2018, 16).

The room for improvement with regard to peacekeeping also seems significant. First of all, since geographic information system technology is becoming increasingly accurate, user-friendly, and less costly, the time seems ripe for the real-time tracking of UN peacekeepers (Karlsrud 2017b, 62–63; Dorn and Semken 2015). As Dorn and Semken (2015, 545) explain, this tracking can help "avoid and respond to ambushes, kidnappings and friendly fire incidents, rapidly send reinforcements and retrieve wounded peacekeepers, ultimately saving lives."

In addition to adopting new technologies, the UN could improve the use of recently acquired technological tools like satellite imagery and drones. Drones represent a new way of seeing and knowing in peacekeeping. As such, they can improve "access to vulnerable populations, providing better information on potential threats to civilians, and increasing access to information in cases where the UN must use force to protect civilians" (Karlsrud and Rosén 2013, 3). While a conventional investigation patrol to a village several kilometers away from a UN peacekeeping base in difficult-to-travel areas like the DRC could take several hours, a drone could arrive in a matter of minutes to send high-quality images of the situation in real time (Karlsrud and Rosén 2013, 5; Dorn 2016, 7). However, the use of drones has not be without problems in UN settings. In Mali, for example, it took several months to get drones operational, as the UN inspectors tried to apply standard UN aviation regulations. A lack of spare parts and training of key personnel also initially prevented the UN from fully using its drone capacity in Mali (Van Dalen 2015, 313; Duursma 2018). Nevertheless the UN is moving in the right direction in terms of its use of drones, and it is likely that drones will soon be an indispensable part of UN peacekeeping missions.

With the increased use of new technologies, UN staff will have to process more and more data. Duursma (2018) explains that high volumes of information, especially when some of the information is contradictory, increases complexity and can thus actually undermine effective decision-making. Read, Taithe, and Mac Ginty (2016, 1314) note with regard to the increasing role played by data in the humanitarian sector that the enthusiasm for "data is vastly outstripped by the capacity to meaningfully analyse it." In the context of UN peacekeeping Karlsrud (2017b, 72) states that a data deluge means that the major challenge for UN information analysts "will be to sift through rapid data streams, analyze them, and then produce actionable information in real-time." The UN has already made a very important step in this regard with the development of the Situational Awareness Geospatial Enterprise (SAGE) event database tool. Developed at the UN Support Base in Valencia, Spain, this tool is a web-based database system that allows UN military, police, and civilians in UN peace operations (both UN peacekeeping operations and special political missions) to log incidents, events, and activities. SAGE includes not only incidents pertaining to armed violence but also troop movements, increased tensions, hijackings, abductions, protests, and many more potentially relevant incidents. Instead of just reporting free text, the information in SAGE is stored as structured data. This means that the event is categorized (type of event, number of victims, ethnicity, number and affiliation of perpetrators, geographical coordinates, and so on).

Over time, the gathering of structured data will enable mission leadership to identify trends and indicators for early warning. Consequently one of the most important new technologies the UN can harness is the use of machine learning to analyze SAGE and other structured data, which, in turn, will enable *predictive peacekeeping*. Duursma and Karlsrud (2019, 1–2) define predictive peacekeeping as "a range of analytic tools and peacekeeping practices that serve to forecast where and when armed violence will take place, combined with changes in peacekeeping leadership decision-making, particularly deployment of peacekeeping staff, based on those forecasts." Predictive peacekeeping could help UN staff in the field to anticipate rather than react to events unfolding. An actionable piece of information on where and when armed violence is likely to take place allows the leadership of a peacekeeping mission to take action. The flow of information within the UN system is often said to be mostly unidirectional: people in the field often complain about the lack of information coming from New York, the "black hole into which their reports descend" (Dorn 2010, 278). Predictive peacekeeping based on SAGE could ensure a two-way street of information, providing those in the field with early warnings based on the data they provide.

Risks and Pitfalls of Implementing Peace Technologies

New peace technologies should not do any harm to communities in countries that experience or recover from civil war. It is therefore important to consider the risks and pitfalls of implementing new peace technologies. Six major concerns stand out. First, even while new technologies make more effective early warning possible, this greater early warning capacity will need to be translated into more effective early action. Edward C. Luck (2010), the special advisor to Ban Ki-moon when he was an assistant secretary-general, pointed out in this regard that early warning is not an end in itself: "Early warning without early and effective action would only serve to reinforce stereotypes of UN fecklessness, of its penchant for words over deeds." Indeed civilian populations will be left in harm's way if early warnings are not acted upon (Duursma 2018; Duursma and Karlsrud 2019).

Second, the overreliance on new communication technologies can be at the expense of contextual knowledge of peacebuilders. Sandvik et al. (2014, 227) point out that the "humanitarian space" has shrunk in recent years, particularly in highly dangerous places like South Sudan and Syria (see also Fast 2014). In response, humanitarians, including peacebuilders, increasingly resort to communication tools like Skype for daily management or use SMS surveys rather than conduct those surveys themselves (Sandvik et al. 2014; Meier 2015). Duffield (2016) argues in this regard that a decrease in face-to-face engagements—due to the adoption of these new technologies—can lead to an erosion of ground truth. Indeed there is a limit to the extent to which information can be transferred via SMS or Skype. As a result, relying on these type of technologies can be detrimental to the effectiveness of peacekeeping and peacebuilding, as it further widens the already existing gap between assessments by locals and assessments by peacekeepers and peacebuilders (Read 2016; Müller and Bashar 2017). In other words, peace technologies should help peacekeepers and peacebuilders to be better informed and thus more connected to the communities with whom they work (see Karlsrud 2014), and at the same time peacekeepers and peacebuilders should continue to engage in face-to-face interactions with locals.

Third, and somewhat relatedly, practitioners should be aware of the potential unequal effects of peace technologies. Larrauri and Kahl (2013, 2) refer to this as the "bias of connectivity." Some groups may be more able than other groups to use a given technology. For instance, smartphones are used more often by young, urban, and relatively wealthy people (2). In addition, the technology gap between men and women is much more pronounced in low-income countries (Sandvik et al. 2014, 229). This means that those organizations that solely rely on new technologies to extract information will likely develop policies that are skewed toward those locals who are connected. For example, Mansfield (2015) shows that data collected in Afghanistan is generally biased toward areas that are relatively secure and that offer little risk to those conducting surveys. Moreover surveys in Afghanistan have typically excluded populations living in remote areas. These are not trivial issues resulting in some missing data points but can result in surveys suggesting that the need for peacebuilding activities is greatest in relatively secure areas where international investment is higher and the population is more attuned to the interests of the state. Indeed the population in remote parts of a country typically has a much more negative perception of the government and international peacebuilding efforts (Mansfield 2015, 112). Hence one important precautionary measure peacebuilding organizations could take to mitigate the undesired effects of technology is to not solely rely on quantitative data generated through surveys to guide their efforts. Existing data collection techniques need to be supplemented with geospatial analysis and mapping in order to determine where data are collected and where not, combined with well-focused fieldwork to generate information on remote and dangerous areas.

Fourth, a fundamental aspect of the use of peace technologies is the storage of sensitive data, which could fall into the wrong hands. Sharing sensitive information over a particular technology always entails risks (Larrauri and Kahl 2013, 2). The UN has already become the target of cyber-attacks in Libya and Geneva (Scott-Railton 2013; Parker 2020). Attacks could aim to retrieve data or even change data to alter the understanding of the reality on the ground. Peace practitioners must therefore continuously check whether the technology they are using is secure, as well as decide where the data will be stored, how to keep the data secure, and how long the data should be stored (Karlsrud 2017b, 75). Exchanges of information over the internet and over mobile phones are relatively easy to check. Hence peacekeepers and peacebuilders who use these means of communication have a responsibility to make sure the systems are secure and private, as well inform the end users of their products that no communication system can be completely secure. A telling example in this regard is the confidential relationship UN information analysts have with local informants. Because information from local informants might reveal their identity, the consequences of this type of information falling into the wrong hands could be catastrophic. Indeed al-Qaeda in the Islamic Maghreb has circulated death lists with alleged local informants of the UN peacekeeping mission in Mali (Duursma 2018). Similarly informants in the DRC, who were previously provided with mobile phones to alert UN peacekeepers of imminent violence, are known to have been targeted in retaliation (Karlsrud 2017b, 64).

Jacobsen (2017, 530) has discussed how the UN High Commissioner for Refugees has put into use biometric technologies (mainly fingerprinting and iris scanning) "as a 'routine feature'" in the registration of refugees. These data are gathered to provide assistance "in a better and more efficient way" (OCHA 2013, in Jacobsen 2017, 537), "which in turn has rendered new domains of refugee existence accessible and open to intervention," Jacobsen argues (540). As may be expected, host states such as Kenya and Lebanon have demanded access to these data, and refugees have expressed strong reservations about biometric registration if these data are to be shared with host states. Also, donor states are keenly interested in such data, and in the context of refugees from conflict countries that are part of the ongoing war on terror there is a palpable risk that this interest may be motivated as much by national security concerns as by protecting refugees' rights (539).

Fifth, the advance of new technologies could potentially prolong the existing interventionary order. Existing interventions embedded in self-interest and geopolitical domination are often seen as a root cause of many international and domestic conflicts. The availability of new technologies like battle drones can already be seen as co-opted by states that want to pursue their interests remotely (Williams 2013). During the past decade, the strategic focus of Western countries has moved from large-scale interventions to counterterrorism and "targeted strikes," structurally enabled by the confluence of drone technology, data convergence, and machine learning technologies (Karlsrud 2019; Lindsay 2013). At the same time, new technological advances have the potential to tame these types of interventions. For instance, the collection of data by the Bureau of Investigative Journalism on US drone warfare in Pakistan, Afghanistan, Yemen, and Somalia leads to greater transparency on targeted killing and potentially holds the White House to account. Moreover, new peace technologies can make interventions more people-oriented. Research on peace formation has shown that the prospects for emancipatory peace commonly have not come from external actors, ethnonationalist elites, or critical resistance movements, but from local and everyday acts of peace and agnostic forms for reconciliation (Visoka 2017). Third parties can facilitate this type of peace formation by promoting intercultural awareness and constructive consensus-building over destructive polarization (Jenny et al. 2018).

Sixth, the use of peace technologies can have unintended consequences that peacekeepers and peacebuilders should try to anticipate and mitigate (Miklian and Hoelscher 2018). Public posts on platforms like Facebook played a central role in the formation of democracy movements in the Arab Spring, but the Syrian and the Egyptian governments have used information collected from these platforms to track down and punish activists (Miklian and Hoelscher, 2018; Comninos 2013). Another possible unintended consequence of adopting technological advances is that it can lead to passivism among those working to build peace (Chandler 2016). Larrauri and Kahl (2013, 2) note that in the context of peacebuilding "sending in information but receiving no feedback, clicking a 'like' button but not changing attitudes, discussing an issue online but failing to take action offline" are all examples of passivity resulting from the use of peace technologies. The use of drones in peacekeeping missions has also been reported to have unintended consequences. In eastern DRC, the local population refers to drones as "loud mosquitos," indicating that people see the drones as an unwanted annoyance rather than a crucial tool to help protect them (Karlsrud 2017b, 75). Worse than merely an annoyance, the use of drones in UN peacekeeping missions could also raise expectations. Accordingly, drones must be accompanied by effective public awareness campaigns. Locals need to know why drones are used but also need to be made aware of what the drones' limits are.

Finally, one of the most shocking unintended consequences of the use of a peace technology is how Amnesty International's Eyes on Darfur project tragically backfired. Amnesty International hoped to deter attacks on villages in Darfur using satellite imagery (Levinger 2009). However, a quantitative assessment of the Eyes on Darfur project shows that villages that were monitored actually became more likely to be attacked. This suggests that the government of Sudan increased violence to retaliate against the monitoring and advocacy efforts of Amnesty International (Grant 2015).

Conclusion

We are living in the information age, and the digital revolution has influenced peacekeeping and peacebuilding efforts. Several UN peacekeeping missions use mobile phone technology to collect early warnings from local populations, as well as drones and satellite imagery to enhance situational awareness. Peacebuilding organizations use new technologies to collect information to prevent voter fraud, for early warning, to determine where to conduct community mediation, to deter potential human rights abuses, to hold accountable human rights abusers, and to collect information that could be used by truth and reconciliation councils when a war has ended. While the UN has made some big steps in terms of the technologies it has adopted, there is room for making more use of available technologies. Drones and satellite imagery are now used in some missions, but they could be adopted in more and could play a more central role in those missions that use them. Crucially the UN will need to find a way to analyze the enormous amount of data it produces every day. Machine learning to detect patterns in these data and produce early warnings holds great promise in this regard.

However, the use of new peace technologies is not without risk. Collected data can fall in the wrong hands. In addition, some have expressed concerns about the use of technologies being at the expense of face-to-face engagements, ultimately resulting in peacekeeping and peacebuilding efforts that are divorced from realities on the ground. While this is a real concern and peacekeepers and peacebuilders should try to avoid this, the use of new technologies also has the potential to make peacekeeping and peacebuilding more people-centric. Indeed the crowdsourcing of information for peacekeeping and peacebuilding is much more bottom-up than traditional practices.

In conclusion, technological advances provide new avenues to develop effective mediation, peacekeeping and peacebuilding practices. The use of peace technologies is particularly helpful when it contributes to the activities of peacekeepers and peacebuilders being more people-centric and helps to make the transition from effective analysis to effective action.

Note

1. For an overview of other technological tools that we could not cover in this chapter due to limited space, see, e.g., Dorn (2016) and United Nations (2014).

References

Bailey, A., and O. Ngwenyama. 2016. "Community Bridging through ICTs: Seeking to Overcome Digital and Community Divides." *Journal of Community Informatics* 12:69–89.

Banbury, A. 2016. "I Love the U.N., but It IS Failing." *New York Times*, 18 March.

Breuer, A., and Y. Welp. 2014. *Digital Technologies for Democratic Governance in Latin America*. London: Routledge.

Bromley, L. 2009. "Eye in the Sky: Monitoring Human Rights Abuses Using Geospatial Technology." *Georgetown Journal of International Affairs* 10:159–168.

Build Up and MISTI. 2018. "The Commons: A Pilot Methodology for Addressing Polarization Online." https://milunesco.unaoc.org/mil-resources/the-commons-a-pilot-methodology-for-addressing-polarization-online/.

Chandler, D. 2016. "How the World Stopped Worrying and Learned to Love Failure: Big Data, Resilience and Emergent Causality." *Millennium* 44:391–410.

Comninos, A. 2013. *The Role of Social Media and User-Generated Content in Post-conflict Peacebuilding*. Washington, DC: World Bank.

Convergne, E., and M. R. Snyder. 2015. "Making Maps to Make Peace: Geospatial Technology as a Tool for UN Peacekeeping." *International Peacekeeping* 22:565–586.

De Ville, G., A. A. Buckley, and B. Butler. 2015. "Pantanî Blog: Using ICT for Safeguarding and Sharing Indigenous Social Memory." *Journal of Research and Didactics in Geography* 2:97–102.

Dorn, A. W. 2010. "United Nations Peacekeeping Intelligence." In *The Oxford Handbook of National Security Intelligence*, edited by L. K. Johnson, 275–295. Oxford: Oxford University Press.

Dorn, A. W. 2016. *Smart Peacekeeping: Toward Tech-Enabled UN Operations*. [Report] New York: International Peace Institute.

Dorn, A. W., and C. Semken. 2015. "Blue Mission Tracking: Real-Time Location of UN Peacekeepers." *International Peacekeeping* 22:545–564.

Duffield, M. 2016. "The Resilience of the Ruins: Towards a Critique of Digital Humanitarianism." *Resilience* 4:147–165.

Duursma, A. 2017. "Counting Deaths While Keeping Peace: An Assessment of the JMAC's Field Information and Analysis Capacity in Darfur." *International Peacekeeping* 24:1–25.

Duursma, A. 2018. "Information Processing Challenges in Peacekeeping Operations: A Case Study on Peacekeeping Information Collection Efforts in Mali." *International Peacekeeping* 25:446–468.

Duursma, A., and J. Karlsrud. 2019. "Predictive Peacekeeping: Strengthening Predictive Analysis in UN Peace Operations." *Stability: International Journal of Security & Development* 8:1–19.

Fast, L. 2014. *Aid in Danger: The Perils and Promise of Humanitarianism*. Philadelphia: University of Pennsylvania Press.

Gallagher, A. 2018. *How "Peace Tech" Is Changing Global Conflict*. Washington, DC: US Institute of Peace.

Galtung, J. 1990. "Cultural Violence.: *Journal of Peace Research* 27:291–305.

Global Pulse. 2014. *Mining Indonesian Tweets to Understand Food Price Crises*. UN Global Pulse Methods Paper. Jakarta: UN Global Pulse.

Grant, G. 2015. "Monitoring Conflict to Reduce Violence: Evidence from a Satellite Intervention in Darfur." Unpublished paper. http://cega.berkeley.edu/assets/miscellaneous_files/Gordon_Monitoring_Conflight_to_Reduce_Violence_WGAPE.pdf.

Hilbert, M., and P. López. 2011. "The World's Technological Capacity to Store, Communicate, and Compute Information." *Science* 332:60–65.

Humanitarian Tracker. 2018. "Syria Tracker." https://www.humanitariantracker.org/syria-tracker.

International Telecommunication Union. 2017. "Measuring the Information Society Report." Geneva: International Telecommunication Union.

Internet Society. 2015. *Internet Society Global Internet Report 2015: Mobile Evolution and Development of the Internet*. Washington D.C.: Internet Society.

Jacobsen, K. L. 2017. "On Humanitarian Refugee Biometrics and New Forms of Intervention." *Journal of Intervention and Statebuilding* 11:529–551.

Jenny, J., R. Greenberg, V. Lowney, and G. Banim. 2018. *Peacemaking and New Technologies: Dilemmas and Options for Mediators*. Geneva: Centre for Humanitarian Dialogue.

Karlsrud, J. 2014. "Peacekeeping 4.0: Harnessing the Potential of Big Data, Social Media and Cyber Technology." In *Cyber Space and International Relations: Theory, Prospects and Challenges*, edited by J. F. Kremer and B. Müller, 141–160. Berlin: Springer.
Karlsrud, J. 2017a. "Towards UN Counter-terrorism Operations?" *Third World Quarterly* 38:1215–1231.
Karlsrud, J. 2017b. *The UN at War: Peace Operations in a New Era.* Basingstoke: Palgrave Macmillan.
Karlsrud, J. 2019. "From Liberal Peacebuilding to Stabilization and Counter-terrorism." *International Peacekeeping* 26:1–21.
Karlsrud. J., and F. Rosén. 2013. "In the Eye of the Beholder? UN and the Use of Drones to Protect Civilians." *Stability: International Journal of Security and Development* 2:1–10.
Larrauri, H. P., and A. Kahl. 2013. "Technology for Peacebuilding." *Stability: International Journal of Security and Development* 2:1–15.
Lederach, J. 2015. *Little Book of Conflict Transformation: Clear Articulation of the Guiding Principles by a Pioneer in the Field.* Intercourse, PA: Good Books.
Levinger, M. 2009. "Geographical Information Systems Technology as a Tool for Genocide Prevention: The Case of Darfur." *Space and Polity* 13:69–76.
Lindsay, J. R. 2013. "Reinventing the Revolution: Technological Visions, Counterinsurgent Criticism, and the Rise of Special Operations." *Journal of Strategic Studies* 36:422–453.
Luck, E. C. 2010. "Statement by Edward C. Luck Special Adviser to the United Nations Secretary-General: Informal Interactive Dialogue on Early Warning, Assessment, and the Responsibility to Protect." United Nations General Assembly, 9 August.
Mancini, F. 2013. *New Technology and the Prevention of Violence and Conflict.* New York: International Peace Institute.
Mansfield, D. 2015. *Effective Monitoring and Evaluation in Conflict-Affected Environments.* Washington, DC: US Institute of Peace.
Martin-Shields, C., and E. Stones. 2014. "Smart Phones and Social Bonds: Communication Technology and Inter-Ethnic Cooperation in Kenya." *Journal of Peacebuilding & Development* 9:50–64.
Meier, P. 2015. *Digital Humanitarians: How Big Data Is Changing the Face of Humanitarian Response.* Boca Raton, FL: CRC Press.
Miklian, J., and K. Hoelscher. 2018. "A New Research Approach for Peace Innovation." *Innovation and Development* 8:189–207.
Müller, T. R., and Z. Bashar. 2017. "'UNAMID Is Just Like Clouds in Summer, They Never Rain': Local Perceptions of Conflict and the Effectiveness of UN Peacekeeping Missions." *International Peacekeeping* 24:756–779.
Parker, B. 2020. "EXCLUSIVE: The cyber attack the UN tried to keep under wraps." *The New Humanitarian*, 29 January.
Read, R. 2016. "Tensions in UN Information Management: Security, Data and Human Rights Monitoring in Darfur, Sudan." *Journal of Human Rights Practice* 8:101–115.
Read, R., B. Taithe, and R. Mac Ginty. 2016. "Data Hubris? Humanitarian Information Systems and the Mirage of Technology." *Third World Quarterly* 37:1314–1331.
Salazar, J. F. 2008. "Indigenous Peoples and the Cultural Constructions of Information and Communication Technology (ICT) in Latin America." In *Information Communication Technologies: Concepts, Methodologies, Tools, and Applications*, edited by C. van Slyke, 1966–1975. Hershey, PA: IGI Global.
Sandvik, K. B., M. Gabrielsen Jumbert, J. Karlsrud and Mareile Kaufman. 2014. "Humanitarian Technology: A Critical Research Agenda." *International Review of the Red Cross* 96:219–242.

Scott-Railton, J. 2013. *Revolutionary Risks: Cyber Technology and Threats in the 2011 Libyan Revolution.* Newport, RI: Center on Irregular Warfare and Armed Groups, US Naval War College.

Tellidis, I., and S. Kappler. 2016. "ICT in Peacebuilding: Implications, Opportunities and Challenges." *Cooperation and Conflict* 51:75–93.

United Nations. 2014. *Performance Peacekeeping: Final Report of the Expert Panel on Technology and Innovation in UN Peacekeeping.* New York: United Nations.

UN General Assembly and Security Council. 2015. *Report of the High-Level Independent Panel on Peace Operations on Uniting Our Strengths for Peace: Politics, Partnership and People.* 17 June. file:///C:/Users/user/Downloads/2015-UNGA-HIPPO-Report.pdf.

Ushahidi. 2011. "Voix des Kivus: A Crowd-Seeding System in DRC." 16 May. https://www.ushahidi.com/blog/2011/05/16/voix-des-kivus-a-crowd-seeding-system-in-drc/.

Van Dalen, J. A. 2015. "ASIFU: Baanbrekend Inlichtingenexperiment in Mali." *Militaire Spectator,* 7 August.

Visoka, G. 2017. *Shaping Peace in Kosovo: The Politics of Peacebuilding and Statehood.* Basingstoke, UK: Palgrave Macmillan.

Williams, B. G. 2013. *Predators: The CIA's Drone War on al Qaeda.* Washington, DC: Potomac Books.

PART III

DISAGGREGATING PEACEBUILDING AND STATEBUILDING

CHAPTER 30

STATEBUILDING

DAVID CHANDLER

AFTER the overextension of international protectorates and the problems of developing "exit strategies" from the Balkan states, Timor-Leste, and Afghanistan, in the late 1990s and early 2000s international statebuilding actors refined and shifted their approach. By the mid-2000s, international statebuilding actors were universally hailing a new policy framework, the so-called bottom-up approach, which was to overcome the limits of overly prescriptive and generic international programs that assumed "one size fits all" (there is an extensive literature, summarized in Bennett, Foley, and Krebs 2016; Collinson 2016; Chandler 2017a). However, this alternative framework—one that presupposed difference as a starting point rather than the uniformity of problems and solutions (see Brigg 2008)—similarly has today reached an impasse. Alternative approaches seem mired in discussion of the problems of "relational sensitivity" (Chadwick, Debiel, and Gadinger 2013), the "local turn" (Randazzo 2017), "hybridity" (Millar 2014; Nadarajah and Rampton 2015), "friction" (Björkdahl 2016), or new forms of representation and inclusion (for example, Peterson 2012). In fact, as long as these alternatives still shared the assumptions of external knowledge and direction, the attention to difference merely multiplied the points of impasse, bringing into focus further limits to external policy practices and forms of knowledge (see, for example, Debiel, Held, and Schneckener 2016; Heins, Koddenbrock, and Unrau 2016).

The focus of this chapter is the analysis of a discursive shift away from these "external" articulations of alternative bottom-up understandings of international statebuilding and toward what are described here as exploratory or "affirmative" forms of knowing and agency. This can be seen as a two-stage process. First, with the opening of the black box of endogenous social processes, there is growing recognition of the limits of bottom-up forms of statebuilding intervention as an attempt to establish a new ground for external problem-solving. Second, there is a shift to "the great outdoors" through imagining alternative ways of perceiving and responding: understanding problems as emergent and interactive processes that are invitations to grasp the world in richer and more complex ways that are much more accepting of diverse contexts and situations.

The Shift from Top-Down to Bottom-Up Approaches

In the 1990s international statebuilding interventions in the cause of sustainable development, peace, and postconflict reconstruction tended to assume that the problems lay in elite blockages to peace and development, to be removed through the export of liberal institutional frameworks of democracy and the market—the so-called liberal peace approach (Newman, Paris, and Richmond 2009; Richmond and Franks 2009; Campbell, Chandler, and Sabaratnam 2011; Tadjbakhsh 2011). In the statebuilding literature, perhaps the first influential critique of this top-down understanding—that the export of liberal institutions was enough to guarantee a transition to liberal forms of peace, development, and democracy—was expressed in Roland Paris's (2004) book *At War's End*. Paris argued that the export of liberal institutional frameworks could not be expected to work as a "quick fix" when societies were not ready for liberalism. For Paris, the export of liberal institutions would not contribute to the construction of "liberal" outcomes, of peace and stability, without further attention to society itself.

Paris argued that Western liberal internationalists had underestimated the societal blockages at play through local agency, which prevented the effective operation of liberal institutional frameworks. In advocating "institutionalization before liberalization," he argued that the introduction of liberal institutions could be highly problematic in societies held to lack the right ideational and cultural agency. In effect, liberal freedoms were held to be problematic and counterproductive in societies where agents were understood to interact in non- or aliberal ways. In these cases, the promotion of liberal peace was held to involve the initial limiting of political and economic freedoms. External interventions would have to act to restrict and regulate the political, social, and economic spheres until behavioral and attitudinal changes allowed local agents to accept the necessary norms of liberal peace.

Rather than starting from universal liberal assumptions of the rational and autonomous subject, international statebuilding theorists increasingly argued that the subject had to be understood as embedded in social practices and relations that needed to be altered before liberal institutional frameworks could operate effectively. The lessons of the Balkans, the Middle East, and Africa were increasingly interpreted as the problem of too much liberal freedom—or too much democracy—rather than too little. The title of Paul Collier's (2010) book *Wars, Guns and Votes: Democracy in Dangerous Places* summed up the increasing awareness that statebuilding had to be done gradually and under the guidance of external interveners pursuing the new international agendas of society-focused understandings of behavioral and ideational practices and the local norms and informal structures reproducing them. Here international relations theorists often posed the need to join with anthropologists to understand the societal self-reproduction of barriers to liberal modes of being (see, for example, Fukuyama 1995).

Bottom-up or society-focused approaches provided an understanding of the limits of international statebuilding, facilitating a new intellectual and policy coherence around a set of policy frameworks that perceived the communities targeted by international statebuilding agencies as having to overcome internal or endogenous barriers to their own sustainable peace and development. Bottom-up approaches thus understood the problems of conflict, underdevelopment, and lack of rights and democracy as being self-generated products of communities or societies themselves. For this reason, the solution was not seen to lie merely in transforming policy governance at the level of elites or installing new formal institutional frameworks. Thus international intervening agencies engaged ever deeper into local interactions and relations on the understanding that "we" cannot fix "their" problems but, equally, that they cannot be expected to break out of the reproduction of these problems or traps without external assistance. Changing or adapting behavior and understandings needed to come from within; sustainable forms of peace, development, and democracy could not be given or produced by outside actors, only facilitated or inculcated by understanding the mechanisms through which problematic social practices are reproduced.

External intervention is legitimized, in fact seen as necessary, but responsibility for the outcomes of intervention is placed squarely on the shoulders of the local actors themselves. The spheres in which statebuilding practice operated were thus those of community and civil society, as the emphasis shifted from state-based to societal understandings of the reproduction of vulnerabilities. The World Bank's (2006, v) report *Civil Society and Peacebuilding* can be seen as presaging this shift by making society central: "Today the main question is no longer whether civil society has a role to play in peacebuilding, but how it can realize its potential, what are the roles of various actors, what are critical factors and pre-conditions for their effectiveness, and how can external actors best provide support?" The UK Department for International Development, ActionAid, Christian Aid, Plan UK, Practical Action, and the Tearfund report *Characteristics of a Disaster Resilient Community: A Guidance Note* (Twigg 2009) highlighted the logic of bottom-up approaches in arguing that rather than responding after the event, international disaster risk–reduction approaches needed to be based on external intervention at the level of community practices and interactions in order to develop capacities and reduce vulnerabilities.

The Difficulty of Grounding

It seems to be logically inevitable that any attempt to start from the perspective of the knowledge and technical mechanisms of international agencies and policy actors will constitute new forms of exclusion and marginalization. Even if not starting from supply-centered approaches, which assume Western superiority, these statebuilding approaches nevertheless assume the objective knowledge of these intervening agencies. In other words, their subject-centered perspectives (of their own role as the active agents, acquiring greater, more varied, or more interactive knowledge) is not (as yet) problematized. Thus failures to know "the" problem or engage with it successfully inevitably continue to expose external actors to accusations of being too Eurocentric or

Western in their views and not being open enough to the systems and societies in which they are engaged (Sabaratnam 2013). These forms of criticism cannot be avoided by seeking to develop and innovate technologically, whether it is through big data, open-source mapping technologies, or other means, as whatever the nature of the innovation and no matter how extensive its application and how efficient it may be in delivering information, real and complex life can never be adequately captured.[1]

The application of new technologies increasingly reveals the nature of the problem to be different from how it was previously imagined: they reveal communities to be much more differentiated and reveal that causal chains are often much more mediated and less linear than previously understood. Acquiring greater knowledge of depth, intricacy, and complexity inevitably questions previous knowledge assumptions as well as bringing attention to the epistemological limitations of external attempts to know societies and processes from the bottom up (see, for example, Finkenbusch 2016). The density is overwhelming. The problem for international actors tasked with policy intervention is that discussion and reflection upon the epistemological limits of knowledge is bound up with their own external, Western positionality (Bargues-Pedreny 2016).

Contemporary debates over the limits to what international statebuilders can achieve thus construct policy interventions as a performative epistemology (Pickering 2010); the failure of policymaking is seen to directly manifest the limits of a Western way of knowing. This failure is driven by the conflation of epistemological limitations with a Western, Eurocentric, or colonial positionality. This positionality is then held to have historically been elitist, hierarchical, and exclusionist. The inability to drill down to the required level of depth, to grasp the rich interactive density of complex relational processes, then gives the lie to statebuilding claims of objectivity or of epistemic superiority. All interventionist actors (who, by definition, are external agents intervening with instrumentalist intentionality) are caught in their inability to see the problem as it may appear to those more closely involved, despite their claim to be objectively knowing and addressing it. Contemporary political, scientific, and philosophical sensitivities necessarily bring international aspirations back down to earth in the knowledge that interveners cannot escape their own socially, politically, and technologically mediated frameworks of understanding. It appears that bottom-up approaches cannot step outside of their positionality, even with the nicest and most generous of intentions (or with the most reflexive awareness of the recursive processual nature of assemblages and emergent causality; see Körppen, Ropers, and Giessmann 2011; Ramalingam 2013; de Coning 2016).

Bottom-up or postliberal statebuilding interventions (Chandler and Richmond 2015), while appreciating nonlinear and emergent causality, appear to be unable to overcome the epistemological limits of international policy intervention. Statebuilding as a problem-solving discourse appears trapped in a modernist deadlock, still reproducing "objective" Western understandings in the attempt to externally resolve problems. However, I argue that this experience has opened the possibilities for these limits to be legitimized or worked around. In response to the problems of legitimizing knowledge claims, policy innovators are increasingly shifting perspective toward a richer posthuman understanding of knowledge generation.[2] Problems are increasingly recast as ones of phenomenology rather than merely epistemology. Access to and the construction of "the problem" is transformed by attention to the understanding or perceptions of other agencies or actors.

This shift begins to go beyond the assumptions that the problems are merely ones of epistemology, of extending the knowledge of external actors themselves. Considering how the world might be perceived and questions articulated in different ways, with different tools and techniques, begins to raise questions about the nature of the problem itself (Barad 2007). It is at this point that the necessary limits to bottom-up approaches of designing problem-solving interventions become much clearer. Attempting to resolve "the problem" is then no longer a purely epistemological concern of extending modernist forms of knowledge deeper into social and cultural processes of interaction by fine-tuning techniques of data gathering and breaking down categories of analysis or speeding up the feedback from digital recording and sensing equipment.

A fundamental gulf opens up between the agency of the international statebuilders and the problem itself. Or rather the understanding that there is "a" problem constitutes a fundamental gap between the statebuilding agency and the society concerned, which is continually apparent when the intervener needs to acquire knowledge in order to address the problem by providing information and assistance or in terms of knowing more about capacities, choices, and needs. Bottom-up interventions emphasized the need for intervention to be bottom up, but increasingly it becomes apparent that there is no bottom to be found, no solid ground for external problem-solving knowledge and expertise. With this shift, inevitably, governing and knowing agency necessarily becomes understood as more widely distributed.

The shift to bottom-up or problem-centered approaches seeking to redesign policy interventions appears to have had the additional implication of making societies and problems much more opaque, or rather infinitely complex, than initially imagined, thus forcing problems to be increasingly recast as ontological rather than merely epistemological. The point of the distinction is the vantage point or positionality of the knowledge that is required. An epistemological problem can be solved by an expansion of existing frameworks of knowledge, from the subject position of an external actor (in this case, the statebuilding agency concerned). An ontological shift in perspectives also makes the problem itself less clear, even knowing what the problem is cannot be resolved by such an extension and requires indirect access through the ways of thinking and relating internal to the policy target or situation itself. The question is then no longer "How can we understand more?" or even "What do they want?" or "What do they need?" but an entirely different presentation of the question, away from the initial assumption of the existence of "a" problem. Instead the starting point is more a question of the ways of knowing of others: "How do they think?" "How do they see the world?" "What language do they use?" "How do they use it?" "What tools do they use?" "What instruments or technologies do they use to make sense of the world?" The knowledge sought is how phenomena appear—how information is processed and things are perceived—to others.

In this way, the barriers revealed by the bottom-up approach appear to be the barriers of the modern or Western episteme itself, and the renegotiation of intervention as a set of knowledge practices begins to formulate the problem in terms that parallel discussions of posthuman or object-oriented ontologies: as the object of analysis seems to be increasingly obscure, to withdraw or recede from the direct or unmediated view of the external actor (see, for example, Harman 2016). The more the external statebuilding agency or actor thinks that it grasps the problem in bottom-up approaches—understands the processes involved, locates the most vulnerable, finds the mechanisms of mediation, interpretation, and translation—the more the problem recedes or disaggregates, and it is clear that what was mistakenly taken as knowledge of "the problem" was merely a self-projection of the categories and understandings of the external actor itself. Rather than coming closer to the problem, to addressing causes and removing barriers, the problems appear to be further away or, more precisely, to have much more relational depth.

The Journey to Acceptance

The critique of earlier top-down or supply-centered policy approaches (UN 2016, 29) as well of those of the alternative bottom-up or demand-driven solutions is precisely that both remain based on projections of Western understandings: of a liberal, modernist, or Eurocentric episteme, which makes "God's-eye view" assumptions that the epistemological barriers to problem-solving can be overcome while ignoring the possibility of ontological barriers to knowledge (see, for example, Chandler 2015). In the work of object-oriented ontology or speculative realism, this problem of ignoring ontological barriers is often termed "correlationism," a problematic, first coined by Quentin Meillassoux (2008, 5–7), which stems from Kant's transcendental idealism. Phenomenological barriers to knowledge are not taken seriously as it is assumed that we never have access to the inner world of experience of other subjects or objects, only to the world as we perceive and experience it, trapped within our own phenomenological world of perception. Thus problems are always understood epistemologically, within our own set of correlations between the world and ourselves.

Problems thus are always framed as "problems for us," never constructed in the ways they may appear for other forms of being or ways of existing. The perceived need to overcome or to bypass these limits has been increasingly raised by decolonial approaches (Mignolo 2011; Mignolo and Escobar 2010; Shilliam 2015; Wynter 2003), and these fit well (in this regard) with the concerns of posthumanist, speculative realist, or object-oriented theorists. The key problematic for bottom-up forms of intervention is thus that of not taking alterity seriously enough (Candea in Carrithers et al. 2010, 175); the study of different local relations and interactions from the God's-eye view of a Western observer or governance agency appears to risk affirming the modernist worldview (of "phenomenology-of") rather than questioning the hegemonic Western assumptions about the objective or scientific nature of knowledge, i.e., that the world is single and uniform and only sociocultural understandings and responses differ (Holbraad in Carrithers et al. 2010, 181).

This reversal of positionality in relation to the problem increasingly links new developments in policy practices with posthuman, speculative, or object-oriented approaches. In international policy discourses, bottom-up approaches are fundamentally challenged by the need to go beyond correlationism, beyond merely the projection of a Western external, or modernist, framing of problems and solutions in order to think outside these mental constraints. As Meillassoux (2008, 7) puts it, this shift can be understood as an exciting challenge of entering "the great outdoors," no longer forced to be constrained by traditional frameworks of gaining access to problems but rather to explore other ways of being and knowing.

The extent of the challenge facing international statebuilding agencies today might be usefully thought through speculatively. To this end, the Southern Reach trilogy makes a good starting point. The trilogy is a series of novels by the American author Jeff VanderMeer first published in 2014: *Annihilation*, *Authority*, and *Acceptance*. A film adaptation of *Annihilation* by the writer-director Alex Garland was released in 2018. Southern Reach is a military establishment sent to explore and understand the strange "Area X." The novels detail some of the experience of sending different missions to the area and how this process of exploration begins to unravel the self-understanding of those involved without ever gaining any knowledge or control over the area itself. In this mysterious and hypnotic landscape the explorers and researchers themselves become "colonized," losing their illusions and their sanity but never discovering anything that could be called traditional knowledge. The allusions to the inversion of modernist or humanist understandings of the relation between the subject and the world—or the bifurcation of culture and nature, which places the human as the knowing active subject and the world as passive object—are clear in that the two main centers of engagement are the Lighthouse (with its allusions to Enlightenment knowledge) and its inversion, the tower/tunnel that goes to subterranean depths (which leads to the disorientation and disappearance of the self).

When the Southern Reach teams cross the border into Area X, their experience of the world becomes "weird" as they lose their compass on reality and their technological instruments become useless. This experience of the loss of power, control, and understanding as leading international institutional experts arrive equipped with their statebuilding expertise and manuals of lessons learned, will be familiar to anyone with experience of statebuilding missions, from Bosnia and Kosovo to Iraq and Afghanistan. The first novel, *Annihilation*, captures the experience of loss of the international statebuilding narrative of progress and its assumptions that gradually more knowledge can be acquired and problems and barriers overcome: "Everyone [mission members] had died or been killed, returned changed or returned unchanged, but Area X had continued on as it always had . . . while our superiors seemed to fear any radical reimagining of this situation so much that they continued to send in knowledge-strapped expeditions as if this was the only option" (VanderMeer 2015, 158–159). The biologist (the leading protagonist) is an expert in "transitional environments," but while

Area X is a transitioning environment, it is not transitioning in the way imagined by international statebuilders. As David Tompkins (2014) in the *Los Angeles Review of Books* and Joshua Rothman (2015) in the *New Yorker* highlight, Area X can be understood as a "hyperobject" in the terminology of contemporary speculative realist philosopher Timothy Morton (2013, 116): "Objects are Tardis-like, larger on the inside than they are on the outside. Objects are uncanny. Objects compose an untotalizable nonwhole set that defies holism and reductionism. There is thus no top object that gives all objects value and meaning, and no bottom object to which they can be reduced."

Perhaps it would be useful to think of statebuilding in these terms, as engaging with something that increasingly seems to make more sense if seen as a "hyperobject." Initially statebuilders assumed that there was a holistic "top object," which gave value and meaning to what they were doing in building peace, building democracy, or building economic capacities; this top object was the state and its institutions. Statebuilding from the top down seemed to make a lot of sense, especially in the mid- and late 1990s, with international powers highly confident in liberal universalism after the collapse of the Soviet challenge (Chandler 2017a). With the perceived failure of statebuilding from the perspective of a top object of meaning from the late 1990s onward, statebuilding from the bottom up assumed that instead of a top object the focus should be upon a bottom object, which could be the (reductionist) key to the statebuilding project, starting from local and contextual knowledges and approaches and working upward to achieve international policy ends.

Thus the annihilation of the statebuilding project does not come with the disastrous experiences of top-down interventions, of regime change and international protectorates, but with the realization that international statebuilders have no handle on reality, that there is nothing to grasp to frame their policy interventions, that there is neither a top to cohere statebuilding as an external project nor a bottom from which alternatives can be launched. This realization and its consequences have been much slower to percolate through the levels of international statebuilding bureaucracy and the groupthink of the academy, where international institutions have been highly reluctant to think through the implications of the lack of mission legitimacy. There has been instead the slow exhaustion of the search for a bottom-up reductionist ground or foundation. Instead of a bottom, international statebuilders found themselves in the disorienting experience of infinite complexity, where none of their knowledge or assumptions appeared to make sense. Very much in line with the speculative "weird" fiction of the Southern Reach trilogy, the inevitable outcome has been acceptance of the limits of external power and knowledge and, in some cases, a welcoming of the release from modernist or humanist responsibilities for problem-solving.

It cannot be emphasized enough that previous approaches to international statebuilding have black-boxed societies, being too little interested in their internal workings and relationships and instead focusing on surface appearances and offering policy advice and assistance on this basis. The opening up of this black box has provided the dynamic that is driving and transforming the design of policy intervention, which increasingly seeks to draw from the rich plurality of the new worlds opened up in the problematization of a narrow bottom-up approach. In fact, as articulated here, it becomes clear that there are two stages of the opening up of the problem. The first stage, external and subject-centered, seeks to drill down, operating within the legacy of the modernist episteme, pluralizing the variables and localizing the factors (as described earlier). The second stage begins to shift to a less modernist framework with a pluralizing ontology, speculating upon multiple ways of knowing or perceiving reality, or of being in the world.

The attempt to move away from addressing a problem to exploring the ways in which it may appear to others transforms the self-understanding of statebuilding actors. This shift from a subject-centered position (which assumes a universal or objective perspective) to a posthuman approach is often unclear in the remaking of international discourses of policy intervention because this means dealing with the alien nature not of objects but of communities constituted as vulnerable or "at risk." Thus Meillassoux's "great outdoors" is recast as an open-ended engagement with the "other," with the "local" or with "grass-roots communities" (see, for example, Mac Ginty and Richmond 2013). The fact that the other can never really be known is not a problem but, on the contrary, positive and enabling, and "expands possibilities for opening to 'new' understandings of difference" where external actors can "value cultural difference independently of claims to have or know culture, attend directly to the process of constituting culture, and open to other ways of knowing human difference" (Brigg and Muller 2009, 136, 138).

Regardless of the bottom-up terminology in which this shift is recast, the phenomenological framing is the same: policy interventions increasingly start from the perceptions of actors closer to the problem itself and its articulation in its concrete local context. The problem is then posed in terms of the ways of knowing and interacting of the local, vulnerable, marginalized, or most at risk. Thus the project design shifts from assuming the problem and looking for its solution to a more open-ended inquiry into understanding the perceptions of the other or the ways in which the problem emerges on its own terms (see, for example, Bahadur and Doczi 2016). Thus interventions seek to see a more affirmative mode of engagement, exploring how other actors and agents see and understand these interactions, grasping the world in its ontological multiplicity (Chandler 2017b). This provides a major challenge to the approach of drilling down to access and open up problems to an external understanding, as these bottom-up approaches are limited by retaining the baggage of the modernist episteme.

Conclusion

This chapter has sought to bring clarity to the discussion of the limits and possibilities of the practices of international statebuilding intervention, which bring to the surface the difficulty of maintaining the legitimacy of the internationalist imaginary of intervention from a universalist, detached, or objective perspective (even if it were possible for statebuilding interventions to be free from the blinkers of power or ideology).

Understanding the conditions of impossibility for traditional or modernist conceptions of international statebuilding—the inability to legitimate the separations and cuts necessary to demarcate a distinct or separate policy sphere—shines an important light on the frameworks through which policy interventions are understood and contested today. It also suggests that to dismiss posthumanist, speculative realist, or object-oriented approaches as somehow not policy relevant would be to miss the broader context in which both academic and policy processes are evolving.

NOTES

1. Critics have argued that new scanning and mapping technologies may distance humanitarian actors even more from these societies (Scott-Smith 2016; Duffield 2016; Meier 2015) or that they may reproduce epistemological blind spots and exclusions in different forms (Read, Taithe, and Mac Ginty 2016; Kitchin 2014; Aradau and Blanke 2015; O'Grady 2016).
2. Posthumanist phenomenology is often seen as starting with Thomas Nagel's (1974) famous essay "What Is It Like to Be a Bat?' or with Deleuze and Guattari's popularization of Jacob von Uexküll's ethology (Deleuze 1988, 124–126), drawing attention to how our perceptions of the world are very different from those of other actors and agencies. This approach pluralizes the world, enabling us to see the world as constituted by many multiple ways of being, decentering the human as an all-knowing actor. A variety of related approaches—such as speculative realism, object-oriented ontology, actor network theory, new materialism, and postphenomenology—have extended the pluralizing perspectives of critical, gender, feminist, black, and decolonial studies to the nonhuman, thus radicalizing perspectivism (see, for example, Bogost 2012; Bryant 2011; Ihde 2009; Morton 2012; Harman 2016).

REFERENCES

Aradau, C., and T. Blanke. 2015. "The (Big) Data-Security Assemblage: Knowledge and Critique." *Big Data and Society* 2 (2):1–12.

Bahadur, A., and J. Doczi. 2016. *Unlocking Resilience through Autonomous Innovation. Overseas Development Institute Working Paper*. London: ODI.

Barad, K. 2007. *Meeting the Universe Halfway: Quantum Physics and the Entanglement of Matter and Meaning*. London: Duke University Press.

Bargues-Pedreny, P. 2016. "From Promoting to De-emphasizing 'Ethnicity': Rethinking the Endless Supervision of Kosovo." *Journal of Intervention and Statebuilding* 10 (2):222–240.

Bennett, C., M. Foley, and H. B. Krebs. 2016. *Learning from the Past to Shape the Future: Lessons from the History of Humanitarian Action in Africa. Humanitarian Policy Group Working Paper*. London: Overseas Development Institute.

Björkdahl, A., K. Höglund, G. Millar, J. van der Lijn, and W. Verkoren, eds. 2016. *Peacebuilding and Friction: Global and Local Encounters in Post-Conflict Societies*. Abingdon, UK: Routledge.

Bogost, I. 2012. *Alien Phenomenology or What It's Like to Be a Thing*. Minneapolis: University of Minnesota Press.

Brigg, M. 2008. *The New Politics of Conflict Resolution: Responding to Difference*. Basingstoke, UK: Palgrave Macmillan.

Brigg, M., and K. Muller. 2009. "Conceptualising Culture in Conflict Resolution." *Journal of Intercultural Studies* 30 (2):121–140.

Bryant, L. 2011. *The Democracy of Objects*. Ann Arbor, MI: Open Humanities Press.

Campbell, S., D. Chandler, and M. Sabaratnam, eds. 2011. *The Liberal Peace?* London: Zed Books.

Carrithers, M., M. Candea, K. Sykes, M. Holbraad, and S. Venkatesan. 2010. "Ontology Is Just Another Word for Culture: Motion Tabled at the 2008 Meeting of the Group for Debates in Anthropological Theory, University of Manchester." *Critique of Anthropology* 30 (2):152–200.

Chadwick, W., T. Debiel, and F. Gadinger, eds. 2013. *Relational Sensibility and the "Turn to the Local": Prospects for the Future of Peacebuilding*. Global Dialogues 2. Duisburg: Käte Hamburger Kolleg/Centre for Global Cooperation Research.

Chandler, D. 2015. "Reconceptualising International Intervention: Statebuilding, 'Organic Processes' and the Limits of Causal Knowledge." *Journal of Intervention and Statebuilding* 9 (1):70–88.

Chandler, D. 2017a. *Peacebuilding: The Twenty Years' Crisis, 1997-2017*. Basingstoke, UK: Palgrave Macmillan.

Chandler, D. 2017b. "Securing the Anthropocene? International Policy Experiments in Digital Hacktivism: A Case Study of Jakarta." *Security Dialogue* 48 (2):113–130.

Chandler, D., and O. P. Richmond. 2015. "Forum: Contesting Postliberalism. Governmentality or Emancipation?" *Journal of International Relations and Development* 18 (1):1–24.

COIC. 2016. *After the World Humanitarian Summit: A Thinkpiece Drawing on Collaboration by OCHA, UNDP, UNHCR, UNICEF, WFP and the World Bank*. New York: Center on International Cooperation.

Collier, P. 2010. *Wars, Guns and Votes: Democracy in Dangerous Places*. London: Vintage.

Collinson, S. 2016. *Constructive Deconstruction: Making Sense of the International Humanitarian System*. Humanitarian Policy Group Working Paper. London: Overseas Development Institute.

Debiel, T., T. Held, and U. Schneckener, eds. 2016. *Peacebuilding in Crisis: Rethinking Paradigms and Practices of Transnational Cooperation*. Abingdon, UK: Routledge.

de Coning, C. 2016. "From Peacebuilding to Sustaining Peace: Implications of Complexity for Resilience and Sustainability." *Resilience: International Policies, Practices and Discourses* 4 (3):166–181.

Deleuze, G. 1988. *Spinoza: Practical Philosophy*. San Francisco, CA: City Lights.

Duffield, M. 2016. "The Resilience of the Ruins: Towards a Critique of Digital Humanitarianism." *Resilience: International Policies, Practices and Discourses* 4 (3):147–165.

Finkenbusch, P. 2016. "Expansive Intervention as Neo-Institutional Learning: Root Causes in the Merida Initiative." *Journal of Intervention and Statebuilding* 10 (2):162–180.

Fukuyama, F. 1995. "The Primacy of Culture." *Journal of Democracy* 6 (1):7–14.

Harman, G. 2016. *Immaterialism: Objects and Social Theory*. Cambridge, UK: Polity Press.

Heins, V. M., K. Koddenbrock, and C. Unrau, eds. 2016. *Humanitarianism and Challenges of Cooperation*. Abingdon, UK: Routledge.

Ihde, D. 2009. *Postphenomenology and Technoscience: The Peking University Lectures*. Albany: State University of New York.

Kitchin, R. 2014. "Big Data, New Epistemologies and Paradigm Shifts." *Big Data and Society* 1 (1):1–12.

Körppen, D., N. Ropers, and H. J. Giessmann, eds. 2011. *The Non-Linearity of Peace Processes: Theory and Practice of Systemic Conflict Transformation*. Opladen: Barbara Budrich Verlag.

Mac Ginty, R., and O. P. Richmond. 2013. "The Local Turn in Peace Building: A Critical Agenda for Peace." *Third World Quarterly* 34 (5):763–783.

Meier, P. 2015. *Digital Humanitarianism: How Big Data Is Changing the Face of Humanitarian Response*. London: CRC Press.

Meillassoux, Q. 2008. *After Finitude: An Essay on the Necessity of Contingency*. London: Continuum.

Mignolo, W. 2011. *The Darker Side of Western Modernity: Global Futures, Decolonial Options*. London: Duke University Press.

Mignolo, W., and A. Escobar, eds. 2010. *Globalization and the Decolonial Option*. Abingdon, UK: Routledge.

Millar, G. 2014. "Disaggregating Hybridity: Why Hybrid Institutions Do Not Produce Predictable Experiences of Peace." *Journal of Peace Research* 51 (4):501–514.

Morton, T. 2012. *Ecological Thought*. Cambridge, MA: Harvard University Press.

Morton, T. 2013. *Hyperobjects: Philosophy and Ecology after the End of the World*. Minneapolis: University of Minnesota Press.

Nadarajah, S., and D. Rampton. 2015. "The Limits of Hybridity and the Crisis of Liberal Peace." *Review of International Studies* 41 (1):49–72.

Nagel, T. 1974. "What Is It Like to Be a Bat?" *Philosophical Review* 83 (4):435–450.

Newman, E., R. Paris, and O. P. Richmond, eds. 2009. *New Perspectives on Liberal Peacebuilding*. New York: United Nations University Press.

O'Grady, N. 2016. "A Politics of Redeployment: Malleable Technologies and the Localisation of Anticipatory Calculation." In *Algorithmic Life: Calculative Devices in the Age of Big Data*, edited by A. Louise and V. Piotukh, 72–86. Abingdon, UK: Routledge.

Paris, R. 2004. *At War's End: Building Peace after Conflict*. Cambridge: Cambridge University Press.

Peterson, J. H. 2012. "Creating Space for Emancipatory Human Security: Liberal Obstructions and the Potential of Agonism." *International Studies Quarterly* 57 (2):318–328.

Pickering, A. 2010. *The Cybernetic Brain: Sketches of Another Future*. Chicago: University of Chicago Press.

Ramalingam, B. 2013. *Aid on the Edge of Chaos: Rethinking International Cooperation in a Complex World*. Oxford: Oxford University Press.

Randazzo, E. 2017. *Beyond Liberal Peacebuilding: A Critical Exploration of the Local Turn*. Abingdon, UK: Routledge.

Read, R., B. Taithe, and R. Mac Ginty. 2016. "Data Hubris? Humanitarian Information Systems and the Mirage of Technology." *Third World Quarterly*, 29 February. doi:10.1080/01436597.2015.1136208.

Richmond, O. P., and J. Franks. 2009. *Liberal Peace Transitions: Between Statebuilding and Peacebuilding*. Edinburgh: Edinburgh University Press.

Rothman, J. 2015. "The Weird Thoreau." *New Yorker*, 14 January. https://www.newyorker.com/culture/cultural-comment/weird-thoreau-jeff-vandermeer-southern-reach.

Sabaratnam, M. 2013. "Avatars of Eurocentrism in the Critique of the Liberal Peace." *Security Dialogue* 44 (3):259–278.

Scott-Smith, T. 2016. "Humanitarian Neophilia: The 'Innovation Turn' and Its Implications." *Third World Quarterly* 37 (12):2229–2251.

Shilliam, R. 2015. *The Black Pacific: Anti-Colonial Struggles and Oceanic Connections.* London: Bloomsbury Academic.

Tadjbakhsh, S., ed. 2011. *Rethinking the Liberal Peace: External Models and Local Alternatives.* London: Routledge.

Tompkins, D. 2014. "Weird Ecology: On the Southern Reach Trilogy." *Los Angeles Review of Books,* 30 September. https://lareviewofbooks.org/article/weird-ecology-southern-reach-trilogy/.

Twigg, J. 2009. *Characteristics of a Disaster Resilient Community: A Guidance Note.* London: University College, London, Hazard Research Centre.

UN. 2016. *One Humanity: Shared Responsibility. Report of the Secretary-General for the World Humanitarian Summit.* New York: United Nations.

VanderMeer, J. 2015. *Annihilation.* London: Fourth Estate.

World Bank. 2006. *Civil Society and Peacebuilding: Potential, Limitations and Critical Factors.* Washington, DC: World Bank.

Wynter, S. 2003. "Unsettling the Coloniality of Being/Power/Truth/Freedom: Towards the Human, after Man, Its Overrepresentation—an Argument." *CR: The New Centennial Review* 3 (3):257–337.

CHAPTER 31

DEMOCRATIZATION AND PEACEBUILDING

CHRISTOPH ZÜRCHER

In the early days of UN peacekeeping, democratization was not considered to be at the heart of the endeavor. Peacekeeping was mainly seen as an activity to separate two warring parties in order to give them the political space for negotiating a peaceful arrangement after war. Since then, peacekeeping has evolved into a much broader activity. With the end of the Cold War, UN peacekeeping became more frequent, more complex, and more ambitious. Building peace became intrinsically linked with building a particular type of postwar polity—a liberal, democratic state organized around democracy and the rule of law. This preference is evident across the peacebuilding universe, as the United Nations and other international organizations, states, and nongovernmental organizations have developed an impressive array of programs for promoting rule of law, good governance, strong civil societies, transparency and accountability, electoral reforms, and other ingredients necessary for a liberal democracy.

There are many reasons why the international community is determined to link peacebuilding with democratic governance. For some, it is a moral imperative in an era when there is no legitimate alternative to democratic governance available; others, most prominently nongovernmental organizations, have long viewed the state as a threat to its citizens and, consequently, want to invest in mechanisms that can constrain its power by strong democratic institutions (Orr 2004; Paris 2004). And finally, echoing the key tenets of democratic peace, many argue that a democratic regime is the most viable bulwark against a relapse into violence (Paris 2004; Sens 2004). Consequently all peacebuilding missions launched by the UN after the Cold War mention that democracy, or its contents, such as good governance, rule of law, free and fair elections, etc., are at the core of the peace that is being built. Only very recently have we seen signs that democracy may feature less prominently in current mission mandates. Both the 2013 mission in Mali and the 2014 mission in the Central African Republic were labeled stabilization missions and prioritized restoring state authority. This slight tuning down of the democracy theme does not necessarily signal a paradigm change, but rather reflects that the UN, in the light of the increasingly difficult and challenging realities on the ground, has become a bit less ambitious. When violence is ongoing and there is no peace to keep, promoting democracy may have to wait, but that does not mean that a democratic liberal peace is no longer the end vision of contemporary peacebuilding.

Defining Democratization

A contemporary political system can be called democratic when "its most powerful collective decision makers are selected through fair, honest, and periodic elections in which candidates freely compete for votes and in which all the adult population [are] eligible to vote" (Huntington 1991, 7). Democratization thus refers to a process during which the political space is opening up, allowing for increased participation in the political process and for more contestation for political influence.

Scholars often distinguish different phases of a democratic transition, such as the opening up of a previously closed authoritarian system, a transition period toward more democracy, and finally a consolidation phase. In the literature on democratization and peacebuilding, most scholars are interested predominantly in whether a postwar society will successfully consolidate democratic institutions. A simple litmus test is the replacing of a government that was not democratically elected by one that is. However, this is a minimal threshold, since elected government can still be powerless, unstable, or corrupt. Also, it may reverse the democratic transition and reinstall authoritarian governance.

A more demanding threshold therefore is whether a postwar government has the political will and the capacity to sustain a free and fair electoral process, an independent judiciary, institutional constraints on the executive power, civil liberties, political rights and the rule of law, and a social contract with the citizens that generates legitimacy for the government. Peacebuilders typically envision a fully developed, liberal democracy as the end state of a peacebuilding operation. Scholars are usually more skeptical about the probability that postwar societies will be able or willing to sustain liberal democracies. What constitutes "success" of democratization in the context of peacebuilding is therefore very much open for debate.

Doing Democratization

This chapter is concerned with democratization in the context of multilateral peacebuilding led by the UN. Typically the UN works through a peacekeeping operation, which is headed by a special representative of the secretary-general. Many, but not all, mandates of recent peacekeeping operations include references to the promotion of democracy, good governance, and the rule of law. Multilateral peacekeeping operations are supported by myriad other actors. Specialized UN agencies and programs play an important role, such as the United Nations Development Program in the field of economic development, the World Food Program in the field of food security, the United Nations High Commissioner for Refugees in the field of refugee protection, and so on. The World Bank and the Asian Development Bank often support postconflict countries with development assistance. Numerous international NGOs and NGOs are also active in postconflict societies. Nation-states also play an important role, as bilateral aid donors or through political dialogue. All these actors typically embrace the overall objective of the peacebuilding endeavor, which is to secure the peace, provide opportunities for economic development, and enable a political transition to a more democratic regime. However, the lack of coordination among these actors is widely recognized as a serious threat to the effectiveness of peacebuilding.

These international actors interact and cooperate with a multitude of domestic actors in order to promote the democratization agenda. We will focus on the priorities and strategies of domestic political elites. But political elites are not the only actors. There are also local NGOs that partner with external actors and are often dependent on the resources that international actors bring. Depending on the context, there may be other civil society actors, such as social movements, political parties, and the media. However, many of the countries where peacebuilding missions take place lack the economic opportunities and the institutional guarantees needed for the emergence of a strong civil society, and the nascent civil society actors are often powerless vis-à-vis the political elites. All of these actors—international and domestic—shape the peacebuilding landscape. It is their interaction that determines, in the end, the outcome of peacebuilding.

Does Peacebuilding Effectively Promote Democratization?

When debating the effectiveness of peacebuilding on democratization, it is useful to take as a starting point the challenges most postwar countries pose. While some authors have argued that the end of civil wars can provide an opening that can be seized for initiating a transition toward more democracy (Wantchekon 2004; Wantchekon and Neeman 2002; Gurses and Mason 2008), many others argue that civil war and its aftermath are not usually conducive to democratic outcomes. At war's end there often is insurmountable hostility among former warring parties; those who have won on the battlefield fear losing their grip on the state at the ballot box and may be tempted to stay in power by undemocratic means, and collective identities have become highly politicized and often exclusionary. War also erodes state capacities, and as a result there are no institutions in place to ensure the acceptance of the rules of the democratic game. This exacerbates the risk that political competition will turn into violent competition. In short, under such circumstances, actors may lack the will or the capacities to engage in a meaningful political process and to accept the bounded uncertainty that comes with democratic rules. It is therefore not surprising that on average countries do not increase their levels of democracy in the years after war ends. Prewar and ten years postwar levels of democracy are roughly the same, indicating that countries do not make democratic gains after war.[1]

Obviously such observations are based on averages and do not imply that all postwar countries are locked in a nondemocratic state. Some countries have indeed emerged after wars as democracies, for example Namibia, East Timor, and, with some limitations, Bosnia. But by and large, democratization is a rare outcome even when a major UN peacekeeping mission is launched to assist the transition process. Since the end of the Cold War, the UN has launched twenty-three major peacebuilding missions. (We define "major mission" here as a mission deployed for six months or longer and with at least five hundred military personnel in the field.) These twenty-three missions were reasonably successful in securing peace, but much less successful in establishing a democratic regime. Five years after the operations began, only two countries were rated "free" by Freedom House[2] that is, is as liberal democracies that extend freedom, fairness, transparency, accountability, and the rule of law from the electoral process into all other major aspects of governance and interest articulation, competition, and representation.

It is unrealistic, perhaps, to expect a *liberal* democracy to emerge from the ashes of war. But even when we apply a lower and less ambitious threshold for success, such as an *electoral* democracy (that is, according to Freedom House's definition, a regime that holds elections but provides less protection for civil liberties than does a liberal democracy), we still find that just nine out of twenty-three countries that hosted major peacebuilding missions qualify.[3] Given the vast amount of resources and hopes that are invested in liberal peacebuilding, these are sobering results.

Peacebuilding for Democracy

As we have discussed, democratization after war is a daunting challenge, yet modern peacebuilding missions are designed precisely to address these challenges. They are launched in order to help domestic elites overcome the many difficulties presented by postwar democratic transitions. Peacebuilders bring tremendous resources: they deploy civilian personnel who assume vital administrative functions and military personnel to guarantee security, and they bring economic aid, which frequently becomes the single most important source of government income. With their military presence and their economic resources, peacebuilders should have some leverage over the government, and this leverage should translate into more democracy. However, this leverage appears rarely enough to nudge a postwar country into a democratic arrangement.

The literature suggests a number of explanations for this. Perhaps the simplest one is that peacebuilders are reaching for the impossible dream, attempting to engineer in years what took centuries for Western European states (Chandler 2006; Kaplan 2008;

Lidén 2009; Lidén, Mac Ginty, and Richmond 2009; Richmond and Mitchel 2011; Tadjbakhsh 2009). Moreover, most peacebuilding missions take place in poor countries that may lack the means to sustain democratic institutions. Also, peacebuilding operations confront highly difficult conditions, including a lack of local assets and high levels of destruction (Collier 2009; Doyle and Sambanis 2006; Orr 2004). Taken together, such explanations suggest that deep-seated structural factors make it difficult for democratic institutions to grow.

Another explanation suggests that peacebuilders often face a dilemma in that they must choose between protecting stability and nurturing a liberal democracy, and stability frequently wins out (Fortna 2008; Jarstad and Sisk 2008). Rather than pressing for more democratic reforms, peacebuilders support the ruling government in order to avoid political tensions that may lead to renewed violence. Still other explanations focus on the shortcomings of the peacebuilding mission itself. The assumption is that peacebuilders would be more successful as democratizers if the missions were better resourced, had a longer time horizon, and could better adapt to the local conditions (Carothers 2002; Paris 2004, 2010; de Zeeuw 2005). Coordination problems among the myriad peacebuilding actors and the peacebuilders' limited ability to learn are also quoted as causes for the modest track record of democratic peacebuilding. Also widespread is the complaint that peacebuilders rarely adapt their strategies to the context but tend to treat postconflict democratization generally as a problem that can be solved by mechanically implementing a series of known tasks (Carothers 2002; Paris 2004).

Peacebuilding Is Interaction

Besides structural impediments in host countries and strategic and material deficits of peacebuilders, by far the most important cause for a compromised peace is the interaction between peacebuilders and the elites of the host country. Peacebuilding is an inherently interactive process in which domestic elites interact with the peacebuilders and thereby resist, revise, and shape the outcome (Richmond and Mitchell 2011; Campbell, Chandler, and Sabaratnam 2011; Barnett and Zürcher 2008; Barnett, Fang, and Zürcher 2014; Bonacker et al. 2009; Hyden 2008; Zürcher 2011; Zürcher et al., 2013; Manning 2008; Autesserre 2010). Importantly, the interests and preferences of peacebuilders and of elites in the host country governments are not always aligned. As a result, the social order that emerges from the interaction between local domestic elites and international interveners may diverge from the blueprints of the international community. Instead of a liberal peace, the emerging social order might resemble a hybrid arrangement, driven by the competing agendas of external and internal actors, resulting in a "compromised poor quality peace" (Mac Ginty 2010) and low levels of democracy.

On the surface, peacebuilders and domestic political elites may seem to have a common objective: building peace after war. However, this objective is very loosely defined, and peacebuilders and local elites can have very different visions. Because their interests are not necessarily aligned, the interaction between peacebuilders and local elites is rarely fully cooperative but is instead a blend of cooperation and resistance. Peacebuilders and domestic elites engage in strategic bargaining that results in some sort of compromise. Clearly the bargaining that takes place in the real world is not a simple game that is concluded by a clearly specified compromise. Instead bargaining takes place constantly, over a long period of time, on many different levels, and between many different players. Peacebuilders and domestic elites interact at the governmental level, for example, when they negotiate the priority areas of the aid package. Representative of aid agencies negotiate with ministries about how to implement the aid package. Field offices of donor agencies negotiate with the subnational administration or even directly with communities. What it is called here in a figurative way "bargaining" and "compromise" reflect aggregated interests, aggregated bargaining power, and the costs of adoption of a certain policy.

Peacebuilders want to secure the peace, but they also seek to establish a democratic political system. Accomplishing this requires in most postconflict countries fundamental social changes, which peacebuilders attempt to foster through a mix of diplomatic pressure and persuasion and typically a large package of ambitious aid programs. Rhetorically the domestic (governmental) elites in the host countries agree with the peacebuilders. In reality the domestic elites may be interested in some but not all aspects of the package that the international community offers.

There may be many different reasons why elites are not always keen to push for democratic standards, but a few stand out. First, introducing democratic rules can create security threats. In some postwar countries, democratization became associated with mass mobilization and ultimately civil war. For example, democratic openings paved the way to ethnic conflict in Rwanda, Tajikistan, and Bosnia. Such experiences do not inspire enthusiasm for democratization. Domestic elites frequently resist calls for democratic reforms by pointing out that further liberalization might empower oppositional forces, potentially leading to renewed civil war. In cases where domestic elites feel highly threatened, not even a very large and intrusive mission may be enough to push for more democracy. For example, the highly intrusive missions in Bosnia and Kosovo were not able to push for a real democratic opening. Paradoxically, the intrusive nature of the mission and the international tutelage generated even more domestic resistance to peacebuilders' policies. This backlash led peacebuilders to abandon part of their democratization agenda and to grant more autonomy to domestic elites.

Second, norms associated with good governance restrict leaders' ability to extort and expropriate resources. Civil wars typically lead to a surge of the informal and illicit economy, at the expense of formal, regulated, and accountable economic activities. The winning side in a civil war is most often also in command of substantial parts of that informal and illegal economy. A push toward better (economic) governance and more accountability will undermine their access to resources.

Third, democratic procedures and good governance will likely threaten the very foundation upon which the precarious and feeble authority (and survival) of many regimes in postconflict states rests: patron-client networks. Clientelistic networks are an endemic feature of weak states; arguably, they are the most basic form of governance practiced in places where infrastructural power is weak. For leaders who cannot rely on a solid governance infrastructure based on widely accepted institutions and staffed by a professional and loyal bureaucracy, clientelistic ties become their primary basis of authority. The ability to co-opt potential rivals into their networks is essential, especially in volatile postwar countries (Kitschelt and Wilkinson 2007; Ilkhamov 2007; Eisenstadt and Roniger 1984; Reno 2000; Bratton and van de Walle 1994). But entrenched clientelistic networks and ruling by patronage is in contradiction to important aspects of the peacebuilders' reform package, which may include anticorruption measures; a call for meritocratic, inclusive appointments to government positions; democratic accountability; and the like.

Fourth, and perhaps most important, domestic elites may resist some elements of the peacebuilding package for the simple reason that those elements contradict their *legitimate* political interests in another policy field. Consider this example from Bosnia, as reported by Peirce and Stubbs (2000). The mayor of a Bosnian community opposed a project dealing with minority groups returning to the central Bosnian town of Travnik. On the national level, the project aimed to strike an ethnic balance so that refugees of all ethnic groups who returned to an area where they were a minority would benefit equally. However, from the mayor's perspective, it would have been "electoral suicide . . . to agree to reconstruction programs targeted solely at minorities, whilst the majority population remains displaced and in poverty" (Peirce and Stubbs 2000, 167). Some international observers interpreted his position as nationalist, undemocratic, and "anti-Dayton," However, the mayor's primary concern was his immediate electorate, and his electorate was concerned only with reforms that impacted their immediate welfare—not that of returning minority groups. This example highlights the fact that resistance to certain aspects of the peacebuilding package is often not a sign of a principled antidemocratic attitude but rather a result of the incentives created by the local political economy. Gross (2017) convincingly shows how local-level actors are often very successful in modifying the peacebuilders' reform proposal in a way that satisfies their economic interests and promotes their political power. They do this not because they oppose in principle the broad democratization agenda but because modifying the reform agenda suits their primary interest, which is not necessarily democratization.

In sum, democratization can entail considerable *adoption costs* for elites. Consequently democratization is often perceived by these elites as a problem rather than a solution; hence they may resist certain aspects of the package. Given the seemingly tremendous power disparity between domestic elites and peacebuilders, it is surprising that peacebuilders ever compromise. But having power is not necessarily the same as effectively exercising power. One reason peacebuilders are inclined to compromise is that it is typically in the peacebuilders' mandate to "align" with local government and to create "local ownership." Furthermore, their resources and time horizons are often limited, since voters at home will not support an expensive peacebuilding mission indefinitely. (Bueno de Mesquita and Downs 2006).

A costly and lengthy standoff between peacebuilders and domestic elites is therefore not something the peacebuilders desire. Consequently peacebuilders may prioritize stability over pushing for structural reforms necessary to achieve the liberal democracy they desire, because such structural reforms may indeed increase tensions and carry the risk of renewed instability. Quite counterintuitively, under certain circumstances domestic elites may even gain bargaining power from their position of relative weakness. Domestic elites who face strong domestic opposition may argue that any attempts to institute liberal reforms may trigger a backlash, endangering not only the government's survival but also national stability. Under such circumstances peacebuilders may be inclined to provide ongoing support without evidence of reform, since stability is considered more important.

In addition, peacebuilders are highly dependent on their domestic counterparts, as their cooperation is essential for the effective, smooth, and stable implementation of the ambitious projects. Without local consent and support, peacebuilders will find it almost impossible to conduct de-mining, to build roads and bridges, to reach out to civil society organizations, or to work with elected representatives and the administration. Also, without a broad buy-in of their local partners, the security for international civilian personnel may be compromised. All of this creates a strong incentive to collaborate with domestic political actors, even when these same actors are skeptical of some vital elements of the peacebuilding package.

Finally, peacebuilders' leverage can quickly depreciate. As peacebuilders deploy resources and become invested in a postwar country, they become hostage to the need to demonstrate success. Domestic elites are aware that mission withdrawal would signal a failure to peacebuilder constituencies at home. Domestic elites may therefore be able to resist or subvert some aspects of the peacebuilders' contract without fear that such behavior will jeopardize political and material support. Because of this, peacebuilder leverage peaks immediately before deployment. As with so many forms of sustained social interaction, problems of compliance and moral hazard also plague peacebuilding missions and contribute to outcomes that are at times far from a liberal peace.

Democratic Successes

Peacebuilding is rarely able to push a host country firmly into democratic rule, but that does not mean peacebuilding is always ineffective. But the question of whether or not peacebuilding is successful at democratization cannot be answered without discussing how "success" should be defined. Various studies have used widely different measurements for success, which contributed to the seemingly contractionary and inconclusive findings. Some studies use absolute gains in democracy *after* war's end (typically a three-point increase on the widely used Polity score) as a measure of success (Gurses and Mason 2008). Other studies measure absolute gains by comparing *prewar* and *postwar* levels of democracy, thus also considering changes that might have occurred during the war. Some studies use a ten-year period, others five or even two years. Other studies use a threshold rather than absolute gains, but these thresholds vary considerably. For example, Zürcher at al. (2013) use two thresholds, a Freedom House score of 2.5 for liberal democracy, and the Freedom House designation of electoral democracy as a less demanding threshold. Doyle and Sambanis (2000) use Polity scores of −7, respectively −4 as two thresholds. These are extremely low thresholds, which are typically considered to denote nondemocratic regimes, but the authors argue that the challenges of postwar peacebuilding warrant such a minimal threshold. Clearly there is no accepted measure of what constitutes democratic success, and it is up to authors to argue for a measure they think is appropriate. For the following discussion, we use "liberal democracy" (a Freedom House score of 2.5 or better) as a strict measure for success and "electoral democracy" as a more lenient threshold. According to this measure, two respectively seven cases qualify as success (see Table 31.1).

Postwar countries are more likely to embrace the democratization package peacebuilders advertise when elites perceive adoption costs to be relatively low. As we have discussed, most important are low threat perceptions. Elites have to be assured that a democratic opening does not strengthen their opponents. If they believe democratization will endanger security, then domestic elites are, quite understandably, reluctant to embrace democracy. Rwanda and Afghanistan exemplify this situation. In both cases, domestic elites felt that democratic openings might empower opposition groups and openly resisted democratic reforms. Similarly Serb and Croat elites in Bosnia were reluctant to embrace democracy since they feared it might endanger the political dominance and exclusionist policies of their respective ethnic groups.

Compare this to the situation in, for example, Namibia and East Timor, both of which qualify as successes. In these countries, political elites faced little or no political competition and were reasonably sure that they would prevail in postwar elections. Furthermore, there was no credible threat from an armed opposition. Both factors drove adoption costs down. In addition, peacebuilders held considerable leverage over political elites. In both countries (Kosovo would be a third example), political elites fought for independence and therefore depended on the support of the peacebuilders and of the international community at large. Since the international community typically looks askance at openly undemocratic policies, the costs of not embracing democracy increased. Countries seeking independence thus find it difficult to avoid a commitment to democracy. In other words, the quest for independence lowers adoption costs.

These examples suggest that democratization through peacebuilding is met with success in cases where conditions on the ground are highly favorable, because domestic elites perceive the costs of adopting democracy as low. But that does not mean that the role of peacebuilders is inconsequential. In East Timor and Namibia, the presence of the peacebuilders safeguarded the success of the first general elections, and their resources helped to build democratic institutions. In Mozambique, the peacebuilders had the trust of both former warring parties because major donors had been working with them for years prior to any peace arrangement; therefore they helped reduce uncertainty about each other's intentions by serving as a guarantor and assisted in compliance after elections. Thus, when local conditions are already favorably disposed to

Table 31.1 Major Multinational Peacebuilding Missions after 1989: Democratic Transition Outcomes

Country	Mission Start Year	Freedom House Score (5 Years after Start)	Electoral Democracy (5 Years after Start)
Croatia	1996	2	Yes
Namibia	1989	2.5	Yes
Mozambique	1992	3.5	Yes
Sierra Leone	1999	3.5	Yes
Liberia	2003	3.5	Yes
Bosnia	1996	4.5	No
Burundi	2004	4.5	Yes
Haiti	2004	4.5	No
Mali	2013	4.5	No
Haiti	1994	5	Yes
Afghanistan	2002	5	No
Central African Republic	2007	5	No
Tajikistan	1997	5.5	No
Kosovo	1999	5.5	No
Côte d'Ivoire	2003	5.5	No
Angola	1995	6	No
Central African Republic	1998	6	No
Democratic Republic of the Congo	2001	6	No
Cambodia	1992	6.5	No
Rwanda	1993	6.5	No
South Sudan	2005	7	No
Darfur UNAMID	2007	7	No
Abyei UNISFA: Sudan and South Sudan	2011	7	No

democratization, the role of the peacebuilders may lie in trust-building and reducing commitment problems. Under such conditions, peacebuilders can, even with relatively low resources, nudge the outcome closer to a democratic outcome.

Finally, an argument could be made that the success of democratization through peacebuilding does not necessarily have to be measured by a concise and universal metric, nor is it necessary that peacebuilding have a direct impact on the immediate prospect of a liberal peace. One might consider peacebuilding successful when it can help to create the prospect of a future opening, laying the groundwork for democratization down the road. This could include less tangible social changes, such as the growing acceptance among segments of the society for democratic rules; an increased interest of the population to engage with the government in negotiating a social contract, where expectations of accountability and the provisions of public services are exchanged for rule acceptance; an increased responsiveness of leaders for the demands of society even in the absence of truly liberal democratic institutions; introducing ceremonies and symbols of democracy (even if the substance is lagging); experimenting with local governance based on elections; or a less regulated circulation of people and ideas. These and other subtle changes, even though they are not easily measured, can make democratic change a possibility. Without the efforts of peacebuilders, many postwar countries today would not have this promise of a possibility.

Conceptual Alternatives to the Peacebuilders' Liberal Vision

Over the past two decades, criticism of the liberal democratic peace has become more widespread. Two related arguments are often made: first, that the liberal peace approach has rarely been successful in the most fragile countries; second, that the liberal peace is a Western concept building on political institutions that have grown under very particular historical conditions. It is therefore neither effective nor appropriate to impose such institutions on non-Western environments. A lively debate has emerged on whether there may be alternatives to, or at least adaptations of, the Western democracy model that are better suited for many postconflict countries and whether supporting such alternatives would lead to better peacebuilding results.

An important theoretical contribution to this debate has been made by Amartya Sen (1999, 2003), who argues that the most essential and indispensable aspect of democracy is the ability of a society to practice what he calls, following John Rawls, "public reasoning" and "government by discussion." Sen maintains that public reasoning, which is closely linked with the tolerance of different points of view and the encouragement of public discussion, has been an important value in many non-Western societies, for example India, China, Japan, Korea, and Iran and in the Arab world and many parts of Africa. Sen's approach therefore decouples particular political Western institutions from the essence of democracy as a universal norm and value. That value, Sen argues, can be achieved in societies that are non-Western.

Youngs (2015) notes that critics of Western democracy have argued that important elements of non-Western democracy are customary or traditional methods of adjudication, less individualism and more communalism, more traditional social values, and more consensual and participatory politics. Many of these elements are in line with Sen's overarching principle of public reasoning as a core value of democracy. Youngs further argues that regimes that emphasize the importance of some or all of these values may be able to generate legitimacy, even if the regime is not (exclusively) based on the kind of procedural legitimacy that flows from Western political institutions. While Youngs does not argue for a complete replacement of Western institutions by non-Western "traditional" procedures, he suggests that improving coordination and learning between different approaches and institutions may increase the quality of governance, thereby creating an opportunity for peace.

International peacebuilders are increasingly aware of the potential value of incorporating non-Western informal institutions in governance systems and occasionally experiment with approaches that give a greater role to such traditional, non-Western institutions and procedures. A few examples from Afghanistan follow. Western peacebuilders have helped to modernize the traditional village council. While leaving the authority of the traditional village council intact, Western donors have incentivized village councils (with the promise of block grants) to select their members by ballot and not based on prestige alone. They have also pushed for inclusion of female members. The hope is to create more inclusive and effective local governance, both preconditions for peace. Western donors have also supported traditional Afghan mediation courts, the *huquqs*, and have helped to integrate them into the formal justice system, hoping that this would make the justice sector more legitimate and more efficient. The Liaison Office (TLO) is another noteworthy initiative in the same vein. TOL's aim is to facilitate the formal integration of communities and their traditional governance structures (such as village councils, elders, and triable structures) within Afghanistan's formal structures in order to improve local governance, stability, and security through systematic and institutionalized engagement with customary structures. TLO is an Afghan initiative, funded by Western peacebuilders.

These examples show how Western donors seek to incorporate elements of more traditional institutions, which often are deeply embedded in society and are widely seen as legitimate, into the more formal, more Western governance structures. However, such experiments are still rather rate and primarily focus on the communal or local level. Also, these are add-ons and modifications rather than alternatives to the formal political institutions that govern the political process at the macro level. There is not (yet) solid evidence about the effectiveness and the sustainability of such hybrid institutional armaments. And most important, such informal arrangements are highly vulnerable to being hijacked by those who have the will and the capability to organize violence. Who, in the absence of a fully functional and accountable state with a legitimate monopoly of violence, would be able to protect these arrangements?

Notes

1. Using the Polity IV index as a measurement of democracy, we see that two years after war ended, only 10% of the countries reached a Polity score of 7 or higher (approximately like Kenya, Moldova, or Mali). About half of all war-affected countries (53%) show a polity score of −5 or lower, that is, a regime type similar to Gambia or Iran; 37% have a polity score of

−7 or lower, similar to Belarus or Uzbekistan. The picture looks equally gloomy five years after the end of civil war: 52% have a Polity score of −5 or lower; 39.4% have a score of −7 or lower; and only 10.6% have a score of +7 or higher. In general, values and distributions two years and five years after war ends are remarkably similar. When comparing the means of the two samples (Polity scores two years and five years after the conflict), we find a high correlation between the two samples of 0.958 (at a significance level of 0.10). Using the same data, we also find that there is hardly a gain in democracy during or after civil wars. When comparing Polity scores five years before with five years after the war, we find that the postwar scores are on average about 3.1 points higher on the scale than the five-year average score before the war. However, these gains in democracy turn out to be temporary and partly driven by a few outlier cases with very low prewar democracy scores. When we compare ten-year Polity averages before the war and ten years after the war, we find that the average Polity scores plunge back to their levels ten years before the war. All data from Doyle and Sambanis (2000).

2. Freedom House rates a country as free and as a liberal democracy when it scores between 2.5 and 1 on a 7-point scale, whereas 7 denotes the least free regime. See https://freedomhouse.org/reports/freedom-world/freedom-world-research-methodology (accessed November 15, 2020).
3. Freedom House defines an electoral democracy as a country that has a competitive multiparty system; universal adult suffrage for all citizens (apart from legal restrictions that may apply to citizens convicted of certain crimes); regularly contested elections by secret ballot with "reasonable ballot security" and the absence of massive fraud that distorts results; and "significant public access" of major political parties to voters through open campaigning and the media. See https://freedomhouse.org/reports/freedom-world/freedom-world-research-methodology (accessed November 15, 2020).

References

Autesserre, S. 2010. *The Trouble with the Congo: Local Violence and the Failure of International Peacebuilding*. Cambridge: Cambridge University Press.

Barnett, M., S. Fang, and C. Zürcher. 2014. "Compromised Peacebuilding." *International Studies Quarterly* 58 (3):608–620.

Barnett, M, and C. Zürcher. 2008. "The Peace-Builder's Contract: How External Statebuilding Reinforces Weak Statehood." In *The Dilemmas of Statebuilding: Confronting the Contradictions of Postwar Peace Operations*, edited by Roland Paris and Timothy D. Sisk, 23–52. London: Routledge.

Bonacker, T., M. Daxner, J. H. Free, and C. Zürcher, eds. 2009. *Interventionskultur: Zur Soziologie von Interventionsgesellschaften*. Wiesbaden: VS Verlag.

Bratton, M., and N. van de Walle. 1994. "Patrimonial Regimes and Political Transitions in Africa." *World Politics* 46 (4):453–489.

Bueno de Mesquita, B., and G. W. Downs. 2006. "Intervention and Democracy." *International Organization* 60 (3):627–649.

Campbell, S, D. Chandler, and M. Sabaratnam. 2011. *A Liberal Peace? The Problems and Practices of Peacebuilding*. London: Zed Books.

Carothers, TT. 2002. "The End of the Transition Paradigm." *Journal of Democracy* 13 (1):5–21.

Chandler, D. 2006. *Empire in Denial: The Politics of State-Building*. London: Pluto Press.

Collier, P. 2009. *Wars, Guns, and Votes: Democracy in Dangerous Places*. New York: Harper.
de Zeeuw, J. 2005. "Projects Do Not Create Institutions: The Record of Democracy Assistance in Post-Conflict Societies." *Democratization* 12 (4):481–504.
Doyle, M. W., and N. Sambanis. 2000. "International Peacebuilding: A Theoretical and Quantitative Analysis." *American Political Science Review* 94 (4):779–801.
Doyle, M W., and N. Sambanis. 2006. *Making War and Building Peace: United Nations Peace Operations*. Princeton, NJ: Princeton University Press.
Eisenstadt, S. N., and L. Roniger. 1984. *Patrons, Clients and Friends: Interpersonal Relations and the Structure of Trust in Society*. Cambridge: Cambridge University Press.
Fortna, V. P. 2008. "Peacekeeping and Democratization." In *From War to Democracy: Dilemmas of Peacebuilding*, edited by Anna Jarstad and Timothy D. Sisk, 39–79. Cambridge: Cambridge University Press.
Gross, L. 2017. *Peacebuilding and Post-War Transitions: Assessing the Impact of External Domestic Interactions*. London: Routledge.
Gurses, M, and T. David Mason. 2008. "Democracy Out of Anarchy: The Prospects for Post-Civil-War Democracy." *Social Science Quarterly* 89 (2):315–336.
Hyden, G, 2008. "After the Paris Declaration: Taking On the Issue of Power." *Development Policy Review* 26 (3):259–274.
Huntington, S. 1991. *The Third Wave: Democratization in the Late Twentieth Century*. Norman: University of Oklahoma Press.
Ilkhamov, A, 2007. "Neopatrimonialism, Interest Groups and Patronage Networks: The Impasses of the Governance System in Uzbekistan." *Central Asian Survey* 26 (1):65–84.
Jarstad, A. K., and Timothy D. Sisk, eds. 2008. *From War to Democracy: Dilemmas of Peacebuilding*. Cambridge: Cambridge University Press.
Kaplan, S. 2008. "Fixing Fragile States." *Policy Review* 152:63–77.
Kitschelt, H, and S Wilkinson. 2007. *Patrons, Clients, and Policies: Patterns of Democratic Accountability and Political Competition*. Cambridge: Cambridge University Press.
Lidén, K. 2009. "Building Peace between Global and Local Politics: The Cosmopolitical Ethics of Liberal Peacebuilding." *International Peacekeeping* 16 (5):616–634.
Lidén, k, R. Mac Ginty, and O. P. Richmond. 2009. "Introduction: Beyond Northern Epistemologies of Peace. Peacebuilding Reconstructed?" *International Peacekeeping* 16 (5):587–598.
Mac Ginty, R. 2010. "Hybrid Peace: The Interaction between Top-Down and Bottom-Up Peace." *Security Dialogue* 41 (4):391–412.
Manning, C. 2008. *The Making of Democrats: Elections and Party Development in Postwar Bosnia, El Salvador and Mozambique*. Basingstoke, UK: Palgrave Macmillan.
Orr, R. 2004. "The United States as Nation Builder: Facing the Challenge of Post-Conflict Reconstruction." In *Winning the Peace: An American Strategy for Post-Conflict Reconstruction*, edited by Robert Orr, 36–52. Washington, DC: CSIS Press.
Paris, R. 2004. *At War's End: Building Peace after Civil Conflict*. New York: Cambridge University Press.
Paris, R. 2010. "Saving Liberal Peacekeeping." *Review of International Studies* 36 (2):337–365.
Peirce, P., and P. Stubbs. 2000. "Peacebuilding, Hegemony and Integrated Social Development: The UNDP in Travnik." In *Regeneration of War-Torn Societies*, edited by Michael Pugh, 157–176. Basingstoke, UK: Palgrave Macmillan.
Reno, W. 2000. "Shadow States and the Political Economy of Civil Wars." In *Greed and Grievance: Economic Agendas in Civil Wars*, edited by Mats Berdal and David M. Malone, 43–69. Boulder, CO: Lynne Rienner.

Richmond, O P., and A. Mitchell. 2011. "Peacebuilding and Critical Forms of Agency: From Resistance to Subsistence." *Alternatives: Global, Local, Political* 36 (4):326–344.

Sen, A. 1999. "Democracy as a Universal Value." *Journal of Democracy* 10 (3):3–17.

Sen, A. 2003. "Democracy and Its Global Roots." *New Republic*, 6 October, 28–35. http://www.columbia.edu/itc/sipa/U6800/readings-sm/sen_democracy.pdf.

Sens, A. 2004. "From Peace-Keeping to Peace-Building: The United Nations and the Challenge of Intrastate War." In *The United Nations and Global Security*, edited by Richard Price and Mark Zacher, 141–160 London: Palgrave Macmillan.

Tadjbakhsh, S. 2009. "Conflicted Outcomes and Values: (Neo)Liberal Peace in Central Asia and Afghanistan." *International Peacekeeping* 16 (5):635–651.

Wantchekon, L. 2004. "The Paradox of 'Warlord' Democracy: A Theoretical Investigation." *American Political Science Review* 98 (1):17–33.

Wantchekon, L, and Z. Neeman. 2002. "A Theory of Post–Civil War Democratization." *Journal of Theoretical Politics* 14 (4):439–464.

Youngs, R. 2015. "Exploring Non-Western Democracy." *Journal of Democracy* 26 (4):140–154.

Zürcher, C. 2011. "Building Democracy While Building Peace." *Journal of Democracy* 22 (1):81–85.

Zürcher, C., C. Manning, K. Evenson, R. Hayman, S. Riese, and N. Röhner. 2013. *Costly Democracy. Peacebuilding and Democratization after War*. Stanford, CA: Stanford University Press.

CHAPTER 32

POWER-SHARING IN DIVIDED SOCIETIES

JOHN DOYLE

Power-sharing models of government as a contribution to peacemaking have dominated constitutional design since the mid-1990s, but they remain highly contested in both the academic literature and policy analysis. From the Dayton Accords on Bosnia and Herzegovina, through the Good Friday Agreement on Northern Ireland, to Iraq, Burundi, and Afghanistan, a wide variety of different types of power-sharing models have been used, at the state level in both unitary and federal states and in regional governments, without being deployed nationally. This level of deployment led McGarry and O'Leary (2006a, 44) to say, "Consociational theory, developed by Arend Lijphart and other scholars, is one of the most influential theories in comparative political science. Its key contention is that divided territories, be they regions or states, with historically antagonistic ethnically, religiously or linguistically divided peoples, are effectively, prudently, and sometimes optimally, governed according to consociational principles."

Power-sharing arrangements of different types have been seen as a means to limit demands for partition and secession without imposing a coercive form of integration on mobilized minorities (O'Leary 2005, 40–44). Wolff (2004) argues that international actors promote power-sharing to assure conflict parties of the relative permanence of the institutions they have agreed upon. They become part of what Barbara Walter (1997) calls the "credible commitment" of international actors to a new regime, that minorities will have some institutional power when the attention of international actors moves to the next crisis.

Power-sharing offers international mediators and local political actors alike a "way out," which at least achieves or maintains a ceasefire and offers the hope, in the absence of armed conflict, that political progress can be made (Zartman 2005). Yet critics of power-sharing fear that it offers a short-term solution only at the cost of longer-term political difficulties. Horowitz (1985, 1991, 2001, 2002, 2014) argues that it promotes the interests of political extremes over more moderate centrist parties. Taylor (1994, 2001) believes it prevents conflict transformation. Wilford (1992) asserts that the privileging of the national or ethnic identities at the heart of most power-sharing models marginalizes other forms of identity and other political interests. Andeweg (2019) questions whether the persistence of elite cooperation is partly responsible for the rise of antisystem parties, as consociationalism seems to deny voters a real choice to change the system itself.

This chapter has the following structure. First of all, the origins of power-sharing as a model of government are explored, through its roots in consociational theory, looking at the strengths and limits of these early debates and the small number of practical examples for contemporary peacebuilding. The chapter then briefly discusses the impact of the end of the Cold War as a relatively unexplored aspect of the emergence of power-sharing as a common model after 1995. The contemporary cases are explored thematically, looking at the claims of power-sharing to offer a way out of armed conflict based on long-term, deeply embedded political divisions and the chapter critiques suggestions that power-sharing locks in specific social identities in a manner that limits future change. Specific aspects of the debate are explored, such as the gendered nature of power-sharing models, the claims that it encourages ethnic outbidding by extreme parties, and the potential impact of the choice of electoral system.

The Origins of Consociational Theory and Its Relevance

The contemporary debates on power-sharing are largely derived from the early work of Arend Lijphart (1969) on consociational theory. Lijphart's original contributions focused on the apparent success of four European cases with (in Lijphart's view) nonmajoritarian consociational forms of government at their core: the Netherlands, Switzerland, Belgium, and Austria. He argued that consociationalism was at the heart of long periods of internal political stability in societies with deep social divisions around language, religion, and ideology. The very premise of whether the Netherlands and Austria were deeply divided societies was, however, contested (Steiner 1981). Austria, Switzerland, and the Netherlands (whatever about Belgium) seemed to be settled liberal democracies, and certainly none of his cases had suffered recent internal civil war. Therefore the lessons that could be drawn from these liberal European states were contested, and the direct application of their governance model to countries at war or to deeply repressive polities did not appear to be strong. The two cases of recent armed conflict where consociationalism had been attempted as part of a peace process prior to the 1990s, Lebanon and Northern Ireland (1973–1974), did not offer compelling narratives of success. Lijphart (1975) argued that Northern Ireland was too deeply divided on the core issue of territory to be a likely case. In the absence of real-world case studies, the research agenda became focused on the conditions under which consociationalism might be possible, called the "favourable factors" debate (Bogaards 1998).

Steiner (1981, 341) argues that identifying societies where consociationalism might be attempted is hindered by the fact that Lijphart was not clear in how a plural society was to be defined operationally. He notes that Lijphart was concerned not simply with cultural diversity but also with subcultural segmentation. Yet Steiner argues that the operational definitions of segmentation often use measures that reflect cultural diversity in itself, and not the level of its political organization. According to Steiner, defining what are, and what are not, "plural" societies is crucial if consociationalism is to be advanced as a form of democracy (344). If Switzerland is not really "plural" at all, but quite homogeneous, as Barry (1975, 487) claims, then it is not very useful as a comparator where all analysts agree that conflict-prone societies are deeply segmented. More generally, even if the Netherlands was deeply segmented and consociational at one point in history, it certainly seems less so today. If consociational government assists those societies to move from being deeply segmented to more unified liberal democracies, then consociationalism may continue to have relevance as a theory and practice for a point in time (even if prolonged) rather than a permanent mode of government.

The challenge of defining deeply divided societies at a social level remains important. If a civil war is analyzed as being primarily an elite-based contest for power or resources, with limited social roots (Collier and Hoeffler 2000), then defining a model of government around those divisions would not be appropriate. However, if a prolonged conflict reflects deeply embedded social divisions with limited cross-cutting political cleavages to soften the primary division, then a model of peacebuilding based on attempting to build a single integrated society, but where clear and obvious minorities will be excluded from power in the short term, is unlikely to be acceptable to highly mobilized minority communities. Steiner (1981, 342), in his critique of Lijphart, when conducting a thought experiment on what proportion of Catholics needed to vote for a Catholic party to be plural, and indeed what policies would constitute Catholic policies, reflects the origins of consociational theory rather than its potential application in the 1980s. His question could be more easily answered if he had chosen an example from national identity and demands for statehood. A policy platform based on demands for statehood, or resistance to demands for secession, is easy to identify. The support base in elections or opinion polls for rival political parties based on demands for devolution or secession resting on social or identity characteristics is also possible to operationalize. Identities will change over time but have stronger social characteristics than being in favor of or against a particular model of delivering a health system, for example. Therefore much of the critique of the definition of plural societies as a starting point for power-sharing can be overcome if its use as a model is confined to cases of deep ideological, ethnonational, or socioreligious communal divisions, which can be analyzed in terms of long-term political mobilization with clear political demands, interacting with deep social divisions. Using this approach, research can clearly identify supporters of a United Ireland or an independent/autonomous Kurdistan based on both electoral/political behavior over considerable time, on broad political demands and on political actors' strong but never absolute interconnections with other forms of social organization and identity.

The Impact of the End of the Cold War

Lijphart's (1969) early work on consociationalism saw external factors as being favorable to the development of internal power-sharing when they were equally threatening to all significant factions in a divided society. He wrote, "There are three factors that appear to be strongly conducive to the establishment or maintenance of cooperation among elites in a fragmented system. The *most* striking of these is the existence of external threats to the country" (217). Where external actors were seen differently by internal actors, or where they supported one political community over another, Lijphart (1975) saw this as impeding the emergence of consociational settlements, and this framed his analysis of the failure of the power-sharing experiment in Northern Ireland in 1973–1974. In both outcomes, however, external actors are seen in a negative light, even if their threats might lead to internal cooperation.

This debate, though useful in exploring the internal dynamics that limited the successful use of consociational theory, was too unaware of the geopolitical context of the Cold War and its impacts on war and peace. During the Cold War there were very few cases where the superpowers were indifferent as to outcome. It was all too easy for them to intervene and support favored actors in civil wars and also attractive for actors to exaggerate their support for one ideological view over another in order to secure diplomatic and military support. The malign effect of external actors in the Lebanon case acted as a further barrier to the conceptualization of external actors having a potentially substantive positive role in consociationalism (McGarry and O'Leary 2009, 41–42). As a result, this context-specific frame, understanding external interventions as negative and threatening, was turned into a core aspect of the theory without substantive critical reflection. The sudden increase in the number of cases in the 1990s had much more to do with the changed geopolitical environment than with any change in the nature of the conflicts themselves or in evidence-based lesson learning by mediators.

Exploring the Contemporary Cases

In response to that sudden increase in the number of power-sharing cases from 1995 onward, the debate shifted away from being mainly about Lijphart's four historic classic examples or theoretically possible situations, to an exploration of case studies focused on the impact of actual power-sharing executive models. The key issues at the heart of contemporary debates are the appropriate use of power-sharing: Which deep divisions are best addressed by power-sharing models of peacemaking? Does power-sharing lead to ethnic outbidding, the strengthening of more militant parties, and increased polarization? Does power-sharing entrench ethnicity or national identity within the political system, pushing out cross-community political organizations, or does it create a space to move beyond armed conflict, where identities can shift in response to new circumstances? How has consociationalism interacted with other political issues such as gender equality?

The case for power-sharing rests on the normative and empirical claim that permanently excluded political communities, which have a strong sense of collective self-identity and a record over some considerable time of political mobilization, in particular if that has resulted in protracted social conflict (Azar 1990), are likely to be successful at resisting political domination by an exclusionary majority force. If long-term political exclusion has not resulted in successful integration, but rather has led to latent or armed conflict, then other forms of governance seem empirically and normatively necessary. Partition or extensive devolution may provide a solution in situations where populations are relatively geographically separate. But in contexts where populations are mixed, forced population movement or hegemonic repression does not provide compelling evidence of a useful approach for peacebuilding. Power-sharing, though utilized in a limited number of cases, and primarily in the relatively recent period since 1995, offers a potential approach to shared governance that avoids the costs of forced movement or permanent exclusion.

Taylor (2001) and other critics argue that power-sharing cements political identities and does not allow for normal and healthy evolution of social identities. The arrangements for Lebanon, which do not reflect demographic changes (Hamdan 2013), and elements of the Dayton Accords that marginalize counterhegemonic groups (McCrudden and O'Leary 2013) are typical examples used to highlight the dangers of power-sharing in general. However, while all identities are socially constructed and all change over time, there was no realistic expectation in the majority of cases where power-sharing was introduced that normal evolution of identities was likely to deal with the consequences of exclusion at the heart of a prior conflict. In a context of group-based discrimination or exclusion over a protracted period there is little evidence from contemporary peace negotiations that political actors representing minorities are willing to put their faith in majority rule by their opponents combined with court-endorsed individual rights (McCrudden and O'Leary 2013).

In cases of deeply divided societies, enduring conflict, and popular political mobilization, majoritarian forms of executive formation have generally failed. Minorities with an experience of discrimination or exclusion that are large enough to prevent hegemonic rule by a majority will typically have both an expectation of some political influence and an ability to resist permanent exclusion. If the nature of electoral behavior consistently reflects deep social divisions and leaves a minority—whether defined by political goals, national identity, ethnicity, or religious identity—in a permanent sate of exclusion, then power-sharing represents a means by which a polity can be governed and an executive formed that includes a structured form of inclusion for all politically significant groups. Without endorsing the wider argument, there is some evidence in Collier and Hoeffler's (2000) work to suggest that attempts to dominate or assimilate large and politically mobilized minorities increases the likelihood of armed conflict. Collier's work does not look at governance models, but its analysis of the impact of the size of minorities suggests that attempts at assimilation or domination of large minorities are problematic.

The more interventionist role of international mediation in the post–Cold War period saw a significant increase in the role of external actors in shaping peace agreements with power-sharing elements. Stedman, Rothchild, and Cousens (2002), Paris and Sisk (2009), and Jarstad and Sisk (2008) addressed the challenges for external actors during the implementation of these agreements, which included tensions between the implementation of reforms focused on democratization and those focused on ending violence. From Bosnia-Herzegovina to Northern Ireland, Burundi, Kosovo, Afghanistan, and Iraq, a variety of external actors used their influence to push power-sharing models of democracy over majoritarian ones, and consociational arrangements became normatively acceptable (O'Leary 2005, 40–44; Visoka 2017). Walsh and Doyle (2017) look at external intervention in the implementation period of the Good Friday Agreement on Northern Ireland, finding that external intervention in a manner consistent with internal consociationalism led to the most positive outcomes in advancing contentious issues such as postconflict policing. The new wave of power-sharing after 1995 rests on a wider normative support by international actors, and also seems to suggest that the best outcomes require an appropriate, locally informed mix of consociationalism and external support. It is not a question of choosing one over the other.

The apparent negative consequences of power-sharing for local representation and peace formation have been much discussed, with claims that power-sharing leads peace processes to become deadlocked or authoritarian based on elite-level pacts (Taylor 2001). However, the causal links between power-sharing in itself and other aspects of either the conflict or the peace process have not been well explored, and in many cases all the negative outcomes of a complex conflict, external intervention, and ultimate power-sharing model are assumed to be related to the choice of consociationalism. Therefore it is necessary to unpick the processes of political deadlock and political progress in different examples and to explore the place of power-sharing among the many other social and political forces leading to any given outcome, much more than the literature has done to date.

The longest-lasting of the contemporary power-sharing models is Bosnia-Herzegovina, and there are few who rush to the defense of that current political system. However, untangling the causal effects of the particular style of power-sharing in use at the center in Bosnia-Herzegovina, as distinct from the highly differentiated model of devolution or the impact of external powers or simply the legacy of a bitter war, is not a simple matter. Sircar (2006) offers an interesting comparison of two early cases, Bosnia-Herzegovina and Northern Ireland, both with active external engagement and nearby kin states. He argues that in the case of Northern Ireland there are strong traits of what he calls "transnational consociation," with strong intergovernmental cooperation from the UK and Ireland acting (at that time) as "reliable long-term guarantors of the settlement" (261). However, he argues that the limited intergovernmentalism between Zagreb and Belgrade has meant that "post-conflict institutions are held together by international agencies that do not have as durable a link to the conflict zone as the 'reference states.' Therefore, a durable transnational consociation with the 'reference states' as guarantors is more likely in Northern Ireland than in Bosnia-Hercegovina" (2). Going beyond individual and dual case studies, Ottmann and Vüllers (2014, 20), in one of the first cross-case studies of power-sharing from their new data set, suggest that even the promise of power-sharing in a peace agreement has a positive impact on reducing the probability of a return to violence. They also suggest that the range of issues that power-sharing impacts—not just politics but security and economic and territorial control—are important factors.

Within the range of liberal critiques of power-sharing there is a strongly emerging debate on the impact of consociational approaches on gender equality and inclusion. The literature, though diverse, draws strongly on the cases of Northern Ireland and Bosnia-Herzegovina and is overwhelmingly critical (Deiana 2016; Hayes and McAllister 2012; Kennedy, Pierson, and Thomson 2016). Some more recent literature has focused on unpicking consociationalism and the case studies and seeking to identify the particular impact of power-sharing compared to other factors that have led to negative outcomes for gender equality and inclusion—such as patriarchy in general, conservative political actors, and negative attitudes to the importance of gendered perspectives among mediators (Kapic 2020; Byrne and McCulloch 2018; Bell 2018). There is no doubt that gender equality and inclusion have been low priorities for the main mediators and most political actors in almost all contemporary cases where power-sharing models were practiced. In Burundi there were quotas for women as well as ethnically defined communities by design. In Northern Ireland women activists set up a new party, and an electoral system designed to allow the tiny loyalist organizations to secure elected representatives had the unforeseen consequence of allowing a party with an explicit gender equality mandate to enter the talks. This had an impact on some of the other parties, at least in forcing them to select more women candidates for election; at the time of writing, three of the five largest parties, including the two biggest, are led by women. However, progress on gender equality has been very modest. Quotas in police recruitment for Irish nationalists rapidly changed the composition of the new police service, but despite explicit recognition of the problem of the tiny proportion of women officers, no quota was introduced and no real progress was made. Likewise in Bosnia-Herzegovina, beyond the formality of quotas in Parliament, there has been no measurable progress on women's rights (Kapic 2020; Deiana 2018). However, as Bell (2018) highlights, there is often no absolute incompatibility between power-sharing and gender equality, and strategies for engagement by feminists are possible. Indeed power-sharing, with its focus on quotas and proportionality, is ideologically suited to an engagement with feminist critiques. There is nothing in consociational theory to prevent gender quotas as well as national quotas, gender balance in negotiating teams, or an agenda for gender equality in ethnonational groups. Ultimately there is little evidence that power-sharing is a causal factor in the poor outcomes in contemporary cases; the internal and external political actors who dominated those processes gave gender equality no priority and thereby reinforced exclusion.

Power-sharing's most compelling defense has been that it offered a potential model to advance from conflict to, at the very least, conflict management and maybe peacebuilding, when other forms of majoritarian democracy and integration were rejected by significant political forces and were widely seen as inadequate by political analysts. Discussion of conflict transformation was nonthreatening, but there were few practical models whereby the shift from armed conflict to conflict transformation could occur without a significant transition of some sort. The Kurds' first political preference, for example, in the aftermath of the US invasion of Iraq may have been independence, but they were well aware of the potential cost of pursuing that option and were at least open to a mix of devolution and power-sharing. They were extremely unlikely to accept Iraq-wide majority rule and what O'Leary and Salih (2005) call the "reextinction" of Kurdistan and Kurdish identity into an undifferentiated and assumed singular Iraqi identity. Kurdish groups also had the political and military power to prevent this. Iraq is far from a successful postconflict state, but a mix of power-sharing at the center and recognition of Kurdistan's de facto separate status within a federation created a more promising polity than any likely alternative for now, and the consociational elements of the Iraqi constitution, insofar as they have been implemented, are hardly responsible for the wider challenges in the country and the region. The Kurdish case is also a good example of the importance of context. Whereas in Bosnia and Herzegovina, the creation of an excessive degree of devolution and the hollowing out of the (small) central state are at the heart of its difficulties, in the Kurdish case an appropriate mix of consociationalism and federalism is central to any positive roadmap for the current situation.

Northern Ireland is the case most often quoted by those who argue that power-sharing encourages support for more militant ideological perspectives. Most analysis of the Northern Ireland case believes that the IRA ceasefires of 1994 and 1997 could be maintained without a short-term move to Irish unity as long as that option could be pursued peacefully, but there was no prospect that the peace process could advance based on majority rule in Northern Ireland (Doyle 1998). Majority rule over fifty years had seen systematic discrimination against the Irish nationalist population in employment, housing, and policing and their exclusion form political life. Given the deep political divisions in Northern Ireland, reflected strongly in voting behavior, there was no prospect, regardless of voting system used, that a political system could be devised acceptable to the 40% of the population of Northern Ireland who self-identify as Irish, based on de facto majority rule by unionists, with some limited individual rights for Irish nationalists (Connolly and Doyle 2015). Like other cases, therefore, the case of Northern Ireland starts with "Nothing else seemed possible."

It is now more than twenty years since the 1998 Good Friday Agreement, a period that has seen the maintenance of the ceasefires and historically low levels of political violence. This gives a reasonable time period to explore the impact of power-sharing. At first glance voter behavior has dramatically shifted over the past twenty-five years. In the 1993 local government elections (based on proportional representation, unlike Westminster), before the ceasefires, the Ulster Unionist Party (UUP) received 29% of the vote; the more moderate of the two Irish nationalist parties, the Social Democratic and Labour Party (SDLP), 22%; the more militantly unionist Democratic Unionist Party (DUP) 17%; Sinn Féin 12%; and the centrist Alliance Party 8%. By the time of the 2017 Assembly elections, the DUP had 28.1% support, with Sinn Féin just behind at 27.9%—significantly overtaking their more centrist rivals, the UUP (13%) and SDLP (12%), while Alliance had recovered somewhat to 9%.

This voting pattern is reasonably consistent, and both Horowitz (2001) and Taylor (2001) take it as evidence of ethnic outbidding. However, this may be only partly true. Both of the two new dominant parties had radically moderated their positions during the peace process (Mitchell, O'Leary, and Evans 2001, 2002). The DUP walked out of the talks in 1997 and campaigned for a no vote in the 1998 referendum on the Good Friday Agreement, preferring a return to a security-focused approach. A vote for Sinn Féin before 1994, whatever the party might say, was seen as an endorsement of some sort for the IRA armed campaign. By 2007 both parties were supportive of a peace process and willing to share power in an Executive. The is also mixed evidence on their ability to share power—notwithstanding many periods of Executive collapse. The first Executive formed under the terms of the Good Friday Agreement was led by what was then the largest and the more moderate party in each community, the UUP and SDLP. The Executive collapsed on four separate occasions during its first term of office, ultimately leading to elections in November 2003. The Executive could not be reformed after the 2003 elections, but after new elections in 2007 a second Executive was formed, led by the largest two parties at the time, the DUP and Sinn Féin, and it continued functioning reasonably well until 2017, through two further elections.

In 2016 the SDLP, UUP, and Alliance Party, though entitled to ministerial posts, chose instead to leave the Executive and form an Official Opposition for the first time—which is permitted under the power-sharing rules. Ministerial positions were proportionally allocated between the DUP and Sinn Féin, with independent unionist Claire Sugden serving as minister of justice by agreement between the parties. This ability to leave government without its collapsing is one of two interesting features of the Northern Ireland case. No party is required to enter government, but no party with the minimum necessary support can be excluded against their will. It is not therefore a mandatory coalition, but inclusionary (McGarry and O'Leary 2006a, 262). In brief, the second useful feature is that Executive posts are allocated proportionally, avoiding the need for postelection negotiations on portfolios. The largest party gets first choice of ministry, followed by others in turn using a D'Hondt system.

Although the Northern Ireland government collapsed in January 2017 in a row over corruption, and the Executive was not reestablished until January 2020, this was an unusual situation, as the DUP for most of this period was supporting a minority British Conservative government in London, with whom they shared a key interest in supporting Brexit (despite the fact that a majority in Northern Ireland voted to remain in the EU). With direct influence in the UK government, there was little incentive for the DUP to share power locally with Sinn Féin (Connolly and Doyle 2019). It was only with the election of a majority Conservative British government under Boris Johnson in December 2019, no longer relying on DUP support, that meaningful talks on restoring devolution resumed.

The alleged negative consequences of ethnic outbidding is not supported by the evidence in Northern Ireland. The two more militant parties shared power for ten years and agreed on some really contentious issues, such as policing. During their time in shared office, both parties clearly grew into the role and seemed to value the opportunity to be in government. The possibility of returning power to London, so-called Direct Rule, remains unpopular in Northern Ireland—though supported by a minority of unionists who continue to oppose power-sharing in principle. While some key legacy issues remain unresolved, including state support for the Irish language and how to deal with official recognition of the past and support for victims, there is no shared position on these issues between moderate nationalists and unionists either. It is also impossible to overstate the negative impact of Brexit on the peace process (Connolly and Doyle 2019).

As a subset of his wider critique, Horowitz (2001) has argued in particular that proportional representation combined with power-sharing leads to more negative outcomes and that an alternative vote system would force candidates to moderate their position in order to win support from more centrist voters. Northern Ireland uses a proportional representation single transferable vote (PRSTV) system for the Assembly and local government, but uses the First Past the Post system for Westminster. Lijphart (1991) did not like PRSTV, as it reduced proportionality and opened the possibility of gerrymandering via district manipulation, but there is no evidence that his fears have been seen in practice in Northern Ireland (Mitchell 2014). Bogaards (2019, 13) remains unconvinced that consociationalism is responsible for the rise of more militant parties in Northern Ireland, saying there is no evidence to support this as yet. Voters in Northern Ireland have by and large remained loyal to parties, and especially to ideological blocs, in both PRSTV and Westminster elections—even at the cost of "wasting" their vote by voting for parties with no prospect of winning in a First Past the Post-election. Even in the immediate post-1998 Agreement election, which used PRSTV, pro-Agreement unionist voters overwhelmingly gave their lower preference votes to anti-Agreement unionists rather than moderate pro-Agreement nationalists (O'Kelly, Doyle, and Boland 2010).

Fiji utilized the alternative vote electoral system, advocated by Horowitz (2001) and other critics of power-sharing, but in Fiji it was responsible for isolating minority moderate parties and in the specific case of Northern Ireland, such a voting system would see the Alliance Party lose most of their seats in Northern Ireland, being reduced from a party in power in their own right to an influencing bloc on the margins between the two large voting blocs. Lebanon is generally seen as the modern consociational system with the least flexibility, given its confessional structures. The two main parties representing the Shia community have called for a move to a proportional representation system, as the current system underrepresents them, as they believe that an open proportional representation system, where the voter in effect to chooses their self -identity, would strengthen their position and not weaken it (Hamdan 2013). Bogaards (2019) in fact concludes that the whole dichotomy of consociationalism versus centripetalism, set up by Horowitz, is not useful and that systems in reality often incorporate elements of both.

Conclusion

Notwithstanding the highly contested debate, there is no evidence that it is possible to move more immediately from armed conflict to a process of conflict transformation without some intermediate phase. The debate then reduces to what sort of "immediate." There is no evidence that de facto integrationist and assimilationist measures based on strengthened individual rights are either attractive to marginalized minorities or have any realistic probability of being implemented, if the previous minority conflict actors are excluded from political power, control of security, and a proportionate share of economic resources.

From a normative perspective, building a power-sharing model based on electoral results overcomes one of the strong objections to power-sharing: the fear that it locks a historically specific set of social divisions into the political system, in the manner of Lebanon. Basing each cycle of executive formation on the results of the previous election can facilitate internal electoral competition within political communities and the emergence of new parties outside of the traditional political divisions. It could also facilitate other forms of representation, such as gender quotas. The experience of Lebanon, however, suggests that once a system is in place, party elites privileged by that system have strong incentives and abilities to prevent reform.

There are of course major continuing problems in almost all of the cases that have sought to implement power-sharing arrangements, from Bosnia-Herzegovina to Iraq. However, unpicking the probable negative impacts of different factors does not offer much evidence that power-sharing is a significant negative factor, and there is growing evidence in single case studies and cross-country comparisons that it is associated with more positive outcomes. The important and growing role of external actors, both in reaching agreements and in their implementation, remains understudied, and research programs could engage with those focused on the "local turn" in peacemaking in order to better understand the positive and negative factors in this important interaction between external support and local agency (Mac Ginty and Richmond 2013).

References

Andeweg, R. B. 2019. "Consociationalism in the Low Countries: Comparing the Dutch and Belgian Experience." *Swiss Political Science Review* 25 (4):408–425.

Azar, E. 1990. *The Management of Protracted Social Conflict: Theory and Cases.* Aldershot, UK: Dartmouth.

Barry, B. 1975. "Political Accommodation and Consociational Democracy." *British Journal of Political Science* 5 (4):477–505.

Bell, C. 2018. "Power-Sharing, Conflict Resolution, and Women: A Global Reappraisal." *Nationalism and Ethnic Politics* 24 (1):13–32.

Bogaards, M. 1998. "The Favourable Factors for Consociational Democracy: A Review." *European Journal of Political Research* 33:475–496.

Bogaards, M. 2019. "Consociationalism and Centripetalism: Friends or Foes?" *Swiss Political Science Review* 25 (4):519–535.
Byrne, S, and A. McCulloch. 2018. "Is Power-Sharing Bad for Women?" *Nationalism and Ethnic Politics* 24 (1):1–12.
Collier, P., and A. Hoeffler. 2000. *Greed and Grievance in Civil Wars.* Policy Research Working Paper 2355. Washington, DC: World Bank. http://documents.worldbank.org/curated/en/359271468739530199/pdf/multi-page.pdf.
Connolly, E., and J. Doyle. 2015. "Ripe Moments for Exiting Political Violence: An Analysis of the Northern Ireland Case." *Irish Studies in International Affairs* 26:147–162.
Connolly, E., and D. Doyle. 2019. "Brexit and the Changing International and Domestic Perspectives of Sovereignty over Northern Ireland." *Irish Studies in International Affairs* 30:217–233.
Deiana, M. 2016. "To Settle for a Gendered Peace? Spaces for Feminist Grassroots Mobilization in Northern Ireland and Bosnia-Herzegovina." *Citizenship Studies* 20 (1):99–114.
Deiana, M. 2018. *Gender and Citizenship. Promises of Peace in Post-Datyon Bosnia-Herzegovina.* London: Palgrave Macmillan.
Doyle, J. 1998. "'Towards a Lasting Peace'? The Northern Ireland Multi-party Agreement, Referendum and Assembly Elections of 1998." *Scottish Affairs* 25:1–20.
Hamdan, A. 2013. "The Limits of Corporate Consociation: Taif and the Crisis of Power-Sharing in Lebanon Since 2005." In *Lebanon: After the Cedar Revolution*, edited by A. Knudsen and M. Kerr, 39–59. Oxford: Oxford University Press.
Hayes, B C., and I. McAllister. 2012. "Gender and Consociational Power-Sharing in Northern Ireland." *International Political Science Review* 34 (2):123–139.
Horowitz, D. 1985. *Ethnic Groups in Conflict.* Berkeley: University of California Press.
Horowitz, D. 1991. "Making Moderation Pay: The Comparative Politics of Ethnic Conflict Management." In *Conflict and Peacemaking in Multiethnic Societies*, edited by J. V. Montville, 451–475. New York: Lexington Books.
Horowitz, D. 2001. "The Agreement: Clear, Consociational and Risky." In *Northern Ireland and the Divided World*, edited by John McGarry, 89–108. Oxford: Oxford University Press.
Horowitz, D. 2002. "Explaining the Northern Ireland Agreement: The Sources of an Unlikely Constitutional Consensus." *British Journal of Political Science* 33 (2):193–220.
Horowitz, D. 2014. "Ethnic Power Sharing: Three Big Problems." *Journal of Democracy* 25 (2):5–20.
Jarstad, A., and T. Sisk, eds. 2008. *From War to Democracy: Dilemmas of Peacebuilding.* Cambridge: Cambridge University Press.
Kapic, T. 2020. "The Impact of Consociational Peace Agreements on the Descriptive Representation of Women in National and Sub-National Political Institutions in Divided Societies: The Case of Bosnia and Herzegovina." PhD dissertation, Dublin City University.
Kennedy, R, C. Pierson, and J. Thomson. 2016. "Challenging Identity Hierarchies: Gender and Consociationalism." *British Journal of Politics and International Relations* 18 (3):618–633.
Lijphart, A. 1969. "Consociational Democracy." *World Politics* 21 (2):207–225.
Lijphart, A. 1975. "The Northern Ireland Problem: Cases, Theories, and Solutions." *British Journal of Political Science* 5 (1):83–106.
Lijphart, A. 1991. "The Alternative Vote: A Realistic Alternative for South Africa?" *Politikon* 18 (2):91–101.
Mac Ginty, R, and O. P. Richmond. 2013. "The Local Turn in Peace Building: A Critical Agenda for Peace." *Third World Quarterly* 34 (5):763–783.

McCrudden, C, and B. O'Leary. 2013. *Courts and Consociations: Human Rights versus Power-Sharing*. Oxford: Oxford University Press.

McGarry, J., and B. O'Leary. 2006a. "Consociational Theory, Northern Ireland's Conflict, and Its Agreement. Part 1: What Consociationalists Can Learn from Northern Ireland." *Government and Opposition* 41 (1):43–63.

McGarry, J., and B. O'Leary. 2009. "Power Shared after the Deaths of Thousands." In *Consociational Theory: McGarry and O'Leary and the Northern Ireland Conflict*, edited by R. Taylor, 15–84. London: Routledge.

Mitchell, P. 2014. "The Single Transferable Vote and Ethnic Conflict: The Evidence from Northern Ireland." *Electoral Studies* 33 (1):246–257.

Mitchell, P., B. O'Leary, and G. Evans. 2001. "Northern Ireland: Flanking Extremists Bite the Moderates and Emerge in Their Clothes." *Parliamentary Affairs* 54 (4):725–742.

Mitchell, P., B. O'Leary, and G. Evans. 2002. "The 2001 Elections in Northern Ireland: Moderating 'Extremists' and the Squeezing of the Moderates." *Representation* 39 (1):23–36.

O'Kelly, M., J. Doyle, and P. J. Boland. 2010. "How Many Ways Can You Look at a Proportion? Cross-Community Vote Transfers in Northern Ireland before and after the Belfast Agreement." *Royal Statistical Society Series A (Statistics in Society)* 173 (1):215–235.

O'Leary, B. 2005. "Debating Consociational Politics: Normative and Explanatory Arguments." In *From Power Sharing to Democracy: Post-conflict Institutions in Ethnically Divided Societies*, edited by S. J. R. Noel, 3–43. Toronto: McGill-Queens University Press.

O'Leary, B., and K. Salih. 2005. "The Denial, Resurrection, and Affirmation of Kurdistan." In *The Future of Kurdistan in Iraq*, edited by B. O'Leary, J. McGarry, and K. Salih, 3–43. Philadelphia: University of Pennsylvania Press.

Ottmann, M., and J. Vüllers. 2014. "The Power-Sharing Event Dataset (PSED): A New Dataset on the Promises and Practices of Power-Sharing in Post-conflict Countries." *Conflict Management and Peace Science* 32 (3):327–350.

Paris, R., and T. Sisk, eds. 2009. *The Dilemmas of Statebuilding: Confronting the Contradictions of Postwar Peace Operations*. London: Routledge.

Sircar, I. 2006. "Transnational Consociation in Northern Ireland and in Bosnia-Hercegovina: The Role of Reference States in Post-Settlement Power-Sharing." PhD dissertation, London School of Economics. https://core.ac.uk/download/pdf/4187804.pdf.

Stedman, S., D. Rothchild, and E. Cousens, eds. 2002. *Ending Civil Wars: The Implementation of Peace Agreements*. Boulder, CO: Lynne Rienner.

Steiner, J. 1981. "The Consociational Theory and Beyond." *Comparative Politics* 13 (3):339–354.

Taylor, R. 1994. "A Consociational Path to Peace in Northern Ireland and South Africa?" In *New Perspectives on the Northern Ireland Conflict*, edited by A. Guelke, 161–174. Aldershot, UK: Avebury.

Taylor, R. 2001. "Consociation or Social Transformation?" In *Northern Ireland and the Divided World: Post-Agreement Northern Ireland in Comparative Perspective*, edited by J. McGarry, 36–52. Oxford: Oxford University Press.

Walsh, D, and J. Doyle. 2017. "External Actors in Consociational Settlements: A Re-examination of Lijphart's Negative Assumptions." *Ethnopolitics* 17 (1):21–36.

Walter, B. F. 1997. "The Critical Barrier to Civil War Settlement." *International Organization* 51 (3):335–364.

Wilford, R. 1992. "Inverting Consociationalism? Policy, Pluralism, and the Post-Modern." In *Northern Ireland: Politics and the Constitution*, edited by B. Hadfield, 29–46. Buckingham, UK: Open University Press.

Wolff, S. 2004. "The Institutional Structure of Regional Consociations in Brussels, Northern Ireland, and South Tyrol." *Nationalism and Ethnic Politics* 10 (3):387–414.
Visoka, G. 2017. *Shaping Peace in Kosovo: The Politics of Peacebuilding and Statehood.* Basingstoke, UK: Palgrave Macmillan.
Zartman, W. I. 2005. *Cowardly Lions: Missed Opportunities to Prevent Deadly Conflict and State Collapse.* Boulder, CO: Lynne Rienner.

CHAPTER 33

STATEBUILDING, SECURITY-SECTOR REFORM, AND THE RULE OF LAW

PAUL JACKSON

INTERNATIONAL statebuilding has evolved into a hydra-like approach to peacebuilding more generally, with security assistance and rule of law (ROL) promotion taking place within larger, comprehensive approaches to security-sector reform (SSR) (Hänggi 2004; Schroeder and Chappuis 2014, 132). The international approach to constructing a state monopoly over the use of force has become reinforced by defining this state monopoly as an integral part of the ROL or a shared international set of norms that "civilized states" conform to (Schroeder and Kode 2012; Delcourt 2015). ROL itself, having been expanded from a largely domestic space to international relations approaches, is now used as a foundation for state authority or even as a substitute for the principle of sovereignty (Delcourt 2015). Pandolfi and McFalls (2010) even go as far as categorizing ROL as a form of "mobile sovereignty" whereby global bureaucracies and armies of experts justify interventions in the name of a transnational ROL that is deterritorialized, standardized, and exportable.

Most contemporary international approaches to SSR and ROL incorporate assumptions about the nature of the state, specifically that particular forms of the state are preferable as outcomes of statebuilding efforts and that SSR and statebuilding agendas aim not just at reconstructing security institutions but also at social engineering by remolding sociopolitical institutional structures. This drives medical-themed approaches that assume interventions to "correct" failed states can make them functioning members of an international network of liberal states leading to security. This challenge has driven discussion among many of those involved in statebuilding where there is a recognition of the explicitly political aspects of these efforts. In particular, there has been a broadening and deepening of approaches to take account of the lack of definable boundaries around concepts like security and recognizing that ROL approaches explicitly affect local societal norms and hierarchies. While this has led to discussion on the incorporation of nonstate actors into formalized systems of security and justice, it has also driven recognition of hybrid approaches that recognize more than one core approach to security provision and the law.

This chapter starts by defining ROL and SSR while emphasizing the political nature of both. It then outlines what we actually know about these processes. The evidence itself is remarkably disparate and unclear, although recent work (Jackson and Bakrania 2018) has identified significant patterns within the extant literature. In particular the field is dominated by evidence that is relatively easy to collect rather than more difficult and intangible aspects. Everyone admits that politics is important, but measuring to what degree remains very difficult. From the evidence, the chapter outlines three core issues identified across examples and then identifies three ways that programming approaches could change to create a more positive trajectory.

Success or Failure?

Most of the evidence regarding SSR in particular exists in the form of case studies. In Africa alone there are significant cases of SSR reform in Liberia, Sierra Leone, Guinea-Bissau, Somalia, Sudan, South Sudan, Mali, Côte d'Ivoire, Burundi, Chad, Libya, the Central African Republic (CAR), and the Democratic Republic of Congo (DRC). A recent paper by Detzner (2017) seeks patterns in these cases and finds an unrelenting list of SSR failure. Detzner shows that the most common failures are more complex than the normal reason given, "poor implementation," and there is a small universe of success to sit alongside the failures.

Lack of local ownership is frequently cited as a core reason for failure. In particular, agreement with the central government is necessary but not sufficient if SSR and ROL reforms are to take root in broader societies. In Guinea-Bissau, for example, consultation with broader society was limited (Kohl 2014), which echoes work in Sierra Leone where justice elements were neglected until long after central government reform (Jackson and Albrecht 2011). The use of private companies to carry out SSR also raises questions about local involvement. In Liberia, for example, training carried out by US private security companies remained unaccountable even to central government, and a comprehensive security review for the US government, which contracted the companies, carried out by the Read Corporation failed to involve local actors to the extent that Liberian security agencies were unaware it had taken place (Detzner 2017). It is also important to note that extensive dialogue does take place but may be significantly undermined if there is no follow-up (Moulaye 2015). In many cases, governments are so unstable that consultations may be launched more than once, reflecting regime change, as in the CAR (Moulaye 2015).

Key elements of SSR can go wrong individually, notably the critical stage of disarmament, demobilization, and reintegration (DDR), where earlier programs in Sierra Leone, Liberia, Mali, and the DRC were subject to problems, including the creation of fighters who were then demobilized in multiple conflicts. The reintegration and rehabilitation of fighters have always been underresourced, and the presence of fighters from previous West African conflicts in Libya after the fall of Gaddafi illustrates the additional security problems created by poorly executed DDR programs. In some instances, like Southern Sudan and the CAR, fighters become so frustrated over delays that they give up and return to conflict, whereas multiple DDR programs in some places, notably the DRC, have produced very few results. In many cases the promise of a DDR program may actually act as a recruiting tool or become an integral part of the conflict cycle without longer-term planning for a transition from fighter to civilian (Debos 2016).

As well as going wrong there are several elements of SSR that are frequently just left out of reform programs either for political reasons or for being too difficult. For many reasons presidential guard units and intelligence organizations are frequently excluded from SSR programs, a notable exception being the SSR program in Sierra Leone, which introduced a very effective intelligence coordination system with considerable success (Jackson and Albrecht 2011). The police, prisons, and judiciary are frequently subject to reform much later in any given process, and yet they are the security institutions that are closest to the population. The problem is exacerbated when one takes into account nonstate actors, which are frequently seen as existing on the boundary of security and justice when for most people in the countryside they *are* security and justice (Jackson 2011; Debos 2016).

Among all of the cases of failure, there has been some success in SSR alongside some success with specific programs that might have failed as an overall program. Sierra Leone, Ethiopia, and South Africa are frequently cited, and to this Somaliland and Puntland add interesting and important local examples of success. Of these both South Africa and Ethiopia were doing SSR before SSR formally existed, and Sierra Leone was instrumental in the ideas behind SSR since the UK led both the discussions in the Organization for Economic Cooperation and Development's Development Assistance Committee and the reforms in Sierra Leone at the same time (Jackson and Albrecht 2011). These three cases illustrate both the importance of government and broader societal buy-in, a transformational process (in Sierra Leone's case there was effectively no formal security structure) and widespread political sharing of the political process. In other words, there was strong political legitimacy for reform that enjoyed broad support. Bryden and Brickhill (2010) examine DDR efforts in southern Somalia, Puntland, and Somaliland. In southern Somalia, externally driven DDR in support of successive interim "governments" have been widely viewed with suspicion and alarm, whereas locally led efforts in Somaliland and Puntland have achieved significant success again, on the back of significant local legitimacy.

As Detzner (2017) and Jackson and Bakrania (2018) suggest, success stories share characteristics of political legitimacy and local ownership that is broader than the formal state institutions; there is a wider societal involvement. The case evidence shows that focusing on the early mobilization of local political demand may be more important for the success of SSR approaches than external, international pressure.

What Do We Know about the Rule of Law and Security-Sector Reform?

ROL is a wide-ranging and essentially contested concept that covers a multitude of sins. While SSR might be relatively new, its ancestor, security assistance, is an old approach to international relations and its modern, systematized version based on "build, train, equip" approaches dates back to the Second World War. Promoting ROL has become part of international orthodoxy only recently, however, and yet, despite difficulties in establishing definitions and boundaries, it has become a central tool of international intervention (Tamanaha 2004). The concept of ROL is contested at least partly because it ranges from "thin" definitions that establish primacy of the law and appropriate regulatory order. In practice, thin definitions are never that thin, as they also incorporate specific assumptions about how law is produced and its role in society (Trebilock and Daniels 2008). "Thick" definitions that explicitly incorporate normative approaches to what a "good" society should look like shift the focus away from basic legal function and toward the capacity to enforce checks and balances on power, access to justice, conflict resolution, protection of rights, and enforcement of human rights. For these functions to be effective, there is also an underlying set of assumptions about the capacity to make these available, including independence of the judiciary, access to justice systems, and institutional capacity.

A basic definition of ROL would incorporate consensual agreement to be bound by a set of prescribed rules and regulations designed to manage social, political, and economic interactions and norms through nonviolent means. In addition, there is sufficient institutional capacity and oversight to enforce there rules. Of course, this is an explicitly political statement about the nature of power within societies and who decides what the law should be. In many societies there are clashes between international, liberal approaches to ROL and local, traditional or customary systems that seek to uphold different local societal norms (Jackson 2011). ROL approaches, then, are not technocratic interventions but directly affect the political characteristic of how societies choose to organize themselves. ROL therefore becomes an area of contestation in itself either as a means of breaking down power structures or preserving them.

SSR is a relatively recent phenomenon but is rooted in earlier debates about civil-military relations (Jackson 2011). SSR was designed as an instrument to achieve both security and development by synchronizing security with development objectives so that both support, not harm, each other. It is different from a number of contemporary academic debates within peacebuilding because it combines an explicitly normative agenda of linking liberal statebuilding with human security. The concept has been driven, at least early in its development, by policymakers on the ground rather than by analytical depth or an explicit theoretical framework, and research has tended to follow policy within SSR. Early development of SSR was driven by the democratization of Eastern

Europe, then South Africa, both of which involved transformation from autocratic or nondemocratic regimes to Western democracy. The 1994 Code of Conduct on Politico-Military Aspects of Security, adopted by the Organization for Security and Cooperation in Europe, required member states to accept democratic control over military and other security forces, a set of principles that were then taken on as security-sector reform by the United Kingdom and then by the Development Assistance Committee of the Organization for Economic Cooperation and Development, chaired at the time by the UK. Since then, SSR has developed rapidly within all major multilateral organizations and is now a common element of policy in postconflict interventions and approaches toward failed states.

SSR has become a critical element of this as it has moved beyond its traditional focus on security institutions and toward a more comprehensive approach focused on legitimacy, democratic control and oversight, sound governance, and ROL. Indeed the inclusion of ROL approaches is part of the international community's approach to building effective security institutions, and SSR programs tend to be multisectoral, crossing security agencies, intelligence, policing, and justice provision as well as incorporating nonstate actors in security and justice provision (Andersen 2011; Schnabel and Born 2011; Schroeder and Chappuis 2014). Such comprehensive approaches have been seen as part of the solution to technocratic approaches to institution-building that have characterized much traditional security assistance (Schroeder and Chappuis 2014; Jackson and Bakrania 2018). SSR literature itself reflects a long-term change from a focus on narrow, technical approaches and concepts to far wider concerns encapsulating traditional justice, gender issues, and the nature of justice (Bakrania 2015; Jackson and Bakrania 2018). Many of these developments have come about as a result of feedback from experience on the ground rather than through conceptual development, resulting in a situation in which SSR has suffered from "benign analytical neglect" (Peake, Scheye, and Hills 2013, vii).

The shift to a more comprehensive approach has not been unproblematic, and the tension between ROL and SSR has led to a wide variety of programming variation but also a recognition that there are many competing sources of security that may exist beyond the state, particularly at the local level, and that these remain poorly understood (Jackson 2011). This has led to a clash between the mobile sovereignty model and hybrid approaches to security provision, particularly around issues of local ownership, sovereignty, and the role of international agencies (Mobekk 2010; Chandler 2013). What is constant within these discussions in the central role of the state and debates around statebuilding are essentially discussions that aim not only at reconstructing security institutions but at social engineering by remolding sociopolitical institutional structures.

SSR links an explicitly normative agenda of liberal statebuilding with human security, designed to construct a particular form of state governed by international norms known as ROL, including democratic control of armed forces, governance of security, and accountability for the actions of state agencies responsible for security. SSR has never overcome the issue identified by Brzoska (2006, 2), that a comprehensive approach to SSR has been "sound in theory but problematic in practice." The history of SSR has been one where theory has also been rather unsound, with very few comprehensive studies on the theory behind SSR or its underlying assumptions (Jackson 2011). In practice the picture is no more encouraging, with very few examples of successful SSR and a singular inability to overcome an issue identified by Chanaa (2002, 16) as a "conceptual-contextual divide." A decade later Sedra (2011) stated that "this divide had become a chasm." As Schroeder and Chappuis (2014 136) note, most writing on SSR focuses on individual cases and, within that, the implementation of existing policies rather than on alternatives or on questioning underlying assumptions. The literature is curiously nontheoretical, generally confirming Schroeder and Chappuis's finding that the growth in case studies has taken place within existing policy approaches rather than by developing new theoretical frameworks within which SSR can develop. SSR remains underpoliticized, lacks understanding of local dynamics, and has been decoupled from debates on statebuilding or liberal peacebuilding more generally (Jackson 2011; Peake, Scheye, and Hills 2013; Schroeder and Chappuis 2014).

There is considerable evidence around capacity building within organizations, particularly government institutions; strategic or statutory frameworks; community-based approaches; and restructuring of the security and justice sector. In other words, the literature tends to reflect SSR programming (Jackson 2011; Jackson and Bakrania 2018). It is perhaps unsurprising that the literature reflects traditional approaches because examples of newer and more politically sensitive approaches are scarce. This is reflected in a dearth of evidence on, for example, combating urban violence and organized crime, nonstate justice and capacity building, and integrated political engagement. The issue of politics and SSR has long been identified within the security literature, but despite its constant feature as an "issue" it has rarely been directly confronted other than as an explanation of why interventions failed (Jackson and Bakrania 2018).

This is further reinforced when analyzing what is missing from much of the literature: the lack of evidence on the practicality of ROL approaches, specifically the application, compliance, and interpretation of laws. While a plan or law may exist on paper, it is by no means certain that its existence reflects the empirical situation. Knowledge about the relationships between state and nonstate systems is underdeveloped and tends to be dominated by normative approaches that either privilege statist solutions and universalist approaches like human rights or reify the local without examining the politics of what that means for those seeking justice (Jackson and Bakrania 2018). Areas of SSR identified as having specific weaknesses include incentives to improve service delivery; crime rates that are frequently difficult to verify; data on legal awareness and choice; gender-based violence; judicial redress in support for rights; economic development; poverty reduction; property rights, particularly access to land, which has been identified as a conflict trigger but appears absent in many analyses of SSR interventions despite its prevalence as a type of court case within both formal and informal justice systems; and women's empowerment and gender equality (Denney and Domingo 2015; Jackson and Bakrania 2018).

Dilemmas within SSR and ROL Approaches

SSR and ROL are subject to significant tensions in terms of practical application, alongside significant resistance to the idea of mobile sovereignty supported by an international idea of ROL that is neither entirely international nor of the state. As Kratochwil (2013, 16) points out, a disconnected idea of international law in this way could be interpreted as an imperial project supported by "enthusiasts who see norms cascading (but strangely enough never degenerating, or giving rise to disregard and circumventions), by an 'activist' judiciary—hardly rooted in constitutional structure and a functioning political process—or by the great powers, perhaps even coalitions of the willing, who feel empowered by the universalist nature of their goals."

The era of global ROL as a justification for intervention and of SSR approaches is not the end of the state, but it does represent a reconfiguration of power. At the international level this may be through empowerment of significant international organizations and actors whose intervention is not governed by a traditional reluctance to intervene precisely because they are not state actors (Delcourt 2015). As Rajkovic (2012, 29) points out, ROL is powerful precisely because it appears to be the rule of no-one in particular; adding that the " . . . institutionalization of 'global' judicial authority can deliver the rule of 'no one' in global governance." This has led to inconsistencies between statebuilding, security, and development. There is an assumption that human security can be best served by creating a functioning state that will provide security as a public good and that development will provide benefits to the general population. However, human security in terms of "freedom from fear" and citizen security in terms of an entitlement to protection by the state remain elusive for many people, and the state's responsibility to protect citizens remains unrealized. This lack, in turn, may lead to claims of legitimization of international intervention in failed states under ROL (Luckham and Kirk 2012).

The transfer of the political architecture of the liberal state to nonliberal societies in the form of statebuilding raises the question of whether the pacific nature of liberalism really is the expression of a local population's political will (Jackson 2011). The core of international approaches to security surrounds a normative approach that assumes a preference for the international community to use the legitimacy of state structures, but also the assumption that local actors have internalized the requisite norms of behavior privileged by the international community itself (Pospisil and Kuhn 2016). The literature on SSR reflects a previous understanding of intervention that rests on the assumption that every person is a rational liberal waiting to be liberated (Duffield 2001, 2002). This then leads to a language that assumes "politics" is what happens when local groups oppose the serene progress of liberal statebuilding, not actually reconfiguring power by building states. This linear approach assumes progress consists of a set path toward liberal peace; the only things standing in the way are lack of resources, capacity, international will, and local predatory elites.

Existing, linear approaches are subject to key sets of tensions, reflected in gaps within the literature and available evidence base: state-centrism, technocentrism, and the nature of ownership (Jackson and Bakrania 2018).

State-Centrism

One of the clearest characteristics of most SSR literature is the tension between state-centric SSR programs and the social and political realities of fragmented polities on the ground. While virtually all analysts accept that there are problems with the nation-state in many of the contexts in which states are failing, there is still a tendency to accept the technocratic parameters of statebuilding. Casting the nation-state as the norm ignores the broadening and deepening of security at all levels, the intrastate nature of much conflict, international conflict actors, and the role of the state itself as participant. There remains a pseudo-medical assumption that the right mixture of policies can create a healthy nation-state.

State-centrism itself changes local power structures among those who benefit from training, equipment, and resources. In situations where the state is one actor among many, this could effectively enhance just another armed faction. For a state to actually exist requires the population to buy in to the idea of the state at some level to provide legitimacy. In a liberal state this is commonly taken as participation in the political processes, such as periodic democratic elections. However, formal legitimacy may not be achievable or even desirable for citizens. A technocratic approach may have the effect of creating a state superstructure with no legitimacy among the general population but that provides avenues to power for local elites, such as Timor-Leste (Lemay-Hébert 2009, 2011), or that leads to an artificial state overlaying existing political systems, like Afghanistan (Hodes and Sedra 2013). Statebuilding is also very uneven within states. Even where states have had a functioning core before, during, or even after conflict, it may only rarely or partially have penetrated into the countryside (Jackson 2007). Many people simply have never received services directly from states. At best, this can produce a political hybrid where local people have a say and also have a choice in terms of accessing services, including security, but also encompassing a variety of plural providers (Denney 2014). However, there is a risk that political hybridity also reinforces the position of local elites and neo-patrimonial rule.

Technocentrism

The history of SSR programming and its literature has been greatly concerned with technical aspects of the reform process. SSR and ROL approaches tend to concentrate on measurable outputs like training, infrastructure, and "train and equip" models, whereas statebuilding itself has tended to concentrate on what is being constructed rather than the politics of what is being constructed (Denney and Domingo 2014).

This decontextualized approach removes politics of reform to a place where it can be categorized as a "spoiler" rather than an integral part of the reform process. Many interventions are frequently carried out by international bureaucrats, police, or military whose concerns are primarily technical rather than political. This is exemplified by the contractual relationships that characterize SSR in Liberia and that shift sovereignty from the government of Liberia to the contractee, the US.

Technocratic approaches to SSR represent a form of antipolitics machinery in terms of their disregard for local political sensitivities, but also in their relentless application and reapplication. As Richmond (2011) points out, the contemporary peacebuilding consensus assumes that there is an agreed underlying set of norms and that all interventions in support of this will receive the support of agents engaged in peacebuilding. This approach creates standards, benchmarks, and frameworks for creating a pacific world that constitutes a specific type of knowledge that can then be transferred to conflict zones as international norms (Jackson and Bakrania 2018). This knowledge is accompanied by a set of practices designated by technical terminology as "tools," "indicators," "templates," or "instruments" that can be considered apolitical and objective ways to represent reality, and yet none of it is actually objective (Körppen 2011, 81).

Local Ownership

SSR and ROL programming is inextricably entwined with statebuilding (Jackson 2011). This matters for who owns this process and what choices those people may exercise in the unfolding of programming priorities. While clearly SSR could privilege compliance with international state norms and rules, and this may be in line with the desires of some potential local owners, it raises the question of how local ownership can be exercised if these local owners do not want to comply with international norms. Local ownership may be contested. The use of customary power structures also facilitates the exercise of hidden transcripts of power that trap the dominant as well as the weak within the same web of socialized roles and behavior (Scott 1990).

The literature is divided on the issue of local ownership. The core question relates to agency within the process, but despite the standard use of the term, there is no consensus on either what it means or how it can be enacted (Donais 2008; Mobekk 2010). As Schroeder and Chappuis (2014) point out, local ownership is either romanticized or seen as a problem to be overcome. This reflects a wariness about local ownership and local owners themselves. They may not be representative, they may not be willing to be inclusive, and they may be elite, but they may be the real owners and unwilling to relinquish control. International involvement in strengthening security institutions may cement the positions of ruling groups, and it may be naïve to assume that such local owners would accept programs designed to change or dilute their own power (Nathan 2007). They may also strengthen formal institutions at the expense of local authorities or exist in complete separation, much like the Special Court for Sierra Leone, where there were virtually no links between the international ROL and local interpretations of what was just (Jackson 2015).

If SSR is to shift from purely internationally led approaches to more inclusive programs, then the literature implies that we lack sufficient knowledge. The socially constructed knowledge of success identified by Bliesemann de Guevara and Kostić (2017) implies that we lack a socially constructed knowledge of failure within the literature. Specifically, there is an absence of empirically informed localized interventions that start with realistic political analysis (Mannitz 2014).The overwhelming reaction to failures and shortcomings in SSR has been to develop new versions of existing technocratic solutions, increase funding, or improve communication, despite evidence of limited impact. Within the literature, however, there has been a reflection of wider approaches, particularly ideas around hybridity and nonstate actors (Mac Ginty and Richmond 2013; Lemay-Hébert and Freedman 2017). The practical approaches and categories deployed by the international community within SSR need to be coupled with a measure of humility in terms of the ambitious aims of some programming allied to a lack of knowledge about exactly how international interventions will affect already existing systems and how their implementation is likely to coexist with those systems. As Körppen (2011) points out, this creates a tautology for liberal intervention because local ownership in practice is supported only if it supports liberal approaches, which is itself illiberal.

Toward a New Approach to ROL and SSR?

It is clear from the extant literature on SSR that practitioners and researchers have recognized a central challenge in how to deal with the perceived failures of linear approaches (Denney and Domingo 2014). Usually the debate is couched in terms of assimilating nonstate actors (Mannitz 2014). These debates are also linked to the development of hybrid approaches within the peacebuilding literature (Mac Ginty and Richmond 2013). This has led to more discussion and advocacy of a move beyond linear approaches to the state and toward recognition of the links between external and domestic actors (Schroeder and Chappuis 2014; Lemay-Hébert and Freedman 2017) and works that try to uncover the hidden discourses behind or in parallel with the dominant public transcripts of international peacebuilding interventions (Bliesemann de Guevara and Kostić 2017). With concepts such as "everyday," "hybrid," and "postliberal" forms of peace, critical peacebuilding scholars have presented alternatives in light of the fundamental criticisms mounted against the adherents of liberal interventionism in both policy and academia.

Liberal statebuilding envisages states being constructed where institutions supported the establishment of societal frameworks in which liberal individuals could flourish.

Liberal peace was therefore an outcome of creating the liberal state, including security and justice institutions at its heart, and incorporating democratic control, ROL, and professionalization of bureaucracies (Lemay-Hébert, 2009). However, a newer approach requires a very different concept of the political, notably that externally imposed liberal structures obscure the real underlying politics of states rather than being neutral arbiters. Nonlinear approaches seek to work at the societal level, understanding the local politics of resistance and adaptation that have been neglected.

As Chandler (2013) rightly notes, this shift from linear to nonlinear approaches partly has its roots within the peacebuilding work of Lederach (1997) and the development of a "process-orientated" understanding whereby the role of external actors is not to impose institutions but to assist in establishing a framework within which local societies can pursue their own peacebuilding processes. In peacebuilding, Lederach (1997, 2005) concluded that understanding politics incorporated not institutions, leaders, and parties but societal spaces, practices, and processes. For Scott (1990), the movements below the surface are what determine the support for visible political action. He emphasizes the whole system, not just the visible elements, and the socially embedded relationships, networks, and practices that form what he terms "infrapolitics."

What, then, are the core concerns that a nonlinear approach to SSR and ROL will have to address? I have picked out four groups here: institutions, institutional power, developing processes rather than structures, and engagement with hidden politics. First, institutions remain important. In the contemporary world, systems of domination and power remain, and they are overwhelmingly states, even if their reach is limited in some places. States require institutional structures to exist, although these may not all be exactly the same. Security institutions, for example, are critical to the maintenance of secure populations, and all states have variations on militaries and/or police. Professionalism within these services remains critical, otherwise they can become the chief threat to populations rather than the chief protector. What is lacking here is the consideration of how those institutions relate to political frameworks. In other words, how far are militaries and police controlled within civil legislation and political frameworks, but also how far are they related to the infrapolitics of ethnicity, corruption, kinship, personal rule, or regional bias? Work on Sierra Leone from 1997 to 2014, for example, showed that while the technical capabilities of the security services had been successfully reconstructed, there remained issues that directly affected the lack of civil and political oversight and also the creeping politicization of recruitment to the police, for example (Jackson and Albrecht 2011; 2016: Denney 2014).

Second, taking up processes rather than institutions implies that instead of starting out with a predetermined view of what a military or police force looks like and then working toward that, the various qualities of a desired and agreed security service would emerge as a result of discussion. There are also several relationships to be negotiated with regard to the provision of justice and policing, not least how far local security providers are able to exercise authority, what offenses should be subject to formal police inquiry, and whether there is access to recourse for those seeking justice. In practice the system across much of the African countryside, for example, shows many more people using customary authorities than formal policing structures and most people facing institutional multiplicity at the local level. What a new generation of SSR requires here is precisely Lederach's idea that external intervention should enable local people and societies to make their own choices within locally existing frameworks. This would, of course, allow a degree of choice for someone who might lose out in the traditional or customary system—usually a woman or youth—who would then be able to pursue justice using a different avenue, including the state.

Third, an engagement with hidden politics, the emphasis on process over constructing institutions alongside the continuing importance of institutions themselves over time, raises some very difficult questions for those wishing to intervene. Emphasis on process requires reconfiguration of a programming architecture that rests on finite time periods and measurable outputs. Working within existing frameworks to develop processes that enhance security and justice and therefore SSR takes considerable understanding, effort, and time. Nonlinear approaches and the incorporation of hidden politics requires effort in understanding the local context, which takes time and is frequently not done well, if at all. Multilayered approaches already advocated within some of the SSR literature also tend to avoid discussions of power within those systems, hence Chandler's (2013) assertion that some of these approaches establish a way of criticizing linear, liberal approaches but end up reifying local solutions instead of recognizing them as power structures in their own right.

Conclusions

The failures of previous SSR programs are all too evident and have prompted a focus on the interaction between the international community and states affected by SSR programming. The lack of clear success in SSR has led to a reconsideration of approaches, not least in terms of the limitations of the liberal state and the incorporation of nonlinear approaches to security. However, in many ways this analysis remains only partial, since there is little actual knowledge of hidden politics by the international community in practice, and in many ways much of this theoretical approach reifies the local without really examining the implications of incorporating traditional systems that bring their own politics and power structures with them. Identifying hidden politics with "resistance," for example, places local political structures in a heroic opposition to the international order, whereas in reality those local structures and the actual hidden politics of states are far more complex. It also fails to recognize that local politics has its own hierarchies and power structures that may be less than heroic in terms of resistance to liberal values. The reality is that a multiplicity of institutions, overlapping political networks, and hidden politics that are not actually "hidden" at all, just not recognized or known by international analysts, constitute an evolving political process of contestation over power and resources

The development of nonlinear approaches to SSR can be seen as a way of incorporating traditional structures into a broader global security system and is a means by which the other can be assimilated into liberal strategies and, by implication, the liberal world-system, thereby overcoming something that is perceived as an obstacle or spoiler in a wider process of modernization. A new-generation SSR and ROL approach needs to take on three groups of issues: recognizing that institutions are important but not necessarily the institutions that have been supported in the past or supported in isolation from infrapolitics; developing processes that are inclusive and may be sustainable in the medium to long term; and engaging with the hidden politics of states, recognizing that they are not blank slates and that institutions of the state do not exist in a political vacuum.

References

Andersen, L. R. 2011. *Security Sector Reform and the Dilemmas of Liberal Peacebuilding*. DIIS Working Paper. Copenhagen: Danish Institute of International Studies.

Bakrania, S. 2015. *Security and Evidence Mapping*. Birmingham, UK: GSDRC, University of Birmingham.

Bliesemann de Guevara, B., and R. Kostić. 2017. "Knowledge Production in/about Conflict and Intervention: Finding 'Facts,' Telling 'Truth.'" *Journal of Intervention and Statebuilding* 11 (1):1–20.

Bryden, M., and J. Brickhill. 2010. "Disarming Somalia: Lessons in Stabilisation from a Collapsed State." *Conflict, Security & Development* 10 (2):239–262.

Brzoska, M. 2006. "Introduction: Criteria for Evaluating Post-conflict Reconstruction and Security Sector Reform in Peace Support Operations." *International Peacekeeping* 13:1–13.

Chanaa, J. 2002. *Security Sector Reform: Issues, Challenges and Prospects*. Oxford: Oxford University Press.

Chandler, D. 2013. "Peacebuilding and the Politics of Non-linearity: Rethinking 'Hidden' Agency and 'Resistance.'" *Peacebuilding* 1 (1):17–32.

Debos, M. 2016. *Combatants and the State in Chad: Governing Chad's Inter-war*. London: Zed Books.

Delcourt, B. 2015. "The Rule of Law as a Vehicle for Intervention." *Journal of Intervention and Statebuilding* 9 (4):471–494.

Denney, L. 2014. *Justice and Security Reform: Development Agencies and Informal Institutions in Sierra Leone*. Abingdon: Routledge.

Denney, L., and P. Domingo. 2014. "*Security and Justice Reform: Overhauling and Tinkering with Current Programming Approaches*." Overseas Development Institute, April. http://www.odi.org/publications/8330-security-justice-fragile.

Denney, L., and P. Domingo. 2015. *Context-Relevant, Flexible and Transnational Programming: The Future of Security and Justice?* London: ODI.

Detzner, S. 2017. "Modern Post-conflict Security Sector Reform in Africa: Patterns of Success and Failure." In "African Peace Missions and Security Sector Governance." Special issue, *African Security Review* 26 (2):1–27.

Donais, T., ed. 2008. *Local Ownership and Security Sector Reform*. Zurich: Lit Verlag for the Geneva Centre for Democratic Control of Armed Forces.

Duffield, M. 2001. *Global Governance and the New Wars: The Merging of Development and Security*. London: Zed Books.

Duffield, M. 2002. "War as a Network Enterprise: The New Security Terrain and Its Implications." *Journal for Cultural Research* 6:153–165.

Hänggi, H. 2004. "Conceptualising Security Sector Reform and Reconstruction." In *Reform and Reconstruction of the Security Sector*, edited by A. Bryden and H. Hänggi, 3–20. Geneva: Geneva Centre for the Democratic Control of Armed Forces.

Hodes, C., and M. Sedra. 2013. *The Search for Security in Post-Taliban Afghanistan*. London: Routledge.

Jackson, P. 2007. "Reshuffling an Old Deck of Cards? The Politics of Decentralisation in Sierra Leone." *African Affairs* 106 (January):95–111.

Jackson, P. 2011. "Security Sector Reform and State Building." *Third World Quarterly* 32 (10):1803–1822.

Jackson, P. 2015. "Whose Justice in Sierra Leone? Power, Security and Justice in Post-conflict Reconstruction." In *Evaluating Transitional Justice: Accountability and Peacebuilding in Post-conflict Sierra Leone*, edited by Chris Mahony, Rebekka Friedman, and Kirsten Ainley, 197–215. Basingstoke: Palgrave.

Jackson, P., and P. Albrecht. 2011. *Reconstructing Security after Conflict: Security Sector Reform in Sierra Leone*. Basingstoke: Palgrave.

Jackson, P., and P. Albrecht. 2016. *Securing Sierra Leone, 1997–2013: Defence, Diplomacy and Development in Action*. Whitehall Papers No. 82. London: Royal United Services Institute.

Jackson, P., and S. Bakrania. 2018. "Is the Future of SSR Non-linear?" *Journal of Intervention and Statebuilding* 12 (1):11–30.

Kohl, C. 2014. *The Reform of Guinea Bissau's Security Sector: Between demand and practice*, PRIF Report No. 126, Peace Research Institute Frankfurt

Körppen, D. 2011. "Space beyond the Liberal Peacebuilding Consensus: A Systemic Perspective." In *The Non-Linearity of Peace Processes Theory and Practice of Systemic Conflict Transformation*, edited by D. Körppen, N. Ropers, and H. J. Giessmann, 77–96. Leverkusen: Barbara Budrich.

Kratochwil, F. 2013. "Politics, Law and the Sacred: A Conceptual Analysis." *Journal of International Relations and Development* 16 (1):1–24.

Lederach, J.-P. 1997. *Building Peace: Sustainable Reconciliation in Divided Societies*. Washington, DC: US Institute for Peace.

Lederach, J.-P. 2005. *The Moral Imagination: The Art and Soul of Building Peace*. Oxford: Oxford University Press.

Lemay-Hébert, N. 2009. "Statebuilding without Nation-Building? Legitimacy, State Failure and the Limits of the Institutionalist Approach." *Journal of Intervention and Statebuilding* 3 (1):21–45.

Lemay-Hébert, N. 2011. "The 'Empty Shell' Approach: The Set Up Process of International Administrations in Timor-Leste and Kosovo, Its Consequences and Lessons." *International Studies Perspectives* 12 (2):188–209.

Lemay-Hébert, N., and R. Freedman, eds. 2017. *Hybridity: Law, Culture and Development*. Law, Development and Globalization Series. London: Routledge.

Luckham, R., and T. Kirk. 2012. *Security in Hybrid Political Contexts: An End-User Approach*. London: Justice and Security Research Programme, LSE.

Mac Ginty R., and O. Richmond. 2013. "The Local Turn in Peace Building: A Critical Agenda For Peace". *Third World Quarterly* 34(5):763–783.

Mannitz, S. 2014. "From Paternalism to Facilitation: SSR Shortcomings and the Potential of Social Anthropological Perspectives." *International Peacekeeping* 21 (2):165–180.

Mobekk, E. 2010. "Security Sector Reform and the Challenges of Ownership." In *The Future of Security Sector Reform*, edited by M. Sedra, 230–243. Waterloo: Centre for International Governance Innovation.

Moulaye, Zeïni. 2015. "Missed Opportunities for Comprehensive Security Sector Reform in Mali." In *Learning from West African Experiences in Security Sector Governance*, edited by Alan Bryden and Fairlie Chappuis, 79–95. London: Ubiquity Press.

Nathan, L. 2007. "No Ownership, No Commitment: A Guide to Local Ownership of Security Sector Reform." Global Facilitation Network for Security Sector Reform, University of Birmingham.

Pandolfi, M., and L. McFalls. 2010. "Global Bureaucracy: Irresponsible but Not Indifferent." In *Conflict, Security and the Reshaping of Society: The Civilization of War*, edited by A. Dal Lago and S. Palidda, 171–184. London: Routledge.

Peake, G., E. Scheye, and A. Hills, eds. 2013. *Managing Insecurity: Field Experiences of Security Sector Reform.* London: Routledge.

Pospisil, J., and F. Kühn. 2016. "The Resilient State: New Regulatory Modes in International Approaches to State Building?" *Third World Quarterly* 37 (1):1–16.

Rajkovic, N. M. 2012. "'Global Law' and Governmentality: Reconceptualizing the 'Rule of Law' as 'Rule through Law.'" *European Journal of International Relations* 18 (1):29–52.

Schnabel, A., and H. Born, eds. 2011. *Security Sector Reform in Challenging Environments.* Münster: Lit Verlag.

Schroeder, U., and F. Chappuis. 2014. "New Perspectives on Security Sector Reform: The Role of Local Agency and Domestic Politics." *International Peacekeeping* 21 (2):133–148.

Schroeder, U., and J. Kode. 2012. "Rule of Law and Security Sector Reform in International State-Building: Dilemmas of Converging Agendas." *Hague Journal on the Rule of Law* 4:31–53.

Scott, J. 1990. *Domination and the Arts of Resistance: Hidden Transcripts.* New Haven, CT: Yale University Press.

Sedra, M. 2011. "Introduction: The future of security sector reform." In M. Sedra (Ed.), *The future of security sector reform*. Ontario: The Centre for International Governance Innovation. http://www.cigionline.org/publications/2010/11/future-security-sector-reform

Tamanaha, B. Z. 2004. *On the Rule of Law. History, Politics, Theory.* Cambridge: Cambridge University Press.

Trebilock, M. J., and R. J. Daniels. 2008. *Rule of Law Reform and Development: Charting the Fragile Path of Progress.* Cheltenham, UK: Edward Elgar.

CHAPTER 34

TRANSITIONAL JUSTICE AND PEACEBUILDING

CATHERINE TURNER

Transitional justice is a relatively new field of inquiry. Centrally concerned with justice in times of political transition, the concept initially emerged from the field of political science and the documentation of how democratic regimes pursue accountability for the crimes of a prior regime. The field has expanded rapidly since the early Latin American transitions and is now broadly interdisciplinary, with scholarship found across disciplines such as law, political science, philosophy, social psychology, and anthropology. Since the early 2000s it has existed as a distinct field of scholarship and practice in the postconflict landscape (Teitel 2000; Bell 2009). The aim here is to provide a basic overview of the aims and objectives of transitional justice as a form of postconflict intervention and to consider the relationship between transitional justice and peacebuilding in that context.

From the outset transitional justice has been linked to the rule of law. Attempts to ensure accountability for the crimes of Latin American dictatorships emphasized the importance of the rule of law for the consolidation of democracy (Kritz 1995). This link was further refined by Teitel's (2003, 69) now authoritative definition of transitional justice as "the conception of justice associated with periods of political change, characterized by legal responses to confront the wrongdoings of repressive predecessor regimes." However, while transitional justice was initially characterized by an emphasis on legal response, the form of that response varied beyond that associated with domestic criminal law. Beyond simply attributing legal responsibility for wrongful acts, transitional justice was concerned with addressing political, ethical, and moral dilemmas associated with widespread human rights abuses (Bell, Campbell, and Ní Aoláin 2004). This definition recognized that justice in the aftermath of conflict was not simply a straightforward application of criminal responsibility but rather was concerned with the much broader project of rebuilding trust in society and its institutions. This required more than basic legal accountability.

In 2004 the United Nations produced a report that defined transitional justice as "[t]he full range of processes and mechanisms associated with a society's attempts to come to terms with a legacy of large-scale past abuses, in order to ensure accountability, serve justice and achieve reconciliation" (paragraph 8). This report was the moment when the concept of transitional justice, which had emerged only a decade previously, was given definite form in international policy (Turner 2016, 1). The report is now widely cited as the definition of justice in the aftermath of war, violence, or repression. It is generally accepted that transitional justice incorporates some of the same goals as peacebuilding. For the United Nations (2004, paragraph 7) justice implies "regard for the rights of the accused, for the interests of victims and for the well-being of society at large." This definition undoubtedly incorporates a legal definition of justice. The rights and interests of victims derive from the normative pillars of "the Charter of the United Nations . . . together with the four pillars of the modern international legal system: international human rights law; international humanitarian law; international criminal law; and international refugee law" (paragraph 9). In its 2011 report the United Nations further highlighted how the normative framework supporting transitional justice had been strengthened to include a "right to justice, truth, and guarantees of nonrecurrence" (paragraph 19). As a result of its location between a number of different legally normative regimes, analysis of the requirements of transitional justice has tended to focus on the legal requirements of accountability and combating impunity (United Nations 2006).

Transitional justice as a field of both scholarship and practice has expanded exponentially to include a range of not only legal but also political and cultural mechanisms whose aim is to support peaceful regime change (Mihr 2017). The UN definition also clearly articulates a broader social function of justice in the form of its concern for the well-being of society. This aspect is less clearly defined and is one with which transitional scholars have grappled since the early years of the development of the field. This concern with social well-being expands the scope of transitional justice beyond strict legal justice toward the goals of peacebuilding. The mutual aims of both transitional justice and peacebuilding include promoting respect for human rights and the rule of law, reforming governance to increase accountability, and transforming relationships necessary to secure peace (Lambourne 2014). While this expansion has enhanced the diversity of transitional justice as a field, it has also brought challenges. The broad interdisciplinary nature of scholarship means that it arguably lacks a clear or coherent foundational premise or agreed outcomes (Ainley 2017). And while they remain largely distinct fields of scholarship and practice, there is increasing interest in the relationship between transitional justice and peacebuilding. Scholars have begun to map the points of commonality and convergence between the two, asking why they remain so distinct (Baker and Obradović-Wochnik 2016; Sharp 2013; Lambourne 2014).

This chapter outlines transitional justice as it has been institutionalized in international law, policy, and practice. As such it focuses on transitional justice as an elite discourse rather than as a grassroots project. The reason for this is the potential of elite discourses to constrain what is considered appropriate at the local level (Turner 2016) and because it is the tension between international frameworks and local peacebuilding that is the subject of most of the critique of transitional justice from a peacebuilding perspective. The chapter provides an overview of the key transitional mechanisms and explores how each is intended to contribute toward the goal of peace. The chapter considers the professionalization of transitional justice as a field (Subotic 2012) and its embeddedness in normative ideas of justice (Turner 2013). In addressing a peacebuilding audience the chapter seeks first to explain the arguments made in favor of legally normative transitional justice mechanisms, before exploring the difficult relationship between law and politics in transitional justice scholarship that contributes to the disciplinary gap that is so evident between transitional justice and peacebuilding.

Defining Transitional Justice

Since the adoption by the United Nations in 2004 of a definition of transitional justice there has been a steady move toward a professionalized field (Sriram 2017; Subotic 2012). The definition rests on four key pillars, or mechanisms, that are available to states in transition: trials, truth commissions, reparations, and guarantees of nonrecurrence (United Nations 2004). The design of each of these mechanisms has, in turn, become codified in a series of "toolkits" that provide best practice guidance for their establishment. The aim has been to create a technical approach to justice in the aftermath of violence that transcends divisive politics (Sriram 2017; Turner 2016). And yet it has proven difficult to remove politics from questions of justice, as discussion of each of the mechanisms demonstrates.

Trials

In the early years trials were the primary means of pursuing justice after war or repression. Whether one traces the origin of transitional justice to the Nuremberg Trials or to the Latin American transitions to democracy, the prosecution of leaders for crimes against humanity underpinned the political transition (Teitel 2003). The aim of trials was to mark a break from the prior regime, to make a statement that the widespread violation of human rights was unacceptable in the new political order. Accountability, it was argued, was necessary to move on. As a result, transitional justice, from its early years, has been characterized as being both backward- and forward-looking. Backward-facing accountability is what secures forward-looking guarantees of nonrecurrence (Teitel 2000). This forward-looking aspect is shared with the practice of peacebuilding in that it aims to prevent the outbreak or recurrence of armed conflict. Trials are important because "they express public denunciation of criminal behaviour, provide a direct form of accountability for perpetrators and ensure a measure of [retributive] justice for victims" (United Nations 2004, paragraph 39). These goals are easily recognizable from domestic criminal law. The function of justice in this context is retributive. It is the straightforward punishment of a criminal for a crime committed. This is also a strongly normative approach to justice in the aftermath of conflict. Trials bring to justice those who have been responsible for "serious violations of human rights and humanitarian law, putting an end to such violations and preventing their recurrence" (paragraph 38). When seen from this perspective, criminal justice is a necessary precondition to peace. The success or failure of trails in this view is measured with reference to strictly legal criteria: Was the process fair? Was a conviction secured? How long was the sentence? (J. Clark 2011).

However, trials have also been interpreted as playing a broader role in relation to political and social transformation (Akhavan 1998). With the establishment of the International Criminal Tribunal for the former Yugoslavia trials began to be linked to aims more commonly associated with the restoration of peace than traditional criminal justice. UN Security Council Resolution 808 (1993, preamble), which established the International Criminal Tribunal for the former Yugoslavia, made an explicit link between the trial of individual perpetrators and the restoration of international peace and security. The prospect of being held individually liable for crimes against humanity or war crimes was intended to have a strong deterrent effect on political and military leaders who might otherwise be tempted to ignore the laws of war (Fletcher and Weinstein 2002). Trials, it was claimed, could help by "establishing detailed and well-substantiated records of particular incidents and events. They could help to delegitimize extremist elements, ensure their removal from the national political process and contribute to the restoration of civility and peace and to deterrence" (United Nations 2004, paragraph 39). According to Daly (2002), without societal transformation, the society stays violent, nationalist, or racist, so there is no way to guarantee nonrecurrence. Sometimes prosecution will work—by putting behind bars those who mobilized the masses and incited violence and making it impossible for them to commit the same wrongs again.

Establishing the guilt of individual perpetrators was also seen as a means of demonstrating that abuses occurred as a result of individual agency and not as an inevitable consequence of ethnic or intercommunal differences. It was hoped that attributing responsibility to individuals could break cycles of collective blame, leaving open the possibility of reconciliation by dispelling the desire for collective retribution. In addition to securing individual convictions and creating a deterrent effect, the Tribunal was imbued with the responsibility for establishing a historical record, creating the conditions for victim catharsis, (re)establishing the rule of law, and contributing to the restoration of peace (United Nations 2004, paragraph 38). For the United Nations, "criminal trials can also contribute to greater public confidence in the State's ability and willingness to enforce the law" (paragraph 38). These socially oriented goals expand the aims of the trials beyond purely retributive justice and toward a more transformational approach. Criminal justice not only undoes past injustice but also serves to promote "the normative transformation" into a new system that respects the rule of law (Teitel 2000).

Beyond punishment, the United Nations (2011, paragraph 7) more recently highlighted the role of trials in building the capacities of national justice and security actors through their involvement in national prosecution initiatives, creating more conducive conditions for ongoing national investigation and prosecution of human rights abuses. This approach addresses the concern that international prosecution alone was insufficient to meet the needs of victims. International trials were costly, they were geographically remote from victim populations, and they applied alien rules of procedure and evidence. An alternative approach was to create so-called hybrid tribunals, such as those created for Sierra Leone, Cambodia, East Timor, and Lebanon. Hybrid tribunals sit in the affected state and comprise both local and international judges. While still adjudicating on the basis of international law, the physical location of the tribunal in the affected state itself is intended to more effectively link international transitional justice policy with local communities. For the United Nations (2011) hybrid tribunals have substantial potential both to build confidence in the state and to build the capacity of national justice systems through the inclusion of national prosecutors and judges in the transitional justice process.

This expansion of the goals of criminal trials continues to underpin justifications for international criminal justice while simultaneously causing tension over the extent to which the criminal law is the best means to pursue these ends. In particular it is disputed whether individual criminal responsibility ensures accountability or simply avoids any scrutiny of the collective aspects of violence and repression in a way that allows cycles of group victimhood and blame to continue unaddressed (Subotic 2011; Fletcher and Weinstein 2002). The narrow and individually focused approach of trials and the limits of this form of reckoning in contexts of widespread violence are addressed by other transitional justice mechanisms.

Truth Recovery

In contrast to the retributive function of trials, truth commissions are intended as a more victim-centered approach to justice (Hayner 2010). In recent years the value of truth in transitional justice has been incorporated into the legal framework in the form of a "right to truth" that victims can assert to force action from governments (Sweeney 2018). Truth commissions and other often informal approaches to truth recovery play a central role in delivering the right to truth enjoyed by victims. By placing victims at the heart of the process the aim of modern truth commissions is to provide a space for social reconciliation through acknowledgment of harms caused (Tutu 1999; Mamdani 1996). Truth commissions are nonjudicial in their function. Their role is not to adjudicate on the guilt or innocence of specific perpetrators but rather to build a broader picture of violations that have occurred during a conflict, particularly where there are contested truths (Kritz and Finci 2001).

Truth commissions (including commissions of inquiry and other mechanisms for information retrieval) can be used for a range of purposes. From this perspective truth commissions also have the potential to avoid some of the limitations of a strictly legalist approach. For example, the selection of testimony and the opportunities afforded to victims to tell their stories to the commission are not, in theory, constrained by rules of procedure and evidence in the same way criminal trials are. These rules determine who gets heard and who—and what—does not. Similarly, the purpose of testimony is to encourage acknowledgment of the harm caused to victims rather than to build a case against an accused. This makes truth commissions less adversarial than prosecutions—a benefit in politically divided societies. Since the South African Truth and Reconciliation Commission (TRC) truth commissions have also been regarded as playing a role in interpersonal reconciliation. The TRC famously promoted truth-telling, acknowledgment of crimes, and forgiveness, even where this meant that punishment would be avoided (Mamdani 2002). In adopting this approach, the South African government at the time prioritized restorative justice over the retributive criminal justice approach favored in transitional justice. Rather than simply pursing accountability and punishment, the TRC was put forward as pursuing an alternative form of what justice means in the aftermath of mass or systemic violence—which was complementary to the goals of trials (Leebaw 2008).

Truth commissions therefore prioritize acknowledgement and dignity rather than the pursuit of strict accountability. For some this is achieved by prioritizing the recognition of the harm caused to the victim (Haldemann 2008). This approach was central to the recent Truth and Dignity Commission established in Tunisia in which restoration of dignity was cited as a priority in a context where victims had been consistently vilified (International Centre for Transitional Justice 2015). A commission, through its report, can make a powerful statement that the violations of human rights that occurred were in no way justified.

More tangibly, truth commissions can also help to address a culture of denial. Truth commissions are usually mandated to produce a report that establishes a historical record of what happened. This report, and the narrative that it contains, is considered important in preventing recurrence not only by naming specific perpetrators but, more broadly, by shaping collective historical memory (Haldemann 2008, 726). This narrative can in turn be used for social education purposes; for example, the findings can be disseminated and included in school curricula, media outputs, and public dialogue, thereby exposing communities to uncomfortable truths and decreasing resistance to acknowledging human rights violations (Walker 2013). The aim is to provide a unified narrative of the conflict that reduces the possibility of denial. The reports also provide the basis for other mechanisms, such as reparations and guarantees of nonrecurrence, by providing an authoritative account of both the extent of victimhood and the legal and institutional failings that allowed the widespread abuse of human rights or humanitarian law to occur.

While truth commissions are the primary institution associated with truth recovery in transitional justice, they are also the most formal. Truth recovery and the curation of historical memory in the aftermath of conflict are of increasing interest in transitional justice. Informal community-based mechanisms can operate either alongside official mechanisms or as an alternative where no agreement has been reached on formal mechanisms. For example, the East Timorese Commission on Reception, Truth and Reconciliation incorporated "community reconciliation proceedings" into its work. These took place in cooperation with indigenous communities. Perpetrators whose crimes did not qualify as serious human rights violations participated in these proceedings, where they had to express repentance and ask the affected community to be readmitted as members. In that way, the commission helped restore relations in communities (Hayner 2010). However, where division and disagreement remain over the need for transitional justice or the approach that should be taken, truth recovery can be as simple as documentation. The collection of factual or forensic information has been a feature of transitional justice since its early days. Early truth commissions in countries such as Chile and El Salvador were created as mechanisms for information recovery where prosecution was not possible for legal political reasons (see Mallinder and McEvoy 2011).

This approach highlights the importance of truth as a goal in itself rather than simply fact-finding as a means to prosecution. Two more recent examples demonstrate this approach to documentation in the absence of formal transitional justice mechanisms. In the Balkans the Kosovo Memory Book has documented those killed or missing as a result of conflict between 1998 and 2000 (see Visoka 2016, 2017). In Northern Ireland the Lost Lives project presented a record of every person killed as a result of the conflict from 1969 to 1998 (McKittrick et al. 1999). By bringing together information from resources such as media reporting and eyewitness testimonies these volumes present a factual record of deaths and disappearances in a way that avoids politicization of truth recovery in what remain deeply divided societies.

While truth recovery still takes diverse forms, which in turn emphasize different goals, the importance of truth underpins transitional justice. Truth recovery has evolved from a pragmatic response to the limitations of transitional justice in fragile democracies to a foundational normative principle underpinned by a legal "right" to effective investigation articulated by the United Nations (2011) and the European Court of Human Rights.[1] This simply reflects the centrality of truth as an intrinsic need rather than as an instrumentalist means to an end in the aftermath of violence.

Reparations

The third pillar of transitional justice is the provision of reparation for victims of human rights abuses. Reparations sit alongside fact-finding mechanisms such as trials and truth commissions, which may recommend reparation for victims of specified abuses or patterns of abuse uncovered by the commission. Reparations as a mechanism of transitional justice "promote justice by redressing gross violations of international human rights law" (United Nations 2006). They comprise various aspects of compensation, restitution, rehabilitation, satisfaction, and guarantees of nonrecurrence. In 2005 the UN General Assembly recognized a "right" to reparations through the adoption of the Basic Principles and Guidelines on the Right to a Remedy and Reparation for Victims of Gross Violations of International Human Rights Law and Serious Violations of

International Humanitarian Law (United Nations, 2005b). The existence of a right for victims gives rise to a correlative duty by the state to give effect to the right. The aim of reparations is the restoration of the victim; as such they take many different forms. On a basic level it may be financial compensation for victimhood or the loss of a family member. For example, the Argentine Indemnification Law (Law No. 24.043, enacted 27 November 1991) establishes the right to monetary compensation to all who were at disposition of the National Executive Authority during the state of emergency or were incarcerated based on military courts' orders (regardless of whether they file claims for damages). But monetary compensation is not the only available option. The aim of restitution, a remedy more commonly found in private law, is to restore the victim to the position they would have been in had the abuse not happened. This might include the restoration of property rights, for example, or the restoration of rights and entitlements within the legal system. For example, the Victims' Law in Colombia (Ley de Victimas, 2011) declares that victims whose land had been taken away from them or who were forced to abandon it have a right to restitution (Summers 2012).

Reparations also aim to rehabilitate individuals and communities that have suffered. For example, the Organic Law on Establishing and Organizing Transitional Justice in Tunisia includes a reparations fund called Fund for Dignity and Rehabilitation of Victims of Tyranny (Republic of Tunisia n.d.). Besides monetary compensation, other envisaged measures are free medical services to those wounded as well as to the family members of those who are considered to be martyrs of the Revolution (International Centre for Transitional Justice 2018). For the United Nations (2011, paragraph 26) "reparations to victims can help to reduce community resentment towards parallel mechanisms such as incentive based disarmament, demobilization and reintegration programmes, which may be seen as favouring perpetrators over victims." They are also increasingly viewed as means of addressing gender inequality and the root causes of violence against women and girls, by linking individual reparations with broader efforts to address economic and social marginalization (paragraph 27). This is achieved through collective measures such as programs that redress the social consequences of conflict through education and access to health (see also Walker 2016).

Finally, reparations may be symbolic, in the form of gestures that are intended to acknowledge harm suffered and to restore dignity (International Centre for Transitional Justice 2015). According to the UN Basic Principles, measures of satisfaction can, inter alia, include public apology or acknowledgment and acceptance of responsibility, commemorations to the victims, and the search for the whereabouts of the disappeared, as well as providing assistance in the recovery, identification, and reburial of the remains. That being said, apology can be mentioned as a measure of satisfaction as well as a form of symbolic reparation. For example, in June 2012 President Mauricio Funes of El Salvador publicly apologized for the El Mazote Massacre of 1982 in which a thousand people were killed. In March 2014 President Uhuru Kenyatta of Kenya offered an apology for "all past wrongs" that the Kenyan government had committed, including the 1984 Wagalla Massacre, as well as the postelection violence in 2007–2008. And in June

2010 Britain's Prime Minster David Cameron issued an apology to the families of those killed in the Bloody Sunday shootings of 1972, describing the killings as "unjustified and unjustifiable." However, while apologies may be of symbolic value, they are often insufficient as a form of satisfaction to victims and may need to be accompanied by other measures of accountability (Carranza et al. 2015).

In addition to the individual and collective benefits reparations offer victims, in international law and policy they are vested with the social function of promoting reconciliation and restoring confidence in the state (United Nations 2004, paragraph 54). "Programmes to provide reparations to victims for harm suffered can be effective and expeditious complements to the contributions of tribunals and truth commissions, by providing concrete remedies, promoting reconciliation and restoring victims' confidence in the State" (paragraph 54). From this perspective, certain forms of pubic commemoration have also been interpreted as a form of reparation (United Nations 2004). The link between memory and transitional justice is explored in much greater depth in nonlegal disciplines, raising questions of the extent to which memory can be articulated as a specific legal duty in response to violence (Fletcher and Weinstein 2017). Commemoration takes transitional justice away from the legally normative and places it more squarely in the peacebuilding terrain. To date, while commemoration is discussed as an aspect of transitional justice, studies from the Balkans, Southern Africa, Lebanon, Northern Ireland, and elsewhere suggest that such acts tend to be focused on single-identity forms of memory that can be politically divisive in that they celebrate different interpretations of past events that can be mutually incompatible (Brown 2019; Subotic 2013a; Jelin 2007). Therefore, while commemoration may play a role in confronting the past, it remains less clear how these approaches can appeal to a more inclusive national identity (Brown 2019).

Guarantees of Nonrecurrence

Although considered to be a pillar of transitional justice, guarantees of nonrecurrence remain largely underresearched (Sandoval 2017; Roht-Arriaza 2015). Initially included within the scope of reparations, there is increasing focus on the role that such guarantees actually play in preventing the recurrence of conflict. The guarantee has yet to be fully articulated in legal terms, leaving transitional justice lagging behind the peacebuilding field (Mayer-Rieckh 2017, 417). The guarantee of nonrecurrence derives from the existence of human rights in law. As Mayer-Rieckh highlights, the breach of an obligation on the part of a state requires action to remedy the breach and to prevent its recurrence (422). States are therefore under an obligation to prevent recurrent breaches of international human rights law. In the context of transitional justice such measures focus primarily on strengthening the rule of law (United Nations 2011, paragraph 35). Initiatives may include the reform of domestic judicial systems, the repeal of emergency laws, and strengthening human rights guarantees in the legal system; reforming institutions; and establishing civilian oversight of the military and police. Where transitional justice is led by the United Nations there will also be a focus on strengthening domestic compliance with international legal standards (paragraph 36).

The goal of these measures from a transitional justice perspective is to prevent the recurrence of violent conflict by enshrining strong human rights guarantees into the national legal system. While policy approaches from the UN Rapporteur stress the importance of guarantees of nonrecurrence (United Nations 2015), knowledge of their application in practice remains scarce. The insights from practice that do exist show that the understanding of guarantees of nonrecurrence remains ambiguous and context-specific. By way of example, in the Guatemalan peace process, the nonrepetition of violent events was connected to the right to truth and the establishment of the Clarification Commission. In Sierra Leone, the TRC Act of 2000 suggests that guaranteeing nonrepetition is one of the commission's functions. Most recently, in Colombia, guarantees of nonrepetition included in the Final Agreement concern establishment of the Truth, Coexistence and Non-Recurrence Commission as well as a comprehensive rural reform, criminal sanctions, and early acknowledgment of collective responsibility.

Gaps between Transitional Justice and Peacebuilding

Having outlined very briefly the definition of each of the core transitional justice mechanisms, attention now turns to some of the gaps that exist between the worlds of transitional justice and peacebuilding. There are three key gaps that bear further exploration. The first is the challenge that transitional justice does not take into account the politics of justice in transition. The second is that transitional justice is not capable of delivering transformative justice because of its preoccupation with retributive justice. The third is that transitional justice does not deliver on its promises.

Transitional Justice: Stuck between Law and Politics

One of the key points to note when considering the relationship between transitional justice and peacebuilding is that despite efforts to conceptualize locally owned transitional justice, the origins of the field in legal concepts of retributive justice mean that analysis and evaluation of the successes and failures of transitional justice remains stubbornly rooted in legal notions of justice. And yet there is a profound tension between the desired impartiality of legal procedure and the deeply political contexts in which it is to be applied.

This itself follows from the perceived value that law brings to transitions. From the outset law has been used in transition as a means of transcending divisive political rhetoric (Teitel 2000; Campbell and Ní Aoláin 2005). The impartiality of the law is intended to stand above the fray of politics and to offer at least the possibility of impartial adjudication between competing parties. Further, the application of clearly defined legal rules is intended to enhance the legitimacy of transitional mechanisms generally, but trials specifically. The impartiality delivered by legal adjudication is the very thing that enables authoritative claims toward creating a historical record, for example, or elaborating a shared narrative of a conflict (Leebaw 2008). The value of law in transition is that it is by its nature different from politics. The law makes absolutist statements about the need to ensure justice that simultaneously raise the expectations of victims while constraining what is possible through the application of stringent rules of procedure and evidence that mean most cases will never be prosecuted in an international court. As a result, the unfulfilled promise of law can create obstacles for peacebuilding. First, it can undermine faith in the legal system, which can act as a destabilizing force in fragile states. Second, it creates tension where the outcome of mass atrocities appears to be impunity. Rather than transcending political divisions, the effect of law can be to entrench them further. This stands in contrast to the approach of peacebuilding and the emphasis on the restoration of relationships. Moving beyond violent political division relies on the ability to establish dialogue and empathy between former enemies. It is here that the ideal of justice remains in tension with the ideal of peace. Law and rights are powerful tools for advocacy. And yet there is a delicate balance to be struck between maintaining a strict position on legal rights and advocating for peacebuilding work that is necessary to build the trust and relationships that enable consensus on justice to emerge.

Where lawyers critique transitional justice with reference to due standards of law, the political science and peacebuilding literature highlights more clearly the political aspects of transitional justice that appear to be overlooked in the legal literature. These critiques include, for example, that transitional justice takes insufficient account of local peacebuilding efforts, that it does not permit local participation, and that the priorities of transitional justice are those of powerful actors—both national and international—who have vested interests in how transitional justice is shaped (Baker and Obradović-Wochnik 2016; Sharp 2013). For example, the International Criminal Court (ICC) is consistently criticized for adopting an approach that is simultaneously too legal and too political (P. Clark 2018). By attempting to maintain the principle of complementarity, whereby the ICC intervenes only where the national government is unable or unwilling to establish effective transitional justice mechanisms, the Court is criticized for remaining remote from victim populations, failing to deliver justice by showing too much deference to local actors. The ICC was established to carry out investigations and prosecutions of the perpetrators who commit the most serious offenses. If it does not prosecute due to complementarity it risks standing accused of undermining justice and exacerbating political tensions. If it does, then it remains open to critique on account of imposing international justice (Roach 2011). The tension between peace and justice, or law and politics, has been a defining dichotomy since the emergence of the field of transitional justice. The tension that surrounds decisions whether to exercise international jurisdiction in relation to the prosecution of serious crimes or not exemplifies the difficult path to be negotiated by transitional justice actors.

Transitional Justice Cannot Be Transformational

A second critique of transitional justice, and one that is gaining traction, is that the narrow focus on the violation of civil and political rights prevents a sustained engagement with questions of economic or social justice (Laplante 2008; Mani 2008; Miller 2008). Trials in particular rely on the ability to identify a specific offense that has been committed by a named individual to establish accountability. Truth commissions offer greater flexibility, but even then their terms of reference tend to reduce the harms for which justice is sought down to harms that mirror legal offenses (murder, torture, rape, etc.); (Ní Aoláin and Turner 2007). With clear terms of reference for what a truth commission can and cannot investigate, it is inevitable that alternative forms of harm will not be captured. There is increasing attention on the inability of transitional justice mechanisms to adequately account for the structural conditions that cause conflict and to recommend measures that would reform systems rather than individual persons or institutions (Muvingi 2009; Gready and Robbins 2014). This is evidenced, for example, in the greater attention to the potential for reparations (as a form of transitional justice) to be transformative by addressing socioeconomic factors.

This idea that reparations can be used as tools for transformation is most prevalent in the scholarship on gender justice. Authors such as Rashida Manjoo (2017) and Ruth Rubio-Marin (2009) argue that transformative reparations operate on three different levels, including the individual, where the female victim is given access to remedies, but also institutional, where institutional reforms are triggered, and structural, where the status quo ante is amended.

However, this is the work of peacebuilding, and it is not comfortably accommodated in legally normative approaches to justice (Parlveliet 2017; Walker 2016). Even where the terms of reference of a truth commission include scrutiny of social and economic harm, as seen, for example, in Sierra Leone, there is no guarantee that recommendation in the report will be enforced (McAuliffe 2017). The exclusion of social and economic harms from transitional justice is for many symptomatic of the close relationship between transitional justice and liberal peacebuilding (Matua 2015). Scholars highlight the preference for market-friendly liberal democracy that underpins transitional justice at the expense of redistributive equality (Laplante 2008). This approach in itself is symptomatic of the embeddedness of transitional justice in West-centric narratives of peace and conflict that largely ignore localized critiques that would challenge the foundations of the field (Dawan 2011; Nesiah 2011). In many cases stabilization, development, and the need to create the conditions for peacebuilding can actively discourage a focus on transitional justice that may be politically divisive. For example, in Uganda the UN Development Assistance Framework for 2016–2020 no longer mentions transitional justice or conflict, which had been included in the previous version. Similarly, the EU's strategy for Uganda makes no mention of the past. The structural inequalities that give rise to conflict remain invisible both to an individually focused model of justice (Lambourne 2014) and to development-focused interventions (Bernath, 2018). To address this limitation a closer engagement with peacebuilding would be needed.

Transitional Justice Does Not Deliver on Its Promises

To add to this, and largely arising from the examples just noted, is the fact that research now suggests that transitional justice is not delivering on its stated aims. For example, despite the claim that transitional justice (and international criminal justice in particular) represents impartial adjudication of conflict-related crimes, the recent threatened withdrawals from the ICC have revealed the fundamentally political nature of the Court and the way it is perceived by those states subject to investigation. The success of the ICC, or any transitional justice mechanism, cannot be measured solely with reference to technical criteria that relate only to the legal operation of the institution. This critique is also linked to the claim that the role of transitional justice mechanisms is to create an agreed historical record of conflict that can help to establish the "truth" of a conflict. In particular a stark division has emerged between the evaluation of the success of transitional justice globally and the way it has been perceived in affected communities. For example, recently published studies of the Balkans have demonstrated that the International Criminal Tribunal for the former Yugoslavia has failed to persuade key target audiences that the findings of its judgments were accurate (Orentlicher 2018; Milanović 2016). Indeed the effect of the Tribunal may have been to foment rather than dispel distrust. This finding is common in more recent work on transitional justice and peacebuilding in the Balkans. For example, Spoerri (2011) argues that the fact that former Yugoslav governments were compelled by the EU through the conditionality principle to arrest and extradite accused war criminals to the International Criminal Tribunal for the former Yugoslavia serves as an obstacle toward reconciliation and revelation of truth in the region. If justice is perceived as "coerced" or "imposed," it becomes *injustice* in the eyes of many, with the risk that indicted accused war criminals become heroes facing persecution at the hands of international courts (Spoerri 2011; Subotic 2013b).

Some literature assessing the success of the South African TRC has cast a similarly critical eye on the extent to which victims of the apartheid regime feel that justice was done by the TRC (Madlingozi 2007). Similar tensions exist in respect of the implementation of transitional justice in Nepal, Colombia, Sri Lanka, and Northern Ireland. Documenting the success of the truth commission in Haiti, Quinn (2009) noted that the commission failed to acknowledge the country's conflicted past due to a wide range of factors; these include the lack of political will and funding, security concerns, shortage of time, and insufficient capacity. Despite the drive for an impartial discourse that transcends political divisions, transitional justice itself remains divisive. While this is the case, transitional justice debates remain more clearly focused on how to overcome technical difficulties in implementation rather than looking more broadly at how justice could contribute to social transformation.

Conclusion

Despite the limitations of transitional justice in delivering reconciliation-oriented goals, there remain potential synergies between the fields of transitional justice and peacebuilding. For example, transitional justice is often contentious because it is regarded as a continuation of conflict. Where transitional justice debates become polarized, victims become collateral damage. If the whole project of transitional justice is seen as political, then there is no space for hearing or understanding the justice demands of victims and no scope for using law to mediate difficult political conversations. There is potential for dialogue to be used to explore understandings and fears about transitional justice in a way that frames justice in a less confrontational and more restorative light. These dialogues are not a replacement for transitional justice, but a process whereby the requirements of justice are negotiated, creating the conditions for agreed mechanisms. This aspect of building consensus on the means and ends of justice appears to be absent from current transitional justice practice but is an area of practice that could usefully be developed by learning from the peacebuilding community. Ultimately the project of transitional justice would be strengthened by developing a more inclusive approach that enabled greater participation in the way in which justice is designed and delivered.

Note

My sincere thanks to Maja Davidovic and to Zyra Edwards for their respective contributions to this chapter.

1. *Jordan v. United Kingdom* (2003) EHRR 2; *McKerr v. United Kingdom* (2002) 34 EHRR 20; *Kelly v. United Kingdom*, App No. 17579/90 (ECHR, 4 May 2001); *McCann, Farrell and Savage v. United Kingdom* (1996) 21 EHRR 1.

References

Ainley, K. 2017. "Evaluating the Evaluators: Transitional Justice and the Contest of Values." *International Journal of Transitional Justice* 11 (3):421–442.

Akhavan, P. 1998. "Justice in the Hague, Peace in the Former Yugoslavia: A Commentary on the United Nations War Crimes Tribunal." *Human Rights Quarterly* 20 (4):737–816.

Baker, C., and J. Obradović-Wichnik. 2016. "Mapping the Nexus of Transitional Justice and Peacebuilding." *Journal of Conflict Intervention and Statebuilding* 10 (3):281–301.

Bell, C. 2009. "Transitional Justice, Interdisciplinarity and the State of the 'Field' or 'Non Field.'" *International Journal of Transitional Justice* 3 (1):5–27.

Bell, C., C. Campbell, and F. Ní Aoláin. 2004. "Justice Discourses in Transition." *Social and Legal Studies* 13 (3):305–328.

Bernath, T. 2018. "Off the Agenda as Uganda Moves toward Development: Uganda's Transitional Justice Process." In *Transitional Justice, International Assistance, and Civil Society*, edited by P. Arthur and C. Yakinthou, 114–143. Cambridge: Cambridge University Press.

Brown, K. 2019. "Political Commemoration and Peacebuilding in Ethnonational Settings: The Risk and Utility of Partisan Memory." *Peacebuilding* 7 (1):51–70.

Campbell, C., and F. Ní Aoláin. 2005. "The Paradox of Transitions in Liberal Democracies." *Human Rights Quarterly* 27 (1):172–213.

Carranza, R., C. Corre, and E. Naughton. 2015. *More Than Words: Apologies as a Form of Reparation*. New York: International Center for Transitional Justice.

Clark, J. 2011. "Peace, Justice and the International Criminal Court." *Journal of International Criminal Justice* 9 (3):521–545.

Clark, P. 2018. *Distant Justice: The Impact of the International Criminal Court on African Politics*. Cambridge: Cambridge University Press.

Daly, E. 2002. "Transformative Justice: Charting a Path to Reconciliation." *International Legal Perspectives* 12 (1–2):73–183.

Dawan, N. 2011. "Transitions to Justice." In *Gender in Transitional Justice*, edited by S. Buckley-Zistel and M. Zolkos, 264–283. London: Routledge.

Fletcher, L., and H. Weinstein. 2002. "Violence and Social Repair: Rethinking the Contribution of Justice to Reconciliation." *Human Rights Quarterly* 24 (3):573–639.

Fletcher, L., and H. Weinstein. 2017. "Transitional Justice and the Plight of Victimhood." In *Research Handbook on Transitional Justice*, edited by C. Lawther, L. Moffett, and D. Jacobs, 244–266. Cheltenham, UK: Edward Elgar.

Gready, P., and S. Robbins. 2014. "From Transitional to Transformative Justice: A New Agenda for Practice." *International Journal of Transitional Justice* 8 (3):339–361.

Haldemann, F. 2008. "Another Kind of Justice: Transitional Justice as Recognition." *Cornell International Law Journal* 41:675–737.

Hayner, P. 2010. *Unspeakable Truths: Transitional Justice and the Challenge of Truth Commissions*. London: Routledge.

International Centre for Transitional Justice. 2015. "ICTJ Program Report: Tunisia." 6 March. https://www.ictj.org/news/ictj-program-report-tunisia.

Kritz, N. J., ed. 1995. *Transitional Justice*. Vol. 1: *General Considerations*. Washington, DC: US Institute of Peace.

Kritz, N, and J. Finci. 2001. "A Truth and Reconciliation Commission in Bosnia and Herzegovina: An Idea Whose Time Has Come." *International Law Forum du droit international* 3(1):50–58.

Jelin, E. 2007. "Public Memorialization in Perspective: Truth, Justice and Memory of Past Repression in the Southern Cone of South Africa." *International Journal of Transitional Justice* 1 (1):138–156.

Lambourne, W. 2014. "Transformative Justice and Peacebuilding." In *Transitional Justice Theories*, edited by S. Buckley-Zistel, T. Koloma, C. B. Beck, and F. Mieth, 19–39. London: Routledge.

Laplante, L. 2008. "Transitional Justice and Peacebuilding: Diagnosing and Addressing the Socioeconomic Roots of Violence through a Human Rights Framework." *International Journal of Transitional Justice* 3 (1):331–355.

Leebaw, B. 2008. "The Irreconcilable Goals of Transitional Justice." *Human Rights Quarterly* 30 (1):95–118.

Madlingozi, T. 2007. "Good Victim, Bad Victim: Apartheid's Beneficiaries, Victims and the Struggle for Social Justice." In *Law, Memory and the Legacy of Apartheid: Ten Years after AZAPO v President of South Africa*, edited by W. Le Roux and K. Van Marle, 107–126. Pretoria: Pretoria University Law Press.

Matua, M. 2015. "What Is the Future of Transitional Justice?" *International Journal of Transitional Justice* 9 (1):1–9.

Mallinder, L. and McEvoy, K. 2011. "Rethinking Amnesties: Atrocity, Accountability and Impunity in Post-Conflict Societies." *Contemporary Social Science* 6(1): 107–128.

Mamdani, M. 1996. "From Justice to Reconciliation: Making Sense of the South African Experience." In *Crisis and Reconstruction: African Perspectives*, edited by C. Leys and M. Mamdani, 17–25. Uppsala: Nordisk Afrikainstitut.

Mamdani, M. 2002. "Amnesty or Impunity? A Preliminary Critique of the Report of the Truth and Reconciliation Commission of South Africa." *Diacritics* 32 (3–4):33–59.

Mani, R. 2008. "Dilemmas of Expanding Transitional Justice, or Forging the Nexus between Transitional Justice and Development." *International Journal of Transitional Justice* 2 (3):253–265.

Manjoo, R. 2017. "Introduction: Reflections on the Concept and Implementation of Transformative Reparations." *International Journal of Human Rights* 21 (9):1193–1203.

Mayer-Rieckh, A. 2017. "Guarantees of Non-Recurrence: An Approximation." *Human Rights Quarterly* 39 (2):416–448.

McAuliffe, P. 2017. "Dividing the Spoils: The Impact of Sharing on Possibilities for Socioeconomic Transformation in Postconflict States." *International Journal of Transitional Justice* 11 (2):197–217.

McKittrick, D et al. 1999. *Lost Lives: The Stories of the Men, Women and Children Who Died As A Result of the Northern Ireland Troubles*. Mainstream Publishing.

Mihr, A. 2017. "An Introduction to Transitional Justice." In *An Introduction to Transitional Justice*, edited by O. Simic, 1–27. London: Routledge.

Milanović, M. 2016. "The Impact of the ICTY on the Former Yugoslavia: An Anticipatory Post Mortem." *American Journal of International Law* 110 (2):233–259.

Miller, Z. 2008. "Effects of Invisibility: In Search of the 'Economic' in Transitional Justice." *International Journal of Transitional Justice* 2 (3):266–291.

Muvingi, I. 2009. "Sitting on Powder Kegs: Socioeconomic Rights in Transitional Societies." *International Journal of Transitional Justice* 3 (2):163–182.

Nesiah, V. 2011. "Missionary Zeal for a Secular Mission: Bringing Gender to Transitional Justice and Redemption to Feminism." In *Feminist Perspectives on Contemporary International Law: Between Resistance and Compliance*, edited by S. Kuovo and Z. Pearson, 137–158. Oxford: Hart.

Ní Aoláin, F., and C. Turner. 2007. "Gender, Truth and Transition." *UCLA Women's Law Journal* 16:229–279.

Orentlicher, D. 2018. *Some Kind of Justice: The ICTY's Impact in Bosnia and Serbia*. Oxford: Oxford University Press.

Parlveliet, M. 2017. "Human Rights and Peacebuilding: Complementary and Contradictory Complex and Contingent." *Journal of Human Rights Practice* 9 (3):333–357.

Quinn, J. R. 2009. "Haiti's Failed Truth Commission: Lessons in Transitional Justice." *Journal of Human Rights* 8 (3):265–281.

Republic of Tunisia. n.d. Organic Law on Establishing and Organizing Transitional Justice. Translated by the International Center for Transitional Justice. https://www.ohchr.org/Documents/Countries/TN/TransitionalJusticeTunisia.pdf.

Roach, S. C. 2011. "The Turbulent Politics of the International Criminal Court." *Peace Review* 23 (4):546–551.

Roht-Arriaza, N. 2015. "Measures of Non-Repetition in Transitional Justice: The Missing Link?" University of California Hastings College of the Law, Legal Studies Research Paper Series, 1–41.

Rubio-Marin, R. 2009. "The Gender of Reparations in Transitional Societies." In *The Gender of Reparations: Unsettling Sexual Hierarchies while Redressing Human Rights Violations*, edited by R. Rubio-Marin, 63–120. Cambridge: Cambridge University Press.

Sandoval, C. 2017. "Reflections on the Transformative Potential of Transitional Justice and the Nature of Social Change in Times of Transition." In *Justice Mosaics: How Context Shapes Transitional Justice in Fractured Societies*, edited by R. Duthie and R. Seils, 166–200. New York: International Center for Transitional Justice.

Sharp, D. 2013. "Beyond the Post Conflict Checklist: Linking Peacebuilding and Transitional Justice through the Lens of Critique." *Chicago Journal of International Law* 14 (1):165–196.

Spoerri, M. 2011. "Justice Imposed: How Policies of Conditionality Effect Transitional Justice in the Former Yugoslavia." *Europe-Asia Studies* 63 (10):1827–1851.

Sriram, C. 2017. "Beyond Transitional Justice: Peace, Governance and the Rule of Law." *International Studies Review* 19 (1):53–69.

Subotic, J. 2011. "Expanding the Scope of Post-conflict Justice: Individual, State and Societal Responsibility for Mass Atrocity." *Journal of Peace Research* 48 (2):157–169.

Subotic, J. 2012. "The Transformation of International Transitional Justice Advocacy." *International Journal of Transitional Justice* 6 (1):106–125.

Subotic, J. 2013a. "Diverging Paths in Western Balkans." *Current History* 112 (752):107–113.

Subotic, J. 2013b. "Remembrance, Public Narratives, and Obstacles to Justice in the Western Balkans." *Studies in Social Justice* 7 (2):265–283.

Summers, N. 2012. "Colombia's Victims' Law: Transitional Justice in a Time of Violent Conflict?" *Harvard Human Rights Journal* 25 (1):219–235.

Sweeney, J. 2018. "The Elusive Right to Truth in Transitional Human Rights Jurisprudence." *International and Comparative Law Quarterly* 67 (2):353–387.

Teitel, R. 2000. *Transitional Justice*. Oxford: Oxford University Press.

Teitel, R. 2003. "Transitional Justice Genealogy." *Harvard Human Rights Law Journal* 16:69–94.

Turner, C. 2013. "Deconstructing Transitional Justice." *Law and Critique* 24 (2):193–209.

Turner, C. 2016. *Violence, Law and the Impossibility of Transitional Justice*. Abingdon, UK: Routledge.

Tutu, D. 1999. *No Future without Forgiveness: A Personal Overview of South Africa's Truth and Reconciliation Commission*. New York: Doubleday.

United Nations. 2004. *Report of the Secretary General on the Rule of Law and Transitional Justice in Conflict and Post-Conflict Societies*. UN Doc S/2004/616.

United Nations. 2005. *Updated Set of Principles for the Protection and Promotion of Human Rights through Action to Combat Impunity*. UN Doc E/CN.4/2005/102/Add.1.

United Nations. 2005b. *Basic Principles and Guidelines on the Right to a Remedy and Reparation for Victims of Gross Violations of International Human Rights Law and Serious Violations of International Humanitarian Law*. UN Doc A/Res/60/147.

United Nations. 2011. *Report of the Secretary General on the Rule of Law and Transitional Justice in Conflict and Post-Conflict Societies*. UN Doc S/2011/634.

United Nations. 2015. *Report of the Special Rapporteur on the Promotion of Truth, Justice, Reparations and Guarantees of Non-recurrence*. UN Doc A/HRC/30/42.

United Nations Security Council. 1993. Resolution 808. 22 February.

Visoka, G. 2016. "Arrested Truth: Transitional Justice and the Politics of Remembrance in Kosovo." *Journal of Human Rights Practice* 8 (1):62–80.

Visoka, G. 2017. *Shaping Peace in Kosovo: The Politics of Peacebuilding and Statehood*. Basingstoke, UK: Palgrave Macmillan.

Walker, M. U. 2013. "Nunca Más." In *Jus Post Bellum and Transitional Justice*, edited by L. May and E. Edenberg, 262–284 Cambridge: Cambridge University Press.

Walker, M. U. 2016. "Transformative Reparations? A Critical Look at the Current Trend in Thinking about Gender Just Reparations." *International Journal of Transitional Justice* 10 (1):108–125.

CHAPTER 35

RECONCILIATION AND PEACEBUILDING

GRÁINNE KELLY

The past three decades have seen an increasing prevalence of the secular application of the term "reconciliation" to societies emerging from violent conflict. This is, perhaps, reflective of the growth and intimacy of intrastate conflicts and interethnic polarization since the early 1990s. Routinely peacebuilding scholars and practitioners include reconciliation in the list of essential tasks associated with societal and political reconstruction in the postsettlement context, and the term is regularly incorporated as either an overarching value or a framework for practical action in peace accords. At the state level, countries emerging from violent conflict have established explicitly described truth and reconciliation commissions, including in Sierra Leone, Chile, and Timor Leste. Specific reconciliation laws have been enacted in Algeria, South Africa, and Mali, and Rwanda and the Solomon Islands, among others, have established government departments with the expressed purpose of making progress in reconciliation. Rafts of international donors now support reconciliation-focused programming, precipitating a reorientation in the portfolios of nongovernmental organizations, and the establishment of new community-based programs to facilitate local processes. And yet, for all this activity and rhetoric, the concept of reconciliation, and how it is operationalized, has swung between contestation, skepticism, and ambivalence, even within these formalized structures.

Typically, interested scholars begin by conceding that the concept of reconciliation is vague and ill-defined, and by acknowledging that differing interpretations exist both across different cultures and within individual societies (Anstey and Rosoux 2017; Hamber and Kelly 2005; Hazan 2009; Long and Brecke 2003; Murphy 2010). Given its Judeo-Christian theological roots, its eclectic application spanning everything from interstate relations to individual healing, and accusations of its misuse for narrow political interests, Simpson's (2014, 4) observation that the concept "lacks universal relevance and remains foreign to too many political cultures, languages and parts of the world" is predicable. The lack of clarity as to what reconciliation entails is observable in policy circles from the United Nations to individual national governments and in the ways in which religious and community leaders, victims and perpetrators articulate their understanding at the local level. It is regularly linked with a range of other concepts, including truth (Asmal, Asmal, and Suresh Roberts 1997; Gibson 2006), trust (Govier and Verwoerd 2002), forgiveness (Amstutz 2004; Minow 1998), and psychosocial identity transformation (Kelman 2004, Bar-Tal and Bennink 2004). Nonetheless it appears that, independent of efforts to devise theoretical conceptions of the term, individuals, communities, and societies do develop and articulate their own versions of reconciliation, based on their sociopolitical context, individual and societal expectations, and vested interests. This has led some to argue that the absence of conceptual clarity has hindered its implementation and that sketching out a broad framework of actions, while supporting a flexible approach, is a more realistic undertaking in a divided society (Schaap 2008).

At a basic level, reconciliation is understood as the pragmatic process of (re)building relationships fractured by conflict. Lederach (2001, 842) emphasizes its human dimension, noting that it is "first and foremost about people and their relationships." Kelman (2010, 3) explains that reconciliation is about "learning to live together in the post-conflict environment," while Hayner (2002, 161) views it as the task of "building or rebuilding relationships today that are not haunted by the conflicts and hatreds of yesterday." This emphasis on affective interpersonal relationships, and the associated language of "rehumanization" used in this context, may obscure a wider view of reconciliation that encompasses political, social, and economic institutions as instruments in its wider apparatus. Hutchison and Bleiker's (2015, 81) observation that reconciliation deals with "the deep societal wounds that inevitably open up after war and other traumatic events" might allow for a more encompassing definition to emerge which recognizes the importance of developing relationships of trust and reciprocity between individuals and groups (horizontal reconciliation), as well as between people and wider social and political institutions (vertical reconciliation). The *quality* of those relationships will depend on a range of factors, including the nature of the relationship between parties prior to the conflict, how the parties came to be in conflict, their conduct during the conflict, and their attitudes and behaviors following settlement of the conflict.

Bloomfield (2006, 16) describes reconciliation as "a term that problematically is still used both to represent the overall peace-building process as well as one constituent part of it." As with the wider concept of peacebuilding, the instigation and encouragement of reconciliation processes within the overall conflict cycle remains contested. Some place reconciliation firmly in the postsettlement phase (Hirsch 2012; Lambourne 2004), while others view it as an ongoing task of addressing damaged relationships throughout the conflict cycle, despite the obvious limitations associated with ongoing violence (Lederach 1997; Bar-Tal and Bennink 2004; Bar-Tal 2009). Bashir (2008) argues that reconciliation should be seen as a precondition rather than an outcome of deliberative democracy, while Anstey and Rosoux's (2017) research indicates that political reconciliation should be viewed as a prerequisite to securing the relationships necessary to push an agreement over the line. Referencing the South African peace process, the authors argue that "[r]econciliatory intent lubricated the negotiated transition from its inception in the mid-1980s" (2) and provided an opportunity for the conflicting parties to build their confidence and courage to enter the process. Significantly, an early focus on the building of working relationships between negotiators is cited as a contributing factor in the ability of the South African process to weather several major political crises and to ultimately reach a multiparty agreement in 1993. Brouneus (2007) cautions, however, that a minimum level of security is required for reconciliation processes to be safely considered, citing the sense of fear of reprisal reported by some victims and witnesses following their testimony-giving to the truth-seeking International Criminal Tribunal for the former Yugoslavia (see also Stover 2004). What is clear is that the identification of suitable entry points for reconciliation processes will be specific to the particular contexts and the existence of enabling environments in which it can take hold.

Resistance to reconciliation has emerged from a range of sources. A common criticism focuses on the implication that it seeks to restore relationships to an imagined, harmonious past, which likely never existed (Moon 2008), although this semantic argument has been addressed by others who accept the ambiguity inherent in the term (Rosoux 2017). It has also been criticized for the instrumental ways in which it can be invoked by political leaders as a way of "managing" the wrangling of a society with past injustices or as an exercise in hollow symbolic gestures (Schaap 2008). The April 2014 statement recognizing the significance of the Armenian Genocide Remembrance Day by Prime Minister Recep Tayyip Erdoğan of Turkey was welcomed by some as having reconciliatory—or at least normalization—intent. However, the statement fell well short of an apology or acknowledgment of the oppression of the non-Turkish population during the Ottoman period and was criticized as a politically expedient attempt to defuse international criticism ahead of the centenary of the Armenian genocide in 2015.

And yet the term continues to have traction and relevance as a specific peacebuilding task in a range of contexts. Thus we must continue to grapple with its meaning. When the elements of reconciliation are broken down, it is clear that it is not one practice but a series of practices, which intersect with a range of other associated peacebuilding tasks. Perhaps the most helpful way to counter the criticism that reconciliation is too vague a concept to form a coherent sociopolitical intervention is to further examine it from three different perspectives: first, to engage with the debate as to whether reconciliation is a process, an outcome, or an end-state; second, to examine the concept as a set of specific yet interdependent and interrelated tasks or component parts; third, to locate the various methods and actors within three societal levels of intervention—political, societal, and individual—and examine their similarities and differences. Finally, this chapter will acknowledge the specificity of context while identifying areas of further research that can illuminate the empirical realities of both addressing the past and building a new future in societies emerging from violent conflict.

Reconciliation as Process or End-State

One of the conceptual stumbling blocks for reconciliation centers on whether it is a process, an outcome, or an aspirational end-state. While peacebuilding implies an ongoing, action-oriented process, the term "reconciliation" could equally suggest a goal to be achieved or a settled state of relationships at the end of a series of interventions. Galtung's (2001, 4) observation that "[r]econciliation is a theme with deep psychological, sociological, theological, philosophical, and profoundly human roots—and nobody really knows how to successfully achieve it" infers an end-state, while Lederach's (2001, 842) definition, that it is a "dynamic, adaptive process aimed at building and healing," focuses attention on ongoing reconciliatory activities. Reconciliation as an end goal has been variously associated with a range of other objectives, including mutual recognition, coexistence, mutual trust, forgiveness, and equality. In this context, individual or societal resistance to the concept of reconciliation may arise due to resistance to a perceived and, for them, disagreeable end-state. Arguably this is a circular debate that is unlikely to be settled, and replacing the language of reconciliation with a narrower conception will not resolve the definitional dilemma.

That said, the dominant discourse encourages a pragmatic understanding of reconciliation as an ongoing process without an idealistic destination (Bloomfield, Barnes, and Huyse 2003; Hamber and Kelly 2005; Little 2017). It also accepts that, given the protracted nature of contemporary conflicts, the histories of accumulated grievances and violence, and territorial proximity, reconciliation is a long-term, even generational process, and a degree of realism as to the pace of change possible and the willingness of parties to engage is required. This measure of reconciliation must also contend with the competing expectations of the international community—often embedded in the texts of the peace settlements reached—and the local pressures associated with translating a normative aspiration into a practical reality. The Arusha Peace and Reconciliation Agreement for Burundi (2000, 3) committed the signatories to the "attainment of genuine unity, reconciliation, lasting peace, security for all, solid democracy and on equitable sharing of resources in Burundi." Despite Vandeginste's (2017, 185) observation that reconciliation has become a "political mantra" in postaccord Burundi, the Peace and Reconciliation Commission envisaged to examine the large-scale impacts of the conflict from 1962 to 2008 was not established until fifteen years later, as a result of political tensions and disagreements between the Burundian state, the United Nations, civil society organizations, and donor countries. While acceptance of its generational nature allows for new opportunities for reconciliation to reveal themselves over time, Cole (2014, 2) adds a note of realism to the expectations associated with a move from violence to relative peace, observing that "sometimes new political and social realities intersect with legacies of conflict in ways that restrict possibilities for new relations."

Due to the ambiguity of the term, reconciliation has been dichotomously interpreted as either a shallow/thin or a deep/thick concept (He 2009; Nagy 2002) and instrumentalized accordingly. Crocker (2000, 6) refers to "nonlethal coexistence" as a thin form of reconciliation, noting that its attainment, "when cease fires, peace accords, and negotiated settlements begin to take hold, can be a momentous achievement" and cautions against a "thicker" form of reconciliation that "requires friendliness and forgiveness." Bar-Tal and Bennink (2004) also advocates for the more modest ambition of coexistence, while Stover and Weinstein (2004) reframe it as social reconstruction.

Given the nature of intrastate political negotiations, which seldom involve a change in territorial boundaries, one might identify coexistence as an essential first stage, on the basis of intercommunal proximity alone. Coexistence itself can be found along a broad definitional continuum; however, it has been faulted as a minimalist form of reconciliation that freezes protagonists to the conflict in their original identities and narratives, without allowing for their possible transformation. Communal coexistence was one of the central tenets of the 1989 Taif Agreement, which established a power-sharing system in Lebanon following years of civil war. Nearly three decades on, while intracommunal social cohesion is reportedly strong, the country has struggled to translate the economic interdependence of the main confessional groups into wider social cohesion across those sectarian divides (Aoun and Zahar 2017). If identities are embedded by the institutional arrangements established to get peace processes over the line—as has also occurred in Bosnia and Northern Ireland—this can have a significant impact on the ability of a society to conceive of new forms of relationships in the postsettlement phase. A thicker form of reconciliation would require individuals and groups to move beyond an absence of direct hostility to recognize hurts caused and damage done, to reverse the structural causes of marginalization, and to develop relationships based on mutual trust and interdependence. That said, reconciliation does not take place within a static postconflict context. While it is accepted that the power dynamics, parties involved, and wider context are in constant flux during the course of the violent conflict, the postconflict setting is also transforming and mutating, with different actors, issues, and opportunities to contend with. As such, peacebuilding processes must be mindful of attending to the relationship-focused elements of intervention—which entails changes in attitudes, behaviors, emotions, and beliefs—and which may require constant recalibration and reorientation.

A Component-Driven Process of Reconciliation

Bloomfield (2006, 4) has accused "reconciliation" of being "a grossly over-packed term" that requires disassembly into its component parts. He attempted to do so by disaggregating reconciliation as justice, truth, healing, and reparations, viewing these as "complementary and interdependent instruments of the overall relationship-building process of reconciliation" (11). Lederach (1997) conceived of reconciliation as both the process and the place where four elements—truth, mercy, justice, and peace—meet. Kriesberg (2004) also identified four essential dimensions of postconflict reconciliation: shared truth, justice, regard (recognizing the humanity of the others and their human rights), and security. He contended, however, that these facets of reconciliation cannot be fully addressed or achieved concurrently, and indeed one aspect can act in contradiction with another at any given time. Lederach (1997, 31) also acknowledged the paradox inherent in the attempt to both deal with "the painful past" and "search for the articulation of a long-term, interdependent future." While reconciliation requires attendance to historical grievances and provision of acceptable forms of redress for actions of the past, its forward-facing elements, which address the need for the transformation of future relationships, provide a direction of travel both for individuals and the society as a whole.

The correlation between the pursuit of justice and reconciliatory outcomes is regularly made, if not always well conceived, given the potential for one to work in opposition to the other. The concept of justice in times of transition has been expanded to include both restorative and reparative mechanisms, which may deliver more effectively than more restrictive, state-focused retributive justice mechanisms. However, there are dangers inherent in conflating the two processes which are neither inextricably linked nor necessarily contradictory. The profile of postaccord reconciliation was significantly raised by the establishment of the South African Truth and Reconciliation Commission (1995–2003) and its many subsequent imitators. The achievement of justice had previously been framed as a prerequisite to reconciliation (Pankhurst 1999) or as incompatible goals (Lambourne 2004), and thus placing them explicitly on the agenda of a formal state-centered mechanism raised expectations for their dual delivery. In the context whereby truth is exchanged for amnesty from prosecution, victims and survivors may perceive this as too high a price to pay for the delivery of the intangible outcome of reconciliation and thus disengage. Increasingly the specific components of transitional justice are agreed as part of the elite-level political negotiations, which involve the main conflict protagonists. Thus there is a need to ensure that, during the process of horse-trading that takes place during political negotiations, the specific needs and expectations of victims and survivors are acknowledged and built in to the institutional mechanisms proposed. The peace agreement reached between the Fuerzas Armadas Revoulcionarias de Colombia–Ejército del Pueblo guerrilla group and the government of Colombia in late 2016 positioned the rights of victims at its center, with explicit details included as to the mechanisms that would be established to provide accountability for serious crimes and fulfill the rights of victims to truth, justice, reparation, and nonrepetition. The comprehensive nature of the agreement reflected the prolonged engagement of victims and their advocates during the several years of peace negotiations, their participation in regional and national forums and "peace roundtables" (*mesas regionales de paz*), which generated proposals for the negotiating parties, and their direct participation as delegates during the political talks in Havana.

Attempts to deal with the legacy of the violent conflict need not be viewed within the narrow confines of criminal justice or truth-seeking mechanisms, however. Reparative processes, including formal acknowledgment of wrongdoing, apologies, memorialization, and material or symbolic reparations, may deliver more effectively on the quality of interpersonal and intercommunal relationships sought. While sincerely articulated apologies from perpetrators and other complicit parties are viewed as contributing to psychosocial healing and reconciliation (Barkan and Karn 2006), most do not consider forgiveness a necessary component (Minow 1998; Dwyer 1999), as it places undue emotional demands on victims (Brudholm 2008). Hutchison and Bleiker (2015, 82) argue for greater attention to be paid to the "deep emotional wounds" that linger after conflict and to the personal reactions that play a wider role in social and political change processes: "Central here is an attempt to legitimize compassion and empathy as active components of the reconciliation process."

Reconciliation is complicated—and thus an understanding of the various "strands" that require attention is necessary (Hamber and Kelly 2005, 2018). A specific focus on interventions and practices that build positive interpersonal relationships is vital, and this necessitates acknowledgment of past injustices as well as a shared vision of an interdependent future. Attending to these strands individually without making fundamental changes to the structures that gave rise to and sustained the conflict is futile, as sustainable relationships flourish in a context of greater economic, political, and social justice and equality. Where cultures of fear, suspicion, and mistrust of the "other" predominate, reconciliation involves both attitudinal and behavioral change at grassroots, societal, and state levels (Hamber and Kelly 2005). It also requires recognition of the complex needs and identities of all individuals, many of whom have been dehumanized and marginalized during the course of the violent conflict. Reconciliation demands that we are attentive to the contradictions and ambivalences inherent in addressing individual strands in isolation and the challenges of managing the paradoxical process of addressing multiple strands of reconciliation simultaneously.

Multilevel Reconciliation

It is widely accepted that reconciliation involves top-down, middle-out, and bottom-up processes in order to maximize its reach across and within divided societies and to address estrangement between individuals, communities, and institutions (Lederach 1997; Charbonneau and Parent 2012). It is also agreed that each level requires tailored modes of intervention, with differing expected outcomes. There is less agreement on the choreography of these multilevel processes—which leads and which follows—or the degree of synchronicity required so that diverse processes become mutually reinforcing rather than potentially undermining. While horizontal reconciliation can create ties of trust between individuals and communities, vertical reconciliation between political institutions and the wider population reinforces the complex web of relationships required to knit the society together. Reconciliation processes are required to address multiple issues—mistrust, trauma, estrangement, suspicion, and fear—and thus require multiple entry points in order to reach the appropriate constituencies and locations. State-led reconciliation processes feature regularly in the text of peace accords, both explicitly in terms of truth, justice, memory, and acknowledgment mechanisms to be established, and implicitly in the visionary forewords that articulate a new future for the state. Following a United Nations–led process, the preamble of the Libyan Political Agreement brokered in 2015 committed the signatories to "building a secure and coherent society in which national reconciliation, justice, respect for human rights and freedom of expression prevail" (UN Support Mission in Libya 2015, 3).

In one sense, it is possible to see state-led institutions as potential agents of reconciliation. However, given state involvement in the majority of contemporary conflicts, the act of designing appropriate and acceptable reconciliatory processes to address the legacy of the conflict is a particular challenge (Murphy 2010). States are required to strike a balance between showing effective leadership in addressing the imperatives of reconciliation at the national level and attempting to control both the spaces in and the processes by which reconciliation can occur. Getting this wrong can result in suspicion, disengagement, and even contempt for such processes. If reconciliation is associated with the wider process of postconflict nation-building, this may be viewed as assimilative rather than self-determining and result in some individuals or groups experiencing greater levels of estrangement (Schaap 2008). When the Sri Lankan Civil War was ended with brutal military force in 2009, the ethnic-majority Sinhalese government called for all citizens to coalesce under a unified, postwar nation-state. The state's insistence that the country was not experiencing ethnic tensions—but instead had a terrorist problem which was addressed by the state-military apparatus—has resulted in the continued alienation of many within the Tamil community, deepening skepticism that the state will overturn decades of impunity on all sides, and persistent claims of discrimination against minority groups (Lewer 2017; Ruwanpura 2016). It should not be assumed that top-level reconciliation processes—particularly those symbolic or rhetorical in nature—will trickle down to the local level or significantly impact the lives of individual victims or combatants.

Community-level processes provide different opportunities to promote reconciliation and take many forms. They can be state-endorsed but locally delivered, such as those associated with local transitional justice or reintegration processes. The often cited example of Gacaca courts to address the overwhelming challenge of impunity following the Rwandan genocide was state-supported but involved local leaders and customary rituals and practices and was devolved to local regions. Following the violent civil conflict in the Solomon Islands, between 1998 and 2003 three dominant approaches to address the demands for acknowledgment and justice were followed. Alongside a strong rule-of-law, criminal justice process instigated by the Australian-led Regional Assistance Mission to the Solomon Islands in 2003, a state-led Truth and Reconciliation Commission was formed in 2009. The third approach was the implementation of customary, indigenous, justice-focused practices, known as *kastom*, which are widely used throughout the region, typically established by community groups, churches, and other civil society structures. Building on long-established, kinship-based methods of conflict resolution, and the underlying assumption that justice requires financial or material compensation, this bottom-up approach typically involves reconciliation ceremonies through which damaged relationships are restored to harmony. These ceremonies can involve rituals such as speech-making, meal-sharing, gift-exchanging, cleansing, and singing and dancing together, often intertwining traditional and Christianity-based elements. Scholars have explored the tensions and disconnections that have arisen between traditional and conventional Western approaches to justice in the region over the past decade (Boege 2019; Jeffery 2013; Quinn 2017). While formal justice mechanisms are resource-intensive and can reach only a small proportion of the population directly, localized processes can be delivered with fewer resources, greater speed, and potentially greater community involvement. However, while locally and culturally focused processes may serve a range of purposes, they can also be predisposed to either romanticization of their reach and impact or dismissal if they do not conform to dominant human rights, gender, or judicial norms.

Community-led, relationship-focused processes can begin prior to peace agreements being reached and can act as conflict dampeners during periods of violent escalation. These often require the specific intervention of middle-level leaders, such as churches, civil society groups, and international NGOs that recognize the need to build a constituency for future peacemaking interventions. Much of the success in developing an inclusive peace negotiation process in Northern Ireland, which culminated in the 1998 Belfast Agreement, was credited to the years of quiet relationship-building and shuttle diplomacy between opposing sides, facilitated by influential church, community, and business leaders from the mid-1980s onward (McCartney 1999; Cochrane 2001). Recent years have seen a shift in resources by the donor community to support locally led dialogue, bridge-building, and reintegration processes, in recognition of the need for local ownership and sustained engagement to embed the wider peacebuilding processes (Assefa 2005; Paffenholz 2010). Daly and Sarkin (2007, 5) acknowledge the "best intentions of the reconciliation promoters" but call for greater interrogation of their practices, suggesting that "their programs are based on unquestioned platitudes, and not on realistic assessments of the attitudes and capacities of the people whom they are meant to benefit."

While reconciliation can address the collective experience, the task of relationship-building is ultimately located within the individual—whether that is a particular political leader of high influence or a rank-and-file citizen whose life was deeply affected by violent conflict. What is possible to achieve at a person-to-person level is highly contextual and contingent on the nature of the conflict and the outworkings of the peace settlement. For some societies, the prospects for face-to-face engagement are possible, albeit still challenging and requiring skillful facilitation. In others, the nature of the conflict is such that the main protagonists are not in physical proximity to one another, and other, less direct—and arguably less optimal—contact would be required. In either case, individual processes focus on the attitudinal, behavioral, and emotional changes that take place and that may, or may not, contribute to wider societal reconciliation. A narrow view of reconciliation places the responsibility on those directly involved in or impacted by the conflict to reconcile. This responsibility has been accused of weighing too heavily on the shoulders of victims and the requirement that they should demonstrate "symbolic hope" (Simpson 2014, 7) to the wider society through acts of acceptance, forgiveness, and bridge-building with perpetrators. Victims can feel that they are trapped by the overarching rhetoric to reconcile as a forward-looking endeavor, while grappling with their own bitterness and feelings of injustice (Hamber 2009; Schaap 2008). Arguably, individual processes of reconciliation are too personal and private to be viewed as a necessary component of the wider peacebuilding imperative, and one should caution against the transference of the language of healing, forgiveness, hurt, pain, and trauma from the interpersonal to the broader national processes (Bloomfield 2006).

Conclusions

While reconciliation following violent conflict has been described as "going through a renaissance" in recent years (Cole 2014, 1), calls for greater clarity on its scope and nature persist. Although it has moved up the agenda of international bodies, including both the UN General Assembly (2015) and UN Development Program (2015), there is either a struggle to find a common language or an unspoken assumption of an agreed approach among the various actors involved. Three factors make conceptual precision difficult. First, the task of reconciliation has been described both as a necessary precursor to peace processes and as a key component of postaccord peacebuilding, arguably stretching its remit too widely to maintain coherence. Second, reconciliation will be experienced differently depending on the specific historical, political, and cultural context, and must be tailored accordingly if it is to maintain relevancy. This means that what is emphasized as reconciliation imperatives in one context may have little meaning in another. Third, reconciliation is paradoxical in nature (Lederach 1997; Hamber and Kelly 2005), and making gains in one area (such as pursuing truth-telling) may have at least short-term negative effects in another area (such as intergroup cohesion). Reconciliation therefore involves the calculation of risk in pursuing or emphasizing one approach over another and the constant recalibration and review of what works and in what context (Kelly and Braniff 2016). Additionally, as postconflict states are often unstable or weak, tensions may arise between the articulated ambitions for reconciliation and the realities of politics, developmental needs, and resources available to deliver on them. While relationship-building should be central to the delivery of reconciliation, attention should be paid to the quality of relationships that are being formed. Not all relationships are benign, and conflict entrepreneurs may seek to exploit weaknesses and opportunities to ensure that new, exclusionary forms of cohesion are created.

With a surprising deficit in empirically based studies of reconciliation in action, three future directions are proposed to address the significant gap that has developed between theory and practice (Brouneus 2007, 5). First, greater clarity is required as to which individual practical interventions work in support of reconciliation processes, in what combinations they might usefully be delivered, and at what stage—in parallel or sequentially? For example, should truth commissions come before the rewriting of historical textbooks? What are the specifics and differences of reconciliation processes on an interpersonal or intergroup level, and how do they relate to a broader national or international political level? Do symbolic acts of reconciliation by political leaders lead to perceptible attitudinal or behavioral change in the population, or vice versa? What categories of practices exist, and are some more effective than others? Can we evaluate the impacts of these practices at individual, societal, and national levels, and how can we establish causation?

Second, there is greater scope for attending to the reconciliatory aspects of political, economic, security, judicial, and social reform and assessing the impact that key decisions taken can have on the horizontal relationships between citizen, community, and state. Institution-building following conflict is often viewed as a technical peacebuilding exercise measurable by degrees of transformation over time. However, establishing new cultural norms of transparency and accountability within institutions and ensuring societal acceptance of and engagement with them will uncover opportunities to build relationships of respect and legitimacy and provide for their longer-term stability.

Third, there are opportunities to further examine and understand the specifics of reconciliation processes in a range of contexts, including those involving negotiated peace versus victor's peace, societies marked by significant power asymmetries, those where the state structures are contested or where the conflict has been frozen and requires continued international presence. This could include an examination of the consequences of exclusion or nonparticipation of individuals or groups in reconciliation processes and its impact on the peace sustainability.

One must caution against the perfunctory or superficial invocation of reconciliation without an examination of the specific contexts, needs, experiences, resources, and aspirations of the local population. To do so risks reconciliation becoming an anodyne concept that is easily invoked and instrumentalized because it is so loosely understood. That being said, policymakers, donors, and practitioners should exercise restraint in advocating for a maximalist conception of reconciliation where it can place undue pressures on a society already traumatized by conflict. If reconciliation is articulated as a long-term, incremental process rather than an unrealistic end-goal, then it has a greater chance of bringing both individuals and societies along with it. Reconciliation has the potential to become a demanding and emancipatory concept, but ultimately this will depend on the legitimacy, authenticity, and quality of the processes undertaken.

References

Aoun, J., and M.-J. Zahar. 2017. "Lebanon: Confessionalism, Consociationalism, and Social Cohesion." In *Peacebuilding in Deeply Divided Societies: Towards Social Cohesion*, edited by F. Cox and T. Sisk, 103–136. Cham, Switzerland: Palgrave Macmillan.

Amstutz, M. 2004. *The Healing of Nations: The Promise and Limits of Political Forgiveness.* Lanham, MD: Rowman & Littlefield.

Anstey, M., and V. Rosoux. 2017. "Introduction: Negotiating Perilous Relations. Quandaries of Reconciliation." In *Negotiating Reconciliation in Peacemaking: Quandaries of Relationship Building*, edited by V. Rosoux and M. Anstey, 1–14. London: Springer.

Arusha Peace and Reconciliation Agreement for Burundi. 2000. 28 August. http://peaceaccords.nd.edu/wp-content/accords/Arusha_Peace_Accord____.pdf.

Asmal, K., L. Asmal, and R. Suresh Roberts. 1997. *Reconciliation through Truth.* London: St. Martin's Press.

Assefa, H. 2005. "Reconciliation: Challenges, Responses, and the Role of Civil Society." In *People Building Peace II: Successful Stories of Civil Society*, edited by P. van Tongeren, M. Brenk, M. Hellema, and J. Verhoeven, 637–645. London: Lynne Rienner.

Barkan, E., and A. Karn, eds. 2006. *Taking Wrongs Seriously: Apologies and Reconciliations.* Stanford, CA: Stanford University Press.

Bar-Tal, D. 2009. "Reconciliation as a Foundation of Culture of Peace." In *Handbook on Building Cultures of Peace*, edited by J. de Rivera, 366–377. New York: Springer.

Bar-Tal, D., and G. Bennink. 2004. "The Nature of Reconciliation as an Outcome and as a Process" In *From Conflict Resolution to Reconciliation*, edited by Y. Bar-Siman-Tov, 11–39. Oxford: Oxford University Press.

Bashir, B. 2008. "Accommodating Historically Oppressed Social Groups: Deliberative Democracy and the Politics of Reconciliation." In *The Politics of Reconciliation in Multicultural Societies*, edited by W. Kymlicka and B. Bashir, 48–69. Oxford: Oxford University Press.

Bloomfield, D. 2006. *On Good Terms: Clarifying Reconciliation.* Berlin: Berghof Research Center for Constructive Conflict Management.

Bloomfield, D., T. Barnes, and L. Huyse, eds. 2003. *Reconciliation after Violent Conflict: A Handbook.* Stockholm: International IDEA.

Boege, V. 2019. "Reconciliations (Melanesian Style) and Transitional Justice." *Global Change, Peace & Security*, online, 22 January. doi:10–1080/14781158.2019.1563776.

Brouneus, K. 2007. *Reconciliation and Development.* Berlin: Friedrich Ebert Stiftung.

Brudholm, T. 2008. *Resentment's Virtue: Jean Amery and the Refusal to Forgive.* Philadelphia: Temple University Press.

Charbonneau, B., and G. Parent. 2012. *Peacebuilding, Memory and Reconciliation: Bridging Top-Down and Bottom-Up Approaches.* New York: Routledge.

Cochrane, F. 2001. "Unsung Heroes? The Role of Peace and Conflict Resolution Organizations in the Northern Ireland Conflict." In *Northern Ireland and the Divided World: Post-Agreement Northern Ireland in Comparative Perspective*, edited by J. McGarry, 137–156. Oxford: Oxford University Press.

Cole, E. 2014. "Reconciliation as a Peacebuilding Practice: New Questions, New Ideas." *USIP Insights*, Fall (2). https://www.usip.org/sites/default/files/Insights-Fall-2014.pdf.

Crocker, D. A. 2000. "Retribution and Reconciliation." *Philosophy and Public Policy Quarterly* 20 (1):1–6.

Daly, E., and J. Sarkin. 2007. *Reconciliation in Divided Societies.* Philadelphia: University of Pennsylvania Press.

Dwyer, S. 1999. "Reconciliation for Realists." *Ethics and International Affairs* 13 (1):81–98.

Galtung, J. 2001. "After Violence, Reconstruction, Reconciliation and Resolution." In *Reconciliation, Justice and Coexistence: Theory and Practice*, edited by M. Abu-Nimer, 3–23. Lanham, MD: Lexington Books.

Gibson, J. L. 2006. "Overcoming Apartheid: Can Truth Reconcile a Divided Nation?" *Annals of the American Academy of Political and Social Science* 603 (1):82–110.

Govier, T., and W. Verwoerd. 2002. "Trust and the Problem of National Reconciliation." *Philosophy of the Social Sciences* 32 (2):178–205. doi:10.1177/004931032002003.

Hamber, B. 2009. *Transforming Societies after Political Violence.* New York: Springer.

Hamber, B., and G. Kelly. 2005. *A Place for Reconciliation? Conflict and Locality in Northern Ireland.* Belfast: Democratic Dialogue.

Hamber, B., and G. Kelly. 2018. "*Northern Ireland: Case Study.*" In *Challenging the Conventional: Making Post-Violence Reconciliation Succeed*, edited by Kofi Annan Foundation and Interpeace, 86–130. Geneva: Kofi Annan Foundation and Interpeace.

Hayner, P. 2002. *Unspeakable Truths: Facing the Challenge of Truth Commissions.* New York: Routledge.

Hazan, P. 2009. "Reconciliation." In *Post-Conflict Peacebuilding: A Lexicon*, edited by V. Chetail, 256–267. Oxford: Oxford University Press.

He, Y. 2009. *The Search for Reconciliation: Sino-Japanese and German-Polish Relations since World War II.* New York: Cambridge University Press.

Hirsch, A. K., ed. 2012. *Theorizing Post-Conflict Reconciliation: Agonism, Restitution and Repair.* Oxford: Routledge.

Hutchison, E., and R. Bleiker. 2015. "Reconciliation." In *Routledge Handbook of Peacebuilding*, edited by R. Mac Ginty, 81–89. Oxford: Routledge.

Jeffery, R. 2013. "Enduring Tensions: Transitional Justice in the Solomon Islands." *Pacific Review* 26 (2):153–175.

Kelly, G., and M. Braniff. 2016. "A Dearth of Evidence: Tackling Division and Building Relationships in Northern Ireland." *International Peacekeeping* 23 (3):442–467.

Kelman, H. C. 2004. "Reconciliation as Identity Change: A Social Psychological Perspective." In *From Conflict Resolution to Reconciliation*, edited by Y. Bar-Siman-Tov, 111–124. Oxford: Oxford University Press.

Kelman, H. C. 2010. "Conflict Resolution and Reconciliation: A Socio-Psychological Perspective on Ending Violent Conflict between Identity Groups." *Landscapes of Violence* 1 (1):1–9.

Kriesberg, L. 2004. "Comparing Reconciliation Actions within and between Countries." In *From Conflict Resolution to Reconciliation*, edited by Y. Bar-Siman-Tov, 81–110. Oxford: Oxford University Press.

Lambourne, W. 2004. "Post-Conflict Peacebuilding: Meeting Human Needs for Justice and Reconciliation." *Peace, Conflict and Development* 1 (4):1–24.

Lederach, J. P. 1997. *Building Peace: Sustainable Reconciliation in Divided Societies.* Washington, DC: US Institute of Peace.

Lederach, J. P. 2001. "Civil Society and Reconciliation." In *Turbulent Peace: The Challenges of Managing International Conflict*, edited by C. A. Crocker, F. O. Hampson, and P. Aall, 841–854. Washington, DC: USIP.

Lewer, N. 2017. "Tensions between Short Term Outcomes and Long Term Peacebuilding in Post-War Sri Lanka." In *Negotiating Reconciliation in Peacemaking*, edited by V. Rosoux and M. Anstey, 277–304. London: Springer.

Little, A. 2017. "Fear, Hope and Disappointment: Emotions in the Politics of Reconciliation and Conflict Transformation." *International Political Science Review* 38 (2):200–212. doi:10.1177/0192512115590635.

Long, W. J., and P. B. Brecke. 2003. *War and Reconciliation: Reason and Emotion in Conflict Resolution.* Cambridge, MA: MIT Press.

McCartney, C. 1999. "The Role of Civil Society." In *Striking a Balance: The Northern Ireland Peace Process*, edited by Conciliation Resources, 44–49. London: Conciliation Resources.

Minow, M. 1998. *Between Vengeance and Forgiveness: Facing History after Genocide and Mass Violence.* Boston: Beacon Press.

Moon, C. 2008. *Narrating Political Reconciliation: South Africa's Truth and Reconciliation Commission.* Lanham, MD: Lexington Books/Rowman & Littlefield.

Murphy, C. 2010. *A Moral Theory of Political Reconciliation*. Cambridge: Cambridge University Press.

Nagy, R. 2002. "Reconciliation in Post-Commission South Africa: Thick and Thin Accounts of Solidarity." *Canadian Journal of Political Science/Revue canadienne de science politique* 35 (2):323–346.

Paffenholz, T. 2010. *Civil Society and Peacebuilding: A Critical Assessment*. Boulder, CO: Lynne Rienner.

Pankhurst, D. 1999. "Issues of Justice and Reconciliation in Complex Political Emergencies: Conceptualising Reconciliation, Justice and Peace." *Third World Quarterly* 20 (1):239–255.

Quinn J. R. 2017. "*Kastom* in Dispute Resolution: Transitional Justice and Customary Law in the Solomon Islands." In *Transitional Justice in Practice*, edited by R. Jeffery, 63–84. New York: Palgrave Macmillan.

Rosoux, V. 2017. "Reconciliation as a Puzzle: Walking among Definitions." In *Negotiating Reconciliation in Peacemaking: Quandaries of Relationship Building*, edited by V. Rosoux and M. Anstey, 15–26. London: Springer.

Ruwanpura, K. N. 2016. "Post-war Sri Lanka: State, Capital and Labour, and the Politics of Reconciliation." *Contemporary South Asia* 24 (4):351–359. doi:10.1080/09584935.2016.1240756.

Schaap, A. 2008. "Reconciliation as Ideology and Politics." *Constellations* 15 (2):249–264. doi:10.1111/j.1467-8675.2008.00488.x.

Simpson, G. 2014. "Reconciliation beyond Conceptual Debates." *USIP Insights*, Fall (2). https://www.usip.org/sites/default/files/Insights-Fall-2014.pdf.

Stover, E. 2004. "Witnesses and the Promise of Justice in The Hague." In *My Neighbor, My Enemy: Justice and Community in the Aftermath of Mass Atrocity*, edited by E. Stover and H. M. Weinstein, 104–120. Cambridge: Cambridge University Press.

Stover, E., and H. M. Weinstein, eds. 2004. *My Neighbor, My Enemy: Justice and Community in the Aftermath of Mass Atrocity*. Cambridge: Cambridge University Press.

UN Development Program. 2015. *Reconciliation as Framework for Preventing Conflict and Sustaining Peace*. July. http://www.ijr.org.za/home/wp-content/uploads/2012/07/Policy-Document-Inclusive-Reconciliation-Final-July-2015.pdf.

UN General Assembly. 2015. *High-Level Thematic Debate of the United Nations General Assembly Promoting Tolerance and Reconciliation: Fostering Peaceful, Inclusive Societies and Countering Violent Extremism*. 21–22 April. https://www.un.org/pga/wp-content/uploads/sites/3/2015/03/160315_hltd-promoting-tolerance-reconciliation-concept-note.pdf.

UN Support Mission in Libya. 2015. *Libyan Political Agreement*. 17 December. https://unsmil.unmissions.org/sites/default/files/Libyan%20Political%20Agreement%20-%20ENG%20.pdf.

Vandeginste, S. 2017. "Burundi's Bumpy Road to Reconciliation without Truth." In *Negotiating Reconciliation in Peacemaking: Quandaries of Relationship Building*, edited by V. Rosoux and M. Anstey, 185–211. London: Springer.

CHAPTER 36

RELIGION AND PEACEBUILDING

JOHN D. BREWER

We know a great deal about what happens to politics when it is infused with religious belief (for example, Norris and Inglehart 2004), and the dynamics of religious extremism are well charted (for example, Juergensmeyer 2000). Religious extremism is rarely conducted under a fairness rule that respects the human dignity of others and treats unbelievers as moral beings. This form of religion is referred to by Appleby (2015, 45–47) as "pathological religion." It is possible to tolerate "true religion" when the limits of intolerance are defined by this fairness rule and religious believers do not morally enervate others and deny their humanity. Religion is not, therefore, inevitably extremist; the three Abrahamic faiths of Christianity, Judaism, and Islam give considerable doctrinal support for peace, justice, and reconciliation. It is more often the case that political extremists use religion for political purposes, giving their conflicts a religious hue even when the substance of the conflict is deeply political, because "true religion" easily mobilizes zealous adherents.

Religious extremism is thus mostly a surrogate for political, social, and ethnic conflicts that are not disputes over religious texts, rituals, symbols, or practices. Religious violence is never just about doctrine; it is primarily about political, cultural, and patriarchal power (Walsham 2006). Religion occasions and facilitates violence; it rarely causes it. It is for this reason that Karen Armstrong (2014) refers to religious violence as a myth; what is mythological is not the violence but that it is religious in substance.

There are, however, some examples where religious resources and religious symbols are central to the conflict. It is difficult to deny the religious form to the Islamic State's political violence, or that of Afrikaners in apartheid South Africa, Ulster Protestants, or Jews in the occupied settlements, where "chosen people" status ties them religiously and politically to the land (Akenson 1992; Brewer 2003). In these cases, the conflict is both political *and* religious, for religion maps onto and represents real material and political differences. The conflict is over the legitimacy of the state and access to its political, economic, and cultural resources, but religious affiliation defines the boundaries of the groups that are in competition. Religion provides some of the cultural resources for drawing moral boundaries between the ethnic groups in political competition; it facilitates ethnic and cultural "othering"; religious symbols become associated with political contestation; and faith-based organizations take sides in the war. Even if the religious affiliations of protagonists no longer have strong theological meaning for most people, cultural religion can survive as a relic of a former religious tradition despite changed political circumstances, giving a religious hue to the battle lines. It is for this reason that some conflicts are *experienced* as religious in their form, even though the substance of the conflict is thoroughly political. Academic disagreements about the extent to which Northern Ireland's conflict is religious revolve around this confusion (for example, Barnes 2005; Mitchell 2006).

However, I want to ask a different question: What happens to *religion* when it elides and overlaps with politics? The answers to this question significantly shape the central theme of this chapter, since religious peacebuilding is constrained and limited by the distorting effects that politics has on religion whenever the two elide. In these circumstances, religion is mostly seen as part of the problem, implicated in the abuse of human rights, especially in transitional societies that are still in, or are emerging out of, conflict (Anderson 2003).

Politicized Religion

Analysts have long overlooked the impact of politics on religion, preferring instead to document the impact of religion on politics. I suggest there are several problems for religious belief, practice, and observance when politics and religion correspond. First, there are hermeneutical problems affecting the interpretation of religious texts. Doctrine is given a political interpretation, and politics shapes the canonical emphases. There is also doctrinal excision, whereby politically inexpedient texts are omitted from mention. In times of war, the words of the gods of peace tend to be neglected. Hermeneutics can even affect hymnody, with hymn choice and worship songs reflecting political preferences. For example, the English Christian hymn "I Vow to Thee My Country," sung to the jingoistic tune from Holst's "Jupiter" from *The Planets* suite, falls in and out of favor according to patriotic fervor.

Second, religious belief and practice are politicized. Religion is mobilized to provide the meaning, motive, and moral justification for political action, giving theological and scriptural bases to courses of political action. For example, scriptural support for apartheid, slavery, and anti-Catholicism (on the last, see Brewer and Higgins 1998) formed part of their respective political contestations. This affects, third, the meaning of religion and religious identification. Faith commitment is wrapped up with political loyalty as a form of cultural religion rather than personal faith. Cultural religion and personal religiosity divide; cultural religion implicates an ethnoreligious identity that carries no obligation to practice; religion becomes as much an expression of cultural and political identity as personal faith. It is the political and constitutional content of cultural religion that matters more to most Northern Irish Protestants, for example, than theology.

A fourth problem is that religion fragments. The chaos of political diversity affects the unity of religion as it fragments under denominational differences and doctrinal schisms, which can represent political as much as religious differences. The different political positions that evangelical churches in the Deep South of the USA take on Islam is a good example. Sunni and Shia differences within Islam reflect nation-state allegiances as much as religious disputes. Differences within Judaism often map onto political differences in Israeli society, particularly over whether or not the protection of human rights is an ancient Talmudic principle, a dispute that is critical for the support some Jews give to Palestinian rights.

A fifth problem is that religion narrows in the search for religious and political authenticity. Fragmentation can cohere around narrower and ever more exclusive forms of religious belief and political position, with religious belief and behavior becoming stricter and the band of "true believers" smaller. Liberalization and secularization have lost religion its monopoly of moral discourse, and fragmentation can result in very exclusive forms of faithfulness and "truth." This is reinforced when the religiously faithful retreat also into narrow political positions. The persistence of religious sects and cults in late modernity is as much for political reasons as matters of faith (Sedgwick 2002), and the political beliefs of sects separate them as much as does their religion, notably Quakers.

A sixth problem emerges in situations of religious plurality (on which, see Banchoff 2008). Where different "true religions" offer competing versions of religious truth, religious diversity is difficult to manage with the usual strategy of the privatization of religion, its separation from public life and retreat into the domestic sphere. In situations of religious plurality where, for political reasons, the elision between religion and culture is also strong and religious observance and identification remain high, it is almost impossible to promote the privatization of religion. Equality of religious practice and belief is easier to enact in settings where religion does not matter, or not enough to want believers to have their beliefs count in public affairs. Cultural religion, however, politicizes religion in the public sphere and makes religious plurality politically problematic. Cultural religion can, for example, proliferate hate crimes linked to religion in settings of very marked religious diversity. Exclusive Buddhist nationalism in Myanmar, for example, fuels religious hate crimes against the Muslim Rohingya, as does Sinhalese Buddhist nationalism against Christians, Hindus, and Muslims in Sri Lanka.

A seventh problem is that religious spaces become divided and partisan: some religious spaces are out of bounds and are not seen as neutral and above the fray. Entering each other's religious spaces can be seen as taking the "road leading to hell." This can limit the capacity for ecumenical and interfaith dialogue, or at least force such dialogue into secular spaces, denuding it of some of its religious form. Finally, religion can become an obstacle, not a facilitator, to reconciliation, seen as part of the problem, not part of the solution (for Northern Ireland, see Brewer, Higgins, and Teeney 2012), with religiously motivated fear, anxiety, and hatred inhibiting what Hutchinson and Bleiker (2013) refer to as religion's capacity for "emotional reconciliation."

In short, politics corrupts religion when they overlap and elide. Because of this distortion, it is more difficult for religion to be a site of reconciliation during and after conflict. This constrains and limits the potential for religious peacebuilding.

Religious Peacebuilding

The literature on religious peacebuilding has expanded rapidly in the past two decades in part because of the increase in religiously motivated violence in the Balkan wars and the rise of Islamic extremism (for overviews of the field, see Abu-Nimer 2013; Brewer, Higgins, and Teeney 2010, 2012; Coward and Smith 2004; Omar, Appleby, and Little 2015; Schlack 2009). Much of this literature is from the USA because it is particularly suited as a cultural space for this kind of work. There is a plurality of religions in the US as part of its racial and ethnic mix, but prior to 9/11 the country never witnessed a holy war and thus had no historical memory of religious hatred and violence of the kind that affects most of Europe. Its separation of church and state ensures no one religion has become the established faith and been accorded privileged political status. It is also a society where religious practice remains high, against the trend toward secularization in the West, which encourages people to take religion seriously. In the US, where church and state are separated, there is an implicit requirement for public figures to articulate their private religious views, resulting in an easy penetration of the public sphere with politicians' private religious beliefs: religious enlightenment is taken to be the right *to* believe and for those beliefs to count in public affairs. In Europe, church and state are closer, with many state religions, and religious enlightenment is taken to be the right *not* to believe or to believe something entirely unorthodox. As a result, the US is a society in which religion is recognized as a rich resource in politics, part of political diplomacy (for example, see Johnson 2003).

What is more, the increasing importance of religion in post–Cold War conflicts has affected US foreign and domestic policy to the point where the focus on religion as a site for reconciliation became the reverse face of the attention on religion as a site of contestation. This is captured in the title of a pioneering text in this genre by Scott Appleby (2000), *The Ambivalence of the Sacred*, by which he refers to religion's capacity for peacemaking within the midst of its warmongering. Religious peacebuilding became part of a US interest in procuring global political stability.

The repositioning of research to address this need was sudden and dramatic. The US Institute for Peace established a research program on religious peacebuilding in the mid-1990s, which has been prolific (see Smock 2001, 2002, 2006, 2008); so did the Woodrow Wilson International Center for Scholars. The US Institute for Global Engagement established a "religion and security" initiative in 2003 designed to explore the intersection between religion and political stability. New journals began to appear, such as the

Review of Faith and International Affairs in 2002, and research centers and institutes sprang up to capture this zeitgeist, whose personnel have published the leading texts in this new intellectual field. Three can be mentioned for illustration. The Center for World Religions, Diplomacy and Conflict Resolution at George Mason University in Washington, DC, is run by Marc Gopin (2002, 2012), an expert on the role that religion and culture play in conflict resolution, particularly in the Middle East. The International Peace and Conflict Reconciliation Program at the American University in Washington, DC, is led by Mohammed Abu-Nimer (2013), an international expert on interfaith dialogue and Muslim contributions to conflict resolution. The Yale Center for Faith and Culture is directed by Miroslav Volf (1996, 2006), author of several pioneering works on faith and the problems of reconciliation following mass violence.

Religious peacebuilding in the US is distinguished by three defining characteristics: (1) an emphasis on interfaith dialogue as the primary form of religious peacebuilding, (2) the commensurate privileging of ecumenism as the chief peace strategy in religious peacebuilding, and (3) the use of the case study method. It also tends to concentrate on positive cases, situations where religious bodies, para-church organizations, and faith-based NGOs bring warring factions together and where religion is above the fray and considered neutral so that religious actors have legitimacy as peacemakers (for example, see Cejka and Bamat 2003; Johnson 2003; Smock 2006). They proffer religion as a particularly significant factor in peace when one or more parties to the conflict have strong religious identities and when religious leaders come from all sides of the dispute. Churches are portrayed as trusted institutions, respected for their values, having moral warrants in opposing injustice and unique leverage in promoting reconciliation, and possessing the capability to mobilize and follow through locally in the wake of political agreement. Smock (2001) listed several key roles for faith-based NGOs in peace processes, among them training, mediation, conflict prevention, dialogue, and reconciliation.

There are some successful examples of religious peacebuilding, such as the short-lived peace deal in Sudan in 1972 and the stable settlement in Mozambique in 1992, but there is no conceptual apparatus that focuses on the more numerous cases where religion is part of the problem arising from what earlier I described as "politicized religion." Brewer, Higgins, and Teeney (2010, 2012), for example, dispute that these claims apply where religion is involved in the political contestation or elides with ethnicity, "race," and structural cleavage, when churches take sides and associate with specific parties to the conflict and form part of regressive civil society, where churches and para-church organizations uphold failed or failing states, or when cultural religion supersedes personal religious practice so that the parties bring no specifically religious virtues to the negotiations.

The intellectual focus has also largely been on praxis, not theory. Brewer, Higgins, and Teeney (2010, 2012) proffer one such theoretical model built up from the Northern Ireland case that has used by Steen-Johnsen (2017) and applied to several East African countries. There is now a need for scholars to theorize the field and to chart where, when, and how religious peacebuilding can work or not.

For example, secret and confidential back-channel dialogue was the primary contribution of religious peacebuilders in Northern Ireland, bringing political and militant groups together who could not be seen to be talking in public. Sacred spaces were admirably suited to this private dialogue because of the great respect in which the churches were held and because sacred spaces exude redemption, signify confidentiality and secrecy, and encourage the moderation of language and position. Ironically the Good Friday Agreement in 1998 made this role superfluous, and churches have not been able to find a role for themselves in dealing with the legacy issues in a society learning to live together. This contrasts markedly with the role of religious peacebuilding in South Africa, where the churches' more high profile antiapartheid activities in the public sphere allowed religious peacebuilders to recoup legitimacy in order to help shape the postapartheid reconciliation process. This was most notably the work of Archbishop Desmond Tutu on the Truth and Reconciliation Commission (see Shore 2009), a role some critics have claimed was too religiously focused on extracting forgiveness from Black apartheid victims (Brudholm 2008).

Religious peacebuilders are not on the whole professionally trained conflict mediators. There are some faith-based charities (for example, Christian Aid and Trocaire) and humanitarian aid agencies with links to religious traditions (for example, the Red Cross and the Red Crescent Movement), with highly trained specialists, but theological training and ordination rarely equip journeying prelates, pastors, and priests with the skill set to know how or when to intervene in local conflict disputes or to know what strategies to adopt to facilitate peacebuilding. This is why religious peacebuilding is largely reduced to interfaith dialogue and ecumenism; they can better understand the need for communication with people of other faiths like themselves and how to conduct it. Their role in conflict transformation is thus very limited. However, religious peacebuilders are more skilled in what elsewhere I refer to as social transformation (on the distinction between conflict transformation and social transformation, see Brewer 2015); the requirement to address the emotional legacy of conflict, such as the practice of forgiveness, mercy, emotional empathy, and reconciliation; and to introduce social justice, equality of opportunity, and fairness. These should be areas of religious expertise. It is for this reason that religious peacebuilding takes on a higher profile once the violence ends and society emerges out of conflict to begin to learn how to live together in tolerance.

Religion and Transitional Justice

This realization is behind a further significant realignment in the intellectual landscape of religious peacebuilding, namely, the interest in religious contributions to transitional justice. Transitional justice studies is in an interdisciplinary space, a meeting place for lawyers, criminologists, sociologists, moral philosophers, and theologians, among others, whereas peacebuilding is still commanded over by international relations and political science (see Brewer 2010, 2013 for a sociological entry into the field; see Mac Ginty 2013, 117–194 for other disciplinary perspectives). Thus contributions to transitional justice are much more disciplinarily diverse. One of the first discussions of political forgiveness, for example, was by a theologian (Shriver 2009). Some social science debates about forgiveness are couched within a Christian perspective (see Amstutz 2004) or use specific scriptural texts to flesh out the meaning of political forgiveness (for example, Satha-Anand 2002). Collaborative research between a sociologist and a theologian has shown that high levels of forgiveness among victims in Sri Lanka can be located in the religious resources that the country's many world religions afford (see Wijesinghe and Brewer 2018).

There are many ways in which this focus on religious contributions to transitional justice is manifest. Ex-combatants' social reintegration after conflict, for example, forms one of many debates in transitional justice concerning perpetrators, others of which include amnesty, reparations, and apologies. There has been work done on the religious beliefs of ex-combatants and the extent to which these beliefs encourage the transition from a military to a political strategy and facilitate subsequent involvement in the peace process (see Brewer, Mitchell, and Leavey 2013, 2016). There is now a volume of writing on the religious roots to restorative justice (Hadley 2001). Attention has been given to the contribution of religion to truth recovery and memory (for example, Hayes and Tombs, 2001), religion's role in managing the aftermath of violence and mass atrocity (Brudholm and Cushman 2009; Rios 2015) and to religious bases to the political ethic of reconciliation (Philpott 2012). In a series of writings, Daniel Philpott, the US-based political scientist working at the University of Notre Dame, which has both a Catholic ethos and considerable expertise in international peace studies, has championed religious contributions to transitional justice (for a selection, see Philpott, 2006, 2007a, 2007b, 2009, 2012).

In an attempt to delineate and classify the field of religious-based transitional justice, Brewer, Mitchell, and Leavey (2013, 2016) draw attention to some key distinctions (what follows is taken from Brewer, Mitchell, and Leavey 2016). To begin to understand the role of religion in transitional justice we need to distinguish two aspects to the relationship: the kind of religious input and the type of transitional justice intervention. We can distinguish these briefly as the foundation blocks to conceptual clarification.

In terms of religious input, this can vary significantly. One way to distinguish religious inputs is to differentiate in terms of religious values, doctrine, organization, and personnel. Religious input can thus involve merely the use of vague religious principles and rhetoric as mobilization and legitimation strategies. Examples of this are Christian love used as a pillar for reconciliation between victim and perpetrator (notably Ure 2008), the search for the spiritual roots to restorative justice, or the use of Christian values to motivate truth recovery (notably Tutu's Christian approach to his chairing of the South African Truth and Reconciliation Commission). The various activities of religious and faith-based NGOs, churches, and para-church organizations in peacebuilding and reconciliation constitute one of the more obvious forms of religious input, emphasized strongly in Smock's (2001) special report on faith-based NGOs. Input can also be more individualized, simply referring to the activities of key religious personnel, acting as particular role models of reconciliation, forgiveness, or hope, a number of which are collated by Little (2007) and Gopin (2012), or when acting in concert as a significant collective group, as emphasized in Brewer, Higgins, and Teeney's (2012) analysis of key church figures in Northern Ireland's peace process. Another form of religious input is interfaith dialogue to assist in relations between the three Abrahamic faiths, where various forms of doctrine in each are explored for their common Abrahamic core as the basis for dialogue.

Some of these forms of religious input utilize the special expertise of religious actors or are exclusively religious in content, but others put religious actors and resources as only one among many secular possibilities, to be used as best fit the requirements. In this latter regard, religious inputs can compete with secular ones, and religious values and doctrine compete with moral frameworks that are secular. In some inputs, therefore, we see religion working to its strengths, in areas where there is relevance, expertise, useful contacts, powerful leverage, and experience; in others it confuses and confounds the problem, perhaps with religious actors and NGOs simply getting in the way as well-meaning but naïve amateurs. This suggests that religious inputs need to be carefully and strategically planned to see where they can best make a difference, rather than just thrown in out of a wish for religious actors to be seen to do something. Where religious inputs are appropriately placed, it follows there should be no hostility from competing contributors to the intrusion of religion.

The key problems around religious inputs are therefore twofold: determining which religious inputs are most appropriate and, where they are thought to be so, trying to persuade the competing contributors not to go it alone but to develop fruitful collaboration between religious and nonreligious inputs.

Distinguishing between kinds of religious input is not the only way to unpack the relationship between religion and transitional justice; another is by the kind of transitional justice intervention in which religion gets involved. Interventions vary, such as truth recovery, managing the past, new forms of memory work, assisting in the social reintegration of ex-combatants, work with victims and the like, reparations, and so on. We might refer to these as modes of transitional justice intervention. In one sense religion might be thought of as a synthesizing and encompassing "sacred canopy" hanging over all of society, competent to work in every field of transitional justice, from peacemaking to prisons. The pioneering ambitions of Philpott veer toward this position. He sees a very broad role for religion in transitional justice, mentioning reconciliation, forgiveness, truth commissions, trials, reparations, apologies, trauma relief, conflict resolution, and peacebuilding (see especially, Philpott 2007b).

However, the notion that religion is a sacred canopy with the competence to intervene in all walks of life is not a view that can be sustained in late modernity. Brewer's (2010, 60) approach to the role of religion in transitional justice, for example, is more circumspect, seeing particular areas of the social peace process where religion is better suited than in other areas, particularly in respect to truth recovery and work with prisoners (see, for example, Brewer, Higgins, and Teeney 2012, 64–67). The role of the churches in "local" bottom-up truth recovery processes, for example, is especially noteworthy as alternatives to state-led, top-down processes (see, for example, Rios 2015), which are often in the control of the former regime, as occurred frequently in Latin America. This happened, of course, precisely in societies where the Catholic Church was the dominant and largest civil society group, and the only civil society group with the authority, legitimacy, and relative political safety to conduct rival truth recovery procedures in opposition to the state. Religion was not an input into truth recovery in Argentina or Rwanda (as noted by Philpott 2007b, 104–105). In the latter case this is because religious actors were eager participants in the slaughter, leaving the churches and religious-based NGOs with no credibility in calling for truth recovery afterward and much to hide from any truth recovery process. The context of the intervention is therefore critical to the feasibility of a religious input in particular modes of intervention; clearly, in addition to its obvious social structural dimensions, "context" here includes the nature and history of the conflict itself.

This brief discussion suggests three variables are critical to understanding the potential contribution of religion to transitional justice: the type of religious input, the mode of intervention, and the context to the relationship, as represented in Figure 36.1.

This conceptual mapping implies that religion has a role in transitional justice when—and only when—the relevant types of input can be distinguished, the appropriate forms of input identified, and the best modes of intervention specified, *for the context* in which it is occurring. This schema suggests there will be contexts in which religious inputs ought to occur when they currently do not, and contexts where the input is inappropriate or ought not to occur at all. Even enthusiasts who claim religious inputs are feasible in all modes of transitional justice and in all contexts still ought to have the realism to caution religious actors to garner the information to understand which kind of religious input is needed and what is the best mode of intervention to make for the specific context if it is not to backfire or prove counterproductive.

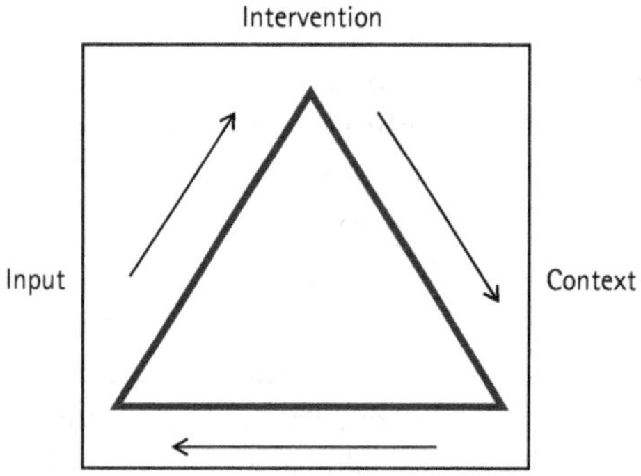

FIGURE 36.1 Religion and transitional justice.

Conclusion

Negotiating across religious differences through interfaith dialogue is a viable form of peacebuilding when religious actors are neutral or when the conflict is about religion, but when the conflict is political or both political and religious, politicized religion blurs the distinction between cultural religion and personal religiosity. Ethnonational group identities can be shaped in part by religion but render very little respect for or practice of religious virtues that promote peacebuilding. In such circumstances all peacebuilders, let alone religious peacebuilders, have more difficult tasks because of the distorting effects of politics on religion. When religion and politics elide, it is difficult for religious peacebuilders to contribute to conflict transformation and the cessation of violence. They may moderate the level of violence, such that the conflict would have been more heinous but for their efforts, but ending it is not primarily their expertise. A far more significant contribution is to social transformation once the violence has stopped. Learning to live together in tolerance once violence has moderated draws on virtues in which religion has expertise, such that religious contributions to transitional justice can be very positive.

References

Abu-Nimar, M. 2013. "Religion and Peacebuilding." In *Routledge Handbook of Peacebuilding*, edited by R. Mac Ginty, 69–80. London: Routledge.

Akenson, D. 1992. *God's People*. London: Cornell University Press.

Amstutz, M. 2004. *The Healing of Nations*. Lanham, MD: Rowman and Littlefield.

Anderson, J. 2003. *Religious Liberty in Transitional Societies*. Cambridge: Cambridge University Press.

Appleby, S. 2000. *The Ambivalence of the Sacred*. Lanham, MD: Rowman and Littlefield.

Appleby, S. 2015. "Religious Violence: The Strong, Weak and Pathological." In *The Oxford Handbook of Religion, Conflict and Peacebuilding*, edited by A. Omar, S. Appleby, and D. Little, 33–60. Oxford: Oxford University Press.

Armstrong, K. 2014. *Fields of Blood*. London: Bodley Head.

Banchoff, T. 2008. *Religious Pluralism*. Oxford: Oxford University Press.

Barnes, L. P. 2005. "Was the Northern Irish Conflict Religious?" *Journal of Contemporary Religion* 20:55–69.

Brewer, J. 2003. "Contesting Ulster." In *Homelands*, edited by R. Robin and B. Strath, 283–304. Brussels: Peter Lang.

Brewer, J. 2010. *Peace Processes: A Sociological Approach*. Cambridge, UK: Polity Press.

Brewer, J. 2013. "Sociology and Peacebuilding." In *Routledge Handbook of Peacebuilding*, edited by R. Mac Ginty, 159–170. London: Routledge.

Brewer, J. 2015. "Peace Processes." In *International Encyclopedia of Social and Behavioral Sciences*, 2nd ed., edited by James Wright, 648–653. Oxford: Elsevier.

Brewer, J., and G. Higgins. 1998. *Anti-Catholicism in Northern Ireland 1600–1998*. Basingstoke, UK: Macmillan.

Brewer, J., G. Higgins, and F. Teeney. 2010. "Religious Peacemaking: A Conceptualisation." *Sociology* 44:1019–1037.

Brewer, J., G. Higgins, and F. Teeney. 2012. *Religion, Civil Society and Peace in Northern Ireland*. Oxford: Oxford University Press.

Brewer, J., D. Mitchell, and G. Leavey. 2013. *Ex-Combatants, Religion and Peace in Northern Ireland*. London: Palgrave.

Brewer, J., D. Mitchell, and G. Leavey. 2016. "Northern Ireland: Religion and Transitional Justice." *Conflict Studies Quarterly* 14:74–91.

Brudholm, T. 2008. *Resentment's Virtue*. Philadelphia: Temple University Press.

Brudholm, T., and T. Cushman. 2009. *The Religious in Responses to Mass Atrocity*. Cambridge: Cambridge University Press.

Cejka, M., and T. Bamat. 2003. *Artisans of Peace*. Maryknoll, NY: Orbis Books.

Coward, H., and G. Smith. 2004. *Religion and Peacebuilding*. New York: SUNY Press.

Gopin, M. 2002. *Holy War, Holy Peace*. Oxford: Oxford University Press.

Gopin, M. 2012. *Bridges across an Impossible Divide*. Oxford: Oxford University Press.

Hadley, M. 2001. *The Spiritual Roots of Restorative Justice*. Albany, NY: SUNY Press.

Hayes, M. A., and D. Tombs. 2001. *Truth and Memory*. Leominster, UK: Gracewing.

Hutchinson, E., and R. Bleiker. 2013. "Reconciliation." In *Routledge Handbook of Peacebuilding*, edited by R. Mac Ginty, 81–90. London: Routledge.

Johnson, D. 2003. *Faith-Based Diplomacy*. Oxford: Oxford University Press.

Juergensmeyer, M. 2000. *Terror in the Mind of God*. Berkeley: University of California Press.

Little, D. 2007. *Peacemakers in Action*. Cambridge: Cambridge University Press.

Mac Ginty, R., ed. 2013. *Routledge Handbook of Peacebuilding*. London: Routledge.

Mitchell, C. 2006. "The Religious Content of Ethnic Identities." *Sociology* 40:1135–1152.

Norris, P., and R. Inglehart. 2004. *Sacred and Secular*. Cambridge: Cambridge University Press.

Omar, A., S. Appleby, and D. Little, eds. 2015. *The Oxford Handbook of Religion, Conflict and Peacebuilding*. Oxford: Oxford University Press.

Philpott, D. 2006. *The Politics of Past Evil*. Notre Dame, IN: University of Notre Dame Press.

Philpott, D. 2007a. "Explaining the Political Ambivalence of Religion." *American Political Science Review* 103:516–528.

Philpott, D. 2007b. "What Religion Brings to Restorative Justice." *Journal of International Affairs* 61:93–110.

Philpott, D. 2009. "When Faith Meets History: The Influence of Religion in Transitional Justice." In *The Religious in Responses to Mass Atrocity: Interdisciplinary Perspectives*, edited by T. Brudholm and T. Cushman, 174–212. Cambridge: Cambridge University Press.

Philpott, D. 2012. *Just and Unjust Peace*. Oxford: Oxford University Press.

Rios, S. 2015. *Religion, Social Memory and Conflict*. London: Palgrave.

Satha-Anand, C. 2002. "Forgiveness in South East Asia." *Global Change, Peace and Security* 22:235–247.

Schlack, A. 2009. *The Role of Religion in Peacebuilding and Conflict Resolution*. Saarbrucken: VDM.

Sedgwick, M. 2002. "Sects and Politics." *Nova Religio* 6:165–173.

Shore, M. 2009. *Religion and Conflict Resolution*. Farnham, UK: Ashgate.

Shriver, D. 2009. *An Ethic for Enemies*. Oxford: Oxford University Press.

Smock, D. 2001. *Faith Based NGOs and International Peacebuilding*. Washington, DC: US Institute of Peace Press.

Smock, D. 2002. *Interfaith Dialogue and Peacebuilding*. Washington, DC: US Institute of Peace Press.

Smock, D. 2006. *Religious Contributions to Peacemaking*. Washington, DC: US Institute of Peace Press.

Smock, D. 2008. *Religion in World Affairs*. Washington, DC: US Institute of Peace Press.

Steen-Johnsen, T. 2017. *State and Politics in Religious Peacebuilding*. London: Palgrave.

Ure, M. 2008. "Introduction." *European Journal of Social Theory* 11:283–298.

Volf, M. 1996. *Exclusion and Embrace*. Nashville, TN: Abingdon Press.

Volf, M. 2006. *The End of Memory*. Grand Rapids, MI: Eerdmans.

Walsham, A. 2006. *Charitable Hatred*. Manchester: Manchester University Press.

Wijesinghe, S., and J. Brewer. 2018. "Peace Religiosity and Forgiveness among War Veterans in Sri Lanka." In *The Sociology of Compromise after Conflict*, edited by J. Brewer, B. Hayes, and F. Teeney, 129–156. London: Palgrave.

CHAPTER 37

FOREIGN AID AND PEACEBUILDING

RACHEL M. GISSELQUIST

FOREIGN aid is a core component of peacebuilding and among the largest external financial flows to fragile states and conflict-affected areas.[1] Nevertheless, troubling critiques have been raised in the literature about its overall impact and effectiveness. Questions are raised not only about whether programs have achieved their core aims but also about whether they have succeeded at the most basic level to do no harm. This chapter considers the contributions of aid to peacebuilding alongside key critiques. While the response from some has been to call for sharp reductions in foreign aid, a starting point of this chapter is that such a response is ill-conceived: critiques notwithstanding, aid has played a positive role in many countries, and improving foreign aid for peacebuilding is a clear priority for global development. Estimates are that almost 25% of the world's population lives in fragile contexts, that 80% of the world's poorest people may live in fragile states by 2030, and that their needs are unlikely to be met with domestic resources and other financial flows alone (OECD 2018). The core challenge at hand is not in further cataloguing aid's failures but in building a base of knowledge that can inform better aid interventions. This implies continuing study not only of which aid interventions have failed but also of those that have worked and the conditions under which success can be replicated or serve as a model elsewhere (see Autesserre 2017; Gisselquist 2015). As explored here, the extant research literature tells a great deal about the precise impacts of particular interventions in particular contexts and the aggregate impact of aid across multiple countries. It offers rich insights about what has and has not worked but leaves major gaps about why, how, and whether findings are transferrable from one context to another. Moreover, even if knowledge is built from multiple cases, the setting of aid priorities, implementation, and evaluation in particular contexts all require continued local engagement and attention to local priorities, needs, and legitimacy.

This chapter begins with consideration of the role of aid in conflict-affected societies and the ways aid may support peacebuilding. This is followed by discussion of why aid does not always realize such positive impact, focusing on the challenges of aid effectiveness in weak states and key (unintended) negative impacts on public accountability, state capabilities, and violence. Discussion then turns to how to improve aid impact and effectiveness, with particular attention to issues of context, local ownership, and building state capabilities.

Foreign Aid and Aid Effectiveness in Conflict-Affected States

"Foreign aid," commonly used interchangeably with "official development assistance" (ODA), refers to bilateral and multilateral government aid provided on a concessional basis (with at least a 25% grant element) to promote economic development and welfare (Findley 2018, 362–363; Lancaster 2006, 9–12). Although colonial states provided what is sometimes called foreign aid, its origins as an institution are generally associated with the adoption of the Marshall Plan in 1948. Foreign economic aid to developing countries since the Second World War is estimated to amount to more than US$7 trillion (Findley 2018, 360). The OECD's *States of Fragility 2018* report estimates external financial flows to fragile contexts at US$240 billion in 2016, up from US$220 billion in 2014. Of this, remittances, which have been the largest flow since 2008, accounted for 45%, ODA 28%, foreign direct investment 22%, and other financial flows 5% (Schiano Lomoriello 2018, 170–171).

As its history suggests, foreign aid since its beginnings has been closely bound with political and strategic considerations—from its historical role in reconstruction after World War II and Cold War competition for influence (Lancaster 2006) to more contemporary consideration of aid to "failed" and "fragile" states as a component of the war on terror (Fleck and Kilby 2010; Woodward 2017). Thus the use of aid in advancing donor country interests and priorities—and all that this implies for its responsiveness to local needs and priorities—deserves careful attention (see Alesina and Dollar 2000; Bermeo 2018; Dreher et al. 2013). More broadly, the influence of Northern actors on Southern actors through aid and peacebuilding activities and the "securitisation of development" should be considered (Duffield 2001, 2007). We will return to some of these issues in the context of aid's negative impacts.

At the same time, the role of aid in peacebuilding can be seen to have evolved within the context of policy and research on international development, in particular discussions on aid effectiveness and the pursuit of global development goals over the past several decades. In 2000 the Millennium Development Goals (MDGs) set out objectives for UN member states to achieve by 2015 in the areas of poverty alleviation, education, gender equality, child mortality, maternal health, disease, environmental sustainability, and a global partnership for development. Improving the effectiveness of development assistance was a core component of efforts to meet the MDGs. It was also clear early on that fragile states as a group lagged behind others on MDG targets and indicators and posed additional challenges for aid effectiveness. In 2005 the Second High Level Forum on Aid Effectiveness noted in the Paris Declaration "the long-term vision . . . to build legitimate, effective and resilient state and other country institutions," alongside the need to adapt the guiding principles of effective aid to "environments of weak ownership and capacity and to immediate needs for basic service delivery" (OECD 2008, 6). In 2007 the OECD's *Principles for Good International Engagement in Fragile States and Situations* set out ten guidelines (OECD/DAC 2007, 1-3):

1. Take context as the starting point.
2. Do no harm.
3. Focus on statebuilding as the central objective.
4. Prioritize prevention.
5. Recognize the links between political, security, and development objectives.
6. Promote nondiscrimination as a basis for inclusive and stable societies.
7. Align with local priorities in different ways and in different contexts.
8. Agree on practical coordination mechanisms between international actors.
9. Act fast but stay engaged long enough to give success a chance.
10. Avoid pockets of exclusion.

In 2008 these guidelines were explicitly built upon in the Third High Level Forum and the Accra Agenda for Action. The Accra forum in turn supported the establishment of the International Dialogue on Peacebuilding and Statebuilding, composed of a network of OECD Development Assistance Committee members and key multilateral agencies (the International Network on Conflict and Fragility), the G7+ group of fragile and conflict-affected states, and the Civil Society Platform for Peacebuilding and Statebuilding.

By 2011 statebuilding was arguably a "new development paradigm" (Marquette and Beswick 2011). The New Deal for Engagement in Fragile States, developed by the International Dialogue and endorsed at the Fourth High Level Forum on Aid Effectiveness, reflected emerging international consensus that "current ways of working in fragile states need serious improvement" (IDPS 2011, 2). It further outlined five Peacebuilding and Statebuilding Goals (PSGs):

Legitimate Politics—Foster inclusive political settlements and conflict resolution
Security—Establish and strengthen people's security
Justice—Address injustices and increase people's access to justice
Economic Foundations—Generate employment and improve livelihoods
Revenues and Services—Manage revenue and build capacity for accountable and fair service delivery. (2)

That same year, *Conflict, Security, and Development* was the title of the World Bank's flagship World Development Report, which highlighted similar goals and focused on breaking cycles of violence and insecurity.

As the MDG era came to close, fragile states continued to lag behind in achieving global goals, making peacebuilding a continuing priority in the 2030 Agenda for Sustainable Development. Sustainable Development Goal (SDG) 16 in particular speaks to "peace, justice and strong institutions." More broadly, SDG commitment to "leave no one behind" has brought attention to gaps in progress in all SDG areas for fragile and conflict-affected states (Samman et al. 2018). It is worth underscoring that, while these discussions point to a significant role for external assistance in peacebuilding and development, they also have spotlighted the importance of local priorities and local needs, with initiatives designed by agreement between aid-recipient countries and "development partners"—internally led processes. Indeed "local ownership" has become a buzzword for international actors over the past decade, despite clear gaps between expressed commitment and practice (Richmond 2012).

From Aid to Peacebuilding

Aid is seen to influence peacebuilding through at least four broad (interrelated) channels. The first operates through effects on various socioeconomic development outcomes, including economic growth, poverty reduction, improvements in health and education, and changes to the structure of the economy, such as reduced reliance on primary commodity exports (e.g., Braithwaite, Dasandi, and Hudson 2016; Collier and Hoeffler 2000). Aid also may support development in postconflict countries indirectly by serving as a signal that can attract foreign direct investment (Garriga and Phillips 2014). In addition, absorptive capacity for aid may be higher in postconflict countries; in a study of seventeen countries in their first decades postconflict, Collier and Hoeffler (2004) find that absorptive capacity is no greater than normal during the first three years but approximately double thereafter.

Both of the links in this chain—from aid to development, and from development to the likelihood and nature of violent conflict—have been subject to debate in the research literature (Mac Ginty and Williams 2016). In terms of the former in particular, a substantial body of work has raised questions about whether aid does indeed have an overall positive impact on growth in the aggregate (see Moyo 2009; Rajan and Subramanian 2008), despite positive effects shown in micro-level studies (Mosley 1986). Notably the "micro-macro paradox" seems to be moving toward resolution in recent cross-national work, which shows an overall positive impact for aid on various indicators of growth, social welfare, and economic transformation (e.g., Arndt, Jones, and Tarp 2016; Dalgaard and Hansen 2017). This literature also suggests that aid's aggregate impact may be mediated through a variety of other factors, including the quality of donors themselves (Minasyan, Nunnenkamp, and Richert 2017).

Second, aid may have an impact on peacebuilding via impact on society, including the motivations and opportunities faced by multiple social groups and actors within the private sector. This is the role highlighted, for instance, in the 2011 World Development

Report, with aid as facilitative of "offering a stake in society" to groups at risk of engaging in violence (World Bank 2011, 8), as well as of attention in the New Deal to legitimate politics and inclusive political settlements in particular (see Yanguas 2017). Work on aid's role in building social cohesion or social capital to prevent conflict also falls within this set (e.g., Fearon, Humphreys, and Weinstein 2009). Along these lines, the 2018 *Pathways for Peace* report emphasizes the importance of addressing grievances related to exclusion from access to power, opportunity, and security, along with a role for external assistance therein (United Nations and World Bank 2018). When aid functions best, it often supports locally initiated, community-supported peacebuilding activities (Autesserre 2017). "Bottom-up" legal empowerment initiatives, such as community-based paralegal programs that receive international support, offer some examples. Studies of Timap for Justice in Sierra Leone and the Judicial Facilitators Program in Nicaragua, for instance, suggest positive impact on access to justice, conflict resolution, and levels of conflict (Dale 2009; Gramatikov et al. 2015).

Third, aid may work through impact on aid-recipient state and governance institutions. As Bräutigam and Knack (2004, 260) note, at its best "aid can release the binding constraint of low revenues for governments committed to development" and, by extension, to the PSGs. Alternatively donors might support better governance by providing aid selectively to governments with better policies or disbursing it alongside actual policy improvements (see Devarajan, Dollar, and Holmgren 2001). Aid also may directly involve efforts to build or rebuild state authority, legitimacy, and capacity to fulfill basic state functions in the areas of security, justice, economic management, and public service provision. Likewise it may fill in service gaps in specific sectors, thus supporting basic welfare and political stability (see Collier 2012), a role that could be extensive during periods of political transition (e.g., Howard 2014). Aid-funding initiatives to support core government functions have shown some success. For instance, in the area of public financial management reform, a World Bank (2012) study of eight postconflict countries identified "substantial progress" in Afghanistan, Kosovo, Sierra Leone, and West Bank and Gaza, noting that these countries also received among the highest levels of aid. Other examples can be seen in initiatives that support external advisors within government ministries, such as the Scott Family Liberia Fellows Program (Lepistö, Gisselquist, and Ojala 2015).

Fourth, aid may support peacebuilding at the individual and household levels, in turn offering micro foundations for some of these approaches. For instance, Fearon, Humphreys, and Weinstein (2009) find that an aid-supported community-driven reconstruction program carried out by the International Rescue Committee in northern Liberia supported new patterns of social cooperation, with effects that persisted after the program's conclusion. Cilliers, Dube, and Siddiq (2018) analyze the impact of truth and reconciliation efforts designed and implemented with aid support in Sierra Leone by Famul Tok, a nongovernmental organization, showing positive impact on forgiveness, levels of trust, and social networks. (Results also show negative effects on levels of anxiety, depression, and posttraumatic stress disorder, suggesting the value of some revisions to these initiatives as well.) Considering disarmament, demobilization, and reintegration efforts in South Sudan, Phayal, Khadka, and Thyne (2015) identify factors that made them successful in terms of ex-combatant satisfaction. Closely connected are studies of why individuals engage in violence (e.g., Gurr 1970; Justino 2009; McDoom 2013), which in turn identify factors influencing individual behavior that could be influenced by aid, such as relative deprivation.

Challenges and Unintended Effects

While aid may support peacebuilding, it does not always do so. Indeed its overall impact appears to be highly mixed (see Donaubauer, Herzer, and Nunnenkamp 2017; Findley 2018). A number of critiques speak to why. We highlight four here.

Aid (In)Effectiveness in "Poor" Policy Environments

Building on the literature in development economics, one of the clearest challenges for aid and peacebuilding is the ineffectiveness of aid in "poor" policy environments. In an influential paper, Burnside and Dollar (2000) found that aid had a positive impact on growth in countries with "good" fiscal, monetary, and trade policies but little effect in countries with "poor" policies. Given that poor policy and state fragility often go together (see Chauvet and Collier 2008), the Burnside-Dollar hypothesis raises particular concerns about aid effectiveness in conflict-affected areas. In addition, a key policy implication of this finding is that aid should be distributed selectively to countries with good policies, an approach affirmed by the Monterrey Consensus in 2002 (Easterly 2003). Focusing on the period 1984–2003, Dollar and Levin (2006) find increasing selectivity in foreign aid in terms of democracy and property rights, particularly for multilateral assistance. Selectivity in this sense implies some considerable international constraints on locally led policies in the effort to promote "get[ting] the policies right" (see Grindle 2007). Selectivity in turn may raise a fundamental challenge for fragile states: their weak state capabilities may have implications for their abilities to implement the good policies needed to secure aid, but external assistance may facilitate building stronger institutions.

Governance and Accountability

A second set of critiques relate to the unintended effects of aid on domestic governance and accountability, which can be particularly problematic for weak and aid-dependent states. One simple illustration of the outsized influence donors have played historically in such environments is seen in estimates of ODA as a share of government expenditure: in 1999, 99% in Rwanda, 63% in the Central African Republic, 51% in Uganda, 45% in Sierra Leone, and 43% in Burundi (Bräutigam and Knack 2004, 258).

In such contexts, aid can serve to orient accountability within recipient states toward donors instead of citizens, and with low reliance on domestic revenues, governments may face weak pressure to maintain popular legitimacy (Moss, Pettersson, and van de Walle 2006). Overly technocratic approaches can lead to aid that props up illegitimate leaders (Easterly 2013). Within government, aid may strengthen the hand of the executive branch that receives aid monies against legislative checks (Bräutigam and Knack 2004). High levels of aid may further weaken state institutions and governance practices over time; soft-budget constraints, for instance, may loosen the costs of poor government budgeting (263–264). A lack of reliance on domestic revenues can mean little investment in building the tax administration and revenue authority, a central pillar of state capacity (Bräutigam, Fjeldstad, and Moore 2008). How extensive is the "aid-institutions paradox"? In cross-national work, Jones and Tarp (2016, 266) find a small *positive* net effect of total aid on political institutions, which is "driven primarily by the positive contribution of more stable inflows of 'governance aid.'" Further research is needed into the conditions under which the aid-institutions paradox operates and is avoided.

State Capabilities

A closely related set of critiques pertains to aid's negative effects on the functioning and capabilities of state institutions. Unrealistic and ill-conceived expectations may place harsh burdens on weak institutions (Duffield 2007; Grindle 2004). Such effects may be exacerbated by donor fragmentation, which may overtax already burdened government institutions with the need to respond to demands from international partners (Bräutigam and Knack 2004). Bringing together a number of critiques in this vein, Andrews, Pritchett, and Woolcock (2017) argue that aid can support "successful failure" of state institutions, allowing them to survive—and even maintain legitimacy—with low and stagnating capability. They find this works through two channels: isomorphic mimicry (i.e., institutions adopt the forms and appearances of well-functioning organizations), which may be particularly effective in environments where functioning is difficult to measure, and "premature load bearing" (expecting too much too soon from weak institutions and causing them to fail under pressure).

Violence and "Do No Harm"

The most fundamental critiques of aid for peacebuilding center on the principle "do no harm" and especially the effects of aid on violent conflict (Anderson 1999; Woodward 2017). In Rwanda, for instance, Uvin (1998, 3) argues, "the way *development* was defined, managed, and implemented was a crucial element in the creation and evolution of many of the processes that led to genocide." In the Democratic Republic of Congo, Autesserre (2010, 11–12) finds that "peacebuilding culture enabled foreign actors to pursue an intervention strategy that permitted, and at times even exacerbated, fighting, massacres, and massive human rights violations during and after the transition."

Research also explores other channels through which aid may be linked to violence. For instance, it may serve as lootable resource itself, which may increase the spoils of war (Grossman 1992). Aid projects that exclude portions of the community may prompt more insurgent activity (Karell and Schutte 2018). Decreases in aid revenues may be linked to violence (Nielsen et al. 2011). Insurgents may sabotage aid programs because they are concerned such aid will weaken their popular support (Crost, Felter, and Johnston 2014). These effects in turn may be mediated by state capacity: Dasgupta, Gawande, and Kapur (2017, 605), for instance, find that India's National Rural Employment Guarantee Scheme had a pacifying impact on Maoist violence "in districts with sufficient pre-existing local state capacity to implement the program effectively."

Studies of aid's effect on violence overall show mixed results. In analysis based on thirty-nine sub-Saharan African countries from 1981 to 1999, de Ree and Nillesen (2009) find a statistically significant and economically important negative effect of foreign aid flows on civil conflict duration and no significant relationship between aid flows and civil war onset. However, in a systematic review of nineteen studies, Zürcher (2017, 506) concludes that "aid in conflict zones is more likely to exacerbate violence than to dampen violence." In his review of this literature, Findley (2018) underscores mixed impacts and the methodological challenges of causal identification in this area. While most such analyses are based on externally developed measures of conflict and violence, an alternative perspective based on new data on local perceptions of peace is offered by Firchow (2018). This analysis suggests little correlation between external interventions after war and higher levels of peacefulness.

How to Improve Aid for Peacebuilding

Within policy circles there has been broad consensus around the ten fragile states "principles" adopted in 2007 (OECD 2011, 47–49). These principles reflect coordinated efforts to address core critiques, but while sometimes treated as guidance for development programming, they offer something much looser in reality. They underscore that aid may cause harm, for instance, but do not provide guidance on what about aid may cause harm or how to prevent it from doing so. This lack of specificity reflects gaps in the research literature about what to do and what works in aid for peacebuilding (see Justino 2019). Consideration of three of these principles illustrates both nuances and tensions, alongside considerable space for future research.

Take Context as the Starting Point

Several decades of development research now critiques one-size-fits-all aid programming and expectations that institutions in Southern countries should follow precisely Northern models. Research on statebuilding expresses similar conclusions (e.g., Englebert and Tull 2008). Along these lines, the 2011 World Development Report emphasized differentiated country strategies, policies based on situation-specific analysis, and a framework taking into account types of violence, transitional opportunities, institutional challenges, stakeholders, and stresses (World Bank 2011). That said, acontextual approaches remain common. For instance, the Marshall Plan is a touchstone for many development thinkers, with claims that similarly massive investment and attention could spur reconstruction elsewhere. Yet comparative consideration, for instance of postwar Japan with contemporary Afghanistan and Iraq, points to fundamentally different statebuilding tasks in many of today's fragile states: "preserving existing state capacity" in the former and "attempting to build state strength where it did not previously exist" in the latter (Monten 2014, 173).

Lack of attention to context is another common feature in standard approaches to aid impact evaluation, particularly the claim by some researchers that "what works" in development is best identified through experimental methods (Banerjee 2007). Indeed if evidence based on a population sample in one country truly does show that an intervention "works" and is thus a good model elsewhere, aid practitioners may be better advised *not* to start with context, but rather with building a toolbox of universally applicable interventions. That said, recent experimental work shows new grappling with context, macro-level theory, and issues of external validity. To better take context into account, a key priority for future research is to identify the aspects of context that matter most in understanding why an aid-supported initiative has worked (or not) and whether it is transferrable elsewhere. Can we narrow down from long laundry lists of political and economic factors? Can we identify peacebuilding-related interventions that have worked similarly across very diverse contexts?

Align with Local Priorities in Different Ways in Different Contexts

Another point emphasized in the aid literature is the importance of local ownership and of linking with local priorities, needs, and agendas (Booth 2012), reflected also in "the local turn" in work on peacebuilding (Leonardsson and Rudd 2015). This principle is explicitly linked to aid programming at all levels—from the expectation that aid priorities are set by agreement between aid-recipient country governments and development partners to efforts like village-level community-driven development—but practice may be very different. Moreover, in situations where state capacity is weak and authority contested, international actors may effectively constrain local ownership to meet what von Billerbeck (2015, 299) describes as key "operational goals—the liberalization of the postconflict state and the delivery of demonstrable output in the short term."

A related challenge concerns the very identification of "local priorities" and who among a plethora of diverse societal groups and actors, with different and sometimes conflicting interests and positions, represents them. This is complicated in any society, but especially so in conflict-affected societies where governments have weak or contested legitimacy. Partly in recognition of such challenges, the concept of ownership has evolved through the high-level forums on aid effectiveness; initially synonymous with domestic government ownership, particularly by the executive branch, it subsequently has expanded to include parliamentarians, civil society, local government, and the private sector. Such expansion is promising but not without risks: it could also effectively place greater power with donors to pick and choose among local partners those whose local priorities best match their own. Consideration of such tensions is another area ripe for future research.

One promising approach is the Everyday Peace Indicators, which uses participatory research methods to measure difficult concepts such as peace, reconciliation, governance, and violent extremism. It thus offers bottom-up, local measures against which to consider interventions, which is a way to link aid initiatives more clearly to local priorities through evaluation (Firchow and Mac Ginty 2017; Mac Ginty 2013). Work based on the Everyday Peace Indicators in South Africa, South Sudan, Uganda, and Zimbabwe shows that bottom-up narratives in conflict situations are articulated in different ways and raise different issues than top-down narratives (Mac Ginty and Firchow 2016).

A closely related question—especially given critiques about the (unintended) negative effects of aid on governance—is precisely how best to support local institutions that legitimately represent local priorities. Community-driven development programs have received considerable attention, but recent reviews suggest mixed impacts (Casey 2018; King and Samii 2014). Social accountability initiatives, including legal empowerment, may be more promising (Goodwin and Maru 2017). Resnick and van de Walle (2013) find that democracy assistance can have a robust positive impact on both vertical and horizontal accountability, for instance through support of the electoral process and of state institutions offering checks on the executive, such as parliament and the judiciary.

Focus on Statebuilding as the Central Objective

The PSGs treat peacebuilding and statebuilding as inextricably intertwined. Troubling, then, is the fact that the international community's track record in statebuilding is mixed at best (Lake and Fariss 2014; Paris and Sisk 2009). Furthermore, while there is a large literature on weak and fragile states, it speaks much more to why states fail than how they exit fragility, much less the role of aid in their doing so (Carment and Samy 2017; Gisselquist 2014).

Whether statebuilding should be the central objective in all peacebuilding situations is not obvious. Foreign aid always involves recipient states at some level, but initiatives vary considerably in their degree of statebuilding engagement. It may be efficient, for instance, for a vaccination campaign to be largely donor-led and implemented, contributing to the PSG goal of service delivery in a conflict-prone environment. Other aid-supported activities, such as public financial management reform, necessarily involve extensive engagement with state institutions. It may be strategically wise for donors to maneuver in terms of sequencing of different types of initiatives and state engagement (e.g., collaboration with diverse local partners instead of national-level state institutions; Gisselquist 2015). Given the challenges of reforming major institutions, it may be best for donors to pursue statebuilding incrementally (and alongside initiatives aimed at other development objectives). One example in terms of aid-supported public financial management is the role of a small, highly skilled "parallel public service" to support high-level policy development, planning, and implementation in postconflict Sierra Leone (Tavakoli, Cessay, and Cole 2015).

Perhaps the clearest statebuilding recommendations in the aid literature concern the importance of "staying long" and minimizing the risks of aid volatility. As the New Deal emphasizes, institution-building is a decades-long process. Yet, historically, aid to postconflict countries has tapered out over the first decade—precisely during the period when it may be most useful (Collier and Hoeffler 2004).

Another approach to building state capability that has received a lot of attention is Problem-Driven Iterative Adaption (Andrews, Pritchett, and Woolcock 2017). A "learning by doing" approach, it is centered around four components:

- Solving "particular problems in local contexts."
- Creating an "authorizing environment" for decision-making that supports experimentation.
- Involving active and ongoing experiential learning and iterative feedback loops.
- Engaging broad sets of agents to ensure that reforms are relevant, politically legitimate, and implementable.

While this approach holds considerable promise in comparison to business as usual, it has not clearly resolved the core challenges outlined, especially in terms of how to identify and align with diverse local priorities.

Conclusion and Future Research

This chapter has considered the role of aid in conflict-affected societies, including the positive and negative effects that may underlie an overall mixed impact on peacebuilding. Assessing aid's impact and effectiveness is not straightforward. It requires distinguishing aid's impact from that of other factors and considering what might have been in its absence. Over the past decade or so, experimental methods and, more broadly, the identification revolution in economics has considerably strengthened the internal validity of causal claims about aid's impact at the project and program level. A key challenge with such methods, however, is that while they have strong internal validity, they tend to have weak external validity—i.e., they have weak leverage on whether impacts identified in one context can be expected elsewhere. In addition, in focusing on establishing causal links between an intervention and various outcomes, they tend to pay less attention than other approaches (such as qualitative case studies) to intermediate processes. Better understanding of such processes could in turn offer useful insight into whether a program could be tweaked for improvement or when being used as a model in another context. Moreover, experimental methods in particular are not well suited to study many of the types of peacebuilding interventions we may expect to have impact, such as national-level institutional reforms. Practical and ethical challenges further complicate such studies in many areas and may be especially pronounced in conflict-affected societies. Unsurprisingly, Cameron et al.'s (2015) mapping of the evidence drawn from impact evaluations relevant to twenty-five categories across the five PSGs shows large evidence gaps: it identifies only two (community-driven reconstruction and psychosocial programs for victims) as having a large enough number of studies to be promising for evidence synthesis.

For some observers, the challenges to aid effectiveness and impact in fragile contexts place a question mark over whether aid should be provided at all. This chapter argues, by contrast, that aid for peacebuilding is—and should be—a priority for global development. Summarizing what we know about how to improve the link between aid and peacebuilding spotlights key gaps with respect to a nuanced understanding of why and under what conditions aid works and does not and how this can inform aid policy. To support better aid, development research in this area should prioritize these tricky questions. Experimental methods in particular are not the best way to identity what works in aid for peacebuilding; they are one tool in a broader toolkit. Much more attention should be paid to local assessment of what works, which is too often defined based on external assessment that may differ from local perceptions. In short, wrestling with the most pressing questions on aid and peacebuilding requires more diverse methodologies than are standard in much of the literature on foreign aid.

Among other core topics for future research on aid and peacebuilding, we highlight three here. First, conflict-linked migration has wide-ranging implications for peace and development, including the spread of conflict across borders. The research literature on South-North and North-North migration is more developed, and the volume of migration across Southern borders and within Southern countries calls for more research attention. Second, the climate crisis has alarming potential implications for conflict. While climate-related aid has increased, more evidence is needed on the role of aid in supporting better climate policy and addressing climate-linked conflict (Kono and Montinola 2019). Third, another area for future research concerns the changing international aid system (Strange et al. 2015). As new development actors and South-South cooperation weaken the influence of Western donors, what will be the implications for aid and peacebuilding?

NOTE

1. The concept of a fragile state is contested and problematic but remains in common use in standard discussions on aid and development (see Brinkerhoff 2014; Grimm, Lemay-Hébert, and Nay 2014; Woodward 2017). "Fragility" here is understood "as the combination of exposure to risk and insufficient coping capacity of the state, system, and/or communities to manage, absorb, or mitigate those risks" (OECD 2016, 21). Fragility and conflict are strongly related; while not all fragile states are experiencing conflict, violence is one of the main factors contributing to fragility, and fragility can lead to violence.

REFERENCES

Alesina,. A., and D. Dollar. 2000. "Who Gives Foreign Aid to Whom and Why?" *Journal of Economic Growth* 5 (1):33–63.

Anderson, Mary B. 1999. *Do No Harm: How Aid Can Support Peace—Or War.* Boulder: Lynne Rienner.

Andrews, M, L. Pritchett, and M. Woolcock. 2017. *Building State Capability.* Oxford: Oxford University Press.

Arndt, A., S. Jones, and F. Tarp. 2016. "What Is the Aggregate Economic Rate of Return to Foreign Aid?" *World Bank Economic Review* 30 (3):446–474.

Autesserre, S. 2010. *The Trouble with the Congo: Local Violence and the Failure of International Peacebuilding.* Cambridge: Cambridge University Press.

Autesserre, S. 2017. "International Peacebuilding and Local Success: Assumptions and Effectiveness." *International Studies Review* 19 (1):114–132.

Banerjee, A, ed. 2007. *Making Aid Work.* Cambridge, MA: MIT Press.

Bermeo, S. B. 2018. *Targeted Development: Industrialized Country Strategy in a Globalizing World.* Oxford: Oxford University Press.

Booth, D. 2012. "Aid Effectiveness: Bringing Country Ownership (and Politics) Back In." *Conflict, Security & Development* 12 (5):537–558.

Braithwaite, A., N. Dasandi, and D. Hudson. 2016. "Does Poverty Cause Conflict? Isolating the Causal Origins of the Conflict Trap." *Conflict Management and Peace Science* 33 (1):45–66.

Bräutigam, D., and S. Knack. 2004. "Foreign Aid, Institutions and Governance in Sub-Saharan Africa." *Economic Development and Cultural Change* 52 (2):255–285.

Bräutigam, D., O. H. Fjeldstad, and M. Moore. 2008. *Taxation and State-Building in Developing Countries: Capacity and Consent.* Cambridge: Cambridge University Press.

Brinkerhoff, D. W. 2014. "State Fragility and Failure as Wicked Problems: Beyond Naming and Taming." *Third World Quarterly* 35 (2):333–344.

Burnside, C., and D. Dollar. 2000. "Aid, Policies, and Growth." *American Economic Review* 90 (4):847–868.

Cameron, D. B., A. N. Brown, A. Mishra, M. Picon, H. Esper, F. Calvo, and K. Peterson 2015. *Evidence for Peacebuilding: An Evidence Gap Map.* 3ie Evidence Gap Map Report 1. New Delhi: International Initiative for Impact Evaluation (3ie).

Carment, D, and Y. Samy. 2017. *Exiting the Fragility Trap: Rethinking Our Approach to the World's Most Fragile States.* WIDER Working Paper. Helsinki: UNU-WIDER.

Casey, Katherine. 2018. "Radical Decentralization: Does Community-Driven Development Work?" *Annual Review of Economics* 10 (1):139–163.

Chauvet, Lisa, and Paul Collier. 2008. "What Are the Preconditions for Turnarounds in Failing States?" *Conflict Management and Peace Science* 25 (4):332–348.

Cilliers, J., O. Dube, and B. Siddiq. 2018. *Can the Wounds of War Be Healed? Experimental Evidence on Reconciliation in Sierra Leone*. New Delhi: International Initiative for Impact Evaluation (3ie).

Collier, P. 2012. *How to Spend It: The Organization of Public Spending and Aid Effectiveness*. WIDER/ReCom Working Paper. Helsinki: UNU-WIDER.

Collier, P., and A. Hoeffler. 2000. *Greed and Grievance in Civil War*. Washington, DC: World Bank.

Collier, P., and A. Hoeffler. 2004. "Aid, Policy and Growth in Post-conflict Societies." *European Economic Review* 48 (5):1125–1145.

Crost, B., J. Felter, and P. Johnston. 2014. "Aid under Fire: Development Projects and Civil Conflict." *American Economic Review* 104 (6):1833–1856.

Dale, P. 2009. *Delivering Justice to Sierra Leone's Poor: An Analysis of the Work of Timap for Justice*. Washington, DC: World Bank.

Dalgaard, C., and H. Hansen. 2017. "The Return to Foreign Aid." *Journal of Development Studies* 53 (7):998–1018.

Dasgupta, A., K. Gawande, and D. Kapur. 2017. "(When) Do Antipoverty Programs Reduce Violence? India's Rural Employment Guarantee and Maoist Conflict." *International Organization* 71 (3):605–632.

de Ree, K., and E. Nillesen. 2009. "Aiding Violence or Peace? The Impact of Foreign Aid on the Risk of Civil Conflict in Sub-Saharan Africa." *Journal of Development Economics* 88 (2):301–313.

Devarajan, S., D. Dollar, and T. Holmgren. 2001. *Aid and Reform in Africa: Lessons from Ten Case Studies*. Washington, DC: World Bank.

Dollar, D., and V. Levin. 2006. "The Increasing Selectivity of Foreign Aid, 1984–2003." *World Development* 34 (12):2034–2046.

Donaubauer, J., D. Herzer, and P. Nunnenkamp. 2017. "The Effectiveness of Aid under Post-Conflict Conditions: A Sector-Specific Analysis." *Journal of Development Studies* 55 (4): 720–736.

Dreher, A., S. Klasen, J. R. Vreeland, and E. Werker 2013. "The Costs of Favoritism: Is Politically Driven Aid Less Effective?" *Economic Development and Cultural Change* 62 (1):157–191.

Duffield, M. 2001. *Global Governance and the New Wars: The Merging of Development and Security*. London: Zed Books.

Duffield, M. 2007., *Development, Security and Unending War: Governing the World of Peoples*. Cambridge, UK: Polity Press.

Easterly, W. 2003. "Can Foreign Aid Buy Growth?" *Journal of Economic Perspectives* 17 (3):23–48.

Easterly, W. 2013. *The Tyranny of Experts: Economists, Dictators, and the Forgotten Rights of the Poor*. New York: Basic Books.

Englebert, P., and D. M. Tull. 2008. "Postconflict Reconstruction in Africa: Flawed Ideas about Failed States." *International Security* 32 (4):106–139.

Fearon, J. D., M. Humphreys, and J. M. Weinstein. 2009. "Can Development Aid Contribute to Social Cohesion after Civil War? Evidence from a Field Experiment in Post-Conflict Liberia." *American Economic Review* 99 (2):287–291.

Findley, M. G. 2018. "Does Foreign Aid Build Peace?" *Annual Review of Political Science* 21 (1):359–384.

Firchow, P. 2018. *Reclaiming Everyday Peace: Local Voices in Measurement and Evaluation after War.* Cambridge: Cambridge University Press.
Firchow, P., and R. Mac Ginty. 2017. "Measuring Peace: Comparability, Commensurability, and Complementarity Using Bottom-Up Indicators." *International Studies Review* 19 (1):6–27.
Fleck, R. K., and C. Kilby. 2010. "Changing Aid Regimes? U.S. Foreign Aid from the Cold War to the War on Terror." *Journal of Development Economics* 91 (2):185–197.
Garriga, A. C., and B. J. Phillips. 2014. "Foreign Aid as a Signal to Investors: Predicting FDI in Post-conflict Countries." *Journal of Conflict Resolution* 58 (2):280–306.
Gisselquist, R. M. 2014. "Aid and Institution-Building in Fragile States: What Do We Know? What Can Comparative Analysis Add?" *Annals of the American Academy of Political and Social Science* 656 (1):6–21.
Gisselquist, R. M. 2015. "Good Aid in Hard Places: Learning from 'Successful' Interventions in Fragile Contexts." *International Peacekeeping* 22 (4):283–301.
Goodwin, L., and V. Maru. 2017. "What Do We Know about Legal Empowerment? Mapping the Evidence." *Hague Journal on the Rule of Law* 9 (1):157–194.
Gramatikov, M., N. Barendrecht, M. Kokke, R. Porter, M. Frishman, and A. Morales 2015. "Impact Assessment of the Facilitadores Judiciales Programme in Nicaragua." *International Peacekeeping* 22 (4):398–418.
Grimm, S., N. Lemay-Hébert, and O. Nay. 2014. "'Fragile States': Introducing a Political Concept." *Third World Quarterly* 35 (2):197–209.
Grindle, M. S. 2004. "Good Enough Governance: Poverty Reduction and Reform in Developing Countries." *Governance* 17 (4):525–548.
Grindle, M. S. 2007. "Good Enough Governance Revisited." *Development Policy Review* 25 (5):533–574.
Grossman, H. I. 1992. Foreign Aid and Insurrection. *Defence Economics* 3 (4): 275–288.
Gurr, T. R. 1970. *Why Men Rebel.* Princeton, NJ: Princeton University Press.
Howard, L. M. 2014. "Kosovo and Timor-Leste: Neotrusteeship, Neighbors, and the United Nations." *Annals of the American Academy of Political and Social Science* 656 (1):116–135.
IDPS. 2011. "A New Deal for Engagement in Fragile States." International Dialogue on Peacebuilding and Statebuilding. https://www.pbsbdialogue.org/media/filer_public/07/69/07692de0-3557-494e-918e-18df00e9ef73/the_new_deal.pdf.
Jones, S., and F. Tarp. 2016. "Does Foreign Aid Harm Political Institutions?" *Journal of Development Economics* 118 (January):266–281.
Justino, P. 2009. "Poverty and Violent Conflict: A Micro-Level Perspective on the Causes and Duration of Warfare." *Journal of Peace Research* 46 (3):315–333.
Justino, P. 2019. "Governance Interventions in Conflict-Affected Countries." *Journal of Development Studies* 55 (7):1364–1378.
Karell, D., and S. Schutte. 2018. "Aid, Exclusion, and the Local Dynamics of Insurgency in Afghanistan." *Journal of Peace Research* 55 (6):711–725.
King, E., and C. Samii. 2014. "Fast-Track Institution Building in Conflict-Affected Countries? Insights from Recent Field Experiments." *World Development* 64:740–754.
Kono, D. Y., and G. R. Montinola. 2019. "Foreign Aid and Climate Change Policy: What Can('t) the Data Tell Us?" *Politics and Governance* 7 (2):68–92.
Lake, D. A., and C. J. Fariss. 2014. "Why International Trusteeship Fails: The Politics of External Authority in Areas of Limited Statehood." *Governance* 27 (4):569–587.
Lancaster, C. 2006. *Foreign Aid: Diplomacy, Development, Domestic Politics.* Chicago: University of Chicago Press.

Leonardsson, H., and G. Rudd. 2015. "The 'Local Turn' in Peacebuilding: A Literature Review of Effective and Emancipatory Local Peacebuilding." *Third World Quarterly* 36 (5):825–839.

Lepistö, R. S., R. M. Gisselquist, and J. Ojala. 2015. "'Embedded' Assistance: Finn Church Aid's Secondment in Somalia." *International Peacekeeping* 22 (4):354–371.

Mac Ginty, R. 2013. "Indicators+: A Proposal for Everyday Peace Indicators." *Evaluation and Program Planning* 36 (1):56–63.

Mac Ginty, R., and P. Firchow. 2016. "Top-Down and Bottom-Up Narratives of Peace and Conflict." *Politics* 36 (3):308–323.

Mac Ginty, R., and W. Williams. 2016. *Conflict and Development.* London: Routledge.

Marquette, H. and D. Beswick. 2011. "State Building, Security and Development: State Building as a New Development Paradigm?" *Third World Quarterly* 32 (10):1703–1714.

McDoom, O. S. 2013. "Who Killed in Rwanda's Genocide? Micro-space, Social Influence and Individual Participation in Intergroup Violence." *Journal of Peace Research* 50 (4):453–467.

Minasyan, A., P. Nunnenkamp, and K. Richert. 2017. "Does Aid Effectiveness Depend on the Quality of Donors?" *World Development* 100:16–30.

Monten, J. 2014. "Intervention and State-Building: Comparative Lessons from Japan, Iraq, and Afghanistan." *Annals of the American Academy of Political and Social Science* 656 (1):173–191.

Mosley, P. 1986. "Aid-Effectiveness: The Micro-Macro Paradox." *IDS Bulletin* 17 (2):22–27.

Moss, T., G. Pettersson, and N. van de Walle. 2006. *An Aid-Institutions Paradox? A Review Essay on Aid Dependency and State Building in Sub-Saharan Africa.* Washington, DC: Center for Global Development.

Moyo, D. 2009. *Dead Aid: Why Aid Is Not Working and How There Is a Better Way for Africa.* New York: Farrar, Straus and Giroux.

Nielsen, R. A., Michael G. Findley, Zachary S. Davis, Tara Candland, and Daniel L. Nielson. 2011. "Foreign Aid Shocks as a Cause of Violent Armed Conflict." *American Journal of Political Science* 55 (2):219–232.

OECD. 2008. *The Paris Declaration on Aid Effectiveness and the Accra Agenda for Action.* Paris: OECD.

OECD. 2011. *International Engagement in Fragile States: Can't We Do Better?* Paris: OECD.

OECD. 2016. *States of Fragility 2016: Understanding Violence.* Paris: OECD.

OECD. 2018. *States of Fragility 2018.* Paris: OECD.

OECD/DAC. 2007. *Principles for Good International Engagement in Fragile States and Situations.* Paris: OECD.

Paris, Roland, and Timothy D. Sisk, eds. 2009. *The Dilemmas of Statebuilding: Confronting the Contradictions of Postwar Peace Operations.* London: Routledge.

Phayal, A., P. B. Khadka, and C. L. Thyne. 2015. "What Makes an Ex-Combatant Happy? A Micro-analysis of Disarmament, Demobilization, and Reintegration in South Sudan." *International Studies Quarterly* 59 (4):654–668.

Rajan, R. G., and A. Subramanian. 2008. "Aid and Growth: What Does the Cross-Country Evidence Really Show?" *Review of Economics and Statistics* 90 (4):643–665.

Resnick, D., and N. van de Walle, eds. 2013. *Democratic Trajectories in Africa: Unraveling the Impact of Foreign Aid.* Oxford: Oxford University Press.

Richmond, O. P. 2012. "Beyond Local Ownership in the Architecture of International Peacebuilding." *Ethnopolitics* 11 (4):354–375.

Samman, E., P. Lucci, J. Hagen-Zanker, T. Bhatkal, A. Telias Simunovic, S. Nicolai, E. Stuart, and C. Caron. 2018. *SDG Progress: Fragility, Crisis and Leaving No One Behind.* London: ODI.

Schiano Lomoriello, R. 2018. "What Sources of External Development Finance Are Available to Fragile Contexts?" In *States of Fragility 2018*, edited by OECD, 169–210. Paris: OECD.

Strange, A. M., Austin M. Strange, Axel Dreher, Andreas Fuchs, Bradley Parks, and Michael J. Tierney. 2015. "Tracking Underreported Financial Flows: China's Development Finance and the Aid-Conflict Nexus Revisited." *Journal of Conflict Resolution* 61 (5):935–963.

Tavakoli, H., I. B. Cessay, and W. Percy Onipede Cole. 2015. "Substantial but Uneven Achievement: Selected Success When Stars Align. Public Financial Management Reforms in Sierra Leone." *International Peacekeeping* 22 (4):337–353.

United Nations and World Bank. 2018. *Pathways for Peace: Inclusive Approaches to Preventing Violent Conflict*. Washington, DC: World Bank.

Uvin, P. 1998. *Aiding Violence: The Development Enterprise in Rwanda*. West Hartford, CT: Kumarian Press.

von Billerbeck, S. B. K. 2015. "Local Ownership and UN Peacebuilding: Discourse versus Operationalization." *Global Governance* 21 (2):299–315.

Woodward, S. L. 2017. *The Ideology of Failed States: Why Intervention Fails*. Cambridge: Cambridge University Press.

World Bank. 2011. *World Development Report: Conflict, Security, and Development*. Washington, DC: World Bank.

World Bank. 2012. *Public Financial Management Reforms in Post-Conflict Countries: Synthesis Report*. Washington, DC: World Bank.

Yanguas, P. 2017. "The Role and Responsibility of Foreign Aid in Recipient Political Settlements." *Journal of International Development* 29 (2):211–228.

Zürcher, C. 2017. "What Do We (Not) Know about Development Aid and Violence? A Systematic Review." *World Development* 98:506–522.

CHAPTER 38

LOCAL OWNERSHIP, LEGITIMACY, AND PEACEBUILDING

TIMOTHY DONAIS

WHATEVER else it may represent, the progressive embedding of the local ownership norm within peacebuilding scholarship and practice is a triumph of ambiguity. Like many international norms, the idea that peace processes need to be owned by, rather than administered to, conflict-affected societies has spread rapidly in large part because it is vague and because it easily lends itself to being appropriated by a range of different actors for a range of different purposes (Krook and True 2010, 104). Local ownership is also difficult, in principle, to oppose—much like equally ambiguous international norms such as democracy or the rule of law—and few serious scholars or practitioners now advocate "foreign ownership" (or equally anachronistic concepts such as trusteeship) as an alternative means of addressing the problems of fragile and conflict-affected states. In this way, the ownership principle has crept to the forefront of current academic debates around peacebuilding and has in many ways become the new orthodoxy of international intervention; in the words of one UN report, ownership is "an imperative, an absolute essential, if peacebuilding is to take root" (United Nations 2010, 9; Lemay-Hébert and Kappler 2016, 896; Ejdus and Juncos 2018, 7). This is despite—or perhaps because of—the contested nature of "the local" as a locus of agency, ongoing imprecision around what it means to "own" a peace process, and the fragmenting of consensus around the very meaning of peacebuilding itself.

In the context of peacebuilding, debates around ownership are also invariably debates about legitimacy, authority, and accountability. The center of gravity around such issues has shifted significantly toward the local as the inherent legitimacy—and applicability—of liberal principles has come under increasing challenge, in large part due to liberal peacebuilding's poor track record. External peacebuilders, in other words, have felt increasingly compelled to justify their programmatic interventions not on the basis of whether they work but rather on the basis of the extent to which they are inclusive of, or at least sensitive to, local preferences, traditions, and cultural norms and are carried out in partnership with relevant local actors or institutions. To a significant extent, "supporting local actors and structures" has replaced "implanting liberal norms and institutions" as a core justification for international intervention in conflict-affected societies (Miklian, Liden, and Kolas 2011, 286). While the bottom-up logic of this approach remains compelling—peacebuilding is more likely to endure when it is actively embraced by those who have to live with its consequences (Donais 2009, 3)—the evidence that external interveners have substantively embraced the local ownership principle, at least at the level of project implementation, remains rather thin. As often as not, local ownership is a legitimizing rhetorical device deployed by external actors who remain reluctant to cede control over their interventions in the name of local ownership, largely because of the fear that "local actors will, if left to their own devices, revert to undemocratic, sectarian, or inhumane practices" (von Billerbeck 2015, 302). While local ownership may represent the new orthodoxy of peacebuilding, therefore, the fundamental question "Who decides?" remains very much in play.

It is not my intention in this chapter to attempt to resolve such dilemmas, since the inherent *unresolvability* of ownership questions, at least as they relate to the quotidian interactions between international and domestic actors, has become part of the very fabric of peacebuilding itself. Rather, I consider two of the most fundamental questions at the core of the local ownership debate—ownership *of what* and ownership *by whom*—in light of ongoing shifts within the broader environment within which peacebuilding unfolds. In doing so, I make two interrelated arguments. First, I suggest that the referent object of ownership—the very idea of a *peace process*—is itself becoming increasingly elusive in the majority of contexts in which international actors intervene in the name of building peace. Even leaving aside the question of whether any single actor, or collection of actors, can ever realistically *exercise ownership* over the content, direction, and pace of war-to-peace transitions, such transitions unfold increasingly in the absence of coherent roadmaps provided by either overarching normative principles or comprehensive peace accords. Thus, in the same way that the distinction between conflict and postconflict has become difficult to sustain, it is becoming ever more challenging to distinguish something we might call a peace process from broader political, economic, or social developments that may or may not contribute to the emergence of a stable, legitimate—and minimally just—social order.

Second, while local ownership may be an empirical reality in the sense that whatever peace is to emerge in Afghanistan or South Sudan or Bosnia-Herzegovina will overwhelmingly be the product of domestic rather than external political forces and dynamics, the specific nature and identity of "the local" remains as contested as ever. Indeed this dimension of the local ownership debate has been marked by a deepening polarization between elite-centric and everyday-centric conceptions of ownership, neither of which has fully come to terms with the question of how to bridge—with or without international facilitation and support—the state-society divide that remains central to the broader peacebuilding *problématique*. Both of these arguments, I suggest, have significant implications for the future of peacebuilding as a coherent project.

Ownership: Discourse and Practice

At the level of discourse, the definitional imprecision inherent in both constituent elements of the local ownership principle has led international norm entrepreneurs increasingly to adopt broad, inclusive conceptualizations of what local ownership might actually mean in the context of peacebuilding. The 2011 New Deal of the International Dialogue on Peacebuilding and Statebuilding, facilitated by the Organization for Economic Cooperation and Development, for example, is built around the notion of "country-led and country-owned" transitions out of fragility and insists that "constructive state-society relations, and the empowerment of women, youth, and marginalized groups, are at the heart of successful peacebuilding and statebuilding." The 2015 Report of the UN Secretary-General's Advisory Group of Experts on the UN's peacebuilding architecture reaches similar conclusions. Couched in the language of "inclusive national ownership," the report emphasizes "the shared national responsibility to drive and direct [peacebuilding] efforts . . . across all key strata and divides, across a spectrum of political opinions and domestic actors, including minorities" (United Nations 2015, 18). While there is widespread (and unsurprising) consensus that sustainable peace processes should be inclusive, constructive, and participatory, however, there is far less consensus on how to go about fostering or facilitating such processes in the fractured and deeply contested environments in which peacebuilding unfolds. As Sarah von Billerbeck (2015, 303) has observed, UN reports in particular are bullish on the positive effects of ownership on both legitimacy and sustainability, even as "they remain silent on how to build ownership in practice."

While formal definitions of local ownership have tended to be expansive and inclusive, the paradoxical nature of local ownership as a top-down policy prescription—in combination with the structural power imbalances inherent in peacebuilding praxis and reservations about whether local actors can be trusted to "do the right thing"—has to date worked rather decisively against the willingness of external interveners to defer to the primacy of the local. In operational terms, then, local ownership has been defined quite restrictively by peacebuilding's "global governors" (Campbell 2018, 5) and has been less about transferring power and authority from outsiders to insiders and more about securing local buy-in for policies, prescriptions, and programs that continue to be generated beyond the borders of conflict-affected states. As Filip Ejdus (2018) suggests with reference to the employment of the ownership principle in EU programming, the central technology of ownership in international intervention is "responsibilization." Ownership, in this sense, is "not designed in the first place to align the intervention to the needs of the locals, but primarily to align the local elites with the objectives of the intervention" (39). As recent literature on resistance has underlined, however, locals—especially local elites—are not so easily "aligned" through the levers of socialization and incentivization and have consistently demonstrated both ingenuity and determination in bending, subverting, and resisting reform efforts in ways that advance their own political agendas at the expense of those of the broader international community (Ejdus and Juncos 2018; Bagayoko, Hutchful, and Luckham 2016; Mac Ginty 2011). Ultimately, then, this fundamental tension—between expansive, inclusive rhetoric and restrictive, contested practice—has remained *the* definitive characteristic of the local ownership debate since it first gained traction in the field of peacebuilding in the late 1990s.

Ownership of What?

Growing attention to questions of ownership in peace processes is, of course, one element of the larger local turn in peace and conflict research. A product of a deepening disillusionment with the poor performance—and neocolonial connotations—of the post–Cold War liberal peacebuilding project, the local turn amounts, at its core, to a reconsideration of agency. One branch of this turn, heavily influenced by critical and postcolonial theory, calls for "a decolonization of knowledge about peace making and peacebuilding" and challenges core Western rationalities of order and governance (Mac Ginty and Richmond 2013, 765). In so doing, it self-consciously seeks to upend conventional peacebuilding logics, making the case that genuinely emancipatory peacebuilding needs to start from the everyday lived realities of those caught up in conflict and needs to focus on creating the space in which a grounded, organic peace can begin to emerge, rooted in the cultural repertoires and social capital of "ordinary" people and communities. Debates around local ownership comprise the second primary branch of the local turn (Leonardsson and Rudd 2015). These debates have straddled the scholarly/policy divide—and have been by extension less critical and more problem-solving in orientation—and have centered around the challenge of effectively incorporating local voices, actors, and perspectives within peacebuilding processes, in the interests of solidarity, legitimacy, and efficacy. The significant overlap between the literatures on ownership and on hybridity underscores that debates around local ownership have been less about transcending liberal approaches and more about creatively blending the liberal and the local (neither of which exist as pure, uncontaminated categories) to produce outcomes that are more legitimate, more just, and more durable than either the liberal or the local are capable of generating on their own. If decolonization is the leitmotif of the local turn's critical variant, then partnership is the problem-solving equivalent.

It is no coincidence that the local ownership norm began to take hold, in the first decade of the new millennium, just as the liberal peacebuilding project itself began to falter. In its earlier iterations, local ownership was viewed as one possible means of saving liberal peacebuilding (Paris 2010), primarily by accommodating local preferences, traditions, and practices in a manner that would secure the collaboration of local actors, and of local elites in particular. Local ownership, in other words, was to be achieved by means of hybridity—defined by Brown and Gusmao (2009, 62) as "the co-existence of introduced (generally liberal institutional) models of governance and local governance practices, rooted in place and culture, and enjoying widespread social legitimacy"—and in ways that didn't require external interveners to suspend or disregard their own normative preferences. While hybridity can be viewed as a natural—unplanned and uncontrolled—outcome of the interaction of competing logics and orders, "prescriptive" hybridity has also been a prominent feature of modern peacebuilding. From Afghanistan's constitutional Loya Jirga to Rwanda's Gacaca courts, hybridity has been viewed as a means of enhancing the legitimacy and depth of international interventions by melding abstract liberal ideas—whether around democracy or justice—to established local traditions and practices.

One of the more important debates to have emerged within the hybridity literature in recent years has in fact revolved around the limits and possibilities of hybridity as a deliberate peacebuilding strategy. In some quarters prescriptive hybridity has been embraced as a means of operationalizing the local turn through a framework of "hybrid peace governance" (Belloni 2012), which strategically combines elements of both international and local to create a robust and broadly legitimate framework for peacebuilding (Paffenholz 2015, 863). This "best of both worlds" scenario—aimed, ultimately, at achieving "co-ownership" of peace processes—has, however, been criticized as both cynical and naïve. As Jenny Peterson (2012, 17) has suggested, it enables hybridity to be deployed as yet one more instrument of legitimization, designed to co-opt local institutions and practices on behalf of what remains a hegemonic liberal project. Hybridity becomes, in this sense, just "one more thing to be planned and administered" by international actors (Millar 2014, 502). As Gearoid Millar has demonstrated with reference to Sierra Leone, however, efforts to graft local institutional forms onto internationally accepted normative structures—around truth-telling rituals in the context of transitional justice, for example—typically overstate the malleability of the conceptual frameworks though which local actors make sense of their worlds, and thus rarely produce predictable, let alone peace-promoting, effects on local populations. At the same time, Thania Paffenholz (2016, 215) notes, hybrid political orders also have a tendency to be instrumentalized by local elites for their own purposes, and therefore "the mere fact of being a hybrid order does not necessarily contribute constructively to peacebuilding."

More broadly, the legitimizing promise of hybridity has proven elusive, in large part because of the challenges of reconciling fundamentally different sources of legitimacy in the unsettled, highly contested normative environments of postconflict contexts. While the liberal norms underpinning conventional approaches to peacebuilding are no longer assumed to carry inherent legitimacy among conflict-affected societies—they are in most cases too abstract, too foreign, and too indifferent to the needs and priorities of local populations (Roberts 2011)—they of course remain deeply embedded within the intervention strategies of external actors. At the level of practice, then, it remains the case that beyond the rhetorical level, the local ownership norm is routinely supported by outsiders only insofar as it legitimizes "higher-order" liberal normative principles (Sending 2009). Conversely, as Willemijn Verkoren and Mathijs van Leeuwen (2013, 164) have found, at the local level those social actors "that are the most deeply rooted, completely trusted, and experienced as legitimate on the ground, are often not the same as those most in step with Western norms." In contexts where there is in fact little overlap between local and international legitimacy, not only do outsiders struggle to differentiate between legitimate and illegitimate forms of local agency, but they actively seek out "local owners" whose structures and worldviews (and accounting practices) most closely resemble, or can be made to resemble, their own. Thus, while much of the force of the local ownership norm lies in its appeal to "grounded legitimacy"—working with the grain of endogenous cultures, institutions, and traditions in the interests of establishing durable political authority (Clements 2014)—almost nowhere have international peacebuilders been willing to invest the time, resources, and political capital required to fully ground their own interventions in nonliberal, contextually specific sources of local legitimacy.

Hybridity, then, offers no easy shortcut for international actors looking for a way around the impasse of liberal peacebuilding. Indeed the realization that hybridity is not a condition that can be easily shaped, crafted, or manipulated (Mac Ginty and Richmond 2016, 220) resonates with growing skepticism about the *tractability* of conflict-affected spaces and has deeper implications for thinking about the very nature of peacebuilding itself. In truth, peacebuilding is more contested than ever, with liberal, postliberal, and hybrid variants competing and coexisting at the level of both theory and praxis. The short-lived (and only ever partial) consensus around peacebuilding's liberal core—including its theory of change and the associated assumptions about the agency required to enact such change—has unraveled, and the prospects for another "master" peacebuilding narrative emerging to take its place remain dim. Whatever its blind spots, the liberal peacebuilding paradigm at least offered an internally consistent and relatively comprehensive vision of the means and ends of peacebuilding processes and provided a coherent answer to the question "What is to be owned?" Unmoored from these liberal premises—including commitments to human rights, democracy, free markets, and the rule of law—it is less clear what peacebuilding stands for, beyond relatively abstract commitments to emancipation and local empowerment. In embracing the inherent nonlinearity of peacebuilding (Debiel and Rinck 2016, 241), emerging postliberal scholarship still struggles to offer a compelling narrative either for how local agential capacities are to be empowered or for how local-level agency leads to national-level peace. Such scholarship, in rejecting the notion that the problems of conflict-affected states are "amenable to political solutions in terms of instrumental governing interventions on the basis of cause-and-effect understandings" (Chandler 2017, 209), also throws into question the very possibility of imagining progressive, deliberate transformations from conflict to peace that are inherent in the very idea of a peace process. In the space of several decades, then, peacebuilding has run the gamut from liberal hubris (outsiders can solve all the problems of war-torn societies) to postliberal humility (outsiders can solve none of them).

While postliberal perspectives are far from hegemonic and continue to carry more weight in academic debates than in operational contexts, shifts in the broader political contexts within which peacebuilding unfolds are also indicative of an unsettling of the ontological premise that the period of separation between war and peace is occupied by a coherent, discernible peace process, recognizable as such even if the specifics shift from context to context. In the first place, fewer contemporary war-to-peace transitions are framed by comprehensive peace accords—akin to Northern Ireland's Good Friday Agreements (1998) or Bosnia's Dayton Peace Accords (1995)—that map out specific parameters of peace processes, specific commitments of relevant parties, and coherent visions of a postconflict political order. Indeed, following a surge in the number of negotiated civil war settlements following the end of the Cold War, this trend appears to have been reversed over the past fifteen years (Walter 2017, 474). As the joint United Nations and World Bank (2018, 37) Pathways to Peace study concluded in 2018, intrastate conflicts of the contemporary era share similar characteristics—including a multiplicity of armed factions and significant involvement of outside states—that make them "particularly resistant to negotiated settlements." Somewhat paradoxically, then, given growing attention to the importance of inclusivity in both the making and the implementation of peace processes (Castillejo 2014), relatively few contemporary peacebuilding contexts are backed by broad-based, comprehensive, and inclusive peace agreements (Colombia's 2016 peace agreement being at least a partial exception). More commonly, as in the UN's most prominent "conflict management" operations in Africa (Gowan 2018), peacebuilding unfolds in the decided absence of either a peace to keep or a peace process to implement, and negotiations—such as those that produced South Sudan's latest fragile accord in 2018—have focused more on elite power-sharing and less on producing coherent roadmaps for peace or addressing deep-rooted conflict causes.

In parallel, the past decade has witnessed a significant scaling back of peacebuilding ambition on the part of the broader international community (the anemic international response to peacebuilding needs in post-Qaddafi Libya being a tragic case in point). For better or worse, the era of large, well-resourced peacebuilding operations appears to have run its course, a casualty of the declining confidence in the transformative potential of liberal ideas (Pospisil 2017, 9), a decline that reached its nadir following the US-led interventions in Iraq and Afghanistan. While this development has opened up crucial space for creative thinking about how to do peacebuilding differently, the downside may very well be diminishing enthusiasm—and commitment—on the part of key donors and international organizations to fully engage with the messy, open-ended, and invariably long-term dynamics implicit in the notion of bottom-up and locally owned peacebuilding. At the very least, a key consequence of the local turn—and the turn away from liberal universalism—in policy terms has been the progressive abandonment of the goal of holism, or the notion that discrete programmatic interventions should "add up" to more than the sum of their parts and coherently nest within a clearly articulated overall peacebuilding strategy. Evidence that a more "destructured" approach is in the process of being embraced—at least in rhetorical terms—can be seen in the recent emergence of the UN's "sustaining peace" agenda. Eschewing the "deterministic design" logic inherent in liberal peacebuilding, and building on theories of both resiliency and complexity, the sustaining peace agenda aims to shift UN peace support efforts toward identifying and supporting local-level political and social capacities able to build peace by strengthening social cohesion (de Coning 2018, 304).

Ultimately, then, both theory and practice appear to be pointing toward a new framework for peacebuilding rooted precisely in such "destructured" dynamics. This has profound implications for thinking about the nature of ownership and about the nature of the relationships between international and domestic actors in such processes. In the first place, it reinforces a broader trend away from thinking of ownership in terms of the imperative of building consensus among a critical mass of a conflict-affected society around a predefined peace process—whether liberal, postliberal, or otherwise—and a method for its operationalization. Indeed by focusing on means rather than ends, such a move effectively collapses the distinction—never easy to sustain—between an ontologically distinct peace process (typically shepherded and/or structured by a coalition of international actors in often uneasy partnership with the national governments of conflict-affected states) and broader political, social, and economic dynamics that may not be encompassed within a formal peace process but which profoundly shape the prospects for a minimally stable and just peace to emerge over much longer timeframes. As Keith Krause (2016, 169) notes, peacebuilding interventions, regardless of their ambitions, "are often just moments in a much longer process of social and political change." More important, perhaps, it implies a recognition that far from an aspiration, local ownership is in fact an unavoidable empirical reality. In other words, ongoing contestation over ownership of particular peacebuilding *projects* should not be conflated with ownership of the kinds of domestic political *processes* that ultimately determine whether or not sustainable peace takes hold; outsiders might have some limited influence over these dynamics, but they cannot control them, let alone engineer particular outcomes. Thus, while tensions between donor and local ownership can be expected to continue to play out at the micro level—where external actors' long-standing preoccupations with control and accountability (especially but not exclusively financial) sit uneasily alongside claims that outsiders should defer to the superior contextual and cultural knowledge of local partners in the design and implementation of peacebuilding projects—at the macro level international actors have progressively retreated from making substantive ownership claims over the overall shape, direction, or content of peacebuilding processes.

In the wider context of the local turn, therefore, contemporary thinking about peacebuilding praxis has in fact (re)turned—after a twenty-year detour down the path of liberal peacebuilding[1]—back to some of the foundational insights of the field. This includes, perhaps most notably, the work of John Paul Lederach, who wrote in the mid-1990s that "the greatest resource for sustaining peacebuilding in the long term is always rooted in the local people and their culture" (cited in Leonardsson and Rudd 2015, 826). It also includes the work of the UN Research Institute for Social Development's Rebuilding War-Torn Societies project, which concluded in a 1995 report that a policy of betting on the local "may in the short term be more laborious, less spectacular, and take more time, but in the long-term may be the only realistic option" (cited in Prendergast 1997, 61). Indeed the gambling metaphor may be apt, since it very much remains to be seen whether the internal dynamics of conflict-affected societies—even with strategic external facilitation—are in fact capable of generating the kinds of just, stable, inclusive, and legitimate outcomes that the liberal peacebuilding project could not. In this, much depends on the dynamics between different categories of local owners, to which this discussion now turns.

Ownership by Whom?

Perhaps the greatest ambiguity of all with regard to making sense of local ownership in the context of peacebuilding surrounds the very nature and character of the local. While internationally supported peacebuilding activities must increasingly be *legitimized* by virtue of unfolding either in support of or in collaboration with local actors, it continues to be the case that "the local" as a conceptual category obscures far more than it illuminates (as does, of course, the conceptual category of "the international"). Indeed there are clear orientalist overtones to the manner in which both scholars and practitioners casually speak of the possibilities for locally owned processes in conflict-affected countries—as if locals were not as vividly diverse, differentiated, and "unmanageable" as in other contexts—in terms they would never contemplate when referring to political dynamics in, for example, Western countries.

While the local turn has focused much-needed attention on the imperative of unpacking "the language of localism" (Mac Ginty 2016, 193), it has also led to a deepening polarization of views around who, precisely, counts as a "relevant local" for the purposes of peacebuilding. On the one hand, despite their nods to participation and inclusion, high-level policy fora, from the International Dialogue on Peacebuilding and Statebuilding to the UN Peacebuilding Commission, still tend towards state-centrism, conflating local with national, and national with governmental, in their conceptions of ownership (Tschirgi 2015). On the other hand, as noted earlier, the more critically inspired versions of the local turn have driven an increased focus on the possibilities for locating emancipatory peacebuilding practice at the level of "the everyday," far removed from the corridors of power. While the two perspectives highlight a crucial social cleavage—that between status quo–oriented political elites and "ordinary citizens," most of whom experience the status quo as unrelenting misery—to date there have been few systematic efforts to understand how the exercise of local agency in the one pole affects the exercise of agency in the other; the gap between these two understandings of the local, in other words, remains largely unbridged.

This disconnect is evident with regard to the ongoing critical project of grounding peacebuilding praxis both in the politics of the everyday and in a progressive project of emancipation. Beyond the enormous difficulties that outsiders have in accessing—let alone constructively engaging with—the complex dynamics that shape life at the level of village and neighborhood (Mac Ginty 2016, 193), there lies the persistent and vexing challenge of how precisely to empower the disempowered. If emancipation, as Karen Fierke (2017, 17) notes, is "the act of freeing, whether from the assumptions that blind us to alternatives, or from the structures of power that constrain human potential," in the context of peacebuilding it is arguably those structures of power *internal* to conflict-affected states that are simultaneously the most constraining of human potential and the least unsettled by critical scholarship. In other words, while critical scholars have directed the bulk of their attention to the imperative of emancipating local agency from the disempowering assumptions and practices of the liberal peacebuilding project (Finkenbusch 2016, 252), there has been comparatively less engagement with the challenge of overcoming the disempowering impacts of elite-level power struggles (or outright power grabs) within conflict-affected states. Typically such dynamics have deeply corrosive impacts on state-society relations and reinforce profound inequalities in the distribution of power and resources, which in turn seriously curtail the political, economic, and social space available to community-level actors. Post-Dayton Bosnia provides an illustrative case in point of these dynamics at play. Notwithstanding the short-lived Bosnian Spring of 2014, the progressive consolidation of power in the hands of the country's three main nationalist power structures—both during and in the aftermath of the intrusive liberal peacebuilding project that lost steam after about 2005—has both neutralized and demoralized the everyday as a potential source, and site, of positive peacebuilding agency. Domestic elites, across a diverse range of conflict-affected contexts, typically represent at least as great an obstacle to the empowerment of grassroots agential capacities as liberal peacebuilders, and it is not clear that even the most enlightened and empathic efforts to facilitate local-level empowerment can effectively capacitate nonelite actors to resist, let alone overcome, predatory elite behavior.

These ongoing tensions between elite and everyday conceptions—and practices—of ownership also have profound implications for the legitimacy of postconflict transitions. Whether an emerging political order is considered to be legitimate by a critical mass of domestic local actors hinges very much on the nature—and the quality—of the relations between governors and governed. In other words, as Kevin Clements (2014, 13) has suggested, "legitimacy is determined by whether the contractual relationship between the state and citizens is working effectively or not." Rebuilding state-society relations on durable, locally legitimate foundations, therefore, is not solely a question of going with the grain of endogenous culture and tradition—however important this might be—but also of (re)establishing a web of mutual rights, obligations, and expectations that enables elites to govern with minimal coercion and with maximal support from citizens (Mcloughlin 2015). From this perspective, the core challenge of peacebuilding—and of any meaningful effort to foster local ownership—lies not just in empowering local-level actors as "carriers" of positive agency for peacebuilding, or restraining the worst impulses of empowered elites, but also in facilitating processes that bring both sets of owners into productive conversation with each other around the necessarily long-term process of defining a new social contract (Pouligny 2005, 499).

All of this leads, finally, back to the issue that has always been at the heart of the local ownership debate: the nature of the relationship between international and domestic actors in conflict-affected settings. For all the attention devoted to the local turn in recent years, few observers have gone so far as to advocate for the logical extension of the local ownership argument, which is that outsiders should simply withdraw from conflict contexts and leave it to the locals to sort matters out for themselves. As Peter

Finkenbusch (2016, 254) has observed, even postliberals have not moved beyond the belief that "endogenous solutions still have to be somehow 'facilitated by external actors.'" Accepting the twin assumptions that (a) outsiders will continue to engage in conflict-affected contexts and (b) the broader objective of peacebuilding remains conflict transformation, the crucial question facing outside actors—in a world in which the means to transform conflict remain both deeply contested and inadequately understood—is how to engage in ways that will enhance the prospects for peace to emerge over the long term. Falling back on the formula of "developing existing local capacities and capabilities" (Chandler 2016, 51) seems inadequate, since in the context of complex social systems with nonlinear dynamics, how can outsiders anticipate which processes, dynamics, and capacities are likely to contribute to peace, and which may in fact undermine it? In making judgments around what kinds of everyday agency should be valorized in the name of peacebuilding, interveners cannot even rely on the distinction between violent and nonviolent forms of local agency (privileging the latter while marginalizing the former), since in practice such a distinction is often both nebulous and unstable (Randazzo 2016, 1355–1356). External actors have in fact struggled across a range of conflict-affected contexts to remain on the right side of the "Do no harm" principle and have all too often, as Krause (2016, 169) suggests, been "complicit by default in the power politics of local actors by providing resources that reinforce existing structures, inequalities, and forms of overt and covert violence." For external actors, ultimately, the shift to the conceptual apparatus of emancipation, facilitation, and empowerment may be read as simply reframing—rather than resolving—the complex questions surrounding how agency, both external and local, may be most effectively mobilized in the name of sustainable peacebuilding.

Conclusion

In a paper published in 1996, Ken Bush captured the essence of the local ownership principle by suggesting that the challenge of peacebuilding, at its core, "is to nurture and create the political, economic and social space within which indigenous actors can identify, develop, and employ the resources necessary to build a peaceful, just and prosperous society" (86). More than two decades on, the challenge remains, even if both the theoretical and empirical terrain upon which peacebuilding unfolds is shifting. And while local ownership has now achieved the status of international norm, its inherent ambiguity continues to leave it poorly placed to prescribe with any precision the manner in which agency and resources—whether internally or externally derived—might be combined in conflict-affected contexts to increase the odds that a stable, sustainable, and minimally just peace will emerge.

As elaborated in this chapter, what is shifting—in rather fundamental ways—are understandings not only of the means and ends of peacebuilding but of the very *status* of peacebuilding as a coherent, self-contained project. The reconceptualization of peacebuilding, from a time-bound, coordinated, and managed process of rebuilding both states and state-society relations to a necessarily long-term and nonlinear mélange of social change processes that are inherently resistant to externally driven social engineering, renders increasingly self-evident the proposition that peacebuilding outcomes will be determined overwhelmingly by dynamics internal to conflict-affected states. While external forces and structures still frame the limits of the possible (as they do to greater or lesser degrees in any national context), the willingness and ability of international actors to collectively exercise meaningful ownership over such dynamics is clearly on the wane.

While local ownership may increasingly be an empirical reality at the macro level (if not necessarily at the micro or project level), recasting external peacebuilders as supporting actors—especially in the context of the simultaneous rise of the local turn and fall of the liberal peacebuilding project—also necessitates a careful reconsideration of the nature of such engagement, the changing (and increasingly digital) nature of the tools and instruments through which it is undertaken, and the legitimacy of the actors who undertake it. For all of the negative lessons captured in recent years about how *not* to do peacebuilding, outside actors still know far too little about the levers at their disposal to "nurture space" within conflict-affected environments, about the nature of the relevant building blocks of peace, justice, and prosperity within any given context, and about which indigenous actors and structures—at which levels of society—might be best positioned to most effectively advance the cause of peace.

Note

1. David Chandler (2017) has recently framed this, in an echo of E. H. Carr, as "the twenty years' crisis."

References

Bagayoko, N., E. Hutchful, and T. Luckham. 2016. "Hybrid Security Governance in Africa: Rethinking the Foundations of Security, Justice and Legitimate Public Authority." *Conflict, Security and Development* 16 (1):1–32.

Belloni, R. 2012. "Hybrid Peace Governance: Its Emergence and Significance." *Global Governance* 18 (1):21–38.

Brown, M. A., and A. F. Gusmao. 2009. "Peacebuilding and Political Hybridity in East Timor." *Peace Review* 21:61–69.

Bush, K. 1996. "Beyond Bungee Cord Humanitarianism: Towards a Developmental Agenda for Peacebuilding." *Canadian Journal of Development Studies* 17(4): 75–92.

Campbell, Susanna. 2018. *Global Governance and Local Peace: Accountability and Performance in International Peacebuilding*. Cambridge: Cambridge University Press.

Castillejo, C. 2014. "Promoting Inclusion in Political Settlements: A Priority for International Actors." Norwegian Peacebuilding Resource Centre Report, March. https://www.files.ethz.ch/isn/179668/e064fd8c68d1be1dd3af802be5a97dd8.pdf.

Chandler, D. 2016. "The Future of Peacebuilding." In *Peacebuilding in Crisis: Rethinking Paradigms and Practices of Transnational Cooperation*, edited by T. Debiel, T. Held, and U. Schneckener, 41–55. New York: Routledge.

Chandler, D. 2017. *Peacebuilding: The Twenty Years' Crisis, 1997–2017*. London: Palgrave Macmillan.

Clements, K. 2014. "What Is Legitimacy and Why Does It Matter for Peace?" *Accord* 25:13–16. https://www.c-r.org/accord/legitimacy-and-peace-processes/what-legitimacy-and-why-does-it-matter-peace.

Debiel, T., and P. Rinck. 2016. "Rethinking the Local in Peacebuilding." In *Peacebuilding in Crisis: Rethinking Paradigms and Practices of Transnational Cooperation*, edited by T. Debiel, T. Held, and U. Schneckener, 240–256. New York: Routledge.

de Coning, C. 2018. "Adaptive Peacebuilding." *International Affairs* 94 (2):301–317.

Donais, T. 2009. "Empowerment or Imposition? Dilemmas of Local Ownership in Post-conflict Peacebuilding Processes." *Peace and Change* 34 (1):3–26.

Ejdus, F. 2018. "Local Ownership as International Governmentality: Evidence from the EU Mission in the Horn of Africa." *Contemporary Security Policy* 39 (1):28–50.

Ejdus, F., and A. Juncos. 2018. "Reclaiming the Local in EU Peacebuilding: Effectiveness, Ownership and Resistance." *Contemporary Security Policy* 39 (1):4–27.

Fierke, K. 2017. "Critical Theory, Security and Emancipation." *Oxford Research Encyclopedia of International Studies*. https://oxfordre.com/internationalstudies/view/10.1093/acrefore/9780190846626.001.0001/acrefore-9780190846626-e-138.

Finkenbusch, P. 2016. "'Post-liberal' Peacebuilding and the Crisis of International Authority." *Peacebuilding* 4 (3):247–261.

Gowan, R. 2018. "Peace Operations in 2017–18: Balancing Conflict Management and Political Approaches as an Era Comes to a Close." In R. Gowan and R. Rappa, *Global Peace Operations Review: Peace Operations 2018*. New York: Center on International Cooperation. https://peaceoperationsreview.org/wp-content/uploads/2018/05/gpor_Peace_Operations_2018_full_final_WEB.pdf.

International Dialogue on Peacebuilding and Statebuilding. 2011. "The New Deal for Engagement in Fragile States." http://www.pbsbdialogue.org/en/new-deal/about-new-deal/.

Krause, K. 2016. "Reforming the Security Sector and Rule of Law." In *Peacebuilding in Crisis: Rethinking Paradigms and Practices of Transnational Cooperation*, edited by T. Debiel, T. Held, and U. Schneckener, 155–174. New York: Routledge.

Krook M., and J. True. 2010. "Rethinking the Life Cycles of International Norms: The United Nations and the Global Promotion of Gender Equality." *European Journal of International Relations* 18 (1):103–127.

Lemay-Hébert, N., and S. Kappler. 2016. "What Attachment to Peace: Exploring the Normative and Material Dimensions of Local Ownership in Peacebuilding." *Review of International Studies* 42 (5):895–914.

Leonardsson, H., and G. Rudd. 2015. "The 'Local Turn' in Peacebuilding: A Literature Review of Effective and Emancipatory Local Peacebuilding." *Third World Quarterly* 36 (5):825–839.

Mac Ginty, R. 2011. *International Peacebuilding and Local Resistance: Hybrid Forms of Peace*. Houndmills, UK: Palgrave Macmillan.

Mac Ginty, R. 2016. "What Do We Mean When We Use the Term 'local'?" In *Peacebuilding in Crisis: Rethinking Paradigms and Practices of Transnational Cooperation*, edited by T. Debiel, T. Held, and U. Schneckener, 193–209. New York: Routledge.

Mac Ginty, R., and O. Richmond. 2013. "The Local Turn in Peace Building: A Critical Agenda for Peace." *Third World Quarterly* 34 (5):763–783.

Mac Ginty, R., and O. Richmond. 2016. "The Fallacy of Constructing Hybrid Political Orders: A Reappraisal of the Local Turn in Peacebuilding." *International Peacekeeping* 23 (2):219–239.

Mcloughlin, C. 2015. "Researching State Legitimacy: A Political Approach to a Political Problem." Research Paper No. 36. Developmental Leadership Program, University of Birmingham. http://publications.dlprog.org/StateLegitimacy_PoliticalApproach.pdf.

Miklian, J., K. Liden, and A. Kolas. 2011. "The Perils of 'Going Local': Liberal Peace-Building Agendas in Nepal." *Conflict, Security and Development* 11 (3):285–308.

Millar, G. 2014. "Disaggregating Hybridity: Why Hybrid Institutions Do Not Produce Predictable Experiences of Peace." *Journal of Peace Research* 51 (4):501–514.

Paffenholz, T. 2015. "Unpacking the Local Turn in Peacebuilding: A Critical Assessment towards an Agenda for Future Research." *Third World Quarterly* 36 (5):856–874.

Paffenholz, T. 2016. "Peacebuilding Goes Local and the Local Goes Peacebuilding." In *Peacebuilding in Crisis: Rethinking Paradigms and Practices of Transnational Cooperation*, edited by T. Debiel, T. Held, and U. Schneckener, 210–225. New York: Routledge.

Paris, R. 2010. "Saving Liberal Peacebuilding." *Review of International Studies* 36:337–365.

Peterson, J. H. 2012. "A Conceptual Unpacking of Hybridity: Accounting for Notions of Power, Politics and Progress in Analyses of Aid-Driven Interfaces." *Journal of Peacebuilding and Development* 7 (2):9–22.

Pospisil, J. 2017. "Pathways to Post-Liberal Peace: Perspectives on the 'Common Good' in Peace and Statebuilding." Working Paper. Political Settlements Research Programme, University of Edinburgh. http://www.politicalsettlements.org/wp-content/uploads/2017/05/201704_WP7_Pospisil_BRIJ5361_Working_Paper_Pospisil.pdf.

Pouligny, B. 2005. "Civil Society and Post-Conflict Peacebuilding: Ambiguities of International Programmes Aimed at Building 'New' Societies." *Security Dialogue* 36 (4):495–510.

Prendergast, J. 1997. "Post-conflict Reconciliation Programming." Occasional Paper. Life and Peace Institute.

Randazzo, E. 2016. "The Paradoxes of the 'Everyday': Scrutinising the Local Turn in Peace Building." *Third World Quarterly* 37 (8):1351–1370.

Roberts, D. 2011. "Post-Conflict Peacebuilding, Liberal Irrelevance and the Locus of Legitimacy." *International Peacekeeping* 18 (4):410–424.

Sending, O. 2009. "Why Peacebuilders Fail to Secure Ownership and Be Sensitive to Context." *Security in Practice* 1. Norwegian Institute of International Affairs. https://www.nupi.no/en/Publications/CRIStin-Pub/Why-Peacebuilders-Fail-to-Secure-Ownership-and-be-Sensitive-to-Context.

Tschirgi, N. 2015. "Bridging the Chasm between International and Domestic Approaches to Peacebuilding." *RCCS Annual Review* 7. https://journals.openedition.org/rccsar/605.

United Nations. 2010. *Review of the United Nations Peacebuilding Architecture*. A/64/868-S/2010/393, 21 July.

United Nations. 2015. *The Challenge of Sustaining Peace: Report of the Advisory Group of Experts for the 2015 Review of the United Nations Peacebuilding Architecture*. A/69/968-S/2015/490, 29 June.

United Nations and World Bank. 2018. *Pathways for Peace: Inclusive Approaches to Preventing Violent Conflict*. Washington, DC: World Bank.

Verkoren, W., and M. van Leeuwen. 2013. "Civil Society in Peacebuilding: Global Discourse, Local Reality." *International Peacekeeping* 20 (2):159–172.

Von Billerbeck, S. 2015. "Local Ownership and UN Peacebuilding: Discourse versus Operationalization." *Global Governance* 21:299–315.

Walter, B. 2017. "The New New Civil Wars." *Annual Review of Political Science* 20:469–486.

CHAPTER 39

ENVIRONMENTAL PEACEBUILDING

FLORIAN KRAMPE AND ASHOK SWAIN

For international and domestic actors, postconflict situations constitute one of the most difficult policy arenas to understand and operate within. In this context, the role of natural resources management has received increasing scholarly attention over the past three decades, as researchers have focused on the ecological foundations for a socially, economically, and politically resilient peace (Krampe 2017). Availability of renewable natural resources is increasingly falling short of meeting human needs. While population growth and economic activities are severely crippling the natural resource base on which human beings are dependent for survival, war and violent conflicts also contribute significantly to the destruction of the environment. Environmental destruction is commonly seen as a repercussion of violent conflict or conflict-induced migration. Besides the direct adverse effects of the armed conflict on the environment, in some cases environmental change is a deliberate objective. Large-scale environmental destruction has taken place in most parts of the developing world due to the civil wars in those regions. This is not only a direct impact of warfare but also a result of more complicated connections. For instance, internal violence makes it impossible to develop sustainable agriculture and leads to massive deforestation and the destruction of wildlife (Swain 2012).

For the most part since the end of the Cold War, postconflict peacebuilding strategies have been dominated by a neoliberal agenda that often favors short-term economic growth solutions over long-term environmental and resource availability concerns (Richmond 2012). The significant overlap between countries with heightened vulnerability to resource availability and countries coming out of war (UN Environment Program 2009) provides a prima facie case for suggesting that integrating environmental concerns in peacebuilding activities needs to be prioritized. However, states recovering from prolonged conflicts might not be interested in pursuing sustainable development policies rooted in environmental and social justice; rather, their newly enhanced political and economic strength and legitimacy might simply yield more effective resource capture. On the one hand, the postconflict setting might increase the state's capacity to seek more control over environmental resources; on the other hand, the postconflict setting facilitates broader societal engagement for environmental resource–induced cooperation. However, there can be no durable and sustainable peace if natural resources that sustain livelihoods and the ecosystem are destroyed or degraded. Analysis of past intrastate conflicts suggests that conflicts associated with natural resources are twice as likely to revert to conflict in the first five years. Nevertheless very few peace negotiations aiming to settle conflicts linked to natural resources have addressed resource management mechanisms (Binningsbø and Rustad 2008).

Machlis and Hanson (2008, 734) see the governance of environmental resources at the core of postwar efforts, as this "could help avert resource conflicts, reduce degradation of war-dominated ecosystems, and increase postwar restoration of ecosystem services, thereby encouraging peace and security." Environmental resources bear the potential for a swift economic recovery, while equally being considered triggers for conflicts if not managed smartly (Lujala and Rustad 2012). Jensen and Lonergan (2012, 9) argue that "integrating natural resource management and environmental sustainability into peacebuilding" is the way to avert uncontrolled exploitation in the aftermath of conflict. Decisions about the restoration, management, and protection of natural resources have vital consequences for short-term stability, long-term sustainable development, and successful peacebuilding. Every state, including the ones recovering from a period of violent conflict, needs to both use and protect its vital natural resources, such as water, land, and forest. The recognition that these environmental resources can contribute to violent conflict accentuates their potential significance as pathways for cooperation and the consolidation of peace in postconflict societies. Thus environmental resources can contribute to peacebuilding through economic development and sustainable livelihoods, while cooperation over their management can also serve as an effective platform or catalyst for enhancing dialogue, building confidence, exploiting shared interests, and broadening cooperation between segmented and conflicting groups.

Consequently international peacebuilding and development actors increasingly work with the complexity of the humanitarian, development, environment, and peacebuilding nexus in areas highly affected by climate change. Achim Steiner, administrator of the UN Development Program (UNDP) and former chief of the UN Environment Program (UNEP), emphasizes the need to connect our understanding of these complex drivers of conflicts and the need to address the broader developmental failures that often underpin insecurity and conflict.[1] Many agencies, including UNDP and UNEP, are increasingly searching for pathways to conflict-sensitive programming, i.e., to minimize the negative impacts of their work in the field. But the interest is moving beyond "do no harm" principles to specifically maximize positive impacts on conflict. For instance, the latest Environment Strategy of the UN Department of Field Support (2017, 3), states that UN peace operations should aim "to seek a positive long-term legacy through the development of specific environment-related projects that may benefit societies and ecosystems over the long term."

Environmental peacebuilding conceptually emerged along two separate pathways—one political and one academic—in the 1990s that have become increasingly merged in recent years. Politically concern for natural resources in the context of postconflict peacebuilding entered the agenda during the 1990s through initiatives of UNEP and its leadership under Klaus Töpfer. Against the backdrop of wars in the 1990s and the increasing concern that resource scarcity, especially water, would fuel conflicts and cause conflict relapses, UNEP started work in this area.[2] Initially this work focused on postconflict environmental assessments in the wake of the Balkan wars (UN Environment Program 2000, 2001). This work subsequently evolved in more substantive policy and research efforts, especially since 2006 led by UNEP's postconflict and disaster management branch (Krampe 2017). We will subsequently explore this pathway as the "resource risk perspective."

Academically, environmental peacebuilding emerged in the late 1990s as a critique of the research focus on environmental conflict. A group of scholars argued that research should study the rise of cooperative solutions in the face of natural resource scarcity rather than be focused only on the associated risks (Wallensteen and Swain 1997, 702). Decisively more broadly conceptualized than postconflict peacebuilding in the peace and conflict debate, this research effort has in the past decades produced rich empirical studies focusing, among others, on cooperation and rivalries over transnationally shared water resources between states (Earle et al. 2015; Ide and Detges 2018; Swatuk 2015; Wolf 1998; Zeitoun and Mirumachi 2008). More recent research looks also at the role of environmental services, climate change adaptation, and mitigation within states (Kashwan 2017; Krampe 2016b; Matthew 2014). We will subsequently discuss this research stream as the "cooperation perspective."

In this chapter we take stock of the current state of the art on environmental peacebuilding. For that, we provide a summary of the most common definitions before looking back at the development of environmental peacebuilding along the two most noticeable perspectives and the remaining challenges and pathways for future research.

Definitions of Environmental Peacebuilding

Environmental peacebuilding is an emerging interdisciplinary field of study, deriving from the environmental security discourse, and comprises environmental studies, geography, and peace and conflict research. Following the research of the past three decades, in its broadest form, environmental peacebuilding encompasses research efforts that link the management of natural resources to conflict prevention, conflict resolution, and peacemaking, as well as postconflict peacebuilding. As such it is conceived more broadly than most contributions in this book, going beyond the level of postconflict societies. The binding element in this research is often "environmental cooperation," which serves as either independent or dependent variable, depending on the specific research focus. As such, overarchingly, environmental peacebuilding can be defined as the sustainable management of natural resources to prevent conflict and build peace—before, during, or after conflict—emphasizing the potential for environmental cooperation to support peace and stability.

Yet on closer review there are notable distinctions among different researchers around the meaning of peace and stability but also relating to the time frame of a conflict. These differences allow grouping previous research in two dominant perspectives that have shaped the environmental peacebuilding discourse over the past two decades (Krampe 2017). The cooperation perspective is driven by the potential of environmental cooperation to contribute to peace through spillover effects. This perspective focuses primarily on the interstate level and often on conflict prevention rather than postconflict peacebuilding. In contrast, the resource risk perspective recognizes resource-induced instability, especially after intrastate conflicts, and stresses the need to mitigate these risks to sustain the absence of violence (negative peace) by facilitating environmental cooperation. Along the two perspectives we find more nuanced and narrow definitions.

Examples of the cooperation perspective derive from bringing together the literature on cooperation theory in international relations and social capital theory in development studies. It is not difficult to find a strong argument for the general proposition that cooperation over environmental resources between rival actors in a postconflict society can have positive spinoffs for peace in other contentious areas (Swain and Öjendal 2018): the basis for achieving this environmentally induced conflict resolution and peacemaking is that cooperation over environmental resources among disputing groups will help to remove mistrust and promote regular interactions, creating norms for reciprocity and infusing a broader understanding of the peace (see, e.g., Krampe and Gignoux 2018). On that premise, there is growing emphasis among the international development policy community in promoting diplomacy over shared water resources. This water diplomacy aims to get disputing riparian countries to build a framework of water sharing and, on that foundation, facilitate and diffuse riparian cooperation to other areas of contention. Pohl and Swain (2017) therefore define water diplomacy as endeavors aimed at receiving political support in nudging governments into cooperative behavior to overcome inertia and vested interests for water sharing and to convince political decision-makers of the necessity and benefits of technical cooperation on shared water resources to realize the full potential of regional peace and development.

Notably these definitions go beyond the narrower peace and statebuilding discourse, yet they serve as an entry point to consider the potential of environmental issues as harbingers of peace in postconflict societies. UNEP, for instance, focuses on the postconflict timeframe, often linked to UN peacebuilding efforts. In this, UNEP recognizes that environmental issues can contribute to violent conflict; about 40% of armed conflicts have been linked to natural resources. As a solution they recommend environmental peacebuilding in the form of cooperation over sustainable natural resource management to aid conflict prevention and prevent a relapse to conflict (UN Environment Program 2009), linking it closely to more negative peace understandings of peacebuilding. In contrast, some define environmental peacebuilding as an opportunity to build positive peace after a conflict, by emphasizing a longer-term development perspective. For instance, Krampe (2017) defines environmental peacebuilding as the conflict-sensitive and sustainable management of natural resources in conflict-affected or postconflict states, by which it provides ecological foundations for a socially, economically, and politically resilient peace.

The Cooperation Perspective

In the late 1990s, criticizing the research focus on environmental and specifically water conflict, scholars argued that research should instead study the rise of cooperative solutions in the face of natural resource scarcity (Wallensteen and Swain 1997, 702). Ken Conca (2001, 227) outlined a first theoretical framework, suggesting that "carefully designed initiatives for environmental cooperation" could facilitate peace. Peace in this case meant "several dependent variables that might be affected by environmental cooperation," such as perceptions of other actors, actors' cost-benefit calculations, and broader societal changes (227). Succeeding empirical assessments of this relationship focused on two mechanisms that affect these variables: "changing the strategic climate," i.e., affecting the cost-benefit calculations of states to make conflict less appealing, and "strengthening post-Westphalian governance," i.e., affecting society broadly by disseminating new transnational norms (Conca and Dabelko 2002, 9). Like earlier studies, the case studies mostly drawn from shared water resources between and among countries confirmed that environmental issues may provide an entry point for cooperation (Conca and Dabelko 2002).

Conflict and cooperation are not necessarily and not even often mutually exclusive. As such, the discussion regarding the relationship between natural resource scarcity, particularly water resources, and conflicts or cooperation has not produced any consensus—and maybe never will. Increasing scarcity of freshwater resources may not generate conflicts or cooperation in itself but does interact with complex societal, political, and geopolitical dynamics, thereby increasing the probability of conflicts. In fact the desire to acquire freshwater resources can be observed at all levels of society, not only between countries but also among different social groups within the basin countries. When several stakeholders depend on the same freshwater system, an individual actor's decision-making on withdrawal and/or pollution of shared resources has the potential to bring about violent conflict (Swain 2004).

However, institutions as well as power asymmetries, geostrategic location, and understanding of a new water user's needs can also lead to cooperation and building of relations (Swain 2012; see also Ide and Detges 2018). Cooperation, as an interactive process, can transform a situation from a potentially destructive conflict into a cooperation-induced productive situation. Cooperation does not only mean that there is an absence of conflict (Zeitoun and Mirumachi 2008); it also implies that there is a mutual desire and willingness to address the conflict through negotiated and peaceful means. States and groups collaborate to protect and manage the scarce water resources, and these collaborations can have positive spinoffs for peace. This "peacemaking" can take place on the foundation of water-induced cooperation. As such, water cooperation has the potential to transform the mistrust and suspicion between and among countries to bring opportunities for shared gains and set up a model of reciprocity (Conca 2001; Krampe and Gignoux 2018). It can also pave the way for greater interaction, interdependence, and societal linkages. The diffusion of bilateral cooperation over the water resources to other areas is not uncommon (Swain 2015).

Some research in this field, especially of water governance in fragile states, focuses on the quality of cooperation, linking it to power asymmetries and hegemony (Zeitoun and Mirumachi 2008). And as indicated, studies of transboundary water management provide abundant examples of environmental cooperation prevailing between nations in the face of water scarcity (e.g., Jägerskog, Swain, and Öjendal 2014). In fact, disputes over water are frequently resolved through interstate cooperation, rather than causing armed conflict (Delli Priscoli and Wolf 2009). The success of these positive examples has given rise to interlinked approaches between science and policy, offering water diplomacy as an approach to managing water issues (Islam and Susskind 2013).

The hope of linking water cooperation and larger regional peacemaking has spawned interest by policymakers in the past decade. The World Bank and other international donors have aimed to bring greater regional peace and cooperation in the Danube Basin, Nile Basin, and Zambezi Basin through river water cooperation. Various international aid agencies are actively engaged in promoting regional peace through transboundary water cooperation in conflict basins such as the Mekong, Salween, Ganges-Brahmaputra, Indus, and Aral Sea (Huntjens et al. 2016). Yet despite a range of theoretical uncertainties and even lack of sufficient empirical proof establishing the link conclusively (Krampe 2017) this has not diminished policy interest in the field.

Some researchers suggest that the real challenge is the lack of a clear strategy for regional peace through water sharing in conflict-affected regions (Grech-Madin et al. 2018). Water cooperation in itself does not promise peace nor sustainability. There are various side effects of water cooperation depending on the form that cooperation takes, the institutional frameworks in place, and the domestic and geopolitical interests that it serves. For instance, the institutional design of treaties and joint river commissions especially determines the character of such dispute resolution (e.g., Mitchell and Zawahri 2015). Others suggest that the history of past conflict and cooperation determines the success of environmental cooperation (Ide and Detges 2018). While institutions matter, agency remains critical. If the individual actors are not striving for sustainable use of water resources, the improved cooperation and joint water exploitation among them will not bring any desirable result in the long run. Rather than bringing a sustainable use of water resources, cooperation between warring countries might facilitate further unsustainable use (Swain and Jägerskog 2016).

As indicated, most of the studies in the cooperation perspective focus solely on interstate relations. Yet there is increasing focus on water issues during and after internal armed conflict. Empirically rich, these studies highlight the potential of natural resources to support postconflict recovery, yet they equally consider them as triggers of conflict livelihoods (Troell and Weinthal 2014). Ignoring water, land, or forest in peacebuilding can be counterproductive because the resultant socioeconomic effects impact community livelihoods (Troell and Weinthal 2014). Sociopolitical factors, such as local discourses and identities, have recently been found to be critical "drivers of the Israeli-Palestinian water conflict" (Ide and Fröhlich 2015, 668), offering new pathways of understanding postconflict water resource management.

The linkage between peacebuilding and natural resources management in postconflict countries can have several dimensions. The management of natural resources, particularly freshwater, might act as a starting point for reestablishing trust and cooperation in a postconflict context. Conflict-affected societies demand a rapid restart of the economy, often focusing on the restoration and sustainable exploitation of natural resources and achieving some broader societal security (Kostić, Krampe, and Swain 2012). Without great care and smart planning, this process of postconflict recovery is likely to lead to poor choices in the early stages, and that can boomerang and intensify mistrust between actors. The need for rapid development has to be balanced with the obvious risk of exploitation and an exceptionally fragile situational context. There is no doubt that environment and water resources cannot be neglected; addressing them as part of peacebuilding is not merely a good idea but a security imperative (Swain 2015).

Aside from water management and more explicit research on postconflict spaces, there is enriching empirical work that studies the potential of transfrontier conservation areas or peace parks. Recognizing demilitarized zones as de facto protected areas, scholars suggest transforming their status to formal protected areas (Swain 2012). This would shift the discourse from border security and division to cooperation and conservation of biodiversity hotspots. Cooperation could then facilitate trust among nations, generate livelihoods for local actors through ecotourism, and potentially provide spillover effects fostering broader cooperation (e.g., Ali 2007). Proposals supporting such peace parks exist for North and South Korea (Healy 2007), India and Pakistan (Swain 2009), and the Balkans (Walters 2015). Nevertheless critics stress the complex politics challenging cooperation in transnational conservation (Duffy 2006), and some scholars caution that the concept supports a neoliberal agenda in developing countries (Swatuk 2014).

A major challenge for peacebuilding in postconflict societies in any part of the world is how to manage the critical natural resource base, such as freshwater, and also how to formulate policies in order to pursue sustainable policies of growth and development. Poor and inappropriate water resource management has the potential not only to throttle poverty reduction measures but also to complicate the whole peacebuilding process itself (Aggestam and Sundell-Eklund 2013; see, e.g., Krampe 2016c; Krampe and Gignoux 2018). Despite existing for almost three decades, the research remains in its infancy on the nexus between natural resource management and sustainable development in war-torn societies that are undergoing processes of peacebuilding.

In spite of the progressive evolution of peacebuilding, learning to avoid the worst pitfalls, development projects commonly omit to take into account aspects of the environment and the sustainable use of natural resources. It is true that peacebuilding strategy cannot eradicate all of the root causes of violent conflict. However, it is also important that careful steps be taken so that state institutions and the international community can address environmental and sustainability issues of natural resource management in postconflict societies to achieve sustainable peace.

The cooperation perspective shows much conceptual promise but features two salient empirical problems. First, the research focuses almost exclusively on interstate relations, yet the majority of contemporary armed conflicts and thus postconflict peacebuilding concerns intrastate relations. Second, empirical measurement of the effects of environmental cooperation on peace has been neglected, thereby disregarding theory development. The positive effects of environmental cooperation on peace therefore are often assumed rather than being empirically verified (Krampe 2017). Indeed research on postconflict Nepal that qualitatively measures the effects of environmental cooperation on political legitimacy—as a proxy for peace—indicates that successful cooperation does not necessarily mean more peaceful relations in the short term (Krampe 2016b).

Most recent studies have taken efforts to focus increasingly on the mechanisms connecting environmental cooperation to peace. On the interstate level, Ide (2018, 351) among others suggests that environmental cooperation "is contingent on a number of scope conditions, such as high environmental attention, internal political stability, wider patterns or traditions of environmental cooperation, and already ongoing processes of reconciliation." In fact, these new efforts suggest that addressing environmental challenges provides chances for peacemaking and peacebuilding. Other works have emphasized the need for peace and conflict research literature to go beyond previous suggestions, mentioned earlier. To explain why the sustainable management of natural resources may facilitate peace, Krampe and Gignoux (2018) suggest three mechanisms: the contact hypothesis, the diffusion of transnational norms, and the service provision–extraction equilibrium.

The Resource Risk Perspective

The second perspective of environmental peacebuilding emerges in tandem with increasing interest in natural resources and conflict by the United Nations. After its establishment in 2005, the UN Peacebuilding Commission emphasized that peacebuilding must pay more attention to natural resources (Islam and Susskind 2013). The subsequent increase in cooperation between the Commission and the UNEP gave rise to a collaborative research project with the Environmental Law Institute, the University of Tokyo, and McGill University. The resulting resource perspective is distinct in three ways: (1) it works against the backdrop of resource-induced conflict as a challenge for postconflict peacebuilding; (2) it focuses on the intrastate level and UN peacebuilding in particular; (3) it broadly includes renewable and nonrenewable resources.

Strongly influenced by the greed and grievance debate, in which the greed-based exploitation of natural resources by rebel groups was seen as a major explanation for the outbreak of civil wars (Collier and Hoeffler 2004; Le Billon 2001), this perspective treats the mismanagement of natural resources as the key cause of conflict relapse (Jensen and Lonergan, 2012); in fact, UNEP research relates 40% of all conflicts to the mismanagement of natural resources (UN Environment Program 2009). Hence UNEP emphasizes the need for capacity development of the UN in terms of natural resource management with the intention to mitigate the risks of resource-induced conflict relapse (7). For example, a range of UNEP initiatives in Sierra Leone has conveyed the need to acknowledge and understand these complex interactions. In this country, which suffered under a decade-long civil war, natural resource management has affected human development, the stability of peace, and prospects for sustainable development (UN Environment Program, n.d.).[3]

UNEP reports are aimed at the securitization of the issue for other UN agencies (Maertens 2019) and are used for the training of peacebuilding staff (Krampe 2016a). These reports stress the sustainable utilization of natural resources in peacebuilding to "capitalize on the potential for environmental cooperation to contribute to peacebuilding" (UN Environment Program 2009). Aiding UNEP's effort to anchor resource issues within UN peacebuilding, the collaborative project undertook extensive independent research. This work produced an impressive array of empirical studies emphasizing the potential of natural resources to support postconflict recovery, while also considering them as triggers of conflict. Jensen and Lonergan (2012, 9) conclude that "integrating natural resource management and environmental sustainability into peacebuilding" is the way to avert uncontrolled exploitation in the aftermath of conflict. Unruh and Williams (2013, 16) argue that issues of land reform have recurrently destabilized postconflict countries; thus they conclude that "aligning land and property interventions" in peacebuilding efforts is essential. Troell and Weinthal (2014) emphasize that water resource management is an important aspect of peacebuilding; ignoring water can be counterproductive because the socioeconomic effects impact community livelihoods.

There is no doubt that decisions about the restoration, management, and protection of natural resources can have vital consequences for short-term stability, long-term sustainable development, and successful peacebuilding in conflict-affected societies (Jensen and Lonergan 2012; Swain and Öjendal 2018). However, political emphasis on faster and time-bound postconflict reconstruction interventions and subsequent peacebuilding can also lead to crude exploitation of natural resources, unsustainable environmental practices, and massive threats to resource-based local livelihoods. These externally promoted peacebuilding processes also provide opportunities for the external actors to exploit natural resources for their own economic and strategic benefit. In fact, it is almost impossible to find even a single case among the several dozens of cases since the end of the Cold War where intervention followed by liberalization and semi-democratization has not produced a period of drastic resource exploitation, often driven by, and for the benefit of, external actors (Jensen and Lonergan 2012).

There are several cases where peace agreements, postconflict reconstruction, and peacebuilding are pursued out of short-term urgency, which has led to long-term unsustainability. Every state, including the ones recovering from a period of violent conflict, needs to sustainably use its vital natural resources, such as water, land, and forest. There is enough evidence to suggest that these environmental resources can contribute to the reemergence of violent conflict. This knowledge demands that the countries recovering from armed conflicts and the international actors must address the resource curse of peacebuilding at an early stage and find strategies, mechanisms, and institutions to prevent the far too common resource exploitation, environmental degradation, and broken livelihood systems that usually follow (UN Environment Program 2009).

In sum, these studies provide an excellent starting point, broaden the research community, and have assembled a unique evidence-based knowledge platform.[4] This research exemplifies the crucial role of natural resources and their management in postconflict countries and has, moreover, raised awareness of the issue among different UN actors, such as UNDP, UN Women, and the Department of Peacekeeping Operations. Yet with all its merits, the current research agenda is not without shortcomings. First, given UNEP's aim to securitize natural resources among UN peacebuilding actors, many of the works are technocratic and policy-oriented and thereby fail—like the cooperation perspective—to develop a thorough theoretical understanding of postconflict natural resource management and how it can facilitate peace. As such, we still do not know if environmental cooperation actually supports the peace process. The studies indicate only that if we neglect postconflict natural resource management, the risk of conflict increases.

Second, the merging of nonrenewable and renewable resources is worth debating. Undoubtedly the conflation of both in Rustad and Lujala's (2012) volume on high-value resources offers important insight into economic mechanisms. However, the underlying greed and grievances debate has moved on considerably over the years (e.g., Cederman, Gleditsch, and Buhaug 2013). The disaggregation of renewable and nonrenewable resources in future studies appears valuable because the dynamics among these resources are different; there is a clear connection between an abundance of nonrenewable natural resources and armed conflict, yet there is only weak evidence that scarcity of renewable natural resources causes armed conflict (e.g., Koubi et al. 2014).

Third, the focus on the risk of resource-induced conflict has unintended consequences, as it stresses risk and securitization of natural resources, even though UNEP, for example, intends to emphasize their potential to facilitate peace. This brings attention to preventing conflict relapse but takes attention from building peace, assuming that peace is more than the absence of violence. These negative impacts have been termed "maladaptation," "backdraft," and "boomerang effects" and theorize and empirically show the security implications of environmental interventions (Dabelko et al. 2013; Mirumachi, Sawas, and Workman 2019; Swatuk and Wirkus 2018). In the specific context of conflict-affected states, environmental interventions can actually be detrimental to building peace as they foster marginalization and affect resource access, piling additional grievances on top of existing conflict-related grievances (Krampe 2019; Swatuk and Wirkus 2018). To date, negligible attention has been given to understanding the effects of climate maladaptation in conflict-affected states. As states affected by conflict are often also impacted by climate change, this is both concerning and problematic.

Remaining Challenges of Environmental Peacebuilding

Environmental peacebuilding as a research field is increasingly gaining traction among a range of interdisciplinary scholars, broadening the field beyond its initial starting points. And it is undeniably a positive development that natural resources and environmental issues have become part of peacebuilding practice and research. This line of inquiry was neglected for too long, especially among "conventional" peacebuilding researchers, who focused primarily on matters pertaining to security-sector reform, political and economic liberalization, and transitional justice.

Yet this research has not led to a cohesive theoretical understanding of the pathways by which environmental cooperation facilitates peace. Most research has focused narrowly on natural resources, thereby ignoring the dynamics and challenges that stem from interactions of natural resource management with the unique social and political processes in postconflict countries (Krampe 2017). Also, UNEP has had difficulties assessing its actual impact on peace, as a suitable matrix and qualitative evaluation methods are missing. Consequently there is very little knowledge of what type of environmental initiatives are in fact contributing to peace in postconflict societies and what are the potential pitfalls (Krampe 2017, 2018).

Future research must increase efforts to offer theoretical explanations for why environmental cooperation would facilitate peacebuilding. Recent work by Ide (2018; Ide and Detges 2018) offers possible explanations on the interstate level that deserve attention and further inquiry. Additionally there is a need to put more emphasis on the intrastate level and conduct more theoretically informed empirical analysis of the long-term interplay between social, political, and ecological processes in postconflict countries and how these processes affect peace. The empirical deep dive by Michael Beevers (2019) on peacebuilding and natural resource governance after the wars in Sierra Leone and Liberia is a long-awaited corrective. Beevers shows how the normative impetus of liberal peacebuilding activities—focusing on the marketization and securitization of conflict resources that drove these wars—rendered issues such as land ownership, environmental protection, and sustainable livelihoods invisible. The consequences, Beevers shows, are prewar governing arrangements in which corruption, exclusion, and exploitation took root.

The link between peacebuilding and natural resources management in fragile postconflict societies has several important dimensions. The ambition for rapid economic development that postconflict peacebuilding often entails must be balanced with the risk of unsustainable exploitation of natural resources. Thus it is important to get a sustainable economic and social policy that takes the limits of natural renewable resources in a postconflict country seriously into account. Such an approach is capable of stabilizing the war-affected country, while delivering lasting conflict resolution and making peace potentially sustainable. Mentioning environmental protection and sustainable use of freshwater resources in a superficial manner is not enough; there is a serious need for sustainable economic policy as a systemic approach in peacebuilding projects. Thus there is a critical need for further studies that offer theorization of environmental peacebuilding. Good examples are emerging (Dresse et al. 2018; Ide 2018; Krampe and Gignoux 2018), but more efforts are necessary to facilitate comparison, careful generalization, and thus theory building of the dynamics and processes related to natural resource management in the context of peacebuilding.

Notes

1. Personal interview by Florian Krampe with Achim Steiner, UNDP, 7 February 2018.
2. Personal interview by Florian Krampe with Klaus Töpfer, UNDP, 23 March 2018.
3. On the impacts, see discussion of Beevers (2019) later in the chapter.
4. The knowledge platform is accessible at www.environmentalpeacebuilding.org.

References

Aggestam, K., and A. Sundell-Eklund. 2013. "Situating Water in Peacebuilding: Revisiting the Middle East Peace Process." *Water International* 39 (1):10–22.

Ali, S. H., ed. 2007. *Peace Parks: Conservation and Conflict Resolution*. Cambridge, MA: MIT Press.

Beevers, M. D. 2019. *Peacebuilding and Natural Resource Governance after Armed Conflict*. Cham, Switzerland: Palgrave.

Binningsbø, H., and S. A. Rustad. 2008. *Resource Conflicts, Resource Management and Post-Conflict Peace*. Oslo: Peace Research Institute Oslo.

Cederman, L.-E., K. S. Gleditsch, and H. Buhaug. 2013. *Inequality, Grievances, and Civil War*. New York: Cambridge University Press.

Collier, P., and A. Hoeffler. 2004. "Greed and Grievance in Civil War." *Oxford Economic Papers* 56 (4):563–595.

Conca, K. 2001. "Environmental Cooperation and International Peace." In *Environmental Conflict*, edited by P. Diehl and N. P. Gleditsch, 225–247. Boulder, CO: Westview Press.

Conca, K., and G. Dabelko, eds. 2002. *Environmental Peacemaking*. Washington, DC: Woodrow Wilson Center Press and Johns Hopkins University Press.

Dabelko, G., L. Herzer, S. Null, M. Parker, and R. Sticklor. 2013. *Backdraft: The Conflict Potential of Climate Change Adaptation and Mitigation*. Vol. 14. Environmental Change and Security Program. Washington, DC: Woodrow Wilson Center.

Delli Priscoli, J., and A. T. Wolf. 2009. *Managing and Transforming Water Conflicts*. Cambridge: Cambridge University Press.

Dresse, A., I. Fischhendler, J. Ø. Nielsen, and D. Zikos. 2018. "Environmental Peacebuilding: Towards a Theoretical Framework. *Cooperation and Conflict* 13 (1): 99–119.

Duffy, R. 2006. "The Potential and Pitfalls of Global Environmental Governance: The Politics of Transfrontier Conservation Areas in Southern Africa." *Political Geography* 25 (1):89–112. http://doi.org/10.1016/j.polgeo.2005.08.001.

Earle, A., A. E. Cascao, A. Jägerskog, A. Swain, and J. Öjendal. 2015. *Transboundary Water Management and the Climate Change Debate*. London: Routledge.

Grech-Madin, C., S. Döring, K. Kim, and A. Swain. 2018. "Negotiating Water across Levels: A Peace and Conflict 'Toolbox' for Water Diplomacy." *Journal of Hydrology* 559:100–109.

Healy, H. 2007. "Korean Demilitarized Zone: Peace and Nature." *Park* 24 (4):61–83.

Huntjens, P., Y. Yasuda, A. Swain, R. de Man, B.-O. Magsig, and S. Islam. 2016. *The Multi-track Water Diplomacy Framework: A Legal and Political Economy Analysis for Advancing Cooperation over Shared Waters*. The Hague: Institute for Global Justice.

Ide, T. 2018. "Does Environmental Peacemaking between States Work? Insights on Cooperative Environmental Agreements and Reconciliation in International Rivalries. *Journal of Peace Research* 55 (3): 351–365.

Ide, T., and A. Detges. 2018. "International Water Cooperation and Environmental Peacemaking." *Global Environmental Politics* 18 (4):63–84.

Ide, T., and C. Fröhlich. 2015. "Socio-environmental Cooperation and Conflict? A Discursive Understanding and Its Application to the Case of Israel and Palestine." *Earth System Dynamics* 6 (2):659–671.

Islam, S., and L. Susskind. 2013. *Water Diplomacy*. New York: Routledge.

Jägerskog, A., A. Swain, and J. Öjendal, eds. 2014. *Water Security*. Vol. 2: *International Conflict and Cooperation*. Los Angeles: Sage.

Jensen, D., and S. Lonergan, eds. 2012. *Assessing and Restoring Natural Resources in Post-Conflict Peacebuilding*. Abingdon, UK: Earthscan from Routledge.

Kashwan, P. 2017. *Democracy in the Woods*. Oxford: Oxford University Press.

Kostić, R., F. Krampe, and A. Swain. 2012. "Liberal State-Building and Environmental Security: The International Community between Trade-Off and Carelessness." In *The Security-Development Nexus: Peace, Conflict and Development*, edited by R. Amer, A. Swain, and J. Öjendal, 41–64. London: Anthem Press.

Koubi, V., G. Spilker, T. Böhmelt, and T. Bernauer. 2014. "Do Natural Resources Matter for Interstate and Intrastate Armed Conflict?" *Journal of Peace Research* 51 (2):227–243.

Krampe, F. 2016a. *Building Sustainable Peace: Understanding the Linkages between Social, Political, and Ecological Processes in Post-War Countries*. PhD Thesis. Uppsala University, Sweden. http://uu.diva-portal.org/smash/record.jsf?pid=diva2%3A945418&dswid=7906.

Krampe, F. 2016b. "Empowering Peace: Service Provision and State Legitimacy in Nepal's Peace-Building Process". *Conflict, Security & Development* 16 (1):53–73.

Krampe, F. 2016c. "Water for Peace? Post-conflict Water Resource Management in Kosovo." *Cooperation and Conflict*, 15 June.

Krampe, F. 2017. "Toward Sustainable Peace: A New Research Agenda for Post-Conflict Natural Resource Management." *Global Environmental Politics* 17 (4):1–8.

Krampe, F. 2019. "Climate Change, Peacebuilding and Sustaining Peace." SIPRI, June. https://www.sipri.org/publications/2019/sipri-policy-briefs/climate-change-peacebuilding-and-sustaining-peace.

Krampe, F., and S. Gignoux. 2018. "Water Service Provision and Peacebuilding in East Timor: Exploring the Socioecological Determinants for Sustaining *Peace*." *Journal of Intervention and Statebuilding* 12 (2):185–207.

Le Billon, P. 2001. "The Political Ecology of War: Natural Resources and Armed Conflicts." *Political Geography* 20 (5):561–584.

Lujala, P., and S. A. Rustad. 2012. *High-Value Natural Resources and Post-Conflict Peacebuilding*. Abingdon and New York: Earthscan.

Machlis, G. E., and T. Hanson. 2008. "Warfare Ecology." *BioScience* 58 (8):729–736.

Maertens, L. 2019. "From Blue to Green? Environmentalization and Securitization in UN Peacekeeping Practices." *International Peacekeeping* 26 (3):302–326.

Matthew, R. A. 2014. "Integrating Climate Change into Peacebuilding." *Climatic Change* 123 (1):83–93.

Mirumachi, N., A. Sawas, and M. Workman. 2019. "Unveiling the Security Concerns of Low Carbon Development: Climate Security Analysis of the Undesirable and Unintended Effects of Mitigation and Adaptation." *Climate and Development* 11 (1):1–13.

Mitchell, S. M., and N. A. Zawahri. 2015. "The Effectiveness of Treaty Design in Addressing Water Disputes." *Journal of Peace Research* 52 (2):187–200.

Pohl, B., and A. Swain. 2017. "Leveraging Diplomacy for Resolving Transboundary Water Problems." In *Water Diplomacy in Action: Contingent Approaches to Managing Complex Water Problems*, edited by S. Islam and K. Madani, 19–36. London: Anthem Publishing.

Richmond, O. P. 2012. *A Post-Liberal Peace*. London: Routledge.

Rustad, S. A., and P. Lujala. 2012. *High-Value Natural Resources and Post-Conflict Peacebuilding*. London: Routledge.

Swain, A. 2004. *Managing Water Conflict*. London: Routledge.

Swain, A. 2009. "The Indus II and Siachen Peace Park: Pushing India-Pakistan Peace Process Forward." *Round Table* 98 (404):569–582.

Swain, A. 2012. *Understanding Emerging Security Challenges*. London: Routledge.

Swain, A. 2015. "Water and Post-conflict Peacebuilding." *Hydrological Sciences Journal* 61 (7):1313–1322.

Swain, A., and A. Jägerskog. 2016. *Emerging Security Threats in the Middle East*. Lanham, MD: Rowman and Littlefield.

Swain, A., and J. Öjendal. 2018. *Routledge Handbook of Environmental Conflict and Peacebuilding*. London: Routledge.

Swatuk, L. A. 2014. *Transforming the Frontier: Peace Parks and the Politics of Neoliberal Conservation in Southern Africa*. Durham, NC: Duke University Press.

Swatuk, L. A. 2015. "Water Conflict and Cooperation in Southern Africa." *WIREs Water* 2 (3):215–230.

Swatuk, L. A., and L. Wirkus. 2018. *Water, Climate Change and the Boomerang Effect: Unintentional Consequences for Resource Insecurity*. London: Routledge.

Troell, J., and E. Weinthal. 2014. "Harnessing Water Management for More Effective Peacebuilding: Lessons Learned." In *Water and Post-conflict Peacebuilding*, edited by E. Weinthal, J. Troell, and M. Nakayama, 405–469. London: Earthscan.

UN Department of Field Support (DFS). 2017. "DFS Environment Strategy". New York: UN Field Support. https://peacekeeping.un.org/sites/default/files/171116_dfs_exec_summary_environment_0.pdf.

UN Environment Program. 2000. *Post-conflict Environment Assessment: FYR of Macedonia*, Nairobi: UN Environment Program.

UN Environment Program. 2001. *Depleted Uranium in Kosovo Post-Conflict Environmental Assessment*. Nairobi: UN Environment Program.

UN Environment Program, ed. 2009. *From Conflict to Peacebuilding: The Role of Natural Resources and the Environment*. Nairobi: UN Environment Program.

UN Environment Program. n.d. *Sierra Leone*. Accessed 25 December 2020. https://www.unep.org/explore-topics/disasters-conflicts/where-we-work/sierra-leone.

Unruh, J., and R. Williams. 2013. "Land: A Foundation for Peacebuilding." In *Land and Post-Conflict Peacebuilding*, edited by J. Unruh and R. Williams, 624). London: Routledge.

Wallensteen, P., and A. Swain. 1997. "Environment, Conflict and Cooperation." In *The Global Environment: Science, Technology and Management*, vol. 2, edited by D. Brune, D. Chapman, M. Gwynne, and J. Pacyna, 691–704. Weinheim: VCH Verlagsgemeinschaft mbH.

Walters, J. T. 2015. "A Peace Park in the Balkans: Crossborder Cooperation and Livelihood Creation through Coordinated Environmental Conservation." In *Livelihoods, Natural Resources, and Post-Conflict Peacebuilding*, edited by Helen Young and Lisa Goldman, 155–166. London: Routledge.

Wolf, A. T. 1998. "Conflict and Cooperation along International Waterways." *Water Policy* 1 (2):251–265. http://doi.org/10.1016/s1366-7017(98)00019-1.

Zeitoun, M., and N. Mirumachi. 2008. "Transboundary Water Interaction I: Reconsidering Conflict and Cooperation." *International Environmental Agreements: Politics, Law and Economics* 8 (4):297–316.

PART IV

POSTLIBERAL PEACE AND PEACE FORMATION

CHAPTER 40

PEACE FORMATION AND THE RESHAPING OF INTERNATIONAL PEACEBUILDING

OLIVER P. RICHMOND

INTERNATIONAL relations theory tends to follow state formation theory and dynamics (Tilly 1985) in that power and hegemony (whether geopolitical or normative in flavor) shapes the state and the international system and has done so since well before World War I. However, there is also a growing amount of evidence that imperial and state power have not been the only influences on the evolution of the state form, or of the international system itself. Indeed, though some of what we might now call an international peace architecture (Richmond, n.d.) has emerged from power perspectives and the roles of states and empires in setting up institutions to mitigate war though diplomacy, trade, norms, and law, this view also overlooks an alternative perspective. Liberal peace thinking focuses on the construction of an international order through cooperation over security and trade to support the checks and balances that have emerged on state formation processes and state power in domestic settings over the past two centuries. However, peace formation is generally overlooked in this process.

How much is the shaping of state and international political community based upon the agency of local actors and their extended networks who are engaged in peaceful forms of politics aimed at emancipation and a positive form of peace rather than war, violence, or extraction? This implies that local forms of peace agency, often ignored, are more influential than generally thought and that state-level and international outcomes, as in constitutional orders brought about by peace treaties and international systems, represent hybrid political orders rather than merely geopolitical or liberal frameworks (Doyle 1983a and 1983b). Working on everyday, local, and hybrid forms of peace emerging from peace formation represents an extension of the spirit and import of much of the critical work in peace and conflict studies over the years, which is beginning to filter through into more mainstream thought and policy (Regan 2013: Diehl 2016). This has endeavored to move the debate from conflict management to resolution, transformation, and peacebuilding, onward to the local level of agency or to frameworks such as hybridity, scales such as the everyday, and related issues of resistance, identity, gender, and justice (Mitrany 1944: Galtung 1969; Burton 1969; Lederach 1997; Paris 2004; Richmond 2005, 2011; Jabri 2010). This represents a challenge to long-standing thinking about the nature of the state and international order (Huntington 1996) as it implies that peaceful political communities are formed through internal but networked and relational processes of legitimation rather than state-level or international processes. Yet, simultaneously, state- and international-level processes block or circumvent such peace formation.

Drawing on critical interests particularly in local agency, resistance, and hybridity in international relations, and the more recent turns to local ownership, micropolitics, networks, and practice debates, as well as poststructuralist interests particularly in power formations, this chapter investigates the positive potential and limits of peace formation, often along the lines suggested by concepts of social and global justice. It extends the arguments often made by those working in the areas of conflict transformation or peacebuilding into a newer, interdisciplinary, and more comparative, transscalar, and networked conceptual framework. Thus the evolving international peace architecture, shaped by a mix of forces from state formation and international hegemony to peace formation, tends also to represent a bottom-up, intensely networked, and mobile phenomenon. Drawing on my earlier work, the chapter places this concept of peace formation parallel to, in opposition to, or alternatively formative of a different type of state and international hegemony. It connects this alternative theory of order formation with critical debates surrounding international relations theory, with implications for governance and the creation of sustainable forms of peace connected to global justice. It does so through an eirenist and critical approach (Richmond 2009), which is salutary in that its conclusions point to the significant limits of such locally and everyday-oriented approaches to the development of postconflict frameworks for political community.

In what follows I summarize the theoretical development of peace formation by asking a range of basic questions about who carries out peace formation, how, and with what results, drawing on the empirical evidence I have amassed through recent research among several international institutions as well as in several peacebuilding and statebuilding spaces around the world (Richmond 2014, 2016).

The Paradox of Peace Formation

Peace formation operates in local contexts and cultural, social, economic, and political spaces and increasingly influences state and international institutions. Though it represents subaltern political claims against elites, power, the state, and international hegemony, it is also closely related and entangled with them through local-to-global networks. However, "local" processes provide contextual and social legitimacy and cannot be imposed externally (Foucault 2003). They seek to determine the terms of emancipation for the community, polis, and eventually for the state. This has historically given rise, via hegemonic states, to both regional and international organization and shaped international law and markets. Private transcripts inherent in peace formation are often networked and multiplied locally and transnationally, through civil society, customary governance, and conflict resolution processes. They may arise in family networks or in professional or social networks; they may be hidden or public. In some cases, they move into the terrain of the state, shaping and modifying institutional processes, governance, and law. They often become internationally connected through international NGOs, the UN system, donors, or family, social, cultural, or labor networks or other forms of association. These processes are often based upon preexisting traditions, institutions, or norms that have been targeted in wartime but recover afterward. They represent significant constituencies and a perspective that spans local, regional, and global historical, material, and social constraints and possibilities. Much of this operates within the older, analog epistemologies of order, focusing on human interactions in organizations and institutions, law, and association.

Peace formation provides a complex positionality through which to understand the realities of the multiple, ongoing attempts to rebuild peaceful social orders. There are two sides to this development. From a small base, peace formation can be seen as examples of "embedded, participatory, and communitarian" spaces in which local and international governance and assistance can converge sensitively. Or they may be seen as creating spaces of parallelism, self-help, and abandonment, where neither the state nor internationals care to tread (Luckham and Kirk 2013).

Thus peace formation offers contextualized, networked agency in support of progressive forms of self-determination and political, social, economic, and cultural rights. It transcends the modern state framework, the norms of liberal peace, and the global political economy. It draws on already existing social capacity, but it is also creative and innovative. Autonomy and human security, networks and associations, formal and informal, are vital for any peace. The more of them there are, the better the mobilization of local agency and the more legitimate and the more stable the peace. In recent times, such agency has shifted into a digital form, able to transcend the usual constraints of sovereignty (authority, territory, distance, power, and knowledge constraints) through networks, mobility, and access to enormous amounts of information upon which to make political calculations. Peace formation may indicate a capacity to bridge the old analog international order with new digital forms of peace (also partly as a response to the ways in which new technologies have also become agents of violence).

These peace formation processes—an assemblage of local, state, transnational, international, and transversal, formal, and informal "peace" processes with a heavy emphasis on the bottom-up perspective—constitute a widening pool of political subjects. They maintain a memory of historical peace practices and institutions related to custom, culture, identity, religion, or Western norms of human rights and representation. They may seek to influence and hold government to account, seek international support to modernize and expand their practices, and set an example to wider society. They slowly insert contextual modes of politics into institution-building processes and legal and constitutional frameworks and influence donors and other international actors. This implies that a local-to-global coalition of actors is required so that a broader legitimacy can emerge. Yet local reconciliation and sustainability appear more feasible if legitimacy emanates from local agreements and processes (not only at the elite or national level). What is important is the way in which, in a minor register, peace formation agency tries to navigate, with varying degrees of success, around power and political deadlock, and whether international processes engage with and support it. It indicates how power is relational, and so can be rejected or modified even from a subaltern position. Peace formation thus endeavors to maintain everyday life where it can. Without it, internationals distribute material and epistemic resources as a simulacrum of peace, or the state operates in an authoritarian mode. With it, the political order may become more representative and also receive guidance for its policies, practices, and institutions. Peace formation offers limited material support and sometimes goes as far as establishing parallel institutions that provide public services. Sometimes local organizations manage to insert themselves into the state formation process, becoming a form of peace infrastructure (Odendaal 2013).

Peace formation actors have emerged not only from direct resistance but as a result of quiet perseverance in view of entrenched and dangerous power structures, drawing on critical social practices. They do so carefully because such "peace work" entails costs, risks, and sanctions aimed at those who take part. This is because peace often threatens local ethnonationalism, patriarchy, and economic and political practices that maintain conflict, political authority, and forms of power. Every engagement peace formation dynamics and networks have with direct, structural, and governmental power inherent in conflict structures puts peace formation at risk of co-optation or complete negation.

Participatory and decentralized systems of democracy assume local agency should not be blocked (Harvey 1998). Peace formation agency is one of several foundations of a legitimate and emancipatory peace (along with the representative state and the international system of law, norms, and institutions) that have emerged from history so far. It is aimed at mitigating direct, structural, and governmental power in favor of the complex subaltern, deemed the font of legitimate political authority. It offers the pragmatic point that subjects of violence and power relations can identify power relations best and probably spent a lot of time thinking about what to do about them in everyday settings, as well as imaging what they might imply about state and international reform in the context of their understanding of emancipatory peace.

The positionality of peace formation is complex because it has to balance local legitimacy, identity, resource inequality, and historical positionality with modernization and liberal theories backed by global institutions—expert knowledge, in other words (Mitchell 2001). It is scalar, networked, and mobile rather than fixed in a territorial space. It is also pitted against the historical accumulation of power and always operates in reaction to violence even though it remains permanently in the background, working for peace. Time and time again, peace actors and networks speak to these problems and have found ways of allowing such differences to coexist in transformative ways, entwining them: ethnonationalism, pluralism, and liberalism in Cyprus, Bosnia and Herzegovina, or Kosovo; custom and indigeneity, self-determination, and the modern state in Timor, Colombia, or Afghanistan; and in all cases, religion, material needs, and the modern political economy; and yet more elements. Local agency sits agonistically between context and structure in a postcolonial world of hybridity. Thus peace formation requires an "ethics of the local": how conflict is caused, how it affects everyday life, how it may be dealt with in context in a positive manner, and how this is related to state and international order, increasing not just in analog form but also now networked, transversal and transscalar, and digital. This is may also require an ethics associated with new digital dynamics. It points to broad questions of justice, resource distribution, and the need for empathy and localized as well as transnational legitimacy. Peace formation may be the basis for legitimacy of any polity aimed at an emancipatory and positive hybrid peace. It transcends power relations at the state level and international frameworks for peace, but is also closely connected to both (Nadarajah and Rampton 2015). It is the basis of concepts like bottom-up peacebuilding, peace infrastructures, and local agency for peace, and it should be closely connected to external processes such as conflict resolution, transformation, peacebuilding, and statebuilding. Indeed it should guide them, as it often has in the production of hybrid forms of political order (Boege, Brown, Clements and Nolan 2008).

Within localized political, cultural, gender, identity, and socioeconomic stratifications, power structures are often based upon discrimination and inequality and their long-term maintenance. This means that any local peace is often negative and relatively unstable. It also means that peace formation groups are often working against the majority in their own contexts, as well as not being fully in line with the goals and preferences of international donors, nor are able to move fully into the open because of security concerns. Both local and international actors, being increasingly aware of the hybridity of the peace framework that emerges from their mutual projects of peace, are also aware that this is a peace in negative hybrid form (Richmond 2015). Thus, in lieu of earlier theories of the social contract, legitimate authority, and state formation, peace formation foregrounds the question of what peace actors perceive as emancipatory, how they attempt to constitute an alternative order, as well the limitations of their capacity to do so. Peace formation is a rapidly evolving and very subtle phenomenon, traveling along a path that might be familiar to scholars from post- or critical development studies (Escobar 2008) but also other, perhaps more liberal thinkers, such as Rawls (1993), Habermas (2001), and Nussbaum (2011) might recognize it.

Who are the local agents of peace formation and what type of peace do they envision?

To understand the local agents of peace formation and their role, we have to comprehend subaltern power, the importance of the local scale in terms of knowledge and legitimacy, their transversal and transnational networks and mobility, but also the marginal nature of peace formers in relation to existing international and local power structures.

So local authorities, peace networks, and advocates of reform situated in specific cultural and material contexts provide crucial signals as to how peace and order can be reconstructed. Yet quite often their messages challenge existing power structures, such as elites, the state, and internationals, and take a long time to percolate through. Peace formation arises nearly always after the event of war and yet is an intergenerational background phenomenon. Peace formation dynamics and actors offer a cultural and ethical site of knowledge about peace and order, which may be translated through tactical and critical agency into political reform. Local or "organic" intellectuals, activists and advocates, some policymakers, politicians, and bureaucrats, some professionals, social, cultural, religious, identity, labor, and leisure networks form everyday associations that provide a three-dimensional vision of peace and order. Often this is deterritorialized, transnational, and transversal, aimed at an emancipatory and empathetic form of peace. Historically such networks, movements, and everyday forms of mobilization drive toward this. Peace formation offers a historical and social positionality necessary for local-scale legitimacy. It increases in scale over time after violence stops and as networks and institutions are built—but peace formers also want autonomy, introducing a dangerous contradiction into the equation between peace and power. Peace formation spans politics, economics, culture, and identity into emotional and creative areas, bridging difference, where power structures and social, political, economic, cultural, and territorial systems accentuate difference as a zero-sum process or, at best, an ugly hierarchy (or "balance"). It is connected to emerging non-Eurocentric thought on global justice (in environmental, security, gender, historical, and distributive forms, among others). Peace formation's new digital capacity expands the rights claims it can make, disguises it, speeds it up, and extends its reach, although it also now confronts the forms of violence new technologies produce.

Are these often marginal peace formation dynamics able to mitigate the violence of local state formation?

Conflict, war, and other forms of violence are mitigated by social, legal, and political institutions and mechanisms in society. However, peace formation is seriously disrupted by violence, and though it works in the long term and in the background, it also requires time to reconstitute after violence. Furthermore, it is often aimed at combating the previous war rather than the next. However, it leads to the possibility that a networked polity, with internal power, exists in society before (or in spite of) the state, which is aimed at maintaining order and responds to expectations for progressive reform, albeit in limited ways (Beck and Grande 2007). This partly represents what may be called the "undercommons" of peace formation (hidden away from the worst excesses of violence, but also spanning a much longer time frame), aimed in part at dealing with local power structures, as well as the newer exclusions that are arising from the contemporary neoliberal global order (Haiven and Khasnabish 2014, 12, 44). In this way peace formation has something in common with new social movement theories. Social actors negotiate their own peace, drawing on long-standing, quiet forms of legitimacy and legitimate authority, and display some, albeit limited agency, even over direct and structural power, and certainly against governmental power. In most cases, such organizations use external concepts such as development, human rights, human security, and democracy upon which to pin their claims. Often less visible is their valorization of mediated identity and their critiques of the inequities of the international system, both past (colonialism) and present (neoliberalism and Eurocentrism).

Indeed peace formation is one of the multiple and entwined historical roots of the state, albeit rarely recognized by elites or internationals, who focus on power and its more negative foundations. Its role has been to balance the state's excesses of power and maintain an emancipatory project both through and despite the state (Chowdhury and Duvall 2014, 218). Peace formation's aim is to expand the possibilities for peaceful everyday life, requiring both transformation and cooperation and a retreat from a methodological exclusion often embedded in practices of power and intervention. Context empowers such agency, even if it cannot countermand structural, direct, or governmental power. Its digital extension provides the knowledge required to extend such claims in a much broader context too.

Many protagonists of peace formation see themselves at the vanguard of international progress and reform, as well as within their own state and community, despite their apparent marginality to state and international political and economic power. They recognize the roles of the UN, EU, World Bank, or international NGOs, but also see them as dependent on increasingly outmoded approaches and norms and subject to hegemonic or dominant state interests.[1] Peace formers often feel they are working for themselves and their own local but also often networked, transnational, and/or transversal communities, in ways that they have defined, and are confident of a slow and hidden multiplier effect emanating from their efforts.[2] They also form networks of their own, sometimes while also formally working for international organizations or NGOs.

Across large parts of the developing and conflict-affected world, it is a normal part of life for citizens to operate in different modes of politics, informal and formal institutions, law, and identity, simultaneously. What has also become clear is that locally sited systems of legitimate authority need to be mirrored by the state if any peace system constructed around statehood is to have internal and international legitimacy. This raises some obvious issues, already noted across the literature, including the self-defeating nature of territorialism, the prioritization of elections (though democratization is a demand emerging broadly from peace formation), and a need to invest in public services in a comprehensive fashion, particularly in education, health, and income support. Inequalities, often the basis of conflict, cannot be dealt with without an active state, supported by international donors, in such areas. This means cutting across local power structures as well as having a strong international consensus, both of which are difficult and fleeting.

Thus peace formation is a slow process (in analog terms at least), which builds upon a number of different and sometimes contradictory systems: the move toward regional integration and international law and global governance (i.e., the liberal international order) is certainly one key aspect, in a formative international community. The reform of the state away from its earlier clash of power and interests into a more social form is another key aspect. But there is also an underground dimension that signifies a different type of political community, both more autonomous but yet far more relational, across time and space, which may not be solely directed toward state reform or international links (especially in the newer digital format). Peace formation seems to be a slow and quiet process of reform, reconciliation, respect, and institutional development rather than a revolutionary process of state or international restructuring. It also appears to be discordant with modern state and international processes, at least superficially, raising concerns that any engagement with peace formation will constitute neocolonialism. This is a paradox, relating the problem of how subaltern claims become politically salient in relation to existing power structures without being co-opted. Both are inevitably mediated, but co-optation is unlikely given these are peace formation rather than state formation forces (i.e., driven by principles rather than material aims). This process is far from a simple dynamic of domination and submission.

In this way, at times peace formation may be open to the criticism that it maintains or merely tweaks a painful and unjust status quo (as in Bosnia and Herzegovina or Israel/Palestine), or is eventually co-opted into the formal power structures of the state without significantly changing them (as in the case of many peace infrastructures), or leads to isolation (as with the intercommunal groups in Cyprus) or a dangerous parallelism in the political order, which threatens the stability of the state if it does not reform to incorporate them (as in Timor Leste). At the same time, it is clear that in every one of these examples, the constituent actors and networks in the peace formation process resolve issues and make some political space for discussions about the nature of peace. In everyday terms, one might say that for as long as they survive, a successful peace process already exists. However, political claims related to the nature of the postconflict state, power-sharing, and the international system (including the international political economy) often overwhelm such small and quiet gains, meaning that peace formation requires transscalar and transversal qualities if it is to have realistic long-term prospects of survival.

Nevertheless one major problem with peace formation is how easily extinguishable it is, or how easily it may drop from view. These dynamics—disappearing to escape sanction and being disrupted or ended by conflict actors—are where the praxis of intervention for humanitarians, development specialists, the various programmers of liberal peacebuilding, and in some cases peacekeeping forces might be reframed.

Peace formation may be stifled by clientelism and neopatrimonialism, which is not in favor of reform. In some cases, however, it appears that peace formation actors are able to mitigate such dynamics even with minimal resources. They may do so within the very same normative or social network that clientelism uses. Dependent upon a deep understanding of local politics, this highlights the need to let local actors take the lead in discursive terms, even if the power required to stop war and make space for diplomacy, mediation, negotiation, and reconciliation (intervention, in other words) must come from outside of the conflict-affected areas. Its global civil and increasingly digital framing collapses the concept of outside/inside to a degree, offering a new perspective of how peace accepts cooperation across different scales.

What are the subaltern critiques of developmental, liberal peacebuilding or statebuilding projects, and how do they respond to their encounter with local actors?

Peace formation points not just to the need to deal with security, rights, and a viable state, but also local historical constructions of identity and authority and international inequality. Justice at local and international levels in general terms is highlighted by a peace formation positionality. Yet it also points to the limits of liberal global justice accounts (Shapcott 2018, 105) and the insertion of alternative traditions. Neoliberalism is widely critiqued by peace formation actors, who argue that it creates significant tensions in a peace process where political and human rights, and justice, are deemed to be key to addressing the roots of a conflict and to building viable and legitimate institutions. Furthermore, any peace process creates new demands for a wider social justice because of the legitimacy of the agency of local actors, even if long latent.

The peace dividend is found in the recommencement and mutual negotiation of improvements in patterns of everyday life disrupted by the conflict. While peace formation offers a progressive rethinking of peace, rights, class, gender, and resource distribution, the state often is controlled by elites who block progress, indirectly utilizing the resources that internationals provide them with. From this arises resistance toward pathway dependency, one-size-fits-all blueprints, standard operating procedures, and postcolonial and anticolonial narratives about intervention. From here also arises the tendency of peace formation to adopt a parallel track, which challenges the state's claims to have become domesticated and oriented toward the interests, rights, and needs of citizens. Parallel track peace formation may also be forced underground in order to survive state or elite repression, to recuperate enough capacity and develop (analog and digital) tactics aimed at reshaping the state through localized strategies or international alliances.

Do peace formation actors modify the liberal, and why? Or do they follow their own peace agendas? Does this lead to a clear alternative or a liberal-local hybrid version?

Peace formation implies a postcolonial version of peacebuilding and statebuilding, which means that its current economic and normative underpinnings need to be expanded, as does their conception of the relationship between society and legitimacy. Peace formation implies concern about inequality in material, gender, and identity terms, as well as with solidarity and empathy. The bottom-up approach drives the formation of legitimate authority if it emerges from peace formation, which will be hybridized and relational, depending upon complex transversal and transnational networks. The subaltern power exercised by peace formation makes unlikely the replacement of the direct, structural, or governmental power of powerful elites or international actors. The dynamics of insurgency, critical agency and the subaltern, resistance, and alterity point to the doggedness of "resistant subjectivities" and "subjugated knowledges" and their focus on peace in everyday life. They all make emancipatory claims, but at best all the examples in existence—Timor Leste, Cyprus, Kosovo, Bosnia, Sierra Leone, Afghanistan, and others—point to the emergence of a negative, hybrid form of peace.

Hybrid political orders are a more significant sign of the capacity of peace formation processes and actors to nudge or reshape power structures, either at the elite level or among those associated with liberal peacebuilding. What tends to arise is a blend of different forms of political authority, connecting peace formation to existing elite power structures, civil society, the state, the international peace architecture, as well as the global economy. A new struggle for dominance of the political order in these terms emerges, and in recent times has been dominated by an alliance between capital and elite power, with peace formation operating from a parallel track, which nevertheless impinges upon the evolution of the state (as might be clearly seen in cases like Timor Leste, Cambodia, or Kosovo in very different ways). A key question for the future will be to engage with the problem of whether the relatively easy shift into a digital terrain for such agency will undermine its capacity to deal with more analog forms of state or global power, but the expectation and the evidence so far points to how peace formation both modifies hegemonic systems and develops its own peace discourses.

How successful have peace formation strategies been?

Four main possibilities emerge from this study for peace formation dynamics.

1. They survive in the informal shadows, away from the risks of crossing powerful actors (as in Kosovo before 1999, where services and self-determination circulated away from the gaze of formal authorities) and avoid the risks of moving into public space.
2. They move in the public space and take this risk (as in Cyprus, where the intercommunal movement stepped out of the shadows with international support in the early 2000s), but they also adopt the liberal peace/neoliberal state international architecture and accept their (lowly) place in the current international hierarchy and system, but without any right or capacity to change it (as perhaps also in Kosovo, Bosnia, or Colombia).
3. They may try to choose only the positive aspects of the liberal peace system and discard the rest (though there are no real examples of this).

4. They may build a hybrid system, in which they have some greater stake in the whole, while also recognizing power relations (as in Timor Leste, Somalia, Somaliland, and others).

These different dynamics indicate that peace formation plays a role in building order, different in every case. This is success of sorts. Peace formation uncovers conflict-sustaining inequality and, at least in discursive terms, indicates the existence of a small but growing network of plural and micro solidarities: it highlights relationality, bridging and translation claims for pluriversality, equaliberty, heterotopias, transscalarity, and transculturalism. It rescues such dynamics from the past, or manufactures them for the future, focusing on their transnational, cultural, and social legitimacy. For many reasons, on paper a positive hybrid peace may thus be imagined.

Peace Formation on the Local Scale

Contextual knowledge is of great tactical value in maintaining peace formation against local and state-level conflict and against inappropriate international blueprints or involvement. It has little capacity against direct violence or war, but it provides pathways through insecurity. It also provides some defense against unrepresentative global governance, though it has little capacity against predatory aspects of global capitalism. Clearly, any legitimate peace and order is unsustainable without the support of subaltern or critical agency and society in general (other than through structural or perhaps governmental power). Peace formation provides a social form of governance, often without formal government. The peace and state that emerge from this process are likely to be contingent, fragmented, decentralized, and capable of supporting peaceful alterity (Axtmann 2004, 259) discursively, if not in practice. Though it cannot respond to direct, structural, or governmental forms of violence, it can work on their delegitimization, digital methods appearing very effective here. It also connects with the emancipatory wing of liberal thought, hybrid notions of law and transitional justice, participatory democracy, human rights, context and dignity, poverty, lack of resources and facilities, and inequality. Concepts such as local ownership, participation, Responsibility to Protect, sustainability, and "do no harm" have emerged because of it, pointing to more critical, postcolonial or anticolonial dimensions. Hybrid approaches to decentralized governance and law, involving local leaders and processes, have also been recognized as important because of it. Expanding rights claims have cascaded from the local scale upward as a result.

Peace formation and its informal critical agencies as well as its collective political mobilization dynamics from within civil society may do little more than mitigate everyday life in material terms, but in the longer term it may be able to move into the open and begin to connect with and shape the state as well as influence donor strategy. It may develop informal governance without a state, following transscalar and transversal networks.

Four key strategies emerge from this analysis. This confluence may start with **parallelism** (as in Afghanistan and Palestine, where peace formation is forced to work under cover and in parallel to existing and new power structures relating to the conflict and the state entities). It may involve **bridging** (as in Timor, where peace formation processes helped translate local forms of legitimate authority and make the state more legible to them). It might involve **co-opting** (as in Bosnia, where is seems that peace formation processes are partly controlled by state formation actors). It could lead to a **takeover** (as in Kosovo, where peace formation actors and state formation actors took control of the state to a large degree). It may influence civil society, the state, or donor strategies. It may be deemed successful if the resultant state or system of governance is more representative and allows for the expansion of rights and justice, donor strategies are more sensitized, and local ownership and autonomy has been upheld, along with democracy. Success may well depend on the way informal processes maintain and improve everyday life and, more ambitiously, form networks that begin to find support within the international community and more formally begin to reshape the state, social norms, as well as donor policy (Odendaal 2013, 23). It may also point to the emergence of scale jumping, where positionalities are altered beyond the canon that can be easily recognized in formal liberal political philosophy (Agnew [1987] 2012). This might seem to overload the potential of peace formation, but it is in line with the autonomy that is often claimed by its proponents.

Methodology and Peaceful Order

Peace formation indicates that "expert knowledge" is subjective and often hegemonic and not well suited to meeting claims made at the local level. Likewise peace formation is limited in terms of scope and power by the very power structures it opposes and by its networked and mobile rather than concentrated nature. A multidimensional methodology and interdisciplinary approach is needed to understand these dynamics, blending social theory, philosophy, anthropology, geography, international relations, and political science. This methodology may also be translated into law and economics at a later moment of consolidation. Peace formation requires a research methodology suited to engaging with local, informal knowledge and creating equal partnerships, decentralized power, and redistributed resources. To avoid re-creating trusteeship and native administration bureaucracies, it needs to be linked to political self-determination and autonomy and through awareness of inequalities relating to postcolonial and gendered analyses, avoid the risk of inadvertently supporting unjust outcomes or processes. The subtleties of peace formation point to how deconstruction assists in the emergence of justice. Subjects have, over the past twenty-five years, uncovered the very power systems that drive conflict as well as how they are perpetuated through intervention (often inadvertently), and now point to how reconstruction may attain justice (Derrida 1992, 15). Peace formation, self-formation, and local knowledges are in a constant struggle for survival against industrialized knowledge and disciplinary power states, the international architecture, and the global economy. Peace formation requires the recovery and assertion of subjugated knowledge. Furthermore, its digital shift to some degree undermines inside/outside, top-down/bottom-up frameworks: it collapses space, speeds up interaction, extends networks, supports the knowledge side of the power-knowledge equation, and offers camouflage in situations of danger. This reframes agency as far more relational than hitherto understood (Qin 2016), and this has significant impact on the micro construction of political order and legitimate authority, whether local, state, or global (Solomon and Steele 2017). New methods are required to engage with these dynamics.

Some Important Caveats

This argument does not constitute a claim that all local, networked, and politicized agency is potentially emancipatory. Indeed peace formation agency is often in a very minor scale. Local actors, communities, and networks are subject to often grotesque power relations around which peace and politics revolve. These power relations are often directly supported by external actors' contacts with military, political, and economic elites and the role of security, economic, political, and other advisors, programmers, and consultants. Global capital also tends to play into the hands of such powerful networks. Peace formation actors are often far outweighed by local nationalists, sectarians, conservatives, or other groups who oppose reconciliation and compromise, and may even support the use of violence. This means it is all the more fascinating that peace actors continue to emerge with clear views on emancipation, peace, the state or form of polity, and international support.

Much of the empirical evidence on peace formation is anecdotal or scattered and impressionistic, partly because of the well-known risks activists and social movements take in responding to entrenched power relations or practices. Any reconstruction of a historical political record, at whichever level, will probably be susceptible to such weaknesses, however. So far a fragmented record and some affiliated theorization has only just been brought into view. It might be said that in a range of literature spanning Marx ([1869] 1969) to Foucault, the sense of political agency has been theorized more strongly than it has been practiced (or at least empirically researched). Nevertheless subaltern agency is not much of a match for governmental, structural, or direct power. Some peace processes exist only (in terms of reconciliation and emancipation) at this level, as the state, regional, or international levels have long been deadlocked (as in the Middle East, Bosnia, Cyprus, and elsewhere). Peace formation's discursive signaling represents a positionality from which local-to-global dimensions of a peace process and settlement can be viewed and tested. Nevertheless this is a highly politicized process: it is predicated upon a struggle for rights and material needs often against entrenched power structures, vested interests, and received wisdom among outside experts.

Conclusion

Foregrounding peace formation, engaging with its signals, and understanding how peace is a political project—one among many—is a crucial next step in the development of an international architecture designed to improve international order rather than just to maintain it, especially in view of the need to reconsider the relationship between universalism and particularism that the tension between peace formation and the international peacebuilding architecture suggests. It is essential for substantive democratization, human rights, understanding the role and nature of the state as well as its legitimacy, and for the legitimacy of intervention to protect vulnerable subjects while also maintaining their political autonomy. Peace formation makes clearer the blockages to peace, who controls them, and how they operate (often with great subtlety).

Finally, foregrounding peace formation also helps to shed light on the growing shift from an analog international system of citizens, states, and international architecture operating within the constraints of sovereignty, territory, human rights, and institutions, to a digital system. In a digital peace framework, rights and justice claims are expanding across dense and far-reaching networks, at a pace never before seen. Distance has collapsed, citizenship is framed in nonterritorial ways, and social legitimacy and responsibility are supplanting the state and the international, at least to a degree. Peace formation is clearly a little-understood dynamic, but one which is influential in reshaping political order.

Notes

1. This was a view I heard repeatedly at NGOs and social organizations in Kosovo, Bosnia, Colombia, Sri Lanka, and Cyprus.
2. For example, the staff at Balkan Sunflowers mentioned this explicitly. Focus Group, Balkan Sunflowers, Pristina, 23 May 2014.

References

Aggestam, K., and A. Björkdahl. 2012. *Rethinking Peacebuilding: The Quest for Just Peace in the Middle East and the Western Balkans* London: Routledge

Agnew, J. (1987) 2012. *Place and Politics.* London: Allen and Unwin.

Ashcroft, B. 2001. *Post-Colonial Transformation.* London: Routledge.

Ashcroft, B, G. Griffiths, and H. Tiffin. 1989. *The Empire Writes Back: Theory and Practice in Post-Colonial Literatures.* London: Routledge.

Axtmann, R. 2004. "The State of the State." *International Political Science Review* 25 (3):259–279,

Beck, U., and Edgar Grande. 2007. *Cosmopolitan Europe.* Cambridge, UK: Polity.

Boege, V., A. Brown, K. Clements, and A. Nolan. 2008. "On Hybrid Political Orders and Emerging States: What Is Failing—States in the Global South or Research and Politics in the West?" *Berghof Handbook Dialogue Series* 8:15–35.

Booth, D., M. Leach, A. Tierney. 1999. *Experiencing Poverty in Africa: Perspectives from Anthropology*. Background Paper 1b. World Bank Poverty Status Report. Geneva: World Bank.

Burton, J. W. 1969. *Conflict and Communication: The Use of Controlled Communication in International Relations*. London: Macmillan and Free Press.

Callinicos, A. 1989. *Against Post-modernism*. Cambridge, UK: Polity.

Chowdhury, A. and R. Duvall. 2014. "Sovereignty and Sovereign Power." *International Theory* 6 (2):191–223.

Connolly, W. E. 2011. *A World of Becoming*. Durham, NC: Duke University Press.

Derrida, J. 1992. "Force of Law." In Deconstruction and the Possibility of Justice, edited by D. Cornell, M Rosenfeld, and D. Gray, 3–65. London: Routledge.

Diehl, P. F. 2016. "Exploring Peace: Looking beyond War and Negative Peace." *International Studies Quarterly* 60 (1):1–10.

Doyle, M. W. 1983a. "Kant, Liberal Legacies, and Foreign Affairs." *Philosophy and Public Affairs* 12 (3):205–235.

Doyle, M. W. 1983b. Kant, Liberal Legacies, and Foreign Affairs." *Part 2". Philosophy and Public Affairs* 12 (4):323–353.

Escobar, A. 2008. *Territories of Difference: Place, Movements, Life, Redes*. Durham, NC: Duke University Press.

Foucault, M. 2003. *Society Must Be Defended*. New York: Picador.

Galtung, J. 1969. "Violence, Peace, and Peace Research." *Journal of Peace Research* 6 (3):167–191.

Habermas, J. 2001. *The Postnational Constellation: Political Essays*. Cambridge, MA: MIT Press.

Haiven, M, and A. Khasnabish. 2014. *The Radical Imagination*. London: Zed Books.

Harvey, Neil. 1998. *The Chiapas Rebellion: The Struggle for Land and Democracy*. Durham, NC: Duke University Press.

Huntington, S. 1996. *The Clash of Civilizations and the Remaking of World Order*. London: Simón & Schuster.

Jabri, V. 2010. *War and the Transformation of Global Politics*. Basingstoke, UK: Palgrave Macmillan.

Kaldor, M. 2003. "The Idea of Global Civil Society." *International Affairs* 79 (3):583–593.

Lederach, J P. 1997. *Building Peace: Sustainable Reconciliation in Divided Societies*. Washington, DC: US Institute of Peace Press.

Levinas, E. 1984. "Peace and Proximity." In *Basic Philosophical Writings*, edited by A. T. Peperzak, S. Critchley, and R. Bernasconi, 161–169. Bloomington: Indiana University Press.

Luckham, E., and T. Kirk. 2013. "The Two Faces of Security in Hybrid Political Orders." *Stability: International Journal of Stability and Development* 2 (2):1–30.

Marx, K. (1869) 1969. *The Eighteenth Brumaire of Louis Bonaparte*. New York: International Publishers.

Mishra, P. 2012. *From the Ruins of Empire: The Revolt against the West and the Remaking of Asia*. London: Penguin.

Mitchell, T. 2001. *Rule of Experts*. Berkeley: University of California Press.

Mitrany, D. 1944. *A Working Peace System: An Argument for the Functional Development of International Organization*. Oxford: Oxford University Press.

Nadarajah, S, and D. Rampton. 2015. "The Limits of Hybridity and the Crisis of Liberal Peace." *Review of International Studies* 41 (1):49–72.

Nussbaum, M. 2011. *Creating Capabilities: The Human Development Approach*. Cambridge, MA: Belknap Press of Harvard University Press.

Odendaal, A. 2013. *A Crucial Link: Local Peace Committees and National Peacebuilding.* Washington, DC: USIP.

Paras, E. 2006. *Foucault 2.0.* New York: Other Press.

Paris, R. 2004. *At War's End: Building Peace after Civil Conflict.* Cambridge: Cambridge University Press.

Qin, Y. 2016. "A Relational Theory of World Politics." *International Studies Review* 18 (1):33–47.

Rawls, J. 1993. "The Law of Peoples." *Critical Inquiry* 20 (1):36–68.

Regan, P. M. 2013. "Bringing Peace Back In: Presidential Address to the Peace Science Society." *Conflict Management and Peace Science* 31 (4):345–356.

Richmond, O. P. 2005. *The Transformation of Peace.* London: Palgrave.

Richmond, O. P. 2009. "Eirenism and a Post-Liberal Peace." *Review of International Studies* 35 (3):557–580.

Richmond, O. P. 2011. *A Post-Liberal Peace.* London: Routledge.

Richmond, O. P. 2014. *Failed Statebuilding.* New Haven, CT: Yale University Press.

Richmond, O. P. 2015. "Dilemmas of a Hybrid Peace: Negative or Positive?" *Cooperation and Conflict* 50 (1):50–68.

Richmond, O. P. 2016. *Peace Formation and Political Order.* Oxford: Oxford University Press.

Richmond, O. P. n.d. *Peace in the 21st Century.* Oxford: Oxford University Press.

Said, E. 2004. *Humanism and Democratic Criticism.* New York: Columbia University Press.

Shapcott, R. 2018. "Global Justice: Shaped Rather Than Found." *International Relations* 32 (1):104–123.

S. Ty, and B. J. Steele. 2017. "Micro-Moves in International Relations Theory." *European Journal of International Relations* 23 (2):267–291.

Tilly, Ch. 1985. "War Making and State Making as Organized Crime." In *Bringing the State Back In*, edited by P. Evans, D. Rueschemeyer, and T. Skocpol, 169–191. Cambridge: Cambridge University Press.

CHAPTER 41

LOCAL RESISTANCE AND HYBRID PEACE

SUNGYONG LEE

THE local-international interaction in postwar peacebuilding has been subject to extensive academic discussion in recent peacebuilding discourse. Studies have paid attention to the issues around the top-down imposition of liberal peace in conflict-affected societies in which cultural and social contexts are incompatible with Western values. The terms "local resistance" and "hybrid peace" soon emerged as buzzwords in conceptual frameworks to capture the dynamic interaction between the stakeholders at the international and local levels, as well as the forms of peacebuilding that developed as a result. With the introduction of the *everyday* concept in the mid-2000s, academic discourse has begun to acknowledge more subtle forms of local resistance, its constructive roles in peacebuilding, as well as the complex nature of hybrid peace.

This chapter introduces and examines the previous academic debates over these questions by focusing on the forms and motivations of local resistance, the patterns of local-international interaction, and the contours of hybrid peace. To achieve this goal, it adopts two sets of typologies regarding motivations of local resistance and local peacebuilders' strategies for external engagement. Moreover it examines local peacebuilding programs promoted in postwar Cambodia. Specifically, this chapter comprises three parts that will explain local actors' resistance against externally driven peacebuilding, the process of hybridization, and the hybrid forms of peacebuilding. Each section starts with a concise introduction to a few major academic debates over the relevant concept. It will then demonstrate how each type of interaction materialized in the context of Cambodian peacebuilding.

It should be noted that hybrid peacebuilding is a lot more than the operational features and organizational structure that come out of the interaction between a few peacebuilding agencies. The dynamics of local-international interaction in postwar peacebuilding are complex and fluid, demonstrating different features according to the local contexts in societies where peacebuilding is promoted. Moreover the dynamics of local resistance are determined in relations to the doings and undoings of international actors (UN peacekeepers, peacebuilders, governments, donors, international NGOs, etc.). Hence a small-scale empirical study on the motivations and actions of local peacebuilders in Cambodia is not meant to represent overall trends in contemporary peacebuilding. Instead the examples in this chapter are selected to offer a concise snapshot of how local resistance and hybrid peacebuilding can be developed in practice in the late 2010s.

"Local" and "local actors" in this chapter denote community residents, traditional leadership, and grassroots organizations that play important roles in planning, implementing, and evaluating peacebuilding programs at substate levels. Hence officials in central government and power elites who have strong influence over national politics are not included in this analysis, although their roles are often critical for local peacebuilding programs. Parts of the ideas included in this chapter were first proposed in the author's previous outputs (especially in Lee 2015, 2019; Lee and Park 2020). In a sense, this chapter is an endeavor to weave together these findings, previously presented in a scattered manner, to trace the transformation of local actors' behavior. The empirical evidence was collected from the author's field studies conducted in 2014, 2016, and 2019.

Local Resistance in Early Phases of Postwar Peacebuilding

Studies on local resistance in postwar peacebuilding during the late 1990s were developed as part of a critical examination of the liberal peacebuilding models. Much discourse was primarily interested in why the liberal peacebuilding models fail to accommodate the unique contexts in local societies and how such insensitivity to context results in failure. Hence the scope of such discussions has been limited, focusing primarily on the incompatibility between (Western) liberal models and local contexts. Frequently from a postcolonial perspective (Bhabha 1994), resistance has largely been understood as reflecting local actors' desire for "an openness to the indefinite possibility that things could be different" (Hoy 2004, 10). Resistance of this type is frequently expressed in overt forms with clear goals and collective forms of violence (Alfred and Corntassel 2005; Newman and Richmond 2006). A few modes of resistance are rebellions, riots, demonstrations, strikes, complications in the implementation process, and media campaigns. Due to its visibility and impact on society, overt resistance has attracted attention from both academic debates and seminal documents on policymaking (Ejdus and Juncos 2018).

The trend in the academic debate began to change in the mid-2000s. Based on the theoretical grounds of Foucault (1978), de Certeau (1984), and Scott (1989), recent studies conceptualize a wide range of everyday activities of local communities as subtle resistance, used by subaltern groups to undermine repressive domination and appreciate the complex features of local resistance (Richmond 2009b, 2010; van der Lijn 2013; Mac Ginty 2011; Galvanek 2013). Resistance is more broadly defined as the actions or nonactions of subnational-level actors that express a refusal to accept certain policies, programs, or themes of peacebuilding. In terms of modes of resistance, these studies categorize communities' subtle actions into several conceptual typologies, including modification, alternation, nonparticipation, foot-dragging, and false compliance (Hellmüller 2013; Dougherty 2004; Mac Ginty 2012; Richmond and Mitchell 2011). Moreover it has been assumed that an actor in resistance is likely to adopt "fluid" forms of resistance, flexibly combining different sources (Visoka 2017).

Why and how did local actors in Cambodia resist the UN-led reconstruction programs? Many examples relevant to this question were reported particularly in the early phase of peacebuilding. A wide range of resistance activities were mobilized against the peace processes led by the UN Transitional Authority, as well as aid programs initiated by bilateral agencies. Table 41.1 roughly categorizes the types of local resistance in Cambodia, focusing on the motivation and the main actors involved in resistance.

Interest-seeking resistance is frequently mobilized by political elites, regional strongmen, business operators, and community leaders in order to create or protect self-interest and power. These actors decide to resist because they believe the introduced peacebuilding programs weaken the advantages that they enjoy in terms of political power, economic benefits, and social privilege. The more interests the local elites have at risk, the stronger the forms of resistance that are likely to be adopted (e.g., systematic sabotage, violent resistance, and foot-dragging). In Cambodia, the Khmer Rouge's withdrawal from the election in 1993 was probably the most well-known example of power-seeking resistance. The Khmer Rouge had been involved in the final stage of the peace negotiation, with an unrealistic expectation of the UN's proposals and a groundless

Table 41.1 Types of Local Resistance in Postwar Cambodia

	Interest-Seeking Resistance	Justice-Seeking Resistance	Resistance in Adherence to Local Contexts	Uninformed Resistance	Involuntary Resistance
Motivation	Threat to power/ interest	Sense of injustice	Cultural incompatibility	Inadequate communication	Practical barriers
Mode	Overt, subtle		Subtle		Nonparticipation
Main Involved Actors	Traditional/power elites ←→	←→			
		Civil society actors ←→	←→	←→	
			Community residents ←→	←→	←→

assumption of China's constant backup. When it realized that its self-centered vision for gaining political power through collaboration with the UN-led peace process was futile, it openly declared the resumption of *revolutionary* campaigns against the UN's demilitarization process and boycotted the first general election (Lee 2013).

Justice-seeking resistance is mobilized when people believe that some of the critical values are negatively affected by externally driven peacebuilding. Some values frequently associated with justice-seeking resistance are nationalism, sovereignty, religious principles, and primary ethics. This type of resistance involves more emotional grievance or anxiety, and external influence tends to be branded as suppression or imperialism. Based on a high level of antiforeign sentiment, local elites or political associations take up collective forms of resistance. For instance, in the early phase of postwar reconstruction in Cambodia, two distinct forms of nationalistic propaganda competed. First, the opposition parties galvanized people's anti-Vietnam zeal and accused the Cambodian People's Party (CPP), which had ruled the country during the civil war with military support from Vietnam, of being a puppet regime of Vietnam. Second, CPP leaders used anti-Western catchphrases and openly expressed their anxiety at overt intervention from the UN Transitional Authority. Many youths in Prey Veng and Svay Rieng provinces were particularly affected by the CPP's antiforeign statements and mobilized campaigns for sabotage and physical violence against the UN's presence, demanding the withdrawal of foreign powers.

Although conceptually distinct, in practice *interest-seeking resistance* and *justice-seeking resistance* are used by local elites who have had practical and cultural benefits from the conventional local systems. Hence the two types of resistance can be closely associated and may be adopted as parts of the same movement. Foreign attempts to exert influence over conventional localities are likely to negatively affect their moral, economic, or political power grounds. The forms of resistance can be proactive and overt (e.g., sabotage, open protest, physical aggression), although more moderate forms are not uncommon (e.g., verbal appeals, avoidance of participation, slowing the implementation process).

In contrast, the next two types of resistance listed in Table 41.1 are more frequently found in grassroots local communities. Moreover, resistance of these kinds is chosen when the local population has difficulty relating to the proposed peacebuilding programs due to incompatibilities with external peace supporters' cultural contexts, communication modes, or priorities in local peacebuilding. This type of resistance does not necessarily involve people's antagonism or ideological opposition to external peacebuilders, and the forms of resistance are likely to be moderate, such as nonparticipation or passive participation. Nonparticipation in this context reflects local actors' conscious abstinence from engagement in proposed peacebuilding programs or public events. The absence of local stakeholders may have a significant impact on the implementation of programs; moreover, the tactic of nonparticipation can be applied as a part of wider resistance strategies (Mac Ginty 2012). From a theoretical perspective, the resistance in these categories is most relevant to recent discourse that illuminates the significance of subtle resistance in local communities' everyday life.

Resistance in adherence to local contexts denotes the situation where local people resist because they consider peacebuilding programs incompatible with their perceptions and daily practice for sustaining their livelihood. Due to the liberal notions that many Western agencies adopt, for example, peacebuilding programs frequently undervalue or ignore the significance of social and cultural contexts in local areas, which include a community-oriented lifestyle, a high-context communication system, mechanisms for social governance (e.g., patron-client relations), and cultural heritage (e.g., local cosmology). Without a high level of anxiety and core agencies for mobilization, these forms of resistance are likely to be moderate and subtle, including individual appeals, silence in public meetings, isolation of peacebuilders from communication, and nonparticipation. One such example in Cambodia can be found during the introduction of private landownership in the postwar period. After the colonial period, the country had gone through a long period of monarchy and communist regimes; thus the vast majority of the population considered land to belong to the monarchy or the state. Some ethnic minorities believed the land belonged to the whole community rather than an individual. With many practical challenges, therefore, many people decided not to bother with land registration, which eventually put them in a vulnerable situation in the consequent landownership clarification.

Uninformed resistance refers to cases where local resistance is generated by insufficient communication and association between external peacebuilders and local communities. A few frequently observed barriers to sound communication between peacebuilding actors and local communities are inappropriate methods and procedures for dissemination (e.g., workshops, work-oriented meetings), a lack of preexisting knowledge of the relevant concepts or themes (e.g., human rights, good governance), and a gap between expectations of different people relating to the functions of the peacebuilding programs and the roles of relevant actors (Mac Ginty 2012). The modes of resistance are normally nonparticipation or passive participation, through which the fundamental objective of the peacebuilding programs is hardly achieved. For instance, community residents feel less interested in the relevant programs when they do not receive sufficient information about the significance of the programs in their lives and ways to participate. In other cases, the community may mobilize resistance based on a misunderstanding of the intention of the programs. In Cambodia during the early 1990s, a large number of public meetings where a few dozen community representatives gathered were major venues for disseminating newly adopted procedures and regulations for local development and governance. Nevertheless many Cambodians who used to consider such public meetings to be ritualistic or perfunctory events were not ready to learn and adopt the information shared in such meetings.

Finally, there are occasions of *involuntary resistance* where people do not collaborate with peacebuilding programs because they are not able to. Resistance action of this type does not reflect local actors' deliberate intentions but still has negative impacts on the implementation of peacebuilding programs. Three main challenges that prevent local populations from participation in peacebuilding are security risks, a lack of material capacity, and the psychological consequences of their exposure to long-term extreme violence. There are occasions when people are not confident in or interested in participating in any type of public activities due to the psychological trauma of long-term violence (Mac Ginty 2012). For instance, the low turnout for voting in some areas, which the Khmer Rouge claimed as resistance, was rather more reflective of community residents' sense of vulnerability. When the Khmer Rouge started a campaign for boycotting the proposed general election in 1993, people in the areas that used to be under the strong influence of the military worried about the negative consequences of their participation in the election. Many victims of Khmer Rouge violence have locked themselves in their houses and refused to come out to any public events for decades (Interview V).

In short, local resistance against the externally driven peacebuilding programs in Cambodia reflects distinct motivations of different local groups. This section roughly categorized the complex motivations into five groups: interest-seeking resistance, justice-seeking resistance, resistance in adherence to local contexts, uninformed resistance, and involuntary resistance. Such a conceptual distinction cannot be clearly delineated in reality, however. One resistance movement may simultaneously incorporate, for example, local elites' interest-seeking resistance and a civil society actor's justice-seeking motivation. Occasionally interest-seeking resistance is manipulated by Cambodian political elites into the form of justice-seeking resistance where civil society actors and/or organized groups of local communities participate. In such cases, widespread antiforeign sentiments turn into more organized and consistent movements, with the backing of local elites. In other cases, community residents' lack of participation may represent their anxiety about cultural incompatibility as well as a lack of material capacity to enable participation. While local elites are more capable of mobilizing overt modes of resistance, local communities' anxiety due to contextual incompatibility occasionally generates explicit reactions (e.g., collective appeal, sabotage).

Hybridization: Transition from Resistance to Engagement

Since it was first proposed by Oliver Richmond, Kevin Clements, Roger Mac Ginty, and other researchers in the mid-2000s, the concept of hybrid peace has been utilized to capture the interaction between different stakeholders in peacebuilding (usually international interveners and local actors) and the mixture of liberal and local models coming out of the interaction. By acknowledging the coexistence of different worldviews, hybridity as an ontological framework moved beyond the binary conceptualization of "international versus local" to explain the peacebuilding models that are open to difference (Richmond 2009b). This approach has generated a range of theoretical debates over the past decade (Richmond 2010; Mac Ginty 2011; Kumar and De la Haye 2012; Lee 2019).

Hybrid peace is a conceptual framework by which to understand both the *process* through which multiple actors interact for hybridization as well as the *outcomes* within which multiple values and perspectives coexist. Both dimensions are important in understanding the nature of hybrid peace. When focusing on process, the significance of hybrid peace has been highlighted as "a constant and multi-sphere series of process" (Mac Ginty 2011, 74) in which there is interplay between different peace agents, networks, and structures. By using the term "hybridization," some studies emphasize the process whereby actors can manage the changes of peacebuilding. Although the details vary, many studies agree that the formation of hybrid peace is determined by the combined impact of the compliance powers of liberal peace; the incentive powers of the liberal peace; the ability of local actors to resist, ignore, or subvert the liberal peace; and the ability of local actors to formulate and maintain alternatives to the liberal peace (Mac Ginty 2010; Lee and Özerdem 2015). Through this process varied forms of hybrid peacebuilding models are developed, reflecting the perspectives, interests, and technical standards of both local actors and international interveners, which will be discussed in more detail in the next section. Hybrid forms of peacebuilding are temporal, dynamic, relational, and contextual in nature in that they are constantly being reshaped and modified according to changes in the social contexts in the conflict-affected countries (Visoka 2011, 2017).

The development of peacebuilding in Cambodia offers a useful example relevant to the process of hybridization. By the time the first general election was conducted in 1993, overt resistance against the international peacebuilding programs had significantly decreased. Instead of denying the foreign influence, many local actors began to seek ways to achieve their goals through coexistence and collaboration with the international peacebuilding agencies. The discussion in this section will highlight some features of this process in Cambodia by focusing on two questions: When and why did local populations that had once engaged in resistance begin to interact and collaborate with their international counterparts? What strategies did local peacebuilders use to push forward their agenda while maintaining interaction with international agencies?

Of the five types of local resistance, the significance of interest-seeking resistance and involuntary resistance was reduced while peacebuilding progressed in the 1990s. Regarding interest-seeking resistance, for instance, the Khmer Rouge—one of the strongest threats to the international peacebuilders—lost its military power and limited its operation to a small area in northeastern Cambodia. Other local elites who had conducted interest-seeking resistance linked themselves with the national political leaders who were constructing political parties in preparation for the first postwar election, in 1993. The new government, established by the first election, maintained a low-key attitude to foreign donors, and local elites abided by this direction. Hence, although local elites' acceptance was an important factor for initiating a peacebuilding program, they tended not to make direct resistance to the operation of these programs.

Regarding involuntary resistance, moreover, the security stabilization and the development of infrastructure removed some of the major challenges that had prevented many of the local people from participating. Multiple sources confirmed that local residents gained confidence from their public participation after seeing the absence of irregular combat, resumption of religious practice, freedom of travel, and recovery of basic state functions. Most immediate and urgent issues that prevented local people's participation in peacebuilding programs were largely addressed by the end of the 1990s (Interviews I, III, VI).

Most relevant to hybridization, the resistance based on other types of motivation—justice-seeking, adherence to local contexts, and miscommunication—transformed its delivery modes due to a few contextual factors. First and foremost, the Cambodian civil society that had been eliminated began to be reconstructed, relying heavily on external financial and advisory support. For example, the number of Cambodian NGOs increased from 12 in 1992 to 407 over a ten-year period. A large number of high-profile social figures and local elites who used to engage in resistance activities participated in Cambodian civil society. Due to the limited resources available within the country, most of these Cambodian NGOs relied for their operational budget on financial support from external funding opportunities (Lee and Park 2020). The need to strengthen collaboration with international actors became more obvious in the late 1990s. By this time, the country had escaped from the social turmoil in the aftermath of the civil war, and the global trend in aid politics had more of a security focus due to the 9/11 terrorist attacks. The combined impact of these two factors resulted in a radical decrease in the amount of international support to Cambodian peacebuilding. A large number of programs were closed down, and many programs sought support from international donors in order to survive (Parks 2008; Interview II).

At the same time, a few major international NGOs, including UN agencies, have made various efforts to incorporate more local perspectives and interests since the mid-1990s. Since recognizing the importance of building national capacities in the conflict-affected countries, themes like capacity building and local ownership became key objectives of their peacebuilding programs. Although many of these programs were inappropriate or insufficient to nurture advanced ownership of local peacebuilding actors, they gradually lowered the anxiety of local people who had previously been involved in resistance against externally driven peacebuilding (Lee 2019).

Another contextual issue was the Cambodian political authority that had been increasingly repressive and authoritarian. Since the successful coup in 1997, Prime Minister Hun Sen and the central government had implemented policies that many local peacebuilding actors believed to be harmful to local populations. A few examples include poor management over land distribution, suppression of basic human rights, inviting uncontrollable foreign investment, and a high level of corruption. The authorities had violently cracked down on the popular appeals of these policies. In contrast, although they had little practical influence, many international agencies openly condemned the government's antihumanitarian actions and lack of transparency. Thus, for many local peacebuilding agencies, the main threat to their programs was the Cambodian government, while they considered many international agencies to be supporters that could protect them from aggressive threats (Lee and Park 2020).

Due to these contextual factors, an increasing number of Cambodian peacebuilding agencies that once engaged in resistance against foreign influence began to accept co-existence with international peacebuilders and see the value and benefits of collaboration. This development decreased the significance of justice-seeking resistance, resistance in adherence to local contexts, and uninformed resistance. Cambodian peacebuilding actors instead explored ways to promote their agenda while maintaining collaboration with international supporters. Although there is variation in the aspects of peacebuilding on which Cambodian peacebuilders aimed to put their ideas, themes that appeared frequently during the interviews were peace and peacebuilding, key operational features (e.g., transparency and accountability), conceptual criteria, and priorities in peacebuilding sectors (Interviews VI, VII, VIII). In pursuit of these goals, local peacebuilding actors developed various strategies that can be roughly categorized into three groups (see Figure 41.1).

Limited engagement was frequently visible in the peacebuilding models developed by traditional/religious leadership who were particularly active in justice-seeking resistance. The local religious/traditional leaders generally enjoyed a relatively strong reputation and solid human network within their respective local communities, which enabled them to deal with the challenges posed by political authorities or economic constraints. The forms of such peacebuilding tended to be less institutionalized and required fewer financial resources to maintain. Hence these local leaders had less need to be dependent on material or advisory support from external donors. Accordingly, many programs developed under such religious/traditional leadership in Cambodia had more opportunities to reflect local perspectives. Although collaboration with external supporters was welcomed, the externals in this collaboration were usually limited to secondary roles such as public relations with international media, crowdfunding, or assistance with logistics. For instance, the peace campaigns mobilized by prominent Buddhist monks adopted traditional folktales and myths as well as Buddhist concepts to

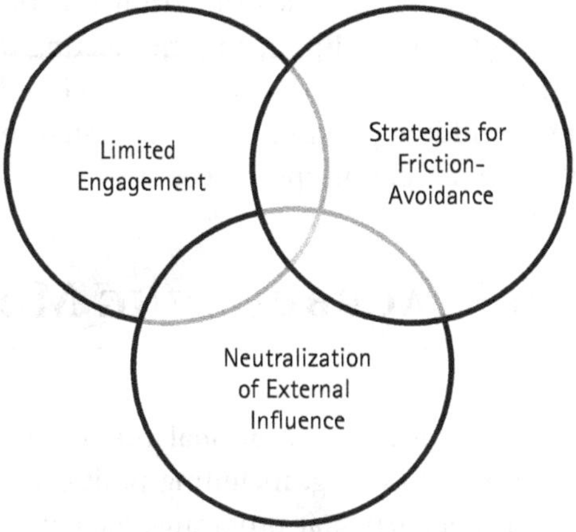

FIGURE 41.1 Strategies for promoting local peacebuilding agendas.

explain and justify various peace-related activities. The procedures and forms of activities also adopted Buddhist rituals and frequently took place in venues with religious or cosmological significance (Soeung and Lee 2017; Interview IV).

Neutralization of external reliance denotes the situations where local peacebuilding agencies attempted to reduce their reliance on their international counterparts since they were aware of various restrictions and challenges raised by international actors. For instance, some local organizations in Cambodia mobilized their operational funding through various income-generation programs, which included community-based tourism, fair trade of local artisans' products, microfinancing, and regular fund-raising campaigns by local communities. In other cases, local peacebuilding agencies created coalitions so they could collectively handle the pressure and demands of their international counterparts. Diversification of international partners was another major strategy. A few preferred types of external partners were civil society associations in the Global North, faith-based peacebuilding agencies, and overseas diaspora groups, who local actors feel are more patient with the challenges in the local communities and more respectful of local needs. It was rare for income-generation programs to raise the funding to cover the full operational budget, and the local coalitions and new peacebuilding partners were unlikely to replace the support from traditional donors. Yet such initiatives offered local actors leverage to negotiate with external partners and continue their core programs regardless of the changes in external funding.

Friction-avoiding approaches denote a range of strategies adopted by local actors in order to push forward their agenda without overtly challenging their international counterparts or creating an alternative form for peacebuilding. Instead a large number of local agencies in Cambodia chose more "quiet" strategies for addressing their anxiety due to factors like the risk of losing the existing partnership, a lack of self-confidence, potential friction with other involved local actors, and internal resistance to radical transformation. For instance, many local organizations used different liberal themes (e.g., human rights, democratic participation, good governance, gender equality) according to the types of donors in order to explain the same programs. When the principles or regulations set by donor agencies seemed unrealistic or undesirable, many local agencies "contextualized" them without communicating the changes to their international partners. On some occasions, local actors disclosed only partial truths in relation to how local peacebuilding programs were operated in order to save face for themselves and their international counterparts (Lee 2019).

Hybrid Peacebuilding Models

What form of hybridity does such local-international interaction bring about? Hybridity can occur in all aspects of peacebuilding, including political orders, governance systems, the operational features of particular programs, the construction of civil society, people's value systems and norms, and the like. One stakeholder's complete assimilation or domination over other actors rarely occurs. In most cases, the transformation is made by reflecting and mingling the values and perspectives of the involved stakeholders. The existing literature pays particular attention to local actors' proactive and subjective actions to adopt, reject, internalize, and modify the imported norms that can be more obviously visible in people's everyday life. Local customary practices surrounding tradition, culture, identity, and subsistence have regulatory power over the extent of liberal elements to be adopted into local peacebuilding or governance. (Out of numerous references, see Boege et al. 2009; Clements 2009; Richmond 2010; Peterson 2012; Shaw and Waldorf 2010; Belloni 2012; Visoka 2017.) Depending on how the mutual interaction is developed, hybrid models can be emancipatory and contributive to locals as well as being utilized to further sustain external domination by co-opting locals (Richmond 2009a). Moreover the nature of peacebuilding programs can constantly transform and develop as the key stakeholders continue mutual interactions, exchanging proposals and resistance between them.

Both development studies and international relations have examined the hybridity of international-local norms. Some studies pay attention to the rules of the international community's regulative functions and propose typologies, such as the norm life cycle and the norm-diffusion spiral (Keck and Sikkink 1998; Risse, Ropp, and Sikkink 1999; Cortell and Davis 2000). Other studies emphasize the roles of local actors in adopting and contextualizing the imported norms and highlight concepts like *norm contestation*, *localization*, and *decoupling*. Regarding local actors' motivations for such contestation and decoupling, a useful framework is the idea of *cultural filter*, which clarifies the normative, structural, institutional, and political constraints embedded in people's daily lives (Acharya 2004; Bettiza and Dionigi 2014; Bonacker, von Heusinger, and Zimmer 2017).

As the academic debates on hybrid peace became theoretically, empirically, and normatively extensive, the number of studies that highlight the conceptual limitations of hybrid peace also increased (Millar 2014; Visoka 2017). For instance, from a normative perspective, studies that highlight the discourse based on the concept of hybridity frequently fail to properly acknowledge the power domination of the externals, and hence may contribute to the domination of liberal models (Chandler 2010). When it is adopted as a descriptive tool without careful acknowledgment of contextual complexities, the hybridity discourse may strengthen the binary conceptualization of local versus international, which the concept aimed to address (Nadarajah and Rampton 2015). Other studies highlight that hybridity has existed in all social activities and does not offer new ideas (Zaum 2012). Reflecting on these limitations, recent studies propose a further development of the discourse of hybridity by addressing the risk of using the concept as a prescriptive tool, diversifying methodological options, and expanding the conceptual basis (Peterson 2012; Mac Ginty 2014; Visoka 2017).

As discussed earlier, many Cambodian peacebuilding actors simultaneously pursued two distinct goals—maintenance of local-international collaboration and development of locally driven peacebuilding—and these efforts have encouraged local peacebuilding agencies to transform their programs in various ways. As a result, local peacebuilders' level of norm adoption varies from integrating Western values into the new modus vivendi to token gestures to acknowledge their familiarity with the norms. The examples relevant to norm diffusion are rare. For instance, few Cambodian agencies fully adopted norms or operational principles imported from the international agencies. However, this type of norm internalization is occasionally visible in faith-based organizations and organizations involved in human rights advocacy. Some activists who participated in the author's interviews confirmed their belief in human rights as a universal value and argued that global standards should be equally applied in Cambodian contexts (Interview VIII).

The majority of the Cambodian peacebuilding agencies instead demonstrate the examples relevant to *norm localization* or *contextualization*. *Norm localization* denotes the cases where local agencies appreciate the importance and utility of imported values and concepts but want to *Cambodianize* them to address their incompatibility with Cambodia's values (e.g., ideas of social hierarchy) or practical contexts (e.g., nonexistence of the civil society defined by European traditions). A development NGO in Phnom Penh, for instance, transformed and redefined its approach toward the principle of "working with communities" at least three times between the mid-1990s and the early 2010s. The director of the NGO mentioned that such frequent modification was unavoidable because the social contexts changed rapidly and the conventional ideas failed to reflect the changes (Interview VI). Since some key liberal concepts (e.g., democracy, freedom of speech) have become widespread, many local agencies incorporate Western views of these themes into their practices without even thinking of the underlying philosophical ground.

In other cases, local actors adopted *contextualization* that denotes local actors' operationalization of the imported principles/values as being more context-sensitive. While local agencies appreciate the key values or themes highlighted by international donor agencies, they modify some elements that seem unrealistic or undesirable for implementation in their local contexts. For instance, while promoting "inclusive participation" in local governance, some organizations facilitate women-only and youth-only meetings since these groups tend to have less of a voice when they are mixed with male elders in the same space. Moreover, while upholding the principle of "transparent and regular evaluation," some local organizations produce their own models for evaluation to challenge international donors' demands for reporting short-term program outcomes in the form of quantifiable figures (Interview VII).

Through this process, there has been constant revision in the adoption of hybridity of local and external norms as a new standard in the local peacebuilding sector and the locality in Cambodia have changed in various directions and occasionally in a more radical and rapid way than many researchers expect. When focusing on fundamental values, for example, the views of Cambodian peacebuilding agencies in relation to concepts like democratic governance, human rights, and peacebuilding in the 2010s became closer to the Western standards than they used to be in the early 1990s. However, at operational levels, a large number of procedural features and institutional requirements that local actors felt irrelevant have been removed or modified over the past decades (Interviews VI, VII and VIII).

Conclusion

This chapter has reviewed conventional academic debates on local resistance, hybridization, and hybrid peacebuilding and examined how such debates are applicable to the local peacebuilding programs in postwar Cambodia. As a concluding discussion, this section will revisit the key findings and present a couple of theoretical and practical implications that are relevant to those findings.

This chapter presented five distinct motivations that encouraged Cambodian actors to mobilize resistance against externally driven peacebuilding programs: interest-seeking resistance, justice-seeking resistance, resistance in adherence to local contexts, uninformed resistance, and involuntary resistance. One resistance movement may reflect multiple motivations of different actors, and the multiple forms of resistance can be mobilized as parts of a wider resistance movement.

From a practical perspective, this means that external agencies should be more nuanced in understanding and responding to local resistance to promote more successful peacebuilding. For instance, while interest-seeking resistance can be dealt with by more strategic institutional arrangements, uninformed resistance can be addressed with relatively simple technical or procedural arrangements for enhancing the mutual communication between local communities and peacebuilding agencies. Nevertheless the identification of complex motivation is a challenging task, especially in the early stages of peacebuilding. In Cambodia, UN agencies and other international organizations failed to recognize this and focused mostly on overt resistance by local elites. They interpreted this to be power-seeking resistance and put most of their effort toward addressing the demands from and coordinating the contradiction between the different elite groups. Hence it was a surprise to them that anti-UN campaigns continued at the local level by youths who were committed to nationalistic propaganda even after the making of political deals between power elites (Interview I).

This chapter demonstrated when and how local agencies that had initially engaged in resistance transformed their attitudes to engage with their international counterparts. Three types of strategies were adopted by local actors to promote their perspective while maintaining external engagement: limited engagement, neutralization of external influence, and strategies for friction-avoidance. Consequently a variety of hybrid forms of peacebuilding have been developed, which demonstrate the incorporation of both local and international perspectives in different extents.

From a theoretical perspective, this process of local-international dynamics in Cambodia demonstrates the dual structure of power in peacebuilding; that is, while the international communities control the official procedures of peacebuilding, local actors determine what and how programs are implemented in practice. At the official level, the transformation in local-international relations may appear to be a result of local actors constantly compromising their views to incorporate international standards. In practice, local actors have found multiple strategies to indirectly reject the imported norms and principles that they find irrelevant or counterproductive. It has always been local actors who ultimately determine how peacebuilding programs will be implemented and how they should be implemented.

References

Acharya, A. 2004. "How Ideas Spread: Whose Norms Matter? Norm Localization and Institutional Change in Asian Regionalism." *International Organization* 58 (2):239–275.

Alfred, T., and J. Corntassel. 2005. "Being Indigenous: Resurgences against Contemporary Colonialism." *Government and Opposition* 40 (4):597–614.

Bhabha, H. K. 1994. *The Location of Culture*. London: Routledge.

Belloni, R. 2012. "Hybrid Peace Governance: Its Emergence and Significance." *Global Governance* 18 (1):21–38.

Bettiza, G., and F. Dionigi. 2014. *Beyond Constructivism's Liberal Bias*. EUI Working Paper 10. Florence: Badia Fiesolana, European University Institute.

Boege, V., A. Brown., K. Clements, and A. Nolan. 2009. "Building Peace and Political Community in Hybrid Political Orders." *International Peacekeeping* 16 (5):599–615.

Bonacker, T., J. von Heusinger, and K. Zimmer. 2017. *Localization in Development Aid*. London: Routledge.

Chandler, D. 2010. "The Uncritical Critique of 'Liberal Peace.'" *Review of International Studies* 36 (1):137–155.

Clements, K. 2009. "Internal Dynamics and External Interventions." *Peace Review* 21 (1):1–12.

Cortell, A., and J. Davis. 2000. "Understanding the Domestic Impact of International Norms: A Research Agenda." *International Studies Review* 2 (1):65–87.

de Certeau, M. (trans. Rendall, S.). 1984. *The Practice of Everyday Life* Vol. 1., Berkeley, CA: University of California Press.

Dougherty, B. 2004. "Right-Sizing International Criminal Justice: The Hybrid Experiment at the Special Court for Sierra Leone." *International Affairs* 80 (2):311–328.

Ejdus, F., and A. Juncos. 2018. "Reclaiming the Local in EU Peacebuilding: Effectiveness, Ownership, and Resistance." *Contemporary Security Policy* 39 (1):4–27.

Foucault, M. 1978. *The History of Sexuality: An Introduction. Vol. I*. New York, NY: Pantheon Books.

Galvanek, J. 2013. *Translating Peacebuilding Rationalities into Practice: Local Agency and Everyday Resistance*. Berlin: Berghof Foundation.

Hellmüller, S. 2013. "The Power of Perceptions: Localizing International Peacebuilding Approaches." *International Peacekeeping* 20 (2):219–232.

Hoy, D.C. 2004. *Critical Resistance: From Poststructuralism to Post-Critique*, Cambridge, MA: The MIT Press.

Keck, M., and K. Sikkink. 1998. *Activists beyond Borders*. Ithaca, NY: Cornell University Press.

Kumar, C., and J. De la Haye. 2012. "Hybrid Peacemaking: Building National 'Infrastructures for Peace.'" *Global Governance* 18 (1):13–20.

Lee, S. 2013. "Lost in Translation: The Problem of Self-Perception in Peace Negotiation." *International Interactions* 39 (2):144–166.

Lee, S. 2015. "Motivations for Local Resistance in International Peacebuilding." *Third World Quarterly* 36 (8):1437–1452.

Lee, S. 2019. *Local Ownership in Asian Peacebuilding*. Basingstoke, UK: Palgrave Macmillan.

Lee, S., and A. Özerdem. 2015. *Local Ownership in International Peacebuilding.* London: Routledge.
Lee, S., and W. Park. 2020. "The Dual Track of Democracy Promotion in Post-war Peacebuilding in Cambodia." *Peacebuilding* 8 (1):78–97. doi:10.1080/21647259.2018.1498671.
Mac Ginty, R. 2010. "Hybrid Peace: The Interaction between Top-Down and Bottom-up Peace." *Security Dialogue* 41 (4):391–412.
Mac Ginty, R. 2011. *International Peacebuilding and Local Resistance.* Basingstoke, UK: Palgrave Macmillan.
Mac Ginty, R. 2012. "Between Resistance and Compliance: Non-participation and the Liberal Peace." *Journal of Intervention and Statebuilding* 6 (2):167–187.
Mac Ginty, R., ed. 2014. *Alternative and Bottom-Up Peace Indicators.* Abingdon, UK: Routledge.
Millar, G. 2014. *An Ethnographic Approach to Peacebuilding.* London: Routledge.
Nadarajah, S., and D. Rampton. 2015. "The Limits of Hybridity and the Crisis of Liberal Peace." *Review of International Studies* 41 (1):49–72.
Newman, E., and O. Richmond, eds. 2006. *Challenges to Peacebuilding.* Tokyo: United Nations University Press.
Parks, T. 2008. "The Rise and Fall of Donor Funding for Advocacy NGOs." *Development in Practice* 18 (2):213–222.
Peterson, J. 2012. "A Conceptual Unpacking of Hybridity." *Journal of Peacebuilding & Development* 7:9–22.
Richmond, O. P. 2009a. "Becoming Liberal, Unbecoming Liberalism." *Journal of Intervention and Statebuilding* 3 (3):324–344.
Richmond, O. P. 2009b. "A Post-Liberal Peace: Eirenism and the Everyday." *Review of International Studies* 35 (3):557–580.
Richmond, O. P. 2010. "Resistance and the Post-liberal Peace." *Millennium: Journal of International Studies* 38 (3):665–692.
Richmond, O. P., and A. Mitchell. 2011. *Hybrid Forms of Peace.* Basingstoke, UK: Palgrave Macmillan.
Risse, T., S. Ropp, and K. Sikkink, eds. 1999. *The Power of Human Rights: International Norms and Domestic Change.* Cambridge: Cambridge University Press.
Scott, J. C. 1989. "Everyday Forms of Resistance." *Copenhagen Papers* 4:33–62.
Shaw, R., and L. Waldorf, eds. 2010. *Localising Transitional Justice.* Redwood City, CA: Stanford University Press.
Soeung, B. and S. Lee. 2017. "The revitalisation of Buddhist peace activism in post-war Cambodia." *Conflict, Security & Development* 17 (2): 141–161.
van der Lijn, J. 2013. "Imagi-Nation Building in Illusionstan." *International Peacekeeping* 20 (2):173–188.
Visoka, G. 2011. "International Governance and Local Resistance in Kosovo." *Irish Studies in International Affairs* 22:99–125.
Visoka, G. 2017. "After Hybridity?" In *Hybridity: Law, Culture and Development*, edited by N. Lemay-Hébert and R. Freedman, 301–324. Abingdon, UK: Routledge.
Zaum, D. 2012. "Beyond the 'Liberal Peace.'" *Global Governance* 18 (1):121–132.

Interviews

Interview I. Community leader in Svay Rieng, Cambodia, November 2014.
Interview II. Peacebuilding practitioner in Battambang, Cambodia, November 2014.

Interview III. Peacebuilding practitioner in Battambang, Cambodia, November 2016.
Interview IV. Buddhist peace activist in Phnom Penh, Cambodia, November 2014.
Interview V. Community leader in Battambang, Cambodia, January 2019.
Interview VI. Peacebuilding practitioner in Phnom Penh, Cambodia, November 2016.
Interview VII. Peacebuilding practitioner in Phnom Penh, Cambodia, November 2016.
Interview VIII. Peacebuilding practitioner in Phnom Penh, Cambodia, November 2016.

CHAPTER 42

HYBRID POLITICAL ORDERS AND CUSTOMARY PEACE

VOLKER BOEGE

For more than two decades, international peacebuilding practice has been informed by what has been called the "liberal peace" approach, and the academic debate has revolved around this approach and its critique.[1] Linked to the discourse on state fragility, liberal peacebuilding was conceptualized as closely linked to statebuilding, with statebuilding following the internationally hegemonic understanding of "the state" as the Western Weberian state, and operationalized as building the institutions of the state and their capacity. Accordingly a core aim of peacebuilding is the establishment of a "proper" state as the guarantor of internal peace, with a liberal market economy and a liberal democratic civil society as underpinnings of such a liberal state. This paradigm of peacebuilding-as-statebuilding (Richmond 2011) is pursued as an exercise of standardization and homogenization.

In the course of numerous international peacebuilding-as-statebuilding interventions, however, it turned out that the intervened-upon societies were not just blank pages on which the international liberal agenda could be written, and that local populations were not just passive recipients of a standardized peace and state delivered to them from the outside. Local people demonstrated agency of their own, often beyond forms of politics familiar to liberal Western views. The concept of peacebuilding-as-statebuilding is thus confronted with the messy realities of governance in the intervened-upon societies.

This reality has been acknowledged by the more recent "turn to the local" (and the concomitant "ethnographic turn"; Millar 2018). Local actors have come to be seen as impacting on the internationals' predetermined peacebuilding/statebuilding agenda and its implementation through various forms of agency, and as capable of appropriating international agendas and resources for their own purposes (Richmond 2011; Richmond and Mitchell 2012; Mac Ginty and Richmond 2013). "The local" is acknowledged as the core arena of peacebuilding, the site of ongoing processes of interchange, entanglement, permeation, reassemblage, and reconfiguration, as co-created and shared by a variety of interacting local, national, regional, and international actors and institutions (Mac Ginty 2015, 2016; Kappler 2014; Mac Ginty and Richmond 2013; Albrecht and Moe 2014).

In this mix, local actors and societal structures and institutions that do not fit into the Western liberal format of "state," "civil society," or "business/economy" warrant particular attention: this is the realm of the "local local" (Richmond 2012, 6), which first and foremost comprises indigenous customary networks and their institutions and actors. Mac Ginty's (2016, 207) observation that international peacebuilders and statebuilders have difficulties engaging with the local holds particularly true for this sphere: internationals are "poorly equipped to see the local," not only because of lack of access, but also because of "epistemological biases." Accordingly, in the "local ownership" discourse, which is so fashionable with international donors and peacebuilders, only a rather narrowly understood local is appealed to: "the local that is part of the Western civil society imaginary while the rest is deemed violent and intolerable" (Richmond 2011, 153).

It is this "rest" that this chapter will focus on. In a first step I'll outline the relevance of customary institutions and traditional authorities for governance and peace in the context of hybridization of political order. In a second step the importance of customary peace(building) and some of its features will be explored. I conclude with short reflections on implications for peacebuilding theory and practice.

Hybridization of Political Order

Traditional[2] societal structures—extended families, lineages, clans, tribes, religious brotherhoods, village communities—and traditional authorities, such as village elders, clan chiefs, healers, big men, tribal kings or queens, and spiritual leaders, and the customary law and indigenous knowledge that comes with them shape the everyday social reality of large parts of the population in postcolonial states even today. Today's customary or traditional actors, institutions, and procedures are of course *not* those of the precontact and precolonial past. Societies everywhere in the world have come into contact with outside influences and have not been left untouched by (originally European) capitalist expansion, colonialism, missionizing, imperialism, and globalization. Therefore there are no clear-cut boundaries between the exogenous "modern" and the endogenous "traditional" or "customary"; rather there are ongoing processes of assimilation, articulation, transformation, and/or adoption. Nevertheless the use of the terms "custom," "customary institutions," "traditional," "traditional authorities," etc. is warranted because they mark specific characteristics, which are rooted in precolonial/precontact times, that distinguish them from introduced institutions that belong to the realm of the state, the market economy, and civil society. It has to be kept in mind, however, that these terms and the social institutions they refer to are not well defined and fixed, but contested, fluid, and constantly changing. Custom is not static; it is in constant flux. It adapts to new circumstances when exposed to external "modernizing" factors, at the same time itself exerting influence on these factors.

Customary institutions of governance that had existed prior to colonial rule and the subsequent declaration of the independence of newly formed postcolonial states have survived the onslaught of colonialism and "national liberation." They have been subject to considerable change and had to adapt to new circumstances. They have, however, shown remarkable resilience. By contrast, attempts to consolidate the introduced form of statehood after it had been formally established as a result of decolonization met with manifold obstacles. The import or imposition of the modern Western state structures in the course of decolonization did not lead to the establishment of the "modern" state (that is, a state along the lines of the Western model)—as might have been expected by the former colonial powers and the indigenous political elites who steered their countries toward independence. At the time of independence the state more often than not was nothing more than an empty shell. Many of the newly independent states had no history of precolonial unitary rule, and people did not have a sense of shared citizenship and national identity.

For many states in the Global South even today it is difficult to exercise sovereign control, to deliver effective governance, to guarantee order, and, based on these services, to gain legitimacy. They have not reached out to and permeated the entirety of their territories and their societies. In certain areas (be it remote rural regions or informal settlements at the fringes of urban centers), only "outposts" of the state can be found. For many people, the state is "far away," not only physically but also mentally. Its institutions are more often than not seen as a threat rather than as providers of security, peace, justice, and other services.

Under such conditions one way to make state institutions work is with the utilization of kin-based and other traditional networks. This, however, allows these networks to imbue (incomplete) state structures with their own logic and their own rules of operating. The state is "infiltrated" by "informal" indigenous societal institutions and social forces. This leads to its aberration from the ideal type of what a "proper" (Western, Weberian) state is supposed to look like (hence the lament about neopatrimonialism, nepotism, corruption, etc.).

On the other hand, the partial intrusion of state agencies impacts nonstate local orders as well. Customary systems are subjected to deconstruction and re-formation as they have to engage with state institutions. In the course of the interchange with the state apparatus, customary institutions and customary authorities adopt an ambiguous position with regard to the state, appropriating state functions and "state talk," but at the same time pursuing their own agenda under the guise of the state surface. This, however, also means changing one's original positionality detached from the state.

So there is change and adaptation in both directions. On the one hand, customary actors and institutions change and adapt in the course of interaction with state institutions and, more generally, in light of the all-powerful, ever-present, taken-for-granted idea of the state. On the other hand, state actors and institutions change and adapt in the course of interaction with the customary sphere. These interactions lead to the hybridization of governance and let hybrid political orders emerge. These hybrid political orders differ fundamentally from (the ideal type of) the Weberian state as the framework for maintenance of peace and order, based on the state monopoly over the legitimate use of physical force. In hybrid political orders, diverse legitimate authority structures, institutions, and actors of governance, sets of rules, logics of order, and claims to power coexist, compete, overlap, intertwine, and blend, emerging from genuinely different societal spheres—spheres, however, which do not exist in isolation but permeate each other (Boege et al. 2009, 2010; Mac Ginty and Richmond 2016; Brown 2018; Hunt 2018; Boege 2018). Political order is not least imbued with the legacies of colonial and precolonial forms of governance. The specific relation between these forms and "the state" differ from place to place and change over time. Hybrid political order is not static, but in constant flux.

While in many countries traditional authority structures just coexist with state structures, without formal regulation of relations, some postcolonial governments deliberately utilize traditional authorities for the purposes of strengthening state capacity and legitimacy by formally incorporating them into state structures, leading to "institutional hybridity" (Mustasilta 2018, 5). This has been the case in sub-Saharan African states in the course of "re-traditionalization" since the beginning of the 1990s (Baldwin 2015; Ubink 2008). Here traditional leaders had been largely discredited in the immediate postindependence era because they often had been incorporated into (indirect) colonial rule as instruments of the colonial masters, and hence the new political elites of the newly independent states often attempted to do away with them as anachronistic and reactionary forces of the past (Baldwin 2015; Ubink 2008). But customary forms of governance persisted, and, sooner or later, the authorities of the independent states—similar to the former colonial rulers—realized that it might be more promising to incorporate them rather than try to suppress and displace them. Hence legislation "(re) incorporated traditional leaders officially into state hierarchies in recognition of their ongoing influence as local players" (Kyed and Buur 2006, 2). In recognition of the relative weakness of state institutions and the relative strength of traditional authorities, governments have come to rely on the latter for performing certain state functions, thus contributing to a resurgence of customary rule, albeit in (partly) new forms and with new functions, e.g., tax collection, policing, registration and census duties (for Mozambique: Buur and Kyed 2006, 848).

Cases of constitutional recognition and integration can be found in African and other postcolonial countries, with different types of institutional coexistence (Mustasilta 2018; Baldwin 2015; Muriaas 2011; Ubink 2008). Ghana, for example, has a firmly institutionalized system of chieftaincy in place, acknowledged by the state. The 1992 Constitution of Ghana and the 2008 Chieftaincy Act clarify the relationship and connection between the chiefly system and the state system. The chiefly system is multilayered, from the village chiefs or head chiefs at the local level up to paramount chiefs or kings of larger ethnically and territorially defined social entities (most notably the king of the Ashanti Kingdom, the Asantehene).[3] This system is reflected in the state context by the formal institutionalization of the National Register of Chiefs, the National House of Chiefs, and Regional Houses of Chiefs. There is a special Ministry of Chieftaincy and Traditional

Affairs. The relationship between state authority and chiefly authority is uneasy, however. The state tries to subordinate the chiefly system, and it has been successful to a considerable extent. Nevertheless even today the "chief is a political and social power center (if even in a circumscribed sense) in the area he rules and ipso facto a microcosm of authority who at times rivals the central government in legitimacy, recognition, and loyalty by subjects" (Boafo-Artur 2003, 134).

In Ghana as elsewhere, state policies aim at the instrumentalization of traditional authorities for state purposes and thus as a means for the reinforcement of the authority of the state (Mustasilta 2018; Ubink 2008). Recognition of traditional leaders is conceptualized as an act that manifests the state's authority. Traditional leaders, on the other hand, might utilize their new position to reinforce their own authority. At the same time, however, they are in danger of losing this authority, exactly because they are now perceived as agents of the state. In Mozambique, for instance, "the obligations placed on chiefs to collect taxes and to police rural communities are greeted with discontent by many rural citizens, with the effect of potentially pitting chiefs against the communities from which, *de facto* and *de jure*, they derive their legitimacy" (Kyed and Buur 2006, 14).

In many other postcolonial countries the situation is similar: the fact that traditional leaders like chiefs are both "state" and "nonstate"—that they are "doubly embedded"—comes with the challenge to have to operate in two sociopolitical contexts (Thelen, Vetters, and von Benda-Beckmann 2014; Holzinger, Kern, and Kromrey 2016; Kyed 2017). Or, to put it in another way, the former CEO of the National Council of Chiefs in Vanuatu compares chiefs to turtles: turtles have to live in two worlds, in the water and on land; they have to swim in the water and to crawl on land; they cannot swim on land or crawl in water; they have to adapt to the environment. Chiefs in their locality act according to *kastom*;[4] however, in the National Council of Chiefs, the Malvatumauri, which is a formal constitutional body, they have to act according to the laws of the state, the Chiefs' Act, etc.[5] Accordingly the Malvatumauri's CEO has to have one leg in kastom, the other leg in the state system. He has a Public Service Commission position, but at the same time has to do kastom work. The Malvatumauri itself is linked to the Ministry of Justice. Similar arrangements can be found in numerous postcolonial states.

Hence chiefs are not "pure" "nonstate" "customary" actors anymore. But they are still around. Contrary to the prophecies (and hopes) of all sorts of modernization theorists, "chieftaincy did not wither and die" (Ubink 2008, 10)—chiefs and other traditional leaders are still with us today, and they won't disappear anytime soon. What I said earlier about "tradition" and "custom" also applies to the institution of chiefs and chiefly systems: they have their roots in precolonial political organization and leadership structures, but they are not manifestations of such precolonial leadership, although they are referred to as "traditional authorities" today (and present themselves as such). Rather, the category chief goes back to colonial times when colonial masters tried to harmonize and instrumentalize the plethora of precolonial leadership types for their own purposes, with "chiefs" emerging as the unifying leadership category (although there were still a variety of types of chiefs). And processes of change and adaptation of chiefly systems continued in the postcolonial period, albeit under fundamentally different conditions. In Vanuatu, for example, the leadership category chief emerged in the course of interaction between local societies and external actors, in particular missions and the colonial administration. Different types of chiefs evolved, and these types overlapped and mixed over time (Lindstrom 1997, 213). Today's "*kastom jifs*, although undoubtedly *kastom*, are not simply customary. The emergence and construction of the popular identity *jif*, from a plethora of local leadership positions, were shaped by the events and interests of postcontact, colonial society" (212).

The institution of chief and chiefly systems—or other traditional leadership categories and systems—are of major importance for the provision of peace, security, and justice in the hybrid political orders of many postcolonial countries today. They fulfill important governance and conflict resolution functions in the everyday life of communities. They are a constant presence in their communities. Conversely, state institutions in charge of maintaining peace and order (the police, courts, corrections, etc.) hardly penetrate large swaths of "their" state territory. They lack capacity and legitimacy; they very much depend on good—more or less formal or informal—working relationships with local customary authorities; and more often than not they have to leave the maintenance of peace and order largely to those authorities.

In Liberia, for example, state security institutions lack effectiveness and legitimacy. Police capacities remain limited, constrained by logistical shortcomings. In large rural areas there is no police presence at all, with nearly 80% of police officers serving in the capital city and its surrounds in Montserrado County (Karim and Gorman 2016, 184). People complain about police inefficiency, corruption, mismanagement, brutality, human rights violations, low morale, and poor discipline (Zanker 2015). There is widespread lack of trust in the state justice institutions. Under these conditions, people put more trust in nonstate security institutions that are embedded in community life; "social networks and community dispute settlement have been key sources for peace" (Herbert 2014, 2). In this context, customary institutions figure prominently. Quarter chiefs and town chiefs, clan chiefs and paramount chiefs, as well as associated councils of elders and advisors are responsible for the maintenance of local peace and order, often in cooperation with religious actors and institutions: Christian and Muslim religious leaders, so-called witch doctors and other representatives of indigenous spirituality, and in particular the secret societies. Their realm of responsibility and the influence of their priests reaches well beyond the religious sphere in a narrow sense. They are in charge of the sociocultural, mental, and spiritual well-being of communities and the relationships between social groups. They also are hybrid authorities, drawing their legitimacy from the invisible world of the spirits, and at the same time they are officially recognized by the state; they come under the assistant minister for cultural affairs within the Ministry of Internal Affairs. In Liberia, as in other hybrid political orders, state and nonstate security institutions do not merely coexist, but there are overlaps, linkages, and interactions. However, there is not much effort to bring these different institutions together in systematic transparent collaboration. Rather, collaboration is largely ad hoc, informal, tacit.

Customary Peace

While the idea of the Weberian state has as its core the state monopoly over the legitimate use of physical force, realities in hybrid political orders are far from that ideal type. In hybrid political orders the provision of peace, security, and justice is not based on such a state monopoly but on the relations and interactions of a plethora of diverse actors and institutions. Customary actors and institutions play a prominent part. They provide what might be called "customary peace."

To speak of customary peace in general and abstract ways, however, is problematic, as the features that we try to capture by this term are first and foremost characterized by the fact that they are locally specific and locally embedded, varying from country to country, from region to region, often from community to community. There are as many different types of customary peace as there are different communities with a specific history and specific kastom/custom. This contextual cultural embeddedness is a decisive feature of customary peace (Boege 2006). It marks a significant difference from "modern" Western concepts, which lay claim to hegemonic universality and universal applicability. So any talk about "customary peace" is somewhat contrary to the very spirit of kastom/custom forms of peace. Nevertheless, in line with Western academic conventions and for the purposes of this text, in what follows I present an ideal type of customary peace, derived from African and Melanesian empirics—though I am aware that this procedure is alien to the spirit of customary peace in the contexts from which this ideal type is distilled.[6]

Customary peace is grounded in and addresses issues of everyday community life. The local peace (and the local understanding of what peace is) is in general embedded in a close net of relations among people and between people and land. Peace is understood in terms of harmony and balance in the life of communities, and "community" also includes the nonhuman natural world (the land, animals, and other aspects of nature) and the nonhuman spiritual world (the spirits of the ancestors, gods, God). Harmony and balance require reciprocity in relationships in all dimensions of everyday life. Relations are not only relations within and between human beings, but also with nonhuman living beings, with spirits, and with God. Disturbance of community harmony and balance can take the form of physical violence, e.g., in conflicts over land, but also other forms, like swearing or sorcery, gossip and verbal insults, destruction of property; more often than not, a variety of conflictual issues are linked. Such disturbances necessitate dispute settlement or peacebuilding activities to restore reciprocity and harmony. Only then can the very survival (both in the material and the spiritual sense) of the community be secured for the future. This implies a restorative and holistic approach to peace and peacebuilding.

Restoration of relationships is key. Accordingly, customary peace hinges very much on the concept of reconciliation; reconciliation is the prerequisite for restoration of relationships between former adversaries and of social harmony within the community.

Reconciliation can take place only on the basis of a common understanding of the history of a given conflict (or other disturbances of the peace); a consensus on that history has to be elaborated. On this basis, perpetrators or members of parties involved in conflict can take responsibility for their deeds, apologize, and ask for forgiveness, and victims or members of the other conflict party can accept apologies and forgive, and then reconciliation can take place. Usually this is a long-term and multifaceted process; it has to aim for overall consensus, and it has to reach an agreement on the conditions of reconciliation that is accepted and adhered to by everybody. It is holistic in the sense that it not only addresses a single conflict or incident or an isolated issue but it also takes into account the overall relationships between conflict parties and their histories and the overall social circumstances in which the conflict took place. And it addresses not only the issues of the visible material world, reason, and the "hard" facts, but also the spiritual (and emotional and psychological) dimension and assures the inclusion of the spiritual world (the ancestors, the gods, God) into any peace settlement. Peace is not only a political or justice or social issue; it cannot be compartmentalized but exists as an all-inclusive whole.

Restoration of peace therefore transcends the material dimension to include the immaterial dimension: emotions, nonverbal communication, ritual and ceremony. Highly ritualized peace ceremonies that align the actors of the visible and the invisible world are means of peacebuilding in their own right, more powerful than mere written or spoken words, appealing to emotions and feelings and the spiritual realm, providing cleansing and purification. Peacemaking through prayer is a powerful form of conflict resolution in many societies of the Global South, where religion plays a much more prominent role in everyday life than in secular Western societies.

The focus on reconciliation in customary peacebuilding encompasses an understanding of justice as basically restorative. Restoration of livelihoods and relationships is more important for the future well-being of communities than punishment of offenders. Accordingly, customary law is guided by ideas of restorative justice (retributive elements can play a part, though).[7] People generally find customary law easier to understand, more accessible, fairer, cheaper, more transparent, faster and more efficient, and also more legitimate than state law because it is embedded in local culture. Customary law and customary means of dispute settlement are seen as providing solutions to most peace and order issues in communities. In Ghana, for example, the chiefs have their own customary courts where they settle cases, together with their councils, according to customary law, which the state acknowledges as a source of justice in its own right. The "primary concern of the chief's court is not to be vindictive or punitive unless circumstances necessarily mandate such a solution. The primary purpose of justice in the chief's tribunal is to promote harmony and reconciliation between parties. The ultimate aim is the restoration of social equilibrium. . . . The adjudicators work to ensure that parties thereafter continue to live and relate to each other as good neighbours, friends and relatives even after the dispute" (Dzivenu 2008, 26). People often prefer the chief's courts to the state legal system because procedures are simple, cheap, and culturally embedded. Decisions are mostly respected, but people have the option to carry cases through the hierarchy of customary courts or to the state justice system.

The interface between customary law and state law is often uneasy, characterized by (partial) contestation, (partial) complementarity, and (partial) incompatibility.[8] Customary law has a specific relationship to the state: it is nonstate, distinct from the laws of the state, but at the same time accepted by the state as a source of law in its own right, besides the statutory law (with variations in the concrete legal regulations in different countries). Formally, customary law is inferior to state law and confined to certain areas of jurisdiction; serious crimes such as murder and rape, for example, are the prerogative of state law. In practice, however, boundaries are fluid, and customary law often reaches beyond its formal restrictions. For example, fieldwork in several West African and Melanesian countries found that in some communities cases of murder and rape were taken directly to the police, while in others even those cases were in the first instance dealt with by the local authorities. Moreover some forms of customary law that are practiced are not sanctioned by the state, and some forms of customary law cover areas that are not (and cannot) be covered by state law but are of major importance for community peace, e.g., sorcery or witchcraft. Often the custodians of customary law do not have any formal enforcement powers; hence they are very much dependent on the willingness of community members to abide by customary law. Naming and shaming seem to be the most powerful instruments to prevent, control, and punish any disturbances of social order.[9]

Today local customary and introduced liberal understandings of justice and peace are in constant dialogue. Customary law is not static; it can be changed, and it changes all the time. Currently there are intense debates in many communities in postcolonial countries about adaptations of customary law in the face of new challenges, in particular with regard to the position of women and youth. In many postcolonial societies there are areas of customary law in which women and youth are disadvantaged. In general, women and young people are nevertheless supportive of customary law, but often they demand reforms to ensure that they are not overlooked, marginalized, or disadvantaged in dispute settlements. In Melanesian countries, for example, women and youth often complain about biased or incompetent chiefs and demand more of a voice in the kastom context, but they are not against the chiefly system and kastom as such. It is their own system, whereas the state system is seen as alien; for example, in Vanuatu courts are often called "kot blong waetman" (the white man's court; Forsyth 2011, 181). These demands indicate that customary peace has its problematic sides.[10] Power imbalances with regard to age, gender, kinship networks, or resourcefulness matter. And there can be tribes, clans, or extended families that are more powerful than others. Power asymmetries can affect the ways customary peace is built and maintained.

Furthermore, customary peace hinges on the existence of a community of relationships and values that are rooted in a common worldview and shared institutions. It works within and between societal groups in the local context. Dealing with "outsiders" or "upscaling" is much more difficult. Conflicts between "us" and "them," who do not share "our" kastom/custom, are much harder to tackle. If "they" do not understand, do not respect, or are unwilling to participate in "our" ways, customary peacebuilding runs into difficulties. In postconflict Bougainville, for example, customary ways of reconciliation played a crucial role for peacebuilding in the local context. Attempts to apply this form of peacebuilding to restoration of relationships with the multinational mining company whose operations had been a major cause of the preceding war so far have failed completely (Boege 2019).

Moreover customary law may include practices that contradict universal standards of human rights; e.g., perpetrators may be subject to severe corporal punishment. In Liberia, for example, trial by ordeal (called "sassywood") is a widespread practice (Isser, Lubkemann, and N'Tow 2009, 57). Though this might be seen as inhumane from a Western perspective, locals might see it as more humane than the Western practice of imprisonment.

Finally, customary peace is not totally nonviolent, and it is not necessarily sustainable over long periods of time. Peace deals might be revoked in the near or distant future when circumstances change. Recourse to violence is a legitimate option for dealing with conflicts in the context of customary peace. The controlled and ritualized conduct of physical violence can be integral to societal order and the maintenance and restoration of peace; such forms of violence are not perceived as being destructive disturbances of peace, but as integral and indispensable elements of peaceful order. Furthermore, violence against weak members of the community, in particular domestic violence against women and children, can be a "normal" feature of customary peace (as it is in "peaceful" Western societies). So there is no reason to romanticize customary peace. But there is every reason to take it seriously and engage with it.

Conclusion: Working across Difference, Engaging with Customary Peace

Beyond the provision of everyday peace and order in the local context, customary ways of building and maintaining peace also play a role in the internationally supported peacebuilding endeavors in the postconflict regions of the Global South.[11] Peacebuilding interventions everywhere have to engage with local customary forms of peace(building). Locals introduce their custom/kastom way of knowing and doing peace, thus derailing or redirecting preplanned peacebuilding interventions (Millar 2018, 4). Customary peace thus contributes to the emergence of new hybrid forms of peace (Richmond and Mitchell 2012; Mac Ginty 2011). It shapes those hybrid forms, and it changes in the course of hybridization.

Customary institutions and actors in many places and on many occasions have proven their capabilities and legitimacy in building peace, preventing and transforming conflicts, and maintaining societal order. Protagonists of liberal peacebuilding and statebuilding, however, have problems engaging with the "local-local," "i.e. what lies beneath the veneer of internationally sponsored local actors and NGOs constituting a 'civil' as opposed to 'uncivil' society" (Richmond 2011, 10). They hesitate to acknowledge and engage with such unfamiliar forms of peace(building) as they cling to their idea of what a proper peace (and state) should look like. People on the ground are much more flexible in comparison. They constantly engage with the interplay of profoundly different norms and forms of governance and peace. They show formidable pragmatism and adaptability when it comes to combining the indigenous customary and the exogenous "modern," exploring what works for their circumstances and incorporating it into their culture and customs—which are far from static "traditional" but are fluid and interculturally adaptive, hybridizing all the time. People are engaged in constantly negotiating the emergence of forms of peace, political order, and belonging that best suit their needs, history, culture, aims, and aspirations, and that can provide the framework for the peaceful conduct of conflicts and relatively stable and sustainable order—beyond Western liberal concepts (of state and peace).

Taking customary peace seriously has fundamental implications for the wider debate about and the praxis of peacebuilding and statebuilding. It is prudent to "refocus peace efforts away from top-down institutionalist approaches and towards a local and experiential approach" (Millar 2018, 4), accordingly to pursue positive mutual accommodation of state-based and customary peace approaches rather than the liberal peace's peacebuilding-as-statebuilding course. Based on the recognition and appreciation of diversity, the task is to work across difference. The challenge is to articulate state, custom/kastom, and civil society so that new forms of governance and peace emerge that are more effective and legitimate than Western models. This not least necessitates acknowledging and working with custom/kastom as a living system instead of rejecting it as an anachronistic residue of premodern times.

Notes

1. As a consequence of the extensive criticism it has been subjected to, some commentators "now argue that the idea of a liberal peace is in major crisis or, indeed, already dead" (Kappler 2017, 32). Its practices, however, are still pretty much alive (33).
2. For discussions of the term "traditional," see Zartman (2000), Huyse (2008), Ubink (2008), Holzinger, Kern, and Kromrey (2016).
3. In northern Ghana chieftaincy is historically less elaborated, with several originally acephalous (chiefless) ethnic groups. This makes the situation of customary governance in the North more complicated today and contributes to the frequency of (violent) chieftaincy disputes.
4. The term "kastom," a Pidgin derivative of the English "custom," is frequently used today in Vanuatu and other Melanesian countries, by both people in the villages and politicians in order to stress their cultural heritage and the difference between their own ways and foreign ways of governance, economics, conflict resolution, peacebuilding, etc., often depicting kastom as rooted in ancient precolonial traditions, a set of rules developed by the ancestors (Keesing 1993, 589). Kastom governance is thus presented as the other of, or even alternative to, introduced state-based governance. At the same time, kastom is

different from the customs of the precontact and precolonial past. It has developed since the times of first contact and colonization due to interaction and exchange with various external influences, and it continues to change today. In the following, I use "kastom" or "custom/kastom" so as to indicate the hybrid quality of the phenomena I refer to.

5. Interview with Selwyn Garu, Port Vila, 10 April 2011.
6. The following is based on several research and practice projects conducted over the past two decades in various countries in Oceania and Africa, which in general included more or less extensive components of field research.
7. Of course, talking about customary law in such general terms is a gross oversimplification. The main characteristic is its specificity, its connectedness to specific places and people. There is no "customary law" in the singular; customary law differs from place to place, within countries, and all the more so in different regions of the Global South. The point I want to make here is that customary law is a crucial source of solving conflicts and maintaining peace in its own right, distinct from the laws of the state. It enables customary peace.
8. The term "rule of law" in a context where custom/kastom continues to play a significant role in maintaining peace and order has a completely different meaning from what is understood by this term in the context of state and liberal peace.
9. There might be other means of enforcement, for instance magic/sorcery or direct physical force, but understandably this is not much talked about to outsiders; magic is forbidden in most societies today on religious grounds, and direct physical force is forbidden due to the proclamation of the state's monopoly over the use of legitimate physical violence.
10. For the following, see in more detail Boege (2006).
11. For case studies, see Richmond and Mitchell (2012) and Aning et al. (2018).

References

Albrecht, P., and L. Wiuff Moe. 2014. "The Simultaneity of Authority in Hybrid Orders." *Peacebuilding* 3 (1):1–16.

Aning, K., V. Boege, M. A. Brown, and C. T. Hunt, eds. 2018. *Exploring Peace Formation: Security and Justice in Post-colonial states.* London: Routledge.

Baldwin, K. 2015. *The Paradox of Traditional Chiefs in Democratic Africa.* New York: Cambridge University Press.

Boafo-Arthur, K. 2003. "Chieftaincy in Ghana: Challenges and Prospects in the 21st Century." *African and Asian Studies* 2 (2):125–153.

Boege, V. 2006. *Traditional Approaches to Conflict Transformation: Potential and Limits.* Berghof Handbook for Conflict Transformation. Berlin: Berghof Research Center for Constructive Conflict Management.

Boege, V. 2018. "The Hybridisation of Peace, Security and Justice: Cases from West Africa and Oceania." In *Exploring Peace Formation: Security and Justice in Post-colonial States*, edited by K. Aning, V. Boege, M. A. Brown, and C. T. Hunt, 207–225. London: Routledge.

Boege, V. 2019. "Reconciliations (Melanesian Style) and Transitional Justice." *Global Change, Peace & Security* 31 (2):139–157.

Boege, V., A. Brown, K. Clements, and A. Nolan. 2009. "On Hybrid Political Orders and Emerging States: What Is Failing—States in the Global South or Research and Politics in the West?" In *Building Peace in the Absence of States: Challenging the Discourse on State Failure*, edited by M. Fischer and B. Schmelzle, 15–35. Berghof Handbook Dialogue Series 8. Berlin: Berghof Research Center for Constructive Conflict Management.

Boege, V., A. Brown, K. Clements, and A. Nolan. 2010. "Challenging Statebuilding as Peacebuilding: Working with Hybrid Political Orders to Build Peace." In *Palgrave Advances in Peacebuilding: Critical Developments and Approaches*, edited by O. P. Richmond, 99–115. New York: Palgrave Macmillan.

Brown, M. A. 2018. "The 'Hybrid Turn': Approaches and Potentials." In *Hybridity on the Ground in Peacebuilding and Development: Critical Conversations*, edited by J. Wallis, L Kent, M. Forsyth, S. Dinnen, and S. Bose, 21–36. Canberra: ANU Press.

Buur, L., and H. M. Kyed. 2006. "Contested Sources of Authority: Re-claiming State Sovereignty by Formalizing Traditional Authority in Mozambique." *Development and Change* 37 (4):847–869.

Dzivenu, S. 2008. "The Politics of Inclusion and Exclusion of Traditional Authorities in Africa: Chiefs and Justice Administration in Botswana and Ghana." *Political Perspectives* 2 (1):1–30.

Forsyth, M. 2011. "The *Kastom* System of Dispute Resolution in Vanuatu." In *Working Together in Vanuatu: Research Histories, Collaborations, Projects and Reflections*, edited by J. Taylor and N. Thieberger, 175–182. Canberra: ANU E-Press.

Herbert, S. 2014. *Conflict Analysis of Liberia*. Birmingham: GSDRC, University of Birmingham.

Holzinger, K., F. G. Kern, and D. Kromrey. 2016. "The Dualism of Contemporary Traditional Governance and the State: Institutional Setups and Political Consequences." *Political Research Quarterly* 63 (3):469–481.

Hunt, C. T. 2018. "Hybridity Revisited: Relational Approaches to Peacebuilding in Complex Sociopolitical Orders." In *Hybridity on the Ground in Peacebuilding and Development: Critical Conversations*, edited by J. Wallis, L. Kent, M. Forsyth, S. Dinnen, and S. Bose, 51–65. Canberra: ANU Press.

Huyse, L. 2008. "Introduction: Tradition-Based Approaches in Peacemaking, Transitional Justice and Reconciliation Policies." In *Traditional Justice and Reconciliation after Violent Conflict: Learning from African Experiences*, edited by L. Huyse and M. Salter, 1–22. Stockholm: IDEA.

Isser, D. H., S. C. Lubkemann, and S. N'Tow. 2009. *Looking for Justice: Liberian Experiences with and Perceptions of Local Justice Options*. Peaceworks No. 63. Washington, DC: US Institute of Peace.

Kappler, S. 2014. *Local Agency and Peacebuilding*. Basingstoke, UK: Palgrave Macmillan.

Kappler, S. 2017. "The Securitization of International Peacebuilding". In *Securitization in Statebuilding and Intervention*, edited by T. Bonacker, W. Distler, and M. Ketzmerick, 31–52. Baden-Baden: Nomos.

Karim, S., and R. Gorman. 2016. "Building a More Competent Security Sector: The Case of UNMIL and the Liberian National Police." *International Peacekeeping* 23 (1):158–191.

Keesing, R. M. 1993. "Kastom Re-Examined." *Anthropological Forum* 6 (4):587–596.

Kyed, H. M. 2017. "Hybridity and Boundary-Making: Exploring the Politics of Hybridization." *Third World Thematics* 2 (4):464–480.

Kyed, H. M., and L. Buur. 2006. *Recognition and Democratisation: "New Roles" for Traditional Leaders in Sub-Saharan Africa*. DIIS Working Paper No. 2006/11. Copenhagen: Danish Institute for International Studies.

Lindstrom, L. 1997. "Chiefs in Vanuatu Today." In *Chiefs Today: Traditional Pacific Leadership and the Postcolonial State*, edited by G. White and L. Lindstrom, 211–228. Stanford, CA: Stanford University Press.

Mac Ginty, R. 2011. *International Peacebuilding and Local Resistance: Hybrid Forms of Peace*. New York: Palgrave Macmillan.

Mac Ginty, R. 2015. "Where Is the Local? Critical Localism and Peacebuilding." *Third World Quarterly* 36 (5):840–856.

Mac Ginty, R. 2016. "What Do We Mean When We Use the Term 'Local'? Imagining and Framing the Local and the International in Relation to Peace and Order." In *Peacebuilding in Crisis: Rethinking Paradigms and Practices of Transnational Cooperation*, edited by T. Debiel, T. Held, and U. Schneckener, 193–209. London: Routledge.

Mac Ginty, R., and O. P. Richmond. 2013. "The Local Turn in Peace Building: A Critical Agenda for Peace." *Third World Quarterly* 34 (5):763–783.

Mac Ginty, R., and O. Richmond. 2016. "The Fallacy of Constructing Hybrid Political Orders: A Reappraisal of the Hybrid Turn in Peacebuilding." *International Peacekeeping* 23 (2):219–239.

Millar, G. 2018. "Introduction: The Key Strengths of Ethnographic Peace Research." In *Ethnographic Peace Research*, edited by G. Millar, 1–19. Cham, Switzerland: Palgrave Macmillan.

Muriaas, R. L. 2011. "Traditional Institutions: A Typology of Co-Existence in Sub-Saharan Africa." *Forum for Development Studies* 38 (1):87–107.

Mustasilta, K. 2018. "Including Chiefs, Maintaining Peace? Examining the Effects of State-Traditional Governance Interaction on Civil Peace in Sub-Saharan Africa." *Journal of Peace Research*, 6 September. doi:10.1177/0022343318790780.

Richmond, O. P. 2011. *A Post-Liberal Peace.* London: Routledge.

Richmond, O. P. 2012. "A Pedagogy of Peacebuilding: Infrapolitics, Resistance, and Liberation". *International Political Sociology* 6:1–17.

Richmond, O. P., and A. Mitchell, eds. 2012. *Hybrid Forms of Peace: From Everyday Agency to Post-Liberalism.* Houndmills, UK: Palgrave Macmillan.

Thelen, T., L. Vetters, and K. von Benda-Beckmann. 2014. "Introduction to Stategraphy: Toward a Relational Anthropology of the State." *Social Analysis* 58 (3):1–19.

Ubink, J. 2008. *Traditional Authorities in Africa: Resurgence in an Era of Democratisation.* Leiden: Leiden University Press.

Zanker, F. 2015. *A Decade of Police Reform in Liberia: Perceptions, Challenges and Ways Ahead.* SSR 2.0 Brief Issue No 4. Kitchener, Ontario: Centre for Security Governance.

Zartman, I. W. 2000. "Introduction: African Traditional Conflict 'Medicine.'" In *Traditional Cures for Modern Conflicts: African Conflict "Medicine,"* edited by I. W. Zartman, 1–11. Boulder, CO: Lynne Rienner.

CHAPTER 43

LOCAL INFRASTRUCTURES FOR PEACE

ANDRIES ODENDAAL

THE approach to local peacebuilding that became known as "local infrastructures for peace" is, at its core, an attempt to ensure that the relevant actors working toward peace within a specific context do so based on a shared vision and in a manner that is synergistic and empowering of local resources. Its intention is to create linkages and enhance productive relationships between various actors for peace, much as roads, bridges, and the internet link people and enable productive collaboration. At first glance, therefore, it is a practical attempt to deal with the "perennial coordination challenge" (Newman 2013) that working for peace at all levels, whether international or local, face. But it is more than a mere coordination strategy. It is also, by the assumptions that inform the approach, a theoretical construct that is open to critique and contestation.

The concept was born from practice. It speaks to practical realities on the ground, inter alia, the competitive practices of many international organizations at the local level (Newman 2013); the scattered, short-term, and fragmented nature of most NGO activities (Paffenholz 2013a); the absence of a coherent and shared vision among peace actors (Lederach 1997); the disconnect between top-down and bottom-up processes (Hellmüller 2018); and the persistence of violence in postagreement contexts due to unresolved local conflicts (Autessere 2010; Van Baalen and Höglund 2017). Given this complex reality, it should not be a surprise that "infrastructures for peace" came to mean different things to different people. The dilemma is partly theoretical, with reference to conflicting understandings of peace (Richmond 2016). It is also very practical. What are legitimate and justifiable efforts to encourage greater synergy and collaboration between peace actors? How do we avoid the Scylla of ceding control of peace processes to hegemonic, faceless bureaucracies, and the Charybdis of futile turf battles between peacebuilders for the superiority of *their* approach and *their* vision of peace? How do we provide flexible yet reliable platforms to facilitate meaningful interaction among those whose conflict it is and relevant external actors? And who are "we" to do this?

What follows is a reappraisal of local infrastructures for peace as a practical response to the reality and complexity of local peace. There is, in the peacebuilding field, a "common tension" (Richmond, Pogodda, and Ramović 2016, 11) or even an "endemic gap" (Hughes, Öjendal, and Schierenbeck 2015, 823) between theory and practice. Furthermore, there is a troubling scarcity of robust research (Autessere 2017). This is the case even though local peacebuilding has over the past decade become a mainstream objective of various international institutions and funders (Paffenholz 2013b, 2015). While this deficit in research is slowly changing, most of the writing on local infrastructures for peace has been based on project reports and reflections of practitioners implementing an "action-learning" approach (Lederach 1997; Odendaal 2010, 2013; Kumar and De la Haye 2012; Chuma and Ojielo 2012; Van Tongeren 2013; Unger and Lundström 2013; Giessmann 2016).

This chapter does not pretend to have broken free from this trend. It instead presents a reinterpretation of the relevance of infrastructures for peace from a practitioner's perspective, considering past experiences and current theoretical debates. It argues for an appreciation of the complex, interlinked nature of global, national, and local conflicts and the necessity of flexible yet sustained and productive dialogue spaces at the points of frictional interactions.

Initial Understandings of Peace Infrastructures

Interpretations of "the local" has decisive implications for local infrastructures for peace. The local is indeed "a social construction" (Mac Ginty 2015, 848) with, inevitably, numerous interpretation possibilities. Two versions, though, have dominated: the local as subsidiary to "the national," thus the recipient of top-down interventions; and the local as sites of resistance against national or international impositions by virtue of local interests, local aspirations for peace, and local agency. In this view the local constitutes essential bottom-up building blocks for any meaningful peace. The consequences of these two dominant understandings were two contrasting ways of conceptualizing and implementing peace infrastructures, the one "top-down," the other "bottom-up."

Top-Down Peace Infrastructures

The earliest peace infrastructures (Nicaragua in 1988; South Africa in 1991; Northern Ireland in 1998; Former Yugoslav Republic of Macedonia in 2001) were established following the achievement of a national peace accord. The main aim with these infrastructures was to assist in the implementation of the national peace accord at the local level. The novelty at the time was the tacit acknowledgment by the collective national leadership that peace would not automatically "trickle down" to the local levels. Peace at the local level required the engagement of local actors. It required that the role played by local histories, conditions, actors, and relationships be acknowledged, and that, for peace to take hold, local communities should forge their own peace, albeit in the context of a larger national framework.

The specific objectives with such infrastructures differed according to the context. In Nicaragua the main objective was to monitor the ceasefire and facilitate local reconciliation; in South Africa to prevent local violence; in Northern Ireland to normalize community-police relationships; and in Macedonia to improve ethnic relationships. The infrastructures consisted, of course with considerable variations, of inclusive local committees or councils composed of either an inclusive group of trusted individuals or representatives of local government, political actors, and civil society. They were linked with national institutions responsible for implementing the peace accord and supported by state bureaucracies (Odendaal 2013). These designs were of a fairly static, top-down nature, guided by the assumption that local leaders, in cooperation with national agencies, had the legitimacy and capacity to resolve local conflicts, but with national support.

International bodies were quick to pick up on these developments. In particular the United Nations Development Program, in collaboration with other UN departments, set out on a program to support the establishment of infrastructures for peace. By 2012 the UN had provided support to infrastructures for peace in more than thirty countries (Ryan 2012; Kumar and De la Haye 2012). The reasons for and implementation histories of these infrastructures varied considerably (Giessmann 2016), as did the outcomes. A key success factor was the extent to which local communities appropriated the structure and made it their own. Whenever there were attempts to tamper with the inclusivity of local structures or to capture direct political control of the structure, results were disappointing. Also, when local structures relied on consensus-seeking as their main strategy they were more effective than when they resorted to coercive measures (Odendaal 2013).

The main advantage of these formal infrastructures—formal in the sense that they were instituted by virtue of an official proclamation or law—was that they enabled dialogue between state actors (local government and the local security services) and representatives of various local social formations. They had the potential to "transform local, state and international patterns of exclusion, bridging informal peace formation with formal politics and the state" (Richmond 2014,154). They also released some funding for peace efforts at the local level and facilitated access to specialist services and resources as required. In some contexts, especially where there had been a civil war, the existence of such a formal structure legitimized the collective search for peace—something that was unthinkable while the war was raging (Odendaal 2013).

However, this approach has been criticized for its statist and neoliberal assumptions (Verzat 2012; Mac Ginty and Richmond 2015); for its excessive bureaucratization (Lederach 2012); and for engaging easily identifiable and accessible local elites at the cost of more relevant but less accessible levels of societal relations (Chandler 2013). There have, indeed, been too many examples of infrastructures that lived in the minds of their designers but did not take root in the experiences of those affected by the conflict. In such cases their contribution to local peace was minimal.

Bottom-Up Peace Infrastructures

In peacebuilding literature there were two "local turns" (Paffenholz 2015), both advocating a bottom-up approach. In the first the name of John Paul Lederach looms large. In fact, he is widely credited with developing the leading theory regarding infrastructures for peace. His "integrated framework for peacebuilding" (Lederach 1997) has had a decisive impact over the past two decades (Paffenholz 2013b). By his own admission, this theory had its roots more in his practical experiences than in theoretical considerations. His main thesis was that the resources for the transformation of a conflict resided within the society whose conflict it was. These resources should be galvanized and empowered through an infrastructure. The infrastructure should be understood as a *process-structure*. "A process-structure is made up of systems that maintain form over time yet have no hard rigidity of structure. . . . The goal is not stasis, but rather the generation of continuous, dynamic, self-regenerating processes that maintain form over time and are able to adapt to environmental changes." Furthermore, "such an infrastructure is made up of a web of people, their relationships and activities, and the social mechanisms necessary to sustain the change sought. This takes place at all levels of the society" (Lederach 1997, 84).

It is no surprise that Lederach subsequently distanced himself from the more institutionalized understandings of infrastructures for peace. In 2012 he expressed two reservations regarding the way some infrastructures for peace had been established. First, there was an excessive focus on institutionalization, which shifted the focus away from "creative and generative processes of peacebuilding and social change" to "institutionalised capacity, functions and roles." Institution-building is a self-obsessed and self-perpetuating process, by implication, while the key challenge posed by peacebuilding is "how to provide for robust creativity and adaptive capacity in the wider system of change" (12). His second concern regarded the "low view of context" that informed external designs of peace infrastructures (10). This "low view" did not mean the absence of context analysis; rather, it meant that realist understandings of political power dominated such analyses. It was essentially neocolonialist because the "deeper histories of lived conflict, narrative and 'thick' understanding of culture remain at the periphery" (10). Consequently, in Lederach's vocabulary, "infrastructure" had been replaced by "platform" (11) or by "web" (Lederach 2005; Paffenholz 2013b; Odendaal 2017).

"Web" actually came to dominate his writing and especially his practice. The metaphor refers to spider webs. A spider covers an open space by anchoring its web on all sides of the space, thus illustrating the need to be inclusive of all "owners" of a conflict. It is the flexibility and adaptability of this structure that distinguishes it from hard or static infrastructure metaphors. Lederach has therefore not turned his back on his initial theory; he simply switched metaphors to correct a misunderstanding of his theory. His recent practice in Nepal offers important clues to what he had in mind and will be discussed in more detail later.

The second "local turn" debate took place primarily in international relations circles. It has been stimulated not only by contrasting intellectual paradigms (critical theory versus "a hybrid of liberalism and realism"; Richmond 2016, 61), but by the practical reality of the failure of much international peacebuilding under the reigning liberal paradigm. The debate provided a substantive critique of the international peacebuilding "industry." Jabri (2016, 164) stated it strongly: "Peacebuilding is now one of the prevailing modes through which 'armies in all but name' come to govern the lands and populations of those still deemed incapable of self-government." It has become, in her view, a neocolonial enterprise determined by two structural forces: the colonial legacy and its continuing impact on the present, and the unequal structure of the global political economy (154).

From a methodological perspective the critique boiled down to the prominence of statebuilding in peacebuilding strategies, especially the manner in which the Western democratic model of the state has been universalized and imposed as the ultimate peacebuilding mechanism, resulting in the lack of local legitimacy (Richmond 2016); the sheer hubris of international peacebuilders (Jabri 2016); and the imposition of prefabricated standardized projects on widely divergent contexts, thus ridiculing their professed respect for "local ownership" (Miklian, Lidén, and Kolås 2011). "The local" emerged in this debate as the alternative to the perceived failures of liberalism and statism. It has been described as the domain where the international liberal and its national instruments do not hold sway; where culture, history, and identity determined behavior; and where local critical agency expressed itself (Mac Ginty and Richmond 2015). "The local" is thus a site of resistance, the space where authentic peace is formed, and where the imposition of a fake peace is rejected. "The local" is also not a geographical area but must be understood as "de-territorialised, networked and constituted by people and activity rather than place" (Mac Ginty 2015, 841).

The understanding of the local as the site of resistance has been criticized for its stereotyping of the international (Da Costa and Karlsrud 2013), for its excessive focus on Western actors within the international (Paffenholz 2015, 858), for portraying the international and the local as binary opposites (Schierenbeck 2015), and for romanticizing the local (Paffenholz 2015). In its starker formulations this understanding has limited value for appreciating the complex layout of the local. Defining the local as a site of resistance to external interventions obscures the nuances and contradictions of local conflict patterns. The local is not a uniform site. The interests and agendas of local actors cannot be squeezed into pro-state and anti-state positions (Autessere 2010; Kalyvas 2006; Van Baalen and Höglund 2017). Portraying the international and the local as binary opposites stereotypes both concepts at the cost of recognizing the complexity, diversity, and nuance of agendas to be found at each of these levels and between them.

There are numerous examples of bottom-up processes. These initiatives have been called by different names, such as innovative social coalitions (Paladini Adell 2012), zones of peace (Hancock 2016), (informal) local peace committees (Van Tongeren 2013 Odendaal 2013), and local peace networks (Connolly and Powers 2018). Prominent among these have also been women's movements (Richmond 2014). Probably the most successful has been Somaliland, a new (yet internationally unrecognized) state formed peacefully by way of an entirely bottom-up approach (Lederach 1997 Paffenholz 2003).

These bottom-up movements are either homegrown or initiated by international or national NGOs. What they share is that they are locally owned peace initiatives forging wider horizontal but limited vertical relationships or networks. They face two risks: their need for resources obliges them to seek increasingly short-term and prescriptive international funding, and political capture or oppression correlates with how effective the network has become. The greater their success, the greater the risk (Connolly and Powers 2018; Mitchell and Hancock 2012). It is therefore not only official peace infrastructures that face the risk of political capture or manipulation.

Beyond the Top-Down/ Bottom-Up Dichotomy

Judged on the starker formulations of both the top-down and bottom-up positions, they seem incompatible. Either you work with national and international state institutions or you resist their role and go it alone. However, the past decade saw developments that point to a clearer understanding of both the relevance of the local for peace as well as its complex interrelatedness with the national and the international. These developments point to a position beyond this dichotomy.

A first important development was the increasing appreciation of the relevance of the local. As Richmond (2014, 23) stated, "The realisation has now dawned that any peace should be local." It is demonstrated, inter alia, by the mainstreaming of "local ownership" in the language and project plans of the main international peacebuilding organizations. In theory, at least, it acknowledges that peace should have "grounded and embedded legitimacy, which connects to the everyday lives of the citizens" (ix). The importance of the local lies in its being the touchstone of legitimacy.

The mainstreaming of "local ownership," however, is not without its dilemmas and complications. By virtue of the bureaucratic need for precision planning and budgeting, local ownership is often interpreted through the predetermined objectives of international peacebuilding institutions and squeezed through their prefabricated project-planning designs. As Miklian, Lidén, and Kolås (2011) claim, organizational structures and agendas of international agencies regularly determine *which* "local" will be acceptable and under what conditions. Foreign aid organizations and funders "have *already* overromanticised the local in their projects, reports and mission statements *while being hamstrung by their very organisational structures against providing the benefits to local community actors that they promise*" (289, italics in original).

This dilemma goes deeper than arrogance or the laziness of merely paying lip service to local ownership. It lies in the inherent contradiction between respect for the local and the larger-scale planning and budgeting regimes of large organizations. Consequently, during the planning phase, *assumptions* about local ownership are projected as reality in order to meet bureaucratic requirements. There is clearly much work still to be done to transform the theoretical commitment to local ownership to the hard work of developing a "thick analysis" in each case.

Closely linked with the concept of ownership is the growing realization that peace is political. In 2016 both the General Assembly and the Security Council of the United Nations (2015) adopted the report of the UN High-Level Independent Panel on UN Peace Operations. The first of four "essential shifts" in approach that the report advised was to acknowledge the fundamental political nature of peace: "Lasting peace is achieved not through military and technical engagements, but through political solutions" (10). It implied that peace could be achieved only by engaging in the complex turmoil of interests, needs, and fears of the various sections of a society at various levels. Peace did not require military superiority or technocratic perfection but the appropriation of processes and outcomes by those whose conflict it was. It thus required a commitment "to open and impartial dialogue with all parties, States and non-state actors" (26). "The legacy of the 'white-SUV culture' must give way to a more human face that prioritizes closer interaction with local people to better understand their concerns, needs and aspirations" (30).

Another important development is the mainstreaming of "complexity." The liberal peacebuilding practice of the past had been dominated by "a linear cause-effect problem-solving model" (de Coning 2018, 302). It entailed an expert analysis of the causes of the conflict, followed by designing programmatic interventions. Now "it is increasingly acknowledged that peacebuilding is about influencing the behaviour of *social systems* that have been affected by conflict" (305, emphasis added).

According to Lederach (2012), these social systems are "complex-adaptive systems," which are different from complex systems. Lederach used the examples of a marching band and a jazz quartet to illustrate the difference. A complex system is like a marching band; their music is complex yet coordinated and predetermined. With a jazz quartet, though, the musical key and starting point is agreed, but as the music unfolds "a dynamic, generative and responsively interactive process emerges *unpredictably*" (11, italics in original).

A systemic understanding of the local has important implications for peace infrastructures or webs. First, the behavior of the system is determined by the nature of interactions within the system. Since the interactions are "rich, dynamic, fed back, and, above all, nonlinear, the behavior of the system as a whole cannot be predicted from an inspection of its components" (Cilliers 2000, 24). It requires, in Lederach's words, a "thick" analysis.

Second, relationships are fundamental. Change takes place when patterns of interaction change (Cilliers 2000), that is, when the nature of relationships change.

Third, every local system is unique (Cilliers 2000). It may share conditions and dynamics with other social systems, yet outsiders to this system will always interact with the particular nature of a system at any given moment. Local systems specifically share "the national" and "the international" with other local systems, but they integrate these into their respective systems in dissimilar ways. Hence the phenomenon that national peace agreements may restore peace in some local areas but not in others (Autessere 2010; Van Baalen and Höglund 2017; Kalyvas 2006).

Fourth, a complex-adaptive system has a robust structure, yet its boundaries cannot be determined accurately. It is inherently an open system, with the capacity to integrate external interventions into itself. Hence the observation that the national and the international, by virtue of being actors at the local level, are integrally part of the local system. There is no local where the global has had no influence, whether in the form of ideas, aspirations, fears, resources, or interventions. With reference to Homi Bhabha, Leonardsson and Rudd (2015, 833) mention that local agencies in processes of peace are "constantly accepting, adopting, resisting and/or mocking ambivalent and unstable foreign interventions, [and] outcomes are redefined." There is also no local where the national, in the form of state structures or political influences, is not an active player. The state is a reality in all local conflicts, whether through its overbearing domination, its dysfunctional absence, or its policies and interventions. The policies, projects, and activities of both the national and the international are subjected to the same pattern of interactions that characterize that local system.

Recent literature has produced two concepts that recognize this interactive relationship between the local, national, and international, namely "friction" and "hybridity." Friction is "the awkward, unequal, unstable and creative qualities of interconnection across differences" (Björkdahl and Höglund 2013, 294, quoting Tsing). The site of peacebuilding is a space where local, national, and global actors, ideas, and practices "rub against each other," producing "new power relations, agencies, ideas and practices that may or may not resemble their originals" (Björkdahl and Höglund 2013, 292).

Hybridity recognizes the unexpected, unintended, and mixed *outcomes* of the interaction between top-down and bottom-up forces and takes into account the levels and dynamics between top and bottom (Mac Ginty and Sanghera 2013, 4). At its most advanced, Mac Ginty and Richmond (2015, 6) explain, "the concept of hybridity amounts to a rejection of conflict scientism, or the notion that conflicts can be 'understood' if only we have enough data and the correct formula. Instead, it is part of a critical epistemological project that takes seriously multiple sources of agency and is prepared to look beyond institutions for explanations of politics and power (and hints of future emancipation)." Richmond (2014), furthermore, distinguishes between positive and negative hybridity. Positive hybridity, in his view, is the outcome of consensual interactions between the three sites or levels of peace: the local and everyday, the state, and the international system. What enables positive hybridity, however, is reliance on what is authentically local—and not what is offered by elite power holders or seekers implicated in conflict, or external actors embedded in a disconnected international system. Peace is enabled when the local is respected for its critical contribution.

In summary, the local is an indispensable part of the peace equation. It is important for its resistance to external imposition and domination, but also for its contribution to the transformation of society. Each local conflict system is unique. It must be engaged in terms of its own complex dynamics. Yet each local system is also part of larger horizontal and vertical systems. Consequently the transformation of the local has implications for these larger systems, and vice versa. From the perspective of local peace infrastructure methodology, it is this interaction within and between the various levels of conflict that matters. The relevance of peace infrastructures or webs is that they encourage and support dialogue at the points of friction. Dialogue, particularly types of dialogue that embrace radical disagreement (Ramsbotham 2010), is what makes friction intelligible. It gives meaning to the encounter and its hybrid outcomes. The core assumption regarding local peace infrastructures is therefore that they enable difficult dialogues. They are platforms or webs with the ability to initiate, encourage, and sustain dialogue within conflict systems.

Lederach in Practice: A Case Study

Given Lederach's importance for the topic discussed here, the approach that he stimulated in Nepal deserves attention. Since 2003 he was deeply involved in processes to consider alternative approaches to peace at a time when the country was in the grip of the Maoist insurgency (Lederach 2015). He worked closely with a team of respected social actors that, importantly, included The McConnell Foundation funding organization. For the best part of a decade they worked at a deeper understanding of the context. One of the conclusions they reached was that, whatever the outcome would be of national peace negotiations ongoing at the time, conflicts over natural resources (water, forest, and land) would continue at the local level. These conflicts were widespread. They were central to the well-being of local communities as they concerned survival in very difficult economic conditions. They were also deeply rooted in Nepal's history of social and political inequality and exclusion. The transformation of these conflicts would therefore contribute substantively to justice and social change in Nepal. In 2013 the Natural Resource Conflict Transformation Centre (NRCTC) in Nepal was established as a vehicle for their vision and to facilitate the practical implementation of transforming local resource conflicts.

In 2017 this author led a process to review the initial years of NRCTC (Odendaal 2017). Two observations are pertinent here. First, NRCTC had registered remarkable success. The process produced significant reconciliation in 94 of 102 local conflict cases taken on during 2013–2017, with publicly signed and celebrated agreements; 45% of these conflicts had been in deadlock for more than a decade, two of them in deadlock for more than forty years, and five between thirty and forty years. The process structure consisted of a funder that had actively participated in the process to develop a thick analysis; a Nepali organization of facilitators who learned from Lederach, but also from local practice (NRCTC); and the creation of dialogue structures wherein the "owners" of the conflict agreed to have them represented at the dialogue interface (the "spider web") by persons resembling Lederach's description of "insider-partials" (Wehr and Lederach 1991). The dialogue was largely self-mediated by members of the group, but with support by NRCTC. It was a slow process, on average taking thirteen months but in some cases much longer. It focused primarily on reaching a shared understanding of the deeper causes of the conflict. During this process a fundamental shift in historically determined relationships invariably took place. A practical solution became possible following these developments.

Second, these processes primarily engaged local actors, often at the infrapolitical level. Yet the nature of their conflicts reflected the social, political, and economic realities of the larger Nepali society, such as unequal access to natural resources as historically determined by caste and ethnicity, the lingering impact of the violence of the civil war, the dysfunctionality of local government, the historic legacy of bonded labor practices, and internal migration patterns. While arriving at their own local agreements, however, they were still faced with the reality of the national condition and the state. Often the state was a silent participant in the discussions because of its neglect, weakness, or reluctance, but at times it played a constructive role. In one specific case the implementation of the local agreement required a change in national legislation that led to a frustrating impasse because, inter alia, of the frequent rotation of national governments. In another case six villages agreed to a new style of shared management of their use of a water canal, which ended two decades of acrimony. However, to address the real structural cause of the conflict—the fact that there was an insufficient water supply—massive infrastructural investments would be necessary. While the local transformation that has taken place has high symbolic relevance for Nepali society, it also demonstrates the limitations of an approach that focuses solely at the local level. In theory Lederach recognized the interlinked nature between levels of society. Some of the local dialogue teams indeed engaged with local and national government, but the process structure has thus far not provided convincing clues regarding the problem of scale. In Lederach's view, the growth of the spider web should take place organically from within. It is therefore still too early to reach a definitive conclusion regarding its possibility in this specific context. The establishment of "vertical webs," however, is clearly a yet unfulfilled objective. It is a desired outcome, yet fraught with practical dilemmas.

Vertical Webs

While the necessity for vertical webs seems, from a methodological perspective, clear and urgent (Hellmüller 2018), the practical reality is the inherent power imbalance between large and small social systems. The power imbalance between local and global, in other words, extends beyond political and economic issues. It also exists in the nature of social systems. It operates independently of grand theoretical or ideological persuasions. The arrogance of liberal power is not greater or more destructive than the arrogance of hybrid or postliberal power. The local is almost doomed to irrelevance in the context of overpowering social and political processes (Mac Ginty 2015).

The larger the social system, the greater the need for formalization and institutionalization. Large bureaucracies require top-down planning, meticulous budgeting, strict timelines and accounting procedures, and formalized meeting structures. The threats of bureaucratization and political control are therefore more pertinent at these levels. This poses a difficult dilemma. The rationale for official or formal infrastructures assumes that the formal engagement of state structures and representatives of alternative social or political movements enables the possibility of sustained dialogue that would otherwise be very difficult to organize. By formalizing such structures, by implication, the state commits itself to sustained dialogue. However, the danger of political capture or "instrumentalized hybridity" (Mac Ginty and Richmond 2015) looms large. There are formal dialogue settings, though, that have achieved enough to encourage continued attention to this possibility (Odendaal 2013; Siebert 2013; Giessmann 2016; Moubarak 2018). The main "success factor" is the shared ownership of these platforms by all the participants. The enduring challenge is how to ensure that the dialogue taking place at these platforms goes deep, is sustained, and succeeds in transforming relationships and perceptions.

Conclusions

The past two decades have witnessed an increasing and welcome attention to the importance of the local in efforts to promote peace. Attention to the local has been caused by, on the one hand, the deepening appreciation of the complexity and tenacity of local conflicts and, on the other hand, by the seeming failure of large-scale international peacebuilding ventures that were largely insensitive to the local. Both in practice and in theory the focus shifted to the relative importance of the local in relation to national and international strategies, veering from understandings of the local as a key ingredient of a national strategy to the local as the site of resistance against top-down peacebuilding.

Several developments during the past decade contributed to an understanding that has the potential to move beyond the top-down versus bottom-up contradiction. At the core of these is the view of local conflict as a complex-adaptive system. This means that each local conflict should be seen and approached in terms of its own unique dynamics. In addition, the local is always nested in the larger national and international systems, and the national and international are integrated within the local system. The hybrid outcome of peacebuilding will always be determined by the nature not only of horizontal interactions but also of vertical interactions between the local, national, and international.

The contribution of infrastructures for peace (called by whatever name) is their focus on creating a network of inclusive platforms for deep dialogue at the points of friction, whether horizontally or vertically. The inclusivity of these platforms implies primarily that those whose conflict it is should be adequately represented. It may also include those who provide meaningful support to peace, including national and international actors, as long as local agency is honored.

The nature of complex-adaptive systems prescribes flexible structures, fine-tuned to the very specific needs at each friction point. Approaches that prefer prefabricated or bureaucratized models should take heed of this requirement. On the other hand, the local cannot be romanticized to the point where the national and international have become, by definition, the enemy of peace. There is no plausible peace where the national and international have been expunged. The challenge is to enhance engagement, to subject seemingly incompatible positions to deep dialogue and to the possibility of transformation. Such transformation, however, is possible only when local voices are truly heard.

While there is evidence regarding the high risk of political capture of formalized dialogue structures, there is also evidence of the benefit that such structures offer. If the formalization of a dialogue structure encourages the possibility of constructive dialogue between state structures and citizens at the points of friction, it cannot be denounced simply because of its formality. It should, rather, be assessed on its relevance to the context, its inclusive nature, and the quality of dialogue it enables.

References

Autessere, S. 2010. *The Trouble with the Congo: Local Violence and the Failure of International Peacebuilding.* Cambridge: Cambridge University Press.

Autessere, S. 2017. "International Peacebuilding and Local Success: Assumptions and Effectiveness." *International Studies Review* 19:114–132.

Björkdahl, A., and K. Höglund. 2013. "Precarious Peacebuilding: Friction in Global-Local Encounters." *Peacebuilding* 1 (3):289–299.

Chandler, D. 2013. "Peacebuilding and the Politics of Nonlinearity: Rethinking 'Hidden' Agency and 'Resistance.'" *Peacebuilding* 1 (1):17–32.

Chuma, A., and O. Ojielo. 2012. "Building Standing National Capacity for Conflict Prevention and Resolution in Kenya." *Journal of Peacebuilding and Development* 7 (3):25–39.

Cilliers, P. 2000. "What Can We Learn from a Theory of Complexity?" *Emergence* 2 (1):23–33.

Connolly, L., and L. Powers. 2018. "Introduction." In *Local Networks for Peace: Lessons from Community-Led Peacebuilding,* edited by L. Connolly and L. Powers, 2–10, New York: International Peace Institute.

Da Costa, D., and J. Karlsrud. 2013. "'Bending the Rules': The Space between HQ Policy and Local Action in UN Civilian Peacekeeping." *Journal of International Peacekeeping* 17:293–312.

de Coning, C. 2018. "Adaptive Peacebuilding." *International Affairs* 94 (2):301–317.

Giessmann, H. 2016. *Embedded Peace. Infrastructures for Peace: Approaches and Lessons Learned.* New York: UNDP.

Hancock, L. 2016. "Agency and Peacebuilding: The Promise of Local Zones of Peace." *Peacebuilding,* 3 November. doi:10.1080/21647259.2016.1253604.

Hellmüller, S. 2018. *The Interaction between Local and International Peacebuilding Actors.* Cham, Switzerland: Palgrave Macmillan.

Hughes, C., J. Öjendal, and I. Schierenbeck. 2015. "The Struggle versus the Song—The Local Turn in Peacebuilding: An Introduction." *Third World Quarterly* 36 (5):817–824.

Jabri, V. 2016. "Post-Colonialism: A Post-Colonial Perspective on Peacebuilding." In *The Palgrave Handbook of Disciplinary and Regional Approaches to Peace,* edited by O. Richmond, S. Pogodda, and J. Ramović, 154–167. Hampshire, UK: Palgrave Macmillan.

Kalyvas, S. 2006. *The Logic of Violence in Civil War.* Cambridge: Cambridge University Press.
Kumar, C., and J. De la Haye. 2012. "Hybrid Peacemaking: Building National Infrastructures for Peace." *Global Governance* 18:13–20.
Lederach, J. P. 1997. *Building Peace: Sustainable Reconciliation in Divided Societies.* Washington, DC: USIP.
Lederach, J. P. 2005. *The Moral Imagination: The Art and Soul of Building Peace.* Oxford: Oxford University Press.
Lederach, J. P. 2012. "The Origins and Evolution of Infrastructures for Peace: A Personal Reflection." *Journal of Peacebuilding and Development* 7 (3):8–13.
Lederach, J. P. 2015. "Memoirs of Nepal: Reflections across a Decade." Redding, CA: The McConnell Foundation.
Leonardsson, H., and G. Rudd. 2015. "The 'Local Turn' in Peacebuilding: A Literature Review of Effective and Emancipatory Local Peacebuilding." *Third World Quarterly* 36 (5):825–839.
Mac Ginty, R. 2015. "Where Is the Local? Critical Localism and Peacebuilding." *Third World Quarterly* 36 (5):840–856.
Mac Ginty, R., and O. P. Richmond. 2015. "The Fallacy of Constructing Hybrid Political Orders: A Reappraisal of the Hybrid Turn in Peacebuilding." *International Peacekeeping*, 5 November. doi:10.1080/13533312.2015.1099440.
Mac Ginty, R., and G. Sanghera. 2013. "Hybridity in Peacebuilding and Development: An Introduction." *Journal of Peacebuilding and Development* 7 (2):3–8.
Miklian, J., K. Lidén, and Å. Kolås. 2011. "The Perils of 'Going Local': Liberal Peace-Building Agendas in Nepal." *Conflict, Security & Development* 11 (3):285–308.
Mitchell, C., and L. Hancock, eds. 2012. *Local Peacebuilding and National Peace: Interaction between Grassroots and Elite Processes.* London: Continuum.
Moubarak, W. 2018. "The National Dialogue of Lebanon: An Experiment beyond External Mediation." Occasional Paper 90. Abu Dhabi: Emirates Center for Strategic Studies and Research.
Newman, E. 2013. "The International Architecture of Peacebuilding." In *Routledge Handbook of Peacebuilding*, edited by R. Mac Ginty, 311–324. London: Routledge.
Odendaal, A. 2010. *An Architecture for Building Peace at the Local Level: A Comparative Study of Local Peace Committees.* New York: UNDP.
Odendaal, A. 2013. *A Crucial Link: Local Peace Committees and National Peacebuilding.* Washington, DC: USIP.
Odendaal, A. 2017. *"Stitching Different Fabrics into One Cloth": A Report on the Effectiveness and Relevance of the "Spider-Approach" of the Natural Resource Conflict Transformation Center in Nepal.* Kathmandu: NRCTC-N.
Paffenholz, T. 2003. *Community-Based Bottom-Up Peacebuilding: The Development of the Life and Peace Institute's Approach to Peacebuilding and Lessons Learned from the Somali Experience (1990–2000).* Kijabe: Life and Peace Institute.
Paffenholz, T. 2013a. "Civil Society." In *Routledge Handbook of Peacebuilding*, edited by R. Mac Ginty, 347–359. Oxford: Routledge.
Paffenholz, T. 2013b. "International Peacebuilding Goes Local: Analysing Lederach's Conflict Transformation Theory and Its Ambivalent Encounter with 20 Years of Practice." *Peacebuilding* 2 (1):11–27.
Paffenholz, T. 2015. "Unpacking the Local Turn in Peacebuilding: A Critical Assessment towards an Agenda for Future Research." *Third World Quarterly* 36 (5):857–874.
Paladini Adell, B. 2012. *From Peacebuilding and Human Development Coalitions to Peace Infrastructure in Colombia.* Berghof Handbook Dialogue Series 10. Berlin: Berghof Foundation.

Ramsbotham, O. P. 2010. *Transforming Violent Conflict: Radical Disagreement, Dialogue and Survival*. Oxford: Routledge.

Richmond, O. P. 2014. *Failed Statebuilding. Intervention and the Dynamics of Peace Formation.* New Haven: Yale.

Richmond, O. 2016. "Peace in International Relations Theory." In *The Palgrave Handbook of Disciplinary and Regional Approaches to Peace*, edited by O. Richmond, S. Pogodda, and J. Ramović, 57–68. Hampshire, UK: Palgrave Macmillan.

Richmond, O. P., S. Pogodda, and J. Ramović. 2016. "Introduction." In *The Palgrave Handbook of Disciplinary and Regional Approaches to Peace*, edited by O. Richmond, S. Pogodda, and J. Ramović, 1–17. Hampshire, UK: Palgrave Macmillan.

Ryan, J. 2012. "Infrastructures for Peace as a Path to Resilient Societies: An Institutional Perspective." *Journal for Peacebuilding and Development* 7 (3):14–24.

Schierenbeck, I. 2015. "Beyond the Local Turn Divide: Lessons Learnt, Relearnt and Unlearnt." *Third World Quarterly* 36 (5):1023–1032.

Siebert, H. 2013. *National Peace and Dialogue Structures: Strengthening the Immune System from Within Instead of Prescribing Antibiotics*. Berghof Handbook Dialogue Series 10. Berlin: Berghof Foundation.

Unger, B., and S. Lundström. 2013. *On Framing, Setting Up and Supporting Peace Infrastructures: Introduction*. Berghof Handbook Dialogue Series 10. Berlin: Berghof Foundation.

United Nations. 2015. *The Challenge of Sustaining Peace: Report of the Advisory Group of Experts for the 2015 Review of the United Nations Peacebuilding Architecture*, A/69/968-S/2015/490.

Van Baalen, S., and K. Höglund. 2017. "'So, the Killings Continued': Wartime Mobilization and Post-War Violence in KwaZulu-Natal, South Africa." *Terrorism and Political Violence*, 19 May, 1–18. doi:10.1080/09546553.2017.1318126.

Van Tongeren, P. 2013. "Potential Cornerstone of Infrastructures for Peace? How Local Peace Committees Can Make a Difference." *Peacebuilding* 1 (1):39–60.

Verzat, V. 2012. *Infrastructures for Peace: A Grass-roots Way to Do State-building?* Berghof Handbook Dialogue Series 10. https://www.berghof-foundation.org/en/publications/handbook/handbook-dialogues/10-peace-infrastructures/.

Wehr, P., and J. P. Lederach. 1991. "Mediating Conflict in Central America." *Journal of Peace Research* 28 (1):85–98.

CHAPTER 44

EMANCIPATORY PEACE

GËZIM VISOKA

The quest for emancipation has emerged as one of the aspirational goals of critical approaches to peacebuilding. As a concept, emancipation is central to most critical political and social theories. It is associated with social change and freedom from domination. It is also associated with the empowerment of individuals and communities to determine freely their political, economic, and social affairs, and with social justice and equality (Welzel 2013). Yet the concept of emancipation in critical theory remains problematic, largely because scholars are divided on whether the purpose of critical knowledge is to offer conceptual and practical alternatives for addressing current problems or for achieving emancipation. This chapter examines the contribution of critical peace and conflict studies to the advancement of different, more practical visions of emancipation. It argues that visions of emancipatory peace represent an important attempt to move critical theory away from an abstract, intellectual field to a more ambitious, consolidated body of knowledge that is more likely to shape practices and improve the conditions for a just and lasting peace.

The project of emancipation emerged first and foremost as a critique of the current liberal interventionism, which has guided international interventions for building peace, reconstructing state institutions, and installing a neoliberal economy in postconflict societies. Because of its association with democracy and human rights, the project of liberal interventionism is often associated with liberal norms and values. Although emancipation is widely associated with liberation and liberalism, it is not the core objective of liberal peacebuilding. Yet building stable, functioning state institutions that ensure order, justice, and security necessarily entails emancipation. This problematic nature of liberal peacebuilding, and its lack of success in practice, has been subject to widespread academic criticism. Early critique mostly engaged in identifying what was missing from the theory and practice of liberal peacebuilding via a negational description of the existing order of things, while avoiding spelling out emancipatory alternatives. This type of critique deconstructed the ethics and politics of international interventions, sought faults rather than solutions, aspired to unmask power relations and the dynamics of dominance and to create space for a politics of resistance.

These critics feared that any prescriptive knowledge about how to achieve emancipation would be co-opted by policymakers who would use it to expand interventionary practices. So, by not offering an alternative, they aimed to see the liberal peace model fall under its own contractions.

As we have seen in many countries, the peace installed by the international community is a victor's peace, rooted in new power struggles, state governmentality, and the reproduction of structural violence in new mutations while ignoring reconciliation, rights, equality, and justice. This peace has ignored local agency, context, culture, and needs and—most important—was not based upon local legitimacy or social consensus. Responding to this, a new type of critique offering an alternative vision of emancipatory peace has emerged as a project for freeing subjects from constraints and removing the barriers and blockages to peace, development, and justice that were either embedded socially and politically prior to the violent episodes of the conflict or that have emerged as a result of international intervention and postconflict peacebuilding and statebuilding efforts. Visions of an emancipatory peace in conflict-affected societies focus on a number of displacements and transformations. They propose building peace from the bottom up, not from the top down, as was the case with liberal peacebuilding. This means that any attempt to devise a social contract and peace agreements should take into account local needs, culture, and context rather than the geopolitical interests and agendas of foreign governments. It means dislocating the ownership of peace processes to local actors and recognizing the agency and role of a much wider range of local stakeholders than merely ethnonationalist elites, warlords' elites, and metropolitan civil society. The site of peacebuilding should be not only institutions, as the epicenters of power, but also the everyday and peripheral sites where the critical mass for peace resides. This rearrangement is seen as a viable source for generating local legitimacy for the peace process and for positively influencing the dynamics of peace at the national, regional, and international levels. This vision of peace does not propose to exclude international actors from the peacebuilding process; rather, it suggests hybridizing the positive and emancipatory aspects found in the local and international structures and actors. However, to allow this vision of a positive hybrid peace to contribute to peace formation, the international community needs to change the logic of engagement: from imposition to enablement.

While this vision of emancipatory peace has opened up new possibilities for redesigning the peace configuration at the local and national levels, and potentially on a regional and international scale, it has also been criticized for lacking practicality and continuing to be embedded in Western epistemologies. Questions also remain about how to measure emancipatory peace and its broader implications, and about how to sustain it under the conditions of a persisting struggle of identity politics, power and governance, and equality and tolerance of difference and freedom. Finding out the value and viability of this vision of emancipatory peace requires further empirical research and an examination of its application in practice. The task of the next phase of critical peace and conflict studies is thus—along with interrogating existing policies and practices—to explore the practical viability of different proposals for emancipation.

This chapter is structured as follows. The first section gives a brief overview of the significance of emancipation for different strands of critical theory. The tensions between deconstructive and reconstructive modes of critique will be examined in order to shed light on the extent to which critical theory can contribute to emancipatory knowledge for the purpose of furthering social and political change. The second section begins by exploring the early critiques of liberal peacebuilding before proceeding to outline some of the emerging visions of an emancipatory peace in conflict-affected societies. The discussion will mostly survey the recommendations made in existing literature and offer critical observations on their conceptual and practical viability. The third and final section traces some of the critiques of and responses to the visions of emancipatory peace. The focus will be on exploring the limits of existing work in critical peace and conflict studies and putting forward ideas for a future direction.

Emancipation and Critical Theory

Critical theory is considered one of the major epistemological approaches for generating normative and practical alternatives, and one of its central themes is emancipation. Etymologically, "emancipation" means "transferring the ownership of a thing or person," referring thus to "the act of setting free, such that one person is no longer the property of another" (Coole 2015, 532). Ontologically, a commitment to emancipation originates from an unsatisfactory view of how the world works and, in particular, dissatisfaction with how state-centric political orders reproduce domination, inequality, and discrimination between different social classes and groups. However, the quest for an emancipatory alternative remains a major challenge for any critical political theory. As the emancipatory project is embedded in the universalist values of equality, justice, and freedom, by default it is involved in promoting identity politics rather than eliminating them, and may therefore cause a power struggle to be reshuffled rather than eliminated. This notwithstanding, as Roger D. Spegele (1996, 8) maintains, the purpose of critical theory is "to assist us to change its practice, and the world with which it is inextricably bound up, for the better." Robert W. Cox (1981, 128) argues that the role of critical theory is to "open up the possibility of choosing a different valid perspective from which the problematic becomes one of creating an alternative world." Similarly, Richard Wyn Jones (1999, 56) maintains that "without the ability to claim that a better world is possible or even conceivable, there is no means by which the present can be criticized." Andrew Linklater (1998, 10) asserts that "normative arguments ... are incomplete without a parallel sociological account of how they can be realized in practice ... and normative and sociological advances are incomplete without some reflection on practical possibilities."

As the purpose of critical knowledge is to promote social and political transformation, when this knowledge is put into practice it is expected to improve individual and community life. At its core, critical theory conceives knowledge as being not objective but influenced by preexisting purposes and interests, predisposed to privileging agency over social structures, and committed to emancipation through dialogue and resistance. Critical knowledge should not be used instrumentally to maintain current power relations—its purpose should be to promote emancipatory social transformation and to pursue justice, equality, and inclusive progress. The test of emancipatory knowledge is not its accuracy of prediction but its successful implementation of change in practice. As Spegele (1996, 9) predicated, emancipatory knowledge "requires that theory be connected up with practice in a certain way, namely, that if the theory is correct and people 'put it into practice,' their lives will go better individually and collectively." Pinar Bilgin (1999, 33) argues that critical perspectives "offer a normative basis to criticise the existing practices and conceive emancipatory alternatives," which are pursued on a "non-statist, non-militarised and non-zero-sum" basis. In this context, critical theory serves the purpose of examining "elements that could be thought of, either separately or collectively, as alternative" to be able to explore "what possibilities exist" for change (Cox 2012, 20). Most of the critical theories engaged in offering emancipatory alternatives suggest that sharing an understanding of emancipation requires individual and collective changes—in particular, the removal of structural and agential blockages to self-realization (Rancière 2009, 13), about "the freeing of someone from the control of another" (Bingham and Biesta 2010, 9), or what Booth (1991, 319) defines as "freeing people from those constraints that stop them carrying out what they would freely choose to do." The structures and relations that need to be transformed are those that "entail systematic inequality, disadvantage and vulnerability" (Nunes 2014, 120). These blockages could be norms, rules, governing authorities, cultural practices, or actors with disproportionate political and material power.

On the other hand, Jacques Rancière (1991, 9, 98) argues that "only individuals can be emancipated" and that "any individual can always, at any moment, be emancipated and emancipate someone else. . . . A society, a people, a state, will always be irrational." Resistance is affiliated with protest, whereas emancipation is closely linked to transformation and liberation. Thus emancipation cannot be achieved through resistance alone, but should be seen as being about "social creativity, introducing new values and aims, new forms of co-operation and action" (Pieterse 1992, 13). In this sense, emancipation is about expanding individual human freedom, about equal participation in society, free from the collective drivers of structural violence (Bargetz 2015). As Mats Alvesson and Hugh Willmott (1992, 432) have argued, emancipation is "the process through which individuals and groups become freed from repressive social and ideological conditions, in particular, those that place socially unnecessary restrictions upon the development and articulation of human consciousness." From this situated perspective, emancipation could take the shape of "the transformation of structures and relationships of vulnerability through localised political action, aimed at the creation of spaces in people's lives so that they are enabled to make decisions and act beyond mere survival" (Basu and Nunes 2013, 69).

Although emancipation remains a central concern for critical theory, it has fallen short of offering a practical solution for how to achieve it (Fierke 2007). In theory, we know quite well that emancipation is about radically improving the lives of oppressed people. Without some sort of practical alternativity and an emancipatory commitment, critique as a political endeavor risks losing its social significance. Cynthia Weber (2017,

54) rightly argues that critical scholars in international relations who tend to endorse antinormative knowledge risk turning critique into a dogmatic form of knowledge with moral certainty and a singular understanding of history, ideology, and political community. Similarly, Daniel L. Levine (2012, 69) argues that there is growing realization in international relations that "critique is a necessary but secondary task; the priority is to return to practical theory as quickly as possible." Practicality is essential for generating emancipatory alternatives. Practicality shifts the focus from abstract criticality and normativity to contextual critiques that account for everyday practices and interactions. This would be essential for rescuing critique and preventing it from becoming a postempirical endeavor. Critical knowledge that engages with policy alternatives "is not only pragmatic, it is also politically enabling: it forces us away from instrumental problem-solving perspectives towards a wider framework of pragmatic thought where narrow instrumental goals are overridden by wider normative and political concerns" (Kurki 2013, 260). According to Aoileann Ní Mhurchú and Reiko Shindo (2016, 5), "critique can help us to develop different ways of talking about, evaluating, doing and interrogating the changing nature of politics, relations and experiences of the international in a globalising world." Hence critique is inevitably implicated in world-making and, with a much clearer understanding of alternativity, can steer the thrust for world-changing in a more emancipatory, just, and inclusive direction. In examining the relationship between critique and emancipation, the remainder of this chapter delves into peace and conflict studies to explore their visions of an emancipatory peace.

Emancipation and Peacebuilding

Since its inception, peace and conflict studies has maintained a normative and practical commitment to justice and emancipation (Galtung and Fischer 2013). In particular, critical approaches to peace and conflict studies have emerged in response to the fallacies of international interventions aimed at peacebuilding in conflict-affected societies—interventions whose primary goal has been to criticize power relations, values, and the impact of liberal interventionist paradigms. While the early critique of liberal peace and its discontents regarded critique as an intellectual exercise concerned with conceptual and philosophical problems, the more recent work of critical peace scholars tends to perceive critique as a social practice that must have a real-world impact and should offer a vision of emancipatory peace. The remainder of this chapter examines the key features of critical peace theories and explores how they propose to achieve emancipation in conflict-affected societies.

The Liberal Peace and Its Discontents

To understand the different visions of emancipatory peace it is important first to unpack the critiques of liberal peace as a framework for guiding international interventions in conflict-affected societies. What we know today as liberal peacebuilding has its roots primarily in the Enlightenment work of Immanuel Kant and his proposal for perpetual peace, transmitted to contemporary international affairs by Michael W. Doyle and his liberalist thinking in the early 1980s. Doyle (1983, 206–207) argues that the key to a liberal peace is the existence of a domestic liberal democratic system in which political and civil rights are respected and advanced, governments are democratically elected and exercise limited powers, economic activity is largely free from state intervention, and the social welfare of citizens is protected. The added value of liberal peace is seen in its establishment of a lasting peace where nations, at the domestic level, calculate the cost of war rationally and then use their influence to avoid conflicting relations with other nations. Moreover, as Roland Paris (2010, 350) argues, "liberalism contains a universalist (and universalising) vision of emancipation through political and economic liberalisation, but it simultaneously embraces an ethic of individual and collective choice or self-government, which can conflict with universalist formulas."

Liberal peacebuilding operates on a number of Western self-mirroring assumptions that would emancipate local subjects (Visoka 2016). First and foremost, the political system in the relevant society is deemed undemocratic, having caused political polarization and, ultimately, violent conflict. As an emancipatory antidote to this, democratization and power-sharing are implanted; they are seen as being more inclusive forms of politics, in which participation and representation and the design of political institutions both encourage societal liberalism and empower individuals of all identities and backgrounds. Democratically elected leaders are perceived as less likely to engage in warfare thanks to political accountability and their fear of losing domestic support (Doyle 2005). The statebuilding process mainly reflects neoliberal principles and seeks to replicate some of the norms, values, and institutions underpinning the Western world. While local subjects call for personal freedom, social recovery, and political emancipation, the new social contract often disappoints. In principle, building democratic institutions is considered the appropriate method for achieving the peaceful resolution of a conflict by channeling political interests through institutionalized processes, encouraging debate, and promoting compromise in accordance with democratic rules (Paris 2004, 159). Democracy is considered a suitable political regime as it improves living conditions and the quality of life and choices, guarantees human rights, and promotes individual and collective responsibility. Most important, promoting democracy is regarded as a long-term investment in the stabilization of war-torn societies and the expansion of geopolitical allies of Western powers.

Under liberal peacebuilding, the former security apparatus is seen as an authoritarian one that should be replaced by a more democratic security system, facilitated by the UN and other bilateral donors, by creating a new police force and army through a democratic reform and consolidation process. The establishment of security infrastructure is considered fundamental for ensuring the stable consolidation of state institutions, as well as for postwar social and economic recovery (Muggah 2009). The rule of law is seen as an essential part of emancipation whereby all individuals are treated equally through independent, impartial, and fair judicial standards. Such a system is meant to free local subjects from customary and traditional forms of dispute resolution, which have been biased against women and other socially disadvantaged groups. Moreover a neoliberal economy is imposed to promote economic growth and overcome centralist and socialist models of economy. A vibrant civil society is seen as crucial for promoting human rights, social inclusion, and ethnic reconciliation and for overcoming radical and nationalist politics. The emancipatory role of civil society is seen as helping to hold government accountable, represent citizens' interests, bolster an environment of pluralism and trust, and create platforms for democratic citizenry where members socialize and promote collective responsibility (Diamond 1999).

This vision of liberal peace has informed most of the engagements by the UN and regional organizations and donors in conflict-affected societies and has served as their ideological tapestry for building a sustainable peace. In 1992 the UN secretary-general's report *An Agenda for Peace* set out the landmark policy strategy on how to intervene in violent conflict and establish a liberal peace. From that point on, postconflict peacebuilding began to be defined in terms of actions that sought to "identify and support structures which will tend to strengthen and solidify peace in order to avoid a relapse into conflict" (UN Secretary-General 1992, paragraph 55). This conceptual scope of liberal peacebuilding was further refined by the UN in 2001, when it stated that peacebuilding efforts should "focus on fostering sustainable institutions and processes in areas such as sustainable development, the eradication of poverty and inequalities, transparent and accountable governance, the promotion of democracy, respect for human rights and the rule of law and the promotion of a culture of peace and non-violence" (UN Secretary-General 2001, 2). Despite extensive investment, in practice this vision of liberal peacebuilding has produced mixed results. Bosnia and Herzegovina, Kosovo, Iraq, and Afghanistan, among other countries that have been subjected to liberal peacebuilding, continue to suffer from a fragile peace, dysfunctional institutions, and social inequality (Visoka 2016). What was intended as a liberal and emancipatory solution to peacebuilding after conflict often gave rise to other problems, which in turn triggered new illiberal responses and reversal effects and generated peace-breaking dynamics, thereby delaying the prospects for establishing a sustainable peace.

As a result of the shortcomings of liberal peacebuilding in practice, a critical scholarship emerged that has examined the performance, effectiveness, and implications of mainstream peacebuilding praxis. For a number of years, critical peace and conflict studies has focused on this type of critique while only hinting vaguely at emancipatory alternatives. The critique has mostly concentrated on "revealing the implicit political underpinnings of international peacebuilding, and [demonstrating] their flaws," such as power imbalance, ideological inconsistency, and flawed sources of legitimacy (Tadjbakhsh 2011, 6). Early critiques argued that the problematic nature of liberal peacebuilding, its adverse effects—such as the exacerbation of social and economic tensions, political conflict, and sectarian division—and the contested value system, top-down governance approach, and misunderstanding of local context have undermined local agency and prevented addressing some of the underlying sources of the conflict (Newman, Paris, and Richmond 2009). Some of these critiques relate to the creation of an environment for delaying sustainable development by imposing good governance and a democratization agenda, and by challenging local processes by imposing external conditionality, which reflects normative incoherence under the conditions of contested legitimacy. In most cases, external interventions have attempted to re-engineer postconflict societies by introducing Western political and economic governance models that have often been unsuited to the local context and culture and applied contrary to the democratic will of the local people (Richmond 2011). External desires for stability, democracy, and a market economy have been converted into political obligations for local subjects, thereby suppressing local agency, needs, and rights as well as leading to different forms of resistance and hybrid modes of governance. This in turn has led to the failure of the liberal peacebuilding model as manifested in the rise of predatory and authoritarian states and deep local dissatisfaction with and resistance to Western statebuilding models.

Moreover, driven by security concerns and economic and strategic self-interest, states and international organizations engaged in postconflict reconstruction often demonstrate an unwillingness to commit adequate resources, to operate in a coordinated and effective manner, or to be accountable for the human rights abuses and local damage directly caused by their international staff (Philpott and Powers 2010). In particular, excluding citizens from democratic processes undermines both the prospects for public deliberation and active support for political processes—elements that are required to flesh out the institutional skeleton with substance. As David Chandler (2010a, 15) argues, statebuilding does not operate on the basis of the principles of human rights, political legitimacy, and the social contract; instead it seeks "to secure stability through balancing internal and external interests and concerns as matters of technical and administrative competences in the formulations of good governance." While ethnic power-sharing arrangements in conflict-affected societies were intended to accommodate all ethnic groups, build consensus, and promote political moderation, in reality they led to encapsulation and the capture of institutions by a handful of ethnonationalist protagonists who continue to rule and use the seats reserved for minorities as a convenient threshold for government formation, thereby rendering political change impossible. Under these conditions, any political moderation on either side was policed by ethnonationalist conglomerates and was either sidelined or incorporated into their political realms. This ambivalence has suited ethnic elites, however, as they have continued to secure popular legitimacy, based not on the quality of their democratic performance but on the extent to which they oppose political compromise and moderation and (their flagship policy) promote ethnic nationalism.

Accordingly, critical scholars have come to the conclusion that, in the aftermath of violent conflicts, international statebuilding has failed to support a consensual conception of the social contract. Instead it has removed political agency from local subjects and transferred it to a handful of ethnonationalist groups. The political longevity of ethnonationalist elites, in turn, has been maintained by tactical subordination in international missions, ruling over local subjects through externally imposed conditionality, neopatrimonialism, the misuse of both peaceful and armed resistance, corruption and coercion, and threatening to destabilize the fragile peace and undermine the credibility of international missions. This gap between the new state institutions and the needs of local subjects has both undermined state legitimacy and exacerbated social and political disengagement. Failed statebuilding has led to the emergence of different forms of local resistance, such as a refusal to obey an international authority or local institutions, to pay taxes, to participate in elections, or to share the common burden (Jahn 2007; Richmond 2011).

Notably, most of the early critiques of peacebuilding avoided proposing any alternative to liberal peace. In fact, some of this early critical research was more inclined to tackle questions relating to the removal of obstacles and the creation of conditions for enforcing the liberal peace than to look for alternative emancipatory possibilities or to engage critically with the legacies of liberal peacebuilding. In response, proponents of liberal peace have argued that the critique is inappropriate and is incongruent with reality on the ground. They point out that, regardless of the shortcomings of liberal peacebuilding, there is no better alternative available to replace it. For instance, Paris (2010, 339–340, emphasis in original) has argued that "much of the critical literature is actually exposing variations within, rather than alternatives to, liberal peacebuilding," and that "there is no realistic alternative to *some form* of liberal peacebuilding strategy." By constantly deconstructing liberal peace discourses and practices, critical scholars came to the realization that they may "unintentionally contribute to the constitution of liberal peace as a unitary discursive framework, making it consistent in its inconsistency, timeless in its temporality, solid in its fluidity, and liberal in its illiberalism" (Visoka 2019, 690). In short, critical scholars have realized that knowledge should not only be about unmasking power relations and the dynamics of dominance and creating space for a politics of resistance, but should also play a role in generating practical knowledge that makes emancipatory transformation possible (Visoka and Richmond 2018).

Visions of an Emancipatory Peace

In response to the futility of early critiques of liberal peacebuilding, a renewed effort to suggest more explicit and emancipatory forms of peacebuilding has emerged. Oliver P. Richmond and Roger Mac Ginty (2015, 183) admit that "while the problem-solving approach is unapologetically prescriptive in its policy recommendations, much of the critique has avoided recommending, let alone imposing, alternatives on those in conflict-affected areas." The most elaborate alternative to liberal peacebuilding to date is the work of Richmond, who advocates for an emancipatory peace. Richmond (2010, 34) argues that "without an engagement with needs and welfare, peacebuilding will not lead to a sustainable outcome because there are few peace incentives for citizens or elites." To open up the space for possible emancipatory alternatives, Richmond (2011) places the focus on what he labels "post-liberal peace," which entails transforming liberal peace in order to engage proactively with the local people, to recognize their needs and seek their support and consent. Richmond argues that this approach is a precondition for coming up with a new social contract—one that recognizes the plurality of peace, remains open to customary governance, and prioritizes social welfare and justice.

While most critical scholars seek to advance the quest for emancipation based on postnational constellations, Richmond suggests reclaiming emancipation from most local levels of society. The everyday is seen as "a site of dynamics including resistance and politicisation, solidarity, local agency, hybridity, and also of passivity and depoliticization" (Richmond 2010, 671). By focusing on the everyday, postliberal peace is able to understand contextual manifestations of peace, where citizenship, rights, and duties have an everyday meaning, rather than introducing peacebuilding models at a global level or constructing virtual peace by relying on institutions or local elites. This new vision of peace—where an everyday form of peace offers care and assistance for the most marginalized but also mediates culture and identity, institutions, and custom—is seen as emancipatory in nature (Richmond 2011, 3). Aware of the ambiguous and sensitive nature of the everyday, Richmond (2010, 672) argues that in a postliberal peace it should not be seen as a binary exclusion where the local prevails over the international, or the nonliberal over the liberal, but should instead be "a site where these meet and are negotiated, leading variously to repulsion, modification or acceptance, and hybridity." The everyday thus represents a balance between the local and the international, and it enables "a better understanding of how local actors fulfill their needs, [and] maintain their institutions and identities, while appearing to conform to the strategic institutions of biopolitical governance" (683).

For critical scholars, any peacebuilding agenda should have a detailed understanding of local cultures, traditions, and needs. Peace is more likely to last if it takes account of local needs, is legitimated by the beneficiary communities, and is not necessarily designed to serve the geopolitical and economic interests of external powers or the agendas of donors. Andries Odendaal (2013, 17) argues that, in postconflict societies, "understanding the importance of local dynamics and the nature of the interactions between national and local dynamics has important consequences for peacebuilding." An emancipatory peace regards local and community approaches to conflict resolution and peace formation as fundamental to a sustainable peace because affected communities play a major role in shaping peace from the bottom up, not from the top down, as was the case with liberal peace and other predatory and interventionist approaches. As Chuck Thiessen (2011, 123) maintains, "emancipatory peacebuilding, in short, broadens the narrow top-down state-building focus of liberal peacebuilding, and holistically redirects the project as a grassroots, bottom-up activity—engaging with the local and the marginalized." It is seen as being conditional on the ability of local actors—with transnational assistance—to tackle context-dependent and local structural violence, promote social justice as a pathway to peace, and promote openness, dialogue, recognition, and difference, while also rejecting the domination of external actors and local ethnopolitical agents.

The vision of emancipatory peace was elaborated around the concept of hybrid peace, which shares the positive elements of both liberal peace and local and indigenous forms of peace. Richmond has never denied sympathy with the emancipatory wing of liberal peace, affiliated with civil peace and a peaceful civil society, which in practice is often disregarded by peacebuilding organizations. He has argued that there are some "emancipatory forms of human security, associated with individual emancipation and social values, that offer a significant opportunity to enhance the process of building peace in postconflict states" (Richmond 2007, 459). Hybridity, he says, represents "a transmutation of both the liberal and the local discourses of peace, even in view of their relatively unequal material relations and, of course, their differences and similarities" (Richmond 2011, 151). Richmond is aware, however, that hybrid peace may have both positive and negative aspects. A positive hybrid peace could represent "complementary practices related to democracy, self-determination, agency, autonomy, solidarity, human rights and needs, and a rule of law, with customary social support networks, customary forms of governance and political order" (18–19). A negative hybrid peace, on the other hand, could be a combination of "liberal institutionalism and market development solutions with patriarchal, feudal, communal, or sexist, practices" (18–19). Looking through the lens of the alternative concept of hybridity, Richmond and other scholars, such as Mac Ginty, seek to overcome the hierarchical order of Western political agencies, cultures, and modes of knowledge and to show how at the same time, in practice, they interact and are mutually constitutive of local, transitional, and non-Western knowledge and practices. As they argue, postliberal and hybrid forms of peace have "tried to highlight the possible alternatives" and also "to utilise more imaginative conceptualisations of power and agency, seeing them as networked, co-constitutive and adaptive" (Richmond and Mac Ginty 2015, 181). In other words, to overcome the ineffectiveness of hegemonic, top-town interventionism, a contextualized, hybridized peace—which would be more "locally sustainable, resilient, and legitimate"—is proposed (Richmond 2009, 572). So, in this regard, hybridity has a performative and emancipatory function, as acts of resistance and coexistence can open up a space for the transformation of power relations, increase equality, address claims for justice and welfare, and liberate local subjects from international domination (Richmond and Mitchell 2012; Visoka 2017).

In the pursuit of an emancipatory peace, the question of local legitimacy has emerged as a central issue. Local legitimacy has come to signify a wide range of everyday and institutional attributes that indicate the subject's acceptance of and support for local or international authorities based on individual or collective norms and values (see Visoka 2020). Similarly, the discourse of local ownership signifies an attempt to transform paternalistic forms of interventionism to more facilitative and consultative forms of engagement that respect local needs, dynamics, and interests. More recently Richmond (2014) has suggested "peace formation" as the derivative product of hybrid forms of postliberal peace, which emphasizes the primacy of local legitimacy and perspectives in conjunction with facilitative and constructive international engagement. According to Richmond (2014, xiv), "peace-formation dynamics lead to the creation of new and contextual forms out of social, political, economic, cultural and historical forms of peace in local, transnational and international terms, while still remaining cognizant of the liberal peace and its norms, technologies, capacities and advantages." Peace formation therefore entails multiple articulations of local agents, who come together to secure legitimacy for peace processes while still remaining open to external involvement, not in interventionist terms but as enablers and facilitators of locally rooted peace networks, agencies, and infrastructures. Semantically, the metaphor of peace "building" symbolizes a peace constructed using certain models, plans, and standards, while peace "formation" focuses on how it is shaped and its ultimate form and signifies peace processes that are more gradual, bottom-up, and natural.

The vision of emancipatory peace presented here also has implications for the logic of international engagement in conflict-affected societies, which may be changed from one that is based on conditionality and intervention to a new mode of engagement that is more conducive to and supportive of localized initiatives and the dynamics of emancipation. Richmond (2011, 110) advocates for "international support for these processes, and guidance on technical aspects of governance and institution building without introducing or endorsing hegemony, inequality, conditionality, and dependency." While emancipation may be considered an antidote to intervention, in most cases it requires some sort of intervention and support from the outside "by someone who is not subjected to the power that needs to be overcome" (Bingham and Biesta 2010, 30). Richmond (2014, 209) accepts that "peace formation requires direct intervention to stop violence . . . but afterwards it requires empathy, sensitivity, listening, enablement and support from external actors rather than trusteeship, while it maintains its own autonomous authority." He has recently coined the term "peace enablement" to signify "a more hands-off, but also more supportive, form of assistance" (224), one that displays an awareness that conflict-affected citizens need greater security, stability, and material support in order to overcome the multiple structures of violence and inequality. For peace enablement to take place, however, the international actors would need to have not only local knowledge and contacts but also local support, consensus, and legitimacy. It would require engaging a more diverse selection of local actors in the peacebuilding and statebuilding process, as well as showing sensitivity to local practices, needs, and culture. International peace enablement is an interesting idea, and it remains to be seen how it will be developed further and how it will be received by the proponents and critics of peacebuilding interventions.

Critiques of Emancipatory Peace

As might be expected, any scholarly effort to offer emancipatory alternatives tends to become the subject of criticism. Scholars have shown skepticism about how much the visions of emancipatory peace discussed here differ from liberal peacebuilding, while others have highlighted the lack of practical strategies for achieving emancipatory peace or the unviability of such alternatives. David Chandler (2010b, 146) criticizes the proposal of an emancipatory peace for "its inability to overcome social, economic and cultural inequalities" in postconflict societies and for creating a consensus whereby the problem of intervention to further statebuilding and peacebuilding is framed as a "problem of the relationship between the liberal West and the non-liberal Other."

Chandler (2013, 24–25, 32) further argues that a return to nonlinearity, to the local, and to an intersubjective understanding of practices moves away from the structuring economic and social relations, which he implies "market inequality" and "structuring asymmetries of intervention." He adds that the nonlinear discourses of hidden local agency neither create the basis of any genuine understanding of the limits to liberal peace nor provide any emancipatory alternative (31). He further argues that nonlinearity is problematic because "peacebuilding interventions can no longer be held to account through highlighting the gap between their legitimating promise and their reality in terms of outcomes" (32), and that failures of liberal peace could be seen as successes from the point of view of hybrid peace. From this point of view, therefore, achieving emancipation by using Western agencies, structures, and institutions is highly problematic and underscores the limits of emancipatory peace.

Adding to this, John Heathershaw (2013, 277, 280) argues that current visions of an emancipatory peace remain "entrapped by the terms of the liberal peace debate where liberals compete against locals," thereby bolstering "the myth that they are naturally oppositional." Similarly, Meera Sabaratnam (2013) regards these local/international binaries as symptomatic of a Eurocentric critique of liberal peace. She suggests, instead, decolonizing the critique of liberal peace by problematizing what is discounted by Eurocentric thinking, repoliticizing postconflict fields and promoting positional critique, and bringing up the subject of the materiality and material impact of peacebuilding processes. This critique owes its relevance to the fact that most of the international scholarship on postconflict societies is based on an unrepresentative body of international scholarship and tries to mediate, deviate from, and reinterpret this knowledge, consequently constructing a different social reality which is then interpreted through different measurements, reference points, and analytical concepts (Visoka and Musliu 2019). On the question of hybrid peace, Gearoid Millar (2014, 501) labels it a "prescriptive hybridity" that "assumes that administering hybrid institutions will foster predictable peace-promoting experiences among local people." He argues that "internationally planned and administered hybrid institutions will not result in predictable experiences and may even result in negative or conflict-promoting experiences" (501). Similarly, Suthaharan Nadarajah and David Rampton (2015, 50) argue that the hybrid peace approach is "a problem-solving tool" that "reproduces the liberal peace's logics of inclusion and exclusion, and through a reconfiguration of the international interface with resistant 'local' orders, intensifies the governmental and biopolitical reach of liberal peace for their containment, transformation, and assimilation." In response, Richmond and Mac Ginty (2015, 175) assert that these arguments "somewhat miss the point" because "[h]ybridity has been adopted as a tool precisely to interrogate categories and combat lazy dichotomies often used by policymakers and mainstream theorists, as well as to understand political tensions/resistance over peacebuilding." In defense of hybridity as a conceptual tool, they argue that "[t]he non-acceptance of established political categories allows us to reassess boundaries and identify the connections, networks and co-constitutive elements that comprise the context for contemporary peace-support intervention" (180).

The second major aspect of the critique of emancipatory peace is its lack of specificity or any elaboration of what such a vision of sustainable peace would look like in practice. While the critical knowledge I have examined here has made a valuable contribution at a conceptual level, so far it has insufficiently linked their proposed conceptual alternatives with the practical aspects of emancipatory politics and possibilities. Paris (2010, 357) criticized the concept of emancipatory peacebuilding for lacking elaboration, claiming that, even when this happens, "we may . . . discover that emancipatory peacebuilding is not really opposed to liberal peacebuilding at all." Recently, Keith Krause (2019, 296) has added that critical approaches to peace and conflict lack a critical praxis, namely an "effective political engagement with existing structures and institutions" that would "provide local actors with understandings of the potential for emancipatory socio-political change". Other scholars have argued that the postliberal peacebuilding model proposed by Richmond does not extend far enough in its attempts to eradicate the causes of war or reduce the structural conditions of political violence. For example, Richard Jackson (2018, 6) points out that "all of these critiques and alternatives leave largely unquestioned the centrality of militarism and legitimate political violence to the ontology and epistemology of peace." He suggests "post-liberal peace plus" as an emancipatory alternative "in which a radically pacifist, locally organised, agonistic politics replaces the Western-oriented, top-down state-building blueprint which is currently central to peacebuilding theory and practice" (1–2). Jackson offers a critical alternative to existing approaches to peacebuilding by arguing that a genuine peace is one which is post-Westphalian and post-Weberian, where the state, society, and politics are disarmed and demilitarized and security is provided by alternative, unarmed civilian defense, and the basis of political interactions is pacifism rather than ethnic antagonism.

The vision of emancipatory peace presented here should of course be subjected to critical interrogation. As it is still a project in its infancy, it is too early, however, to draw conclusions about its viability or drawbacks (see Aning, Brown, Boege and Hunt 2018; Piccolino 2019). Early evidence suggests that the agenda for an emancipatory peace has influenced debates on human security and transitional justice as well as the work of local, regional, and international organizations and think tanks. From an emancipatory human security perspective, peace is for "people whose security is at stake and who need to feel secure, as opposed to that of external actors in a peacebuilding situation, or of hegemony, power, and institutions" (Tadjbakhsh 2010, 129). From this perspective, the most suitable peace is one that "responds to emancipation from insecurities in everyday life, be they basic threats to survival and bare life, functional threats that hamper livelihoods, or those concerned with dignity" (129). Thiessen (2014, 40) adds, "[E]mancipation, as the chief aim of security, requires bottom-up approaches where individuals are empowered to voice, negotiate and develop forms of human security tailored to their particular situation." Similarly, it is suggested that emancipatory approaches to transitional justice "side with the powerless," that they expose "the structures and conditions that create victims, and in turn [empower] victims to enable social change" (Robins and Wilson 2015, 211). Such emancipating "processes involve shifting the locus of transitional justice from the metropolis, where those most able to access a global rights discourse are located, to the communities where violations occur" (221–222). The top-down agenda for transitional justice in many conflict-affected societies tends to focus mainly on the institutional level, often ignoring aspects such as truth-seeking, victim support, apologies, reparation, and community reconciliation. Accordingly, emancipatory transitional justice should ground "transition in the everyday lives of the marginalized and [drive] transitional processes away from efforts to enforce a status quo towards a politically transformative practice respondent to human needs" (222). In their programs, NGOs such as Peace Direct and Conciliation Resources have enshrined visions of emancipatory peace focusing on local peace formation structures, actors, and knowledge. And think tanks such as the Stockholm International Peace Research Institute have expanded their research to look at everyday peace and locally focused interventions in conflict-affected societies (see O'Driscoll, 2019). As the crisis of the liberal international order deepens, the quest for alternative approaches to peacebuilding will most likely turn to searching for postliberal forms of peace—a search in which the vision of emancipatory peace presented here could be one of the sources of inspiration.

Conclusion

This chapter has given an overview of the role of critical theory in shaping the different visions of emancipatory peace for conflict-affected societies. The role of critical theory is not only to criticize existing discourses, policies, and practices and identify contradictions in them, but also to offer alternatives for enhancing human emancipation. In light of this, the chapter explores the relationship between critical theory, peacebuilding, and emancipation. The critical debates in peace and conflict studies have gone through two major phases. The first was mostly a critique of liberal peacebuilding interventions: it focused on identifying the flaws and failures of peacebuilding and statebuilding in policy and practice. This phase avoided searching for alternative solutions or making suggestions for how to remedy the dominant approaches to peacebuilding in conflict-affected societies. Despite its limited scope and its failure to offer emancipatory alternatives, this first phase paved the way for a more elaborate critique to emerge—one that would offer a more detailed vision of how to achieve a more emancipatory peace. Proponents of the second phase of critique have taken up the challenge of providing emancipatory knowledge which has practical relevance for vulnerable societies across the world. As discussed here, the solutions for emancipation, justice, and lasting peace offered by visions of emancipatory peace have focused mainly on those aspects of local agency, culture, and knowledge that are overlooked and ignored by the liberal peacebuilding approach. Rather than offering prescriptive solutions, they propose broad conceptual and structural parameters that could be more useful in achieving emancipation. They entail giving primacy to local agency and bottom-up approaches to conflict resolution, peacebuilding, and social transformation, where local culture, needs, and knowledge play a vital role in shaping policies and designing solutions to social, political, and economic problems. It is a vision that seeks to empower local subjects to own peace and the social contract and to pursue development with justice, and governance with accountability. The vision of emancipatory peace does not seek to replace the international with the local; it proposes changing the dynamics of engagement from intervention to enablement. This vision of emancipation means emancipation from the oppressive structures that are either installed by the international community or are a byproduct of international intervention and from the subsequent impositions on the relevant state and society. Primarily it entails freeing local subjects from international intervention, and it is put forward to be used also as a basis for tackling both the root causes and the immediate triggers of violent conflict. As this vision of emancipatory peace is in its infancy, however, it needs further substantiation and application in practice to assess its viability and its success in achieving its goals. All forms of critique and theorizing intended to produce emancipatory knowledge inevitably have to understand the structures of violence, inequality, and injustice so that they can offer alternative practical, ethical, and political knowledge to counter the contemporary causes, drivers, and consequences of power, violence, conflict, and intervention—in the mode not of intervention but of enablement, in the form not of conditionality but of facilitation, not as a peace that is imposed but as peace that is bottom-up, contextual, and constitutive.

References

Alvesson, M., and H. Willmott. 1992. "On the Idea of Emancipation in Management and Organization Studies." *Academy of Management Review* 17 (3):432–464.

Aning, K., M. A. Brown, V. Boege, and Ch. T. Hunt, eds. 2018. *Exploring Peace Formation: Security and Justice in Post-Colonial States.* London: Routledge.

Bargetz, B. 2015. "The Distribution of Emotions: Affective Politics of Emancipation." *Hypatia* 30 (3):580–596.

Basu, Soumita, and João Nunes. 2013. "Security as Emancipation." In *Critical Approaches to Security: An Introduction to Theories and Methods*, edited by L. J. Shepherd, 63–76. Abingdon, UK: Routledge.

Bilgin, P. 1999. "Security Studies: Theory/Practice." *Cambridge Review of International Affairs* 12 (2):31–42.

Bingham, C., and G. Biesta. 2010. *Jacques Rancière: Education, Truth, Emancipation.* London: Continuum.

Booth, K. 1991. "Security and Emancipation." *Review of International Studies* 17 (4):313–326.

Chandler, D. 2010a. *International Statebuilding: The Post-Liberal Governance.* London: Routledge.

Chandler, D. 2010b. "The Uncritical Critique of 'Liberal Peace.'" *Review of International Studies* 36 (S1):137–155.

Chandler, D. 2013. "Peacebuilding and the Politics of Non-linearity: Rethinking 'Hidden' Agency and 'Resistance.'" *Peacebuilding* 1 (1):17–32.

Coole, D. 2015. "Emancipation as a Three-Dimensional Process for the Twenty-First Century." *Hypatia: A Journal of Feminist Philosophy* 30 (3):530–546.

Cox, R. W. 1981. "Social Forces, States and World Orders: Beyond International Relations Theory." *Millennium: Journal of International Studies* 10 (2):126–155.

Cox, R. W. 2012. "For Someone and for Some Purpose: An Interview with Robert W. Cox." In *Critical Theory in International and Security Studies*, edited by S. Brincat, L. Lima, and J. Nunes, 16–34. London: Routledge.

Diamond, L. 1999. *Developing Democracy: Towards Consolidation.* Baltimore: Johns Hopkins University Press.

Doyle, M. W. 1983. "Kant, Liberal Legacies, and Foreign Affairs." *Philosophy and Public Affairs* 12 (3):205–235.

Doyle, M. W. 2005. "Three Pillars of the Liberal Peace." *American Political Science Review* 99 (3):463–466.

Fierke, K. 2007. *Critical Approaches to International Security.* Cambridge, UK: Polity.

Galtung, J., and D. Fischer. 2013. *Johan Galtung: Pioneer of Peace Research.* New York: Springer.

Heathershaw, J. 2013. "Towards Better Theories of Peacebuilding: Beyond the Liberal Peace Debate." *Peacebuilding* 1 (2):275–282.

Jackson, R. 2018. "Post-Liberal Peacebuilding and the Pacifist State." *Peacebuilding* 6 (1):1–16.

Jahn, B. 2007. "The Tragedy of Liberal Diplomacy: Democratization, Intervention, Statebuilding (Part II)." *Journal of Intervention and Statebuilding* 1 (2):211–229.

Krause, K. 2019. "Emancipation and Critique in Peace and Conflict Research." *Journal of Global Security Studies* 4 (2):292–298.

Kurki, M. 2013. *Democratic Futures: Re-Visioning Democracy Promotion.* London: Routledge.

Levine, D. J. 2012. *Recovering International Relations: The Promise of Sustainable Critique.* Oxford: Oxford University Press.

Linklater, A. 1998. *The Transformation of Political Community.* Cambridge: Polity Press.

Millar, G. 2014. "Disaggregating Hybridity: Why Hybrid Institutions Do Not Produce Predictable Experiences of Peace." *Journal of Peace Research* 51 (4):501–514.

Muggah, R., ed. 2009. *Security and Post-Conflict Reconstruction: Dealing with Fighters in the Aftermath of War.* London: Routledge.

Nadarajah, S., and D. Rampton. 2015. "The Limits of Hybridity and the Crisis of Liberal Peace." *Review of International Studies* 41 (1):49–72.

Newman, E., R. Paris, and O. P. Richmond, eds. 2009. *New Perspectives on Liberal Peacebuilding.* Tokyo: UN University Press.

Ní Mhurchú, A., and R. Shindo. 2016. "Introduction: Being Critical and Imaginative in International Relations." In *Critical Imaginations in International Relations*, edited by A. Ní Mhurchú and R. Shindo, 1–10. London: Routledge.

Nunes, J. 2014. *Security, Emancipation and the Politics of Health: A New Theoretical Perspective.* London: Routledge.

Odendaal, A. 2013. *A Crucial Link: Local Peace Committees and National Peacebuilding.* Washington, DC: US Institute of Peace.

O'Driscoll, D. 2019. *Building Everyday Peace in Kirkuk, Iraq: The Potential of Locally Focused Interventions.* SIPRI Policy Paper No. 52. Solna, Sweden: Stockholm International Peace Institute.

Paris, R. 2004. *At War's End: Building Peace after Violent Conflict.* Cambridge: Cambridge University Press.

Paris, R. 2010. "Saving Liberal Peacebuilding." *Review of International Studies* 36 (2):337–365.

Philpott, D., and G. F. Powers. 2010. *Strategies of Peace: Transforming Conflict in a Violent World.* Oxford: Oxford University Press.

Piccolino, G. 2019. "Local Peacebuilding in a Victor's Peace: Why Local Peace Fails without National Reconciliation." *International Peacekeeping* 26 (3):354–379.
Pieterse, J. N. 1992. "Emancipations, Modern and Postmodern." *Development and Change* 23 (3):5–41.
Rancière, J. 1991. *The Ignorant Schoolmaster: Five Lessons in Intellectual Emancipation.* Stanford: Stanford University Press.
Rancière, J. 2009. *The Emancipated Spectator.* Translated by G. Elliott. London: Verso.
Richmond, O. P. 2007. "Emancipatory Forms of Human Security and Liberal Peacebuilding." *International Journal* 62 (3):459–477.
Richmond, O. P. 2009. "A Post-liberal Peace: Eirenism and the Everyday." *Review of International Studies* 35 (3):557–580.
Richmond, O. P. (2010). "Resistance and the Post-Liberal Peace." *Millennium: Journal of International Studies* 38 (3):665–692.
Richmond, O. P. 2011. *A Post-Liberal Peace.* London: Routledge.
Richmond, O. P. 2014. *Failed Statebuilding: Intervention, the State, and the Dynamics of Peace Formation.* New Haven, CT: Yale University Press.
Richmond, O. P., and R. Mac Ginty. 2015. "Where Now for the Critique of the Liberal Peace?" *Cooperation and Conflict* 50 (2):171–189.
Richmond, O. P., and A. Mitchell, eds. 2012. *Hybrid Forms of Peace: From Everyday Agency to Post-Liberalism.* Basingstoke, UK: Palgrave Macmillan.
Robins, S., and E. Wilson. 2015. "Participatory Methodologies with Victims: An Emancipatory Approach to Transitional Justice Research." *Canadian Journal of Law and Society / Revue Canadienne Droit Et Société* 30 (2):219–236.
Sabaratnam, M. 2013. "Avatars of Eurocentrism in the Critique of the Liberal Peace." *Security Dialogue* 44 (3):259–278.
Spegele, R. D. 1996. *Political Realism in International Theory.* Cambridge: Cambridge University Press.
Tadjbakhsh, Sh. 2010. "Human Security and the Legitimisation of Peacebuilding." In *Palgrave Advances to Peacebuilding: Critical Developments and Approaches*, edited by O. P. Richmond, 116–136. Basingstoke, UK: Palgrave Macmillan.
Tadjbakhsh, S., ed. 2011. *Rethinking the Liberal Peace: External Models and Local Alternatives.* London: Routledge.
Thiessen, C. 2011. "Emancipatory Peacebuilding: Critical Responses to (Neo)Liberal Trends." In *Critical Issues in Peace and Conflict Studies*, edited by T. Matyók, J. Senehi, and S. Byrne, 115–142. New York: Lexington Books.
Thiessen, C. 2014. *Local Ownership of Peacebuilding in Afghanistan: Shouldering Responsibility for Sustainable Peace and Development.* Lanham, MD: Lexington Books.
UN Secretary-General. 1992. *An Agenda for Peace.* UN Doc. A/47/277-S/24/24111, 17 June.
UN Secretary-General. 2001. *Prevention of Armed Conflict: Report of the Secretary-General.* UN Doc. A/55/985-S/2001/574, 7 June.
Visoka, G. 2016. *Peace Figuration after International Intervention.* London: Routledge.
Visoka, G. 2017. *Shaping Peace in Kosovo: The Politics of Peacebuilding and Statehood.* Basingstoke, UK: Palgrave Macmillan.
Visoka, G. 2019. "Critique and Alternativity in International Relations." *International Studies Review* 21 (4):678–704.
Visoka, G., and V. Musliu, eds. 2019. *Unravelling Liberal Interventionism: Local Critiques of Statebuilding in Kosovo.* London: Routledge.

Visoka, G., and O. P. Richmond. 2018. "After Liberal Peace? From Failed State-Building to an Emancipatory Peace in Kosovo." *International Studies Perspectives* 18 (1):110–129.
Visoka, G. 2020. "International Intervention and Relational Legitimacy." In *Local Legitimacy and International Statebuilding*, edited by O. P. Richmond and R. Mac Ginty, 44–68. Edinburgh: Edinburgh University Press.
Weber, C. 2017. "What's the Point of IR? Or, We're so Paranoid, We Probably Think This Question Is about Us." In *What's the Point of International Relations?*, edited by S. L. Dyvik, J. Selby, and R. Wilkinson, 46–56. London: Routledge.
Welzel, C. 2013. *Freedom Rising: Human Empowerment and the Quest for Emancipation.* Cambridge: Cambridge University Press.
Wyn Jones, Richard. 1999. *Security, Strategy and Critical Theory*. Boulder, CO: Lynne Rienner.

Index

Tables, figures and boxes are indicated by *t*, *f* and *b* following the page number.

www.ingramcontent.com/pod-product-compliance
Lightning Source LLC
Chambersburg PA
CBHW082133060326
40795CB00005B/390